SIPRI Yearbook 1991
World Armaments and Disarmament

sipri

Stockholm International Peace Research Institute

SIPRI is an independent international institute for research into problems of peace and conflict, especially those of arms control and disarmament. It was established in 1966 to commemorate Sweden's 150 years of unbroken peace.

The Institute is financed mainly by the Swedish Parliament. The staff, the Governing Board and the Scientific Council are international.

The Governing Board and the Scientific Council are not responsible for the views expressed in the publications of the Institute.

sipri

Stockholm International Pece Research Institute
Pipers väg 28, S-171 73 Solna, Sweden
Cable: SIPRI
Telephone: 46 8/655 97 00
Telefax: 46 8/655 97 33

SIPRI Yearbook 1991

World Armaments and Disarmament

Stockholm International Peace Research Institute

OXFORD UNIVERSITY PRESS
1991

Oxford University Press, Walton Street, Oxford OX2 6DP
Oxford New York Toronto
Delhi Bombay Calcutta Madras Karachi
Petaling Jaya Singapore Hong Kong Tokyo
Nairobi Dar es Salaam Cape Town
Melbourne Auckland
and associated companies in
Berlin Ibadan

Oxford is a trade mark of Oxford University Press

Published in the United States
by Oxford University Press, New York

Yearbooks before 1987 published under title
'World Armaments and Disarmament:
SIPRI Yearbook [year of publication]'

British Library Cataloguing in Publication Data
Data available
ISSN 0953–0282
ISBN 0–19–829145–0

Library of Congress Cataloging in Publication Data
LC card number 83-643843
ISSN 0347–2205

Typeset and originated by Stockholm International Peace Research Institute
Printed and bound in Great Britain by
Biddles Ltd., Guildford and King's Lynn

Contents

Part II. Military expenditure, the arms trade and armed conflicts

Part IV. Special features

Annexes

Acknowledgements

This year, in the midst of changing world developments which necessitated a permanent reappraisal of 1990, SIPRI's *Yearbook* work was a special challenge. We are proud to have secured the co-operation of four distinguished international authors outside the SIPRI staff and aside from the regular contributors to the *Yearbook*: Dr Harald Müller, David Fischer, Professor Barry R. Posen and Sir Brian Urquhart. We are also grateful to those who provided us with valuable suggestions as to how to improve the *Yearbook* to the benefit of a globally dispersed readership.

A dedicated staff produced SIPRI's twenty-second *Yearbook,* the last *Yearbook* to appear during my service as SIPRI Director. I feel greatly indebted to Connie Wall, who again displayed great skill and leadership in shouldering the tremendous responsibility of editing the *Yearbook* and piloting it through all the thorny stages of production.

Special thanks are due in particular to the remarkably committed editors—Billie Bielckus, Paul Claesson, Jetta Gilligan Borg, Don Odom and Gillian Stanbridge—who also set the book in final camera-ready format.

We also extend thanks to the conscientious secretaries—Joan Allison, Fatima Asp-Barreto, Cynthia Loo, Marianne Lyons and Ricardo Vargas-Fuentes—and to Gerd Hagmeyer-Gaverus for programming and other computer assistance.

My special gratitude is due to my secretary, Bella Kjellgren, who again supported me with her noble patience and striking energy.

Dr Walther Stützle
Director of SIPRI

GLOSSARY AND CONVENTIONS

Acronyms

ABM	Anti-ballistic missile
ACE	Allied Command Europe (NATO)
ACM	Advanced cruise missile
ACV	Armoured combat vehicle
ADM	Atomic demolition munition
AFAP	Artillery-fired atomic projectile
AFB	Air Force Base
AIFV	Armed infantry fighting vehicle
ALCM	Air-launched cruise missile
AMF	Allied Mobile Force
ANC	African National Congress
ASAT	Anti-satellite
ASEAN	Association of South-East Asian Nations
ASLCM	Advanced sea-launched cruise missile
ASM	Air-to-surface missile
ASUW	Anti-surface warfare
ASW	Anti-submarine warfare
ATBM	Anti-tactical ballistic missile
ATC	Armoured troop carrier
ATTU	Atlantic-to-the-Urals (zone)
AWACS	Airborne warning and control system
BMD	Ballistic missile defence
BW	Biological warfare/weapons
BWC	Biological Weapons Convention
CAS	Committee on Assurances of Supply
CBM	Confidence-building measure
CBW	Chemical and biological warfare/weapons
CD	Conference on Disarmament
CEP	Circular error probable
CFE	Conventional Armed Forces in Europe
CMEA	Council for Mutual Economic Assistance (as COMECON)
COCOM	Coordinating Committee
COMECON	Council for Mutual Economic Assistance (as CMEA)
CONUS	Continental USA
CORRTEX	Continuous reflectometry for radius versus time experiments
CPC	Conflict Prevention Centre
CSBM	Confidence- and security-building measure
CSCE	Conference on Security and Co-operation in Europe
CTB(T)	Comprehensive test ban (treaty)
CTOL	Conventional take-off and landing
CW	Chemical warfare/weapons
CWC	Chemical Weapons Convention
CWFZ	Chemical weapon-free zone
DEW	Directed-energy weapon
DOD	Department of Defense
DOE	Department of Energy
DST	Defence and Space Talks
EC	European Community
ECOWAS	Economic Community of West African States
ECU	European Currency Unit
EFA	European Fighter Aircraft
ELINT	Electronic intelligence
ELV	Expendable launch vehicle

EMP	Electromagnetic pulse
Enmod	Environmental modification
ERW	Enhanced radiation (neutron) weapon
EUCLID	European Cooperative Long-term Initiative on Defence
FBS	Forward-based system
FEL	Free electron laser
FOC	Full operational capability
FOFA	Follow-on forces attack
FOST	Force Océanique Stratégique
FOTL	Follow-on to Lance
FROD	Functionally related observable difference
FROG	Free-rocket-over-ground
FY	Fiscal year
GAO	General Accounting Office
GBR	Ground-based radar
GCC	Gulf Co-operation Council
GDP	Gross domestic product
GLCM	Ground-launched cruise missile
GNP	Gross national product
HACV	Heavy armoured combat vehicle
HLTF	High Level Task Force
IAEA	International Atomic Energy Agency
ICBM	Intercontinental ballistic missile
ICJ	International Court of Justice
ICO	Islamic Conference Organization
IEPG	Independent European Programme Group
IFV	Infantry fighting vehicle
INF	Intermediate-range nuclear forces
IOC	Initial operational capability
IRBM	Intermediate-range ballistic missile
JCG	Joint Consultative Group
JSG	Joint Strategy Group
JVE	Joint verification experiment
KEW	Kinetic-energy weapon
KTO	Kuwait Theatre of Operations
Laser	Light amplification by simulated emission of radiation
LDDI	Less developed defence industry
MAD	Mutual assured destruction
MARV	Manœuvrable re-entry vehicle
M(B)FR	Mutual (and Balanced) Force Reduction (Talks)
MD	Military District
MIRACL	Mid-Infrared Advanced Chemical Laser
MIRV	Multiple independently targetable re-entry vehicle
MLRS	Multiple launcher rocket system
MOD	Ministry of Defence
MOU	Memorandum of Understanding
MRD	Motor rifle division
MRV	Multiple re-entry vehicle
MSC	Military Staff Committee
MSOW	Modular stand-off weapon
MTCR	Missile Technology Control Regime
MTM	Multinational technical means (of verification)
NATO	North Atlantic Treaty Organization
NMP	Net material product
NNA	Neutral and non-aligned (states)
NPG	Nuclear Planning Group

NPT	Non-Proliferation Treaty
NRRC	Nuclear Risk Reduction Centre
NST	Nuclear and Space Talks
NSWTO	Non-Soviet WTO
NTI	National trial inspection
NTM	National technical means (of verification)
NTS	Nevada test site
NWFZ	Nuclear weapon-free zone
ODA	Overseas development assistance
OECD	Organization for Economic Co-operation and Development
OMG	Operational Manœuvre Group
O&M	Operation and maintenance
OOV	Object of verification
OPANAL	Agency for the Prohibition of Nuclear Weapons in Latin America
OSI	On-site inspection
OSIA	On-Site Inspection Agency
PLO	Palestine Liberation Organization
PNE(T)	Peaceful Nuclear Explosions (Treaty)
POMCUS	Prepositioned Organizational Material Configured to Unit Sets
PSDN	Packet Switched Data Network
PTB(T)	Partial Test Ban (Treaty)
R&D	Research and development
RMA	Restricted Military Area
RPV	Remotely piloted vehicle
RV	Re-entry vehicle
SACEUR	Supreme Allied Commander, Europe
SALT	Strategic Arms Limitation Talks
SAM	Surface-to-air missile
SCC	Standing Consultative Commission
SDI	Strategic Defense Initiative
SDIO	SDI Organization
SICBM	Small ICBM
SLBM	Submarine-launched ballistic missile
SLCM	Sea-launched cruise missile
SLV	Space launch vehicle
SNF	Short-range nuclear forces
SSD	Special Session on Disarmament (UN)
SS(M)	Surface-to-surface (missile)
SRAM	Short-range attack missile
SRBM	Short-range ballistic missile
SSBN	Nuclear-powered, ballistic-missile submarine
SSGN	Guided-missile submarine, nuclear-powered
SSN	Nuclear-powered attack submarine
START	Strategic Arms Reduction Talks
SVC	Special Verification Commission
SWS	Strategic weapon system
TASM	Tactical air-to-surface missile
TEL	Transporter–erector–launcher
TLE	Treaty-limited equipment
TNF	Theatre nuclear forces
TTB(T)	Threshold Test Ban (Treaty)
V/STOL	Vertical/short take-off and landing
WEU	Western European Union
WHO	World Health Organization
WTO	Warsaw Treaty Organization (Warsaw Pact)

Glossary

Anti-ballistic missile (ABM) system	Weapon system for intercepting and destroying ballistic missiles and their warheads in flight.
Anti-Ballistic Missile (ABM) Treaty	Treaty signed by the Soviet Union and the United States in 1972 in the SALT I process which prohibits the development, testing and deployment of sea-, air-, space- or mobile land-based ABM systems.
ATTU zone	The Atlantic-to-the-Urals zone of the 1990 Treaty on Conventional Armed Forces in Europe (CFE). Ceilings on NATO and WTO treaty-limited equipment holdings in the ATTU zone are set in the Treaty. *See also:* Conventional Armed Forces in Europe (CFE) Negotiation, Treaty-limited equipment (TLE).
Ballistic missile	A missile which follows a ballistic trajectory (part of which may be outside the earth's atmosphere) when thrust is terminated.
Binary chemical weapon	A shell or other device filled with two chemicals of relatively low toxicity which mix and react while the device is being delivered to the target, the reaction product being a supertoxic chemical warfare agent, such as nerve gas.
Biological weapon (BW)	Living organisms, whatever their nature, or infective material derived from them, which are intended for use in warfare to cause disease or death in man, animals or plants, and which for their effect depend on their ability to multiply in the person, animal or plant attacked, as well as the means of their delivery.
Charter of Paris for a new Europe	*See:* Paris Documents.
Chemical weapon (CW)	Chemical substances—whether gaseous, liquid or solid—which might be employed as weapons in combat because of their direct toxic effects on man, animals or plants, and the means of their delivery.
Circular error probable (CEP)	A measure of missile accuracy: the radius of a circle, centred on the target, within which 50 per cent of the weapons aimed at the target are expected to fall.
Conference on Disarmament (CD)	Multilateral arms control negotiating body, based in Geneva, which is composed of 40 states, including all the nuclear weapon powers. The CD reports to the UN General Assembly.
Conference on Confidence- and Security-Building Measures and Disarmament in Europe	The Stockholm Conference, part of the CSCE process, was held in 1984–86. The Stockholm Document was signed on 19 September 1986. *See also*: Confidence- and Security-Building Measures (CSBM) Negotiations.

Conference on Security and Co-operation in Europe (CSCE)	Conference of all the European states except Albania plus the USA and Canada, which began in 1973 and in 1975 adopted a Final Act (also called the Helsinki Declaration), containing, among others, a Document on confidence-building measures and disarmament. Follow-up meetings were held in Belgrade (1977–78), Madrid (1980–83) and Vienna (1986–89). The Concluding Document of the Vienna Meeting of the Conference on Security and Co-operation in Europe, adopted in Vienna in January 1989, established the CFE and the CSBM Negotiations. On 19–21 November 1990, a summit meeting of heads of state and government of the CSCE was held in Paris. *See also:* Conventional Armed Forces in Europe (CFE) Negotiation, Confidence- and Security-Building Measures (CSBM) Negotiations, Paris Documents.
Confidence- and Security-Building Measures (CSBM) Negotiations	The CSBM Negotiations, part of the CSCE process and with the participation of all the CSCE states, were held in Vienna from March 1989 to November 1990 and built upon the results of the Stockholm Conference. The Vienna Document 1990 was included in the set of Paris Documents. The Negotiations were rejoined in Vienna on 28 November 1990. *See also:* Vienna Document 1990.
Conventional Armed Forces in Europe (CFE) Negotiation	Negotiation between the member states of NATO and the WTO on conventional force reductions in Europe, held in Vienna from March 1989 to November 1990. Part of the CSCE process. The CFE Treaty, which sets ceilings on treaty-limited equipment in the ATTU zone, was signed in Paris on 19 November 1990. The follow-up to the CFE Negotiation, CFE IA, formally opened on 26 November 1990, with the same mandate and same states, and is to be completed before the 1992 CSCE follow-up meeting in Helsinki. The objective of CFE IA is to limit the personnel strength of conventional armed forces in the ATTU zone. *See also:* ATTU zone, Paris Documents, Treaty-limited equipment (TLE).
Conventional weapon	Weapon not having mass destruction effects. *See also:* Weapon of mass destruction.
Cruise missile	Unmanned, self-propelled, guided weapon-delivery vehicle which sustains flight through aerodynamic lift, generally flying at very low altitudes to avoid radar detection, sometimes following the contours of the terrain. It can be air-, ground- or sea-launched and deliver a conventional, nuclear, chemical or biological warhead.
Defence and Space Talks	Talks between the USA and the USSR, conducted since 1985 parallel to START under the Geneva Nuclear and Space Talks (NST), on ballistic missile defences and on means of preventing an arms race in space. *See also:* Nuclear and Space Talks.

European Community (EC)	The EC was created in 1951–57 by six governments—Belgium, France, the Federal Republic of Germany, Italy, the Netherlands and Luxembourg—based on the 1951 Treaty of Paris, which established the European Coal and Steel Community (ECSC), and on the 1957 Treaties of Rome, which established the European Economic Community (EEC) and the European Atomic Energy Community (Euratom). The three Communities, intended to promote economic expansion, growth of employment and a rising standard of living in the member states through establishment of a common market and common economic policies, are now regarded as a single entity. The 12 EC members in 1990 also included Denmark, Ireland, Greece, Portugal, Spain and the UK.
First-strike capability	Theoretical capability to launch a single attack on an adversary's strategic nuclear forces that nearly eliminates the second-strike capability of the adversary.
Flexible response	The NATO doctrine for reaction to an attack with a full range of military options, including the use of nuclear weapons.
Helsinki Declaration	*See*: Conference on Security and Co-operation in Europe.
Initial operational capability (IOC)	Date by which a weapon system is first operationally deployed, ready for use in the field.
Intercontinental ballistic missile (ICBM)	Ground-launched ballistic missile with a range in excess of 5500 km.
Intermediate-range nuclear forces (INF)	Theatre nuclear forces with a range of from 1000 up to and including 5500 km. *See also*: Theatre nuclear forces.
International Atomic Energy Agency (IAEA)	With headquarters in Vienna, the IAEA is endowed by its Statute, which entered into force in 1957, with the twin purposes of promoting the peaceful uses of atomic energy and ensuring that nuclear activities are not used to further any military purpose.
Kiloton (kt)	Measure of the explosive yield of a nuclear weapon equivalent to 1000 tons of trinitrotoluene (TNT) high explosive. (The bomb detonated at Hiroshima in World War II had a yield of about 12–15 kilotons.)
Launcher	Equipment which launches a missile. ICBM launchers are land-based launchers which can be either fixed or mobile. SLBM launchers are missile tubes on submarines.
Launch-weight	Weight of a fully loaded ballistic missile at the time of launch.

Megaton (Mt)	Measure of the explosive yield of a nuclear weapon equivalent to 1 million tons of trinitrotoluene (TNT) high explosive.
Multiple independently targetable re-entry vehicle (MIRV)	Re-entry vehicle, carried by a nuclear missile, which can be directed to separate targets along separate trajectories (as distinct from MRVs). A missile can carry one or several RVs. *See also:* Re-entry vehicle (RV).
Multiple re-entry vehicle (MRV)	Re-entry vehicle, carried by a nuclear missile, directed to the same target as the missile's other RVs. *See also:* Re-entry vehicle (RV).
Mutual assured destruction (MAD)	Concept of reciprocal deterrence which rests on the ability of the nuclear weapon powers to inflect intolerable damage on one another after receiving a nuclear attack. *See also:* Second-strike capability.
National technical means of verification (NTM)	The means used to monitor compliance with treaty provisions which are under the national control of individual signatories to an arms control agreement.
Neutral and non-aligned (NNA) states	The group of 12 European states (Austria, Cyprus, Finland, Holy See [Vatican City], Ireland, Liechtenstein, Malta, Monaco, San Marino, Sweden, Switzerland and Yugoslavia) which work together in the CSCE.
North Atlantic Treaty Organization (NATO)	Established in 1949 by a treaty between 12 states: Belgium, Canada, Denmark, France, Iceland, Italy, Luxembourg, the Netherlands, Norway, Portugal, the UK and the USA. The 16 member states in 1990 also included the Federal Republic of Germany, Greece, Iceland, Spain and Turkey. (France and Greece are not in the military structures of NATO.)
Nuclear and Space Talks (NST)	Negotiations between the USA and the USSR on strategic nuclear weapons (START) and space weapons and defence issues (the Defence and Space Talks), held in Geneva since March 1985. The INF negotiations of 1985–87 were also included in the NST. *See also:* Nuclear and Space Talks.
Nuclear Risk Reduction Centres (NRRC)	Established by the 1987 US–Soviet NRRC Agreement. The two centres, which opened in Washington and Moscow in 1988, exchange information by direct satellite link in order to minimize misunderstandings which might carry a risk of nuclear war. Notifications concerning exchange of information about nuclear explosions under the 1974 Threshold Test Ban Treaty, the 1976 Peaceful Nuclear Explosions Treaty and the 1990 protocols to the two treaties shall also be submitted through the US and Soviet NRRCs.

Organization for Economic Co-operation and Development (OECD)	Established in 1961 to replace the Organization for European Economic Co-operation (OEEC). With the accession of Canada and the USA, it ceased to be a purely European body. OECD objectives are to promote economic and social welfare by co-ordinating policies. The 24 members in 1990 were Austria, Australia, Belgium, Canada, Denmark, Finland, France, the Federal Republic of Germany, Greece, Iceland, Ireland, Italy, Japan, Luxembourg, the Netherlands, New Zealand, Norway, Portugal, Spain, Sweden, Switzerland, Turkey, the UK and the USA. Yugoslavia participates with a special status.
Paris Documents	A set of five documents adopted at the November 1990 Paris CSCE summit meeting. They include the CFE Treaty, the Joint Declaration of Twenty-Two States, the Charter of Paris for a new Europe, the Supplementary Document to give new effect to certain provisions contained in the Charter, and the Vienna Document 1990. Several new CSCE institutions were set up in the Paris Documents. *See also:* Conventional Armed Forces in Europe (CFE) Negotiation, Paris Documents, Vienna Document 1990.
Peaceful nuclear explosion (PNE)	Application of a nuclear explosion for non-military purposes such as digging canals or harbours or creating underground cavities.
Re-entry vehicle (RV)	That part of a ballistic missile which carries a nuclear warhead and penetration aids to the target and re-enters the earth's atmosphere and is destroyed in the terminal phase of the missile's trajectory. A missile can have one or several RVs; each RV contains a warhead.
Second-strike capability	Ability to receive a nuclear attack and launch a retaliatory blow large enough to inflict intolerable damage on the opponent. *See also:* Mutual assured destruction.
Short-range nuclear forces (SNF)	Nuclear weapons with ranges up to 500 km; not limited by the INF Treaty. *See also*: Theatre nuclear forces.
Special Verification Commission (SVC)	US–Soviet consultative body established in accordance with the 1987 INF Treaty, to promote the objectives and implementation of the Treaty.
Standing Consultative Commission (SCC)	US–Soviet consultative body established in accordance with the SALT agreements, to promote the objectives and implementation of the agreements.
Stockholm Conference	*See*: Conference on Confidence- and Security-Building Measures and Disarmament in Europe.

Strategic Arms Limitation Talks (SALT) — Negotiations between the Soviet Union and the United States which opened in 1969 and sought to limit the strategic nuclear forces, both offensive and defensive, of both sides. The SALT I Interim Agreement and the ABM Treaty were signed in 1972. The negotiations were terminated in 1979, when the SALT II Treaty was signed. *See also:* Strategic Arms Reduction Talks (START).

Strategic Arms Reduction Talks (START) — Negotiations between the Soviet Union and the United States, initiated in 1982, which seek to reduce the strategic nuclear forces of both sides. Suspended in December 1983 but resumed under the Nuclear and Space Talks that opened in Geneva in March 1985. *See also*: Nuclear and Space Talks.

Strategic Defense Initiative (SDI) — The programme announced by President Reagan in his 1983 'Star Wars' speech for research and development of systems capable of intercepting and destroying nuclear weapons in flight and rendering the USA safe from the threat of a nuclear strike by another state.

Strategic nuclear weapons — ICBMs, SLBMs and bomber aircraft carrying nuclear weapons of intercontinental range (over 5500 km) which allows them to reach the territories of the other strategic nuclear weapon powers.

Terminal guidance — Guidance provided in the final, near-target phase of the flight of a missile.

Theatre nuclear forces (TNF) — Nuclear weapons with ranges of up to and including 5500 km. In the 1987 INF Treaty, nuclear missiles are divided into intermediate-range (over 1000 km) and shorter-range (500–1000 km). Also called non-strategic nuclear forces. Nuclear weapons with ranges up to 500 km are called short-range nuclear forces. Those with ranges of 150–200 km are often called battlefield nuclear forces. *See also*: Short-range nuclear forces.

Throw-weight — The sum of the weight of a ballistic missile's re-entry vehicle(s), dispensing mechanisms, penetration aids, and targeting and separation devices.

Treaty-limited equipment (TLE) — The five categories of NATO and WTO equipment on which numerical limits are established in the 1990 CFE Treaty: battle tanks, armoured personnel carriers, artillery, combat aircraft and attack helicopters. *See also:* Conventional Armed Forces in Europe (CFE) Negotiation, ATTU zone.

Toxins — Poisonous substances which are products of organisms but are inanimate and incapable of reproducing themselves as well as chemically induced variants of such substances. Some toxins may also be produced by chemical synthesis.

Vienna Document 1990	The Vienna Document 1990 on new CSBMs, adopted in Vienna and included in the set of Paris Documents, repeats many of the provisions in the 1986 Stockholm Document and expands several others. It established a communications network and the Conflict Prevention Centre. *See also:* Confidence- and Security-Building Measures (CSBM) Negotiations, Paris Documents.
Warhead	That part of a weapon which contains the explosive or other material intended to inflict damage.
Warsaw Treaty Organization (WTO)	The WTO, or Warsaw Pact, was established in 1955 by a treaty of friendship and collaboration between eight countries: Albania, Bulgaria, Czechoslovakia, the German Democratic Republic, Hungary, Poland, Romania and the USSR. Albania ceased to participate in 1961 and formally withdrew from the Treaty in 1968. There were six WTO member states after the October 1990 unification of Germany. On 31 March 1991 the military organs and structures of the WTO were dismantled.
Weapon of mass destruction	Nuclear weapon and any other weapon which may produce comparable effects, such as chemical and biological weapons.
Western European Union (WEU)	Set up in 1948 by the 50-year Treaty of Brussels on Economic, Social and Cultural Co-operation and on Collective Defence, and established in London in 1955. In 1950 the WEU defence organization functions were transferred to the NATO command. Four protocols to the Brussels Treaty were signed in 1954 and came into effect in 1955. In one protocol, the German Federal Republic and Italy were accepted under special conditions, which were annulled in 1980. The nine members in 1990 were Belgium, France, Germany, Italy, Luxembourg, the Netherlands, Portugal, Spain and the UK.
Yield	Released nuclear explosive energy expressed as the equivalent of the energy produced by a given number of tons of trinitrotoluene (TNT) high explosive. *See also:* Kiloton and Megaton.

Conventions

. .	Data not available or not applicable
—	Nil or a negligible figure
( )	Uncertain data
m.	million
b.	billion (thousand million)
$	US $, unless otherwise indicated

Introduction: From the known past to the unknown future

WALTHER STÜTZLE

In 1990 the dynamic of events was greater than in any other year in post-war history. It was a year of revolution and hope, and of triumph and disappointment. 1990 was also the first year of the post-cold war era.

In the context of the *SIPRI Yearbook*, three developments stand out: the formal end of the cold war, marked by the overcoming of the division of Europe and the unification of Germany; developments in arms control and arms reductions; and the Persian Gulf conflict.

I. The end of the cold war: Germany and Europe in transition

The unification of Germany and the signing of the Charter of Paris for a new Europe and the Joint Declaration of the Twenty-Two States of the North Atlantic Treaty Organization (NATO) and the Warsaw Treaty Organization (WTO) are the most remarkable events.

On 28 November 1989 Federal German Chancellor Helmut Kohl stated in the German Bundestag: 'Nobody knows at the present time what a united Germany will look like. I am, however, sure that unity will come, if it is wanted by the German people'.[1] Less than 10 months later, the terms of unification were negotiated and agreed: on 31 August 1990 the Federal Republic of Germany and the German Democratic Republic signed the Treaty on the Establishment of the Unity of Germany.[2] On 12 September 1990 the Four Powers—France, the Soviet Union, the United Kingdom and the United States—together with the FRG and the GDR signed the Treaty on the Final Settlement with respect to Germany.[3] On 3 October 1990 German unification became effective. As a result of this process, the community of states witnessed a number of 'firsts' in European history, all of them relevant as developments proceed from creating unification to actually absorbing the unified state into the European system.

[1] See 'Rede von Bundeskanzler Helmut Kohl im Deutschen Bundestag, Bonn, 28 November 1989', *Bulletin der Bundesregierung*, no. 134 (29 Nov. 1990); see the English text in Adam Daniel Rotfeld and Walther Stützle (eds), SIPRI, *Germany and Europe in Transition* (Oxford University Press: Oxford, 1991), pp. 120–23.

[2] Rotfeld and Stützle (note 1), pp. 182–83.

[3] For the full text of the Treaty, see appendix 17C in this volume. With the ratification also by the Soviet Union on 4 Mar. 1991 and the deposit of the instrument of ratification on 15 Mar. 1991, the Treaty became effective and Germany regained full sovereignty.

1. For the first time, German unity resulted from a process of peace, not from acts of war.

2. The united Germany of 1990 is a completely new German state—one which never before existed in European history, neither within these boundaries nor with the proven democratic record of its larger part, the former FRG.

3. As the new German state acquired full sovereignty, it explicitly did so with a firm commitment to exercise its economic and military power only through the integrative systems of the European Community and the Atlantic Alliance.

4. While the country grew larger, its armed forces were diminished by 40 per cent. Prior to, although linked to, forthcoming limitations for other CSCE (Conference on Security and Co-operation in Europe) countries, Germany accepted a unilateral troop ceiling of 370 000[4] and accepted on its territory a special military zone, which is identical to the territory of the former GDR and which is free from 'foreign armed forces and nuclear weapons or their carriers'.[5] Thus, in the context of unification and in marked difference to what used to be the traditional policy, both NATO in general and the FRG in particular accepted the concept of establishing 'special zones'.

As 1990 drew to a close and 1991 began to unfold, it also of course became very clear that unification presented the German Government and people with a very complex agenda, at the centre of which is the promise to bring the standard of living in the former GDR up to the West German level. Soon it became obvious that the promise of the Bonn Government not to increase taxes had no solid basis and had to be dropped. As the perception grew that the gap between promises and economic reality was widening, severe social unrest in the former GDR became a real prospect.

Only seven weeks after German unification, heads of CSCE states and governments signed the Charter of Paris for a new Europe.[6] More than any other document, this Charter reflects the end of the post-war period. In a number of ways some very important 'firsts' in European history were established in the Charter of Paris, thus rendering Paris 1990 distinctly different from events of a similar dimension, for example, the Peace of Westphalia of 1648 that brought

[4] See 'Declaration on the German contribution to the reduction in conventional armed forces in Europe, Statement by Hans-Dietrich Genscher, Minister for Foreign Affairs of the Federal Republic of Germany, Vienna, 30 August 1990', in Rotfeld and Stützle (note 1), p. 126. The troop-ceiling commitment is incorporated in the Treaty on the final settlement with respect to Germany, Art. 3; see appendix 17C in this volume (pp. 611–13): 'The Government of the Federal Republic of Germany undertakes to reduce the personnel strength of the armed forces of the united Germany to 370,000 (ground, air and naval forces) within three to four years. This reduction will commence on the entry into force of the first CFE agreement [Vienna Negotiation on Conventional Armed Forces in Europe]. Within the scope of this overall ceiling no more than 345,000 will belong to ground and air forces . . . The Federal Government assumes that in follow-on negotiations the other participants in the negotiations, too, will render their contribution to enhancing security and stability in Europe, including measures to limit personnel strength'.

[5] Article 5, para. 3, of the Treaty on the final settlement with respect to Germany; see appendix 17C in this volume, pp. 611–13.

[6] For the Charter of Paris for a new Europe, see appendix 17B in this volume, pp. 603–10.

the Thirty Years' War to an end or the Paris Peace Conference of 1918–19 that settled World War I.

1. Paris 1990, unlike Paris 1918–19, was not a conference at which the winners imposed the rules on the losers; Paris 1990 brought together sovereign states with equal rights and established equal responsibilities for the future security of Europe.

2. With the Charter of Paris, all European states (except Albania) and the two North American states with immediate responsibility in Europe, the United States of America and Canada, have stipulated the major elements of a peace-keeping structure without entering into a peace treaty.

3. The Charter of Paris testifies to the fact that the 'national unity of Germany' is a product of peace and 'an important contribution to a just and lasting order of peace for a united, democratic Europe aware of its responsibility for stability, peace and co-operation'.[7]

4. Paris 1990 is not the diplomatic finish of hectic immediate post-war diplomacy but rather the result of a long and patient East–West process, full of set-backs.

5. For the first time, all participants established common, pan-European security structures and institutions.

6. For the first time in European history, agreement was reached among all participants on the major principles which should govern future European affairs.

The first set of principles applies to the polity of individual states. This set of principles requires states to adhere to human rights, democracy, social justice and economic welfare, to political pluralism and, most fundamental, to the rule of law. The Charter of Paris recognizes the individual human being as the centre of all political endeavour. Thus it puts an end to the totalitarian communist philosophy of substituting the power of both one party and the state for the individual human being, and hence considerably broadens the Helsinki Final Act of 1975.

The second set of principles has to do with relationships among states. European states have pledged no longer to seek security *against* each other but rather *in co-operation with* each other. Both the Charter of Paris and the Joint Declaration of Twenty-Two States express the intention to substitute co-operative security for security through deterrence. This is no small step in Europe's war-ridden history. It is in recognition of this crucial principle that two major documents—the Joint Declaration of Twenty-Two States[8] and the Treaty on Conventional Armed Forces in Europe (CFE)[9]—were signed by heads of state or government.

While the Joint Declaration formally declares an end to the adversarial relationship between NATO and the WTO and recognizes that 'security is indivis-

[7] See appendix 17B in this volume, pp. 603–10.
[8] For the full text, see appendix 17A in this volume, pp. 601–602.
[9] For the text of the CFE Treaty, see appendix 13A in this volume, pp. 461–74.

ible', the CFE Treaty applies this principle to the cold war military set-up and provides for the reduction of forces to a considerably lower level.[10] And it is in consequence of this new philosophy of co-operative security that a number of measures were agreed in Paris to provide for transparency in European defence and to prohibit distrust from producing rules of secrecy. The arms reductions and transparency rules attained at Paris reflect the effort of European and North American states to take concrete steps to operationalize, for the first time in European history, the non-use-of-force principle. Only someone who is not conscious of the many severe dangers that Europe has experienced since 1945 could underestimate the value of this achievement.[11]

The third major principle established in the Charter of Paris is simply this: the notion of 'Europe' no longer stands for Western Europe alone and for its effort to turn the European Community into a Political Union. Europe actually means Europe. The Charter of Paris formally constitutes a responsibility of all CSCE states for each other. Paris 1990 made the European train change from West European narrow-gauge tracks to all-European wide-gauge tracks.

This principle will prove to be the most difficult one for political leaders to work with in the future. For Western industrialized countries, the Charter of Paris constitutes the duty to accept responsibility for helping to overcome the economic backwardness of East European neighbours and of the Soviet Union. With it goes the duty to explain to the respective Western domestic audiences that new sacrifices need to be accepted, for example, higher taxation in order to generate the financial means for East European countries. Political leaders in the East need to explain to their people that economic reform in general and the transition to a market economy in particular require of the individual no less than to change course by 180 degrees: trained to wait for state directives, it is now the readiness of the individual to take the initiative on which success will ultimately depend.

As developments further unfold, however, it needs to be remembered that it is mostly principles that were established in Paris and only the first steps taken on some long, distant road. The European agenda ahead is still a rather tall one: it is a very long way from a commitment to democracy to creating a working democratic system. This is particularly true for countries that lack democratic experience and institutions and are now confronted with the dual task of carrying out political change and economic change simultaneously. From promising social justice and economic welfare to actually establishing it will take years rather than months; and rising nationalism will render the process of exercising the right of self-determination more complicated, not less. To go from an arms surplus to truly deep arms reductions, based on new and purely defensive strategies, doctrines and training handbooks, takes more than just a year or two. First and foremost it requires changed mind-sets—ready to understand the new situation and the new political context—and the ability to

[10] For an evaluation of the CFE Treaty, see Jane M. O. Sharp, 'Conventional arms control in Europe', chapter 13 in this volume, pp. 407–60.

[11] For a more detailed evaluation of the CSCE process, see Adam Daniel Rotfeld, 'New security structures in Europe: concepts, proposals and decisions', chapter 17 in this volume, pp. 585–600.

turn all this into new political guidelines for security and defence. In the London Declaration on a transformed North Atlantic Alliance, issued in July 1990, the NATO member countries have promised to do so.[12] In June 1990 member states of the Warsaw Treaty Organization announced their intention to transform the WTO 'into a treaty of sovereign states with equal rights, formed on a democratic basis'.[13] On 25 February 1991 the WTO states signed an agreement in Budapest to dissolve the military structure of the WTO, effective on 31 March 1991.[14]

Concerning the future of Europe, a great number of questions remain to be answered. The success or failure of the Soviet reform effort is obviously of overriding importance. To engineer a peaceful transformation of the Soviet Union—from a centrally governed, one-party state with a basically defunct economic system to a democratically organized Union of Sovereign Republics, run on the principles of a market economy—amounts to one of the most difficult political tasks faced by a political leadership in Europe in this century. Not only will it require every skill and statesmanship to convince the people of the Soviet Union of the worthiness of this objective and to make them understand that there is no better alternative. Political leadership in Moscow is simultaneously required to constantly convince the neighbours of the Soviet Union that this process of domestic reconstruction can be managed without posing threats to the security of other states. Developments in the Baltic republics in general and the use of Soviet military force against the will of the Lithuanian Government in particular have shown how delicate the balance is. And despite all its shortcomings, the Soviet Union is still a formidable military power with a large arsenal of nuclear weapons. Lest the European states fall victim to misperceptions, it is crucial to remember that the Soviet Union, for the first time in all its history, is trying to adopt a democratic system.

With the military structure of the WTO dissolved, all the members of the Atlantic Alliance need to adjust their work on a new political and military strategy. Since the 22 NATO and WTO countries agreed on 19 November 1990 in Paris 'that security is indivisible and that the security of each of their countries is inextricably linked to the security of all States participating in the Conference on Security and Co-operation in Europe',[15] time may be ripe for governments to think about transforming NATO into a new European–American Treaty Organization (EATO),[16] open to every European country as principally provided for in the NATO Treaty of 1949, hence also the Soviet Union.[17] Although the difficulties of constructing such a new entity must not

[12] For the full text, see Rotfeld and Stützle (note 1), pp. 150–52.

[13] See the Declaration of the Consultative Political Committee of the WTO states, Moscow, 7 June 1990, in Rotfeld and Stützle (note 1), pp. 152–55.

[14] See *International Herald Tribune*, 27 Feb. 1991, p.1 (excerpts).

[15] The Joint Declaration of Twenty-Two States, appendix 17A in this volume, pp. 601–602.

[16] The EATO idea was first presented and developed in Walther Stützle, 'West und Ost in einem Bündnis. Plädoyer für eine europäisch-amerikanische Allianz unter Einschluß der Sowjets', *Die Zeit*, no. 22 (25 May 1990), p. 4.

[17] Article 10 of the NATO Treaty stipulates: 'The Parties may, by unanimous agreement, invite any other European State in a position to further the principles of this Treaty and to contribute to the security

be underestimated, at least four immediate advantages of political–strategic importance should make the idea attractive:

1. NATO has always stressed its character as a West European–American alliance based on common values. With the signing of the Charter of Paris, the alliance reaches from Vladivostok to San Francisco. EATO would palpably reflect this new situation and thus commit the Soviet Union very directly to its responsibility for the security of all Europe, much like that of the United States.

2. Subjects such as arms reductions, progress in arms control, defence planning and planning for CSCE peace-keeping operations in Europe, should the need arise, in any case require an all-European approach if they are to be successfully dealt with in the spirit of the Charter of Paris.

3. Soviet membership in EATO should render meaningless Soviet concerns that some Soviet republics or some former WTO allies might not only seek independence from Moscow but might also want to join a military alliance against the Soviet Union. In the context of transforming the USSR into a Union of Sovereign Republics, the future of the Baltic republics is the obvious case in point. In light of the dissolution of the WTO, the search of Czechoslovakia, Poland and Hungary for a new security framework, including military security, illustrates the problem. The point here is that an EATO-like structure could prove to be a helpful instrument to achieve both: protection of newly acquired independence without generating a new Soviet security trauma.

4. The future of the US military presence in Europe is going to rank high on the agenda of the Atlantic Alliance. This very issue and the need for credible arrangements—reduced presence in Europe, assured reinforcement capability and redeployment questions[18]—are of as much interest to the security of the USSR and its former WTO allies as to the security of all other CSCE countries.

In summary, at the end of 1990 Europe was a very different place. Remarkable alterations were made. Germany was unified, NATO was in the process of transforming, the WTO was in the process of disappearing, the CSCE process had acquired a substantially new quality, and difficulties associated with the transformation of the Soviet Union had become very apparent.

of the North Atlantic area to accede to this Treaty. Any State so invited may become a party to the Treaty by depositing its instrument of accession with the Government of the United States of America. The Government of the United States of America will inform each of the Parties of the deposit of each such instrument of accession.'

[18] See Jane M. O. Sharp (ed.), SIPRI, *Europe after an American Withdrawal: Economic and Military Issues* (Oxford University Press: Oxford, 1990). See also the more recent report: *The United States & NATO in an Undivided Europe: A Report by the Working Group on Changing Roles and Shifting Burdens in the Alliance* (Johns Hopkins Foreign Policy Institute: Washington, DC, 1991). The group, comprised of a great number of influential and experienced people such as former SACEUR General Andrew Goodpaster, former US Defense Secretary Harold Brown as well as Senator Sam Nunn, recommends *inter alia* that US forces in Europe should be reduced to less than 100 000. The group also observes that 'the United States serves vital functions for European security, reassuring allies and former adversaries alike'; p. 2.

II. Arms control and reduction of armaments

Compared with previous years, arms control negotiations were remarkably successful in 1990, although not entirely free from disappointment.

In the long history of arms reduction and control efforts, the CFE Treaty of 19 November 1990 represents the first major international agreement on the reduction of conventional armaments. The Treaty covers battle tanks, armoured combat vehicles, artillery, combat aircraft and attack helicopters, deployed by the 22 NATO and WTO countries on the land territory between the Atlantic Ocean and the Ural River/Caspian Sea (the Atlantic-to-the-Urals, or ATTU, zone). 'According to data released at its signature, the Treaty requires 44 829 TLE [Treaty Limited Equipment] items to be removed from the ATTU zone' and '27 000 and 7 000 items to be destroyed for the WTO and NATO, respectively'.[19] The fact that the CFE Treaty was agreed by the hitherto and still most heavily armed alliances of our times, NATO and the WTO, comprising the vast majority of the world's strongest industrial powers, makes the Treaty a unique step forward. Together with the CSBM (Confidence- and Security-Building Measures) Document signed in Vienna,[20] the 22 states succeeded in providing for reduction and transparency and thus in putting some military substance to the political commitment of letting military reality reflect the obligations under the renunciation-of-force principle.

However, since the residual forces are still on a very high level, much work remains to be done in the context of further CFE negotiations as well as by way of unilateral measures. This includes the completion of the withdrawal of Soviet forces from both the Czech and Slovak Federal Republic and Hungary by mid-1991, the conclusion of an agreement with Poland about complete withdrawal from Poland and complete withdrawal from Germany before 1994.

Concerning chemical weapons the United States, ahead of the promised schedule, withdrew all its stockpile from the Federal Republic: *circa* 7000 ammunition tons of nerve agent (sarin and VX) were removed to the US facility at Johnston Atoll, in the Pacific, in order to be destroyed there.

On 1 June 1990 the United States and the Soviet Union concluded an agreement on destruction and non-production of chemical weapons and on measures to facilitate the multilateral convention on banning chemical weapons.[21] Both countries pledge to begin the destruction no later 'than by 31 December 1992' (Article VI) and to 'reduce and limit [their] chemical weapons' so that 'its aggregate quantity' does not 'exceed 5,000 agent tons' by 'no later than 31 December 2002' (Article VI). The agreement also provides for a substantial exchange of data and verification.[22]

[19] See Sharp (note 10).
[20] See the Vienna Document 1990 in appendix 13B in this volume, pp. 475–88.
[21] See the full text of the Agreement in appendix 14A in this volume, pp. 536–39.
[22] For an evaluation of CBW, see J. S. Lundin and Thomas Stock, 'Chemical and biological warfare: developments in 1990', chapter 4 in this volume, pp. 85–112.

One of the disappointments of 1990 was that, again, no Soviet–US treaty on the reduction of strategic nuclear armaments (START) was achieved. It is difficult to understand why it was not completed. The tremendous political agenda that political leaders had to deal with as much as actual START issues may account for this failure. The list of unresolved issues has been narrowed down substantially as a consequence of intensive involvement on the leadership level, and the basic structure of the treaty is well established. Neither side will have more than 6000 accountable warheads on no more than 1600 accountable nuclear delivery vehicles.[23]

The Defence and Space Talks are also continuing. Disagreement centres around different positions *vis-à-vis* the 1972 Anti-Ballistic Missile (ABM) Treaty: while the Soviet Union insists on the traditional narrow interpretation of the Treaty, the United States wishes to tailor the ABM interpretation to the goal of incorporating strategic defences into security strategies. The link between START and strategic defence is the single most important issue to the future of the US–Soviet nuclear relationship.

Some important progress is to be reported with regard to the control of the proliferation of missile technology.[24] When first established in 1987, the Missile Technology Control Regime (MTCR) had only seven members: Canada, France, FR Germany, Italy, Japan, the UK and the USA. Spain joined in 1989. In 1990 Australia became a member. Belgium, Luxembourg, the Netherlands and, most important, the Soviet Union, in June 1990, announced adherence to the restrictions on technology exports under the MTCR. With Denmark, Portugal, New Zealand and Norway deciding to join soon, the MTCR was further strengthened. Of course, as the Gulf War in general and the firing of Iraqi Scud missiles in particular demonstrated, the currently existing MTCR is insufficient for at least two reasons: it lacks the membership of important ballistic missile technology producers, for example, China, India and Brazil; and the regime is void of a compliance-enforcement mechanism. Still, as 1990 has borne out, 'ballistic missile proliferation is actually slowing down' and, apart from economic constraints, which account for this development, 'the most important common factor, however, appears to be the MTCR'.[25]

Although the fourth Review Conference on the Non-Proliferation Treaty, held from 20 August to 14 September 1990, failed to produce a final document, this last review conference before the 1995 extension conference gave reason for hope. For the first time France and China, neither a state party to the Treaty, decided to send observer delegations. And the USA accepted a 'clear cut link between the future of the NPT and the CTBT [comprehensive test ban treaty] and endorsed early action on the latter'.[26]

[23] See Regina Cowen Karp, 'US-Soviet nuclear arms control', chapter 11 in this volume, pp. 383–402.
[24] See Aaron Karp, 'Ballistic missile proliferation', chapter 9 in this volume, pp. 317–43.
[25] See Karp (note 24).
[26] See David Fischer and Harald Müller, 'The fourth review of the Non-Proliferation Treaty', chapter 16 in this volume, pp. 555–84.

Concerning the implementation of the US–Soviet INF Treaty, signed in December 1987, both parties were approaching the final elimination of all intermediate-range and shorter-range missiles and launchers. Both sides fulfilled the elimination target for 1990 'comfortably'. As of 1 December 1990, of a total 1846 Soviet missiles only 66 remained for elimination, while the equivalent figures for the USA are 180 out of 846 missiles.[27] It is also worth mentioning that the USA, in 1990, ceased the production of plutonium for its nuclear arsenals.[28]

In summary, the hallmark of 1990 clearly was the as yet unratified CFE Treaty, bolstered by the Vienna Document on Confidence- and Security-Building Measures and the ongoing withdrawal of Soviet troops from Eastern Europe.

III. The Persian Gulf War

On 2 August 1990 a United Nations member state, Iraq, invaded another UN member, Kuwait, and declared it its 19th province. In the midst of smooth developments in Europe and swift negotiations about German unification, the world was reminded of the fact that peace was a global issue, not a regional one. Iraq's attempt to end the existence of a sovereign UN member state represented the first challenge to the UN system after the end of the cold war and 'provided the Security Council with a major challenge which is also the first full-scale test of collective action against aggression by the United Nations through a united Security Council'.[29] Twelve UN Security Council resolutions[30] demanding the immediate and unconditional withdrawal of Iraq, imposing sanctions and, finally, authorizing the use of force after 15 January 1991 did not cause the Iraqi leadership to change its mind or course of action. Nor did the largest and speediest projection of military power into the region, led by the USA and the UK and joined by over 30 more countries,[31] lead Iraq to comply with the UN resolutions. Hence, the Gulf War began on 17 January 1991 and ended with a cease-fire on 28 February—after the complete defeat of Iraqi forces, the destruction of Kuwait, very severe damage to the economic infrastructure of Iraq, casualties among the civilian population throughout the conflict area as well as an ecological disaster which will affect at least this entire region.

The UN's overall performance in the Iraq–Kuwait conflict has indicated the potential this world organization has for dealing with conflicts as long as the

[27] See Stephen Iwan Griffiths, 'The implementation of the INF Treaty', chapter 12 in this volume, pp. 403–406.

[28] See Robert S. Norris, Richard W. Fieldhouse, Thomas B. Cochran and William M. Arkin, 'Nuclear weapons', chapter 1 in this volume, pp. 3–40.

[29] See Brian Urquhart, 'The role of the United Nations in the Iraq-Kuwait conflict in 1990', chapter 18 in this volume, pp. 617–26.

[30] For the full texts of the resolutions and the voting record, see appendix 18A in this volume, pp. 627–35.

[31] See Barry R. Posen, 'Military mobilization in the Persian Gulf conflict', chapter 19 in this volume, pp. 639–54.

five permanent members of the Security Council are no longer paralysed by East–West rivalry and thus find it possible to act jointly. A few characteristics, however, deserve special mention:

1. Sanctions were not given a specified period of time to prove their effectiveness or ineffectiveness, nor was the transition from sanctions to military action made dependent on a mandatory review procedure under UN Security Council authority and voting rules.
2. The UN Security Council authorized the peace enforcement action but did not control it since 'the very idea of a UN command under the Security Council, although traditionally accepted for peace-keeping operations, was not seriously considered for enforcement operations in the Persian Gulf'.[32]
3. To translate the UN mandate, as enshrined in UN Security Council Resolution 678, into military operations was completely left to the coalition countries, hence also the decision when and under what conditions to settle for a cease-fire and, consequently, to negotiate an armistice.

In summary, the Gulf War was conspicuous for its UN authorization as much as for the absence of UN control.

In view of the great number of conflicts throughout the world, the United Nations will have to address how to best improve on the positive side of the Gulf experience and how to remedy the deficiencies. If this task is properly discharged, hope may no longer be unfounded that the international community will move away from the role of onlooker at the destruction of countries such as Lebanon or Liberia, the oldest African democracy, which was torn to pieces in 1990.

On many accounts 1990 was a remarkable if not unique year. Only future historians, however, will be able to pass a sufficiently informed judgement, based on knowledge of the entire context. SIPRI is acutely aware of this. The *SIPRI Yearbook* annual accounts and analyses of events do not have this advantage of time and historical hindsight, but are restricted to the artificial limits of calendar years. The job ahead for SIPRI and other researchers and for historians is to continue research, based on the facts as they become known.

IV. Context and agenda: SIPRI's anniversary and the future

In May 1991 SIPRI looks back on the first 25 years of its history. It is a period in history during which international efforts have achieved a lot to improve the conditions for peace. However, developments also suggest that a lot remains to be done. A few examples should help to illustrate the point.

In June 1968, two years after SIPRI had begun its work, NATO countries launched the signal of Reykjavik, calling for negotiations on mutual (and balanced) reductions of conventional forces in Europe (the MBFR Talks). Negotiations about the mandate did not begin until 1972. Eighteen years later,

[32] Urquhart (note 29).

in November 1990, the CFE Treaty was signed. No shorter was the time needed from the first preparatory meetings for the CSCE, held in Finland in 1972, to the signature of the Charter of Paris in 1990. The agenda ahead is tall: further cuts in armaments, the introduction of manpower ceilings for all participating states, further refined verification schemes, conversion and, eventually, the problems of naval arms control are only some of the issue areas to which states will have to turn. And as the agenda becomes more complex, the negotiating format will become less handy: with the dissolution of the Warsaw Treaty Organization, effective on 31 March 1991 and as provided for in the Charter of Paris, after the Helsinki Follow-Up Meeting in 1992 the CFE Negotiation will develop into negotiations open to all 34 CSCE states.[33]

In 1961, five years before SIPRI opened its doors, a UN General Assembly resolution called on all states to conclude an international agreement to refrain from the transfer or acquisition of nuclear weapons. In 1970 the Non-Proliferation Treaty entered into force. Twenty years later, it is still an instrument far from being as effective as was hoped at the time. Two nuclear powers, France and China, have not yet become party to this crucial instrument of international security. The number of threshold countries has grown larger, not smaller. And in fewer than five years the NPT is up for extension or non-extension. A great number of difficult issues remain to be solved: for example, the extension of safeguards to the entire nuclear fuel cycle, protection against misuse of new technologies for the uncontrolled production of fissile material, more binding rules for nuclear safety policy in light of the 1986 Chernobyl accident, and the link between non-proliferation and reaching a comprehensive test ban treaty.[34]

In 21 years of negotiating a control and reduction scheme for their strategic nuclear inventory, beginning in 1969, the United States and the Soviet Union have not been able to produce more than the ABM Treaty and the Interim Agreement on the limitation of strategic offensive arms, known as the SALT I Agreement of 1972. Should the START negotiations be concluded as planned, reductions will be less than substantial (for the USA only 11 per cent of the warheads, and for the Soviet Union 21 per cent) and the established nuclear deterrence structure (sea-, land- and air-based systems) will remain unchanged. Hence, the future agenda will list important and complex issues such as the content of and eventual transition to minimum deterrence, the offensive–defensive strategic defence equation, the inclusion of China, France and the UK in the START effort and the conclusion of a comprehensive test ban treaty. All these questions must be researched if the policy task is to help 'to recognize emerging opportunities that permit the evolution of security strategies that might make nuclear weapons increasingly less relevant in the conduct of states'.[35]

[33] See the Charter of Paris, section 'Security', in appendix 17B in this volume, pp. 603–10.

[34] See Fischer and Müller (note 26).

[35] Regina Cowen Karp (ed.), SIPRI, *Security With Nuclear Weapons? Different Perspectives on National Security* (Oxford University Press: Oxford, 1991), p. 18.

In 1966 world military expenditure was about US $568 billion (in 1990 prices); in 1990 world military expenditure amounts to US $950 billion. This represents an increase of almost 70 per cent over the past 25 years.[36] To recognize the dimension of the problem will help to understand that the decline in world military expenditure, as observed during the past three years, is a very relative one and one that proceeds from very high levels. It is also true that the modest reduction is mainly due to structural features, such as technological over-sophistication and high budgetary costs. The fundamental change in East–West relations has yet to filter through towards substantial cuts in defence spending. Therefore, the prospect for a disarmament dividend is very limited at present.[37] Much of the research work on when the dividend is to be realistically expected has yet to be done before political decisions can be taken on how to reallocate the 'dividend'.

In 1967 the Charter of Algier, for the first time, established that developed countries should devote 1 per cent of their gross national product (GNP) to the net transfer of resources to developing countries, including private flows. In 1970 the UN reduced the target to 0.7 per cent of the GNP for official development assistance (ODA). Notwithstanding the remarkable development effort produced over the past 24 years, facts and figures speak a sobering reality: major donor countries are still far from meeting the 0.7 per cent GNP target, and most of the receiving countries have not resisted the temptation to give 'defence' a priority position in terms of resource allocation over economic development and social services.[38] In 1988 total ODA from Western nations (Australia, Canada, Japan, New Zealand, the USA and Western Europe) was $48.1 billion, amounting to 0.35 per cent of the combined GNP of these countries.[39] SIPRI estimates show that the aggregate Third World external debt has now surpassed $1400 billion. For many Third World countries, debt-servicing and military expenditure take up 40–80 per cent of all government revenue, 'leaving very little for other essential needs. The most startling feature of the late 1980s is the emergence of *negative net transfers* whereby the debtors are paying more to the creditors than they are receiving in new money'.[40] Thus, the need to recognize the relationship between development in Third World countries, financial flows such as ODA, and debt as an extremely important issue of international security is greater than ever, as is the need to shed further scientific light on it.

[36] For this figure I am indebted to Dr Somnath Sen, Senior Researcher on the SIPRI military expenditure project.

[37] For a detailed assessment, see Saadet Deger, 'World military expenditure', chapter 5 in this volume, pp. 115–63.

[38] See, e.g., Saadet Deger and Somnath Sen, *Arms and the Child, A SIPRI–UNICEF Report on the Impact of Military Expenditure in Sub-Saharan Africa on the Survival, Protection and Development of Children* (SIPRI: Stockholm, 1990).

[39] See Somnath Sen, 'Debt, financial flows and international security', in SIPRI, *SIPRI Yearbook 1990: World Armaments and Disarmament* (Oxford University Press: Oxford, New York Toronto, 1990), chapter 6, pp. 203–17.

[40] See Somnath Sen, 'Debt, financial flows and international security', chapter 6 in this volume, pp. 181–95.

Both the Iraqi occupation of Kuwait and the ensuing Gulf War have again demonstrated the risk for all the international community if armaments export is pursued unrestrained. Eighty per cent of Iraq's import of major equipment in 1980–89 came from three of the permanent members of the UN Security Council—the USSR (53), France (20) and China (7)[41]—substantially financed by Iraq's victims—Kuwait and Saudi Arabia. The argument is not to substitute restrictions on trade in weapons for the solution of the underlying fundamental political issues. The point, however, is not to complicate political solution by fuelling respective areas such as the Middle East with the export of weapons. As a lesson of the Gulf War, this has even been accepted by France, a traditionally very unrestricted weapon exporter. In a television message entirely devoted to France's lessons from the Gulf War, President François Mitterrand stated that control over the trade in weapons must be accepted by both, 'aussi bien les pays qui vendent des armes que ceux qui les achètent'.[42]

The list of problems to be addressed by both research and politics is a long one. It contains, for example, foremost the need to design and accept schemes that would provide for transparency. Mechanisms to ensure control of international trade are to be based on transparency and require a common understanding of some key principles. With the East–West conflict no longer paralysing the UN, perhaps it has an important role to play, such as to register and monitor the international arms trade. The European Community countries will have to answer the question of whether the search for a common foreign and security policy makes a common policy on trade in weapons imperative. It is 100 years now since the first attempt was made to regulate the international trade in arms. The so-called Brussels Act of 1890 (General Act for the Repression of African Slave Trade) prohibited the introduction into Africa between latitudes 20° North and 22° South of fire-arms and ammunition other than flint-lock guns and gunpowder, except under effective guarantees.[43] The record of success to date, 100 years later, is very limited, and the list of unsolved problems might well hold work for a few more decades to come.

As SIPRI, in May 1991, looks back on 25 years of research work, more than 30 armed conflicts are being waged throughout the world.[44] As much as progress in Europe gives us reason to rejoice, developments in other parts of the world must protect us against turning Eurocentric or euphoric. The 43-day Gulf War, that broke out as this *Yearbook* was in preparation, is only the most salient case in point. There is no 'outbreak of peace' within reach that would cause SIPRI's research to stop or even pause. The need to inform the public to make an informed opinion possible is greater than ever.

[41] See the SIPRI Fact Sheet On Military Expenditure and Iraqi Arms Imports, 8 Aug. 1990 (mimeo).

[42] See the text of President Mitterrand's speech of 3 Mar. 1991 in *Le Monde*, 5 Mar. 1991, p. 5.

[43] See Jozef Goldblat, SIPRI, *Agreements for Arms Control: A Critical Survey* (Taylor &Francis: London, 1982), p. 4.

[44] See Karin Lindgren, Birger Heldt, Kjell-Åke Nordquist and Peter Wallensteen, 'Major armed conflicts in 1990', chapter 10 in this volume, pp. 345–80.

Part I. Weapons and technology

1. Nuclear weapons

Prepared by the *Nuclear Weapons Databook* staff, Washington, DC*

I. Introduction

During 1990, all five acknowledged nuclear weapon nations (the USA, the USSR, the UK, France and China) began the process of transition to the new political, economic and military circumstances. The USSR and the USA each moved towards a less nuclear-armed future as they approached completion of a START (Strategic Arms Reduction Talks) treaty, descended from their peak nuclear deployments of the late 1980s, and took steps (especially the USSR) towards major military reductions and restructuring.

Adjustment to the post-cold war world by all the major powers took place in fits and starts. Military spending habits proved hard to break, and many difficult decisions were postponed. The pace of world events overwhelmed the ability of entrenched bureaucracies, institutions and economies to adapt and fashion new plans and strategies. While there was full recognition that the cold war was over, it became evident that it will take some time to create a new and less nuclear-dependent security framework. Many nations spent the year deliberating how to adjust their budgets, military forces and deployments, and their engrained attitudes to enter this new era in East–West relations. Fundamental questions were raised, but only partial and uncertain answers were offered. How can military budgets be cut? Which weapons and programmes are needed, and which can be stopped? How many US forces should be withdrawn from Europe? What will NATO's future role be? What are tomorrow's probable security threats, and how can they be addressed? The year was one of uncertain transition and change, but one clearly marked by movement away from massive stockpiles of nuclear weapons.

The US Congress debated and deferred most difficult decisions on military programmes, although budget problems and political developments ensured a decreasing military budget. The USA is gearing up for a major reduction and reorganization of its military forces and strategy, including nuclear weapons. Environmental, health and safety problems with the Department of Energy (DOE) nuclear weapon complex resulted in the virtual cessation of new nuclear weapon production and a continuing shutdown of most facilities. With the deployment of the first two Trident submarines, the USA has acquired the beginning of the first sea-based counterforce capability. It was revealed during

* Robert S. Norris, Richard W. Fieldhouse and Thomas B. Cochran, Natural Resources Defense Council, Inc., Washington, DC; and William M. Arkin, Greenpeace USA, Washington, DC.

1990 that many US nuclear weapons have serious safety problems, and some weapons were removed from 'alert' duty.

The Soviet Union experienced extraordinary internal difficulties during 1990, while simultaneously undertaking a major reduction and reorganization of its military forces. Both these phenomena are affecting Soviet nuclear weapons, and a significant public debate emerged on nuclear weapon and related issues for the first time—'nuclear *glasnost*'. Soviet nuclear forces reached a peak in 1989 and are now declining. The Typhoon submarine programme was terminated, SS-24 missile deployment was completed, and the Blackjack bomber programme continued to suffer serious problems.

The new British Prime Minister, John Major, showed no signs that he would significantly change the Conservative Government's defence policy. Like the USA and the USSR, the UK conducted a review of military force structure and strategy to consider possible changes, reductions and reorganization. One possible component of a new British security strategy might be greater co-operation with France on several nuclear weapon programmes, but these remained undecided in 1990. There were no major developments in the British nuclear weapon programme.

France, like the other nuclear weapon powers, was forced to deliberate over how the changed global situation would affect its defence budget in general and certain nuclear weapon programmes in particular. As a result, it is anticipated that in 1991 the French Government will announce some changes to its nuclear forces.

More so than in previous years, in 1990 the Chinese Government provided very little public information about its current nuclear forces, although a few relevant items are known. China and the USSR concluded a border demilitarization agreement that should help reduce Sino-Soviet nuclear tensions. China conducted two nuclear tests during the year.

The tables showing the nuclear forces of all five nations as of January 1991 (tables 1.1–1.7) appear on pages 16–24 of this chapter. Table 1.8 (page 25) provides historical figures for the strategic forces of the five nations.

II. US nuclear weapon programmes

In the United States there was full recognition by virtually everyone that the cold war was over and that fundamental positive changes were taking place in the Soviet Union.[1] There were many tangible signs that the cold war had

[1] The following is representative: 'During the last year, the military threat in Europe from the Warsaw Pact has been virtually eliminated. At the same time, the tensions and open warfare that threatened Western interests in some, but not all, other regions have also abated. These changes have fundamentally altered many of the basic assumptions on which our national security policy, our military strategy, and our defense budgets have been based for the last four decades'; US Congress, Senate Armed Services Committee (SASC), *National Defense Authorization Act for Fiscal Year 1991*, Report 101-384, 20 July 1990 (US Government Printing Office: Washington, DC, 1990), p. 8. See also US Congress, SASC, *Threat Assessment; Military Strategy; and Operational Requirements*, Hearings, 101st Congress, 2nd session (US Government Printing Office: Washington, DC, 1990), S. Hrg 101-780; US Congress, House Armed Services Committee (HASC), *The Fading Threat: Soviet Conventional Power in Decline*,

passed. For example, the US Looking Glass nuclear command post aircraft were removed from continuous airborne alert, certain NATO exercises were cancelled, and most notably, two and one-third US armoured divisions were removed from Germany and sent to the Persian Gulf as part of Operation Desert Shield (see also chapters 13 and 19 in this volume).

In deliberations about nuclear weapon programmes, the major questions were what kinds and numbers of programmes to have in response to a vastly different Soviet Union. After a very contentious year of wrangling over the national defence budget, Congress cut it by a modest 6 per cent, from President Bush's amended request of $307 billion, to $288.3 billion. No major programmes were cancelled. Before Iraq's invasion of Kuwait the expectation was that the final figure would be closer to the $283 billion arrived at by the House of Representatives, rather than the higher Senate figure of $288 billion. In the end the Senate figure prevailed, with the Gulf crisis possibly the cause of the additional $5 billion. The so-called 'peace dividend' disappeared in the face of the new threats and uncertainties in the Middle East. This is also reflected in the US nuclear weapon programme.

The Strategic Defense Initiative (SDI) programme continued to decline in importance as congressional budget cutters reduced the $4.7 billion request to $2.9 billion. The 'Star Wars' programme was hobbled by several technical failures and by Senate action to reorient the programme away from early deployment towards a modest research effort.[2]

There were indications that the Pentagon was taking action to revise fundamentally its strategy and force structure.[3] The degree and the speed of those revisions will depend in part on what transpires in the Persian Gulf crisis, but major changes and reductions are virtually certain. Given the changes in the USSR and Europe, US nuclear targeting and war plans will undergo considerable review and revision—possibly including major revisions to the Single Integrated Operational Plan (SIOP), the US strategic nuclear war plan.

ICBMs

During 1990 the US operational intercontinental ballistic missile (ICBM) force remained at 1000 missiles with 2450 warheads, unchanged over the past

Report, July 9, 1990, 101st Congress, 2nd session (US Government Printing Office: Washington, DC, 1990), Committee Print No. 11.

[2] Broad, W. J., 'Crown jewel of "Star Wars" has lost its luster', *New York Times*, 13 Feb. 1990, p. C1; Broad, W. J., 'Technical failures bedevil star wars', *New York Times*, 18 Sep. 1990, p. C1; Rasky, S. F., 'Senate approves shift in strategy for missile shield', *New York Times*, 5 Aug. 1990, p. 1; Schmitt, E., 'Change on missile shield is resisted', *New York Times*, 6 Aug. 1990, p. B6; Henderson, B. W., 'X-ray laser research slashed as Congress cuts SDI funding', *Aviation Week & Space Technology*, vol. 133, no. 20 (12 Nov. 1990), p. 29.

[3] Gordon, M. R., 'Pentagon drafts new battle plan', *New York Times*, 2 Aug. 1990, p. A1. The outlines of the plan include reducing the armed forces by the mid-1990s from $2.1 million to $1.6 million, cutting 6 of the 18 active Army divisions, reducing the number of deployable aircraft-carriers from 14 to 12, reducing the number of tactical air force wings from 36 to 25, and reorganizing the forces into four basic components: the Atlantic Force, Pacific Force, Contingency Force and Strategic Force. President Bush announced parts of the plan in a speech held in Aspen, Colorado, on 2 Aug., the day Iraq invaded Kuwait.

two years. Early in the year the Air Force proposed retiring 450 Minuteman II missiles as a cost-saving measure but was overruled by the White House.[4]

Controversy continued over whether and how to proceed with the rail-based MX and the Small ICBM (SICBM).[5] In the final fiscal year (FY) 1991 budget, $680 million in research and development (R&D) funds were provided for the two programmes. Congress said that R&D should continue as a prudent and necessary hedge; that the two-missile modernization programme has failed to achieve the political consensus necessary for deployment of both systems; and that, with defence budgets likely to continue to decline, deployment of both missiles will be unaffordable.[6] Congress reversed its view of the SICBM programme from that of a mobile system to one deployed in silos.

An MX test-flight took place on 16 May from Vandenberg AFB, California, the 19th since June 1983. The Rockwell International company's Defense Electronics Division delivered the first MX rail-garrison launch-control test car to the Air Force on 4 October. The command and control car will be tested at Vandenberg AFB and the Association of American Railroads Transportation Test Center in Pueblo, Colorado.[7]

The second SICBM (MGM-134A) test-flight was scheduled for the autumn but was postponed until the spring of 1991. The original test programme had called for 22 test-flights, but six were dropped during the year to save $200–300 million.[8]

Strategic submarine programmes

During 1990 two more submarines carrying Poseidon submarine-launched ballistic missiles (SLBMs) were withdrawn from service, and two new Trident submarines entered service. The *USS Henry Clay* (SSBN-625) was deactivated on 12 March and the *USS Daniel Webster* (SSBN-626) on 30 August. This brings to eight the number withdrawn from service since September 1985. It was also reported that the remaining 11 submarines which carry Poseidon SLBMs would be retired several years earlier than originally scheduled.[9]

The test-flight problems experienced in 1989 during the firing of the Trident II missile from submerged submarines appear to have been

[4] Smith, R. J. and Moore, M., 'Air force calls for scrapping Minuteman II', *Washington Post*, 13 Jan. 1990, p. A1; Fulghum, D., 'Scrapping old ICBMs rejected', *Air Force Times*, 29 Jan. 1990, p. 27.

[5] Medalia, J. E., 'MX, "Midgetman," and Minuteman Missile Programs', Congressional Research Service Issue Brief IB77080, 15 Oct. 1990. It was reported that the Air Force had tentatively decided to kill the rail-garrison system; Capaccio, T., 'Air Force ponders terminating the MX rail garrison missile', *Defense Week*, 10 Dec. 1990, p. 1.

[6] *Congressional Record*, 23 Oct. 1990, pp. H11935–H12275. The Conference Report was also published as House Report 101-923.

[7] 'Aerospace world', *Air Force Magazine*, vol. 73, no. 12 (Dec. 1990), p. 25.

[8] US Congress, House Appropriations Committee (HAC), *Department of Defense Appropriations for Fiscal Year 1991*, Hearings before a Subcommittee of the Committee of Appropriations, House, 101st Congress, 2nd session (US Government Printing Office: Washington, DC, 1990), Part 7, p. 495; Lynch, D. J., 'Midgetman's second flight set this week', *Defense Week*, 29 Oct. 1990, p. 8.

[9] Steigman, D., 'Last Poseidon submarines to end strategic role by 1992', *Navy Times*, 29 Oct. 1990, p. 22.

corrected.[10] The final three (of nine) Performance Evaluation Missile tests were held on 15 and 16 January and 13 February 1990. All nine were conducted from the *USS Tennessee* (SSBN-734). There were also three Demonstration and Shakedown tests, on 12 February from the *Tennessee*, and on 11 March and 26 September from the *USS Pennsylvania* (SSBN-735). The *Tennessee*, loaded with Trident II SLBMs, undertook its first patrol on 29 March. This milestone gave the USA an SLBM counterforce capability for the first time.[11] The second submarine to carry Trident II missiles, the *Pennsylvania*, was prepared for initial deployment. Since 1982 the Trident submarine force has conducted over 140 patrols (as of April 1990).[12] In the 30 years since deployment of the first submarine, the *USS George Washington*, on 15 November 1960, US ballistic missile submarines have conducted over 2850 patrols.

Because of safety problems at the DOE plutonium component plant at Rocky Flats, outside Denver, Colorado, there was speculation that there might not be enough W88 warheads for the second submarine. However, the Navy said that there were enough.[13] The third Trident II submarine, the *USS West Virginia* (SSBN-736), was commissioned on 20 October at Kings Bay submarine base in Georgia and fired its first test missile on 28 November. The *West Virginia*'s initial patrol may be delayed, or its missiles may be armed with W76 warheads borrowed from Trident I missiles, due to the problems at Rocky Flats.[14]

Safety problems with the W88 warhead for the Trident II missile emerged during 1990, including concern about the decision not to use Insensitive High Explosive (IHE) in the warhead.[15] Because conventional high-explosive (HE) used in nuclear warheads can burn or explode in an accident or fire, the USA uses IHE in most of its modern nuclear weapons as a safety measure (see page 10). More IHE is required in a warhead to achieve the same compression as conventional high explosive. Inclusion of IHE increases the weight which in turn decreases the range, and so the Navy decided not to incorporate it in the W88.[16] Because of the way the warheads encircle the third stage of the Trident II, if the volatile high-energy propellant were to explode, plutonium

[10] Kolcum, E. H., 'Three successful launches verify design fixes to Trident 2 D5 ICBM', *Aviation Week & Space Technology*, vol. 132, no. 2 (8 Jan. 1990), pp. 50–51.

[11] Spinardi, G., 'Why the U.S. Navy went for hard-target counterforce in Trident II', *International Security*, autumn 1990, pp. 147–90.

[12] US Congress, HAC, *Department of Defense Appropriations for Fiscal Year 1991*, Hearings before a Subcommittee of the Committee of Appropriations, House, 101th Congress, 2nd session (US Government Printing Office: Washington, DC, 1990), Part 6, p. 368.

[13] US Congress, SASC, *Department of Defense Authorization for Appropriations for Fiscal Year 1991*, Hearings, Senate, 101st Congress, 2nd session (US Government Printing Office: Washington, DC, 1990), Part 7, p. 125; Broder, J. M. and Abramson, R., 'Trident subs may lack warheads', *Los Angeles Times*, 15 Mar. 1990 , p. A32.

[14] US Congress, HAC (note 12), pp. 307–308.

[15] Smith, R. J., 'Trident's mix of propellant with warheads raises safety questions', *Washington Post*, 29 May 1990, p. A6.

[16] US Congress, Senate Appropriations Committee, *Energy and Water Development Appropriations for Fiscal Year 1988*, Hearings before a Subcommittee of the Committee of Appropriations, Senate, 100th Congress, 1st session (US Government Printing Office: Washington, DC, 1988), Part 2, pp. 1127–29.

would be scattered about whether there was IHE in the warheads or not. The Pentagon and the Navy will face a difficult decision in 1991 on whether to incorporate IHE in the W88 and whether to use a safer third-stage propellant.[17]

The Navy requested money for the 18th Trident submarine in the FY 1991 budget and for long-lead items for the 19th and 20th submarines. By the end of the budget process in October, Congress recommended terminating the programme at 18 submarines and prohibited the obligation of funds for any more. For many years the Navy avoided declaring how many Trident submarines it wanted. A fleet of 18 SSBNs will mean 3456 SLBM warheads towards the START treaty ceiling of 4900 ballistic missile warheads. The budget also provided funding for 52 Trident II missiles, bringing the number procured thus far to 247.[18]

The Navy operates two Fleet Air Reconnaissance Squadrons to perform the TACAMO ('take charge and move out') mission of communicating with its SSBNs (nuclear-powered ballistic missile submarines). One TACAMO plane of the Atlantic Fleet based in Maryland, and one of the Pacific Fleet based in Hawaii, are airborne at all times. New Boeing E-6A Hermes aircraft have begun to replace the EC-130s which date from 1964.[19] The first operational mission was completed on 31 October 1989. All 16 Hermes aircraft are scheduled to be in service by June 1991.

Strategic bomber programmes

The B-2 bomber programme was involved in another year of controversy as changing world events seemed to undermine its rationale.[20] On 26 April the Secretary of Defense cut the programme from 132 production aircraft[21] to 75 and reduced the number to be bought in FY 1991 from five to two, and five in FY 1992. The total programme cost diminished from $75.4 billion to $61.1 billion, which caused the unit cost to rise from $571 million to $815 million per plane. The larger force of 120 operational planes, with 12 spares, was to have had 40 per cent of the aircraft (about 48 planes) on ground alert. The smaller force of 75 (60 operational planes) would increase the alert rate to 55 per cent or about 33 planes. The Senate upheld the Pentagon's plan to buy two

[17] US Congress, HASC, *Nuclear Weapons Safety*, Report of the HASC Panel on Nuclear Weapons Safety, Committee Print No. 15, 101st Congress, 2nd session (US Government Printing Office: Washington, DC, 1990).

[18] Medalia, J. E., *Trident Program*, Congressional Research Service Issue Brief IB73001, 15 Oct. 1990.

[19] Andrews, H., 'E-6A', *Naval Aviation News*, July–Aug. 1990, pp. 18–19.

[20] Reports and articles during the year supportive of the B-2 include: Department of the Air Force, 'B-2 survivability against air defense systems', Mar. 1990; Department of the Air Force, 'The case for the B-2', June 1990; Department of the Air Force, 'Bomber modernization: deterrence at the crossroads', June 1990; Rice, D., 'The manned bomber and strategic deterrence: the U.S. Air Force perspective', *International Security*, summer 1990, pp. 100–28. Articles critical of the B-2 include: Brown, M. E., 'The case against the B-2', *International Security*, summer 1990, pp. 129–53; Brower, M., 'The B-2 bomber: impossible cost, dubious mission', Union of Concerned Scientists, Apr. 1990.

[21] The Air Force had planned to procure a total of 133 B-2 aircraft (6 development aircraft and 127 production aircraft), of which 132 would be operational. One of the development aircraft will remain at Edwards AFB, California, as a test aircraft.

planes in FY 1991 while the House voted to kill the programme outright, leaving the 16 planes (6 development and 10 production) authorized in prior years to constitute the programme. While the final compromise authorized $2.35 billion in the procurement account (plus $1.75 billion for R&D), it was unclear whether any more bombers would actually be bought. Resolution of the different House and Senate interpretations, and the future of the programme itself, was left to be decided in 1991.

By late November the first B-2 bomber, which first flew in the summer of 1989, had completed 87 hours of flight-time in 22 flights.[22] On 23 October the plane began the first of the Block 2 'low observable' or 'stealth' test-flights. The first flight of the second B-2 bomber occurred on 19 October.[23] This aircraft is the only one of the six in the full-scale development fleet that will not join the operational force after the test programme ends.

Older model B-52G bombers continue to be retired or removed from the SIOP and assigned conventional missions. The Air Force plans to retire 19 B-52Gs in FY 1990 and 16 in FY1991.[24] It was reported that approximately two dozen B-52Gs were sent to Diego Garcia, armed with Israeli-built Have Nap missiles, in response to the Iraqi invasion of Kuwait.[25] About one-third of the 95 B-52Hs of the Strategic Air Command (SAC) have been fitted with the Common Strategic Rotary Launcher system that is able to launch any of the bomber nuclear weapons.

The Strategic Air Command began to turn over its FB-111A bombers to the Tactical Air Command (TAC) in June where they are redesignated F-111Gs. All 29 aircraft of the 509th Bomb Wing at Pease AFB, New Hampshire, were delivered to Cannon AFB, New Mexico, by the end of the year. The 29 additional FB-111As with the 380th Bomb Wing at Plattsburgh AFB, New York, will be turned over to TAC by the first quarter of 1992.

Further details emerged about the Advanced Cruise Missile (AGM-129A) programme. During the year the missile was publicly displayed, the test programme was completed, and the first operational ACMs reached SAC in June, three and a half years late.[26] The first unit to receive the missile is the 410th Bombardment Wing, at K. I. Sawyer AFB, Michigan. Deployment at three additional bases is planned. The Air Force plans to achieve initial operational

[22] 'No. 1 B-2 completes first round of low observable testing', *Aviation Week & Space Technology*, vol. 133, no. 22 (26 Nov. 1990), p. 30.

[23] 'B-2 starts stealth test series: second bomber makes first flight', *Aviation Week & Space Technology*, vol. 133, no. 18 (29 Oct. 1990), p. 77.

[24] US Congress, Senate Appropriations Committee, *Department of Defense Appropriations for Fiscal Year 1991*, Hearings, Part 3, 101st Congress, 2nd session (US Government Printing Office: Washington, DC, 1990), p. 253.

[25] Atkinson, R., 'U.S. to rely on air strikes if war erupts', *Washington Post*, 16 Sep. 1990, p. A1.

[26] 'Air Force displays advanced cruise missile for first time', *Aviation Week & Space Technology*, vol. 132, no. 20 (14 May 1990), p. 30; Dudney, R. S., 'Strategic modernization in the shakedown', *Air Force Magazine*, vol. 73, no. 10 (Oct. 1990), p. 45; US Congress, HAC, *Hearing on Department of Defense Appropriations for FY 1991*, Hearings before a Subcommittee of the Committee of Appropriations, House, 101st Congress, 2nd session (US Government Printing Office: Washington, DC, 1990), Part 5, pp. 404–407, 420–22; General Accounting Office, *Strategic Missiles: Logistics Support for Advanced Cruise Missile Based on Outdated Plans*, GAO/NSIAD-90-178, Sep. 1990, p. 9.

capability in March 1991.[27] Close-up pictures of the missile and a mockup were made public during the year, showing a large, unusually shaped missile, with longer range and greater stealth characteristics than the ALCM.

The ACM programme calls for 1461 missiles (plus 25 R&D models) at a total cost of $6.8 billion, or $4.6 million apiece. The missile will use the same W80-1 warhead that is currently on the air-launched cruise missile (ALCM). According to the General Accounting Office, the warhead costs add another $979.9 million to the programme. This would mean that an individual W80-1 warhead for the ACM costs about $650 000. Given the problems at Rocky Flats it will be difficult to meet the production schedule. Due to revised force structure plans, the Air Force may cut the programme to 1000 or fewer missiles.[28]

On 8 June, Secretary of Defense Dick Cheney announced that he had ordered the temporary removal of the Boeing short-range attack missile (SRAM-A) from ground alert aircraft, pending the outcome of safety studies. The 14-ft (4.25-m) long, 2200-lb (998-kg) supersonic missile is carried on B-1B, B-52G/H and FB-111A bomber aircraft and has a range of 35–100 miles (55–160 km). An estimated 1100 SRAMs are deployed at 13 SAC bases in 10 states.[29] Approximately 30 per cent of SAC's bombers, currently some 75 planes, are placed on ground alert, which means they are loaded with nuclear bombs and missiles and are capable of being airborne within 15 minutes. Of the operational SRAMs, an estimated 600 were removed from ground alert status. In December Secretary of Defense Cheney decided that the SRAM-As should be permanently removed from bombers on ground alert status although they would be retained in the SAC inventory.[30]

The safety problem arises with the SRAM's 275-lb W69 warhead. A chemical high explosive surrounds a plutonium core or 'pit' inside the warhead. The HE could ignite if there were an accident involving the missile propellant or the bomber fuel. If the HE exploded the plutonium might scatter, posing a severe public health hazard. The likelihood of there being a nuclear explosion is virtually nil.

The W69 warheads were manufactured between October 1971 and August 1976. Since then, the Insensitive High Explosive has been developed and used in all modern warheads and bombs produced since May 1979, except for the W79 artillery shell and the W88 Trident II warhead. About 25 per cent of the estimated 20 000 warheads in the current US stockpile contain the IHE, the rest do not.

The new SRAM II (AGM-131A) will have IHE in its 325-lb W89 warhead with yields up to 200 kt. The supersonic missile is 14 ft (4.25 m) long, 15.6 inches (40 cm) in diameter, weighs 1920 lbs (870 kg) and will have a greater

[27] General Accounting Office, *Strategic Weapons: Long-Term Costs Are Not Reported to the Congress*, GAO/NSIAD-90-226, Aug. 1990, p. 48.

[28] Tyler, P. E., 'Military chiefs detail plans to cut troops, weapons', *Washington Post*, 12 May 1990, p. A1.

[29] These are California, Michigan, New York, Louisiana, Arkansas, Washington, North Dakota, South Dakota, Texas and Kansas.

[30] Smith, R. J., 'Cheney acts to preclude plutonium accident', *Washington Post*, 8 Dec. 1990, p. A8.

range than the SRAM-A. The missile continued in development throughout the year.[31] Live launch test-flights were scheduled to begin in September 1990, but because of engine problems they will not begin until April 1991.[32] The Air Force plans to buy 1633 ACMs at a cost of $2.3 billion, exclusive of the warhead costs. Of the total, 1225 will be deployed in 10 squadrons. Initial and full operational capability (IOC and FOC) are scheduled for April 1993 and October 1998, respectively. There are currently no plans to put the ACM on B-52s. With a much smaller B-2 programme likely, the total number of SRAM IIs may be reduced as well.

Non-strategic nuclear forces

US non-strategic nuclear weapons continued to decline in numbers and in types during the year, as some systems were retired, several more were cancelled, and yet others were withdrawn and eliminated to comply with the 1987 INF (Intermediate-range Nuclear Forces) Treaty. Two US Army nuclear warhead types were completely retired from the stockpile: the B54 warheads for the Special Atomic Demolition Munition (atomic land mines) and W31 warheads for the Nike Hercules air-defence missile.

INF missiles were withdrawn from Europe at a steady pace. Approximately 100 ground-launched cruise missiles (GLCMs) were removed during the year, leaving 68 GLCMs deployed at Comiso air base in Sicily, and 52 at RAF Greenham Common, UK. On 22 August, the 38th Tactical Missile Wing at Wüschheim in the Federal Republic of Germany was deactivated, the fourth wing to be deactivated under the terms of the INF Treaty (see also chapter 12).

The last components of the 4th Battalion, 9th Field Artillery, left Camp Redleg, Heilbronn, Germany, on 26 April. This is the first of the three US Pershing II bases to have all of its missiles removed. The unit was deactivated in October, and the deployed Pershing II inventory in Germany shrank to 27 missiles. It was revealed during the year that the W85 Pershing II warheads were being converted and reused as B61-10 bombs for possible redeployment in Europe.[33]

On 4 October, the day after German unification, the Luftwaffe decommissioned its Pershing 1A force and removed the nuclear-armed missiles from NATO command. The 72 missiles, in two wings at Landsberg am Lech and Geilenkirchen/Tevren, ended 25 years of nuclear alert duty. The US W50 warheads will be returned to the United States and retired, and the US–German Program of Cooperation concerning nuclear weapon arrangements will be terminated.

[31] 'Aerospace world', *Air Force Magazine*, vol. 73, no. 1 (Jan. 1990), pp. 24–25.

[32] Morrocco, J. D., 'Problems with rocket motor delay initial flight of SRAM 2', *Aviation Week & Space Technology*, vol. 132, no. 5 (29 Jan. 1990), p. 31; US Congress, HAC, *Department of Defense Appropriations for Fiscal Year 1991*, Hearings before a Subcommittee of the Committee of Appropriations, House, 101st Congress, 2nd session (US Government Printing Office: Washington, DC, 1990), Part 7, p. 505.

[33] Norris, R. S. and Arkin, W. M., 'Beating swords into swords', *Bulletin of the Atomic Scientists*, vol. 46, no. 9 (Nov. 1990), pp. 14–16.

The FY 1991 budget request included funds for three new short-range non-strategic nuclear weapons: the W82 warhead for the 155-mm artillery shell, the Follow-on to Lance (FOTL) missile, and the tactical variant of the short-range attack missile (SRAM-T). During the first few months of the year arguments were proposed by various NATO and US officials about why these short-range weapons were needed. The arguments lost validity in the face of the changes in Europe. On 3 May President Bush announced that he had decided to terminate the FOTL programme and to cancel any further modernization of nuclear artillery shells.

An artillery shell already in Europe was discovered to be defective.[34] Through computer simulations it was determined that the W79 8-inch artillery shell did not meet certain important safety criteria related to accidental nuclear detonation. The artillery shells were shipped back to the Pantex assembly plant in Texas, and the problem was corrected.

The SRAM-T (AGM-131B) programme came under close scrutiny during the year because it is primarily destined for deployment on US and allied aircraft in Europe, such as the F-15E, F-16, F-111F/G, and German and Italian Tornados. The SRAM-T is 95 per cent common with its strategic counterpart, the SRAM II, although it has a longer range of up to 400 km due to different flight software and because the missile is lofted from the fighter's wing before the engine fires. The 310-lb (140-kg) warhead offers a choice of 10 or 100 kt. Previously designated the tactical air-to-surface missile (TASM), the SRAM-T is under consideration by the UK for its Tornado and naval Sea Harrier aircraft. Congress cut authorization of the $119 million request to $35 million.[35] On 5 November the first 'captive carry' test was conducted successfully at Edwards AFB, California, using an F-15E aircraft with a missile attached.

At the London NATO meeting of heads of state and government in July, Western leaders declared their intent to transform NATO's role and to revise its nuclear strategy through modification of the doctrine of flexible response.[36] In the aftermath of the NATO defence ministers' meeting in Brussels in December, the expectation was that negotiations might start in 1991 with the Soviet Union to eliminate all European land-based nuclear missiles and nuclear artillery. What would remain would be nuclear-capable tactical aircraft with gravity bombs and the SRAM-T.

[34] Smith, R. J., 'Defective nuclear shells raise safety concerns', *Washington Post*, 23 May 1990, p. A1.

[35] Although in manoeuvres by Appropriation Committee conferees the money was restored; US Congress, House of Representatives, *Making Appropriations for the Department of Defense, Conference Report*, 24 Oct. 1990, Report 101-938, p. 108; Morgan, D., 'Defending jobs with a cold war relic', *Washington Post*, 8 Nov. 1990, p. A29.

[36] For the full text of the London Declaration, see Rotfeld, A. D. and Stützle, W. (eds), SIPRI, *Germany and Europe in Transition* (Oxford University Press: Oxford, 1991), pp. 150–52. See also US Department of State, Bureau of Public Affairs, *NATO Transformed: The London Declaration*, Selected Document no. 38, July 1990.

Tactical aircraft programmes

Two aircraft previously planned to be nuclear-capable will not be. The Navy's SH-60F Ocean Hawk helicopter, intended to replace the SH-3D/H Sea King, and the Marine Corps's AV-8B Harrier II were earmarked for nuclear missions during development in the early 1980s, but it has now been decided to make them conventional only.

The Navy released some details and an artist's concept of the A-12A Avenger attack aircraft on 17 August.[37] The plane is a delta-shaped flying wing manufactured by General Dynamics and McDonnell Douglas, and is planned to replace the Navy A-6E and later the Air Force F-15 and F-111. While precise details have not been released, preliminary estimates are that the plane is roughly 37 ft (11.3 m) long with a wingspan of 72 feet (21.9 m) open and 34 ft (10.4 m) folded, and a wing area of 1300 sq. ft (117 sq. m). The plane is estimated to have a payload of 25 000 lbs (11 250 kg) with a range of 1800 miles (2880 km), significantly greater than the A-6E. The Navy plans to deploy them on aircraft-carriers for the attack role, possibly in wings of 16 aircraft rather than 20.[38] The first six A-12s, funded in FY 1990, were under construction in 1990. Roll-out was scheduled for late 1991 with the first test-flight in March 1992. After a three-year test programme, sea trials and crew training, the plane would enter service in mid- to late 1996 with the first squadron at either Lemoore, California, or Whidby Island, Washington.

These milestones will probably slip because the programme is in serious difficulty. In April Secretary Cheney announced that the number of naval A-12s to be procured would be decreased from 858 to 620 and that the Air Force's purchase had been deferred beyond FY 1997.[39] The congressional conferees stated that the 'aircraft is seriously overweight, far behind schedule, increasingly complex in design and more difficult to manufacture, and suffering from management deficiencies'.[40] The eight planes the Navy wanted in the FY 1991 budget were cut. In December three naval officers in charge of the programme were punished for concealing information from Pentagon officials about the delays and extra costs.[41] In January 1991, Secretary Cheney cancelled the programme. Unless Congress reverses the decision, the programme will stop at the original six aircraft.

Plans to relocate the 401st Tactical Fighter Wing from Torrejon Air Base, Spain, to Crotone Air Base, Italy, were derailed pending a re-evaluation of the

[37] Office of the Assistant Secretary of Defense for Public Affairs (OASD/PA), 'Navy releases details of A-12 "Avenger"', News Release no. 390-90, 17 Aug.1990.

[38] Morrocco, J. D., 'Funding cuts may limit carrier air wings to 16 A-12s', *Aviation Week & Space Technology*, vol. 133, no. 14 (1 Oct. 1990), pp. 18–19. See also Sweetman, B., 'The A-12 Avenger; stealth for tactical strike', *International Defense Review*, no. 10 (1990), pp. 1157–59; Richeson, J., 'Avenger', *Naval Aviation News*, Nov.–Dec. 1990, pp. 14–16.

[39] For a review of the problems and controversies surrounding the A-12, see Morrison, D. C., 'Pentagon charade', *National Journal*, 20 Oct. 1990, pp. 2516–21.

[40] *Congressional Record*, 23 Oct. 1990, p. H 12077.

[41] Schmitt, E., 'Admiral is ousted over plane delay', *New York Times*, 5 Dec. 1990, p. A1; Bond, D. F., 'A-12 cost overruns misjudged; Navy removes top program officials', *Aviation Week & Space Technology*, vol. 133, no. 24 (10 Dec. 1990), pp. 26–27.

future need for such a force. The wing's 72 F-16C/D aircraft provide a nuclear capability.

Naval nuclear forces

By the end of 1990, there were 36 surface ships and 50 submarines certified to use the Tomahawk sea-launched cruise missile (SLCM). While the number of platforms capable of firing the Tomahawk cruise missile continues to increase, it does so at a slower rate than originally planned.[42] The overall number of Tomahawk-capable platforms, originally set at 198, will not be reached because of early retirements or programme cuts. Current projections put the number at 90 surface vessels and 85 submarines.

All 367 nuclear Tomahawk land-attack missiles (TLAM/Ns) procured between FY 1982 and FY 1989 have been delivered to the Navy. Of the 400 nuclear and non-nuclear versions authorized in the FY 1991 budget, the Navy had not determined at the end of the year how many of them would be TLAM/Ns. At the beginning of the year it was announced that the total programme of all types was reduced by 400 missiles, from 4030 to 3630. In May, US and Soviet START negotiators decided to limit the number of nuclear SLCMs to 880, a number in excess of the 758 planned in the US programme. Exactly how many more TLAM/Ns may be bought is still to be determined.

The Navy had wanted a new B90 Nuclear Strike/Depth Bomb (NSDB) to replace its old B57 nuclear depth bombs and gravity bombs starting in 1993. The Los Alamos National Laboratory-designed, multi-purpose, thermonuclear B90 entered engineering development (Phase 3) on 23 June 1988. The bomb is 9.8 ft (3 m) long, 13.3 in. (0.3 m) in diameter and weighs 780 lbs (355 kg). It will have a yield in the kiloton range. Congress was not impressed with the Navy's arguments for the B90 bomb and expressed doubts about the need to maintain a nuclear anti-submarine warfare (ASW) capability for naval aircraft. Funds were cut, the programme was delayed, and the Navy will have to report on the mission and need for the bomb in 1991.

The B90 is designed for ASW and strike warfare (against land targets) missions. It is planned to be carried by A-6 Intruder, A-12 Avenger and F/A-18 Hornet strike aircraft on aircraft-carriers, and by US S-3 Viking, P-3 Orion, Italian Atlantique, Dutch NP-3 and British Nimrod ASW aircraft. It is unclear whether the B90 would be carried by carrier-based ASW helicopters. The SH-3D/H Sea King is currently certified to carry the B57 nuclear depth bomb, but the SH-60F Ocean Hawk replacement will not be nuclear-certified. The B90 would be the Navy's first nuclear bomb solely built to its specifications in a long time. The B57 is also used by the Air Force for low-yield, land-attack missions.

[42] Handler, J. and Arkin, W. M., *Nuclear Warships and Naval Nuclear Weapons 1990: A Complete Inventory*, Neptune Papers no. 5 (Greenpeace USA: Washington, DC, Sep. 1990), pp. 3–4, 11.

Problems with the nuclear weapon production complex

The serious environmental and safety problems at the DOE nuclear weapon production facilities that first came to public attention in 1988 continued to be of concern in 1990 (see *SIPRI Yearbook 1989*, pages 10–11).[43] Many key facilities were idle throughout the entire year with little prospect of early resumption of activities.[44] The three (tritium production) reactors at Savannah River were shut down during the period April–August 1988. No new tritium has been produced in over two years. The Rocky Flats plant was shut down in November 1989 and may be closed well into 1991 or beyond. This stoppage caused the new production workload at the Pantex final assembly plant in Texas to be much reduced. Funds for a new plutonium recovery facility at Rocky Flats were cut. The planned Special Isotope Separation plant in Idaho was cancelled in January. Funds for the Special Nuclear Materials Laboratory at Los Alamos were deferred for at least a year. Operations at the Fernald uranium processing plant in south-western Ohio will cease. On 16 October the DOE announced that the PUREX (plutonium–uranium extraction) plant, closed since December 1988, would not reopen. It is evident that the USA will no longer produce plutonium for its nuclear arsenal.

III. Soviet nuclear weapon programmes

1990 was a tumultuous year in the Soviet Union. In the midst of a massive reduction and restructuring of its entire military, including nuclear forces, the USSR experienced internal political problems that spawned public debate on and opposition to nuclear weapons in an unprecedented manner. Several republics openly opposed Soviet nuclear weapons or demanded that they be transferred to republic control. *Glasnost* was finally extended to include nuclear wapons in 1990.

The Soviet nuclear arsenal declined in number during the year, and all signs pointed towards a significantly reduced investment in nuclear weapons in the future. This is in accord with Soviet preparations for the pending START treaty and anticipated follow-on negotiations.

Problems with the Soviet nuclear weapon testing and production complex may force unilateral cutbacks much sooner than will future arms control agreements. In March, it was revealed that the Semipalatinsk nuclear test site in Kazakhstan would be closed by 1993, largely as a result of public protest.[45] The Soviet Union held only one nuclear test in 1990, at the distant

[43] General Accounting Office, *Nuclear Health and Safety: Long-Term Plans to Address Problems of the Weapons Complex Are Evolving*, GAO/RCED-90-219, Sep. 1990.

[44] Lippman, T. W., 'Atomic arms plants face new delays', *Washington Post*, 24 Nov. 1990, p. A11.

[45] Smith, R. J., 'Soviets to close major site of underground atomic tests', *Washington Post*, 10 Mar. 1990, p. A1; '1992 testing halt announced', Moscow World Service, 30 June 1990, in Foreign Broadcast Information Service, *Daily Report–Soviet Union* (hereafter referred to as *FBIS-SOV*), FBIS-SOV-90-127-S, 2 July 1990, p. 51.

Table 1.1. US strategic nuclear forces, January 1991

Weapon system				Warheads		
Type	No. deployed	Year deployed	Range (km)	Warhead x yield	Type	No. deployed
ICBMs						
Minuteman II	450	1966	12 500	1 x 1.2 Mt	W56	450
Minuteman III (Mk 12)	200	1970	13 000	3 x 170 kt	W62	600
Minuteman III (Mk 12A)	300	1979	13 000	3 x 335 kt	W78	900
MX	50	1986	11 000+	10 x 300 kt	W87	500
Total	**1 000**					**2 450**
SLBMs						
Poseidon (11 SSBNs)	176	1971	4 600	10 x 50 kt	W68	1 760
Trident I (20 SSBNs)	384	1979	7 400	8 x 100 kt	W76	3 072
Trident II (2 SSBNs)	48	1990	7 400	8 x 475 kt	W88	384
Total	**608**					**5 216**
Bombers[a]						
B-1B	90	1986	9 800	ALCM	W80-1	1 600
B-52G/H	154	1958/61	16 000	SRAM	W69	1 110
FB-111A	24	1969	4 700	Bombs	[b]	1 600
Total	**268**					**4 300**
Refuelling aircraft						
KC-135 A/R/E	615	1957	. .	. .	. .	. .
KC-10A	60	1981	. .	. .	. .	. .

[a] Numbers reflect Primary Authorized Aircraft. An additional 7 B-1Bs, 25 B-52s and 5 FB-111s are in the total inventory. B-52Gs at Castle AFB, California; Loring AFB, Maine; and Barksdale AFB, Louisiana, some 47 aircraft, have primarily conventional missions. Bombers are loaded in a variety of ways, depending on mission. B-1Bs normally carry up to 16 weapons (SRAMs and either B83 or B61 bombs). B-52s can carry a mix of 8–24 weapons. FB-111s can carry up to 6 weapons (SRAMs or B61 or B43 bombs).

[b] Bomber weapons include four different nuclear bomb designs (B83, B61-0, -1, -7, -9, B53, B43) with yields from low-kt to 9 Mt, ALCMs with selectable yields from 5 to 150 kt, and SRAMs with a yield of 170 kt.

Sources: Cochran, T. B., Arkin, W. M. and Norris, R. S., *Nuclear Weapons Databook, Volume I: US Forces and Capabilities*, 2nd edn (forthcoming); authors' estimates.

Arctic site on Novaya Zemlya, but even there the public outcry, both domestic and foreign, was immediate. Reports have differed about plans to use the Novaya Zemlya site over the next two years, but Soviet television reported in

Table 1.2. US non-strategic nuclear forces, January 1991

Weapon system				Warheads		
Type	No. deployed	Year deployed	Range (km)	Warhead x yield	Type	No. in stockpile
Land-based systems						
Aircraft[a]	1 300	..	1 060–2 400	1–3 x bombs	Bombs[a]	1 800
Missiles						
Pershing II	27[b]	1983	1 790	1 x 0.3–80 kt	W85	0[c]
GLCM	120[b]	1983	2 500	1 x 0.2–150 kt	W84	0[c]
Lance	100	1972	125	1 x 1–100 kt	W70	1 282
Other systems						
Artillery[d]	4 700	1956	30	1 x 0.1–12 kt	·[d]	1 540
Naval systems						
Carrier aircraft[e]	850	..	1 000–1 800	1–2 x bombs	Bombs[e]	1 350
Tomahawk SLCM	325	1984	2 500	1 x 5–150 kt	W80-0	325
ASW aircraft[f]	500	..	1 160–3 800	1 x bomb <20 kt	B57	850

[a] Aircraft include the US Air Force F-4D/E, F-16A/B/C/D and F-111A/D/E/F/G. F-15Es will become nuclear-certified in 1992. Bombs include three types (B43, B57 and B61) with yields from sub-kt to 1.45 Mt.

[b] Remaining missiles will be withdrawn and eliminated by 1 June 1991.

[c] Pershing warheads are being converted into B61-10 bombs. GLCM warheads will likely be placed in inactive reserve.

[d] Total inventory of US Army and Marine Corps nuclear-capable artillery. There are two types of nuclear artillery (155-mm and 203-mm) with four different warheads: a 0.1-kt W48, 155-mm shell; a 1- to 12-kt W33, 203-mm shell; a 0.8-kt W79-1, enhanced-radiation, 203-mm shell; and a variable-yield (up to 1.1 kt) W79-0 fission warhead. The enhanced-radiation warheads will be converted to standard fission weapons.

[e] Aircraft include the US Navy A-6E, A-7E, F/A-18A/C and Marine Corps A-6E and F-18A/C. Bombs include three types with yields from 20 kt to 1 Mt.

[f] Aircraft include US Navy P-3B/C, S-3A/B and SH-3D/H helicopters. Some US B57 nuclear depth bombs are allocated for British Nimrod, Italian Atlantic and Netherlands P-3 aircraft.

Sources: Cochran, T. B., Arkin, W. M. and Norris, R. S., *Nuclear Weapons Databook, Volume 1: US Forces and Capabilities*, 2nd edn (forthcoming); Collins, J. M. and Rennack, D. E., *US/Soviet Military Balance*, Library of Congress/Congressional Research Service, Report no. 90-401 RCO, 6 Aug. 1990; International Institute for Strategic Studies, *The Military Balance 1990–1991* (Brassey's: Oxford, 1990); authors' estimates.

Table 1.3. Soviet strategic nuclear forces, January 1991

Weapon system					Warheads	
Type	NATO code-name	No. deployed	Year deployed	Range (km)	Warhead x yield	No. deployed
ICBMs						
SS-11 Mod. 2	Sego	100	1973	13 000	1 x 1.1 Mt	100
Mod. 3		210	1973	10 600	3 x 350 kt (MRV)	210[a]
SS-13 Mod. 2	Savage	30	1973	9 400	1 x 750 kt	30
SS-17 Mod. 3	Spanker	50	1979	10 000	4 x 750 kt (MIRV)	200
SS-18 Mod. 4/5/6	Satan	308	1979	11 000	10 x 550/750 kt (MIRV)	3 080
SS-19 Mod. 3	Stiletto	250	1979	10 000	6 x 550 kt (MIRV)	1 500
SS-24 Mod. 1/2	Scalpel	36/50	1987	10 000	10 x 550 kt (MIRV)	860
SS-25	Sickle	300	1985	10 500	1 x 550 kt	300
Total		**1 334**				**6 280**
SLBMs						
SS-N-6 Mod. 3	Serb	176	1973	3 000	2 x 1 Mt (MRV)	176[a]
SS-N-8 Mod. 1/2	Sawfly	286	1973	9 100	1 x 1.5 Mt	286
SS-N-17	Snipe	12	1980	3 900	1 x 1 Mt	12
SS-N-18 Mod. 3	Stingray	224	1978	6 500	7 x 500 kt (MIRV)	1 568
Mod. 1			1978	6 500	3 x 500 kt (MIRV)	
SS-N-20	Sturgeon	120	1983	8 300	10 x 200 kt (MIRV)	1 200
SS-N-23	Skiff	96	1986	9 000	4 x 100 kt (MIRV)	384
Total		**914**				**3 626**
Bombers						
Tu-142	Bear H	85	1984	12 800	8 AS-15 ALCMs or bombs	680
Tu-160	Blackjack	21	1988	14 600	6 AS-15 ALCMs, 4 AS-16 SRAMs and 4 bombs	294
Total		**106**				**974**
Refuelling aircraft	..	140–170	..	..	..	..
SAMs[b]	..	6 050	1958–80	50–300	1 x low kt	3 000
ABMs						
ABM-1B	Galosh Mod.	32	1986	320	1 x unknown	32
ABM-3	Gazelle	68	1985	70	1 x low yield	68
Total		**100**				**100**

[a] SS-11 and SS-N-6 MRV warheads are counted as one.

[b] Nuclear-capable land-based surface-to-air missiles probably include SA-2 Guideline, SA-5 Gammon and SA-10 Grumble.

Table 1.3 *cont.*

Sources: Authors' estimates derived from: Cochran, T. B., Arkin, W. M., Norris, R. S. and Sands, J. I., *Nuclear Weapons Databook, Volume IV: Soviet Nuclear Weapons* (Harper & Row: New York, 1989); US Department of Defense, *Soviet Military Power*, 1st–9th edns; DIA, *Force Structure Summary–USSR, Eastern Europe, Mongolia, and Afghanistan*, DDB-2680-170-90, Feb. 1990; Collins, J. M. and Rennack, D. E., *U.S./Soviet Military Balance*, Library of Congress/Congressional Research Service, Report no. 90-401 RCO, 6 Aug. 1990.

---

early December that 19 tests would be carried out at the testing ground before January 1993.[46]

In addition to problems with nuclear testing, nuclear weapon production has slowed, and resources devoted to the nuclear weapon enterprise are declining. In early 1990, a Defence Ministry official stated that spending for nuclear warheads would decrease by 43.2 per cent in 1990, a figure that evidently does not include the budget of the Ministry of Atomic Industry.[47] The USSR is continuing to phase out its older nuclear materials production facilities—the production of highly enriched uranium for military purposes has ceased,[48] the fourth plutonium reactor at Chelyabinsk ceased operation on 17 July, and the fifth reactor ceased operation on 7 November.[49]

Budget cuts and changing times continue to affect the level and intensity of Soviet military activity. In 1989, the number of strategic nuclear force missile test launches was halved.[50] Offensive training missions by the strategic bomber force against North America continued to decline, and general air activity in and around the USSR plummeted. Naval activities and out-of-area operations and deployments continued to shrink.

Politics and nuclear weapons

With the flare-up of ethnic violence between the Soviet republics of Armenia and Azerbaijan in January and the declaration of independence by Lithuania in March, greater attention was focused in 1990 on the security and disposition of Soviet nuclear weapons in the restive republics. As politics in the republics

[46] 'Semipalatinsk nuclear tests to end in 1993', Moscow Television Services, 4 Dec. 1990, FBIS-SOV-90-234, 5 Dec. 1990, p. 1.

[47] 'According to Col. Gen. V. N. Babyev of the Central Finance Directorate of the Ministry of Defence, Soviet spending for nuclear warheads will decrease 43.2% in 1990, from 1,306.2-million rubles in 1989 to 993.8-million rubles in 1990'; Zaloga, S., 'Red hot news', *Armed Forces Journal International*, vol. 127, no. 9 (Apr. 1990), p. 28. According to Viktor Mikhaylov, Deputy Minister of the Ministry of Atomic Energy and Industry, 'Last year, by an "arbitrary" decision, they gave us a 20 per cent cut in resources'; FBIS-SOV-90-140, 20 July 1990, p. 4.

[48] 'Moiseyev responds to readers on Perestroika in armed forces', *Voyenno-Istoricheskiy Zhurnal*, no. 2 (1990), in *JPRS Soviet Union–Military Affairs*, 4 June 1990, p. 24.

[49] 'Plutonium reactor's shutdown threatens layoffs', Moscow Television Service, 17 July 1990, FBIS-SOV-90-140, 20 July 1990, p. 74.

[50] US Department of Defense (DOD), *Soviet Military Power 1990* (US Government Printing Office: Washington, DC, 1990), (hereafter referred to as DOD, *SMP 1990*), p. 29.

Table 1.4. Soviet non-strategic nuclear forces, January 1991[a]

Weapon system					Warheads	
Type	NATO code-name	No. deployed	Year first deployed	Range[b] (km)	Warhead x yield	No. deployed
Land-based systems						
Long-range bomber						
Tu-95	Bear G	60	1984	12 800	2 AS-4	120
Aircraft						
Tu-26	Backfire A/B/C	180	1974	4 000	1–3 x bombs or ASMs	360
Tu-16	Badger A/G	60	1954	3 100	1–2 x bombs or ASMs	60
Tu-22	Blinder A/B	60	1962	2 400	1–2 x bombs or 1 ASM	60
Tactical aircraft[c]		2 200	. .	700–1 300	1–2 x bombs	2 500
Missiles						
SS-1c	Scud B	661	1965	300	1 x 1–10 kt	1 370
. .	FROG 3/5/7	370	1965	70	1 x 1–25 kt	1 450
SS-21	Scarab	300	1978	70	1 x 10–100 kt	310
Other systems						
Artillery[d]	. .	7 000	1973–80	10–30	1 x low kt	2 000
Naval systems						
Aircraft						
Tu-26	Backfire A/B/C	130	1974	4 000	1–3 x bombs or ASMs	260
Tu-16	Badger A/C/G	125	1955	3 100	4 x bombs or ASMs	500
Tu-22	Blinder A	25	1962	2 400	4 x bombs	80
Su-24	Fencer C/D	50	1989	1 300	2 x bombs	100
Su-20	Fitter C	70	1973	630	1 x bomb	70
ASW aircraft[e]	. .	330	1966–82	. .	1 x depth bombs	350
Anti-ship cruise missiles[f]						
SS-N-3 a,c/b	Shaddock/Sepal	144	1960	450	1 x 350 kt	90
SS-N-7	Starbright	64	1968	65	1 x 200 kt	32
SS-N-9	Siren	248	1969	280	1 x 200 kt	92
SS-N-12	Sandbox	248	1976	550	1 x 350 kt	106
SS-N-19	Shipwreck	180	1980	550	1 x 500 kt	72
SS-N-22	Sunburn	126	1981	100	1 x 200 kt	42
Land-attack cruise missiles						
SS-N-21	Sampson	136	1987	3 000	1 x 200 kt	136
ASW missiles and torpedoes						
SS-N-15	Starfish }	400	1973	37	1 x 10 kt }	400
SS-N-16	Stallion }		1979	120	1 x 10 kt }	
FRAS-1		25	1967	30	1 x 5 kt	25

Table 1.4 *cont.*

Weapon system					Warheads	
Type	NATO code-name	No. deployed	Year first deployed	Range[b] (km)	Warhead x yield	No. deployed
Torpedoes[g]	Type 65	520	1965	16	1 x low kt	520
	ET-80		1980	>16	1 x low kt	
Naval SAMs						
SA-N-1	Goa	47	1961	22	1 x 10 kt	200
SA-N-3	Goblet		1967	37	1 x 10 kt	

[a] Assumes remaining SS-20s and SS-N-5 missiles are not operational. The number and existence of SSC-1bs, ADMs and sea mines are unknown.

[b] Range for aircraft indicates combat radius, without refuelling.

[c] Nuclear-capable tactical aircraft models include 725 MiG-27 Flogger D/J, 600 Su-17 Fitter C/D/H, and 875 Su-24 Fencer A/B/C/D/E.

[d] Nuclear-capable artillery include systems of the three calibres: 152-mm (D-20, 2A36/M-1976, 2S3, 2S5 and possibly a new M1986), 203-mm (M55, 2S7 and M-1980) and 240-mm (2S4 and M-240). Some older systems may also be nuclear-capable.

[e] Includes 75 Be-12 Mail, 45 Il-38 May and 60 Tu-142 Bear F patrol aircraft. Land- and sea-based helicopters include 150 Ka-25 Hormone and Ka-27 Helix models.

[f] Number deployed is total launchers on nuclear-capable ships and submarines. Warheads based on an average of 2 nuclear-armed cruise missiles per nuclear-capable surface ship, except for 4 per Kiev and Kirov Class ships, and 4 per nuclear-capable cruise missile submarine, except for 12 on the Oscar Class.

[g] The two types of torpedo are the older and newer models, respectively, with the ET-80 probably replacing the Type 65.

Sources: Cochran, T. B., Arkin, W. M., Norris, R. S. and Sands, J. I., *Nuclear Weapons Databook, Volume IV: Soviet Nuclear Weapons* (Harper & Row: New York, 1989); Polmar, N., *Guide to the Soviet Navy*, 4th edn (US Naval Institute: Annapolis, Md., 1986); Department of Defense, *Soviet Military Power*, 1st–9th edns; DIA, *Force Structure Summary–USSR, Eastern Europe, Mongolia, and Afghanistan*, DDB-2680-170-90, Feb. 1990; Collins, J. M. and Rennack, D. E., *US/Soviet Military Balance*, Library of Congress/Congressional Research Service, Report no. 90-401 RCO, 6 Aug. 1990; International Institute for Strategic Studies, *The Military Balance 1990–1991* (Brassey's: Oxford, 1990); Handler, J. and Arkin, W. M., *Nuclear Warships and Naval Nuclear Weapons: A Complete Inventory*, Neptune Paper no. 5 (Greenpeace USA: Washington, DC, 1990); authors' estimates.

became more heated—varying declarations of sovereignty and independence have included nuclear-free pledges—the deployment and control of nuclear weapons become even more controversial.

Throughout the year, rumours circulated in the Western press that nuclear stocks had been removed from the areas of greatest unrest, but by the end of the year it appeared that few, if any, major deployment changes had been made. None the less, as the Union began to be challenged and the republics

Table 1.5. British nuclear forces, January 1991[a]

Weapon system				Warheads		
Type	No. deployed	Year deployed	Range (km)[b]	Warhead x yield	Type	No. in stockpile
Aircraft						
Tornado GR-1	108[c]	1982	1 300	1–2 x 400/200 kt bombs[d]	WE-177A/B	175[e]
Buccaneer S2B	40	1962	1 700	1 x 400/200 kt bomb	WE-177A/B	
SLBMs						
Polaris A3-TK	64	1982[f]	4 700	2 x 40 kt	MRV	100[g]
Carrier aircraft						
Sea Harrier FRS.1	42	1980	450	1 x 10 kt bomb	WE-177C	25[h]
ASW helicopters						
Sea King HAS 5	46	1976	–	1 x 10 kt depth bomb	WE-177C	
Lynx HAS 2/3	70	1976	–	1 x 10 kt depth bomb	WE-177C	

[a] British systems certified to use US nuclear weapons include 11 Nimrod ASW aircraft based in the UK, 1 Army regiment with 12 Lance launchers and 4 Army artillery regiments with 120 M109 howitzers in Germany.

[b] Range for aircraft indicates combat radius, without refuelling.

[c] The Royal Air Force operates 9 squadrons of dual-capable strike/attack Tornados in Germany and Britain. Total inventory of strike variants, including those for training and spares, is approximately 220.

[d] The US Defense Intelligence Agency (DIA) has confirmed that the RAF Tornados 'use two types of nuclear weapons, however, exact types are unknown'. The DIA further concludes that each RAF Tornado is capable of carrying two nuclear bombs, on the two outboard fuselage stations.

[e] The total stockpile of WE-177 tactical nuclear gravity bombs is about 200, of which 175 are versions A and B. All three weapons use the same basic 'physics package', and the yield is varied by using different amounts of tritium.

[f] The 2-warhead Polaris A3-TK (Chevaline) was first deployed in 1982 and has now completely replaced the original three-warhead Polaris A-3 missile (first deployed in 1968).

[g] It is now thought that Britain produced only enough warheads for three full boat-loads of missiles, or 48 missiles, with a total of 96 warheads. In Mar. 1987 French President Mitterrand confirmed that Britain had '90 to 100 [strategic] warheads'.

[h] The C version of the WE-177 bomb is believed to be assigned to selected Royal Navy (RN) Sea Harrier FRS.1 aircraft and ASW helicopters. The WE-177C exists in both a free-fall and depth bomb modification, by varying the fuzing and casing options. There are an estimated 25 WE-177Cs, each with a yield of approximately 10 kt (possible variable yield).

Sources: Cochran, T. B. *et al.*, *Nuclear Weapon Databook*, *Volume V: British, French and Chinese Nuclear Weapons and Nuclear Weapons Proliferation* (forthcoming); British Ministry of Defence, *Statement on the Defence Estimates, 1980–90* (Her Majesty's Stationery Office: London, annual).

Table 1.6. French nuclear forces, January 1991

Weapon system				Warheads		
Type	No. deployed	Year deployed	Range (km)[a]	Warhead x yield	Type	No. in stockpile
Aircraft						
Mirage IVP/ASMP	18	1986	1 500	1 x 300 kt	TN 80	18
Mirage 2000N/ASMP[b]	45	1988	1 570	1 x 300 kt	TN-81	45
Jaguar A	30	1974[c]	750	1 x 6–8/25 kt bomb[d]	AN-52[e]	30
Refuelling aircraft						
C-135/FR	11	1965	..	..	..	..
Land-based missiles						
S3D	18	1980	3 500	1 x 1 Mt	TN-61	18
Pluton	44	1974	120	1 x 10/25 kt	AN-51[e]	70
Submarine-based missiles						
M-20	32	1977	3 000	1 x 1 Mt	TN-61	32
M-4A	16	1985	4 000–5 000	6 x 150 kt (MIRV)	TN-70[f]	96
M-4B	48	1987	6 000	6 x 150 kt (MIRV)	TN-71	288
Carrier-based aircraft						
Super Etendard/ASMP[g]	36	1978[c]	650	1 x 300 kt ASMP	AN-52[e]	24

[a] Range for aircraft indicates combat radius, without refuelling, and does not include the 90- to 350-km range of the ASMP air-to-surface missile (where applicable).

[b] The Mirage 2000N/ASMP has completely replaced the Mirage IIIE in the tactical nuclear role and the final Jaguar A squadron (15 aircraft) in July 1990. 75 Mirage 2000N aircraft are planned, the last to be delivered in 1992.

[c] The Jaguar A and Super Etendard aircraft were first deployed in 1973 and 1978, respectively, although they did not carry nuclear weapons (the AN-52) until 1974 and 1981, respectively.

[d] Two-thirds of the AN-52 stockpile reportedly consists of the low-yield variant, and one-third the high-yield variant. The AN-52 has an estimated weight of 455 kg, length of 4.2 m, diameter of 0.6 m and span of 0.8 m.

[e] The same nuclear device is used for both the AN-52 warhead (gravity bomb) and the AN-51 warhead (Pluton). Both warheads have the same higher yield of 25 kt (thus said to have the MR-50 charge in common), yet have lower yields of 6–8 kt and 10 kt, respectively.

[f] The *Inflexible* was the only SSBN to receive the TN-70. All subsequent refits of the M-4 into Redoutable Class SSBNs will incorporate the improved TN-71 warhead.

[g] The Super Etendard used to carry 1 AN-52 bomb. At full strength the AN-52 equipped 2 squadrons (24 aircraft) of Super Etendard: Flottilles 11F and 17F, based at Landivisiau and Hyères, respectively. From mid-1989 these two squadrons began receiving the ASMP missile. By mid-1990, all 20 aircraft (to be configured to carry the ASMP) were operational. Although originally about 50–55 Super Etendard aircraft were to receive the ASMP, because of budgetary contraints the number of aircraft so configured dropped to 20.

Sources: Cochran, T. B. *et al.*, *Nuclear Weapon Databook, Volume V: British, French and Chinese Nuclear Weapons and Nuclear Weapons Proliferation* (forthcoming).

Table 1.7. Chinese nuclear forces, January 1991

Weapon system				Warheads	
Type	No. deployed	Year deployed	Range (km)	Warhead x yield	No. in stockpile
Aircraft[a]					
H-6 (B-6)	120	1965	3 100	1–3 x bombs[b]	200+
H-5 (B-5)	30	1968	1 200	1 x bomb	
Q-5 (A-5)	30–50	1970	400	1 x bomb	
Land-based missiles					
DF-3 (CSS-2)	70–100	1970	2 800	1 x 1–3 Mt	85–125
DF-4 (CSS-3)	15–20	1971	4 800–7 000	1 x 1–3 Mt	20–30
DF-5 (CSS-4)	4–10	1979	13 000	1 x 4–5 Mt	10–20
M-9/SST 600[c]	. .	1990	600	1 x ?	. .
Submarine-based missiles[d]					
JL-1 (CSS-N-3)	24	1986	2 800–3 300	1 x 0.5–1 Mt	26–38

[a] All figures for these bomber aircraft refer to nuclear-configured versions only. Hundreds of these aircraft are also deployed in non-nuclear versions.

[b] Yields of bombs are estimated to range from below 20 kt to 3 Mt.

[c] The nuclear capability of the M-9 is unconfirmed.

[d] Two missiles are presumed to be available for rapid deployment on the Golf Class submarine. Additional missiles are being built for new Xia Class submarines.

Sources: Cochran, T. B. *et al.*, *Nuclear Weapon Databook, Volume V: British, French and Chinese Nuclear Weapons and Nuclear Weapons Proliferation* (forthcoming); Lewis, J. W. and Xue, L., *China Builds the Bomb* (Stanford University Press: Stanford, Calif., 1988).

began to create their own military forces, the question of control of the nuclear stockpile was again raised. The sensitive discussions came to a head at the end of November, when Russian Federation President Boris Yeltsin discussed the weapons on the territory of the Russian Republic, stating that about 80 per cent of the Soviet nuclear arsenal was based there.[51]

On 27 November, Defence Minister Dmitri Yazov took the extraordinary step of making a public address on Soviet television:

I am appealing to you at the instruction of the country's President in connection with the unlawful actions in a number of republics that place the country's defence capacity at risk . . . Voices are being heard advocating handing over nuclear weapons to some republics . . . The creation of nuclear-free zones is being declared without taking into account the interests of the defence capacity of our state . . . In this

[51] 'Yeltsin on inter-republic treaties, nuclear arms', Agence France Press (Moscow), 20 Nov. 1990, FBIS-SOV-90-226, 23 Nov. 1990, p. 43.

Table 1.8. Strategic nuclear weapon arsenals of the USA, the USSR, the UK, France and China, 1946–90

	USA		USSR		UK		France		China[b]	
Year[a]	L	W	L	W	L	W	L	W	L	W
1946	125	9	–	–	–	–	–	–	–	–
1947	270	13	–	–	–	–	–	–	–	–
1948	473	50	–	–	–	–	–	–	–	–
1949	447	200	–	–	–	–	–	–	–	–
1950	462	400	–	–	–	–	–	–	–	–
1951	569	569	–	–	–	–	–	–	–	–
1952	660	660	–	–	–	–	–	–	–	–
1953	720	878	–	–	–	–	–	–	–	–
1954	1 035	1 418	–	–	–	–	–	–	–	–
1955	1 260	1 755	–	–	8	–	–	–	–	–
1956	1 470	2 123	22	84	48	–	–	–	–	–
1957	1 605	2 460	28	102	73	–	–	–	–	–
1958	1 620	2 610	56	186	88	40	–	–	–	–
1959	1 551	2 496	108	283	96	70	–	–	–	–
1960	1 559	3 127	138	354	120	105	–	–	–	–
1961	1 532	3 110	187	423	120	163	–	–	–	–
1962	1 653	3 267	235	481	144	180	–	–	–	–
1963	1 812	3 612	302	589	144	207	–	–	–	–
1964	2 012	4 180	425	771	128	204	4	4	1	–
1965	1 888	4 251	463	829	88	199	32	32	1	1
1966	2 139	4 607	570	954	88	194	36	36	7	10
1967	2 268	4 892	947	1 349	88	189	36	36	12	15
1968	2 191	4 839	1 206	1 605	80	232	36	36	21	25
1969	2 109	4 736	1 431	1 815	48	144	36	36	38	45
1970	2 100	4 960	1 835	2 216	64	144	36	36	68	75
1971	2 087	6 064	2 075	2 441	64	144	45	45	97	102
1972	2 167	7 601	2 207	2 573	64	144	70	70	115	120
1973	2 133	8 885	2 339	2 711	64	144	86	86	136	145
1974	2 106	9 324	2 423	2 795	64	144	86	86	158	165
1975	2 106	9 828	2 515	3 217	64	144	102	102	175	180
1976	2 092	10 436	2 545	3 477	64	144	98	98	186	200
1977	2 092	10 580	2 562	4 242	64	144	114	114	201	220
1978	2 086	10 832	2 557	5 516	64	144	114	114	231	250
1979	2 086	10 800	2 548	6 571	64	144	114	114	258	260
1980	2 022	10 608	2 545	7 480	64	144	130	130	280	270
1981	1 966	10 688	2 593	8 296	64	144	130	130	288	285
1982	1 921	10 515	2 545	8 904	64	128	130	130	305	300
1983	1 905	10 802	2 543	9 300	64	112	126	126	320	315
1984	1 943	11 500	2 540	9 626	64	112	126	126	329	329
1985	1 965	11 974	2 538	10 012	64	96	142	222	331	336
1986	1 957	12 386	2 506	10 108	64	96	138	218	320	325
1987	2 001	13 002	2 535	10 442	64	96	138	298	309	319
1988	1 926	13 000	2 553	10 834	64	96	132	292	313	323
1989	1 903	12 100	2 448	11 320	64	96	132	372	302	317
1990	1 876	11 966	2 354	10 880	64	96	132	452	304	324

L: Launchers; W: Warheads

[a] Figures are given as at the end of each year.

[b] Figures for China are for deployed systems only.

Sources: Cochran, T. B., Arkin, W. M. and Norris, R. S., *Nuclear Weapons Databook, Vol. I,* forthcoming (for the USA), *Vol. IV,* 1989 (the USSR) and *Vol. V,* forthcoming (the UK, France and China).

situation I consider it necessary to announce . . . 1. Under no circumstances will the dispersal of nuclear weapons be permitted . . .[52]

Yazov went on to state that the Soviet Army would use armed force to defend military facilities and that military commands would also use force to ensure continuation of utilities, water, food supplies, and so forth, to military garrisons. A week later, Gorbachev's principal defence adviser Marshal Sergey Akhromeyev stated that strategic offensive weapons 'are deployed in the Russian Soviet Federated Socialist Republic, the Ukranian Republic, and the Kazakh Republic. They were also deployed in Belorussia and the Baltic republics until quite recently, when we eliminated medium-range missiles, in keeping with a treaty. Tactical nuclear weapons are deployed in practically all Union republics'.[53]

As if internal problems were not enough, the deployment of nuclear forces in the Baltic region and on the Kola peninsula continued to be an irritant in Soviet foreign relations. In October 1989, President Gorbachev stated in Helsinki that the USSR had withdrawn all of its short-range nuclear missiles to sites beyond range of the countries of northern Europe.[54] The exact meaning of Gorbachev's statement was not immediately apparent, but NATO and Norwegian military officers claimed that tactical nuclear weapons were still deployed on the Kola peninsula and stated that nuclear-armed SS-21 short-range missiles were stationed within sight of the Norwegian border. In response, Deputy Foreign Minister Gennady Gerasimov stated in January that at 'no point on Soviet territory are there tactical nuclear charges capable of reaching the countries of northern Europe'. The Deputy Commander of the Leningrad Military District then stated in February that 'not a single medium-range missile is serving combat duty in the district any longer', and that 'there is no nuclear fleet' in the Baltic Sea.[55] By September, the Soviet Government announced that the sixth and final Golf Class ballistic missile submarine in the Baltic Fleet had been taken out of service and additionally that the Baltic Fleet had been denuclearized. However, the disposition of the ship- and land-based nuclear weapons still remained unclear.

Perhaps to eliminate a similar potential problem with Japan, in July Ground Forces Chief of Staff General Dmitri A. Grinkevich stated that there were no nuclear weapons on the four disputed Kurile islands north of Japan.[56]

The disintegration of central authority, the effects of years of secrecy and safety scandals, together with the benefits of *glasnost*, have now fed a growing local protest movement against military installations and nuclear projects. Most well known is the so-called Nevada–Semipalatinsk–Mururoa movement

[52] 'Yazov announces authorized military measures', Moscow Television Service, 27 Nov. 1990, FBIS-SOV-90-229, 28 Nov. 1990, p. 68.

[53] 'Akhromeyev on union, republic defense', Moscow Domestic Service, 3 Dec. 1990, in FBIS-SOV-90-233, 4 Dec. 1990, p. 24.

[54] Keller, B., 'Gorbachev plans to destroy his A-armed subs in Baltic', *New York Times*, 27 Oct. 1989, p. A10.

[55] 'Effects of cuts on Leningrad military district', *Trud* (Moscow), 13 Feb. 1990, in *JPRS-Soviet Union-Military Affairs*, 3 Apr. 1990, pp. 28–29.

[56] 'Kurils "nuclear free"', *Jane's Defence Weekly*, vol. 14, no. 3 (21 July 1990), p. 79.

in Kazakhstan, which has worked to close the Semipalatinsk nuclear testing ground, precipitated protests against nuclear testing at the Arctic site on Novaya Zemlya, and appears to be expanding into an all-union 'not-in-my-backyard' movement.

Spurred on by the continuing effects of the Chernobyl accident, numerous groups, and even local governments, have taken action against a variety of nuclear power stations.[57] Residents in the Far East towns of Vanino and Sovetskaya Gavan have protested plans to dismantle, and/or dispose of, nuclear reactors from retired submarines of the Pacific Fleet. Protests against contamination at the previously secret Chelyabinsk ('Mayak') nuclear weapons and materials production complex have also become common.[58]

In addition to anti-nuclear protests, ecological groups have taken action against other military projects. For example, protests resulted in an August order to halt construction of the Mukachevo early-warning radar in the Ukraine, and resulted in the closure of a radar in Ulyanovsk.[59] Greater public attention on radiation, contamination and the effects of accidents has also forced local military commanders and garrisons to address these dangers in their public announcements. The commander of the Transcaucasus Military District, for instance, stated in October that 'the troops of the Transcaucasus Military District have neither nuclear, chemical, nor any other weapon which could cause ecological harm'.[60]

Strategic offensive forces

Soviet strategic offensive forces reached peak strength in 1989. With anticipated modernization and the conclusion of a START treaty, they will decline significantly in the coming years. At the end of 1990, Soviet strategic forces comprised 1334 ICBMs with 6280 warheads, 914 SLBMs with 3626 warheads, and 106 bombers with 974 warheads. This is an overall reduction of 440 nuclear weapons since 1989.

The Director of the US Central Intelligence Agency, William Webster, testified in March 1990 before the House Armed Services Committee that 'the Soviets are pursuing a broad-based strategic modernization effort. They are protecting, and in some ways improving, the overall capabilities of their strategic forces'.[61] However, in 1990 the effects of budget cuts, reorganiza-

[57] See, for example, Marples, D., '"Chernobyl"—Summer 1990', Report on the USSR (Radio Liberty), 29 June 1990, p. 14; 'Nuclear power industry situation eyed', Moscow Domestic Service, 28 Nov. 1990, FBIS-SOV-90-231, 30 Nov. 1990, pp. 64–66.

[58] 'Supreme Soviet committees, commissions meet 5 Oct.: committee views nuclear pollution', FBIS-SOV-90-195, 9 Oct. 1990, p. 35.

[59] Lepingwell, J. W. R., 'Soviet early warning radars debated', Report on the USSR (Radio Liberty), 17 Aug. 1990, pp. 11–15; see also Report on the USSR (Radio Liberty), 24 Aug. 1990, p. 30; 'Ulyanovsk radar to be moved for health reasons', Moscow World Service, 16 Jan. 1990, FBIS-SOV-90-011, 17 Jan. 1990, p. 133.

[60] 'Commander claims troops not nuclear-armed', *Izvestia* (Moscow), 3 Oct. 1990, FBIS-SOV-90-194, 5 Oct. 1990, p. 81.

[61] 'Statement of the Director of Central Intelligence before the Armed Services Committee, House of Representatives, 1 Mar. 1990', (mimeo), p. 4.

tions, developmental problems and programme completions began to be felt in the nuclear force. With completion of deployment of the SS-24 missile in 1990, termination of the Typhoon submarine programme and continued problems with the Blackjack bomber, strategic modernization will effectively cease in 1991, both quantitatively and qualitatively.

ICBMs

The Soviet ICBM force declined slightly in 1990, from 6450 to 6280 warheads, as new SS-24 and SS-25 missiles were deployed and older ICBMs were retired. In early August, Soviet officials announced that deployment of the new 10-warhead SS-24 missile would end by 1 January 1991 and that missile production would cease in 1991.[62] At full deployment, the force will likely consist of 36 SS-24 Mod. 1 missiles on 12 railroad trains at three sites,[63] and some 60 SS-24 Mod. 2 missiles will be deployed in converted SS-19 silos. The operational deployment of the SS-24 missile is now virtually complete; at mid-year over 50 SS-19 ICBMs silos had been converted to the new SS-24 Mod. 2 configuration.[64]

The new single-warhead, road-mobile SS-25 missile reached a strength of 300 missiles in 1990, and the missile remained in production. SS-25 deployments are occurring at former SS-20 bases, which have been eliminated as such under the INF Treaty, and it is estimated that the SS-25 force will eventually reach a strength of about 500 missiles and warheads.[65]

The SS-18 missile also remained in production in 1990, as older silo-based Mod. 1/3 and Mod. 4 missiles were replaced with new Mod. 5 and Mod. 6 missiles.[66] The multi-warhead Mod. 5 has 'substantially more accuracy and warhead yield' than the Mod. 4, and the Mod. 6 is a new single-warhead missile.[67] The USSR will be obligated by the START treaty to dismantle half of the SS-18 force, so the number of missiles will decline from 308 to 154 at most in the early 1990s. As new SS-24 and SS-25 missiles are produced, and the remaining SS-18s are upgraded, SS-11, SS-13 and SS-17 ICBMs are being drawn down, and the SS-19 force is being eliminated.[68] Eventually, under START numerical constraints on ballistic missile warheads, it is estimated that the entire force of SS-11s, SS-13s, SS-17s and SS-19s will be eliminated.

[62] DOD, *SMP 1990*, p. 36; 'Scalpel near end of line', *Jane's Defence Weekly*, vol. 14, no. 6 (11 Aug. 1990), p. 181; Smith, R. J., 'Soviets to end production of railroad-based missiles', *Washington Post*, 3 Aug. 1990, p. A31.

[63] This is 10–20 missiles fewer than US intelligence organizations reportedly predicted by the end of 1991; Gertz, B., 'Soviets to double mobile missile force', *Washington Times*, 30 July 1990, p. A6.

[64] DOD, *SMP 1990*, p. 52.

[65] DOD, *SMP 1990*, p. 52.

[66] A small number of missiles in the SS-18 force have always been armed with single warheads, although all of the missiles are normally counted as carrying 10 warheads.

[67] DOD, *SMP 1990*, p. 52.

[68] DOD, *SMP 1990*, p. 52. Over 50 SS-19 silos have been converted to house the new SS-24 Mod. 2. 'Eventually, the Soviets probably will destroy the remaining 300 SS-19 silos not converted to support the SS-24 Mod 2'.

Strategic submarine programmes

The Soviet SLBM force declined slightly in 1990, from 3642 to 3626 warheads and, like the ICBM force, peaked in 1989. The sixth and final boat of the large Typhoon Class was added to the inventory in 1989, and the submarine became operational in 1990.[69] The sixth boat of the Delta IV Class also became operational in 1990, and a seventh Delta IV was launched early in the year.[70]

The rate of ballistic missile submarine production will likely proceed at one Delta IV boat per year for the next few years,[71] and US intelligence believes that 'a new class of SSBN could replace the older Deltas'.[72]

The Soviet Navy continues to retire older Yankee Class submarines, and the force of 34 submarines built in 1967–74 has now been drawn down to 11 at the end of 1990. START considerations will force the USSR to reduce the level of Soviet strategic warheads at sea to some 2000 by the late 1990s, and this will likely be accomplished with the retirement of 22 Delta I and Delta II Class submarines, and the eventual phase-out of the Typhoon forces. Only the SS-N-20 and SS-N-23 SLBMs were in production in 1989 and 1990, and a new missile that was reported under development in the mid-1980s has been cancelled or slowed.[73]

Strategic bomber programmes

A decline in bomber output in 1989 and 1990 will result in a smaller bomber force than was previously expected. Only seven new Bear H bombers were built in 1989, and a similar or smaller number is assumed for 1990.[74] Conversion of Bear B/C bombers to Bear G configurations appear to have ceased at about 60 aircraft in 1990, and the converted bombers are now assumed to have shed their strategic offensive role. The number of nuclear bomber weapons declined from 1228 to 974, as the new AS-15 Kent ALCM became the most prominent air-delivered weapon.

After many years of continuing development and testing problems, the Blackjack bomber programme appears completely stalled. In addition to design and production defects, operating complexity, lack of support equipment and engine problems, the high cost of the new bomber is affecting programme acquisition.[75] Production of the Blackjack continued at the low rate of

[69] DOD, *SMP 1990*, pp. 37, 56.
[70] DOD, *SMP 1990*, p. 53.
[71] 'Statement Presented by Mr. Dennis M. Nagy, Acting Deputy Director for Foreign Intelligence, Defense Intelligence Agency, to the Technology and National Security Subcommittee of the Joint Economic Committee, Congress of the United States, 20 April 1990', (mimeo).
[72] 'Remarks by RADM Thomas A. Brooks, USN, to USNI [US Naval Institute] ASW Seminar, 27 Feb. 1990', (mimeo), p. 6.
[73] DOD, *SMP 1990*, p. 36; DOD, *SMP 1989*, p. 47; DOD, *SMP 1988*, p. 48.
[74] 'Statement Presented by Mr. Dennis M. Nagy' (note 71); DOD, *SMP 1990*, p. 36.
[75] 'Problems of Tu-160 program surveyed', *Krasnaya Zvezda*, 1 May 1990, in FBIS-SOV-90-086, 3 May 1990, pp. 61–66.

about 10 aircraft in 1990,[76] and it is now assumed that the force of 100–140 originally envisaged will be scaled back to roughly 30 aircraft.[77] The Bear H force will likely reach full strength at about 90 aircraft in 1991, before new production on that bomber is also halted.

In a major organizational development, the long-range aviation component of the Air Forces, which had been abolished in 1982, was reinstated in 1990. This new command takes on control of long-range bombers and fighter-bombers assigned to five air armies, consolidating assets in the Soviet Union with the withdrawal of forces from Eastern Europe.

Maritime-oriented Bear G bombers assigned to the 39th Air Army have been removed from the strategic bomber force in a similar fashion to US B-52Gs in the Strategic Air Command losing their nuclear functions. As older Bear A/B/C bombers complete retirement or conversion, virtually all of the long-range bomber force oriented towards strategic offensive missions is able to deliver the new RK-55 (NATO designation: AS-15 Kent) ALCM and the AS-16 Kickback (or Kingpost) short-range attack missile (SRAM).[78] It is estimated that 800 AS-15s have been deployed on Bear H and Blackjack bombers (690 AS-15s were estimated to be deployed at the end of 1989). The AS-X-19 long-range supersonic cruise missile reportedly remains under development, and the US DOD believes that it may be deployed in the early 1990s.[79]

Non-strategic nuclear forces

By the end of 1990, all of the Soviet missiles covered under the INF Treaty—SS-20 Saber, SS-4 Sandal, SS-12M Scaleboard B and SS-23 Spider missiles—were either eliminated or removed from operational duty. The last missile type to be eliminated, the SS-20, is assumed no longer to be operational. By the end of 1990, approximately 600 of the total of 654 SS-20 missiles had been eliminated; a total of about 50–60 SS-20 missiles remained, with a requirement to complete elimination by 31 May 1991.

In addition to the removal of some 2000 warheads from the active Soviet nuclear arsenal under the INF Treaty, Soviet unilateral reductions in non-strategic nuclear weapons continued in Europe in 1990. On 5 June, Foreign Minister Eduard Shevardnadze announced that the Soviet Union would withdraw 'a very substantive' part of its nuclear arsenal stationed in Eastern Europe, and would remove 1500 nuclear warheads in 1990.[80] Shevardnadze stated that 60 tactical missile launchers and more than 25 nuclear artillery

[76] DOD, *SMP 1990*, pp. 36, 38.

[77] Pritchard, C. G., 'Soviet bomber forces boosted by disarmament?', *International Defense Review*, no. 8 (1990), p. 837; Gordon, M. R., 'U.S. says Soviets will field fewer of its latest bombers', *New York Times*, 5 June 1990, p. 20.

[78] Zaloga, S. J., 'Current trends in the Soviet strategic bomber force', *Jane's Soviet Intelligence Review*, vol. 2, no. 8 (Aug. 1990), pp. 338–42.

[79] DOD, *SMP 1990*, p. 53.

[80] Goshko, J. M., 'Shevardnadze announces withdrawal of A-arms', *Washington Post*, 6 June 1990, p. A6; Max, A., 'Soviets plan nuclear withdrawal', *Washington Times*, 6 June 1990.

guns would be withdrawn. The 1990 package followed President Gorbachev's announcement on 11 May 1989 that 500 nuclear warheads would be withdrawn from Eastern Europe (287 missiles, 166 aircraft bombs and 50 artillery projectiles).

The INF and unilateral withdrawals resulted in the removal of some 2500 nuclear warheads from Eastern Europe by the end of 1990—virtually all of the weapons so deployed in Czechoslovakia, Germany, Hungary and Poland. With the agreements to withdraw Soviet forces from Czechoslovakia and Hungary by 1991 and from Germany by 1994, and with Soviet withdrawal from Poland before 1994,[81] thousands of non-strategic nuclear delivery systems will be removed from Europe and returned to the Soviet Union. By the end of 1990, Shevardnadze stated, the various packages of withdrawals would have already removed 140 missile launchers and 3200 nuclear artillery pieces from Eastern Europe to the Soviet Union.[82]

These unilateral withdrawals, together with defensive restructuring of the armed forces, have left the disposition and composition of much of the non-strategic land- and air-based nuclear arsenal uncertain.

The 25-year-old SS-1c Scud medium-range missile remains the primary ground force nuclear weapon, but it is reaching the end of its service life. There has been some speculation of a possible replacement for the Scud missile, but this remains unconfirmed.[83] The SS-21 short-range missile continues to be produced, replacing older FROG rockets. The SS-21 is also being organized into brigades of 18 launchers each at the Army level, replacing FROG battalions of four launchers at the division level.[84] There is no evidence that the large number of older FROG missiles and launchers have been eliminated, although most have been removed from Eastern Europe.[85] There have been reports of increased production of SS-21 missiles.[86]

While modernization of artillery assets continues, defensive restructuring is reducing the size of the artillery force, as artillery battalions are being decreased from six or eight guns to four guns.[87] This will reduce the Soviet nuclear artillery capability. Large numbers of artillery guns are being eliminated throughout the Soviet Army, as a result of both the unilateral reductions and the 1990 Conventional Armed Forces in Europe (CFE) Treaty, and it is estimated that over 10 000 guns are in storage in the Soviet Union. Older guns will likely be eliminated altogether, as newer self-propelled and towed guns continue to replace older towed guns in the force.

[81] 'Soviets to delay Poland pullout', *International Herald Tribune*, 13 Feb. 1991.

[82] 'Text of Shevardnadze's speech', Moscow TASS, 5 June 1990, FBIS-SOV-90-109, 6 June 1990, p. 5.

[83] McCausland, J. D., 'Soviet short-range nuclear forces', *International Defense Review*, no. 9 (1990), p. 941.

[84] DOD, *SMP 1990*, p. 55.

[85] DOD, *SMP 1990*, pp. 77–78. The missile is still deployed with front-line forces; at mid-year, nearly half of all short-range ballistic missiles and rockets remaining in Soviet forces in Eastern Europe were FROG missiles.

[86] DOD, *SMP 1990*, p. 36.

[87] DOD, *SMP 1990*, p. 56; 'Leningrad military district officer interviewed', *DANAS* (Zagreb), 25 Sep. 1990; in FBIS-SOV-90-198, 12 Oct. 1990, p. 93.

The withdrawal of nuclear-capable aircraft from Eastern Europe also continues, as does the general reorganization and resubordination of much of the tactical air force.[88] It is far too early to tell what will eventually happen to the withdrawn aircraft, or to their nuclear capabilities, but it is known that fighter-bomber production fell by about 10 per cent in 1989,[89] to a level half as high as in 1980. Production of the Backfire medium bomber remained the same at about 30 bombers per year in 1990, as Badger and Blinder bombers continued to be retired.[90] A new short-range attack missile, tentatively identified in the West as the AS-9 Kyle, was reported deployed on the Backfire in 1989.[91] A large number of aircraft, including nuclear-capable aircraft, assigned to the Soviet Air Forces have been transferred to Soviet Naval Aviation, where they have taken on maritime duties.

Naval nuclear forces

The Soviet Navy has so far avoided the extreme defence cuts affecting the other armed forces, although modernization programmes have slowed and reductions in force are taking place. Defence Minister Yazov stated in September that, under the current plan for military reform, 'Major changes will take place in the composition, structure, equipment and bases of the navy'.[92] It is estimated that the number of nuclear weapons in the non-strategic naval force remained about the same (about 2600) in 1990.[93]

In 1989, 21 major surface warships and submarines of all types were produced, an increase in the number of naval vessels compared with the average annual production rate of 18 units in the preceding eight years.[94] Twenty new units were started in 1989, an increase of three units compared to 1988.[95]

In April 1989, the fourth Kirov Class nuclear-powered guided cruise missile (CGN) was launched, but work on the fifth Kirov at Leningrad's Baltic Works shipyard ceased soon thereafter, and further construction of the class has been terminated.[96] The fourth Slava Class cruiser was launched in August 1990, and construction of that class has been terminated.[97] The last of 12 Udaloy Class destroyers were also completing construction in 1990.[98] Of the large surface combatants, only the Sovremennyy Class destroyer continues in large-scale production; as of September 1990, the Soviet Navy had received

[88] DOD, *SMP 1990*, pp. 79–80, 96, 97.

[89] 'Statement Presented by Mr. Dennis M. Nagy' (note 71).

[90] DOD, *SMP 1990*, p. 36.

[91] DOD, *SMP 1990*, p. 79.

[92] 'Assessments of armed forces reforms, reductions: Yazov interviewed on cuts', 14 Sep. 1990, FBIS-SOV-90-183, 20 Sep. 1990, p. 53.

[93] For a detailed description of changes in Soviet Navy nuclear capabilities, see Handler and Arkin (note 42).

[94] DOD, *SMP 1990*, pp. 38–39. In 1989, a single Slava Class cruiser, 3 Udaloy and Sovremennyy Class destroyers, and a total of 6 Krivak III Class frigates and Grisha V Class corvettes were deployed, with similar ships entering the force in 1990.

[95] DOD, *SMP 1990*, pp. 35–36.

[96] DOD, *SMP 1990*, p. 84.

[97] DOD, *SMP 1990*, p. 84.

[98] DOD, *SMP 1990*, p. 85.

12 of them, and 10 more were in construction. The first unit of a modified version of the Udaloy Class is under construction, to enter Soviet forces in 1992.[99]

In 1990 the DOD revealed new details about the Soviet conventional aircraft-carrier *Tbilisi*. The *Tbilisi* airwing will probably consist of 20–40 aircraft, including Su-27 Flanker, MiG-29 Fulcrum, vertical/short take-off and landing aircraft, and helicopters. In addition to its airwing, the *Tbilisi* carries 12 SS-N-19 antiship cruise missiles in vertical launchers. Soviet sources had indicated that the *Tblisi* will not be nuclear-armed,[100] but with revelations in 1990 of the inclusion of the nuclear-capable SS-N-19 cruise missile as part of the ship's armament, these reports look more doubtful.

The *Tbilisi* is expected to be deployed with the Northern Fleet in late 1990 or 1991, and to complete fitting out and testing before it becomes fully operational in 1993 or 1994.[101] The second aircraft-carrier, the *Varyag* (formerly *Riga*), is currently fitting out in the Black Sea, while the third carrier, the *Ulyanovsk*, a larger 70 000- to 75 000-metric ton ship, is being built.[102] The USSR has indicated that the aircraft-carrier programme will end when the third unit is finished.

The attack and cruise missile submarine production rate for 1990 remained steady, with the deployment of Victor III, Sierra, Kilo and Akula Class attack submarines, and a single Oscar II Class cruise missile boat. All these submarines are nuclear-capable. The production output of these submarines will probably continue at about six or seven per year after 1990.

Deployment of the long-range nuclear SS-N-21 Sampson SLCM remains somewhat mysterious. According to the 1990 edition of *Soviet Military Power*, 'The SS-N-21 probably can be launched from any appropriately modified modern nuclear-powered general purpose submarine, and probably would be used primarily against Eurasian theatre strategic targets. Specific candidates for employment are Yankee-Notch, Akula, and possibly Victor III and Sierra Class nuclear-powered attack submarines (SSNs)'.[103] This statement changes the description of deployment by *Soviet Military Power* last year, adding the Sierra Class, and specifying the Victor III Class rather than all Victor classes.[104]

The total number of potential SS-N-21 delivery platforms is approximately 30 submarines, and the operational inventory of SLCMs in 1990 is estimated at a total of 136. However, according to the US DOD, a total of some 1000 SS-N-21 missiles have been produced since 1985.[105] The USA reported that a two-year SS-N-21 'improvement program' was completed in 1990, which,

[99] DOD, *SMP 1990*, p. 85.
[100] TASS, Moscow, 'Aircraft take off from new Soviet Tblisi carrier', 22 Nov. 1989.
[101] DOD, *SMP 1990*, p. 89.
[102] DOD, *SMP 1990*, p. 84.
[103] DOD, *SMP 1990*, p. 53.
[104] DOD, *SMP 1989*, p. 47.
[105] DOD, *SMP 1990*, pp. 35, 38. In the pre-Gorbachev years, production of long-range (greater than 600 km) SLCMs averaged 35 missiles per year and has remained at the 200 missile per year level since 1985.

along with extensive flight-testing, could help to explain the otherwise excessive estimates of missile production rates.[106]

Soviet Naval Aviation (SNA) began to experience 'dramatic expansion and modernization' during 1989, a process that continued in 1990.[107] In addition to continued deployment of the Backfire bomber (offset by Badger retirements), approximately 290 Fencer, Fitter, Frogfoot and MiG-27 Flogger aircraft were transferred from the Air Forces to SNA, largely to avoid being counted in the CFE Treaty (see also chapter 13 in this volume).[108] The Northern and Baltic Fleets, in particular, have been strengthened with the formation of new units using aircraft previously assigned to the Soviet Air Forces.

Military reform in the Soviet Union

Gorbachev's push for 'new thinking' and his doctrine of 'reasonable sufficiency' in defence have now become firmly rooted in the armed forces. The shape and capabilities of the military have already been fundamentally altered. As one Soviet analyst working for the US Army wrote in December 1989, 'as we enter a new decade, in a number of respects the Soviet armed forces that we studied and assessed in 1988 no longer exist'.[109] These changes will affect all Soviet nuclear forces.

In late November, Chief of the General Staff General Mikhail Moiseyev unveiled a new, three-phase, 10-year reform and restructuring plan for the armed forces, which was forwarded to the Supreme Soviet. The components of this long-range reform plan are:[110]

1. Reworking of all operational and mobilization plans;
2. Completion of the withdrawal of Soviet troops from Czechoslovakia (1991), Germany (1994), Hungary (1991), Mongolia (1992),[111] and from Poland about 1994;
3. Creation of a new 'strategic grouping of armed forces on Soviet territory' including nuclear forces, and possible creation of separate territorial units for the 15 republics;

[106] 'Statement of RADM Thomas A. Brooks, Director of Naval Intelligence, before the Seapower, Strategic, and Critical Materials Subcommittee of the HASC on Intelligence Issues, 14 Mar. 1990', (mimeo), p. 32.

[107] Brooks (note 106), p. 29.

[108] DOD, *SMP 1990*, p. 79.

[109] Turbiville, G. H., Jr, 'Restructuring the Soviet ground forces: reduction–mobilization–force generation', *Military Review*, vol. 59, no. 11 (Dec. 1989), p. 17.

[110] 'Chief of Staff Moiseyev on military reforms', *Krasnaya Zvezda*, 20 Nov. 1990, FBIS-SOV-90-225, 21 Nov. 1990, pp. 49–54; 'Soviet military to cut arms, size of forces', *Washington Post*, 19 Nov. 1990, p. A29; Iams, J., 'Soviet military', Associated Press, 18 Nov. 1990; 'Assessments of armed forces reforms, reductions: Yazov interviewed on cuts', Moscow, 14 Sep. 1990, FBIS-SOV-90-183, 20 Sep. 1990, p. 53; Keller, B., 'Gorbachev vows to reorganize military', *New York Times*, 18 Aug. 1990; 'Yazov article on military reform plan', *Krasnaya Zvezda*, 5 June 1990, FBIS-SOV-90-109, 6 June 1990, pp. 61–65.

[111] 'Troop withdrawal from Mongolia', Moscow TASS, 17 May 1990, in FBIS-SOV-90-097, 18 May 1990, p. 1. By 1992, 50 000 men, 11 000 pieces of equipment, including 850 tanks and 1100 armoured vehicles, 820 artillery guns, 190 planes and 130 helicopters will be withdrawn from Mongolia.

4. Reorganization of the central command structure;[112]

5. Realignment of military districts,[113] and elimination of staff directorates;[114]

6. Reduction of the number of military educational and scientific research institutions;

7. Reduction of the number of military units, including a 30 per cent reduction in formations and units in the Strategic Rocket Forces, 10–12 per cent in the Ground Forces, 18–20 per cent in the Air Defence Troops, and 6–8 per cent in the Air Forces;[115]

8. Removal of the civil defence, military road building units, and military construction formations and organs from the armed forces;[116]

9. Reduction of the size of the armed forces from 4 million to 3–3.2 million personnel by the year 2000;

10 'Reorganize and amalgamate the branches of the armed forces', that is, eliminate some of the five branches of the armed forces;[117] and

11. Pay increases, to allow military personnel some competition in the new market economy.

IV. British nuclear weapon programmes

Reassessments

The easing of East–West relations, as well as budget constraints, have forced the UK to reassess its security policy. On 6 February Secretary of State for Defence Tom King notified the House of Commons that a review of Britain's armed forces was under way.

On 2 April the Defence White Paper was presented.[118] It focused on the events in Eastern Europe and proposed a budget of £21.2 billion. On 18 June Secretary King announced that the defence budget would be cut by £600

[112] Reductions have already been made 'in command and control agencies of the central administrative staff'; 'Moiseyev responds to readers on *perestroika* in armed forces', *Voyenno-Istoricheskiy Zhurnal*, no. 2 (1990), in *JPRS-Soviet Union-Military Affairs*, 4 June 1990, p. 23.

[113] The number of military districts is already declining. In 1989, the Central Asian and Turkestan MDs were merged, and the Ural and Volga MDs were merged (as were the Ural and Volga MDs). Reductions were also made in the staff of the military districts; see note 112.

[114] An Army directorate has already been abolished; see note 112.

[115] According to General Moiseyev in Feb. 1990, 'More than 50 combined units, units, and subunits of the Missile Forces and the Ground Forces, Air Defence Forces, and Air Forces stationed on the territory of our country have been inactivated' (note 112). 'It is also planned to cut by 30 per cent the number of administrative organs, large strategic formations, combined units, units, and institutions that are not part of branches of the Armed Forces and are not connected with the implementation of combat missions . . .'; 'Chief of Staff Moiseyev on military reforms', *Krasnaya Zvezda*, 20 Nov. 1990, FBIS-SOV-90-225, 21 Nov. 1990, p. 52.

[116] A directive from the Presidium of the Supreme Soviet already removed the KGB Border Troops, Internal Affairs Ministry (MVD) internal troops, and the MOD railroad construction troops from the armed forces in Mar. 1989. This reduced the size of the armed forces by 600 000 men. See Turbiville (note 109), p. 18.

[117] 'Chief of Staff Moiseyev on military reforms', *Krasnaya Zvezda*, 20 Nov. 1990, FBIS-SOV-90-225, 21 Nov. 1990, p. 52.

[118] British Ministry of Defence, *Statement on the Defence Estimates 1990* (Her Majesty's Stationery Office: London, Apr. 1990), 2 vols, CM-1022-I, CM-1022-II.

million. This included the cancellation of 33 Tornado aircraft (26 GR Mk 1 and 7 F Mk 3) and major reductions in army and air force units in Germany.

On 25 July Secretary King announced the results of the *Options for Change* review, whose purpose was to identify how to restructure Britain's armed forces in light of the political and military changes in Europe.[119] It was generally held that the review was thorough and realistic and that the recommended cuts were deeper than expected.

For the most part British nuclear forces were spared. Earlier in the year the idea was floated that perhaps one Trident submarine could be cut from the scheduled four to save £500 million. But the review recommended to keep the programme at four. Overall there will be a regular service personnel cut of 18 per cent by the mid-1990s. While the pace of the reductions would depend on many factors, the Army would be reduced from 160 000 to 120 000, the Royal Air Force from 90 000 to 75 000, and the Royal Navy from 63 000 to 60 000. Civilians employed by the MOD could also be reduced by the same proportion. The Government also proposed that two squadrons of dual-capable Tornados replace the Buccaneer aircraft in the anti-shipping role.

Trident

In early October the Ministry of Defence announced that its planned purchase of 14 Trident II missiles from the USA would be postponed until FY 1992.[120] The MOD said it would not affect the schedule or the budget of the Trident programme. A contract, valued at £500 million, for the third submarine (*Vigilante*) was awarded to Vickers Shipbuilding and Engineering Ltd at the end of the year. Work continued on *HMS Vanguard*, the first boat due in 1994, and on the second boat, *HMS Victorious*.

Tactical air-to-surface missile

No decision was made during the year to replace the WE-177 free-fall bombs with an air-to-surface missile. Various options continued to be studied. Two US candidates under consideration are the SRAM-T made by Boeing Aerospace and a derivative of the Supersonic Low-Altitude Target (SLAT/AQM-127A) made by Martin Marietta. The third possibility is joint development with France of the Aérospatiale Air-Sol-Longue-Portée missile (see below).

[119] *Optims for Change, Statement by the Secretary of State for Defence, Mr Tom King*, CD 92/3 (Her Majesty's Stationery Office: London, 25 July 1990).

[120] Pringle, P., 'Britain "has delayed deal for Trident"', *The Independent*, 29 Sep. 1990, p. 2; Riddell, P. and Atkins, R., 'Defence budget squeeze delays Trident purchase', *Financial Times*, 1 Oct. 1990, p. 10.

V. French nuclear weapon programmes

At the start of the year French defence officials asserted that despite the events in Europe they were not going to make any serious changes to their military budget or forces. The first budget projections for 1991 actually called for a 5.5 per cent increase. By the end of the year, however, decisions had been taken to reassess fundamentally French military forces and budgets.[121] One review, to be completed next year, could overturn decades-long assumptions that underlie French security policy. In the past budgets were modestly reduced and programmes stretched out but never cancelled. With fiscal pressures and the new East–West climate, officials will be forced to set priorities and make hard decisions.

Prime Minister Michel Rocard stated on 22 October that French defence policy will continue to be based 'on a strategy of an autonomous and sufficient nuclear deterrent', and four days later President François Mitterrand announced that the Defence Council would meet by the end of the year to decide how nuclear forces should be modernized. It is possible that Mitterrand may cancel the S4 missile programme, which would mean that the IRBM leg of the *force de dissuasion* would end at the turn of the century after retirement of the S3 missiles.[122]

Force Oceanique Strategique

Le Terrible, the fourth submarine to be refitted with M4 SLBMs, was completed in the summer. This leaves only one more submarine, *Le Foudroyant*, which will complete its refit in 1993. The number of warheads on French strategic submarines has grown more than fourfold with these conversions, from 96 to 416 warheads.

S4 IRBM and ASLP

A programme to upgrade the French land-based IRBM force had been planned, at a cost of 30 billion francs. The S4 missile was to have replaced the S3 at the turn of the century. President Mitterrand is apparently considering dropping the programme and thus abandoning one leg of the French triad when the S3 is eventually retired.[123] The alternative is to develop a 1500-km range missile known as the Air-Sol-Longue-Portée (ASLP). Discussions have been going on for three years with the UK to jointly develop the ASLP, but no final decision had been made by the end of the year.

[121] de Galard, J., 'Nucleaire strategique: la fin de la triade?', *Air & Cosmos*, no. 1308 (Dec. 1990), p. 44; de Briganti, G., 'French will improve military ties in Europe', *Defense News*, 3 Dec. 1990, p. 8.

[122] de Galard (note 121).

[123] de Galard (note 121).

Rafale

The Rafale fighter-bomber aircraft moved closer towards it first flight, scheduled for February 1991.[124] The Rafale D is designed to replace the Mirage IVP beginning in 1998 in the air-to-ground role. The plane is also designed for air defence, air superiority and reconnaissance missions. The French Air Force wants 250 of the planes. The Navy plans an additional 86 of the M version to replace the Crusader aircraft currently on their aircraft-carriers. Naval versions will be delivered first. The first pre-production aircraft was rolled out on 29 October.[125] The first flight of the naval version is scheduled for November 1991, with deliveries to begin in 1996 and service entry in late 1998 to coincide with commissioning of the *Charles de Gaulle* aircraft-carrier. The Rafale will initially carry the nuclear-armed Air-Sol-Moyenne-Portée (ASMP) air-to-surface missile, which may be replaced or supplemented by the ASLP that may be developed with the UK.

Hadès

In July President Mitterrand decided to begin production of the Hadès short-range nuclear missile. A fifth test-flight took place at the Landes Testing Centre. The ballistic missile, with a range of 350–480 km, will have a variable-yield warhead of up to 80 kt. Plans to use an enhanced radiation ('neutron bomb') warhead were dropped.[126] The original programme called for 120 missiles. This was cut to 90, and the final number may be as few as 30 missiles.

VI. Chinese nuclear weapon programmes

During 1990, following the June 1989 massacre in Tiananmen Square, the Chinese Government provided less information than usual about its nuclear weapon programme developments, which is to say very little indeed.[127] Thus, while we do not know of any specific development in the Chinese nuclear weapon programme, it is assumed that previous nuclear modernization decisions are being implemented at a modest pace. A new generation of Chinese solid-fuelled ballistic missiles has ben under development for some years, and details may appear in the near future. China conducted two nuclear tests in 1990, on 25 May and 16 August, with yields estimated at 15–65 kt and 50–200 kt, respectively. China had not held two tests in one year since 1983.

[124] Lenorovitz, J. M., 'Rafale debut opens way for flight test program', *Aviation Week & Space Technology*, vol. 133, no. 19 (5 Nov. 1990), pp. 20–22; Finance, R., 'The Rafale—combat aircraft for Europe', *NATO's Sixteen Nations*, vol. 35, no. 5 (Sep. 1990), pp. 81–84.

[125] 'Rafale C: lever de rideau', *Air & Cosmos*, 3 Nov. 1990, pp. 13–15.

[126] Guisnel, J., 'Hadès missile on production line', *Liberation*, 13 July 1990, p. 6, trans. in FBIS-WEU-90-154, 9 Aug. 1990, pp. 9–10.

[127] Sun, L. H., 'Perils of China-watching', *Washington Post*, 17 Dec. 1990, p. A14.

Sino-Soviet relations

On 23–26 April 1990, Chinese Premier Li Peng visited Moscow to meet with President Gorbachev and other Soviet leaders. The trip reciprocated Gorbachev's historic trip to Beijing in May 1989. On 24 April the two nations signed an agreement for reducing troop levels along their common border 'to a minimum corresponding to good-neighborly relations'.[128] Along with Soviet nuclear weapon reductions under the US–Soviet INF Treaty, the pending START treaty and unilateral Soviet reductions, the border demilitarization agreement will reduce the nuclear aspect of the Sino-Soviet relationship and could help minimize Chinese nuclear modernization incentives. Foreign diplomats in Beijing reportedly acknowledged that China is even seeking military technology from the USSR, particularly aviation technology.[129]

Other developments

There were two reports in 1990 of untoward Chinese nuclear behaviour: one shedding light on China's 1988 detonation of a neutron bomb, the other concerning the export of chemicals to Iraq. In November 1990 it was reported in the USA that China had illegally obtained secret data for its experimental neutron bomb design from the Lawrence Livermore National Laboratory, California, which hosted 41 Chinese visitors over a 21-month period from January 1986 to September 1987.[130] Besides adding evidence to previous reports that the 29 September 1988 test was a neutron bomb design, the news accounts quote a US Federal Bureau of Investigation official as saying that China had the 'most aggressive' foreign spying operation against the USA. It is not known whether China will continue development of tactical warheads such as neutron bombs, particularly in light of improving relations with its neighbours, especially the USSR.

In September and October 1990 *The Independent* reported that the China Wanbao Engineering Company, a subsidiary of the huge state arms export company North China Industries Corp. (NORINCO), had secretly agreed to provide Iraq with seven tons of lithium hydride, a chemical that can be used in the manufacture of nuclear weapons, missile propellant or chemical weapons.[131] It is not clear what purpose Iraq intended for the chemical, but the

[128] Dobbs, M., 'Chinese, Soviets sign troop-cutback pact', *Washington Post*, 25 Apr. 1990, p. A29.

[129] Southerland, D., 'China seeks technology from Soviet military', *Washington Post*, 17 July 1990, p. A12.

[130] Stober, D., 'Lab secrets stolen: Chinese allegedly used data for bomb', *San Jose Mercury News*, 21 Nov. 1990, p. 1A; Wines, M., 'Chinese atom-arms spying in U.S. reported', *New York Times*, 22 Nov. 1990, p. 5. In 1988 the US GAO reported that 118 Chinese nationals visited the three US nuclear weapon laboratories (Lawrence Livermore, Los Alamos and Sandia) in this 21-month period, two of them at LLNL discussing manufacturing details of high-speed cameras used for nuclear weapon diagnostic purposes; US General Accounting Office, *Nuclear Nonproliferation: Major Weaknesses in Foreign Visitor Controls at Weapons Laboratories*, GAO/RCED-89-31, 12 Oct. 1988.

[131] Kelsey, T., 'China ships vital nuclear cargo to Iraq', *The Independent*, 30 Sep. 1990, p. 1; Higgins A. and Kelsey, T., 'Peking's arms undo its good work', *The Independent*, 30 Sep. 1990, p. 19; Kelsey T. and Higgins, A., 'Pressure grows on China over nuclear sale to Iraq', *The Independent*, 7 Oct. 1990, p. 1; Sun, L. H., 'Chinese said to sell chemical to Iraq', *Washington Post*, 1 Oct. 1990, p. A18.

majority of the sale (five tons) was apparently halted. It appears that the company was not under Beijing's control when it made the agreement in violation of the trade embargo against Iraq. This raises the question of whether the Chinese Government can exercise control over the large military–industrial bureaucracy that makes and sells military technology with a nuclear and missile proliferation potential.

2. Nuclear explosions

RAGNHILD FERM

I. Introduction

The total number of nuclear tests conducted in 1990 was 18, the lowest for 30 years. The USSR conducted only one test, and the USA and France carried out fewer tests than in any previous year in the past decade. China, on the other hand, carried out two tests, its highest annual number since 1983.

II. Nuclear explosions in 1990

US and British explosions

The USA carried out eight nuclear tests in 1990, one of which—on 6 April—was unannounced. The British test on 14 November, conducted (as usual) in co-operation with the USA at the Nevada Test Site, was somewhat delayed because of the presence at the test site of British anti-nuclear activists, trying to stop the test.[1] Even before the British test in December 1989 the Governor of Nevada had complained that the continued testing programme caused damage to the environment and to the health of the inhabitants of the state.[2]

Soviet explosions

Until recently the Soviet Union has not published information on the nuclear explosions it conducted before 1985. SIPRI records on Soviet explosions have been based on information from various non-Soviet sources, for example the Swedish National Defence Research Institute (FOA), the US Department of Energy (DOE), the US Geological Survey and the Natural Resources Defense Council (Washington, DC). According to information based on these and other sources, the USSR conducted 649 nuclear explosions in the period 1949–90. In September 1990, however, the Deputy Minister of the Soviet Ministry of Atomic Energy and Industry revealed that, as of that date, as many as 714 Soviet nuclear explosions had been carried out since 1949, when the Soviet Union started its testing activities.[3] It is the number of atmospheric tests before 1963, when the Partial Test Ban Treaty (PTBT) prohibiting testing in the atmosphere was signed, that is considerably higher than was known before, but the figure for underground tests is also somewhat higher. Many experts in the West have long suspected that the total number of Soviet explosions was higher than the accepted figure.

The two main test sites in the USSR are at Semipalatinsk in East Kazakhstan and on Novaya Zemlya in the Barents Sea. At the Semipalatinsk

[1] *The Times*, 15 Nov. 1990.
[2] *The Guardian*, 12 Nov. 1990.
[3] *Krasnaya Zvezda*, 13 Sep. 1990; *Pravda*, 24 Oct. 1990.

test site no nuclear explosions have been conducted for more than a year, mainly because of the growing opposition to the testing activities from environmentalists, medical experts, politicians and peace movements. The environmental conditions in the Semipalatinsk area have been bad for a long time. The test site was set up in 1948, and (according to the new Soviet information) more than 100 atmospheric nuclear tests were conducted there before 1963. A scientific commission, appointed in 1989 by the Soviet Council of Ministers to examine health conditions and the ecological situation in Semipalatinsk, came to the conclusion that people in the area still suffer from the effects of fall-out from these explosions, and children of the second generation of nuclear test victims have inefficient immune systems.[4]

The great majority of Soviet tests after 1963—all conducted underground—were carried out at the Semipalatinsk test site, and obviously even these tests have been hazardous: scientists claim that seismic electromagnetic effects as well as leakage of radioactive gases into the atmosphere have had negative medical consequences for people in the region.[5] In 1989 the Supreme Soviet urged the Soviet Defence Ministry and the Ministry of Atomic Energy and Industry to consider terminating nuclear testing at Semipalatinsk,[6] and in 1990 a decision was taken to stop all testing there by January 1993. Nineteen more explosions are planned before the closure,[7] but the Kazakh parliament maintains that all tests in Semipalatinsk must be stopped immediately.[8]

Novaya Zemlya has been used for Soviet nuclear testing since 1957. About 100 atmospheric nuclear tests have been carried out there, the largest, in October 1961, with a yield of approximately 58 Mt (nearly four times larger than the largest US test).[9] Since the conclusion of the PTBT, the test site has been used only a few times each year, for the largest explosions.

The USSR conducted its first and only nuclear test in 1990 on 24 October at this test site, bringing the total to 715. The explosion attracted much attention, not only because it was the first Soviet test for a year and the first conducted on Novaya Zemlya for almost two years, but also because it was regarded as a confirmation that the USSR had now definitely decided to close the Semipalatinsk test site and start a testing programme at Novaya Zemlya. The Supreme Soviet claimed that it had not approved any decision to carry out the test in October, and the Soviet Department of the Environment protested that it

[4] Soviet Commission of People's Deputies, *Report to the Supreme Soviet's Subcommittee on Power Engineering and Nuclear Ecology of the Committee on Ecology and Rational Use of Natural Resources*, 8 Feb. 1990 (translated by the Soviet Committee of Physicians for the Prevention of Nuclear War); International Physicians for the Prevention of Nuclear War and Institute for Energy and Environmental Research, Radioactive Heaven and Earth: The Health and Environmental Effects of Nuclear Weapons Testing in, on, and above the Earth (Apex Press: New York, and Zed Press: London, 1991).

[5] See note 4.

[6] See SIPRI, *SIPRI Yearbook 1990: World Armaments and Disarmament* (Oxford University Press: Oxford, 1990), chapter 2.

[7] 'Semipalatinsk nuclear tests to end in 1993', Foreign Broadcast Information Service, *Daily Report–Soviet Union (FBIS-SOV)*, FBIS-SOV-90-234, 5 Dec. 1990.

[8] *Soviet Weekly*, 10 Jan. 1991, p. 1.

[9] In this chapter the term 'yield' is used to describe the size of the explosions. In appendix 2A, table 2A.1, only the body wave magnitude (m_b) is given. To be able to give a reasonably correct estimate of the yield it is necessary to have detailed information, for example on the geological conditions of the area where the test is conducted. Therefore, giving the m_b figure is an unambiguous way of listing the size of an explosion.

had not been informed by the Government or the military. A Soviet Foreign Ministry spokesman said that the test had to be conducted because the year-long Soviet test moratorium had affected the country's security.[10] The Supreme Soviet adopted a resolution urging the Government in the future to notify all concerned authorities in good time that a test is to be conducted.[11]

The Nordic countries as well as the inhabitants of the most northern parts of the USSR have expressed concern and protested to the Soviet authorities about possible expansion of testing activities on Novaya Zemlya, pointing out that the fragile Arctic environment should not be exposed to nuclear experiments.[12] In the past, leaks have been reported; the most well known in recent years occurred in August 1987, when radiation spread beyond the boundaries of the USSR. Iodine-131 was detected in Scandinavia and short-lived fission products were recorded in the air at ground level throughout Sweden.[13] Soviet authorities maintain that the Novaya Zemlya area is ideal for testing from the geological and meteorological viewpoints.[14]

The Soviet Council of Ministers announced in January 1991 that the USSR would observe a unilateral moratorium on nuclear tests for the following four months.[15]

French explosions

In 1989 the French Government announced that mainly because of budget restraints the number of nuclear tests in the future would be reduced to six per year.[16] Since 1984 France has conducted eight tests per year, and over the past 10-year period it has accounted for some 20 per cent of all nuclear explosions. Four of the six French tests in 1990 were conducted at Mururoa[17] and two at Fangataufa, a nearby atoll used for larger explosions.

In June 1987 the French marine biologist Commander Jacques-Yves Cousteau and his team visited the Mururoa atoll and its surroundings and took water and sediment samples after a 15-kt nuclear explosion. They discovered damage to the base of the atoll down to 230 m but judged that the short-term risks of the explosions were negligible. However, the team was allowed to stay for only a few days and the sampling zones were restricted.[18]

The Cousteau expedition found significantly elevated levels of caesium-134 in samples from two separate locations. The French Government argued that

[10] 'Explains nuclear test', FBIS-SOV-90-209, 29 Oct. 1990, p. 3.

[11] 'Resolutions adopted on Novaya Zemlya nuclear test', FBIS-SOV-90-211, 31 Oct. 1990, p. 30.

[12] *Press Release*, Ministry for Foreign Affairs, Stockholm, 27 Nov. 1990.

[13] Bjurman, B., *et al.*, *The Detection in Sweden of Short-lived Fission Products Probably Vented from the Underground Nuclear Test at Novaya Zemlya on 2 August 1987*, FOA Rapport C.20673–9.2 (FOA: Stockholm, Sep. 1987).

[14] Litovkin, V., 'The North's cold "secret"', interview with Vice Admiral G. Zolotukhin, *Soviet Weekly*, 21 June 1990.

[15] *Soviet Weekly*, 17 Jan. 1991.

[16] *Le Monde*, 9 June 1989.

[17] The Polynesian spelling of the name of the atoll is Moruroa (big secret). The French misunderstood the name and used the spelling Mururoa. Since this version has been used in official documents, issued by the International Court of Justice, the United Nations, etc., SIPRI has adopted it although it is not the correct spelling.

[18] Fondation Cousteau, Mission Scientifique de la Calypso, *Sur le Site d'Experimentations Nucleaires de Mururoa*, Paris, Nov. 1988.

this was caused by the atmospheric explosions carried out between 1966 and 1974. The Cousteau report has now been re-evaluated by a US oceanographer and physicist, Norm Buske,[19] who claims that since caesium-134 has a half-life of only two years, its abundance in the Cousteau samples cannot be explained by atmospheric explosions carried out more than 15 years ago. Buske argues that the isotopes must have leaked from the more recent underground nuclear explosions. Furthermore, two New Zealand researchers, Professor Manfred P. Hochstein and Michael J. O'Sullivan from the Auckland University Geothermal Institute,[20] using a computer simulation model have come to the conclusion that radioactive leakage from the tests could come to the surface in about 30 years, not in 1000 years as the French Commissariat à l'Energie Atomique (CEA) maintains.

During his visit to the South Pacific Council (set up by France in 1985 to give greater autonomy to French territories in the region), President François Mitterrand pointed out that there was now greater transparancy regarding the testing.[21] Announcements of tests carried out are now made by the French authorities. (In the past such information has been obtained from the New Zealand observatory on Rarotonga, Cook Islands and the Australian Seismological Centre, Canberra, among other sources.) Accordingly the test on 2 June, the first in 1990, was announced by the High Commissioner in French Polynesia; a yield estimate was also given—less than 15 kt. All subsequent French tests in 1990 were also officially reported. Mitterrand reaffirmed that the testing posed no threat to human health but said that, if it ever did, the French Government would make as much information available as possible.

None the less, the continued French testing in Polynesia attracts increasing criticism from scientists and experts as well as from other countries in the region and environmental organizations, who argue that any indication of leakage from underground explosions would require that the French nuclear testing programme in the South Pacific be discontinued.

Chinese explosions

China carried out two tests in 1990: on 26 May and 16 August. This was the first year in which China had conducted more than one test since 1983. According to newspaper reports the first explosion, with a yield of 15–65 kt, was a test of a hydrogen bomb.[22] The yield of the second explosion was estimated at 50–200 kt, probably 150 kt, which is considered remarkably high.[23]

[19] Buske, N., SEARCH technical services, *Cesium-134 at Moruroa—Review of the Calypso Water Samples* (SEARCH: Davenport, Wash., Sep. 1990).

[20] Hochstein, M. P. and O'Sullivan, M. J., 'Geothermal systems created by underground nuclear testing: implications for long-term, direct effects of underground testing', revised version of paper presented at the International Scientific Symposium on a Nuclear Test Ban, Las Vegas, Nev., 15–16 Jan. 1988 (mimeo).

[21] *Le Monde*, 20–21 May 1990.

[22] *Dagens Nyheter*, 14 Nov. 1990, quoting *Wen Wei Po* (Hong Kong).

[23] Swedish National Defence Research Institute (FOA), personal communication, Nov. 1990.

Appendix 2A. Nuclear explosions, 1945–90

Table 2A.1. Registered nuclear explosions in 1990

Date	Origin time (GMT)	Latitude (deg)	Longitude (deg)	Region	Body wave magnitude[a]
USA					
10 Mar.	160000.0	37.113 N	116.055 W	Nevada	5.4
6 Apr.	1700	37. N	116. W	Nevada	
13 June	160000.0	37.262 N	116.420 W	Nevada	5.8
21 June	181500.0	36.993 N	116.004 W	Nevada	
25 July	150000.0	37.207 N	116.214 W	Nevada	4.8
20 Sep.	171500.0	37. N	116. W	Nevada	
27 Sep.	180200.0			Nevada	
12 Oct.	173000.0	37.248 N	116.494 W	Nevada	5.6
USSR					
24 Oct.	145758.0	73.364 N	54.827 E	Novaya Zemlya	6.3
UK					
14 Nov.	191700.7	37.227 N	116.371 W	Nevada	5.7
France					
2 June	172958.7	21.877 S	138.918 W	Mururoa	
7 June	173000.0	21. S	138. W	Mururoa	
26 June	175958.2	22.215 S	138.841 W	Fangataufa	
4 July	175958.4	21.866 S	139.046 W	Mururoa	5.1
14 Nov.	181158.3	22.147 S	138.852 W	Fangataufa	
21 Nov.	165958.0	21.906 S	138.960 W	Mururoa	
China					
26 May	075957.8	41.566 N	88.688 E	Lop Nor	5.8
16 Aug.	045957.7	41.586 N	88.799 E	Lop Nor	6.7

[a] Body wave magnitude (m_b) indicates the size of the event. To be able to give a reasonably correct estimate of yield it is necessary to have detailed information, for example on the geological conditions of the area where the test is conducted. Therefore, to give the m_b figure is an unambiguous way of listing the size of an explosion. m_b data for the US, Soviet and British tests were provided by the Hagfors Observatory of the Swedish National Defence Research Institute (FOA) and data for the French tests by the Australian Seismological Centre, Bureau of Mineral Resources, Canberra.

Table 2A.2. Estimated number of nuclear explosions 16 July 1945–5 August 1963 (the signing of the Partial Test Ban Treaty)

a = atmospheric; u = underground

	USA		USSR		UK		France		
Year	a	u	a	u	a	u	a	u	Total
1945	3	0							**3**
1946	2[a]	0							**2**
1947	0	0							**0**
1948	3	0							**3**
1949	0	0	1	0					**1**
1950	0	0	0	0					**0**
1951	15	1	2	0					**18**
1952	10	0	0	0	1	0			**11**
1953	11	0	4	0	2	0			**17**
1954	6	0	7	0	0	0			**13**
1955	17[a]	1	5[a]	0	0	0			**23**
1956	18	0	9	0	6	0			**33**
1957	27	5	15[a]	0	7	0			**54**
1958	62[b]	15	29	0	5	0			**111**
1949–58, exact years unknown			18[e]						**18**
1959	0	0	0	0	0	0			**0**[d]
1960	0	0	0	0	0	0	3	0	**3**[d]
1961	0	10	50[a]	1[c]	0	0	1	1	**63**[d]
1962	39[a]	57	43	1[c]	0	2	0	1	**143**
1 Jan.– 5 Aug. 1963	4	25	0	0	0	0	0	2	**31**
Total	**217**	**114**	**183**[e] **(215)**[f]	**2**[c]	**21**	**2**	**4**	**4**	**547** **(578)**[f]

[a] One of these tests was carried out under water.
[b] Two of these tests were carried out under water.
[c] These underground tests are not confirmed by the new Soviet information.
[d] The UK, the USA and the USSR observed a moratorium on testing, Nov. 1958–Sep. 1961.
[e] The total figure for Soviet atmospheric tests includes the 18 additional tests conducted in the period 1949–58, for which exact years are not available.
[f] The totals in brackets include the explosions revealed by Soviet authorities in Sep. 1990, the exact years for which are not known.

Table 2A.3. Estimated number of nuclear explosions 6 August 1963–31 December 1990

a = atmospheric; u = underground

	USA[a]		USSR		UK[a]		France		China		India		
Year	a	u	a	u	a	u	a	u	a	u	a	u	Total
6 Aug.–31 Dec. 1963	0	15	0	0	0	0	0	1					**16**
1964	0	38	0	6	0	1	0	3	1	0			**49**
1965	0	36	0	10	0	1	0	4	1	0			**52**
1966	0	43	0	15	0	0	5[b]	1	3	0			**67**
1967	0	34	0	17	0	0	3	0	2	0			**56**
1968	0	45[c]	0	15	0	0	5	0	1	0			**66**
1969	0	38	0	16	0	0	0	0	1	1			**56**
1970	0	35	0	17	0	0	8	0	1	0			**61**

Table 2A.3. *cont.*

	USA[a]		USSR		UK[a]		France		China		India		
Year	a	u	a	u	a	u	a	u	a	u	a	u	Total
1971	0	17	0	19	0	0	5[b]	0	1	0			**42**
1972	0	18	0	22	0	0	3	0	2	0			**45**
1973	0	16[d]	0	14	0	0	5	0	1	0			**36**
1974	0	14	0	18	0	1	7[b]	0	1	0	0	1	**42**
1975	0	20	0	15	0	0	0	2	0	1	0	0	**38**
1976	0	18	0	18	0	1	0	4	3	1	0	0	**45**
1977	0	19	0	18	0	0	0	8[e]	1	0	0	0	**46**
1978	0	17	0	27	0	2	0	8	2	1	0	0	**57**
1979	0	15	0	29	0	1	0	9	1[f]	0	0	0	**55**
1980	0	14	0	21	0	3	0	13	1	0	0	0	**52**
1981	0	16	0	22	0	1	0	12	0	0	0	0	**51**
1982	0	18	0	32	0	1	0	6	0	1	0	0	**58**
1983	0	17	0	27	0	1	0	9	0	2	0	0	**56**
1984	0	17	0	29	0	2	0	8	0	2	0	0	**58**
1985	0	17	0	9[g]	0	1	0	8	0	0	0	0	**35**
1986	0	14	0	0[g]	0	1	0	8	0	0	0	0	**23**
1987	0	14	0	23	0	1	0	8	0	1	0	0	**47**
1988	0	14	0	17	0	0	0	8	0	1	0	0	**40**
1989	0	11	0	7	0	1	0	8	0	0	0	0	**27**
1990	0	8	0	1	0	1	0	6	0	2	0	0	**18**
Total	**0**	**598**	**0**	**464** (500)[h]	**0**	**20**	**41**	**134**	**23**	**13**	**0**	**1**	**1 294** (1 330)[h]

[a] See note *a* table 2A.4.
[b] One more test was conducted this year, but it did not cause any detonation.
[c] Five devices used simultaneously in the same test are counted here as one explosion.
[d] Three devices used simultaneously in the same test are counted here as one explosion.
[e] Two of these tests may have been conducted in 1975 or 1976.
[f] This explosion may have been conducted underground.
[g] The USSR observed a unilateral moratorium on testing, Aug. 1985–Feb. 1987.
[h] See note *f*, table 2A.2.

Table 2A.4. Estimated number of nuclear explosions 16 July 1945–31 Dec. 1990

USA[a]	USSR[b]	UK[a]	France	China	India	Total
929	649 (715)	43	183	36	1	1 841 (1 907)[b]

[a] All British tests from 1962 have been conducted jointly with the United States at the Nevada Test Site. Therefore, the number of US tests is actually higher than indicated here.
[b] The figures in brackets include additional tests announced by the Soviet authorities in Sep. 1990.

Sources for tables 2A.1–2A.4

Swedish National Defence Research Institute (FOA), various estimates; Norris, R. S., Cochran, T. B. and Arkin, W. M., 'Known US nuclear tests July 1945 to 31 December 1988', *Nuclear Weapons Databook*, Working Paper no. 86–2 (Rev. 2C) (Natural Resources Defense Council: Washington, DC, Jan. 1989); Reports from the Australian Seismological Centre, Bureau of Mineral Resources, Geology and Geophysics, Canberra; Cochran, T. B., Arkin, W. M., Norris, R. S. and Sands, J. I., *Nuclear Weapons Databook, Vol. IV, Soviet Nuclear Weapons* (Harper & Row: New York, 1989), chapter 10; Burrows, A. S., *et al.*, 'French nuclear testing, 1960–88', *Nuclear Weapons Databook*, Working Paper no. 89–1 (NRDC: Washington, DC, Feb. 1989); 'Known Chinese nuclear tests, 1964–1988', *Bulletin of the Atomic Scientists*, vol. 45, no. 8 (Oct. 1989), p. 48, see also vol. 45, no. 9 (Nov., p. 52); and various estimates.

3. Military use of outer space

JOHN PIKE

I. Introduction[1]

US military space and strategic defence systems, focused for three decades on the USSR, underwent a profound reorientation in 1990 towards Third World contingencies, embodied in Iraq. Operations Desert Shield and Desert Storm displayed an unprecedented integration of military space systems in support of ground operations, amounting to a revolution in the conduct of warfare.[2] These space support functions were provided to a broader range of forces, in a more direct and timely manner, and over a more extended period of time than in any previous conflict. The lessons learned will inform military planning for decades to come. Planners in the USSR, Europe and in other countries will increasingly have to take into account these developments.

The Persian Gulf conflict also occasioned a transformation of the US Strategic Defense Initiative (SDI). Conceived in 1983 as a counter to a massive Soviet nuclear missile attack, by early 1990 the fading of the cold war had vitiated the initial impetus of the programme. Congress imposed significant reductions on the 1991 SDI budget, slashing the Bush Administration's request by 34 per cent. Although developed and tested before the start of the SDI research and development (R&D) programme, the success of the Patriot missile intercept system against Iraqi Scud missiles gave the SDI a new reason for being, now increasingly oriented to meet tactical and theatre as well as limited strategic missile threats. Research, development, testing and evaluation (RDT&E) of current and planned SDI systems continue, although budget cuts have led to the elimination of some of the more cost-intensive programmes.

In the USSR, upgrades to the existing anti-missile system continue. The completion of the Pechora radar early-warning network is expected in the early 1990s. The dismantlement of the Krasnoyarsk phased-array radar has begun, in compliance with agreements reached with the USA over interpretation of the 1972 Anti-Ballistic Missile (ABM) Treaty. This progress notwithstanding, in the area of arms control and military space activities the USA and the USSR are still divided over a number of fundamental issues, both with regard to the ABM Treaty and to the Defence and Space Talks.

[1] Many aspects of the discussion of Soviet military space activities are based on discussions with Nicholas Johnson, as well as Michael Cassut, Geoffrey Perry and Saunders Kramer. Their works have provided the essential core of insight into the Soviet space effort. Although their individual contributions are not adequately recognized in the following footnotes, many of the sources cited here are based on their careful monitoring of Soviet flight activity (most of the *Aerospace Daily* articles on the Soviet space effort cite Perry, and many of the *Defense Daily* articles on Soviet activities cite Kramer).

[2] The deadline for the preparation of this chapter has mandated a cut-off in the discussion of military space systems at the point of transition from Desert Shield to Desert Storm. See also chapter 19 in this volume.

While these developments dominated the agenda in 1990, other events made it increasingly clear that military space is no longer the sole domain of the USA and the USSR. The year witnessed an increasing proliferation of space systems and space technology, with China, Israel, Pakistan and the United Kingdom each launching a satellite into orbit.

II. US strategic defence programmes

After more than seven years of effort devoted to the development of components and systems oriented towards defending against a concerted attack by Soviet missiles, the Strategic Defense Initiative was reoriented in late 1990 into a Global Protection Against Limited Strikes (G-PALS) system, to defend against tactical and theatre missile threats, as well as limited ICBM strikes against the USA. The G-PALS system is intended to intercept up to 200 long-range ICBM or SLBM warheads aimed at the USA, as well as to defend US forces and allies against tactical and theatre missiles.[3] The new system will be deployed in three stages: a Transportable Protection Against Limited Strikes (T-PALS), which would be an air-transportable system to defend against theatre missiles; a Continental US system (C-PALS), with ground-based interceptors deployed at multiple sites and Brilliant Eyes sensors; and the global system (G-PALS), with space-based Brilliant Pebbles (BP) interceptors.[4]

The order-of-magnitude reduction in the scope of the mission of defending the northern USA has not led to a corresponding reduction in the size of the programme: the space-based components and the ground-based interceptors, totalling 1000 each, constitute 25 per cent and 50 per cent of the previous number, respectively (see table 3.1). The total estimated cost of deploying the G-PALS system was in the range of about $40 billion.

The proliferation of missile technology as a new rationale for SDI found growing support in the Bush Administration and among its backers in 1990, and was also likely to revive congressional endorsement of the programme.[5]

Strategic defence system components and deployment plans

Space-based interceptors and Brilliant Pebbles

The hallmark of the SDI since 1983 has been an initial layer of space-based interceptors that home in on the hot exhaust plume of hostile missiles during their first minutes of flight. The aim is to destroy them before they can deploy multiple and decoy warheads that would stress the performance of subsequent layers of the defence. Originally, plans for this layer of the system called for Space-Based Interceptor (SBI) rockets. A major change in these plans came in

[3] 'SDIO retools for limited threats', *SDI Monitor*, 21 Dec. 1990, pp. 281–82.

[4] 'SDIO works up three limited-strike protection plans', *SDI Monitor*, 18 Jan. 1991, p. 21.

[5] See Defense Secretary Richard Cheney's *News Briefing on FY92 Defense Budget*, 4 Feb. 1991; 'Text of President Bush's State of the Union Message to the nation', *New York Times*, 30 Jan. 1991, p. A12; 'Administration's SDI refocus supported by Nunn', *Defense Daily*, 31 Jan. 1991, pp. 14–15.

Table 3.1. SDI system architecture evolution, 1987–90

	1987	1988	1989	1990
System	Phase One			G-PALS
Cost (FY1988 US $b.)	115	69	55	41
Warheads intercepted	3 000	3 000	3 000	200
Boost-phase space-based interceptors				
Space-Based Interceptors	3 000	1 500	. .	. .
Garages	300	150		
'Brilliant Pebbles'	. .	. .	4 600	1 000
Late mid-course ground-based interceptors				
Ground-based interceptors	1 000	2 000	2 000	750 or
Exoatmospheric-Endoatmospheric Interceptor	. .	. .	. .	1 000
Space-based sensors				
BSTS boost surveillance	12	8	none	none
SSTS mid-course tracking	40	20	18	none
'Brilliant Eyes' mid-course tracking	. .	. .	200?	50–80
Ground-based sensors				
GSTS ground surveillance	. .	some	some	some?
Ground Based Radar	. .	4	4	some?

Sources: Finnegan, P., 'SDIO shifts focus, prepares for cuts', *Space News*, 8 Oct. 1990, pp. 1, 16; 'Ground-based radar moved to demonstration/validation', *SDI Monitor*, 20 July 1990, pp. 163–64; Strobel, W., 'Limited SDI program might cost $9 billion', *Washington Times*, 1 Feb. 1991, p. A3.

early 1989 with adoption of the Brilliant Pebbles concept (the name implying improved capabilities compared with the SBI 'Smart Rocks').[6]

Despite the enthusiastic claims of advocates,[7] a review of the project by the Pentagon's Defense Science Board took a more restrained view, concluding that BP is not a ready replacement for SBI, as it still lacks final design and a programme acquisition strategy.[8] A review of the BP concept by the JASONs (a high-level group of Defense Department scientists) also raised doubts about the project,[9] as did other analysts who raised concerns about the problems posed by controlling such a large number of satellites.[10]

[6] The development, function and capabilities of SBI are discussed in further detail in Pike, J., 'Military use of outer space', SIPRI, *SIPRI Yearbook 1990: World Armaments and Disarmament* (Oxford University Press: Oxford, 1990), pp. 61–63. See also Bennet, R., 'Brilliant Pebbles', *Reader's Digest*, Sep. 1989, pp. 128–32, which provides a useful though uncritical background.

[7] 'Brilliant Pebbles: the revolutionary idea for strategic defense', *Heritage Foundation Backgrounder*, no. 748 (25 Jan. 1990).

[8] Defense Science Board, *SDIO Brilliant Pebbles Space Based Interceptor Concept* (Office of the Under Secretary of Defense for Acquisition: Washington, DC, Dec. 1989).

[9] Perlman, D., 'Critical secret report on star wars', *San Francisco Chronicle*, 19 Feb. 1990.

[10] Garwin, R., 'Are Brilliant Pebbles all that brilliant?', *Aerospace America*, Dec. 1990, pp. 6, 8; 'BM/C3 researchers scramble to add Brilliant Pebbles', *SDI Monitor*, 30 Mar. 1990, p. 83.

The SDI Organization (SDIO) claims that BP would be capable of intercepting missiles with ranges as short as 600–800 km,[11] but the system would face a number of challenges in meeting the Third World missile threat, since these shorter-range missiles have much shorter boost-phases and flight-times, and can fly in depressed trajectories that bring them below the minimum intercept altitude of BP. The system would have to be modified, adding ultraviolet and radar sensors to their previously planned infra-red sensors, in order to intercept theatre missiles prior to their re-entry into the atmosphere.[12]

Ground-based interceptors

The second layer of defence would intercept missile warheads during the mid-course or terminal phases of their flight, just before or after they re-enter the atmosphere. The re-introduction in 1990 of terminal-phase interceptors into SDI deployment planning was a major departure from recent thinking, which had focused exclusively on late mid-course intercepts above the atmosphere.

Until quite recently, work on ground-based interceptors was based on the approach used in the Homing Overlay Experiment (HOE) that successfully intercepted a warhead in 1984. This was followed by the Exoatmospheric Reentry-vehicle Interception System (ERIS), which incorporated a much smaller and lighter kill vehicle. The relative progress among these three generations of interceptor is indicated by the mass of the kill vehicle, which dropped from near 1200 kg with the HOE, to less than 200 kg with ERIS, and to about 25 kg with GBI-X, a smaller and more sophisticated version of ERIS.

GBI-X faces new competition in the G-PALS system in the form of the Exoatmospheric–Endoatmospheric Interceptor (E^2I) which, according to SDIO Director Henry Cooper, would intercept warheads at very high altitudes 'where you can still get a very large footprint but sufficiently low that light decoys would be decelerated by the upper atmosphere'.[13] The name of this system is derived from its trajectory, which takes it outside the atmosphere, and then back into the atmosphere, permitting intercepts at ranges of up to 2000 km.[14] The interceptor would be launched in the general direction of an incoming threat cloud of warheads and decoys, on a trajectory that would take it well above the atmosphere. The interceptor then would re-enter the atmosphere, homing in on targets using an optical target acquisition for a hit-to-kill intercept.[15] Testing of this system, which will use a 40-kg kill vehicle, is slated to run from 1994 through 1996, following contractor selection in late

[11] US Department of Defense, *New Strategic Defense Initiative Program Focus: Global Protection Against Limited Strikes (GPALS)*, Fact Sheet (DOD: Washington, DC, 30 Jan. 1991).

[12] Finnegan, P., 'Brilliant Pebbles program may be redesigned to face Third World, mobile threats', *Defense News*, 10 Dec. 1990, p. 42.

[13] Bates, K., 'SDIO's Cooper says US could deploy Strategic Defense System for $40 billion', *Inside the Pentagon*, 20 Dec. 1990, pp. 10–11.

[14] Broad, W., 'As antimissile era dawns, planners eye panoply of weapons', *New York Times*, 5 Feb. 1991, pp. C1, C10.

[15] 'SDIO to choose between GBI-X, E2I for GPALS' ground-based tier', *Aerospace Daily*, 6 Feb. 1991, p. 210.

1991.[16] This effort is a follow-on to the High-altitude Endo-atmospheric Interceptor (HEDI), which would have intercepted warheads shortly after they begin to re-enter the atmosphere, at ranges of up to 200 km.[17]

Between 750 and 1000 ground-based interceptors would be deployed at five bases under the C-PALS. These will be either the GBI-X or E^2I; competitive selection will take place in the mid-1990s.

The Booster Surveillance and Tracking System

One factor that led to the adoption of Brilliant Pebbles was the claim that the sensors on these interceptor satellites would eliminate the need for large sensor satellites in geosynchronous orbits, such as the Booster Surveillance and Tracking System (BSTS). The elimination of the anti-missile mission requirement for BSTS in SDI led to a decision in 1990 to transfer budget authority for this programme back to the Air Force,[18] which sought to justify continuation of the programme, renamed the Advanced Warning System, on the basis of its improved early warning of missile attack, and enhanced intelligence collection and verification capabilities. The future of this project remains in doubt, since BSTS grew out of the Advanced Warning System which was rejected for deployment in 1983.[19] Compared with the current capabilities of Defense Support Program (DSP) early-warning satellites, the greater sensitivity of the BSTS sensors could improve the ability to track smaller Third World missiles.[20]

Mid-course sensors: Brilliant Eyes

The Brilliant Eyes concept marks a further step in the SDI evolution away from discrimination of real warheads from decoy warheads during the mid-course phase of their flight, as they coast through space prior to re-entering the earth's atmosphere. These 50–80 spacecraft would orbit at altitudes about twice that of the Brilliant Pebbles, or somewhat less than 1000 km.[21] Each spacecraft would be equipped with a combination of long-wavelength infra-red, visible light and laser radar sensors, for tracking targets in mid-course.[22] The 'Brilliant Eyes' constellation of satellites could replace some or all of the three mid-course sensors conceived for the Phase One SDI system (the Space Surveillance and Tracking System, SSTS; the Ground-based Surveillance and

[16] 'Strategic Defense Command begins new interceptor program', *Defense Daily*, 17 Dec. 1990, p. 429.

[17] Adams, P., 'Warhead interceptor will undergo eye checkup in 1990 test', *Defense News*, 28 Nov. 1988, p. 42.

[18] Lawler, A., 'Pentagon revamping BSTS; project moving to Air Force', *Space News*, 14 May 1990, pp. 1, 20.

[19] US General Accounting Office, *DOD Acquisition: Case Study of the Air Force Advanced Warning System*, Report no. GAO/NSIAD-86-45S-14 (GAO: Washington, DC, 31 July 1986).

[20] 'BSTS is in a "time of peril": hard', *SDI Monitor*, 6 July 1990, p. 150.

[21] Strobel, W., 'Limited SDI program might cost $9 billion', *Washington Times*, 1 Feb. 1991, p. A3.

[22] For a discussion of these Phase One sensors, see Pike (note 6), pp. 63–65. For their replacement by BP, see 'SDI constellation grows in brilliance', *Military Space*, 14 Jan. 1991, pp. 3–4.

Tracking System, GSTS; and the Ground Based Radar, GBR).[23] The initial test of mid-course tracking technology will come in the Midcourse Sensor Experiment (MSX), a $400 million 3000-kg spacecraft planned for launch on a modified Titan 2 booster in early 1993.[24]

Anti-tactical missile systems

The visible success of the Patriot anti-tactical missile in destroying Iraqi Scuds in flight over Israel and Saudi Arabia has sparked renewed interest in active defences against tactical missiles, which had waned in the wake of the 1987 INF Treaty. Current US programmes have been grouped under the new so-called Theater Missile Defense Initiative (TMDI).

These projects generally fall into three categories. First are the improved versions of the Army's Hawk and Patriot anti-aircraft missiles, modified to give them the ability to intercept ballistic missiles. The SDI budget includes $130 million in 1990 and $143 million in 1991 for development of more capable systems. Second are missile development efforts, such as the Extended Range Interceptor (ERINT) and the Israeli Chetz (Arrow), which have been under development in the SDI programme, although they were not part of the Phase One SDI deployment plan. Third are new operational requirements for missiles, the Theater High Altitude Area Defense (THAAD) system and the Corps Surface-to-Air Missile (Corps SAM), which may be met by either some of the missiles mentioned above, or newly developed systems.

The Phase Two SDI system

The SDI programme's new focus on near-term deployments of defences against tactical and theatre missiles has been accompanied by reduced emphasis on more advanced follow-on technologies.

The Ground-Based Free Electron Laser (GB-FEL) programme was long regarded as the centrepiece of the SDI programme. The Relay Mirror Experiment (RME) and Low-power Atmospheric Compensation Experiment (LACE) satellites were launched in February 1990 to demonstrate the technical feasibility of the system. In 1989 it was decided that a $500 million Technology Integration Experiment for this project would use a multi-megawatt radio frequency-driven beam generator developed by Boeing and the Los Alamos National Laboratory.[25] Work on this contract and project was terminated in late 1990 and the FEL programme was reduced to a research effort.[26]

The Zenith Star space-based chemical laser space test is intended to demonstrate the availability of a directed-energy weapon that can cope with future

[23] See Grossman, E., 'Small and light "brilliant eyes" could replace three SDI surveillance systems', *Inside the Army*, 28 May 1990, p. 15.

[24] 'Budget cuts push midcourse test into 1993', *SDI Monitor*, 16 Feb. 1990, p. 46.

[25] Gilmartin, P., 'Boeing Aerospace wins SDI contract for RF-driven free electron laser', *Aviation Week & Space Technology*, 23 Oct. 1989, p. 21.

[26] 'SDIO pulls in horns on FEL research', *SDI Monitor*, 21 Dec. 1990, pp. 285–86.

Soviet missiles.[27] Building on ground testing of the Alpha laser, the space-based 2-MW test, initially planned for 1994, has been further deferred to the late 1990s. A smaller Complementary Space Experiment orbital test of a chemical laser is planned for 1993.[28]

The Starlab pointing, tracking and fire control experiment is the third major directed-energy project to be cancelled in 1990. Initiated in 1978 as the Talon Gold, prime contractor Lockheed received several hundred million dollars for this experiment, which was slated to fly on the Shuttle in 1991.[29]

SDI testing

1990 marked a major transition for the SDI programme, as a number of early development projects finally culminated in laboratory or field tests of pre-prototype or 'breadboard' hardware. Indicating both the immaturity of the SDI test programme and the intrinsic challenge of the anti-missile mission, most of these initial tests experienced failures (see table 3.2).

Congressional reductions in the 1991 budget

In 1990, with the end of the cold war, Congress came to question the salience of the Soviet nuclear threat,[30] and as a result voted for significant changes in the 1991 SDI budget.[31] The final budget of $2.9 billion represents a significant reduction from the Administration's request of $4.4 billion.[32] Funding for the nuclear-pumped X-ray laser was eliminated as a separate item, although some research will continue.[33] The SDIO claimed that 'the severe FY 91 budget cut imposed at least a two year delay in deploying any defense for the US people'.[34] Perhaps more significant than the overall funding level was the decision that the SDI budget should be organized by mission areas[35] rather than by technology inputs.[36]

Much of the reduction was achieved by transferring projects from the SDIO to other agencies, rather than through outright cancellation. Over $800 million was appropriated for 1991 for projects which were originally part of the SDI, compared to about $200 million appropriated for projects in 1989 and 1990.

[27] US General Accounting Office, *Strategic Defense Initiative Program: Zenith Star Space-Based Chemical Laser Experiment*, Report no. GAO/NSIAD-89-118 (GAO: Washington, DC, Apr. 1989).

[28] Kiernan, V., 'Scaled-down test of Zenith star concept under DOD review', *Space News*, 11 Dec. 1989.

[29] Foley, T., 'Starlab to engage satellites, rockets, ground-based laser', *Aviation Week & Space Technology*, 26 Oct. 1987, pp. 58–59.

[30] See the Senate floor debate on SDI in *Congressional Record*, 4 Aug. 1990, pp. S12350–99.

[31] McDonald, B., 'Falling star: SDI's troubled seventh year', *Arms Control Today*, vol. 20, no. 7 (Sep. 1990), pp. 7–11.

[32] Kiernan, V., 'SDI handed big cuts, new rules', *Space News*, 22 Oct. 1990, pp. 1, 36.

[33] Henderson, B., 'X-ray laser research slashed as Congress cuts SDI funding', *Aviation Week & Space Technology*, 12 Nov. 1990, p. 29.

[34] O'Neil, M., *FY 92–93 Budget Brief*, SDI Organization Fact Sheet, 2 Feb. 1991.

[35] 'Restructuring the SDI program', *Congressional Record*, 23 Oct. 1990, pp. H-2169–70.

[36] Grossman, E., 'SDIO blueprints reveal plans for phase I funding under other program elements', *Inside the Army*, 24 Dec. 1990, pp. 1, 8–10.

Table 3.2. SDI systems testing in 1990

Date	System	Experiment	Success/ failure	Comment
26 Jan.	HEDI[a]	First test flight	Failure	Premature activation of Flight Destruct System
7 Feb.	HPB[b]	First test flight	Success	Modified Aries launcher
14 Feb.	SDI test satellites[c]	RME[d] LACE[e]	Limited success	Faulty navigation software Poor ground laser performance
29 Mar.	Firepond laser radar	Decoy discrimination	Success	Tracked target throughout flight at range over 800 km
19 Apr.	Alpha chemical laser	Power generation	Failure	45% of rated power; wrong prediction of exhaust flow
25 Apr.	Brilliant Pebbles	BSUVE[f]	Success	Gathered UV target and background signature data
27 Apr.	. .	EXCEDE-3[g]	Success	Nuclear explosion simulation
Early June	AOA[h]	Systems integration	Failure	Initial night flight; data processor failure
25 July	. .	SPEAR[i]	Failure	Aborted; guidance failure
9 Aug.	Chetz (Arrow)	First test flight	Failure	Initially reported as success
25 Aug.	Brilliant Pebbles	First test flight	Failure	Accidental termination of telemetry
19 Sep.	AOA[h]	PBV[j]/warhead tracking	Limited success	Successful track; satellite link communication failure
30 Nov.	Alpha chemical laser	Power generation	Success	100% of rated power achieved (2.2 MW)
17 Dec.	Starbird sounding rocket	First test flight	Success	Reached altitude over 100 km
21 Dec.	Chetz (Arrow)	Second test flight	Success	May have included intercept of a Jericho missile target

[a] High Endoatmospheric Defense Interceptor; the test demonstrated the cooling system's ability to cool the infra-red seeker faceplate, a main goal, but persistent problems with the Flight Destruct System led to delays, eliminating HEDI from consideration for deployment.

[b] High Performance Booster, a new booster system to be used in a number of future tests.

[c] Two satellites launched into orbit as hosts for RME and LACE experiments.

[d] Relay Mirror Experiment; tested the relay of a laser beam from a ground laser to an orbiting mirror to a target; delays eventually traced to faulty navigation software.

[e] Low-power Atmospheric Compensation Experiment; evaluated techniques for compensating laser beam for atmospheric distortion; experiment unsuccessful as result of unexpectedly dim retro-reflections, partially attributed to poor ground-based laser performance.

[f] Bow Shock Ultra-Violet Experiment.

[g] Excitation by Electron Deposition Experiment; payload used an electron accelerator to simulate the effects of nuclear explosions on the upper atmosphere at altitudes above 80 km.

[h] Airborne Optical Adjunct.

[i] Space Power Experiment Aboard Rocket (originally planned for 1989); destroyed after straying off course 35 seconds after launch owing to failure in the guidance system.

[j] Post-boost vehicle.

Sources: 'Connector failure seen cause of HEDI detonation', *Aerospace Daily*, 13 Mar. 1990, p. 442; Gilmartin, P., 'Delay of first HEDI test launch caused by flight destruct system replacement', *Aviation Week & Space Technology*, 30 Oct. 1989, p. 21; 'SDI testing', *SDI Monitor*, 21 Dec. 1990, p. 291; 'SDIO cancels Kwajalein radar for HEDI tests', *SDI Monitor*, 15 May 1989, p. 132; 'Initial Aries run a success: Air Force payload deployed', *Space News*, 19 Feb. 1990, p. 16; Kiernan, V., 'Satellites ready to begin SDI laser tests over Hawaii', *Space News*, 26 Feb. 1990, p. 9; 'RME bounces beam; LACE still has problems', *SDI Monitor*, 6 July 1990, pp. 147–48; 'SDI satellite scores once more', *Space News*, 17 Sep. 1990, p. 2; 'SDIO: RME proves directed energy weapon pointing feasible', *Aerospace Daily*, 12 Dec. 1990, p. 423; 'LACE/RME experiencing some difficulty', *Defense Daily*, 27 Mar. 1990, p. 477; 'Relay mirror successfully bounces beam', *SDI Monitor*, 28 Sep. 1990, p. 222; 'SDIO experiment demonstrates that decoys can be spotted, tracked', *Aerospace Daily*, 11 Apr. 1990, pp. 60–61; 'Firepond test boosts laser radar discrimination', *SDI Monitor*, 13 Apr. 1990, pp. 88–90; Kiernan, V., 'Alpha laser yet to produce full power', *Space News*, 20 Aug. 1990, p. 19; Kiernan, V., 'SDI experiment reveals new detectable ICBM signals', *Space News*, 30 Apr. 1990, p. 17; 'Bow shock experiments finds UV stronger than expected', *SDI Monitor*, 11 May 1990, pp. 108–10; Kiernan, V., 'Sounding rocket simulates nuclear explosion for Star Wars test', *Space News*, 14 May 1990; 'AOA has problems on first night flight', *SDI Monitor*, 8 June 1990, p. 129; 'AOA passes first test, gets ready for night flight', *SDI Monitor*, 25 May 1990, pp. 122–23; 'Milspace testing', *Military Space*, 28 Aug. 1989, p. 8; Kiernan, V., 'Gaffe by guidance experts blamed for SPEAR 2's in-flight destruction', *Space News*, 12 Nov. 1990; 'Israel tests defensive missile', *Washington Post*, 10 Aug. 1990, p. A29; 'Arrow destroyed early in test', *Aviation Week & Space Technology*, 1 Sep. 1990, p. 177; 'Test of missile called partial success', *Washington Post*, 28 Aug. 1990, p. A6; Kiernan, V., 'First Brilliant Pebbles launch marred by telemetry loss', *Space News*, 3 Sep. 1990, p. 9; Kiernan, V., 'Space tests scheduled for Brilliant Pebbles', *Space News*, 19 Mar. 1990, p. 4; 'Airborne surveillance testbed tracks Minuteman 3 missile in flight', *Defense Daily*, 20 Sep. 1990, p. 456; 'PM says AOA/AST generated target data in first operational test', *Aerospace Daily*, 25 Sep. 1990, p. 491; 'Alpha laser powered up for first time in Nov.', *Space News*, 7 Jan. 1991, p. 2; 'Starbird booster passes first flight test', *SDI Monitor*, 21 Dec. 1990, pp. 289–90; 'Israelis fire second Arrow in anti-ballistic missile tests', *Flight International*, 2–8 Jan. 1991, pp. 4–5.

III. Soviet strategic defence programmes

The USSR possesses the world's only operational anti-missile system. Despite recent upgrades to this system, 'US intelligence has no evidence [that] Soviet developments in strategic defences will change the relative survivability of the SLBM or bomber leg of the US triad'.[37] While admitting the high cost of the Soviet project, one of the system's operators stated that: 'Across the ocean, the production line is putting out increasingly improved ballistic missiles, MX, Trident 2, Midgetman . . . Of course, there must be a shield. We are not immune [to] a provocative launch of a missile, either, judging from the development of the situation in the Near East. We eliminate one such mad missile—all expenditures will pay for themselves . . .'.[38]

[37] *Department of Defense Appropriations for Fiscal Year 1991*, Hearings before the Committee on Appropriations, US Senate, 101st Congress, 2nd Session (US Government Printing Office: Washington, DC, 1990), Part 2, p. 371.

[38] Dokuchayev, A., 'ABM system's role in deterrence viewed', *Krasnaya Zvezda*, 5 Oct. 1990, p. 5, in Foreign Broadcast Information Service, *Daily Report–Soviet Union* (*FBIS-SOV*), FBIS-SOV-90-201, 17 Oct. 1990, pp. 41–43.

Construction of the Pechora phased-array early-warning radar network began in the late 1970s, with a total network of 11 planned for completion in the early 1990s.[39] However, construction of the Space Objects Tracking Centre radar at Mukachevo in the Trans-Carpathian region has been halted by local protests over the project's environmental impact.[40] An older Hen House radar near this site will continue to perform the early-warning function in this sector.[41] Dismantlement of the radar at Krasnoyarsk has begun, in compliance with agreements reached with the USA over interpretation of the 1972 ABM Treaty,[42] at a reported cost of about 50 million roubles.[43] According to one Soviet observer, the original construction of the Krasnoyarsk radar was intended to avoid the need to spend 1–3 billion roubles building two radars at more northern locations.[44] The decision to dismantle this radar has been controversial, with some arguing that this will result in a blind spot in the Soviet early-warning network.[45]

IV. Anti-satellite weapon systems

After a flurry of activity in 1989, the US anti-satellite (ASAT) weapon programme turned to the long-haul task of developing, testing and deploying a new ground-based kinetic-energy ASAT system by the late 1990s. Plans for directed-energy weapons have been placed on hold. As for the Soviet ASAT effort, there have been no visible developments of either operational or new systems.

US anti-satellite developments

Despite reductions in the scope of near-term plans, the Bush Administration continues to express strong support for the development of an anti-satellite system. While Chairman of the Joint Chiefs of Staff General Colin Powell reasserted the traditional argument that a US ASAT system was needed to counter the Soviet ASAT system,[46] Secretary of the Army William Stone went further, asserting that an ASAT system also served to deny and negate

[39] Gertz, B., 'CIA warns of verification woes in future treaty', *Washington Times*, 21 Dec. 1988, p. A3.

[40] Badurkin, V., 'Mukachev radar facility prompts local protests', *Trud*, 25 Feb. 1990, FBIS-SOV-90-045, 7 Mar. 1990, pp. 2–3.

[41] 'Transcarpathian Oblast radar project mothballed', *Izvestia*, 15 Aug. 1990, FBIS-SOV-90-163, 22 Aug. 1990, p. 51.

[42] Gertz, B., 'Report late on Soviet arms curbs', *Washington Times*, 1 Jan. 1991, p. A4.

[43] 'Krasnoyarsk radar dismantling "in full swing"', Moscow World Service (radio transcript), 9 Oct. 1990, FBIS-SOV-90-196, 10 Oct. 1990, p. 1.

[44] Zaloga, S., 'Soviet radars draw opposition', *Armed Forces Journal International*, June 1990, p 21.

[45] Surikov, B., 'Krasnoyarsk radar station's future considered', *Moscow News*, 18 Mar. 1990, FBIS-SOV-90-059, 27 Mar. 1990, pp. 2–3. For a candid Soviet view of the development of the Soviet ABM system, see the interview with Grigoriy Kisunko, General Designer of the Moscow ABM system, in 'ABM designer sees squandering of resources', *Sovetskaya Rossiya*, 5 Aug. 1990, FBIS-SOV-90-151, 6 Aug. 1990, pp. 1-3.

[46] *Department of Defense Appropriations for Fiscal Year 1991* (note 37), Part 1, pp. 179 and 188.

the threat posed by Soviet space-based reconnaissance and surveillance systems.[47] Deputy Assistant Defense Secretary for Strategic Defense, Space and Verification Policy Douglas Graham summed up the arguments:

Space Control—the ability to assure such freedom of action, and, when so directed by the National Command Authorities, to deny it to the enemy—has become as important to the USA as sea control capabilities are to the exercise of maritime strategy and air power is to is to land and air warfare. Space control requires an integrated combination of anti-satellite (ASAT) capabilities, space surveillance, and enduring space assets. In view of our continued need to project power, deter war and control escalation during conflict, it is essential for the USA to develop and deploy an operational ASAT system to counter Soviet exploitation of their present space control and space-based targeting capabilities.[48]

Kinetic-energy ASAT

A ground-based kinetic-energy (KE) ASAT system continues to be the primary focus of US ASAT planning.[49] In early 1989 the DOD established a tri-service Joint Program Office to manage ASAT development,[50] with the Air Force responsible for tracking and battle management functions,[51] and the Army responsible for development and operation of the interceptor system.[52] Under this plan, two initial demonstration/validation tests of the interceptor are planned for 1992, with nine full-scale development tests against targets in space planned for 1995.[53] The total cost for this system is estimated to include $1305 million for R&D, $606 million for procurement of 72 missiles, and $341 million for operations and support over 20 years.[54]

The 72 missiles would be deployed by 1996 at a single Army base (in contrast to previous plans for as many as 300 interceptors at several sites),[55] able to intercept several dozen satellites over a period of less than eight hours.[56] Candidate sites for ASAT deployment include the Kwajalein Missile Range and Hawaii in the Pacific (to intercept Soviet satellites soon after launch), as well as Vandenberg Air Force Base in California, Ft Stewart in

47 *Department of Defense Appropriations for Fiscal Year 1991* (note 37), Part 3, p. 124.

48 Address by Deputy Assistant Secretary of Defense for Strategic Defense, Space and Verification Policy, Douglas Graham, to the annual meeting of the American Institute of Aeronautics and Astronautics, Arlington, Va., 2 May 1990.

49 Olson, P., 'ASAT', *Air Defense Artillery*, Mar.–Apr. 1990, pp. 24–31.

50 *US Department of Defense Authorization for Appropriations for Fiscal Years 1990 and 1991*, Hearings before the Armed Services Committee, US Senate, 101st Congress, 1st Session (US Government Printing Office: Washington, DC, 1989), Part 6, pp. 271–305.

51 'Ground sensors to cue ASAT', *Military Space*, 23 Oct. 1989, pp. 1, 8.

52 'DAB selects ground-based ASAT in kinetic-kill milestone one review', *Aerospace Daily*, 15 Dec. 1989, pp. 425–26.

53 'Army studies whether to put new ASAT on military bases or in trucks', *Inside the Army*, 16 Oct. 1989, pp. 1, 6.

54 *Department of Defense Appropriations for 1991*, Hearings before a Subcommittee of the Committee on Appropriations, US House of Representatives, 101st Congrees, 2nd Session (US Government Printing Office: WAshington, DC, 1990), Part 7, p. 169.

55 Finnegan, P., 'Army bolsters ASATs with additional $79 million', *Defense News*, 1 Jan. 1990, p. 3.

56 Grossman, E., 'White House drops ASAT lobby, leaves DOD on its own to fight for funds', *Inside the Army*, 11 June 1990, pp. 1, 5–8.

Georgia, Cape Canaveral in Florida, Wake Island, and Puerto Rico.[57] In July 1990 Rockwell was selected as the prime contractor to develop and build the ASAT interceptor, which would be similar in concept to the Army's GBI-X and ERIS anti-missile interceptors.[58]

Directed-energy ASAT

The US directed-energy ASAT system experienced a significant slow-down in 1990, with selection of the primary concept for development delayed from 1991 to 1994 or 1995 (delaying an operational capability from 1996 to around 2000).[59] None the less, the DOD continues to attach high importance to this undertaking.[60] Near-term planning has focused on the 2.2-MW Mid-Infrared Advanced Chemical Laser (MIRACL) at the White Sands Missile Range in New Mexico, although initial plans for testing in 1991 against inactive US satellites and space debris have been dropped. Other candidates for the ASAT mission include the Army's Ground-Based Free Electron Laser,[61] and the Air Force's ground-based Excimer laser[62] and chemical oxygen-iodine laser (COIL),[63] as well as new low-power Carbon Dioxide and Free Electron Lasers to be installed at White Sands in 1991 and 1993, respectively.[64] Pointing and tracking experiments, as well as low-power atmospheric compensation tests, will be conducted with the 3.5-metre diameter beam-director telescope at the Starfire Optical Range in New Mexico.[65]

Soviet anti-satellite developments

The USSR continues to maintain an operational anti-satellite system, which has not been tested since 1982. Based at the Baikonur Cosmodrome, the system consists of 16 SL-11 Cyclone boosters, which are maintained on alert, capable of launching into orbit radar-guided kill vehicles estimated to be capable of intercepting 10 US satellites in a campaign lasting two days.[66]

[57] Gilmartin, P., 'Defense Department to launch design competition for new antisatellite weapon for the 1990's', *Aviation Week & Space Technology,* 24 July 1989, p. 30; and 'Army trims ASAT design to cut dosts', *Military Space,* 26 Mar. 1990, pp. 4–5.

[58] Asker, J., 'Rockwell selected as sole contractor for $100 million ASAT design effort', *Aviation Week & Space Technology*, 23 July 1990, p. 30.

[59] Kiernan, V., 'Lengthy delay hits laser ASAT work', *Space News*, 20 Aug. 1990, pp. 1, 20.

[60] *Department of Defense Appropriations for 1991*, Hearings before a Subcommittee of the Committee on Appropriations, US House of Representatives, 101st Congress, 2nd Session (US Government Printing Office: Washington, DC, 1990), Part 7, pp. 498–99.

[61] Nelson, R., 'Laser lab braces for slowdown', *Military Space,* 27 Aug. 1990, pp. 1, 4.

[62] *US Department of Defense Appropriations for 1988*, Hearings before a Subcommittee of the Committee on Appropriations, US House of Representatives, 100th Congress (US Government Printing Office: Washington, DC, 1987), Part 6, p. 716.

[63] 'Air Force laser uses simple chemistry', *Military Space*, 30 July 1990, p. 6.

[64] Kiernan, V., 'Battle brewing over laser ASAT space test', *Space News*, 29 Oct. 1990, p. 6.

[65] 'Air Force mirror to test laser ASAT pointing', *Military Space*, 30 July 1990, pp. 3–4.

[66] *Department of Defense Appropriations for 1990*, Hearings before a Subcommittee of the Committee on Appropriations, US House of Representatives, 101st Congress, 1st Session (US Government Printing Office: Washington, DC, 1989), Part 6, p. 212.

V. Arms control and military space activities

The ABM Treaty[67]

During 1989 substantial progress was made in compliance with the 1972 ABM Treaty, with Soviet agreement to dismantle the Krasnoyarsk radar,[68] as well as Pawn Shop and Flat Twin radars.[69] This trend was reversed in 1990, with the initiation of US testing of the Airborne Optical Adjunct (AOA).[70] The AOA is a modified Boeing 767 aircraft that carries an infra-red telescope for tracking and identifying re-entry vehicles while they are still above the atmosphere for mid-course and terminal interception.[71] Article V of the ABM Treaty bans the development or testing of air-based ABM components, and the AOA appears to be inconsistent with this provision.[72] The case that the AOA complies with the ABM Treaty includes seven separate points. These, and the arguments against them, may be summed up as follows:

1. The Boeing 767 cannot stay aloft for a sufficient period of time to be an effective ABM component. However, the aircraft currently has a maximum airborne endurance of about 10 hours, comparable to that of the E-3 Airborne Warning and Control System (AWACS), which performs an air defence function analogous to the AOA's ABM function. If needed, the endurance could be extended to several days through the use of aerial refuelling.

2. The AOA is compliant with the Treaty as long as tests do not involve the transfer of data in real-time to an ABM interceptor. However, AOA tests have in fact involved real-time data transfer to ground stations, which could then relay this data to interceptors.[73] Verification of this capability would require a detailed understanding of the computer software and communications capabilities of AOA, which is clearly beyond the capabilities of the national technical means (NTM) that are the verification standard of the Treaty.

3. The sensor focal plane array (the electronic chip that forms the target image) does not contain a full complement of sensor elements. Again, this characteristic cannot be monitored by NTM, and can thus not be used to determine whether a device has ABM capabilities.

4. The size of the window on the AOA restricts the field of view of the sensor to the point that it is not ABM-capable. However, this is not a measure contained in the Treaty and has not been the subject of bilateral discussions. Furthermore, the AOA sensor telescope is mounted on rails that enable it to look through the window at varying angles. This gives the sensor a field of

[67] Bunn, M., *Foundation for the Future: The ABM Treaty and National Security* (Arms Control Association: Washington, DC, 1990) is now the standard work on this subject.

[68] 'The Kremlin apology: excerpts from speech', *New York Times*, 25 Oct. 1989.

[69] 'Soviets dismantle Gomel radars', *Defense News*, 6 Nov. 1989, p. 2.

[70] Henderson, B., 'Army begins flight tests of airborne optical adjunct', *Aviation Week & Space Technology*, 28 May 1990, p. 113.

[71] Kiernan, V., 'May tests set for trouble-plagued AOA', *Space News*, 5 Mar. 1990, p. 3.

[72] Broad, W., 'Critics say star wars test may be a treaty violation', *New York Times*, 12 May 1990.

[73] 'PM says AOA/AST generated target data in first operational test', *Aerospace Daily*, 25 Sep. 1990, pp. 491–92.

view on the order of 90°, greatly in excess of the approximately 5° of mechanically scanned dish radars considered to be ABM components, and comparable to the 120° typical of phased-array radars.

5. The AOA sensor will initially require external cueing to tell it where to look for incoming targets. However, there is no way to determine this using NTM, since this is a function of software commands to the sensor. Also, limitations on sensor acquisition capabilities are not an inherent definition of ABM capabilities. The Soviet Pawn Shop and Flat Twin radars require cueing from other radars, and the Bush Administration considers these to be ABM components. Similarly, the US Missile Site Radar (MSR), part of the Safeguard system, required cueing from the Perimeter Acquisition Radar, and the MSR was recognized as an ABM component.

6. The AOA does not provide information on the range of targets. However, by using the aircraft's motion and tracking the apparent position of targets over a period of time, the AOA is able to calculate range by triangulation.

7. The AOA is not a 'stand alone' system, able to perform the complete function of an ABM component as defined in Article II, and is therefore not Treaty-accountable. However, most ABM systems have more than one sensor component, each of which plays some role in the management of the battle.

Several of the other tests that pose significant challenges to the traditional interpretation of the ABM Treaty, the BSTS and the Zenith Star chemical laser, have been delayed to the later 1990s. The testing of these space-based devices, in many respects capable of substituting for ABM radars or interceptors, could be inconsistent with Articles V and VI, which ban such testing.

The Defence and Space Talks

There were few new developments in the Defence and Space Talks during 1990, with the USA and USSR continuing to disagree fundamentally about the future role of strategic defences.[74] The core of the US negotiating position is that the purpose of a new defence and space agreement (separate from the ABM Treaty and not linked to the prospective Strategic Arms Reduction Talks (START) agreement[75]) is to facilitate a co-operative stable transition to increased reliance on strategic defences.[76] The US proposal would exclude all space-based sensors from limitation under the new defence and space treaty[77] and continues the policy of the Reagan Administration to endorse virtually unlimited testing of space-based anti-missile components under the so-called broad interpretation of the ABM Treaty. The USSR continues to reject this approach[78] and instead proposes a new agreement to specify a list of threshold

[74] The negotiating positions of the two parties were reviewed in Pike (note 6), pp. 72–75.

[75] For a discussion of START, see chapter 11 in this volume, section V.

[76] For a detailed review of the US position, see Smith, D., 'The Defense and Space Talks: moving towards non-nuclear strategic defenses', *NATO Review*, vol. 28, no. 5 (Oct. 1990), pp. 17–21,

[77] Adams, P., 'US, Soviets edge closer to rewritten ABM Treaty at Defense and Space Talks', *Defense News*, 21 Aug. 1989.

[78] 'Soviets reject transition to strategic defenses—Hadley', *Defense Daily*, 22 Mar. 1990, p. 458.

limits to distinguish between permitted and prohibited activities under the ABM Treaty.[79]

A major development in the Defence and Space Talks occurred at the September 1989 meeting of Soviet Foreign Minister Eduard Shevardnadze and US Secretary of State James Baker,[80] at which the Soviet side indicated a willingness to ratify the START agreement, despite disagreement on the ABM issue, as long as the parties continued to observe the ABM Treaty as it was signed in 1972.[81] At the Washington summit meeting between Presidents Gorbachev and Bush which concluded on 2 June 1990, the parties agreed that:

> Within the existing negotiating framework on Nuclear and Space Arms in Geneva, the two sides will continue negotiations on ABM and space without delay. Thus, in the future talks the two sides will discuss strategic stability issues of interest to them, including the relationship between strategic offensive and defensive arms, taking into account stabilizing reductions in strategic offensive arms and development of new technologies.[82]

Leading US proponents of SDI increasingly insist that there is growing Soviet support SDI, and that this will eventually be reflected in Soviet endorsement of the US negotiating position.[83] This issue has been the focus of growing discussion in 1990.[84] Chief US Defense and Space negotiator David Smith claims that 'commencing last year, dissident Soviet voices have begun publicly to offer positive views on advanced defences'.[85] According to Deputy Assistant Defense Secretary Graham, there are 'some indications that the Soviets may be willing to discuss some form of defensive transition'.[86] However, this view is based on a selective reading of Soviet statements.[87] Even SDIO Director Henry Cooper, former chief US Defense and Space negotiator, concedes that 'the Soviets have only changed the form of their linkage demand'.[88]

[79] Strobel, W., 'US, Soviet arms negotiators edge nearer pact', *Washington Times*, 7 Aug. 1989, p. A8.

[80] 'Excerpts from statement on arms', *New York Times*, 25 Sep. 1989, p. 1.

[81] Gordon, M., 'An arms obstacle falls', *New York Times*, 24 Sep. 1989, pp. 1, 16.

[82] 'Documents from the US–Soviet summit', *Arms Control Today*, vol. 20, no. 5 (June 1990), p. 23.

[83] 'Statement of Steven Hadley, Assistant Secretary of Defense for International Security Policy', *Department of Defense Authorization for Appropriations for Fiscal Year 1991*, Hearings before the Armed Services Strategic Forces Subcommittee, US Senate, 101st Congress, 2nd Session (US Government Printing Office: Washington, DC, 20 June 1990), Part 7, pp. 431-41.

[84] This topic is the subject of a major exchange between Keith Payne and Michael MccGwire in the 5 Nov. 1990 edition of *Inside the Army*, pp. 10–14, which constitutes a definitive rehearsal of this issue. Claims that the Soviet position has changed are also made in *Congressional Record*, 1 Aug. 1990, pp. E2574–77, and 12 Sep. 1990, pp. H7451–52.

[85] Smith, David, 'Soviets view SDI in new light', *Defense News*, 5 Nov. 1990, p. 24.

[86] 'US tries to persuade Soviets to accept limited strategic defense', *Inside the Army*, 22 Oct. 1990, p. 10.

[87] The author regrets that the discussion of this topic in last year's *Yearbook* chapter fell prey to this campaign in suggesting that Alexei Arbatov took a favorable view toward a limited SDI system. Arbatov has always been and remains in opposition to ABM and SDI, and is against even the limited systems permitted under the ABM Treaty. That his views on this matter could be so misrepresented is testimony to the growing confusion of this debate.

[88] Cooper, H., 'Soviets haven't tempered opposition to SDI', *Washington Times*, 13 June 1990, p. E2.

Much attention was given to the statement by Major General V. Belous that 'we should hardly deny the possibility of reasonable compromises in the future and the development of defences for US and Soviet territory against accidental missile launches or blackmail attempts and threats made by third countries'.[89] However, this observation concludes an article that begins by noting:

[SDI's] strategic goals and schemes have not changed substantially in any way. There have only been changes in tactics . . . Western and Soviet specialists have often hastily concluded that this programme has allegedly 'gone out of fashion', and is gradually losing significance . . . SDI, like all other US military programs, is designed to break the military parity that has taken shape and secure strategic superiority over the Soviet Union . . . if the 'space threshold' is crossed, this will result in the unrestrained escalation of the arms race, which no one will succeed in halting.

The official Soviet attitude towards the SDI remains negative. Leading Soviet arms adviser Marshal Sergey Akhromeyev noted that the ABM and prospective START Treaties 'are objectively interconnected, and START cannot be effective outside the effectiveness of the ABM Treaty'.[90] Chief Soviet negotiator Yuri Nazarkin warned that 'if the ABM Treaty is violated, we will probably use our right of withdrawal' from the START agreement.[91] Responding to the Bush Administration's request for a major increase in the SDI budget, Soviet Foreign Ministry representative Vadim Perfilev, noting the impending START agreement, asked: 'why in this light, is it necessary to build up Star Wars? . . . [Such a move] can only lead to destabilization of the strategic balance'.[92]

VI. Soviet and US military space systems

Military space systems have assumed a central place in US military planning for the post-cold war security environment. In early 1990 Deputy Assistant Defense Secretary Graham asserted that:

Space—like the land, the seas and the air—must be viewed as a medium within which military operations may be required . . . Assured access to space and freedom of action in space are increasingly important to America's ability to deploy and employ military forces worldwide . . . Our global responsibilities remain vital to U.S. national security, and space will be an increasingly critical link to our forces and friends overseas . . .[93]

89 Belous, V., 'The SDI syndrome', *Sovetskaya Rossiya*, 23 Mar. 1990, p. 5, FBIS-SOV-90-057, 2 Mar. 1990, pp. 1-3.

90 Adams, P., 'Soviet: SDI still a problem for START pact', *Defense News*, 16 Apr. 1990, p. 7.

91 Palmer, R., 'Soviets say SDI test could endanger pact', *Washington Times*, 1 Mar. 1990, p. A9.

92 Oberdorfer, D. 'Soviets object to Bush administration's request for more SDI spending', *Washington Post*, 31 Jan. 1990, p. A16.

93 Graham (note 48).

These observations took on new significance with the Iraqi invasion of Kuwait.[94] Shortly after the initiation of Operation Desert Shield, one senior US military planner noted that: 'While the cold war may have edged off the front page, the emergence of regional conflicts—as predicted—are dominating our foreign policy'.[95]

The USSR concluded 1990 with a total of 75 launches, one more than the total for 1989, but in marked contrast to the 90 launches of 1988, or the peak of 101 flights in 1982.[96] Leading Soviet space observer Geoffrey Perry noted that the 'launch rate is down but the primary reason is because Soviet satellites now have greater longevity and are replaced only as needed', concluding that despite this slow-down 'the military program is as strong as ever'. Perry suggested that the USSR has begun a shift from launching intelligence satellites to maximize collection capabilities at all times to launching based on need and efficiency.[97] (For data on individual satellites discussed under the programmes below, see table 3.3 and appendix 3A.)

Soviet photographic reconnaissance satellites

The continued slow-down in the Soviet space programme in 1990 was accompanied by a significant reduction in the pace of photoreconnaissance operations . The USSR launched a total of 21 photoreconnaissance satellites in 1990, far fewer than the 31 sent up in 1989 or the 32 in 1988, and fewer than the previous low of 25 in 1987. The pace picked up toward the later part of the 1990, in response to the Gulf crisis.[98] The total number of reconnaissance days (the cumulative number of days each satellite was in orbit, measuring effort and capability) in 1990 was 886, up from 833 in 1989 and 669 in 1988, although still less than the peak of 1128 in 1986.[99] About one-third of the imaging satellites launched in 1990 (6 out of 21) were devoted to military mapping or civil remote sensing missions, the same proportion as in 1989 (10 out of 31), in contrast to the quarter of the launches in 1988 (8 out of 32).

In 1990 three medium-resolution third-generation satellites were launched, compared with two launched in both 1988 and 1989. Of the high-resolution third-generation satellites, four were launched in 1990 as compared with 10 in 1989 and 13 in 1988. The overall pace of third-generation launch activity in 1990 matches the six launches in both 1986 and in 1987, far fewer than the high of 18 launches in 1978. Cosmos 2099, launched on 31 August, was placed in an orbit optimized for coverage of the Persian Gulf area.[100]

[94] 'Space supports Mideast build-up', *Military Space*, 27 Aug. 1990, pp. 1–2.

[95] Hard, D., USAF Maj.-Gen., Director of Air Force Space Acquisition, *Space Systems Status*, Northern Virginia Armed Forces Communications and Electronics Luncheon Address, 28 Aug. 1990.

[96] 'Soviets finish year with 75 launches', *Defense Daily*, Special Supplement, 9 Jan. 1991, p. S-1.

[97] Covault, C., 'Soviet military space operations developing longer life satellites', *Aviation Week & Space Technology*, 9 Apr. 1990, pp. 44–49.

[98] de Selding, P., 'Launch of Gulf satellites delays group's experiments', *Space News*, 17 Sep. 1990.

[99] 'Soviets increased satellite photo/recon levels in 1990', *Defense Daily*, 28 Jan. 1991, pp. 133–34, citing the analysis of Saunders Kramer.

[100] 'Remote sensing, reconnaissance satellites launched by Soviets', *Aerospace Daily*, 11 Sep. 1990, p. 413.

Table 3.3. Operational US military satellites, 1984–94

Numbers for 1991–94 are projected.

Satellite[a]	1984	1985	1986	1987	1988	1989	1990	1991	1992	1993	1994
Imaging intelligence satellites											
KH-11	2	1	1	2	3	3	3	3	2	1	..
KH-12	..	..	..	..	..	1	3	3	3	3	3
Lacrosse	..	..	..	..	1	1	1	2	3	3	3
Boren Add-On	..	..	..	..	..	..	..	..	1	2	3
Electronic intelligence satellites											
Rhyolite	2	2	..	..	..	..	..	..	..	..	..
Jumpseat	2	2	2	2	2	2	2	2	..	..	..
Chalet	2	2	2	2	2	1	1	1	1	1	1
Magnum	..	1	1	1	1	2	3	4	4	4	4
NOSS	16	16	16	16	16	16	16	16	16	16	16
Early-warning satellites											
DSP	5	5	4	5	5	5	5	5	5	5	5
NATO	3	3	3	3	3	2	2	3	3	3	3
Communications satellites											
DSCS2	7	7	6	6	5	3	2	1	1	1	1
DSCS3	1	3	3	3	3	4	4	6	7	8	9
SDS	2	2	2	2	2	2	2	2	2	1	..
Milstar	..	..	..	..	..	..	..	..	1	2	3
FLTSAT	4	4	5	5	5	6	6	5	2	2	2
LEASAT	2	3	3	3	3	4	4	4	5	5	2
UFO	..	..	..	..	..	..	..	..	1	2	5
Navigation satellites											
Transit	5	7	7	9	14	9	7	4	4	4	4
Navstar	6	7	7	7	7	12	16	20	21	21	21
Weather satellites											
DMSP	2	2	2	2	2	2	2	2	2	2	2
Total	**61**	**67**	**64**	**68**	**72**	**72**	**84**	**88**	**88**	**90**	**91**

[a] Acronyms are spelled out in the text.

Of the fourth-generation satellites, which typically remain in orbit for about eight weeks, six were launched in 1990, continuing a downward trend with seven launched in 1989, eight launched in 1988 and nine launched in both 1986 and 1987. Cosmos 2089, launched the day after the Iraqi invasion of Kuwait, was placed in an orbit optimized to cover events in that region.[101] Cosmos 2102, launched on 16 October, was placed in an unusually low orbit from 2 October through 2 November, to facilitate observation of Operation Desert Shield through early November.[102] This satellite was destroyed in orbit

[101] Rains, L., 'Soviets orbit photo satellite 48 hours after Iraq invasion', *Space News*, 13 Aug. 1990.

[102] 'Soviet recon satellites image Persian Gulf area', *Aviation Week & Space Technology*, 19 Nov. 1990, p. 24.

on 20 November 1990, marking the third year in a row that a malfunction on one of these spacecraft resulted in its intentional destruction by on-board explosives. A single fourth-generation satellite dedicated to military mapping and remote sensing was launched in 1990 (Cosmos 2078),[103] consistent with the two launches in 1989 and a single launch in 1988.[104]

1990 marked a second year of uneventful operations of the newest, fifth-generation reconnaissance satellites (in contrast to the poor showing of 1988). Cosmos 2072, launched on 13 April, replaced Cosmos 2049,[105] and was joined at the end of the year by Cosmos 2113. These satellites use electronic transmission to return images in near-real time, while the third- and fourth-generation systems use film returned to earth in re-entry capsules.

US imaging intelligence satellites

US imaging intelligence capabilities continued to expand in 1990, as two new satellites joined the five already in orbit.[106] This increase in the number of satellites in orbit is expected to continue. In contrast to prior years, in which the expanded budget for imaging intelligence satellites was marked by vocal political opposition,[107] there was no public dispute.[108]

The USA continued operations of three KH-11 photographic intelligence satellites through 1990. The oldest KH-11, launched in December 1984, remained in orbit at the end of the year. Other imaging intelligence spacecraft in orbit included the first KH-12 launched in August 1989, as well as the first Lacrosse, launched in 1988. Continuing the expansion of the number of US low-altitude intelligence satellites begun in 1988, on 28 February 1990 the Space Shuttle Discovery deployed on flight STS-36 what appeared to be the second new generation of photographic reconnaissance satellites, popularly referred to as the Advanced Keyhole or the KH-12.[109] This spacecraft was placed in a roughly circular orbit at an altitude of 811 km, with an inclination of 65°. Despite initial Soviet reports suggesting that this satellite had exploded in orbit,[110] ground observers visually tracked this spacecraft in November

[103] Clark, P., 'Soviet spacecraft launches in 1990: 20 Apr.–15 May', *Jane's Soviet Intelligence Review*, vol. 2, no. 7 (July 1990), p. 333.

[104] Clark, P., 'Soviet topographic satellites', *Zenit*, no. 45 (Nov. 1990), pp. 12–16.

[105] 'Soviets launch three satellites, boost reconnaissance abilities', *Aerospace Daily*, 20 Apr. 1990, p. 121.

[106] Richelson, J., *America's Secret Eyes in Space: The US Keyhole Spy Satellite Program* (Harper & Row: New York, 1990) is by far the most definitive and authoritative treatment of this subject ever published, containing an unprecedented wealth of programmatic detail.

[107] Rasky, S., 'Senators balking over verification', *New York Times*, 29 Apr. 1988; Munro, N., 'House Senate committees battle over funding spy system', *Defense News*, 18 Sep. 1989, p. 39.

[108] For details of this plan and the debate over it , see Pike (note 6), pp. 78–79.

[109] The most appropriate nomenclature for the new satellite is probably Advanced Keyhole, as it is reliably suggested that the US intelligence community no longer uses the KH designation system, and thus there is properly no such satellite as the KH-12. Unfortunately, the designated code name for this new spacecraft has not been publicly compromised (the KH-11 was Kennan, KH-9 was Hexagon, and so forth). Since most published reports over the years have referred to this new satellite as the KH-12, that nomenclature will be adhered to here, until the proper designation for the satellite is understood.

[110] Leary, E., 'Problems are reported with new spy satellite', *New York Times*, 18 Mar. 1990.

1990.[111] The third KH-12 was launched on 8 June 1990.[112] A manœuvre on 19 June raised the inclination of the orbit to 63.4°, and following this manœuvre three sub-satellites, generally regarded as performing a signals intelligence function, were released. In October the primary satellite was again manœuvred; its new orbital elements have not yet been established.

Imaging intelligence satellites were widely used in Operation Desert Shield in 1990. Intelligence reports provided warning of the Iraqi invasion nearly a week before it occurred, both on the timing and magnitude of the assault.[113] A few days after the invasion, satellite photography showing the Iraqi military buildup on the Kuwaiti–Saudi border was instrumental in convincing Saudi King Fahd to permit the introduction of US troops.[114] By late October Iraqi forces were shifting position frequently to evade satellite intelligence.[115] Imaging intelligence systems were also used to monitor the effectiveness of the embargo, and by early December satellite images showed a steady stream of trucks entering Iraq from Iran.[116] Civilian LANDSAT and SPOT images were also used to develop up-to-date maps of the theatre of operations.[117]

Soviet electronic intelligence satellites

The Soviet electronic intelligence (ELINT) capability consists of three complementary systems.[118] Six low-altitude satellites comprise the third generation of Soviet ELINT satellites, and Cosmos 2058 was the first of these to be launched since Cosmos 1975 in 1988, suggesting improved operational lifetimes for these satellites.[119] The absence of launches during 1989 had raised doubts about the future of the newer 12-tonne fourth-generation ELINT satellites.[120] These questions were largely dispelled by the launches of Cosmos 2082 on 22 May 1990.[121] However, on 4 October a second was lost due to a malfunction in its SL-16 Zenit booster, which damaged its launch pad when it exploded five seconds after lift-off.[122] No launches were conducted in support of the new fifth-generation ELINT system in geosynchronous orbit.

[111] 'Satellite, believed lost, is spotted in orbit', *Washington Post*, 13 Nov. 1990, p. A11.

[112] Kolcum, E., 'Second Titan 4 carries secret surveillance satellite into orbit', *Aviation Week & Space Technology*, 18 June 1990, p. 27.

[113] Jehl, D., 'Iraq standoff exposes gap in US intelligence force', *Los Angeles Times*, 25 Aug. 1990, pp. A1, A15.

[114] Offley, E., 'Vital space assets', *Seattle Post- Intelligencer*, 23 Aug. 1990, p. 1. These systems were not infallible, since the USA lost track of four Iraqi divisions for a 24-hour period on 7–8 Aug. See 'Dugan orders review of limited SR-71 revival', *Defense Daily*, 20 Aug. 1990, p. 276.

[115] Gertz, B., 'Oil fire in Gulf is Iraqi tactic', *Washington Times*, 1 Nov. 1990, p. A12.

[116] Horwitz, T., 'Wily smugglers keep embargoed supplies flowing into Iraq', *Wall Street Journal*, 5 Dec. 1990, p. 1.

[117] Kiernan, V., 'Satellite data boosts map quality for US troops', *Space News*, 15 Oct. 1990, pp. 1, 28.

[118] Clark, P., 'Soviet worldwide ELINT satellites', *Jane's Soviet Intelligence Review*, vol. 2, no. 7 (July 1990), pp. 330–32.

[119] Clark, P., 'Soviet spacecraft launches in 1990: Part 1', *Zenit*, no. 3 (Apr. 1990).

[120] Clark, P., 'Economic changes in the Soviet space program', *Jane's Soviet Intelligence Review*, vol. 2, no. 5 (May 1990), p. 236.

[121] Rains, L., 'Soviets launch first ELINT spy satellite since 1988', *Space News*, 29 May 1990.

[122] Rains, L., 'Zenit explodes over launch pad', *Space News*, 25 Oct. 1990, pp. 1, 28.

US electronic intelligence satellites

The USA operates several constellations of signals intelligence satellites in geostationary, elliptical and low Earth orbits. Signals intelligence provided one of the first warnings that the Iraqi invasion of Kuwait was likely, when a Soviet-built Tall King radar resumed operation on 29 July 1990. The radar had been out of service for a number of months prior to the invasion.[123] By early October US electronic intelligence had some success in monitoring Iraqi military communications, but the Iraqi Army was also using underground cables to communicate, making it difficult to determine Iraqi military intentions.[124] The geostationary ELINT constellation consists of three or four Magnum satellites and one Chalet/Vortex. The National Reconnaissance Office (NRO) and the National Security Agency (NSA) launched a third Magnum satellite on the Space Shuttle on 14 November 1990.

Soviet ocean surveillance satellites

Following the problems with the nuclear-powered Cosmos 1900, which malfunctioned on 12 April 1988,[125] there were no Radar Ocean Reconnaissance Satellite (RORSAT) launches in 1990, and it is questionable whether this system will again be flown in peacetime. In apparent response, the Electronic Ocean Reconnaissance Satellite (EORSAT) constellation continues to experience significant expansion. Through 1988 this network consisted of two spacecraft flying in a single orbital plane. Additional launches in 1989 led to a brief period during which five EORSATs were operating simultaneously in two distinct orbital planes.[126] This buildup continued in 1990, with the launch of Cosmos 2060 into the same plane as Cosmos 2046 and Cosmos 2051.[127] Cosmos 2103 initiated a new plane, operating in tandem with Cosmos 2033, Cosmos 2046, Cosmos 2060 and Cosmos 2096.[128]

US ocean surveillance satellites

The White Cloud Naval Ocean Surveillance System (NOSS) is the US counterpart to the Soviet EORSAT.[129] No launches under this programme were conducted in 1990.

[123] 'Invasion tip', *Aviation Week & Space Technology*, 6 Aug. 1990, p. 15.

[124] Gertz, B., 'US breathes easier as it spots Iraq's jamming gear', *Washington Times*, 9 Oct. 1990, p. A8.

[125] 'Soviets confirm Cosmos 1900 difficulties', *Aerospace Daily*, 16 May 1988, pp. 252.

[126] Covault, C., 'Soviet military space operations developing longer life satellites', *Aviation Week & Space Technology*, 9 Apr. 1990, pp. 44–49.

[127] Clark, P., 'Soviet spacecraft launches in 1990: 15 Feb.–20 Mar.', *Jane's Soviet Intelligence Review*, vol. 2, no. 5 (May 1990), p. 237.

[128] 'Soviet EORSAT seen going into new orbital plane', *Aerospace Daily*, 16 Nov. 1990, p. 293.

[129] Richelson, J., *The US Intelligence Community* (Ballinger: Cambridge, Mass., 1985), pp. 140–43.

Soviet military communications satellites

The Soviet military communications network includes three classes of satellite that operate in low-altitude orbits. The first-generation spacecraft are launched eight at a time on the SL-8 booster into a single orbital plane. The number of satellites active in this constellation is thus impossible to determine, but the three most recently launched octuplets are usually thought to constitute the bulk of the nominal constellation of two dozen satellites. Most of the two octuplets of these satellites, launched in March 1988 and March 1989, were probably operational in 1990, supplemented by the launch of an additional octuplet on 4 April 1990. The dispersion of the orbital elements of these satellites suggested a possible malfunction in the upper stage or deployment mechanism used on this flight.[130] The second-generation, low-altitude communications satellites, significantly larger than the first-generation satellites, consists of three satellites, each in a unique orbital plane separated by 120°. Following the three launches of 1989, there were two launches of this system in 1990: Cosmos 2056 in January, and Cosmos 2112 in December. The third-generation low-altitude satellites are launched in groups of six. As was the case in 1989, in 1990 there were also two launches in this series, maintaining a total of 12 satellites operating in two planes.

In spite of the growing use of geosynchronous systems in recent years, there is no reason to anticipate the phase-out of the Molniya-1 constellation in the near term. Three replacement Molniya-1 satellites were launched in 1990, maintaining the full complement of eight satellites. A fourth generation of Soviet military communications satellites, designated Potok, also operates in geosynchronous orbit, providing data relay support to the Soviet fifth-generation photographic reconnaissance satellite. Cosmos 2085 was the sole launch under this programme in 1990.[131]

US military communications satellites

The USA maintains several geostationary communications satellite networks. The Defense Satellite Communications System (DSCS), is used by all four military services and a number of government agencies.[132] Two DSCS II satellites, launched in the late 1970s and the 1980s, remain in service, along with four of the more capable and survivable DSCS III spacecraft launched in the 1980s. Beginning in 1991, DSCS III satellites will be launched singly on upgraded Atlas II boosters, with 10 launches planned through 1997.[133]

[130] Clark, P., 'Soviet spacecraft launches in 1990: Part 2', *Zenit*, no. 39 (June 1990), p. 12.

[131] Clark, P., 'Soviet spacecraft launches in 1990: 19 June–1 Aug.', *Jane's Soviet Intelligence Review*, vol. 2, no. 10 (Oct. 1990), pp. 476–77.

[132] Finney, A.T., 'Tactical uses of the DSCS III communications system', NATO Advisory Group for Aerospace Research and Development (AGARD), *Tactical Applications of Space Systems*, Proceedings of Avionics Panel Symposium, Colorado Sorings, Colo., 16–19 Oct. 1990, Report no. AGARD-CP-460 (AGARD: Neuilly-sur Seine, 1990).

[133] 'General Dynamics wins MLV II competition', *Aerospace Daily*, 4 May 1988, p. 185.

The Navy uses a wide range of military communications satellites. The Gapfiller transponders on three Marisat satellites, in service since 1976, were finally taken out of service in 1990. The Fleet Satellite Communications (FLTSATCOM) constellation consists of three back-up satellites, and FLTSATCOM 4, FLTSATCOM 6, and FLTSATCOM 8 launched in 1980, 1986 and 1989 respectively, in front-line service. The Navy's other major system is the Leased Satellite (LEASAT) system, which consists of four Syncom IV spacecraft. The final launch of the LEASAT programme was completed on 9 January by the Space Shuttle.[134] The Navy's new satellite project, the Ultra-High Frequency (UHF) Follow-On (UFO) programme, will launch 10 satellites beginning in 1992.[135]

The two satellites of the Satellite Data System (SDS) support near-real time communications between low-altitude photographic intelligence satellites and ground control stations. NASA's Tracking and Data Relay Satellite System (TDRSS), supports near-real time data transmission from the Lacrosse low-altitude imaging intelligence satellites.[136]

The Milstar satellite programme underwent a major reorientation in 1990, away from support of strategic nuclear warfighting with the USSR,[137] towards support of conventional forces in the Third World.[138] The constellation will be limited to no more than six satellites, rather than the 10 originally planned,[139] and the system will be focused on support of tactical users.[140] The first Milstar launch aboard a Titan 4 from the Eastern Test Range is still anticipated in late 1992,[141] although subsequent flights will be delayed.[142]

Soviet early-warning satellites

The Soviet ballistic missile early-warning satellite network includes a network of nine satellites in Molniya-type orbits. Six launches were conducted in 1990. The first two flights were uneventful, with Cosmos 2063 replacing Cosmos 1793, and Cosmos 2076 replacing Cosmos 1849, but on 22 June the launch of Cosmos 2084 was marred by the failure of the Molniya booster's upper stage, which left the spacecraft stranded in low earth orbit. The TASS announcement of this failure was a mark of the new openness of the Soviet space

134 'Shuttle-deployed Syncom IV-5 arrives on station, begins testing', *Aerospace Daily*, 19 Jan. 1990, p. 110.

135 'Navy satellites approach critical replacement stage', *Aviation Week & Space Technology*, 21 Mar. 1988, pp. 46, 51.

136 Charles, D., 'Spy satellites: entering a new era', *Science*, vol. 243 (24 Mar. 1989), pp. 1541–43.

137 Kiernan, V., 'Officials: changing world heightens demand for Milstar', *Space News*, 8 Oct. 1990, p. 8.

138 Kiernan, V., 'US Congress slashes Milstar funding, orders shift of system to tactical users', *Space News*, 22 Oct. 1990, pp. 3, 37.

139 Langberg, M., 'Lockheed fights for Milstar as cold war thaw threatens', *San Jose Mercury News*, 14 Jan. 1991, pp. 1C, 6C.

140 US House of Representatives, *Conference Report Making Appropriations for the Department of Defense*, 101st Congress, 2nd Session, Report 101-938, 24 Oct. 1990, p. 97.

141 Kiernan, V., 'Electrical tests bring Air Force close to MILSTAR deployment', *Space News*, 28 May 1990, p. 10.

142 Foley, T., 'Slow funding to delay second, third Milstars', *Space News*, 17 Dec. 1990, pp. 4, 21.

programme.[143] The failure left one of the constellation's planes vacant, as Cosmos 1977 reached the end of its operational life.[144] Three more launches ended the year's activity, with the unexplained positioning of Cosmos 2105 between the orbital planes of Cosmos 2001 and Cosmos 2097.[145]

US early-warning satellites

The US Satellite Early Warning System (SEWS) consists of five Defense Support Program spacecraft.[146] Three of these provide frontline operational service, with two additional spacecraft available as back-up units should problems emerge with the primary satellites.[147] Two DSP satellites were used to track Iraqi Scud missile launches.[148] Although the system was slow to provide warning of initial Iraqi test launches in early December,[149] by the end of 1990 the system had been greatly improved.[150] The second Improved DSP (DSP-I) was launched on Titan 4 on 12 November 1990. The DSP-I satellites, of which spacecraft 14 through 22 were on order as of 1990, with options for 23 through 25 under negotiations,[151] will incorporate upgraded sensors and improved resistance to laser attack.[152]

Soviet navigation satellites

The USSR navigation satellite network consists of two systems. The Tsikada constellation of small satellites of modest capabilities is similar to the US Transit system. The GLONASS network of semi-synchronous satellites which provides higher accuracy fixes is similar to the US Navstar system.

In contrast to the US Transit system, which is used by both civilian and military operators, the Soviets use similar satellites in separate military (Cosmos designation) and civilian (Tsikada) networks. The military system consists of a six-satellite constellation, and there were two launches in 1990 to maintain this constellation, in contrast to the four launches of 1989.[153]

GLONASS launches in 1990 included Cosmos 2079, 2080 and 2080 in May 1990 and Cosmos 2109, 2110 and 2111 December. Along with the

143 'Soviets announce failure of early warning satellite', *Aerospace Daily*, 28 June 1990, p. 518.

144 'Two cosmonauts to be launched today', *Aerospace Daily*, 1 Aug. 1990, p. 177.

145 'Soviets launch missile early warning satellite', *Aerospace Daily*, 27 Nov. 1990, p. 335.

146 Ball, D., *A Base for Debate* (Allen & Unwin: London, 1987), provides perhaps the most comprehensive discussion of the DSP system.

147 Kenden, A., 'Military manœuvres in synchronous orbit', *Journal of the British Interplanetary Society*, vol. 36, no. 2 (Feb. 1983), pp. 88–91.

148 Covault, C., 'USAF missile warning satellites providing 90-sec. Scud attack alert', *Aviation Week & Space Technology*, 21 Jan. 1990, pp. 60–61.

149 Toth, R., 'Iraqi missile test had US thinking war had started', *Los Angeles Times*, 21 Dec. 1990, pp. A1, A11.

150 Diehl, J., 'Jordan's troop shifts raise questions in Israel', *Washington Post*, 2 Jan. 1991, pp. A17, A23.

151 'RFP for two more DSP satellites to be released Jan. 31', *Aerospace Daily*, 23 Jan. 1991, p. 125.

152 Covault, C., 'New missile warning satellite to be launched on first Titan 4', *Aviation Week & Space Technology*, 20 Jan. 1989, pp. 34–40. This article is an excellent review of the history and status of this programme.

153 'Soviet Union launches military navigation satellite', *Aerospace Daily*, 20 Sep. 1990, p. 471.

cessation of operations of satellites already in orbit, these launches left the operational GLONASS complement at 11 satellites,[154] one more than at the end of 1989 but still well short of the 15 spacecraft planned for 1995. The ultimate goal is a constellation of 21 satellites.

US navigation satellites

The Transit navigation satellite network continued operations in 1990, with 12 operational and spare Transit satellites in orbit. Most of the military users of Transit, such as the Navy's ballistic missile submarines that were the original impetus for Transit, will soon shift to Navstar. The Transit constellation will remain in service to civilian users at least through the year 2000.[155]

The launch of five Navstar/Global Positioning System (GPS) satellites in 1990, in addition to the five launched in 1989, brought the total constellation to 15 active satellites by the end of 1990. The total of 21 active plus three spare GPS satellites is planned for implementation by 1993.[156]

Traditionally Navstar has been regarded as providing navigation support to vehicles and platforms rather than weapons, but this is changing.[157] Small and inexpensive GPS receivers will be added to the air-launched version of the Navy's Harpoon cruise missile (known as the Standoff Land Attack Missile, SLAM), the Block III version of the non-nuclear Tomahawk long-range cruise missile (to be tested in 1991 for a 1993 initial operational capability), as well as free-fall conventional gravity bombs. The shifting sands and lack of visible landmarks in the Desert Shield/Desert Storm theatre of operations have accentuated the utility of Navstar for ground forces as well.[158]

Soviet weather satellites

The Soviet low-altitude weather satellite network supports both civilian and military users, in contrast to the separate systems operated by the USA. The Soviet military presumably uses data from the several Meteor 2 and Meteor 3 satellites which are usually operational. Two Meteor 2 spacecraft were launched in 1990.[159] Reports from the USSR suggest that these may be the final launches of the Meteor 2 system.[160]

[154] 'Soviets launched 17 spacecraft in December', *Aerospace Daily*, 10 Jan. 1991, p. 56.

[155] Danchik, R. *et al.*, 'The Navy navigation satellite system (TRANSIT)', *Johns Hopkins APL Technical Digest*, vol. 11, nos 1 and 2 (1990), pp. 97–101.

[156] 'Magnavox prepares for GPS buildup', *Military Space*, 25 Sep. 1989, pp. 3–5.

[157] AGARD (note 132) contains several papers that provide an excellent review of the status of Navstar users.

[158] Moore, M., 'US training, tactics shift with desert sand', *Washington Post*, 25 Nov. 1990, pp. A1, A25.

[159] 'Meteor 2-20, after being stored on orbit, begins transmission', *Aerospace Daily*, 19 Nov. 1990, p. 302.

[160] 'Soviets launch Mir resupply vehicle, two satellites', *Aerospace Daily*, 2 Oct. 1990, p. 5.

US weather satellites

The primary US military weather satellite system is the Air Force constellation of two Defense Meteorological Support Program (DMSP) satellites.[161] The fifth DMSP 5D-2 was launched on 1 December 1990. However, due to an upper stage malfunction, the spacecraft entered an orbit about 100 km lower than planned, sharply reducing its operational utility.[162] Weather satellites provided Desert Shield forces with sandstorm predictions, as well as information to predict the dispersal of chemical weapon clouds.[163]

VII. Military programmes of other countries

Just as the proliferation of nuclear and missile technology is of growing concern in the post-cold war world, proliferation of access to space systems (such as Navstar),[164] as well as proliferation of space technology itself, increasingly complicate the international security environment.[165] Leading intelligence analyst Jeffrey Richelson has suggested that India, Israel, Japan and South Africa may have imaging intelligence satellites by the late 1990s, while Brazil may develop such capabilities early in the next century.[166]

1990 marked China's return to space, following an unusually quiescent period in 1989, with no launch attempts. The 33rd Chinese launch, on 23 October, was the twelfth FSW-1 Recoverable Satellite Program photographic intelligence satellite. These satellites, whose development was initiated in 1966, are soon to be replaced by a larger FSW-2 model.[167] Chinese studies of ASAT systems in the early 1980s were halted in the face of technical problems, including satellite tracking difficulties.[168]

Despite growing budget pressures,[169] France continues work leading to a 1991 launch of the Syracuse II military communications satellite system.[170] The 1994 launch of the Helios photographic reconnaissance satellite, which is being developed with Italian and Spanish participation, will culminate a

[161] Several papers in AGARD (note 132) provide informative discussions of the operational utility of the system.

[162] 'DMSP misses intended mark', *Space News*, 17 Dec. 1990, p. 2.

[163] Kiernan, V., 'DMSP satellite launched to aid troops in Middle East', *Space News*, 10 Dec. 1990, p. 6.

[164] Starr, B., 'NBC nightmare threatens to spread', *International Defense Review*, vol. 23, no. 11 (Nov. 1990), p. 1225.

[165] Adams, P., 'New group to examine proliferation of satellite, EW technology', *Defense News*, 5 Feb. 1990, p. 33.

[166] Richelson, J., 'The future of space reconnaissance', *Scientific American*, vol. 264, no. 1 (Jan. 1991), pp. 38–44. According to Richelson's estimates, Iraq's space potential was in 1990 comparable to that of Brazil. However, as a result of the material destruction and political and economic disorder in Iraq following the successful conclusion of Operation Desert Storm, any aspiration on the part of Baghdad to pursue an active space policy has effectively been blocked for the foreseeable future.

[167] Hua-bao, L., 'The Chinese recoverable satellite program', Paper presented at the 40th Congress of the International Astronautical Federation, Malaga, Spain, 7–12 Oct. 1989, Report no. IAF-89-426.

[168] 'Foreign milspace', *Military Space*, 28 Jan. 1991, p. 4.

[169] de Briganti, G., 'Budget reveals slower growth for military space programs', *Defense News*, 3 Dec. 1990, p. 14.

[170] 'French milspace', *Military Space*, 5 Dec. 1988, p. 5.

planning process that started in the early 1970s.[171] France is also studying development of a more advanced Syracuse III communications system for the late 1990s,[172] but development of a 1-metre resolution imaging radar system to complement Helios has been dropped due to high cost.[173]

Israel successfully launched its second satellite, Offeq 2 (Horizon), on 3 April 1990, using a Shavit (Comet) booster based on the Jericho II ballistic missile.[174] The small size of the satellite indicates that it had no intelligence collection capability, despite continuing press reports to the contrary.[175]

Italian military satellite programmes under development in 1990 include the SICRAL (Satellite Italiano Communicazioni Riservate e Allarme, Italian Alarm and Classified Communications Satellite), the French-led HELIOS, and the US-led Navstar/GPS. Under the auspices of a new military space plan, the Italian Defence Staff has proposed an ambitious 10-year plan with a budget of over $13 billion. At least four satellites would be stationed in orbit at all times, with more satellites on the ground for deployment in the event of a military crisis. Seven space systems are proposed for development, including three communications and data relay networks, two observation satellite programmes, one theatre navigation system, and an electronic intelligence system.[176]

The failure of an Ariane booster on 22 February 1990 led to the loss of the Japanese Superbird 1B satellite, which included a military X-band transponder. Reports have surfaced over the years that Japan is also studying the development of an imaging intelligence satellite capability.[177]

Pakistan's first satellite, BADR-A, was launched by a Chinese Long March rocket on 16 July 1990.

The UK achieved a major expansion of its military communications space capabilities in 1990, with the launch of Skynet 4-A aboard a US Titan 3 on 1 January, and Skynet 4-C on an Ariane on 30 August. The UK is already studying options for a Phase II for the Skynet 4 military communications system, with new satellites in the series to be launched in 1996 and 1997.[178] The status of the Zircon signals intelligence satellite programme remains obscure. The growing European interest in military space systems, coupled with increasing European integration, has led to calls for close collaboration on future military space systems, but the UK has reacted negatively to these proposals, preferring to continue intelligence co-operation with the USA.[179]

171 Helios to deliver imagery to 3 nations', *Military Space*, 21 Nov. 1988, pp. 1–3; 'French study military recon satellite', *Aviation Week & Space Technology*, 22 Jan. 1973, p. 15.

172 'Allied milspace', *Military Space*, 19 Nov. 1990, p. 5.

173 de Selding, P., 'Defense Minister says no to French radar spy satellite', *Space News*, 12 Mar. 1990.

174 Brinkley, J., 'Israel puts a satellite into orbit a day after threat by Iraqis', *New York Times*, 4 Apr. 1990.

175 'A new spy in the sky', *Time*, 2 Apr. 1990, p. 33.

176 Politi, A., 'Italy plans military satellite network for early warning, reconnaissance', *Defense News*, 7 Jan. 1991, pp. 3, 31.

177 'Japan plans satellite', *Jane's Defence Weekly*, 16 Sep. 1989.

178 Furniss, T., 'UK studies new military satellite plan', *Flight International*, 7 Oct. 1989, p. 4.

179 de Selding, P., 'UK minister balks at call for European spy satellite', *Space News*, 16 July 1990, pp. 1, 20.

Appendix 3A. Military satellites launched in 1990

Type/Country/ Spacecraft name	Alternative name (Host spacecraft)	Designation	Launch date	Booster	Facil-ity	Mass (kg)	Apogee (km)	Perigee (km)	Inclin. (deg)	Period (min)	Comments
Imaging intelligence											
USSR											
THIRD GENERATION—HIGH RESOLUTION											
Cosmos 2055	SU PHOTO 3H-274	1990-003A	17 Jan.	SL-4	TT	6 300	249	253	62.81	89.5	First since C-2048, a long hiatus
Cosmos 2062	SU PHOTO 3H-275	1990-024A	22 Mar.	SL-4	PL	6 300	250	194	82.30	89.7	Arctic breakup, recovered 5 Apr.
Cosmos 2073	SU PHOTO 3H-276	1990-035A	20 Apr.	SL-4	PL	6 300	240	177	82.30	88.7	Recovered 7 May
Cosmos 2104	SU PHOTO 3H-277	1990-098A	16 Nov.	SL-4	PL	6 500	283	232	62.81	90.6	Re-entered 4 Dec.
THIRD GENERATION—MEDIUM RESOLUTION											
Cosmos 2083	SU PHOTO 3M-100	1990-053A	19 June	SL-4	PL	6 300	412	297	82.50	91.6	Recovered 3 July
Cosmos 2099	SU PHOTO 3M-101	1990-080A	31 Aug.	SL-4	PL	6 300	341	185	82.32	89.3	Recovered 14 Sep.
Cosmos 2120	SU PHOTO 3M-102	1990-115A	26 Dec.	SL-4	PL	5 500	253	188	82.60	88.6	. .
FOURTH GENERATION											
Cosmos 2057	SU PHOTO 4-86	1990-009A	25 Jan.	SL-4	TT	6 500	335	180	64.90	89.5	Re-entered 19 Mar.
Cosmos 2077	SU PHOTO 4-87	1990-042A	7 May	SL-4	PL	6 500	346	195	62.90	89.6	1 orbit behind C-2072, down 4 July
Cosmos 2089	SU PHOTO 4-88	1990-069A	3 Aug.	SL-4	TT	6 500	351	187	62.84	89.8	Over Kuwait 5–6 Aug., recov. 1 Oct.
Cosmos 2101	SU PHOTO 4-89	1990-087A	1 Oct.	SL-4	TT	6 500	335	180	64.90	89.5	Deliberately exploded on 30 Nov.
Cosmos 2102	SU PHOTO 4-90	1990-092A	16 Oct.	SL-4	PL	6 500	339	185	62.85	89.7	175-km perigee 25 Oct.–2 Nov. for Kuwait
Cosmos 2108	SU PHOTO 4-91	1990-109A	5 Dec.	SL-4	PL	6 500	339	196	62.80	89.6	. .
FIFTH GENERATION											
Cosmos 2072	SU PHOTO 5-11	1990-033A	13 Apr.	SL-4	TT	6 800	307	182	64.80	89.4	Replaced C-2049, in orbit during Kuwait occupation
Cosmos 2113	SU PHOTO 5-12	1990-113A	21 Dec.	SL-4	TT	6 500	307	189	64.80	89.2	. .

MILITARY MAPPING AND CIVIL REMOTE SENSING											
Cosmos 2078	SU PHOTO 4T-13	1990-044A	15 May	SL-4	TT	6 800	283	200	69.99	89.0	Recovered 28 June
Cosmos 2086	Resurs-F1 50	1990-062A	20 July	SL-4	PL	5 500	351	183	82.33	89.9	USSR released photos, recov. 3 Aug.
Resurs-F1 51	Resurs-F 8	1990-073A	16 Aug.	SL-4	PL	5 500	272	179	82.34	89.8	. .
Resurs-F1 52	Resurs-F 9	1990-082A	7 Sep.	SL-4	PL	5 500	238	180	82.50	88.7	. .
Resurs-F2 4	Resurs-F 6	1990-047A	29 May	SL-4	PL	5 500	272	259	82.34	88.5	. .
Resurs-F2 5	Resurs-F 7	1990-060A	17 July	SL-4	PL	5 500	238	176	82.20	88.8	Lifetime of 30 days, recovered 16 Aug.
USA											
KH-12 /2 USA 53	AFP 731	1990-019B	28 Feb.	STS	ETR	17 625	808	801	65.00	100.9	Reported lost 7 Mar. but seen 9–14 Oct.
KH-12 /3 USA 59	Meridian?	1990-050A	8 June	Titan 405A	ETR	17 625	455	455	51.00	92.2	. .
China											
FSW-1 12 China 33	. .	1990-089A	23 Oct.	CZ-2C	JI	2 000	311	208	56.98	89.6	. .

Electronic intellligence systems

USSR											
Cosmos 2058	SU ELINT 3-33	1990-010A	30 Jan.	SL-14	PL	4 375	665	634	82.51	97.7	1st since C-1975, coplanar with C-1812
Cosmos 2082	SU ELINT 4-9	1990-046A	22 May	SL-16	TT	12 500	880	852	71.00	102.0	. .
SU ELINT 4-10	. .	Failure	4 Oct.	SL-16	TT	12 500	. .	. .	. .	. .	Pad damaged in explosion
USA											
KH-12A /3 ESS 1	ELINT Sub Sat	1990-050C	8 June	Titan 405A	ETR	300	455	455	51.00	92.2	Deployed 3 ELINT sub-satellites
KH-12A /3 ESS 2	ELINT Sub Sat	1990-050D	8 June	Titan 405A	ETR	300	455	455	51.00	92.2	. .
KH-12A /3 ESS 3	ELINT Sub Sat	1990-050E	8 June	Titan 405A	ETR	300	455	455	51.00	92.2	. .
Magnum 3 USA 67	. .	1990-097B	14 Nov.	STS	ETR	2 275	35 780	35 780	0.00	1 436.0	. .

Naval intelligence systems

USSR											
Cosmos 2060	SU EORSAT 1-32	1990-022A	14 Mar.	SL-11	TT	4 250	417	404	65.03	92.7	Plane of C-2046, C-2051
Cosmos 2096	SU EORSAT 1-33	1990-075A	23 Aug.	SL-11	TT	4 250	418	403	65.02	92.7	. .
Cosmos 2103	SU EORSAT 1-34	1990-096A	14 Nov.	SL-11	TT	4 250	450	410	65.00	92.8	New plane, ops C-2033, –46, –60, –96
Cosmos 2107	SU EORSAT 1-35	1990-108A	4 Dec.	SL-11	TT	4 250	442	414	65.00	92.9	. .

Type/Country/ Spacecraft name	Alternative name (Host spacecraft)	Designation	Launch date	Booster	Facil- ity	Mass (kg)	Apogee (km)	Perigee (km)	Inclin. (deg)	Period (min)	Comments
Military communications											
USSR											
Cosmos 2064	SU COM 1-337	1990-029A	4 Apr.	SL-8	PL	45	1 491	1463	73.98	115.5	Dispersion due to upper stage problem?
Cosmos 2065	SU COM 1-338	1990-029B	4 Apr.	SL-8	PL	45	1 476	1 462	73.98	115.3	. .
Cosmos 2066	SU COM 1-339	1990-029C	4 Apr.	SL-8	PL	45	1 463	1 387	73.98	114.3	. .
Cosmos 2067	SU COM 1-340	1990-029D	4 Apr.	SL-8	PL	45	1 463	1 401	73.98	114.5	. .
Cosmos 2068	SU COM 1-341	1990-029E	4 Apr.	SL-8	PL	45	1 463	1 415	73.98	114.6	. .
Cosmos 2069	SU COM 1-342	1990-029F	4 Apr.	SL-8	PL	45	1 463	1 430	73.98	114.8	. .
Cosmos 2070	SU COM 1-343	1990-029G	4 Apr.	SL-8	PL	45	1 463	1 444	73.98	115.0	. .
Cosmos 2071	SU COM 1-344	1990-029H	4 Apr.	SL-8	PL	45	146	1 460	73.98	115.1	. .
Cosmos 2056	SU COM 2-45	1990-004A	18 Jan.	SL-8	PL	750	810	776	74.04	100.7	. .
Cosmos 2112	SU COM 2-46	1990-111A	10 Dec.	SL-8	PL	750	818	774	74.10	100.7	. .
Cosmos 2090	SU COM 3-44	1990-070A	8 Aug.	SL-14	PL	400	1 414	1 391	82.56	113.8	. .
Cosmos 2091	SU COM 3-45	1990-070B	8 Aug.	SL-14	PL	400	1 414	1 413	82.58	114.1	. .
Cosmos 2092	SU COM 3-46	1990-070C	8 Aug.	SL-14	PL	400	1 414	1 407	82.57	114.0	. .
Cosmos 2093	SU COM 3-47	1990-070D	8 Aug.	SL-14	PL	400	1 414	1 401	82.57	114.0	. .
Cosmos 2094	SU COM 3-48	1990-070E	8 Aug.	SL-14	PL	400	1 414	1 395	82.57	113.9	. .
Cosmos 2095	SU COM 3-49	1990-070F	8 Aug.	SL-14	PL	400	1 414	1 384	82.57	113.8	. .
Cosmos 2114	SU COM 3-50	1990-114A	24 Dec.	SL-14	PL	400	1 442	1 338	82.60	114.1	. .
Cosmos 2115	SU COM 3-51	1990-114B	24 Dec.	SL-14	PL	400	1 442	1 338	82.60	114.1	. .
Cosmos 2116	SU COM 3-52	1990-114C	24 Dec.	SL-14	PL	400	1 442	1 338	82.60	114.1	. .
Cosmos 2117	SU COM 3-53	1990-114D	24 Dec.	SL-14	PL	400	1 442	1 338	82.60	114.1	. .
Cosmos 2118	SU COM 3-54	1990-114E	24 Dec.	SL-14	PL	400	1 442	1 338	82.60	114.1	. .
Cosmos 2119	SU COM 3-55	1990-115F	24 Dec.	SL-14	PL	400	1 442	1 338	82.60	114.1	. .
Molniya 1-77	. .	1990-039A	26 Apr.	SL-6	PL	1 250	40659	612	62.78	736.4	. .
Molniya 1-78	. .	1990-071A	10 Aug.	SL-6	PL	1 250	40616	630	62.84	735.9	. .
Molniya 1-79	. .	1990-101A	23 Nov.	SL-6	PL	1 250	39750	525	62.90	735.0	. .
Cosmos 2085	Potok 7	1990-061A	18 July	SL-12	TT	2 120	35820	35 775	1.53	1436.5	PHOTO-5 data relay satellite

USA											
AFSATCOM D-10	On DMSP 5D-2 /5	1990-105A	1 Dec.	Atlas E	WTR	. .	849	732	98.86	100.7	. .
GLOMAR 2 SECS	USA 55	1990-028A	5 Apr.	Pegasus	EAFB	75	680	495	94.15	96.5	Small experimental comm. satellite
MACSAT 1 S/F	Multisat	1990-043A	9 May	Scout G-1	WTR	68	769	613	89.89	98.6	. .
MACSAT 2 S/F	Multisat	1990-043B	9 May	Scout G-1	WTR	68	769	612	89.89	98.6	Failed to deploy antenna
Leasat 5	Syncom IV F-5	1990-002B	9 Jan.	STS	ETR	1 320	36 363	34858	1.41	1426.9	. .
United Kingdom											
SKYNET 4-A	. .	1990-001A	1 Jan.	Titan 3	ETR	795	35 704	33676	3.37	1436.0	10 weather delays starting 7 Dec. 1989
SKYNET 4-C	. .	1990-079A	30 Aug.	Ariane 4	KO	795	35871	34721	4.41	1410.9	. .
Japan											
Superbird-X 1B	(On SCS 1B)	Failure	22 Feb.	Ariane 4	KO	. .	. .	. .	. .	. .	Booster failed, did not orbit

Ballistic missile early warning

USSR											
Cosmos 2063	SU BMEWS 1-62	1990-026A	27 Mar.	SL-6	PL	1 500	38 975	626	62.80	709.0	Replaced C-1793
Cosmos 2076	SU BMEWS 1-63	1990-040A	28 Apr.	SL-6	PL	1 500	39 319	571	63.01	708.3	Replaced C-1849
Cosmos 2084	SU BMEWS 1-64	1990-055A	22 June	SL-6	PL	1 500	758	585	62.80	98.2	TASS announced upper stage failure
Cosmos 2087	SU BMEWS 1-65	1990-064A	25 July	SL-6	PL	1 500	39 146	587	62.82	717.9	. .
Cosmos 2097	SU BMEWS 1-66	1990-076A	28 Aug.	SL-6	PL	1 500	39 192	607	62.87	706.0	Replaced C-1966, not C-1922
Cosmos 2105	SU BMEWS 1-67	1990-099A	20 Nov.	SL-6	PL	1 500	39 355	576	63.00	709.0	Positioned between C-2001 and C-2097
USA											
DSP-I 15 F-15	USA 65	1990-095A	12 Nov.	Titan 402A	ETR	2 370	35 780	357 80	1.00	1436.0	Delays (payload, weather) 21, 27 Sep.

Military navigation

USSR											
Cosmos 2098	SU NAV 3-68	1990-078A	28 Aug.	SL-8	PL	750	1 977	396	92.96	109.1	Rare eliptical orbit, C-2016 still op.
Cosmos 2100	SU NAV 3-69	1990-083A	14 Sep.	SL-8	PL	750	1 014	961	82.94	104.9	Coplanar with C-2016
Cosmos 2079	GLONASS 44	1990-045A	19 May	SL-12	TT	900	19 131	18 565	64.88	664.5	. .
Cosmos 2080	GLONASS 45	1990-045B	19 May	SL-12	TT	900	19 127	18 910	64.87	671.3	. .
Cosmos 2081	GLONASS 46	1990-045C	19 May	SL-12	TT	900	19 160	19 099	64.89	675.7	. .
Cosmos 2109	GLONASS 47	1990-110A	8 Dec.	SL-12	TT	900	19 142	19 142	64.80	676.0	. .

Type/Country/ Spacecraft name	Alternative name (Host spacecraft)	Designation	Launch date	Booster	Facil- ity	Mass (kg)	Apogee (km)	Perigee (km)	Inclin. (deg)	Period (min)	Comments
Cosmos 2110	GLONASS 48	1990-110B	8 Dec.	SL-12	TT	900	19 142	19 142	64.80	676.0	. .
Cosmos 2111	GLONASS 49	1990-110C	8 Dec.	SL-12	TT	900	19 142	19 142	64.80	676.0	Total of 11 now operational
COSPAS 5	Nadezhda 2	1990-017A	27 Feb.	SL-8	PL	750	1 020	956	82.95	104.9	. .
Cosmos 2061	Tsikada 19	1990-023A	20 Mar.	SL-8	PL	750	1 017	973	82.94	105.0	. .
Cosmos 2074	Tsikada 20	1990-036A	20 Apr.	SL-8	PL	750	1 015	982	83.00	104.8	Replaced C-1904
USA											
Navstar 2A-17	NDS 19 USA 50	1990-008A	24 Jan.	Delta 6925	ETR	818	20 560	20 088	54.64	717.9	Delayed by booster problems
Navstar 2A-18	NDS 20 USA 53	1990-025A	26 Mar.	Delta 6925	ETR	818	20 085	19 769	54.94	707.6	Scrubbed 21 Mar. due to high winds
Navstar 2A-19	NDS 15 USA 63	1990-068A	2 Aug.	Delta 6925	ETR	818	20 435	19 932	54.70	718.0	. .
Navstar 2A-20	NDS 21 USA 64	1990-088A	1 Oct.	Delta 6925	ETR	818	20 392	19 972	54.89	717.9	Activated in 22 days (vs 30 normal)
Navstar 2A-21	NDS 26 USA 66	1990-103A	26 Nov.	Delta 7925	ETR	930	20 293	20 073	54.91	717.5	First with all-up W-sensor

Weather

USSR											
Meteor 2-19	. .	1990-057A	27 June	SL-14	PL	2 750	961	940	82.55	104.1	. .
Meteor 2-20	. .	1990-086A	28 Sep.	SL-14	PL	2 750	975	953	82.53	104.2	Began transmitting 15 Nov.
USA											
DMSP 5D-2 /5 S-10	. .	1990-105A	1 Dec.	Atlas E	WTR	755	849	732	98.86	100.7	Perigee 100 km lower than planned
China											
FY-1 2 China 30	Feng Yun 2	1990-081A	3 Sep.	CZ-4	TY	910	900	885	98.94	102.8	Feng Yun = Wind and Cloud

Nuclear explosion detection

USSR *Soviet nuclear explosion detection sensors are probably mounted on satellites launched for other primary missions, such as early warning or navigation, but these satellites have not been identified.*

USA *US nuclear explosion detection sensors are mounted on satellites launched for other primary missions.*

ARD-1/2 15	(On DSP-I F-15)	1990-095A	12 Nov.	Titan 402A	ETR	. .	35 780	35 780	1.00	1436.0	Advanced Radiation Detector
NDS 10	(On Navstar 2A-17)	1990-008A	24 Jan.	Delta 6925	ETR	. .	20 560	20 088	54.64	717.9	NDS (X-ray & Optical)
NDS 11	(On Navstar 2A-18)	1990-025A	26 Mar.	Delta 6925	ETR	. .	20 085	19 769	54.94	707.6	NDS (X-ray & Optical)

NDS 12	(On Navstar 2A-19)	1990-068A	2 Aug.	Delta 6925	ETR	..	20 435	19 932	54.70	718.0	NDS (X-ray & Optical)
NDS 13	(On Navstar 2A-20)	1990-088A	1 Oct.	Delta 6925	ETR	135	20 392	19 972	54.89	717.9	First NDS to have EMP W-Sensor
NDS 14	(On Navstar 2A-21)	1990-103A	26 Nov.	Delta 7925	ETR	135	20 293	20 073	54.91	717.5	NDS (EMP, X-ray anf Optical)
NUDETS DMSP-10	(On DMSP 5D-2 /5)	1990-105A	1 Dec.	Atlas E	WTR	..	849	732	98.86	100.7	..

Other military missions

USSR											
RADAR CALLIBRATION											
Cosmos 2059	SU RADCAL 2-20	1990-012A	6 Feb.	SL-8	PL	950	2 292	191	65.80	110.1	Deployed 7 sub-satellites in first week
Cosmos 2075	SU RADCAL 2-21	1990-038A	24 Apr.	SL-8	PL	950	515	484	74.02	94.6	Dispensed 13 test objects on 13 August
Cosmos 2106	SU RADCAL 4-3	1990-104A	28 Nov.	SL-14	PL	4 375	550	526	82.50	95.2	Minor military? Similar to C-2053
GEODETIC											
Cosmos 2088	GEO-IK 11	1990-066A	30 July	SL-14	PL	1 500	1 525	1 483	73.60	116.1	GEO-IK 3, C-1950 & C-2037 still op.
USA											
TECHNOLOGY DEVELOPMENT											
STP P87-2 POGS	USA 56	1990-031A	11 Apr.	Atlas E	WTR	68	750	750	90.00	..	..
STP P87-2 SCE	USA 57	1990-031B	11 Apr.	Atlas E	WTR	68	750	750	90.00	..	Selective com./transceiver experiment
STP P87-2 TE	USA 58 MAESTRO	1990-031C	11 Apr.	Atlas E	WTR	68	750	750	90.00	..	..
SPACE SCIENCE											
CRRES-A	..	1990-065A	25 July	Atlas 1	ETR	8 150	3 3604	334	18.13	591.1	..
CRRES-PEGSAT 1	USA 55	1990-028A	5 Apr.	Pegasus	EAFB	192	680	495	94.15	96.5	2 CRRES barium release cannisters
LAUNCH VEHICLE DEVELOPMENT											
Titan 4	SRMU	Explosion	7 Sep.	Titan 4B	EAFB	..	..	..	..	..	Rocket motor dropped and exploded, one person killed
Truax M-3	SEALAR	Ground	9 May	Truax M-3	USA	..	..	..	..	..	3rd drop test in Monterey Bay
Pakistan											
BADR-A	..	1990-059A	16 July	CZ-2E	..	70	989	204	28.49	96.4	Experimental communications
Israel											
Offeq 2	..	1990-027A	3 Apr.	Shavit 1	Pal.	160	1582	207	143.23	102.7	Re-entered 9 July 1990

Type/Country/ Spacecraft name	Alternative name (Host spacecraft)	Designation	Launch date	Booster	Facil- ity	Mass (kg)	Apogee (km)	Perigee (km)	Inclin. (deg)	Period (min)	Comments
Strategic ballistic missile defence											
USA											
LACE	USA-51	1990-015A	14 Feb.	Delta 6925	ETR	1 400	550	531	43.10	95.3	..
RME	USA-52	1990-015B	14 Feb.	Delta 6925	ETR	1 050	479	457	43.11	93.8	..
BSUVE 1	..	Ballistic	25 Feb.	Terrier	WI	63	721	..	..	15.0	3.5 km/sec UV phenomenology test
EXCEED 3	..	Ballistic	27 Apr.	Aries 1	WSMR	2 275	115	..	6.5	..	..
HEDI KITE 1	..	Ballistic	24 Jan.	Sprint	WSMR	..	..	..	..	..	Shroud removal and vehicle separation
Brilliant Pebbles	..	Ballistic	25 Aug.	Black Brant 10	WI	..	..	..	..	13.0	Telemetry failed, tracked by LACE
SPEAR 2	..	Ballistic	25 July	Aries 1	WSMR	1 000	..	350	..	0.6	Spacecraft problems delayed, range safety failed
AOA—flight 1	..	Aircraft	11 May	..	Seattle	..	10	..	..	450.0	Aerodynamic test, computer problems
AOA—flight 2	..	Aircraft	June	..	Seattle	..	10	..	..	..	Sensor gathering star/background
AOA—flight 3	..	Aircraft	June	..	Seattle	..	10	..	..	..	Daytime satellite track
AOA—flight 4	..	Aircraft	June	..	Seattle	..	10	..	..	..	Night satellite tracking
AOA—flight 5	..	Aircraft	June	..	Seattle	..	10	..	..	..	Full end-to-end system checkout
AOA—flight 6	..	Aircraft	July	..	Seattle	..	10	..	..	..	Full end-to-end system checkout
AOA—flight 7	..	Aircraft	Aug.	..	Seattle	..	10	..	..	..	..
AOA—flight 8	..	Aircraft	Aug.	..	Seattle	..	10	..	..	..	..
AOA—flight 9	..	Aircraft	Aug.	..	Seattle	..	10	..	..	..	Final US test, KMR software rehearsal
AOA/AST	..	Aircraft	14 Sep.	..	KMR	..	10	..	..	..	Rehearsal at KMR
AOA/AST	..	Aircraft	19 Sep.	..	KMR	..	10	..	..	330.0	Minuteman 3 target, comms failed
Anti-tactical ballistic missile tests											
USA											
Hawk ATBM 2	Phase III	Ballistic	1 Nov.	Hawk	WSMR	..	8	..	..	..	2 simultaneous launch
Hawk ATBM 3	Phase III	Ballistic	1 Nov.	Hawk	WSMR	..	8	..	..	..	Intercepted Patriot target
Hawk ATBM-T 3	Phase III	Ballistic	1 Nov.	Patriot	WSMR	..	8	..	..	..	Intercepted TBM
Patriot PAC-2 IOC	PAC-2 ATBM	Milestone	31 Aug.	Patriot	..	..	..	..	..	..	New warhead and fusing, 12-km range
Patriot PAC-2 9	PAC-2 9	Ballistic	1 Sep.	Patriot	WSMR	..	..	..	..	..	13th intercept test, Patriot target

Patriot PAC-2 11	PAC-2	Ballistic	Sep.	Patriot	WSMR	..	..	..	..	..	14 tests 1 Sep. through 20 Dec.
Patriot PAC-2 12	PAC-2	Ballistic	Sep.	Patriot	WSMR	..	..	..	..	..	14 tests 1 Sep. through 20 Dec.
Patriot PAC-2 13	PAC-2	Ballistic	Sep.	Patriot	WSMR	..	..	..	..	..	14 tests 1 Sep. through 20 Dec.
Patriot PAC-2 14	PAC-2	Ballistic	Sep.	Patriot	WSMR	..	..	..	..	..	14 tests 1 Sep. through 20 Dec.
Patriot PAC-2 15	PAC-2	Ballistic	Sep.	Patriot	WSMR	..	..	..	..	..	14 tests 1 Sep. through 20 Dec.
Patriot PAC-2 16	PAC-2	Ballistic	Oct.	Patriot	WSMR	..	..	..	..	..	14 tests 1 Sep. through 20 Dec.
Patriot PAC-2 17	PAC-2	Ballistic	Oct.	Patriot	WSMR	..	..	..	..	..	14 tests 1 Sep. through 20 Dec.
Patriot PAC-2 18	PAC-2	Ballistic	Oct.	Patriot	WSMR	..	..	..	..	..	14 tests 1 Sep. through 20 Dec.
Patriot PAC-2 19	PAC-2	Ballistic	Oct.	Patriot	WSMR	..	..	..	..	..	14 tests 1 Sep. through 20 Dec.
Patriot PAC-2 20	PAC-2	Ballistic	Nov.	Patriot	WSMR	..	..	..	..	..	14 tests 1 Sep. through 20 Dec.
Patriot PAC-2 21	PAC-2	Ballistic	Nov.	Patriot	WSMR	..	..	..	..	..	14 tests 1 Sep. through 20 Dec.
Patriot PAC-2 22	PAC-2	Ballistic	Nov.	Patriot	WSMR	..	..	..	..	..	14 tests 1 Sep. through 20 Dec.
Patriot PAC-2 23	PAC-2	Ballistic	Nov.	Patriot	WSMR	..	..	..	..	..	Dual launch, 1 successful intercept
Patriot PAC-2 24	PAC-2	Ballistic	Nov.	Patriot	WSMR	..	..	..	..	..	Dual launch late Nov.
Israel											
Arrow 1 (Chetz 1)	Dem Val	Ballistic	9 Aug.	Arrow	Pal.	10	..	..	..	0.2	Destroyed due to telemetry loss
Arrow 2 (Chetz 2)	Dem Val	Ballistic	20 Dec.	Arrow	Pal.	10	..	..	..	..	2nd test, intercepted Jericho I target
Strategic anti-missile ground tests											
USA											
NTB National Testbed		Ground	19 Oct.	..	Falcon	..	..	..	..	..	Full SATCOM, Compartmented Security
Alpha 5	..	Ground	19 Apr.	..	SJC	..	..	..	..	..	5th lasing test, exhaust flow failure
Alpha 6	..	Ground	30 Nov.	..	SJC	..	..	..	..	..	6th lasing test, full power test
FEL MCTD	RF FEL	Ground	23 Mar.	..	Seattle	..	..	..	..	..	..
FEL HIBAF 2	RF FEL	Ground	Feb.	..	LLNL	..	..	..	..	..	..
FEL HIBAF 3	RF FEL	Ground	July	..	LLNL	..	..	..	..	..	..
Firepond SDI 1	..	Ground	Mar.	..	West.	..	..	..	..	..	1st LIDAR test
Firepond SDI 2	..	Ground	Oct.	..	West.	..	..	..	..	..	2nd LIDAR test
SBI 1	AHIT	Ground	24 July	..	EAFB	18	..	..	..	0.3	Light-weight KKV, tracked satellite
RME test	..	Ground	25 June	..	Hawaii	..	..	..	..	0.1	1st successful test
RME test 2	..	Ground	12 Sep.	..	Hawaii	..	..	..	..	..	2nd successful test

[a] These Space Test Program experiments were carried on the Shuttle Orbiter, and were not free-flying satellites.

Spacecraft/experiment abbreviations

AHIT	Advanced Hoover Interceptor Test
AOA	Airborne Optical Adjunct
ARD	Advanced Radiation Detector
AST	Airborne Surveillance Testbed
BSUVE	Bow Shock UV (ultraviolet) Emission
CRRES	Combined Release Radiation Effects Satellite
EXCEED	Excitation by Electron Deposition
FEL	Free Electron Laser
HEDI	High-Endoatmospheric Defense Interceptor
HIBAF	High-Brightness Accelerator FEL
KITE	Kinetic Intercept Test
KKV	Kinetic Kill Vehicle
LACE	Low-Power Atmospheric Compensation Experiment
LIDAR	Light detection and ranging (laser radar)
MAESTRO	Multiple Autonomous Experimental Spacecraft for Telecommunnications, Recording and Observation
MCTD	Modular Component Tecnical Development
MACSAT S/F	Multiple Access Communications Satellite Store/Forward
NDS	Nuclear Detection System
POGS	Polar Orbiting Geomagnetic Survey
RME	Relay Mirror Experiment
SPEAR	Space Power Experiment Aboard Rocket
SRMU	Solid Rocket Motor Upgrade
TBM	Tactical Ballistic Missile

Launch facility abbreviations

EAFB	Edwards Air Force Base, Calif., USA
ETR	Eastern Test Range, Cape Canaveral, Fl.a, USA
Falcon	Falcon Air Station, Colo., USA
JI	Jichuan Space Center, China
KMR	Kwajalein Missile Range, USA
KO	Kourou, French Guinea
LLNL	Lawrence Livermore National Laboratory, Calif., USA
Pal.	Palmachim, Israel
SJC	San Jose Capistrano, Calif., USA
TT	Tyuratam (Baikonur), Kazakhstan, USSR
TY	Taiyun, China
West.	Westford, Mass., USA
WI	Wallops Island, Va., USA
WSMR	White Sands Missile Range, N. Mex., USA
WTR	Western Test Range, Vandenberg Air Force Base, Calif., USA

4. Chemical and biological warfare: developments in 1990

S. J. LUNDIN and THOMAS STOCK*

I. Introduction

In 1990 two events related to chemical and biological warfare (CBW) were particularly important. The threat of use of chemical weapons (CW) in the Persian Gulf crisis made the possibility of chemical warfare more likely. On the other hand, there was the positive action of the US–Soviet bilateral agreement to reduce their large CW stockpiles, which was signed at the June 1990 summit meeting. These were not the only events of significance in 1990; the others include the following:

1. Before its invasion of Kuwait, Iraq threatened to use chemical weapons against Israel; after the arrival of the coalition forces in Saudi Arabia, Iraq threatened to use chemical weapons against them also, particularly against the US forces.
2. Threats of Iraqi CW use led to counter threats by the coalition countries and to intensified debate about the coupling of CW use to the use of nuclear weapons. The spread of missiles and missile technology in the Persian Gulf region had a significant impact on this discussion.
3. Allegations that a number of countries may be in the process of acquiring biological weapons (BW) continued to be made. Iraq is alleged to have a BW capability.
4. The crisis in the Persian Gulf led to questions about the effectiveness of protective equipment against CW use and increased procurement of such equipment in the crisis area.
5. Efforts to abolish chemical weapons continued. The US–Soviet agreement to substantially reduce their CW stocks also includes a provision to stop CW production. These provisions are of great importance as are others to destroy the majority of US and Soviet CW stockpiles, to exchange data on CW stockpiles and facilities, and to co-operate in finding reliable destruction technologies.

* Fredrik Wetterqvist of the SIPRI Chemical and Biological Warfare (CBW) Programme assisted in preparing references and data for this chapter. The references were gathered from the SIPRI CBW Programme Data Base and were also kindly provided by J. P. Perry Robinson, Science Policy Research Unit, University of Sussex, UK, from the Sussex–Harvard Information Bank. The evaluations made and developments reported are as of 31 December 1990, and do not take into account events after the outbreak of war in the Persian Gulf on 17 January 1991.

6. The successful removal of all US CW stockpiles from Clausen in the Federal Republic of Germany was evidence of the reduced tension in Europe and proof that, with careful planning and safety precautions, it is possible to accomplish the difficult task of moving chemical weapons through densely populated civilian areas.

7. Chemical weapons are extremely costly to destroy. This was demonstrated by the removal of US chemical weapons from the Federal Republic of Germany. Acquisition of chemical weapons inevitably leads to the need to destroy them and to costs greatly out of proportion to the questionable value of acquiring these weapons.

8. It is obvious, technical arguments aside, that the removal and destruction of chemical weapons can create a conflict between disarmament efforts and environmental concerns.

9. Efforts continued to hinder the spread of chemicals and technology which have CW application. The most visible were national efforts in the FRG and the USA to institute legislation and sanctions against national and international trade with chemicals, and against technology which can be used to make chemical or biological weapons, and prosecution of individuals involved in such activities.

The following discussion focuses on Iraq's CW threats, the efforts of the USA and the USSR to abolish chemical weapons, and the removal of US chemical weapons from the FRG. International CBW negotiations are discussed in chapter 14.

II. Developments related to the Persian Gulf crisis

Both before and after Iraq's August 1990 invasion of Kuwait and in the ensuing international crisis, Iraq threatened to use chemical weapons against Israel, the USA and the coalition forces if Iraq were attacked by those states or if nuclear weapons were used against Iraq.[1] Iraq claimed that it also possessed binary chemical weapons (see below).[2] Several countries warned Iraq not to use chemical weapons and indicated that other weapons could be used in retaliation.[3]

[1] See, for example, Morris H., and Sheridan, M., 'Iraq vows to use chemical weapons if Israelis attack', *The Independent*, 3 Apr. 1990, p. 1; Reuters, 'Iraq free to hit back at Israel', *The Independent*, 17 Apr. 1990, p. 10; Reuters, 'We'd use gas, Iraqi says', *International Herald Tribune*, 10 Aug. 1990, p. 1; see also 'Saddam Husayn addresses visiting U.S. Senators', JN1604161390, Baghdad Domestic Service, 1400 GMT, 16 Apr. 1990 (in Arabic) in Foreign Broadcast Information Service, *Daily Report–Near East & South Asia* (*FBIS–Near East & South Asia*), FBIS-NES-90-074, p. 7; Gordon, M. R., 'Iraq prepares toxic gas sites', *International Herald Tribune*, 26 Sep. 1990, p. 6.

[2] 'President warns Israel, criticizes Israel', JN0204132990, Baghdad Domestic Service, 1030 GMT, 2 Apr. 1990 (in Arabic) in FBIS-NES-90-064, 3 Apr. 1990, p. 34.

[3] See, for example, Brown, C., 'Britain warns Saddam against gas war', *The Independent*, 10 Aug. 1990, p. 1; Reuters, 'Cheney warns Iraq on chemical arms', *International Herald Tribune*, 15 Aug. 1990, p. 4; Conference on Disarmament document CD/PV.574, 16 Aug. 1990, p. 21.

Israel stated early in the crisis that it did not fear CW attack by Iraq.[4] Just prior to the invasion of Kuwait, a statement was made by the Israeli Minister for Science and Technology which was interpreted by the press as implying that Israel could retaliate with chemical weapons if attacked.[5] However, the following day the minister denied any knowledge of whether or not Israel possesses chemical weapons.[6] There is thus still no official Israeli confirmation of Israeli CW possession. On other occasions Israeli ministers expressed the conviction that Israel could retaliate tenfold or even a hundredfold against an Iraqi CW attack.[7]

Preparation by Israel to meet an Iraqi CW attack led to the argument that distributing protective equipment to the civil population could be misunderstood as preparation for an *Israeli* CW attack.[8] None the less, in October Israel issued protective masks to its citizens.[9] The Palestinian population was later able to obtain masks by purchasing them.[10] The measures taken were intended to be seen only as protective and not as preparation for war.[11] By November 1990, 3.4 million protective masks had been handed out (approximately 85 per cent of the number to be distributed).[12]

During the early stages of the conflict in the Persian Gulf, US intelligence sources maintained that Iraqi forces had moved chemical weapons into Kuwait and had practised loading and unloading them onto aircraft, possibly for training purposes.[13] Speculation occurred about whether or not Iraq could successfully destroy oil fields by using chemical weapons.[14] There was also discussion of the utility for chemical warfare of Scud ballistic missiles, which Iraq bought and then modified (for a further discussion of Iraqi ballistic missiles, see chapter 9). In view of the small amount of CW payload that can be put into the warhead of a ballistic missile, the relatively low precision of these missiles and the limited number of them available, it is doubtful if chemical weapons used in this way are militarily useful or merely serve to create

[4] See, for example, 'Schamir zeigt keine Furcht: Israel setzt nach Iraks Drohung Spionage-Satelliten ein', *Frankfurter Rundschau*, 4 Apr. 1990, p. 1.

[5] 'Science minister on deterrence to chemical threat', TA2707060490, Jerusalem Domestic Service, 0405 GMT, 27 July 1990 (in Hebrew) in FBIS-NES-90-145, 27 July 1990, p. 30; see also, for example, 'Israel says gas attack would be met by gas', *International Herald Tribune*, 28–29 July 1990, p. 5.

[6] 'Ne'eman defends remarks on chemical weapons', TA2807104590, Jerusalem Domestic Service, 1000 GMT, 28 July 1990 (in English) in FBIS-NES-90-146, p. 24; see also 'Shamir did not support Ne'eman's CW remarks', TA0208045090, Jerusalem Domestic Service, 0400 GMT, 2 Aug. 1990 (in English) in FBIS-NES-90-149, 2 Aug. 1990, p. 33.

[7] TT/Reuters, 'Israelreaktion på irakiskt hot', *Svenska Dagbladet* (Stockholm), 4 Apr. 1990, p. 4; 'Husayn, Israeli figures on region, Iraqi power', PM1605153690, FBIS-NES-90-096, 17 May 1990, p. 1.

[8] Sheridan, M., 'Israeli experts believe chemical weapon attack unlikely', *The Independent*, 24 Aug. 1990; see also note 6.

[9] Brinkley, J., 'All Israelis to get gas masks', *International Herald Tribune*, 2 Oct. 1990, p. 1.

[10] Black, I., 'Israeli gas mask distribution wrapped in discretion', *The Guardian*, 2 Oct. 1990, p. 4; see also Sheridan, M., 'Not enough gas masks for Palestinians', *The Independent*, 25 Aug. 1990, p. 8; Diehl, J., 'Israel starts mass handout of gas masks', *International Herald Tribune*, 8 Oct. 1990, p. 1.

[11] Black, I., 'Gas masks are a precaution, not a provocation, Shamir maintains', *The Guardian*, 8 Oct. 1990, p. 5.

[12] '85 per cent of gas masks distributed', TA0311202290, Jerusalem Domestic Service, 2000 GMT, 3 Nov. 1990 (in Hebrew).

[13] See, for example, Reuters, 'Iraq reported to prepare gas', *International Herald Tribune*, 9 Aug. 1990, p. 5.

[14] 'Iraqi vows to destroy oil fields', *International Herald Tribune*, 21 Sep. 1990, pp. 1–2.

fear when launched at civilians.[15] It did not become clear whether or not Iraq has been able to produce chemical warheads for its missiles.[16] However, Iraqi missile tests at the end of 1990 were judged to be simulations of CW attack.[17]

Varying estimates of Iraqi CW production capability and CW production facilities were made during 1990. The figures given for production capacity differ and range from estimates of 1000 tonnes of mustard gas and the nerve gases sarin and tabun (estimated to suffice for 500 000 projectiles), to specific figures of 720 tonnes of mustard gas and 48 tonnes of nerve gases annually.[18] It has been claimed that hydrogen cyanide is capable of penetrating the filters in protective masks and that, because of this, Iraq has added hydrogen cyanide to its CW arsenal.[19]

Various alleged CW production sites are said to exist in Iraq at *inter alia* Ramadi, Akashat, Al Fallujah, Samarra, Salman Pak and near Mosul.[20] It is claimed that Samarra produces approximately 400 kg of tabun and mustard gas per day and possibly also sarin.[21] A munition-loading facility is allegedly located at Al-Iskandriyah.[22] President Saddam Hussein has claimed that Iraq has 'dual' chemicals (i.e., binary chemical weapons),[23] but independent corroboration of this has not appeared in the technical literature. Some observers doubt whether Iraq can have developed reliable binary chemical weapons and note the technical difficulties which the USA had in developing its system.[24] At the end of 1990 the Director of the US Central Intelligence Agency (CIA) stated that Iraq had a stockpile of approximately 1000 tonnes of CW agent.[25]

Many claims were made during 1990 about the intentional or inadvertent help which Iraq may have had from various countries in building up its chemical warfare capability.[26] A substantial amount of information linked companies in the *Federal Republic of Germany* to Iraq.[27] In August West

[15] Levran, A., 'Threats facing Israel from surface-to-surface missiles', *IDF Journal*, no. 19 (winter 1990), pp. 37–44.

[16] Bellamy, C., 'Chemical strike prowess of Iraq over-estimated', *The Independent*, 23 Aug. 1990, p. 7.

[17] Press Association, 3 Dec. 1990, as reported by Gertz, B., 'Cheney: Iraqi missiles soon could traverse Mideast', *Washington Times*, 4 Dec. 1990, p. 3; see note 15, p. 37.

[18] Institute for Defense and Disarmament Studies, *Arms Control Reporter* (IDDS: Brookline, Mass.), sheets 704.E-2.11–12, Sep. 1990.

[19] 'Treffer mit Roland', *Der Spiegel*, vol. 44, no. 39 (24 Sep. 1990), p. 32.

[20] See note 18; see also Carus, S. W., *The Genie Unlashed: Iraq's Chemical and Biological Weapons Program*, Policy paper no. 14 (Washington Institute for Near East Policy: Washington, DC, 1989); O'Dy, S., 'Chimique: la grande menance', *l'Express*, 14 Sep. 1990, p. 21; Rathmell, A., 'Chemical weapons in the Middle East: Syria, Iraq, Iran and Libya', *Marine Corps Gazette*, vol. 74, no. 7 (July 1990), pp. 59–67.

[21] 'Gezisch am Nachmittag', *Der Spiegel*, vol. 44, no. 48 (26 Nov. 1990), pp. 29–31.

[22] See note 18.

[23] See note 2; another option might be a mixture of two chemical warfare agents.

[24] See, for example, 'Iraqis surprise analysts', *Jane's Defence Weekly*, vol. 13, no. 15 (14 Apr. 1990), p. 686.

[25] Lardner, G., Jr, 'CIA chief says only perception of imminent peril would budge Iraq', *International Herald Tribune*, 17 Dec. 1990, p. 5.

[26] Frankel, G., 'Iraq's war machine: made in the West', *International Herald Tribune*, 18 Sep. 1990, p. 7; 'Study details Iraq's sources of chemical arms', *International Herald Tribune*, 4 Oct. 1990, p. 7; see also note 18.

[27] 'Senfgas von Ahmed', *Der Spiegel*, vol. 44, no. 41 (8 Oct. 1990), pp. 152–53; 'Schlimmer als die Pest', *Der Spiegel*, vol. 44, no. 33 (13 Aug. 1990), pp. 80–85; Nolan, J. E., 'The Iraq syndrome: cutting

German authorities arrested seven process technicians and accused them of having helped the Iraqi State Enterprise for Pesticide Production (SEPP) construct what the Iraqis said was a pesticide factory, and what the West German authorities alleged was a facility for producing chemical warfare agents.[28] In August 1990, it was announced that West German authorities were investigating companies said to have delivered to Iraq material and chemicals which could be used for CW production.[29] In September West German authorities intercepted a sodium cyanide shipment to Iraq in a Turkish harbour. It was claimed that the sodium cyanide was to be used for metallurgic purposes, but this substance can also be used to produce hydrogen cyanide and tabun.[30]

A number of other countries were also implicated in this context. A company in *France* was accused of providing Iraq with CW material and technology.[31] Reports that the former *German Democratic Republic* provided Iraq with information about defensive training against chemical warfare and exported technology for that purpose were confirmed.[32] *Spain*, *Austria* and a number of other countries were also involved in similar activities.[33] The export of chemicals which can be used as CW precursors from the *USA* to Iraq was attempted.[34] The *USSR* denied that it had contributed to Iraqi CW production.[35] *India* has delivered many chemicals which can be used for CW production.[36]

Increased uncertainty about possible future CW use by Iraq in a military conflict heightened concern about protection against these weapons. Shielding troops against chemical weapons in a hot climate and new developments in CW protection are of particular interest. Recently developed protective suits and other garments appear to allow for more effective combat activities.[37]

off technology can buy time', *International Herald Tribune*, 4 Apr. 1990, p. 6; Leyendecker, H. and Rickelmann, R., *Exporteure des Todes: Deutscher Rüstungsskandal in Nahost* (Steidl Verlag: Göttingen, 1990).

[28] See, for example, 'Sieben Haftbefehle nach Giftgasgeschäft mit Irak', *Frankfurter Rundschau*, 18 Aug. 1990, p. 1.

[29] See, for example, Marsh, D., 'W Germans take a close look at chemicals groups', *Financial Times*, 24 Aug. 1990, p. 2.

[30] 'Deutsches Gift für Irak', *Frankfurter Rundschau*, 1 Oct. 1990, p. 2; 'Der Hinweis traf ins Schwarze', *Der Spiegel*, vol. 44, no. 37 (10 Sep. 1990), pp. 112–18.

[31] Ibrahim, Y. M., 'French firm said to aid Iraq on toxic gas', *International Herald Tribune*, 21 Sep. 1990, p. 6.

[32] 'Deutsche Schlamperei half Teufel am Golf', interview with Kh. Lohs in *Neue Berliner Illustrierte*, no. 41 (1990), pp. 56–57; 'Im ersten Kreis der Hölle', *Der Spiegel*, vol. 44, no. 43 (22 Oct. 1990), pp. 97–101; 'Irak: NVA übte Gaskrieg', *Der Spiegel*, vol. 44, no. 34 (20 Aug. 1990), pp. 14–15.

[33] See note 18; see also 'Two firms suspected of Iraqi poison plant deal', AU3008102190, *Vienna Kurier*, 30 Aug. 1990, p. 6 (in German) in Foreign Broadcast Information Service, *Daily Report–West Europe* (*FBIS–WEU*), FBIS-WEU-90-169, 30 Aug. 1990.

[34] 'Woher kommen Iraks Waffen?', *Frankfurter Rundschau*, 7 Sep. 1990, p. 1; 'Angeblich US: Hilfe für den Irak', *Süddeutsche Zeitung*, 3 Aug. 1990, p. 7; Oram, R., 'US charges two with planning to export arms', *Financial Times*, 16 Aug. 1990.

[35] Permanent Mission of the Soviet Union (Geneva), History of USSR's military contacts with Iraq described', TASS (Moscow), 15 Aug. 1990, *Press Bulletin*, no. 156 (2236), (17 Aug. 1990), pp. 4–5.

[36] Clad, J., 'Indian shipments to Iraq underlie its diplomatic ambivalence: chemical reaction', *Far Eastern Economic Review*, 6 Sep. 1990, pp. 10–11.

[37] 'A worst case scenario', *Jane's Defence Weekly*, vol. 14, no. 13 (29 Sep. 1990), p. 568; Ember, L., 'Persian Gulf crisis: U.S. hones chemical arms defense', *News of the Week*, vol. 68, no. 35 (27 Aug. 1990), pp. 4–5; Browne, M. W., 'If the Iraqis use poison gas: new no-man's-land in desert', *International Herald Tribune*, 10 Aug. 1990, p. 6; Bruce, L., 'Desert fighting has particular horrors, experts say', *International Herald Tribune*, 9 Aug. 1990, pp. 1–5.

However, concern was voiced that, in practice, it would be impossible to fight a chemical war in the Saudi desert.[38] A number of tanks and reconnaissance vehicles protected against BW and CW were moved to the area.[39] In 1990 reports increased concerning the purchase of protective equipment by a number of countries in the Gulf region with Iraq being given special mention. Egypt, Israel, Saudi Arabia, Syria and Turkey were among those named.[40] In the USA the CW and BW defence programme received additional funding for protective clothing and equipment because of the Persian Gulf crisis.[41]

The use of chemical weapons in the 1980–88 Iraq–Iran War deserves study because of the threat of CW use in the Persian Gulf.[42] However, there have been few analyses published of the role which chemical weapons played in the Iraq–Iran War, and the conclusions drawn vary. One Israeli study points out that Iraq initially used chemical weapons only against mass attacks, but that it took advantage of the tactical effect of chemical weapons at a later stage of the war.[43] Chemical weapons did not appear to affect the outcome of the war. The CW attacks against the Kurds in Iraq were considered effective. A controversial observation made in the Israeli study is that, with few exceptions, Iraqi use of chemical weapons took place in combat on its own territory and might therefore be perceived as protective. This concept has no support in international agreements. Other observers stress the psychological impact—rather than the tactical effect—which the use of chemical weapons had on the unprotected Iranian troops.[44]

A US Department of Defense reconstruction of the final stages of the Iraq–Iran War alleged that *Iran* used chemical weapons at Halabja during the war. This conclusion was based on the facts that the UN investigative team found victims who showed signs of cyanide poisoning and that Iraq had not used hydrogen cyanide during the war with Iran.[45] Iran has denied the use of

[38] Reuters, 'Blickpunkt: C–Schutzanzüge: Rasch 50 Grad heiss', *Frankfurter Rundschau*, 11 Aug. 1990, p. 2; Horsnell, M., 'Britain supplies protective suits', *The Times*, 23 Aug. 1990; see Browne, M. W., note 37.

[39] Baker, C., 'U.S. asks Germany for combat vehicles used to detect chemicals', *Defense News*, vol. 5, no. 35 (27 Aug. 1990), p. 7; Reuters, 'Bonn provides Fuchs tanks', *International Herald Tribune*, 18–19 Aug. 1990, p. 5; Press Association, 'UK order for chemicals antidotes', *Financial Times*, 21 Aug. 1990, p. 2.

[40] 'Iraq sought anti-CW kit', *Jane's Defence Weekly*, vol. 14, no. 7 (18 Aug. 1990), p. 217; Authers, J. and Capstick, A. B., 'Gas antidote sales increase as fears of war increase', *Financial Times*, 24 Aug. 1990, p. 2; Witt, M. J., 'Chemical weapon concerns boost business for U.K. firms', *Defense News*, vol. 5, no. 37 (10 Sep. 1990), p. 35; Bjerke, D., and Hygstedt, B., 'Gasmasker krigsklassas inte: leverans till Saudiarabien strider inte mot lagen', *Svenska* Dagbladet (Stockholm) 22 Sep. 1990, p. 8 Höhler, G., 'Gasmasken sind in der Türkei begehrt', *Frankfurter Rundschau*, 7 Sep. 1990, p. 2.

[41] Roos, J. G., 'Chemical gear hot issue on hill', *Armed Forces Journal International*, vol. 128, no. 4 (Nov. 1990), pp. 14, 68.

[42] See, for example, Waters, L., 'Chemical weapons in the Iran/Iraq war', *Military Review*, vol. 70, no. 10 (Oct. 1990), pp. 57–63.

[43] Bar, M., 'Strategic lessons of chemical war: historical approach', *IDF Journal*, no. 20 (summer 1990), pp. 48–55.

[44] See, for example, Smith, R. J., 'Quick trigger on poison gas: Iraq gradually shifted tactics during the war with Iran', *International Herald Tribune*, 11–12 Aug. 1990, pp. 1, 4.

[45] Tyler, P. E., 'Iran faulted, too, in gas attack on Kurds', *International Herald Tribune*, 4 May 1990, pp. 1, 4.

chemical weapons during the war.[46] It has also been questioned whether the alleged Iraqi attack on the Kurds at Amadiyyah in the northern part of Iraq actually took place since no gas victims were said to have been reported.[47]

Some countries may wish to acquire chemical weapons because of the perception that these weapons constitute the 'poor man's nuclear weapon'. This is an old concept, but it was given particular emphasis at the 1989 Paris Conference on the Prohibition of Chemical Weapons, where a number of countries from the Middle East wanted to make nuclear weapon disarmament a prerequisite for CW disarmament. The concept continued to be discussed.[48] Iraq has strongly advocated it and has seen its CW capability as a deterrent against nuclear attack, particularly by Israel.[49]

In 1990 chemical (and biological) weapons tended to be regarded as weapons of political terror rather than as militarily useful weapons. However, if these weapons are used against the civilian population of an adversary who possesses nuclear weapons, the consequences could be quite grave. The concept of chemical weapons as a deterrent is meaningless if an adversary uses nuclear weapons in response, and adopting such a policy could lead to the risk of a dangerous lowering of the nuclear threshold.[50] The most threatening chemical warfare situation might be a conflict in the Middle East between Arab nations in which one nation possessed chemical weapons and the other neither chemical nor nuclear weapons.

Developments in Iraq brought chemical warfare to the forefront of public awareness. It is also clear that segments of the chemical industry put commercial interests above ethical considerations and took advantage of the current situation, in which CW production is not yet internationally prohibited.

III. Libya

In 1990 Libya was the only other Arab country which received significant international attention because of its CW capability. Debate about Libyan CW production capacity continued and became more complicated, but after the Persian Gulf crisis began, little further information appeared.

In February 1990, allegations were made of CW production at Rabta,[51] but Libya maintained that the facility produced pharmaceuticals and invited other

[46] 'UN mission denies use of chemical weapons', LD0405114990, *Tehran Domestic Service*, 1030 GMT, 4 May 1990 (in Persian) in FBIS-NES-90-087, 4 May 1990, p. 45.

[47] Pelletiere, S. C., Johnson, D. V., II and Rosenberger, L. R., *Iraqi Power and US Security in the Middle East* (US Army War College Strategic Studies Institute: Carlisle Barracks, Pa., 1990); Associated Press, 'U.S study funds no proof of Iraqis gassing Kurds', *International Herald Tribune*, 18 Dec. 1990, p. 2.

[48] For a short review see SIPRI, *SIPRI Yearbook 1990: World Armaments and Disarmament* (Oxford University Press: Oxford, 1990), pp. 534–35.

[49] See 'Saddam Husayn addresses visiting U.S. Senators' (note 1).

[50] Evron, Y., 'Israel', ed. R. Cowen Karp, SIPRI, *Security with Nuclear Weapons? Different Perspectives on National Security* (Oxford University Press: Oxford, 1991), pp. 277–97.

[51] Institute for Defense and Disarmament Studies, *Arms Control Reporter* (IDDS: Brookline, Mass.), sheet 704.E-2.2, Mar. 1990; Reuters, 'U.S. says Libyan plant produces mustard gas', *International Herald Tribune*, 8 Mar. 1990, p. 1.

states and companies to join in the production.[52] In March there were reports that the alleged CW production facility at Rabta had burned.[53] On the basis of *inter alia* SPOT satellite images, it was later claimed that the fire was a fake.[54] Two West Germans were said to be under investigation by Libyan authorities, and accusations were made by Libya that agents from the FRG, Israel and the USA were responsible for the fire story.[55] Shortly before the fire incident, Libya offered to dismantle the Rabta facility and build a pharmaceutical facility with foreign participation.[56] If, in fact, there was a fire at Rabta it may have been an attempt to draw attention from a facility under construction at the Sebha oasis, 650 km south of Tripoli, which was alleged to be a new CW production facility.[57] It was said to have been built with the help of West German companies under the project code-name Pharma 200,[58] and investigations were started in the FRG about possible West German involvement.[59] Allegations were also made that China had contributed to Libya's CW buildup, which China denied.[60]

It is interesting to note the outcome of the legal action taken against the West German company that helped Libya build a factory in Rabta under the code-name Pharma 150, which was deemed to be 'clearly intended' for chemical warfare agent production. The director of the company was prosecuted not only for tax evasion but also for having given assistance to Libya. He was sentenced to five years' imprisonment.[61] The company was also accused of having prepared plans for the other, larger Pharma 200 facility in Libya, but no additional legal action was taken.[62] The Bundestag later decided not to further investigate West German involvement in the Rabta affair.[63] In

[52] Conference on Disarmament document CD/970, 20 Feb. 1990.

[53] Reuters, 'U.S. says Libyan plant appears incapacitated by heavy fire damage', *International Herald Tribune*, 16 Mar. 1990, pp. 1, 6; Walker, T. and Al-Tahri, J., 'Libya threatens West Germany over blaze at Tripoli chemicals factory', *Financial Times*, 16 Mar. 1990, p. 24.

[54] Darwish, A., 'Libya fire "almost certainly a hoax"', *The Independent*, 10 Apr. 1990, p. 12; Feinstein, L., 'Mystery surrounds "fire" at Libyan chemical plant', *Arms Control Today*, vol. 20, no. 3 (Apr. 1990), pp. 26–27; 'Libya: small fire, much smoke', *The Economist*, vol. 314, no. 7648 (31 Mar. 1990), p. 52; 'Another Libyan chemical plant?', *Arms Control Today*, vol. 20, no. 6 (July/Aug. 1990), p. 27.

[55] 'Aufruf zum Boykott: Khadhafi nennt Bonn: "satanische Kraft"', *Süddeutsche Zeitung*, 21 Mar. 1990, p. 9.

[56] Parmelee, J., 'Libya made offer to dismantle chemical plant', *International Herald Tribune*, 4 May 1990, pp. 1, 4.

[57] Gertz, B., '2nd chemical arms plant spied in Libya', *Washington Times*, 18 June 1990, pp. 1, 6; Henkel, P., 'Zweite Giftgas-Fabrik im Bau? "Report" hat Beweise für erneuten deutsch-libyschen Deal', *Frankfurter Rundschau*, 27 June 1990.

[58] 'Zweite Giftgasfabrik geplant? Imhausen auch wegen Verdacht auf Subventionsbetrug durchsucht', *Frankfurter Rundschau*, 9 Aug. 1990, p. 2.

[59] Reuters, 'Baut Libyen neues Rabta?', *Frankfurter Rundschau*, 7 May 1990, p. 2; Associated Press, 'Rätsel um Chemiefabrik bleiben', *Frankfurter Rundschau*, 21 June 1990, p. 4.

[60] Reuters, 'China will Libyen Chemikalien "für Giftgasproduktion verkaufen"', *Frankfurter Allgemeine Zeitung*, 8 June 1990, p. 7; 'China denies chemical sale to Libya', *International Herald Tribune*, 12 June 1990, p. 2.

[61] See, for example, Behr, A., 'Fünf Jahre Haft für Hippenstiel-Imhausen: Wegen Beteiligung an einer Giftgasfabrik in Libyen und Steuerhinterziehung', *Frankfurter Allgemeine Zeitung*, no. 147 (28 June 1990), p. 4.

[62] 'Kein neuer Prozess', *Der Spiegel*, vol. 44, no. 38 (17 Sep. 1990), pp. 16–17.

[63] 'Bundestag lehnt Rabta-Untersuchungsausschuss ab', *Frankfurter Allgemeine Zeitung*, 21 June 1990, p. 6.

December 1990, it was reported that the Rabta facility was producing chemical weapons and that the new facility at Sebha was nearing completion.[64]

As with Iraq, Libya was also able to buy equipment and chemicals from companies acting out of commercial interest. Governmental and public interest about Libyan CW production seemed to disappear with the start of the Persian Gulf crisis.

IV. Destruction of chemical weapons

The efforts to destroy chemical weapons attracted special attention in 1990. The most significant events were: (*a*) the US–Soviet agreement to begin destruction of the greater part of their CW stockpiles (see chapter 14 and appendix 14A), (*b*) the US destruction programme at the Johnston Atoll facility in the Pacific, (*c*) the Soviet difficulties to set up and implement a destruction programme, and (*d*) the concerns voiced in the USA and the USSR about possible damage to the environment that might result from CW destruction.

The US destruction programme

In conjunction with its authorization of the development of modernized weapons in November 1985, the US Congress directed the Department of Defense to destroy existing US CW stockpiles.[65] The principal approach of the US stockpile disposal programme is to build a destruction facility near each stockpile on the US mainland. In 1988 construction began on a high-temperature incineration facility for CW destruction located on Johnston Atoll.[66] The same year the Army announced that it intended to build similar disposal facilities at each of eight chemical munition storage sites in the continental United States (CONUS stockpiles),[67] and construction of a plant at Tooele, Utah, began in 1989.[68] Construction was planned to start on a destruction facility at the Anniston, Alabama, stockpile site in 1991.[69] In February 1990 it was announced that destruction of stockpiles and munitions of the incapacitating chemical warfare agent BZ had been completed.[70] In August the US Ambassador to the Conference on Disarmament (CD) in Geneva provided information that destruction had actually been completed in June.[71] The US Administration's request for the Chemical Weapons

[64] Institute for Defense and Disarmament Studies, *Arms Control Reporter* (IDDS: Brookline, Mass.), sheet 704.E-2.22, Dec. 1990.

[65] SIPRI, *SIPRI Yearbook 1986: World Armaments and Disarmament* (Oxford University Press: Oxford, 1986), p. 166.

[66] SIPRI, *SIPRI Yearbook 1989: World Armaments and Disarmament* (Oxford University Press: Oxford, 1989), p. 105.

[67] *Chemical Weapons: Obstacles to the Army's Plan to Destroy Obsolete U.S. Stockpile*, United States General Accounting Office, Report to Congressional Requesters, GAO/NSIAD-90-155, May 1990.

[68] SIPRI, *SIPRI Yearbook 1990* (note 48), p. 126.

[69] See note 67.

[70] Conference on Disarmament document CD/PV. 538, 27 Feb. 1990, p. 18.

[71] Conference on Disarmament document CD/PV. 574, 16 Aug. 1990, p. 20.

Demilitarization Program for fiscal year 1991 was $354.4 million.[72] The US Senate approved the 1991 Defense Appropriations Act in October 1990.[73] Only $292.7 million was appropriated for chemical demilitarization (chemdemil), with $5.3 million earmarked for cryofracture research. There were also significant changes in the programme management, scope and reporting.

During a hearing of the Senate Armed Service Subcommittee on Strategic Forces and Nuclear Deterrence, the Assistant Secretary of the Army for Installations, Logistics and Environment stated that, without circumventing safety and environmental requirements, the Army would be unable to meet the previous congressionally mandated date of 30 April 1997, for completion of destruction of the US stockpiles.[74] His view was supported by a report from the General Accounting Office (GAO) which stated that the Army could probably not complete its stockpile destruction by the deadline owing to: (*a*) stringent environmental regulation of the operation of the first US continental incineration plant, (*b*) programme budget cuts, and (*c*) operational delays in testing the first disposal plant on Johnston Atoll.[75] The Army therefore wanted to extend the disposal date to December 1998.[76] Table 4.1 illustrates the consequences which the proposed delay would have on the US destruction programme's previously planned destruction operations.[77]

In a GAO report published in May 1990,[78] the cost of the US chemdemil programme was listed as having doubled between October 1985 and March 1988 and was expected to continue to grow. A March 1988 estimate that construction of all the planned destruction facilities would cost $3407 million had, by September 1989, increased by $66 million. The cost of equipment for all of the planned sites had increased by $197.5 million. The study also indicated that most of the $123 million requested by the Army for use in fiscal year (FY) 1991 may not be needed until FY 1992 because of delays in the chemdemil programme. The Army's Deputy Program Manager for Chemdemil said on another occasion that the entire programme might eventually cost another $200 billion to complete.[79]

[72] Institute for Defense and Disarmament Studies, *Arms Control Reporter* (IDDS: Brookline, Mass.), sheet 704.E-1.1, Mar. 1990; 'Chemical arms funds could double', *Chemical & Engineering News*, vol. 68, no. 9 (26 Feb. 1990), p. 27; 'Defense agencies procurement summary', *Defense News*, vol. 5, no. 7 (12 Feb. 1990), p. 55.

[73] '26 October', *Chemical Weapons Convention Bulletin*, no. 10 (Dec. 1990), p. 11.

[74] 'Chemical weapon burning delay asked', *Defense News*, vol. 5, no. 16 (16 Apr. 1990), p. 22; '5 April in Washington,*Chemical Weapons Convention Bulletin*, no. 8 (June 1990), p. 14; Institute for Defense and Disarmament Studies, *Arms Control Reporter* (IDDS: Brookline, Mass.), sheets 704.E-1.3–1.4, June 1990.

[75] See note 67, p. 11.

[76] See 'Chemical weapon burning delay asked' (note 74); 'Chemical destruction behind, over budget', *Defense News*, vol. 5, no. 25 (18 June 1990), p. 35.

[77] Institute for Defense and Disarmament Studies, *Arms Control Reporter* (IDDS: Brookline, Mass.), sheet 704.E-1.4, June 1990; see note 67, p. 11.

[78] See note 67.

[79] See note 67; '13 June', *Chemical Weapons Convention Bulletin*, no. 9 (Sep. 1990), p. 12.

Table 4.1. The US Chemical Weapons Demilitarization Program

Location	Construction start as of		Disposal operation as of	
	1989	1990	1989	1990
Tooele, Utah 42% (H, HD, HT, L, GA, GB, VX)	Aug. 1989	Oct. 1989	Feb. 1993– Apr. 1997	Oct. 1993– Dec. 1998
Pine Bluff, Arkansas 12% (HD, HT, GB, VX, (BZ))	Sep. 1991	June 1992	Mar. 1995– Dec. 1996	Mar. 1996– Nov. 1998
Umatilla, Oregon 11.6% (HD, GB, VX)	Sep. 1991	June 1992	Mar. 1995– Nov. 1996	Mar. 1996– Oct. 1998
Anniston, Alabama 7.1% (HD, HT, GB, VX)	Sep. 1991	Sep. 1991	Mar. 1995– Apr. 1997	June 1995– July 1998
Pueblo, Colorado 9.9% (HD, HT)	May 1992	June 1993	Nov. 1995– Feb. 1997	Mar. 1997– Nov. 1998
Newport, Indiana 3.9% (VX)	May 1992	June 1993	Apr. 1995– July 1996	Sep. 1996– Aug. 1997
Aberdeen, Maryland 5 % (HD)	May 1992	June 1993	Apr. 1995– July 1996	Sep. 1996– Sep. 1997
Lexington-Blue Grass, Kentucky 1.6% (H, GB, VX)	Sep. 1992	June 1993	Mar. 1996– Feb. 1997	Mar. 1997– Aug. 1998
JACADS training facility[a] 6.6 %(GB, VX, mustard gas)	1986	May 1990– July 1994	June 1989	Nov. 1990– Sep. 1996

[a] The Johnston Atoll Chemical Disposal System (JACADS) number already includes 1.6 % of the US CW stockpiles from the FRG (GB, VX); these JACADS stockpiles also include US CW stocks withdrawn from Okinawa in 1971.

Sources: *Chemical Weapons: Obstacles to the Army's Plan to Destroy Obsolete U.S. Stockpile*, United States General Accounting Office, Report to Congressional Requesters, GAO/NSIAD-90-155, May 1990, p. 11; Institute for Defense and Disarmament Studies, *Arms Control Reporter* (IDDS: Brookline, Mass.), sheet 704.E-1.4, June 1990.

The only CW destruction site outside the continental United States, the Johnston Atoll Chemical Disposal System (JACADS) which was completed in 1988, is located some 1100 km from its nearest Hawaiian neighbour. It was designed and constructed as a model for the eight mainland facilities. The destruction process used is combustion by incineration of the chemical warfare agents; the choice of methods is based on a decision taken by the Army in

1982 to employ high-temperature incineration to destroy CW munitions.[80] The implementation plan for the facility stated that a 16-month operation verification test (OVT) programme should be conducted, which was planned to start in August 1989 with the destruction of rockets—the most unstable munitions on Johnston Atoll. The test-run included four stages: (*a*) disposal of M55 rockets containing the non-persistent nerve gas sarin; (*b*) disposal of M55 rockets filled with the persistent nerve gas VX; (*c*) disposal of bulk containers of the blistering agent mustard gas; and (*d*) destruction of mustard gas artillery shells. The OVT was first delayed until October 1989, and then re-scheduled to start in March 1990.[81] Another delay occurred, and by May 1990, the OVT was behind schedule by nine months.[82] The process again appeared ready to start up on 30 June, but was stopped the same day after only two hours to monitor some anomalies; 15 M55 rockets containing the nerve gas sarin had been taken through the system.[83] In close to one month of test-runs, the facility was shut down twice. By the end of July 1990, fewer than 1000 rockets had been processed through the system, far fewer than planned.[84] By the end of August, nearly 1800 kg of nerve agent and 800 M55 rockets had been destroyed.[85] Information given at the end of October stated that 3319 M55 rockets containing nerve agents had been safely destroyed (approximately 1 per cent of the munitions stored).[86] The delays in the JACADS programme have made it impossible to keep to the schedule for construction of the other eight US mainland sites.[87] A GAO report stated that the JACADS programme was 32 months behind schedule; costs were estimated at $190–$561 million more than budgeted.[88]

The US Congress waited for an environmental impact statement certifying the safe storage and disposal of the US CW stockpiles from the FRG (see below) before it approved removal of chemical weapons to Johnston Atoll. It

[80] That technology was later endorsed in a report by the National Research Council's Committee on Demilitarizing Chemical Munitions & Agents, *Disposal of Chemical Munitions and Agents* (National Academy Press: Washington, DC, 1984).

[81] Ember, L. R., 'Chemical weapons disposal: daunting challenges still ahead', *Chemical & Engineering News*, vol. 68, no. 33 (13 Aug. 1990), pp. 9–19.

[82] See note 67

[83] See note 81; Associated Press, 'Giftgasofen auf Hawaii erprobt', *Frankfurter Rundschau*, 14 July 1990, p. 6; Shulman, S., 'First test incineration', *Nature*, vol. 346, no. 6279 (5 July 1990), p. 5.

[84] Anderson, I., 'Protests grow over nerve gas disposal', *New Scientist*, vol. 127, no. 1729 (11 Aug. 1990), p. 18; 'CW destruction re-starts', *Jane's Defence Weekly*, vol. 14, no. 4 (28 July 1990), p. 105; Rademacher, H., 'Nur Wasser und Kohlendioxyd sollen den Giftgasofen verlassen', *Frankfurter Allgemeine Zeitung*, 7 Aug. 1990, p. 6.

[85] 'Atoll dispute', *Jane's Defence Weekly*, vol. 14, no. 11 (15 Sep. 1990), p. 459; 'Giftgas-Vernichtung verzögert', *Frankfurter Rundschau*, 31 Aug. 1990, p. 2.

[86] 'Johnston Atoll: what the press didn't tell you', *Pacific Research*, vol. 3, no. 4 (Nov. 1990), pp. 21–22.

[87] The schedule for the continental sites is dependent on the Army's certification to Congress that JACADS has successfully destroyed chemical weapons on an operational basis; see also note 81.

[88] '29 August', *Chemical Weapons Convention Bulletin*, no. 9 (Sep. 1990), p. 19; *Chemical Weapons: Stockpile Destruction Delayed at the Army's Prototype Disposal Facility*, United States General Accounting Office, Report to Congressional Requesters, GAO/NSIAD-90-222, July 1990; 'Probleme der USA bei der C-Waffen-Vernichtung', *Neue Zürcher Zeitung*, 1 Sep. 1990, p. 4; Institute for Defense and Disarmament Studies, *Arms Control Reporter* (IDDS: Brookline, Mass.), sheet 704.E-1.9, Sep. 1990.

was published in June[89] and concluded that the Johnston Atoll facility could cope with the European stocks in a safe and environmentally acceptable manner by using its incineration technique. However, in July the organization Greenpeace published a detailed commentary on the environmental impact statement, recommending that 'existing chemical weapons stockpiles remain segregated in above-ground, monitored retrievable storage . . . until such time as a safe and suitable detoxification or treatment method becomes available' and that 'no material whatsoever should be released to the environment in gaseous, liquid or solid phase during the "decommissioning" process; by-products resulting from the process must be non-hazardous'.[90] Based upon the report, Greenpeace tried to stop the withdrawal operation from the FRG and to re-schedule the chemdemil plan on Johnston Atoll, but the request was not granted.[91] A public debate began about the advantages and disadvantages of incineration of the US CW stockpiles at the Johnston Atoll facility.[92] The debate also concerned the possible threat that the Pacific region might in the future be used as an experimental or toxic waste dumping area.[93] This concern about CW destruction on Johnston Atoll was expressed, for example, during a meeting of the Pacific Heads of Government on Vanuatu on 31 July–11 August, but the USA stated that there was no need for such concern.[94] The Pacific nations affected expressed apprehension about the possible use of Johnston Atoll for long-term US CW stockpile storage, particularly taking into account the probable delay of the planned destruction operations.[95]

[89] US Department of the Army, Program Manager for Chemical Demilitarization, *Johnston Atoll Chemical Agent Disposal System (JACADS): Final Second Supplemental Environmental Impact Statement for the Storage and Ultimate Disposal of the European Chemical Munition Stockpile*, June 1990, 2 vols (US Army: Aberdeen Proving Ground, Edgewood, Md., 1990).

[90] Greenpeace International, Pacific Campaign, *Greenpeace Review of Johnston Atoll Chemical Agent Disposal System (JACADS): Final Supplemental Environmental Impact Statement (June 1990) for the Storage and Ultimate Disposal of the European Chemical Munition Stockpile, Addendum to Greenpeace Comments on Previous JACADS Environmental Impact Statements, and Supplements*, 9 July 1990 (Greenpeace: Washington, DC, 1990), pp. 16–17.

[91] 'Greenpeace: Antrag abgelehnt', *Frankfurter Rundschau*, 11 Aug. 1990, p. 5; Grabenströer, M., 'USA scheuen Verbrennung der Giftgaswaffen im eigenen Land', *Frankfurter Rundschau*, 28 May 1990, p. 4.

[92] Findlay, T., 'Green vs peace?', *Pacific Research*, vol. 3, no. 2 (May 1990), pp. 3–7; Findlay, T., 'Chemical disarmament and the environment', *Arms Control Today*, vol. 20, no. 7 (Sep. 1990), pp. 12–16; Hawkins, S., 'Johnston Atoll: the Greenpeace riposte', *Pacific Research*, vol. 3, no. 3 (Aug. 1990), pp. 9–11; *Action Alert: Kalama Island (Johnston Atoll): Chemical Weapons Incineration* (Pacific Concerns Resource Centre: Auckland, 1990); Wilkes, O., 'Chemical weapon burnoff in Central Pacific', *Peacelink*, issue 83 (July 1990), pp. 5–10; Richardson, M., 'U.S. plan for arms worries the Pacific', *International Herald Tribune*, 22 May 1990, p. 1; 'Johnston Atoll: grim news on nerve gas', *Pacific Islands Monthly*, vol. 60, no. 5 (May 1990), p. 20.

[93] Robie, D., 'Dangerous playground', *Pacific Islands Monthly*, vol. 60, no. 5 (May 1990), pp. 10–13; Anderson, I., 'Destruction of chemical arms comes under fire', *New Scientist*, vol. 127, no. 1728 (4 Aug. 1990), p. 21; 'Toxic waste ship under investigation for illegal dumping', *Peace Courier*, no. 4/5 (May 1990), p.13.

[94] 'Pacifikforum besorgt über Chemiewaffen-Verbrennung', *Frankfurter Rundschau*, 2 Aug. 1990, p. 2; 'Protest gegen Giftgastransport', *Franfurter Rundschau*, 18 Sep. 1990, p. 4; 'Pazifik soll kein Mülleimer sein', *Frankfurter Rundschau*, 1 Aug. 1990, p. 5; 'Agreement on chemical weapons: concern over destruction', *Trust and Verify*, no. 11 (June 1990), p. 2; see also *Wireless File*, no. 208, 'Bush promises safety in destroying chemical weapons' (United States Information Service, US Embassy: Stockholm, 26–28 Oct. 1990), pp. 15–17.

[95] 'USA planen Arsenal für C-Waffen im Pazifik', *Süddeutsche Zeitung*, 7 Aug. 1990, pp. 1, 7.

One of the eight CONUS destruction sites under construction is the Army Depot at Anniston, Alabama, where about 7 per cent by weight of the total US CW agent stockpile is located. A draft environmental impact statement about projected on-site incineration at Anniston was distributed in November.[96]

The Soviet destruction programme

In 1990 the Soviet Union provided information that its 'stockpiles of toxic substances exceed US stockpiles by 10 000 tonnes' and that these 10 000 tonnes are CW stockpiles which were accumulated prior to and during World War II.[97] During an official visit in 1987 to the CW testing ground at Shikhany by representatives of the CD delegations, the USSR also demonstrated a mobile destruction unit.[98] The unit is suitable for small-scale destruction tasks only, and the destruction process is based on thermochemical neutralization with optional subsequent incineration of the neutralization products.

In 1989 it was announced that a CW destruction facility had also been built at Chapayevsk. However, owing to the resulting public concern and protests about environmental damage, the facility could not be taken into use.[99] In April 1990, before the May–June US–Soviet summit meeting, the general situation of Soviet destruction was characterized as follows: 'Unfortunately, as yet the Soviet side cannot put forward any definite timetable for the start and end of the process of destroying its chemical weapons within the framework of the bilateral agreement'.[100] According to a Soviet source, some 438 tonnes of various types of chemical weapon had been destroyed between 1970 and 1990. The USSR was said to have a 'certain quantity of mobile installations for the destruction [Shikhany type] of chemical agents', but these were only for destroying 'insignificant quantities of chemical weapons'.[101]

The large Chapayevsk destruction facility was converted into a training centre after a 5 September 1989 decision by the Soviet Council of Ministers.[102] Among other reasons given for the decision was mentioned the fact that it had been considered extremely important to take into account 'the psychological

[96] '14 November 1990', *Chemical Weapons Convention Bulletin*, no. 10 (Dec. 1990), p. 12.

[97] See 'Petrov views destruction of chemical weapons', PM0808094190, Moscow, *Izvestia*, 7 Aug. 1990, morning edn, p. 6 (in Russian) in Foreign Broadcast Information Service, *Daily Report–Soviet Union (FBIS-SOV)*, FBIS-SOV-90-153, 8 Aug. 1990, pp. 56–57; see also Associated Press, 'Schwierigkeiten mit Vernichtung, Moskau: In der UdSSR 40 000 Tonnen C-Waffen', *Süddeutsche Zeitung*, 4 Aug. 1990, p. 8.

[98] Conference on Disarmament document CD/789, 16 Dec. 1987.

[99] SIPRI, *SIPRI Yearbook 1990* (note 48), p. 127.

[100] 'Problems of destroying chemical weapons cited', LD1704084790, *Moscow Domestic Service*, 0630 GMT, 17 Apr. 1990 (in Russian) in FBIS-SOV-90-074, 17 Apr. 1990, p. 1.

[101] 'Chemical troops chief on destroying weapons', PM0408165690, Moscow, *Pravda*, 2 Aug. 1990, 2nd edn, p. 5 (in Russian) in FBIS-SOV-90-151, 6 Aug. 1990, pp. 3–5, .

[102] 'Chemical troops officer on efforts for CW pact', PM1209154790, Moscow, *Krasnaya Zvezda*, 11 Sep. 1990, 1st edn, p. 3 (in Russian) in FBIS-SOV-90-178, 13 Sep. 1990, pp. 1–2; Evstafiev, I. B., 'Control of the chemical industry in the USSR', eds T. Stock and R. G. Sutherland, *National Implementation of the Future Chemical Weapons Convention*, SIPRI Chemical & Biological Warfare Studies, no. 11 (Oxford University Press: Oxford, 1990), pp. 108–19; 'Chemical weapons plant becomes training center', PM 2308112090, Moscow, *Pravda*, 22 Aug. 1990, 2nd edn, p. 2 (in Russian) in FBIS-SOV-90-165, 24 Aug. 1990. p. 1.

factor—people's views and feelings'. The Chapayevsk facility has a relatively small capacity and is only able to destroy up to 500 tonnes per year. It cost 50 million roubles to build and was constructed over a three-year period (60–70 per cent of the cost was accounted for by monitoring and ecological safety measures). A statement was also issued that the USSR had agreed to create 'a limited number of automated facilities—one or two, or a maximum of three—where chemical weapons could be brought from their storage sites to be destroyed . . . such facilities to be sited in sparsely populated areas away from zones of intensive agriculture'.[103] In 1989 an interdepartmental commission for elaborating a state programme for CW destruction was created,[104] and in April 1990 the programme was submitted to the Supreme Soviet. Five options for the elimination of chemical weapons were suggested; they differed in terms of the number of destruction facilities, their locations, the amount of capital investment to be made, and the timing of the beginning and completion of destruction. More than 300 scientists from various scientific institutions were scheduled to work on the programme. The technology to be used has been developed by the Soviet Ministry of Chemical and Petroleum Refining Industry and is based on a two-stage process,[105] which is probably identical to the one demonstrated at Shikhany. At the end of 1990, a final decision about the programme had not yet been taken by the Supreme Soviet.

Unlike the USA, the Soviet Union has large stockpiles of lewisite. The USSR is investigating a method by which it may be possible to reduce lewisite to arsenic trichloride and then to highly purified arsenic, which could be used in the electronics industry.[106] Information was also released that during the 1980s three 'stationary installations' were built for destroying mustard gas and lewisite. Two of these installations have completed processing; a third is being utilized to extract arsenic from lewisite.[107]

At the end of 1990 when a Soviet official was asked if the USSR would use US experiences and elements of US destruction technology, he stated that the destruction of chemical weapons had been somewhat easier for the USA. First, the United States tackled the problem much earlier; second, there was a substantial legislative base making it possible to rapidly decide questions related to the elimination of chemical weapons; third, destruction was to take place at the storage sites, and almost half the US CW stockpiles were concentrated at the Tooele base.[108] The USA and the USSR agreed in bilateral meetings to visit destruction sites at Chapayevsk in the USSR and at Denver, Johnston Atoll and Tooele in the USA (see also chapter 14).

[103] See 'Chemical troops officer on efforts for CW pact' (note 102).

[104] See 'Petrov views destruction of chemical weapons' (note 97).

[105] See note 104.

[106] See note 104; 'U.S. Delegation visits chemical arms plant', LD2708211590, Moscow Television Service, 1430 GMT, 27 Aug. 1990 (in Russian) in FBIS-SOV-90-167, 28 Aug. 1990, p. 1.

[107] 'Problems of eliminating chemical weapons in USSR', *APN Military Bulletin*, no. 12/13 (June/July 1990), pp. 38–42; Petrov, S., 'Chemical weapons: destruction formula', *Pravitelstvenny Vestnik*, no. 29, translated in *Soviet News*, no. 6539 (15 Aug. 1990), p. 275.

[108] See 'Chemical troops officer on efforts for CW pact' (note 102), p. 3.

Soviet estimates indicated that CW destruction costs for the USSR will total 3 billion roubles. Former Foreign Minister Eduard Shevardnadze is said to have expressed dismay about previous Soviet acquisition of useless chemical weapons which now prove extremely costly to destroy.[109]

Other destruction technologies

It seems worthwhile to recall that the CW munitions which constitute the existing stockpiles were not designed to ever be destroyed. Early disposal efforts were primarily accomplished by open-pit burning, atmospheric dilution, burial and ocean dumping. The last US dumping operation was in 1970. After World War II chemical munitions were dumped in the sea at several locations by both Soviet and Allied forces.

In December 1989, the USA presented a paper dealing with the technology used on Johnston Atoll for the demilitarization and disposal of chemical warfare agents and munitions.[110] The method is based on incineration of the chemical agent and decontamination of the metal parts of the munitions.[111] The USSR, on the other hand, favours a destruction method based on a two-step thermochemical neutralization process,[112] but is searching for a suitable alternative for the conversion of its large stockpiles of lewisite.[113]

Alternatives to current destruction technologies continue to be sought.[114] The USA spent $22 million in FY 1990 for design of a process and demonstration facility for 'cryofracture destruction'[115] in Tooele.[116] A full-scale cryofracture plant would cost roughly $175 million to build. Among the techniques under investigation are enzymatic splitting of the carbon-to-phosphorus bond in organophosphorus nerve agents (biodegradation),[117] photochemical

[109] Remnick, D., 'Shevardnadze rebukes "'McCarthyites"', *International Herald Tribune*, 27 May 1990, p. 2.

[110] Conference on Disarmament document CD/CW/WP. 265, 11 Dec. 1989; Rife, R. *et al.*, 'Chemical demilitarization: disposing of the most hazardous wastes', *Environmental Progress*, vol. 8, no. 3 (Aug. 1989), pp. 167–75.

[111] Program Executive Officer, Program Manager for Chemical Demilitarization, *Chemical Stockpile Disposal Program: Final Programmatic Environmental Impact Statement,* 3 vols, Jan. 1988 (US Army: Aberdeen Proving Ground, Edgewood, Md., 1988); Program Manager for Chemical Demilitarization, *Disposal of Chemical Agents and Munitions stored at Tooele Army Depot, Tooele, Utah: Draft Environmental Impact Statement*, Mar. 1989 (US Army: Aberdeen Proving Ground, Edgewood, Md., 1989).

[112] See notes 98 and 101; 'Petrov recounts chemical disarmament progress', LD0408164590 in FBIS-SOV-90-151, 6 Aug. 1990, p. 3.

[113] See note 101.

[114] Lohs, Kh., Lundin, S. J. and Stock, T. (eds), *The Destruction of Chemical Weapons and Chemical Warfare Agents* (SIPRI: Solna, Sweden, 1990).

[115] 'Cryofracture is the controversial method General Atomics is developing to replace the front-end disassembly of munitions now used in the Army's complex process for destroying chemical arms. In the cryofracture process the munition—metal casing, chemical agent and explosive charge—is frozen in liquid nitrogen (at about –200°F [–129°C]), crushed by a 1000-ton hydraulic press, and dropped into a furnace. The high temperature (1400°F [760°C]) in the furnace completely consumes the agent and explosive, and decontaminates the metal pieces', quoted from Ember (note 81), p. 13.

[116] See Institute for Defense and Disarmament Studies, *Arms Control Reporter* (IDDS: Brookline, Mass.), sheet 704.E-110, Dec. 1990.

[117] See Findlay, T. 'Green vs peace?'; and Findlay, T., 'Chemical disarmament and the environment' (note 92); see also 'Declaration of Dr. Wayne Landis', Greenpeace USA versus Michael P. Stone,

degradation,[118] application of chemical catalysts (chemical degradation)[119] and low-temperature electrochemical oxidation (degradation).[120] In Australia a plasma arc furnace is being developed which is capable of completely destroying dioxins and other toxic material at very high temperatures by direct use of electric current.[121] None of the new technologies under development will be available on a large scale in the near future, but these costly investigations must be conducted in order to find alternative destruction technologies which pose a minimal risk to man and the environment.

Environmental implications

Public environmental concern about the transportation of chemical weapons and the possible risks of destruction has increased. Thus what is probably the least acceptable 'destruction technology'—dumping at sea—was again in 1990 a matter of public debate. For example, it is assumed that some 35 000 tonnes of old munitions from World War I were dumped off Zeebrugge, Belgium, in the North Sea.[122] Additional information appeared that an estimated 150 000 tonnes of old CW munitions were dumped in the Skagerrak off the Swedish west coast after World War II.[123] However, there are currently no plans to retrieve and destroy these munitions.[124] Dumping of chemical warfare agents and munitions generally only leads to postponing destruction and creates future environmental risks.

The question of the destruction of old, buried chemical weapons may pose new environmental problems and may also require new destruction technologies. There are many such examples. In the former German Democratic Republic remnants of old CW munition which were buried in the soil continue to be discovered despite attempts in the 1950s to find and remove them.[125] Two barges which are buried in a field near Hamburg in the FRG, are

Secretary of the Army; Richard Cheney, Secretary of the Department of Defense, United States District Court for the District of Hawaii, Civil no. 90-00588 DAE, 20 Aug. 1990, pp. 7–10. There has been discussion that bio-degradation should be used as an alternative disposal technique, a process in which enzymes break down the compound into non-toxic components. A number of enzymes have been identified such as organophosphorous acid (OPA). See, for example, Landis, W. G., 'Initial characterization of a nerve agent hydrolyzing enzyme', French delegation to CRDC, Aberdeen Proving Ground, Edgewood, Md., 19 June 1984.

118 In photochemical degradation the requisite energy is provided by photons of light; see also note 90, pp. 59–61.

119 Chemical degradation includes the use of specific chemical catalysts in the degradation process; see also note 90, pp. 61–62.

120 In electrochemical degradation the chemicals are broken down by the use of electric current; see also note 90, p. 61.

121 See Findlay, T., 'Chemical disarmament and the environment' (note 92), p.16.

122 '19 December', *Chemical Weapons Convention Bulletin*, no. 8 (June 1990), p. 5.

123 Laurin, F., 'Dumpad stridsgas hotar Skagerak', *Svenska Dagbladet* (Stockholm), 16 Oct. 1990, p. 8.

124 'Sänkt stridsgas undersöks inte', *Svenska Dagbladet* (Stockholm), 28 Oct. 1990, p. 10.

125 Lohs, Kh., 'Die Schwerter sind vergiftet: Die Waffen von gestern bedrohen unsere Umwelt', *Die Zeit*, no. 29 (13 July 1990), p. 39; Lohs, Kh., 'Giftige Neuigkeiten: Tiefenprüfung in Halle brachte alte Kampfstoffe zu Tage', *Wochenpost*, vol. 37, no. 32 (10 Aug. 1990), p. 8; Reiner, J., 'Zeitbomben im Adlergrund und in manch anderen Gefilden', *Neues Deutschland*, 31 Aug. 1990, p. 9.

suspected to contain mustard gas and phosgene from World War II.[126] Old chemical weapons, explosives and munitions were buried between 1942 and 1952 at Dehtlinger Teich near Munster in the FRG, and the old chlorine- and arsenic-containing chemical warfare agent Clark was found to be leaking into a nearby well. The environmental department of Lower Saxony has taken measures to investigate the situation.[127] These problems are not only related to old CW munitions. Large stockpiles of conventional weapons, munitions and other toxic material may eventually pose great technological and environmental problems when they have to be destroyed.[128]

In Czechoslovakia, a US company is cleaning up waste dumps left after the Soviet military withdrawal. The Czechoslovak Environmental Minister suspected that such dumps might contain chemical weapons despite earlier official denials.[129] When Soviet forces withdrew from Hungary, an allegation was made that Soviet troops had buried barrels of tear-gas and chemical weapons,[130] but a Soviet general officially denied the allegation.[131]

V. Withdrawal of US chemical weapons from the Federal Republic of Germany

As reported in the *SIPRI Yearbook 1990*, agreements were reached to remove the US chemical weapons stockpiled in the FRG. Information was given that the stockpiles were situated in the small town of Clausen and not, as had earlier been assumed, in Fischbach,[132] and that transport of the munitions from Germany to the USA would take place from the port of Nordenham in Lower Saxony in the summer of 1990.[133] A public discussion immediately started about possible security risks which could be caused by the operation,[134] and conflict arose about when and how to inform the public.[135] Authorities of the Federal Government of the FRG, the Army of the FRG, the Rhineland-Palatinate local government and the US Army invited the press to a 'media day' in Pirmasens on 8 March 1990, to provide information about the with-

[126] 'Immer wieder Giftalarm in Hamburgs Südosten', *Frankfurter Rundschau*, 28 Sep. 1990, p. 32; Toro, T. 'Unwholesome Hamburg unearths a poisonous past', *New Scientist*, vol. 128, no. 1742 (10 Nov. 1990), p. 18.

[127] 'Giftgas "Clark" im Brunnen', *Frankfurter Rundschau*, 24 Mar. 1990, p. 6.

[128] 'Sorgen um Munitionsbestände der NVA', *Frankfurter Allgemeine Zeitung*, 8 Nov. 1990, p. 7; see also Charles, D., 'Counting the cost of the cold war cleanup', *New Scientist*, vol. 128, no. 1738 (13 Oct. 1990), p. 11.

[129] Kamm, H., 'Prague chore: cleaning up after Soviets', *International Herald Tribune*, 25 July 1990, p. 5.

[130] '30 June', *Chemical Weapons Convention Bulletin*, no. 9 (Sep. 1990), p. 14.

[131] 'General denies leaving shells behind in Hungary', LD1007075890, Moscow, TASS, 0728 GMT, 10 July 1990 (in English) in FBIS-SOV-90-132, 10 July 1990, p. 22.

[132] 'Tödlich in 20 Kilometern', *Der Spiegel*, vol. 43, no. 52 (25 Dec. 1989), pp. 58–63.

[133] 'Umschlaghäfen für C-Waffen künftig in Niedersachsen?', *Frankfurter Rundschau*, 7 Dec. 1989, p. 1.

[134] 'Protest gegen Giftgaslager', *Frankfurter Rundschau*, 15 Jan. 1990, p. 4; Reinhardt, C., 'Bonn hält C-Waffen für sicher', *Frankfurter Rundschau*, 13 Jan. 1990, p. 4.

[135] 'Unklarheiten über Abtransport von C-Waffen', *Frankfurter Allgemeine Zeitung*, 9 Dec. 1989, p. 5; Krummenacker, T., 'Wie nun weg mit dem Dreck?', *Tageszeitung*, 9 Mar. 1990, p. 3; Grabenströer, M., 'Geheimniskrämerei um Giftgas', *Frankfurter Rundschau*, 19 Jan. 1990, p. 6.

drawal.[136] The citizens of the Clausen region were given the opportunity to take part in an information seminar the day before.

The press reported that 435 tonnes of the nerve agents sarin and VX were stockpiled in the US munition depot at Clausen, and that withdrawal would take place between August and December 1990.[137] The nerve agents, later reported to equal 400 tonnes, were filled in approximately 102 000 artillery shells,[138] consisting of 155-mm and 8-inch (203-mm) howitzer projectiles,[139] and weighing approximately 7000 tonnes.[140] The 102 000 artillery shells stored in Clausen constituted about 1 per cent of the total US supply of chemical weapons. It was also stated that no other CW storage sites were located in the FRG.[141] The Minister of Defence of the FRG, Gerhard Stoltenberg, announced that the US chemical weapons would be withdrawn from Clausen between July and September. At the end of March, the USA confirmed that withdrawal would be finished by the end of September 1990.[142]

The plans and preparation for withdrawal included the following: containers with the CW munition were to be transported by road to a railway depot near Miesau and from there to be carried by rail to the port of Nordenham. The removal operation was planned to take place in six phases: pre-positioning, site operations, road convoy, railhead operations, rail movement and port operations.[143]

A number of safety precautions were taken during the withdrawal: the ammunition shells were stored in more than 5000 airtight, specially designed vapour-proof secondary steel containers (SSC) to provide added protection.

[136] On media day, 8 Mar. 1990, a number of fact sheets were handed out, the content of which was reported in: 'Giftgasabzug: Koalition der Vernunft', *Die Rheinpfalz*, 8 Mar. 1990; Kling, A., 'Fischbach "entlastet": Clausen bald befreit', *Pirmasenser Zeitung*, 8 Mar. 1990, p. 15; Müller, J. and Schlicher, R., 'Clausener Bürger fordern Vertrauensbeweise ein', *Die Rheinpfalz*, 8 Mar. 1990; Grabenströer, M., 'Was bitte lagert denn nun in Fischbach?', *Frankfurter Rundschau*, 9 Mar. 1990; Halbig, H., 'Der Abtransport der C-Waffen ist "todsicher"', *Der Tagesspiegel*, 9 Mar. 1990, p. 3.

[137] Grabenströer, M., 'Giftgas-Abzug ab August', *Frankfurter Rundschau*, 7 Mar. 1990, p. 4.; Smith, R. J.,'U.S. remove nerve gas from W. German base', *Washington Post*, 6 Mar. 1990.

[138] 'Amerikaner ziehen C-Waffen ab', *Süddeutsche Zeitung*, 8 Mar. 1990, p. 1; 'Chemische Waffen werden abgezogen', *Frankfurter Allgemeine Zeitung*, 9 Mar. 1990, p. 7; 'U.S. chemical arms pullout to begin', *International Herald Tribune*, 8 Mar. 1990, p. 2; Schulte, H., 'USA to speed up CW withdrawal', *Jane's Defence Weekly*, vol. 13, no. 11 (17 Mar. 1990), p. 483; Stengel, E., '"Risikobewertung" zum Giftgas-Transport: Wahrscheinlichkeit von 0,004 Unfällen', *Süddeutsche Zeitung*, 23 Aug. 1990, p. 5.

[139] *Removal of Chemical Weapons from Germany: Chemical Munitions in the FRG*, Fact Sheet 3, 8 Mar. 1990.

[140] See 'Chemische Waffen werden abgezogen' (note 138).

[141] Naumann, K. and Kreibohm, G., *Summary of Federal Republic of Germany Remarks*, 8 Mar. 1990.

[142] 'Giftgas bis September weg', *Frankfurter Rundschau*, 30 Mar. 1990, p. 5.

[143] *Removal of Chemical Weapons from Germany: Overview*, Fact Sheet 1, 8 Mar. 1990; *Removal of Chemical Weapons from Germany: Command and Control*, Fact Sheet 2, 8 Mar. 1990; *Removal of Chemical Weapons from Germany: Secondary Steel Containers (SSC)*, Fact Sheet 5, 8 Mar. 1990; *Removal of Chemical Weapons from Germany: Single Round Containers*, Fact Sheet 7, 8 Mar. 1990; *Removal of Chemical Weapons from Germany: On-site Emergency Response Team*, Fact Sheet 11, 8 Mar. 1990; *Removal of Chemical Weapons from Germany: Technical Escort*, Fact Sheet 13, 8 Mar. 1990; *Removal of Chemical Weapons from Germany: M915 Truck Tractor and M872A3 Trailer Technical Data*, Fact Sheet 17, 8 Mar. 1990; *Removal of Chemical Weapons from Germany: Loading Equipment and Procedures*, Fact Sheet 18, 8 Mar. 1990; *Removal of Chemical Weapons from Germany: Container Handling Systems*, Fact Sheet 19, 8 Mar. 1990; *Removal of Chemical Weapons from Germany: Construction Projects*, Fact Sheet 20, 8 Mar. 1990.

These were then sealed and loaded into transportation containers (MILVANs) which were designed for rail, road or water transport and which met the standards of the International Maritime Dangerous Goods Code (IMDGC). Routes for the road convoys were selected daily by the police. Each railway transport convoy included two separate train convoys separated by a train with US and West German armed forces, medical personnel, police and fire brigades. US Army and West German forces maintained constant air monitoring of the area with gross-level detectors and low-level detectors.[144]

Before the transport operation began, a number of tests were performed on the CW munitions and containers including a fire test, road test, shipboard transportation simulation, pressure test and drop test.[145] US armed forces and police forces of the FRG practised the withdrawal operation in April.[146]

The withdrawal operation, code-named Lindwurm (dragon), started on 26 June with the placing of ammunition in steel-containers.[147] The first convoy departed from Clausen to the temporary storage area at Miesau, a distance of 50 km,[148] on 26 July. This phase of the operation ended on 1 September, two days earlier than planned. Only 28 convoys had been needed instead of the 30 originally planned.[149] Transportation by rail from Miesau to Nordenham began on 12 September.[150] The date for the final phase of the operation, the loading of the 102 000 ammunition shells in the transport containers on two US special vessels, was set for 20 September,[151] but owing to bad weather,[152] the vessels could not sail from Nordenham until 22 September.[153] The sea route

[144] *Removal of Chemical Munitions from Germany: Air Monitoring,* Fact Sheet 9; 8 Mar. 1990; *Removal of Chemical Munitions from Germany: Air Monitoring Systems,* Fact Sheet 10, 8 Mar. 1990; The following air monitoring systems were used: Chemical Agent Monitor System (CAM); Automatic Chemical Agent Detector (M43A1); Depot Area At Monitoring System (DAAMS) and Automatic Continuous Air Monitoring System (ACAMS); see Fact Sheet 10.

[145] See note 141; *Removal of Chemical Munitions from Germany: Testing of the Secondary Steel Container,* Fact Sheet 6, 8 Mar. 1990.

[146] 'Übungen zum Abtransport der C-Waffen', *Frankfurter Allgemeine Zeitung,* 6 Apr. 1990, p. 5; 'Truppen üben den Abtransport von Giftgas', *Frankfurter Allgemeine Zeitung,* 25 Apr. 1990, p. 4; Herl, M., 'Die Höllenfahrt', *Stern Magazin,* no. 20 (10 May 1990), pp. 18–24.

[147] 'US-Armee packt C-Waffen ein', *Frankfurter Rundschau,* 27 June 1990, p. 4; 'Giftgasgranaten zum Abtransport bereit', *Frankfurter Allgemeine Zeitung,* 27 June 1990, p. 6.

[148] Gautsche-Lindner, J., 'Dass das Zeug weggeschafft wird, ohne weitere Verzögerung', *Rheinpfalz,* 26 July 1990, 'US begins to withdraw its chemical weapons', *Financial Times,* 27 July 1990, p. 2; 'Erster Giftgas-Konvoi angekommen', *Frankfurter Rundschau,* 27 July 1990, p. 1; Grabenströer, M., 'Ein hoffentlich erfolgreiches Stück: Der sichere Konvoi', *Frankfurter Rundschau,* 27 July 1990, p. 3; 'U.S. Army starts to withdraw chemical weapons', LD 3007093990, Moscow, TASS, 0927 GMT, 30 July 1990 (in English) in FBIS-SOV-90-147, 31 July 1990, p. 1,

[149] 'Giftgas-Konvois enden früher', *Frankfurter Rundschau,* 29 Aug. 1990, p. 4; 'Chemical arms leave West Germany', *International Herald Tribune,* 3 Sep. 1990, p. 2; Grabenströer, M., 'Luftschlangen am Ende der Giftgastransporte', *Frankfurter Rundschau,* 3 Sep. 1990, p. 4.

[150] 'Chemiewaffentransport mit der Eisenbahn beginnt', *Süddeutsche Zeitung,* 5 Sep. 1990, p. 2; 'Giftgas-Züge rollen an', *Frankfurter Rundschau,* 5 Sep. 1990, p. 4; 'Giftgas-Transport planmässig', *Frankfurter Rundschau,* 15 Sep. 1990, p. 4; Spoo, E., 'Erste Giftgaszüge am Ziel', *Frankfurter Rundschau,* 14 Sep. 1990, p. 1; Riedel, A., 'Stanitzek: Eine fast sympathische Zugfahrt', *Frankfurter Rundschau,* 14 Sep. 1990, p. 17; 'Abtransport amerikanischer C-Waffen aus Deutschland', *Neue Zürcher Zeitung,* 15 Sep. 1990, p. 9.

[151] 'Jetzt zu Schiff in die Südsee', *Süddeutsche Zeitung,* 20 Sep. 1990, p. 6.

[152] 'Sturm verzögert Giftgas-Abzug', *Frankfurter Rundschau,* 20 Sep. 1990, p. 1; 'Giftgas-Schiffe liegen wegen Sturm weiter fest', *Frankfurter Rundschau,* 21 Sep. 1990, p. 1.

[153] 'Bundesrepublik frei von Giftgas', *Süddeutsche Zeitung,* 24 Sep. 1990, p. 6; 'Giftgas-Transport verlässt deutsche Hoheitsgewässer', *Frankfurter Allgemeine Zeitung,* 24 Sep. 1990, p. 1; 'Kriegsschiffe

was kept secret, and the two container vessels were escorted by US Navy vessels. On 22 November it was announced that the ships had arrived safely at the Johnston Atoll facility.[154]

The cost of the withdrawal operation was shared by the USA and the FRG and was estimated at $50 million and 40 million Deutsch Mark, respectively. The final estimate of the cost of the operation from its beginning to the departure of the vessels from Nordenham was 100 million DM.[155] The Clausen depot is now intended to be made available for civilian use.[156]

The debate following public disclosure of information about the withdrawal operation is similar to the discussion during the past 20 years about US stockpiling of chemical weapons in the FRG. The question of possible stockpiling of new binary chemical weapons on the territory of the FRG in a crisis situations, the so-called second part of the 1986 Reagan–Kohl agreement, has thus still not been clarified.[157] Complaints were heard about the information policy of the FRG, which chose not to provide information to the public until all details of the withdrawal plan were settled. Misgivings were also expressed about possible safety risks to residents of the region during transportation of the chemical weapons.[158] Greenpeace, other non-governmental organizations and local citizen groups proposed that an investigation be made of the possibility of destroying the chemical weapons on-site at Clausen using chemical neutralization.[159] In the Bundestag the 'Greens', in co-operation with some Social Democratic Party of Germany (SPD) parliamentarians, petitioned to postpone the start of the withdrawal operation and to conduct additional safety investigations,[160] but this proposal was rejected.[161] There was also discussion about possibly forbidding low-flying air traffic over the area around Clausen and the routes for the CW ammunition convoys during the entire operation, particularly flights to the nearby Ramstein military airport,[162] since the temporary storage area in Miesau is located in its air corridor.[163] In fact, on the

begleiten den Giftgastransport', *Frankfurter Allgemeine Zeitung*, 24 Sep. 1990, p. 6; 'Giftgas-Schiffe legten ab', *Frankfurter Rundschau*, 24 Sep. 1990, p. 1.

154 'US: Giftgastransport beendet', *Frankfurter Rundschau*, 22 Nov. 1990, p. 2.

155 'Abtransport der Granaten kostet 100 Millionen', *Süddeutsche Zeitung*, 21 Sep. 1990, p. 6.

156 Reuters, 'Früheres Giftgas-Depot kann zivil genutzt werden', *Süddeutsche Zeitung*, 19 Oct. 1990, p. 6.

157 See Deutscher Bundestag, 11 Wahlperiod, 202nd session, Bonn, 15 Mar. 1990, pp. 15732–45.

158 'Mainz gegen Vernichtung von C-Waffen in der Pfalz', *Frankfurter Rundschau*, 8 June 1990, p. 4.

159 Badelt, J., 'Wie kommt der Geist zurück in die Flasche?', *AMI*, vol. 20, no. 4 (Apr. 1990), pp. 15–21; Knipp, D. J., 'Die chemische Neutralisation ist leichter kontrollierbar', *Frankfurter Rundschau*, 28 June 1990.

160 Grabenströer, M., 'USA scheuen Verbrennung der Giftgaswaffen im eignen Land', *Frankfurter Rundschau*, 28 May 1990, p. 4.

161 'Giftgas-Transport Ende August?', *Frankfurter Rundschau*, 2 June 1990, p. 1.

162 Deutscher Bundestag, 11 Wahlperiode, 215th session, Plenarprotokoll 11/215, 1 June 1990; Deutscher Bundestag, 11 Wahlperiode, Drucksache 11/7213, Antwort der Bundesregierung auf die Kleine Anfrage der Abgeordneten Frau Beer und der Fraktion Die Grünen, Drucksache 11/6917, 23 May 1990; 'Während C-Waffen-Transport bleibt Ramstein in Betrieb', *Frankfurter Rundschau*, 1 June 1990, p. 4; see note 158.

163 Grabenströer, M., 'Unfall bei Giftgasverladung', *Frankfurter Rundschau*, 3 Aug. 1990, p. 4.

night of 29–30 August a US 'Galaxy' transport airliner crashed at the Ramstein military airport only 12 km from the Clausen depot.[164]

With respect to the possible risk of accidents during the withdrawal operation, it was estimated by an FRG expert commission that the statistical probability of an accident occurring was 0.0004 for the railroad transportation operation.[165] The commission also found that the suggestion for on-site destruction at Clausen[166] using chemical neutralization would have required the construction of a completely new industrial plant near the CW stockpile depot, entailing some risk to the environment and substantial cost.

During the Lindwurm operation only a few minor incidents occurred. One accident took place in Clausen during the loading operation, but no damage resulted.[167] An incendiary munition was discovered on 3 August near the route of the convoy during that stage of the operation, which resulted in a route change.[168] During the penultimate railway convoy, it was reported that security forces had found what appeared to be a bomb,[169] but it turned out to be harmless. The withdrawal operation went according to plan and none of the feared accidents occurred. This is possibly an argument for moving CW stockpiles to one or more destruction facilities rather than building a destruction facility at each stockpile site.

VI. Measures to hinder the spread of chemical weapons

Restrictive legislation and sanctions

In 1990 a number of countries passed legislation to hinder the spread of chemical weapons. In the *Federal Republic of Germany* export regulations for chemical weapons, chemicals related to CW production and know-how focused mainly on three issues: (*a*) legal proceedings against a West German chemical company and its director; (*b*) investigation of and legal action against other companies and individuals accused of providing Iraq and Libya with expertise and construction assistance; and (*c*) legislation enacted by the Bundestag to strengthen the laws prohibiting such activities or to make them illegal. The wording of this legislation was discussed extensively early in 1990 in terms of the type of activities that should be considered punishable, and there was discussion of whether the law should cover research and develop-

[164] Roll, E., 'Absturz nur zwölf Kilometer von Giftgas-Lager entfernt', *Süddeutsche Zeitung*, 30 Aug. 1990, p. 60; Grabenströer, M., 'US-Lufttransporter bei Ramstein abgestürzt', *Frankfurter Rundschau*, 30 Aug. 1990, pp. 1–2.

[165] See Stengel (note 138).

[166] Lutz, B., 'Fragwürdige Eile', *Die Zeit*, 15 June 1990, reproduced in *Die grünen im Bundestag*, 26 May 1990, pp. 57–58.

[167] See note 163.

[168] 'Giftgas-Konvoi nach Munitionsfund umgeleitet', *Frankfurter Rundschau*, 4 Aug. 1990, p. 4; 'U.S. chemical arms are detoured', *International Herald Tribune*, 4 Aug. 1990, p. 2; 'Änderung der Route für Giftgas transport', *Frankfurter Allgemeine Zeitung*, 4 Aug. 1990, p. 1; 'Granaten-Herkunft im Nebel', *Frankfurter Rundschau*, 7 Aug. 1990, p. 4.

[169] Grabenströer, M., 'Bombenattrappe, defekte Lok und eine Drohung', *Frankfurter Rundschau*, 19 Sep. 1990, p. 21.

ment.[170] The law, which was passed on 1 June, prohibits assisting in the development or construction of biological, chemical and nuclear weapons either in the FRG or outside it. It imposes heavy penalties, up to life sentences, for violation of the law, which entered into force on 1 July. However, opposition politicians felt that the law contained too many loopholes.[171]

In February, *Japan* introduced restraints on the export of chemicals which can be used for CW production.[172] *Sweden* worked on legislation to control the import and export of certain chemicals which can be used for weapons of mass destruction; final decisions on the legislation will be taken in 1991.[173] If the legislation is passed, it would be quite similar to the regulations recommended by the Australia Group, a group of 21 countries which meet semi-annually to discuss which chemicals ought to be subject to various national regulatory measures. *Switzerland* began work on legislation which would prohibit the export of chemicals and know-how related to both chemical and biological weapons. A working group has been appointed to work out the text of the legislation.[174] The *United Kingdom* added 15 more chemicals to its precursor list, which now comprises 37 chemicals.[175] Prior to this, a seminar was held in London which was attended by 26 members of the Australia Group and Leipzig Group[176] countries to co-ordinate work on the lists of chemicals which should be subject to national export control measures.[177]

CW and BW export control measures constitute part of the *USA*'s general policy of denying militarily or economically 'hostile' nations access to US high technology and know-how.[178] CW and BW technology, equipment and know-how are regulated in the USA, as are missiles. This policy has been adopted because of the particularly abhorrent nature of these weapons of mass destruction and their political significance in international conflicts. A number of developments related to national and international control of chemicals which might be used for CW production occurred in the USA in 1990. On 17 May, the Senate passed a House bill (HR3033) which imposes sanctions on

[170] See, for example, 'Streit über ABC-Waffenexport', *Frankfurter Rundschau*, 14 Feb. 1990, p. 9.

[171] See, for example, 'Der Bundestag verabschiedet neue Exportkontrollgesetze gegen Waffenhandel', *Frankfurter Allgemeine Zeitung*, 2 June 1990, p. 4.

[172] 'MITI to monitor chemical exports for arms use', OW1402224090, *Tokyo KYODO*, 1347 GMT, 14 Feb. 1990 (in English) in Foreign Broadcast Information Service, *Daily Report–East Asia* (*FBIS-EAS*), FBIS-EAS-90-032, 15 Feb. 1990, p. 6.

[173] Utrikesdepartementet [Swedish Foreign Ministry], *Kontroll av vissa Produkter som kan Användas i Massförstörelsesyfte*, Ds 1990:4 (Allmänna Förlaget: Stockholm, 1990).

[174] Bern, U. M., 'Exportkontrollen für C-Waffen-Technologie: Bundesrat befürwortet Spezialgesetzgebung', *Neue Zürcher Zeitung*, 2 Feb. 1990, p. 25.

[175] Montagnon, P., 'Export controls placed on 15 chemicals', *Financial Times*, 21 Dec. 1990, p. 8.

[176] SIPRI, *SIPRI Yearbook 1988: World Armaments and Disarmament* (Oxford University Press: Oxford, 1988), p. 103.

[177] UK Foreign & Commonwealth Office, Arms Control and Disarmament Research Unit, *Notes on Arms Control*, Jan. 1991; see also George, A., 'Curbs sought on chemical technology', *The Independent*, 10 Dec. 1990; Mallet, V., 'Crisis speeds up drive for chemical weapons curbs', *Financial Times*, 18 Dec. 1990, p. 4.

[178] See, for example, Richardson, M., 'Wary of sharing military technology, U.S. lags in East Asia Market', *International Herald Tribune*, 28 Feb. 1990, p. 8; *Wireless File,* 'Halt weapons spread (editorial)', EUR-515 (United States Information Service, US Embassy, Stockholm, 30 Mar. 1990), p. 25.

countries which use chemical and biological weapons.[179] After lengthy debate in Congress, including discussion of whether or not the new legislation would infringe on executive prerogatives,[180] President George Bush signed the bill in December. It imposes unilateral export control measures on licences for missile technology and on nuclear, biological and chemical weapons and takes effect on 16 February 1991. Fifty precursors for chemical weapons are among the items subject to world-wide export control.[181] As a result of US regulations, two US chemical companies (one of them a subsidiary of a West German chemical company) refused to deliver approximately 7200 kg of thionyl chloride to the US Department of Defense, stating that it was company policy not to sell chemicals which could be used for the production of chemical weapons. (Thionyl chloride is necessary for the production of a key precursor for the nerve gas sarin and is also used in binary chemical weapons.) This may have jeopardized the time-scale which the US Army needed to meet in order to get funds released for further production of binary artillery shells.[182] The West German company declared that it would take every legal measure to hinder the sale of thionyl chloride by its US subsidiary.[183] However, this development may have been overtaken by the US–Soviet agreement not to continue production of chemical weapons.

While efforts to control the proliferation of BW and CW material and know-how continued during 1990, it was still possible for countries to acquire chemical weapons from commercial companies. Even if national anti-proliferation legislation is passed, developments are being made in the fields of commercial biology and chemistry in Third World countries which will render these efforts futile as long as no international agreement exists which prohibits the acquisition and production of chemical weapons.

VII. Developments related to biological weapons

Alleged acquisition, possession and disposal of biological weapons

Allegations continued to be made, particularly by the USA, that approximately 20 countries have acquired or are in the process of acquiring biological

[179] See, for example, 'Chemical weapons: Senate approves sanctions bill', *Defense & Foreign Policy*, vol. 48, no. 20 (19 May 1990), p. 1572.

[180] *Wireless File*, no. 210, 'Administration opposes Export Act CW provisions' (United States Information Service, US Embassy: Stockholm, 30 Oct. 1990), p. 1.

[181] White House, Office of the Press Secretary, Statement by the Press Secretary with *Fact Sheet on Export Control Procedures*, and *Fact Sheet on Enhanced Proliferation Control Initiative*, 13 Dec. 1990; Farnworth, C. H., 'US moves to cut chemicals' spread', *New York Times*, 15 Dec. 1990, p. 7; Silverberg, D., 'US arms proliferation proposal will rise or fall on world support', *Defense News*, 17 Dec. 1990, pp. 4, 36; see also 'Administration acts to assuage defense industry concerns: Bush plan to curb exports to Third World will be in place by mid-February', *Inside the Pentagon*, 20 Dec. 1990, p. 12.

[182] See, for example, Ember L., 'Chemical weapons: firms deny sale of chemical weapons to Army', *Chemical & Engineering News*, vol. 68, no. 14 (2 Apr. 1990), p. 4.

[183] 'Bayer wehrt sich gegen US-Regierung', *Frankfurter Rundschau*, 30 Mar. 1990, p. 5.

weapons.[184] As with the CW potential of Iraq, a number of companies in various countries are said to have contributed to this buildup, particularly mentioned are West German and US companies.[185]

On one occasion President Saddam Hussein declared that Iraq did not possess biological weapons and was aware of the risks of their use.[186] However, PLO leader Yasser Arafat is said to have ascertained that, in the event of war, Iraq would use not only chemical but also biological weapons against Israel.[187] Allegations have been made that Iraq is able to use anthrax as a biological weapon,[188] and that a facility at Salman Pak is the centre for Iraqi biological warfare development.[189] At the end of 1990, plans were announced to vaccinate the US soldiers deployed in the Persian Gulf area against anthrax.[190] Experts are uncertain about Iraq's actual BW capability and which agents it may possess. Typhoid and cholera have been mentioned in addition to anthrax.[191]

Warnings continued to be issued about the risk of development of new biological warfare agents or techniques for developing them. New hybrid-DNA and other biological techniques were seen to be of concern, and several books on this issue appeared in 1990.[192] No new applications of biological weapons were announced; such activities by a state party to the Biological Weapons Convention (BWC) would be a violation of it. A September symposium at Kühlungsborn in the former GDR dealt with the risks to which the misuse of new biological techniques might lead.[193]

The British Ministry of Defence officially returned Gruinard Island to its owners on 24 April 1990. The island was then deemed to be decontaminated from the anthrax spores which had been spread there in 1940 for test purposes, and which had made the island uninhabitable until a large decontamination operation was undertaken in 1986.[194]

[184] 'About 20 countries may be developing chemical weapons and approximately 10 have BW programs', Institute for Defense and Disarmament Studies, *Arms Control Reporter* (IDDS: Brookline, Mass.), sheet 701.B.60, Nov. 1990.

[185] See note 28; 'Wir haben Überraschungen', *Der Spiegel*, vol. 44, no. 41 (8 Oct. 1990), p. 148; 'US-Firmen sollen Irak Mittel für B-waffen geliefert haben', *Frankfurter Rundschau,* 8 Dec. 1990, p. 1.

[186] See 'Saddam Husayn addresses visiting U.S. Senators' (note 1), p. 9.

[187] 'Israel targeted', *Jane's Defence Weekly,* vol. 14, no. 20 (17 Nov. 1990), p. 975.

[188] *Hansard* (House of Commons), vol. 173, no 114 (5 June 1990), col. 432; Center for Strategic & International Studies, 'Iraqi chemical and biological weapons: embargo vs. war', statement by Brad Roberts to the Committee on Armed Forces, US House of Representatives (CSIS: Washington, 6 Dec. 1990).

[189] See notes 44 and 18.

[190] Gordon, M. R., 'Germ warfare: U.S. plans vaccinations', *International Herald Tribune,* 29–30 Dec. 1990, pp. 1, 5.

[191] See, for example, Charles, D. and Nowak, R., 'The poisonous power of chemical warfare', *New Scientist*, vol. 127, no. 1731 (25 Aug. 1990), pp. 22–24.

[192] For a comprehensive overview see particularly Wright, S. (ed.), *Preventing a Biological Arms Race* (MIT Press: Cambridge: Mass., 1990), which also deals with the efforts to strengthen the Biological Weapons Convention; see also *Politics and the Life Sciences*, Special issue: Biotechnology and International Conflict, vol. 9, no. 1 (Aug. 1990), pp. 47, 72–73.

[193] 'Warnung vor Missbrauch: Kolloquium in Kühlungsborn gestern beendet', *Norddeutsche Zeitung,* 19 Sep. 1990, p. 1.

[194] News Release from the Ministry of Defence, London, 'Mod to hand back Gruinard Island', no. 35 (18 Apr. 1990).

New information appeared about another old BW issue, the so-called Sverdlovsk incident, which occurred in 1979 when an anthrax epidemic broke out in the city of Sverdlovsk in the USSR. The incident created serious mistrust between the USA and the USSR, with the USA claiming that the incident was caused by an accident in a local facility alleged to be producing biological weapons in violation of the BWC. The USSR claimed that the epidemic was the result of the consumption of contaminated meat. In an August 1990 article in *Literaturnaya Gazeta,* a local correspondent gave an account of the events which took place in the city at the time, which he felt created additional uncertainty about what might have happened during the outbreak, particularly because of the military interest in the event.[195] Two other articles on the subject in a Minsk newspaper also raised similar questions.[196] For comparison, it is interesting to read a report of the inadequate handling of a recent outbreak of anthrax in Kirghizia.[197]

In an effort to stop the proliferation of biological weapons, in May 1990 President Bush signed the Biological Weapons Anti-Terrorism Act of 1989. It imposes criminal penalties on those who use or help to spread biological weapons.[198] In May the USA also announced its intention to destroy its remaining stock of live smallpox virus, a small amount of which exists at the US Center for Disease Control. (Smallpox was declared eradicated by the World Health Organization (WHO) in 1980.) The destruction was to take place after further scientific work had been concluded, and the USSR was asked to do likewise.[199] Ultimately, both the USA and the USSR agreed to destroy their stocks of smallpox virus, supposedly the only ones in the world, by the end of 1993.[200]

In 1990 the crisis in the Persian Gulf led to increased fear that biological weapons might be used there. As with chemical weapons, more countries may be trying to secure a BW option in response to the possible existence of these weapons in other countries. Efforts continued to hinder the spread of biological weapons, including attempts to improve the BWC and to prevent damaging confidence in it by pointing to the risk which new technical developments, such as genetic engineering, may pose (see also chapter 14).

VIII. Conclusion

Chemical weapons, and to a lesser extent biological weapons, currently play a significant role in international relations. Allegations of use, threats of use,

[195] 'Urals bacteriological accident suspected in 1979', 90SV0013A, Moscow, *Literaturnaya Gazeta,* no. 34 (Aug. 1990), p. 12 (in Russian) in FBIS-SOV-90-172, 5 Sep. 1990, p. 87.

[196] Parfenov, S., 'Consequences of alleged 1979 Sverdlovsk anthrax outbreak explored', *Znamya Yunosti* (Minsk), 24 and 25 Oct. 1990, (in Russian).

[197] 'Veterinarians reproached after anthrax outbreak', PM2908121590, Moscow, *Izvestia,* 29 Aug. 1990, morning edn, p. 2 (in Russian) in FBIS-SOV-90-169, 30 Aug. 1990. p. 76.

[198] Wireless File, no. 99, 'Bush signs law against biological weapons' (United States Information Service, US Embassy: Stockholm, 22 May 1990), p. 6.

[199] Reuters, 'U.S. would destroy smallpox virus', *International Herald Tribune,* 9 May 1990, p. 1.

[200] Reuters, 'Last live smallpox virus to be destroyed by 1993', *Daily Telegraph,* 15 Dec. 1990, p. 12.

acquisition and allegations of acquisition of these weapons increasingly occur. Perhaps this is because chemical weapons appear to be assuming a status where they are regarded as equal to conventional weapons. If such is the case, this might weaken the supposed link between chemical and nuclear weapons. However, it is possible that chemical weapons will become 'political' weapons and, like nuclear weapons, not be used at all in future military conflicts. The developments which have taken place have occurred despite awareness of and general support for efforts to abolish these weapons by international agreement.

Events in the Persian Gulf overshadowed the BW and CW developments of 1990. The use of chemical and possibly biological weapons seemed highly probable in the Persian Gulf conflict. Protective gear against such weapons was needed in the conflict area and in countries in the vicinity of it. Questions have also been raised about the extent to which it would be possible to use personal protective equipment in the hot desert climate in war. The link between CW production capability and the means of delivering chemical weapons, particularly missile capability, needs to be further explored and possibly re-evaluated.

The CBW situation did not change dramatically in 1990 other than that the CW threat became even more real. Efforts continued to detect and prosecute violations of the CW export regulations in various countries resulting in one widely publicized conviction in the FRG. The debate created by these events highlighted the fact that not only Iraq but also other countries in the Middle East, such as Libya, have for many years been regularly and discreetly provided with CW technology by a number of companies. The USA has passed legislation which addresses the breaking of US export rules regulating chemical and biological weapons and their precursors. The FRG, the UK and other European countries have continuously tried to strengthen existing regulations.

The bilateral agreement signed by the USA and the USSR in June 1990 resulted in a pledge to stop CW production, to exchange information about chemical weapons and CW production and storage facilities, to inspect the other's relevant facilities, and in the assumption of an obligation to destroy the greater part of their CW stockpiles. In so doing, the two demonstrated a clear realization of the diminishing return that possession of chemical weapons constitutes. This is particularly evident when the need to destroy old stockpiles approaches and turns out to be both risky and costly, far more costly than acquiring chemical weapons in the first place. Problems related to the destruction of chemical weapons emerged clearly during 1990, especially under the bilateral agreement. Not only were previously accepted techniques questioned but, while clearly in favour of the destruction of chemical weapons, citizens nevertheless objected to such destruction taking place in their own backyards. It was also evident that disarmament and environmental concerns may clash and leave the public unsure about which approach should be taken.

There are a number of conclusions about the transportation and destruction of chemical weapons that can be drawn from the events of 1990. Although much work has been done to find effective and safe destruction methods, much remains to be done. The cost of destruction for any country with a sizeable amount of chemical munitions will be enormous, of the order of billions of dollars. It is worth asking whether those countries which now produce chemical weapons, or which are in the process of acquiring them, have considered this. One reason for the increasing cost is also increased public awareness of the risks involved in destruction operations and the need for adequate safety measures. Finally, it must be noted that the debate about both the destruction and transportation of chemical weapons has created a contradiction for some between a desire for disarmament measures on the one hand and environmental concerns on the other. It might be well to consider what would constitute *acceptable* risk rather than trying to address every conceivable risk. The successful removal of the US CW stockpiles from the FRG may lead to re-evaluation of the possibility to move CW stockpiles to destruction sites rather than to build destruction facilities at each stockpile.

The environmental problems relate particularly to the unearthing of old CW munitions and the discovery of old CW stockpiles. In a larger context this constitutes one aspect of the problem of contamination of the environment by chemical waste products and other chemical pollution. Similar problems exist for old conventional munitions.

The prospect of the possible use of biological weapons in the Persian Gulf crisis was such a matter of concern that protective measures (vaccination against anthrax) were taken by the UN coalition forces stationed in Saudi Arabia. This is an ominous development.

The year 1990 may thus have been a critical year for efforts related to slowing down the proliferation of chemical and biological weapons and attempts to abolish these weapons of mass destruction. The events of 1990 led to a situation of higher risk that a new chemical and biological arms race might start. On the other hand, the clear signal given by the USA and the USSR of starting bilateral destruction of the greater part of their CW stockpiles could constitute hope for a positive reversal of such developments.

Part II. Military expenditure, the arms trade and armed conflicts

5. World military expenditure

SAADET DEGER*

I. Introduction

The decline in world military expenditure, observed for the past two years, accelerated in 1990, mainly as a result of reductions in US and Soviet defence spending. These two countries together account for around 60 per cent of the world total. Military expenditure fell by approximately 6 per cent in the USA and by almost 10 per cent in the USSR, contributing to a decline in world military spending of over 5 per cent in 1990. In terms of the potential 'disarmament dividend', the reduction was modest, however. Most of the resources released were absorbed by the respective economies to fill domestic needs rather than transferred abroad. Nevertheless, the possibility for a trend decline in aggregate world military expenditure is now more definite. Unless there is a dramatic change in the *domestic* political climate of the major powers, which cannot be ruled out given developments in the USSR at the outset of 1991, the process of military expenditure reduction—as distinct from demilitarization—is likely to continue.

US defence spending in 1990 is discussed in section II. The defence budgets for the European NATO countries, with particular emphasis on the Federal Republic of Germany, France and the United Kingdom, are discussed in section III, followed, in section IV, by an analysis of economic and military data for the European Community (EC). Economic, political and structural developments pertaining to Soviet military expenditure are critically analysed in section V, while data for Eastern and Central Europe are discussed in section VI. Section VII provides a brief review of developments in the Asia–Pacific region, with special focus on China and Japan.

The crisis and subsequent war following the Iraqi invasion of Kuwait on 2 August 1990 seemed to confirm the pessimistic view that the proliferation of regional conflicts will be characteristic of the post-cold war era.[1] After a sustained fall in Third World military expenditure since the mid-1980s, the trend was reversed in 1990. Still, there is no indication that defence spending will rise further in the Third World. Economic problems remain the overriding concern for most poor countries, and it was only the rise in defence expenditure of a few nations that offset the aggregate decline. Section VIII discusses the 1990 data on military expenditure in the Third World.

[1] See also chapter 10 in this volume.

* I am indebted to Somnath Sen for research collaboration on the Soviet Union (section V).

II. The United States

In 1990, for the first time since the Viet Nam War, an incumbent President presented a defence budget which proposed a *sustained* reduction in military spending. President George Bush's fiscal year (FY) 1991 budget proposed that real (or inflation-adjusted) expenditure on defence be reduced continuously for the period 1991–95. It was also the second budget since 1971 in which the current year budget request entailed a reduction from the previous year's appropriations.

The Administration's proposals are as yet modest. In terms of cutting the budget, Congress still wields greater power, and can do a better hatchet job, than the President, particularly since the the Department of Defense (DOD) is still cautious about the future. However, overall perceptions are so different in 1990 compared to previous years that it is difficult to maintain the costly equipment and weapon programmes required to fight a high-intensity war. The threat of an all-European war is non-existent; while Soviet military power is of continuing concern to the DOD, Moscow is clearly not the threat it used to be, given the political and infrastructural disintegration in the USSR.

As for future subregional conflicts, prior to the Iraqi invasion of Kuwait, these were expected to follow past patterns of low intensity. The conflict, with the prospect of a war, changed perceptions somewhat. Coming at the peak of the budgetary cycle, it diluted demands for more savage cuts—particularly in manpower and air- and sealift forces—and congressional reductions were probably less than expected. The final appropriations (passed only in October) are closer to the original Bush request than to what the House and Senate had passed separately. However, it should be noted that the cost of the war against Iraq is being calculated separately, and is not part of the regular FY 1991 budget. Most of it is expected to be recovered from allies in the conflict.[2]

US military expenditure could by 1995 be cut to $200–250 billion. Even this reduction is not harsh—the Reagan Administration's first budget was about $225 billion (figures in constant 1990 prices).

The budget

The overall budget debate was more complicated in 1990 because of a number of problems, none of which explicitly related to military expenditure. When President Bush presented his FY 1991 budget in January 1990, he forecasted an aggregate deficit of $61 billion—well within the Gramm–Rudman–Hollings Act (G–R–H) ceiling.[3] The Congressional Budget Office (CBO), which has a better prediction record than the Administration (although its independent forecasts are not legally binding), estimated an overall deficit of $131 billion. It was immediately clear that the perennial problem of the US

[2] See also chapter 19 in this volume.

[3] For background on the G–R–H ceiling, see Deger, S., 'World military expenditure', SIPRI, *SIPRI Yearbook 1989: World Armaments and Disarmament* (Oxford University Press: Oxford, 1989), p. 135.

Table 5.1. US military budgets, FYs 1989–91

Figures are in US $b., current prices.

	Budget authority			Outlays		
Category	1989	1990	1991	1989	1990	1991
Military	290.8	291.4	295.1	294.9	286.8	292.1
Energy, defence	8.1	9.7	11.0	8.1	8.9	10.4
Others	0.6	0.6	0.8	0.6	0.6	0.7
Total[a]	**299.6**	**301.6**	**306.9**	**303.6**	**296.3**	**303.3**

[a] Totals may not add up due to rounding.

Source: *Congressional Quarterly* (various issues).

budget deficits, and its adverse effect on the international economy, would continue. During the year the Savings and Loan (S&L) bail-out[4] meant that the Government would have to increase expenditure much more than anticipated. Only in late October, after the relevant fiscal year had already begun and Government activity was being halted due to lack of appropriated funding, were the final bills passed by Congress and the President.

The aggregate budget deficit in FY 1991, counting new types of spending such as the S&L funding, is expected to be a staggering $253 billion.[5] The G–R–H ceiling has effectively been abandoned. An ambitious deficit reduction package has been worked out such that the budget will be in surplus in FY 1994; the maximum brunt of the deficit reduction scheme falls on future military expenditure reductions. In the 1990s, barring unforeseen circumstances, defence spending and new military capability will depend more on economic factors than on strategic and security concerns. However, like previous plans, there is no way of predicting whether it will work or not.

As regards the *initial* budget request for 'national defense', which comprises mainly the DOD budget and the share of the Department of Energy (DOE) budget earmarked for nuclear weapon research and production, the details of both budget authority and outlays for FYs 1989–91 are given in table 5.1. Budget authority refers to obligations requested; outlays refer to expenditures incurred during the relevant fiscal years. As regards authority, the modest nominal increases are cancelled by inflation. The real reduction of over 2 per cent in FYs 1990–91 continues the trends of the recent past. The only difference is that Congress in the late 1980s forced the real cuts while in 1990 the President started the process himself. As regards outlays, the real reduction is even more modest—less than 2 per cent for FY 1991. However, the cut for outlays in 1990 was rather high, and expenditure cycles, as well as the time profile of spending, means that there is more erratic variation in this

[4] For background on the Savings and Loan bail-out, see the International Monetetary Fund, *World Economic Outlook*, Oct. 1990 (IMF: Washington, DC, 1990).

[5] United States Information Agency, 'New budget plan: is it credible?', *Washington Economic Reports* (USIA: Washingon, DC, 31 Oct. 1990, p. 2. A more recent (Feb. 1991) estimate gives a figure of around $318.3 billion, corresponding to a 5.7 per cent share of the GNP. This is projected to fall to $280.9 billion in 1992. See *Financial Times*, 5 Feb. 1991.

Table 5.2. US active forces, selected categories, 1989–91

Category	1989	1990	1991
Strategic forces			
ICBMs	1 000	1 000	1 000
Strategic bombers	263	244	244
SLBMs	576	608	656
General-purpose forces			
Army divisions	18	18	16
Conventional bombers	61	33	33
Air Force tactical aircraft	1 769	1 743	1 746
Navy tactical aricraft	730	698	684
Aircraft carriers	14	14	14
Battleships	4	4	2
Nuclear attack submarines	96	91	86

Source: *An Analysis of the President's Budgetary Proposals for Fiscal Year 1991* (Congressional Budget Office: Washington, DC, Mar. 1990).

category. It is clear that the fundamental change was in perceptions and form rather than in substance. The Administration believes that a real decline in military expenditure is inevitable but is in no hurry to implement it.

The force structure priorities of the FY 1991 budget are: more qualified and better paid military manpower; technological superiority; more efficient procurement; increased investment in strategic nuclear forces; maritime superiority; increased mobilization through reserve forces; and a more productive military–industrial base. Each of these items has a clear significance for US military power and influence in the post-cold war period.

These priorities are reflected in the data on active forces given in table 5.2 (data for 1991 are forecasts). The strategic forces make net gains over the 1989–91 period; a notably large increase is sought for sea-launched ballistic missiles (SLBMs). The general-purpose forces are targeted for relatively large cuts, particularly for tactical aircraft. As the dangers of a European conflict recede, more reductions are certain. However, with prospects of progress in the Strategic Arms Reduction Talks (START), it is difficult to envisage why strategic forces are emphasized so much. The planned expansion of SLBMs explains the reluctance to participate more fully in naval arms control.

A particularly acrimonious debate between Congress and the Administration continued throughout the year regarding various programmes and costing. The final Appropriations were passed and authorized in late October. The cuts imposed by Congress on the original budget authority request are shown in table 5.3.

Maximum reductions have been imposed on the procurement budget, most of which is allocated to major weapon systems. The cut, in real terms, is the largest annual reduction in 10 years. Expenditure on military personnel is relatively protected, with some manpower decline but also modest pay rises.

Table 5.3. US defence budget authority for FY 1991, comparison of the President's request and final Congress authorization

Figures are in US $b.

	President's request	Congress authorization	Reduction (%)
Military personnel	79.1	78.1	*1.3*
Procurement	77.3	67.2	*13.1*
Operations and maintenance	88.8	83.5	*5.9*
Research and development	38.1	36.0	*5.5*
Construction	9.1	8.4	*7.7*
Energy, defence	11.0	10.9	*0.9*
Others	5.3	5.2	*1.9*
Total	**308.7**	**289.3**	***6.5***[a]

[a] Percentage average.

Source: *Congressional Quarterly* (various issues).

The DOE nuclear programme maintains its funding level, but mainly to cover costs accrued through more stringent environmental protection measures. The most surprising change is the relatively large cut in the President's request for military research and development (R&D), a real reduction after 20 years of continuous rise. However, it should be stressed that these are budget authorities; it will take some time for the changes to filter through to actual expenditure. As a result, actual spending on R&D is not only expected to resist decline in FY 1991, but is even expected to increase in coming budgets.

Table 5.3 provides details of new force structures extracted from the final Authorization Bill, approved in Congress and signed by the President in October 1990 for FY 1991.[6] Troops will be reduced by 80 000 rather than by 100 000 as requested by Congress. Secretary of Defense Richard Cheney's original budget called for a limited reduction of 37 605. Cuts notwithstanding, however, average pay rises of 4.1 per cent, rather than the 3.5 per cent suggested by the DOD, mean that the personnel budget will not decline much. A provision already exists, therefore, to reduce active duty manpower by 22 per cent (by over 200 000 troops) by FY 1995. This represents quite dramatic cuts, and will certainly be contested in the future.

As regards strategic forces, both the rail-mobile 10-warhead MX missile and the smaller single-warhead Midgetman continue to be funded but at much reduced levels relative to request. The Administration will clearly not be able to justify the cost of deploying both, particularly if nuclear arms control efforts are successful and perceptions of the Soviet threat are altered. The MX programme has been limited to development and testing, with the expectation that it ultimately will be mothballed. While the Midgetman will be based in existing silos, the option for subsequent mobile basing will also be maintained. Of the Navy's request, $1.34 billion has been allocated to the purchase of

[6] For details on the Authorization Bill, see *Congessional Quarterly*, vol. 48, no. 42 (20 Oct. 1990), pp. 3524–31; *Congessional Quarterly*, vol. 48, no. 43 (27 Oct. 1990), pp. 3619–24.

52 missiles for the Trident II submarine (initial request $1.5 billion), but due to increasing unit costs, a result of the British Government's delay in placing its order for 1990, fewer missiles than planned will be purchased.

Conventional forces bear the brunt of procurement cuts. The Army had requested funding for 163 current model M-1A1 tanks and 62 improved but untested M-1A2s. After these purchases the plants are expected to be mothballed. Congress reduced funding for procurement, allowing the purchase of 225 tanks, as requested, but instead of allocating funds for production of the new model it limited financing to the modification of M-1A1s to meet M-1A2 specifications. Modest funding ($176 million) was also given for development of a future joint family of armoured vehicles that would share chassis and components.

The Air Force's request for new F-16 fighter aircraft was cut from 150 to 108, with funding cut from $2.9 billion to $1.9 billion. The DOD is thereby forced to break its contract with the supplier, probably incurring penalty costs. The request for F/A-18 fighter-bombers was cut from 66 to 48, with funding reduced from the requested $2.1 billion to $1.5 billion. Minor cuts were made in the procurement of Air Force F-15E and Navy F-14 fighters, but numbers were maintained. More important, development funds for the Advanced Tactical Fighter (ATF, to replace the F-14 and F-15) were largely maintained. Defense Secretary Cheney had already reduced the Navy's request in August, and events in the Persian Gulf made little difference to the final outcome. One new Seawolf Class submarine ($1.46 billion) and four Arleigh Burke Class destroyers equipped with the Aegis system ($3.2 billion) have been authorized for FY 1991.

As for air and sea transport, Congress showed concern about continuing practical problems with the costly C-17 cargo aircraft. The request to purchase six planes (requested funding $2.1 billion) was turned down. Only two planes have been authorized, with FY 1991 procurement limited to $0.4 billion—probably sufficient to buy only one. Restrictions also apply as to how funding may be used for the C-17. The future of the Marine Corps' tilt-rotor V-22 Osprey, a hybrid helicopter/fixed wing transporter, remains uncertain. The DOD wants to cancel it, but Congress continues funding, allocating $0.8 billion in FY 1991, mainly for development but allowing some procurement.

The final result of these deliberations leaves budget authority at $288.3 billion and outlay at $297 billion. While authority has fallen by a hefty 8.7 per cent in FY 1991, compared to the CBO baseline for FY 1990, the fall in outlay is just 3.2 per cent. Reducing defence spending in the short term is clearly difficult, as immediate costs tend to offset long-term savings.

The past

US military expenditure grew rapidly from 1980 to 1986, after which it gently declined. SIPRI estimates are based on calendar year adjustments to fiscal year data according to the NATO definition. In addition to 'national defense'

Table 5.4. US national defence expenditure outlays, FYs 1981–90

Figures are in US $b., current prices.

	1981	1982	1983	1984	1985	1986	1987	1988	1989	1990
Personnel	47.9	155.2	60.9	64.2	67.8	71.5	72.0	76.3	80.7	75.3
O&M[a]	51.9	59.7	64.9	67.4	72.4	75.3	76.2	84.5	87.0	86.1
Procurement	35.2	43.3	53.6	61.9	70.4	76.5	80.7	77.2	81.6	80.9
RDT&E[b]	15.3	17.7	20.6	23.1	27.1	32.3	33.6	34.8	37.0	36.5
Energy, defence	3.4	4.3	5.2	6.1	7.1	7.4	7.5	7.9	8.1	8.9
Other	3.8	5.1	4.7	4.7	7.9	10.4	12	9.7	9.2	8.6
Total	**157.5**	**185.3**	**209.9**	**227.4**	**252.7**	**273.4**	**282.0**	**290.4**	**303.6**	**296.3**

[a] Operations and maintenance

[b] Research, development, testing and evaluation.

Sources: *United States Budget in Brief, Fiscal Year 1990* (US Government Printing Office: Washington, DC, 1989); *Budget of the United States Government* (US Government Printing Office: Washington, DC, 1990).

figures the data also include figures for military aid. The SIPRI estimates show that the change during the period of expansion was of the order of 47.7 per cent; this implies a per annum growth of 6.7 per cent. The fall in 1986–89 was about 5.2 per cent overall, or 1.8 per cent per annum.

At least 12 factors can be presented to explain the evolution of US military expenditures and force structures in the 1980s. Some of them overlap.

1. International factors; for example, the Gulf conflict will create demand for sealift forces.
2. Changing administrations; for example, President Bush has been more responsive to military cuts than President Reagan was even in his second term.
3. Public opinion; for example, by the mid-1980s it was a common view that social expenditures were being cut to accommodate defence expansion.[7]
4. Congressional attitudes; for example, the 'deep cuts' in FY 1991 have been forced by a Democrat-dominated Congress, while the Administration's proposals relate to reductions from previous inflated plans.
5. Domestic economic constraints; for example, the twin deficits of budget and trade, rather than a dramatic change in security perceptions, prompted the first Reagan cuts after FY 1986.
6. Changing perceptions of the Soviet threat; for example, President Reagan's first State of the Union Address, in 1981, identified the 'military expenditure gap' as the central cause of future defence expansion.
7. Developments in arms control; for example, the 1987 INF Treaty allowed for reductions in costs required for modernization.
8. Elections; for example, base closures and procurement expenditure reductions have been opposed by Congressmen nearer elections in 1990.

[7] Schneider E., 'Causal factors in variations in US postwar defense spending', *Defense Analysis*, vol. 4, no. 1 (1988), pp. 53–79.

Table 5.5. US potential annual savings after 1993 from proposed CFE reductions

Figures are in US $b., at constant (1990) prices.

	Operations and support	Procurement	Total	Reduction from FY 1990 budget (%)
CFE reductions	5.1	1.2	6.3	*2.2*
Reductions based on Government plans	12.5	3.0	15.5	*5.4*
Reductions proportioned to WTO reductions	25.3	6.6	31.9	*11.1*

Sources: *Budgetary and Military Effects of a Treaty Limiting Conventional Forces in Europe* (Congressional Budget Office: Washington, DC, 1990); author's calculations.

9. Inter-service competition; for example, the Navy's share in the total budget has increased from 30 per cent in the early 1970s to over 34 per cent in the late 1980s as a result of its drive for a '600-ship navy'.

10. Allied expenditure; for example, the USA has called on all countries affected by the Gulf conflict to contribute men and money in the joint effort.

11. Burden-sharing; for example, the debate over European NATO's response to aggregate alliance spending, and, in the USA, over how much it should spend on its European commitments, remains unsettled.

12. Macro-economic policy; for example, in the early 1980s defence spending was used as a fiscal stabilizer when growth rates turned negative and unemployment reached record levels (9.7 per cent of the labour force in 1982).

Table 5.4 gives military spending outlays for 1981–90 to show the disaggregated trends. The investment part of the expenditure (procurement and R&D) expanded much faster than the operational part (military personnel and operations and maintenance, O&M). The former increased in nominal terms almost threefold; the latter doubled during the period. The share of procurement plus R&D is currently over 40 per cent compared to 32 per cent in 1981. It will not be easy to bring this percentage down quickly.

The future

As indicated, numerous political, economic and military variables will affect the evolution of US defence expenditure in the future. The specific security variables are: (*a*) the impact of arms control, (*b*) the Administration's plans, (*c*) current perceptions that in the post-cold war period some form of military expenditure limitations are necessary and, in particular, (*d*) military R&D.

The impact of arms control will be felt first by the reductions achieved through the Conventional Armed Forces in Europe (CFE) Treaty, signed in November 1990.[8] The first row of table 5.5 gives possible budgetary implica-

[8] See also chapter 13 in this volume.

Table 5.6. US budgetary requests for the FY 1991–95 Five-Year Plan

Figures are in US $b., current prices.

	Budget authority					Outlays				
Category	1991	1992	1993	1994	1995	1991	1992	1993	1994	1995
The President's 1991 budget	307.0	312.5	317.5	321.6	325.7	304.0	309.2	312.3	319.1	319.9
CBO[a] baseline	315.8	328.4	341.6	355.3	369.7	306.9	317.6	328.3	344.8	355.4
Reduction (%)	*2.8*	*4.8*	*7.1*	*9.5*	*11.9*	*0.9*	*2.6*	*4.9*	*7.5*	*10.0*

[a] Congressional Budget Office.

Source: *An Analysis of the President's Budgetary Proposals for Fiscal Year 1991* (Congressional Budget Office: Washington, DC, Mar. 1990).

tions as calculated by the CBO.[9] The effect will mainly be on Army and Air Force (tactical) units, which currently account for about one-third of the total budget. Operations and Support (O&S) costs (including personnel and O&M), will fall by over $5 billion. Cuts in procurement are marginal, since NATO reductions are small in some categories (tanks and armoured combat vehicles, ACVs), negligible in one category (artillery), and non-existent in others (aircraft and helicopters). The total direct effect is just above 2.2 per cent of the DOD FY 1990 budget, and even less for aggregate military expenditure.

The CBO has also calculated the effect of further cuts, over and above CFE limits, given changed political perceptions. The second row of table 5.5 shows the impact of reductions following the Government's plans to withdraw some forces from Europe. Although O&S costs fall the most, the greatest proportionate reduction is in procurement. The total potential fall in spending is about 5.4 per cent of the DOD budget. The final option presented shows the impact of reductions equivalent to those carried out by the Warsaw Treaty Organization (WTO). Savings of over 11 per cent, from current FY funding levels, are anticipated—almost equal to the O&S costs of US troops in Europe. This option of course increases risks, since force balance is destroyed between NATO and the WTO. However, if it is believed that war on a continental scale is not possible then such risks can be justified.

The Administration's plans can be judged by looking at the budgetary programme for FYs 1991–95. These values are usually presented in Then Year (TY) prices, which are based on forecasted inflation for the relevant five years. The CBO then estimates the 'baseline' values, based on relevant 1990 appropriations with increases in each category by an inflationary factor. For any year, comparing the budget figure with that of the CBO baseline provides an index of non-inflationary change *as compared to 1990*. Table 5.6 gives the data for FY 1991–95 for both authority and outlay.

According to Administration plans formulated in 1990, there are to be sustained and increasing reductions in military expenditure in 1991–95.

[9] Congressional Budget Office, 'Budgetary and military effects of a treaty limiting conventional forces in Europe', CBO Paper (CBO: Washington, DC, Sep. 1990).

Table 5.7. Real changes in total US defence expenditure for 1980–90, and per annum cuts required to attain the 1980 level by 1995: budget authority

Figures are percentages.

	Real change			Per annum cuts required
	1980–85	1985–90	1980–90	1990–95
Military personnel	+ 3	+ 1	+ 4	1
Procurement	+ 115	– 29	+ 52	8
Operations and maintenance	+ 38	– 4	+ 32	5
Research and development	+ 82	– 1	+ 80	11
Construction	+ 97	– 21	+ 56	9
Total DOD	+ 52	– 14	+ 31	5
Energy, defence	+ 93	+ 10	+ 113	14
Average	**+ 53**	**– 13**	**+ 33**	**6**

Source: Author's estimates.

However, the reduction remains modest, compared both to past levels and to current expectations. The decline in authority is faster than that of outlay, since it is more difficult to cut down on obligations incurred in the past. On average, the Government believes that a 10–12 per cent cut is to be effected in the next five years—equivalent to approximately 2 per cent per annum.

Current perceptions are that in the post-cold war period some form of military expenditure limitations are necessary. Thus it may be useful to analyse, within hypothesized scenarios, the nature of the cuts required. First, consider the real (net of inflation) change in US military expenditure allocations between 1980 and 1990. The period can be roughly divided in two halves: the first saw dramatic expansion, the second modest decrease. Table 5.7 therefore gives real change, in all categories of defence spending, for the periods 1980–85, 1985–90 and 1980–90. Military personnel expenditure increased very little in real terms, reflecting only increases in manpower. Budgetary procurement expenditure doubled (over 115 per cent change) in 1980–85 and then fell by around 29 per cent in 1985–90. By 1990 military procurement (which for the USA means weapons alone) had increased by over 50 per cent compared to 1980. There was no fall in funding for defence-related R&D, which increased by 80 per cent over the 10 years. The DOE's budget on nuclear weapons, although a small proportion of the total military budget (3 per cent), has had the most spectacular growth.

Since the USA is experiencing an economic recession, a key problem is the scope and feasibility of the financial cuts. The last column of table 5.7 gives estimates of *annual* reductions necessary to attain the 1980 level for each category by 1995. Note that these are estimated for *authority* figures, which usually can be changed faster than outlay or actual expenditures. The figures are therefore optimistic. It will take somewhat longer (two to three years) for actual spending to attain the targets presented here.

Table 5.8. US Government defence R&D expenditure 1970–90, conduct of R&D, obligations

Figures are in US $b., current prices.

Fiscal year	Current prices	Constant prices (1988)	Share of defence in total (%)
1970	8.0	23.0	*52.3*
1971	8.1	23.7	*52.3*
1972	8.9	25.1	*53.9*
1973	9.0	23.9	*53.6*
1974	9.0	21.6	*51.7*
1975	9.7	21.3	*51.1*
1976	10.4	21.6	*50.0*
1977	11.9	23.2	*50.6*
1978	12.6	22.9	*48.8*
1979	13.6	22.1	*48.4*
1980	15.1	21.7	*50.7*
1981	17.8	23.2	*53.8*
1982	22.1	27.1	*60.7*
1983	24.5	29.1	*63.8*
1984	28.3	32.2	*65.5*
1985	33.4	36.7	*67.5*
1986	36.5	39.4	*69.4*
1987	38.4	39.9	*68.5*
1988	39.5	39.5	*67.2*
1989	41.3	39.3	*65.6*
1990	44.0	39.9	*65.4*

Source: *Special Analyses, Budget of the US Government 1991* (US Government Printing Office: Washington, DC, 1990).

The figures in the last column of table 5.7 show that considerable effort will be required to reach the hypothetical target of returning to 1980 expenditure levels by 1995. Aggregate defence spending needs to be reduced by 6 per cent per annum, while the needed annual cut in weapon procurement and R&D is 8 per cent and 11 per cent, respectively. These hypothesized cuts go far beyond the Administrations plans (which assume 2 per cent annual reductions in the total), but are close to congressional reductions made in 1990 for FY 1991.

During the 1980s military-related R&D expanded much faster than aggregate defence spending in the USA. While economic difficulties, budgetary constraints, developments in arms control and political changes in Europe have, over the past few years, resulted in reductions in all other categories of military expenditure, defence R&D remains stubbornly high. After a period of stagnation in the 1970s, R&D expenditure increased fast in real terms between 1980 and 1987. The share of research, development, testing and evaluation (RDT&E) in the defence budget grew from 9.3 per cent in 1980 to 13.0 per cent in 1989, while federal military research activity, in constant 1988 prices, increased from $21.7 billion in 1981 to $39.3 billion in 1989; the increase is roughly 81 per cent in volume terms. Resource transfer and conversion prospects in the military sector are least evident for R&D. In sum, the annual rate of growth of military R&D spending, after adjusting for inflation, was for the 1980s as a whole 6.3 per cent on average. Table 5.8 gives figures for 1970–90.

III. European NATO

In May 1990 the Defence Planning Committee of NATO finally abandoned the rule that since 1977 had set an annual target of 3 per cent real increase in military expenditure. No country had consistently followed this target, since defence spending rise is a function not only of perceived threats but also of economic and aggregate budgetary growth. To expect otherwise—that defence ministries can convince their governments that a continuous and sustained rise of such a magnitude is possible—is unrealistic.

However, aggregate European NATO military expenditure did rise almost continuously from 1980 to 1987, although the rate was lower than the postulated 3 per cent. Starting from a level of almost $140 billion in 1980, defence spending reached over $157 billion in 1987 (all in 1988 dollars). This constitutes an annual growth rate of 1.8 per cent. Since 1987 the level has stabilized.

Of more current interest is the question of how fast defence spending will fall now that political conditions have changed. As the figures show, there was no decline whatsoever until 1989. In 1990 the decline is expected to be less than 2 per cent overall, to a level of around $155 billion. The aggregate hides substantial country variations. Northern and southern flank countries have not reduced budgets significantly and have in some cases increased them to finance force modernization. Other countries announced budgetary reductions in the early 1990s, and most are rethinking their military strategy. France, the FRG and the UK are particularly important, in terms of size and possible reorganization. These countries are discussed separately.

Since military spending of the European NATO countries increased relatively modestly in the 1980s, it will be easy to make cuts which will bring down the level of defence expenditure to that of 1980. SIPRI estimates show that a 2.1 per cent annual decline in 1991–95 will by the end of the period bring European NATO military expenditure down to 1980 levels.

The impact of the CFE process, both on current reductions and future procurement, will be modest. If the assets acquired from the former German Democratic Republic in 1990 are discounted, NATO will be required to eliminate about 3000 tanks (13 per cent of holdings as of 1 January 1990) and 100 ACVs (0.3 per cent of holdings). For artillery destruction, GDR stocks can be used. NATO combat aircraft and helicopter assets in the Atlantic-to-the-Urals (ATTU) zone are below the CFE ceiling. Some individual country data also show how small the impact could be in general. According to British data provided at the signing of the Treaty, 183 of the UK's 1198 tanks are to be eliminated (15 per cent cut), which means cutting about three armoured regiments, and 17 of 3193 ACVs are to be destroyed (0.5 per cent). The UK can also keep all its artillery and increase stocks of aircraft and helicopters.[10]

Manpower is not affected by the CFE Treaty, but it is probable that ceilings will be the topic of the next round of negotiations. European NATO's total armed forces have increased from about 3.5 million to 3.6 million during the 1980s. The increase in spending has outstripped the increase in numbers, since pay and other amenities have been improved. The conditions of service have generally improved over the past decade, reflecting the greater emphasis on increased capability and improvements in the quality and morale of technologically more sophisticated armed forces. The armed forces in aggregate represent 2.7 per cent of the total labour force. A cut in manning by one-third, for example, could increase the unemployment rate by a full 1 per cent.

The maximum direct impact of the CFE process is expected to be on procurement. Upper ceilings on inventories mean that production of tanks, ACVs and artillery will need to be reduced. However, the major expenditures are on other equipment and platforms (such as naval assets), and prospects for deep cuts here are not bright. It should be noted that intrusive verification has been allowed for storage facilities but not for production facilities. This is because the major arms producers, France and the UK, have objected to inspection of plants and firms where production takes place, on the grounds that the capacity of US and Soviet defence industries, being outside the ATTU zone, will not be subject to verification. With R&D expenditure falling the least among all categories of defence spending, as noted above, there are as yet few constraints on procurement except long-term, indirect ones.

Rapid reductions in assets, and substantial cuts in military spending, will take a considerable time to appear. The best hope is that spontaneous arms control, growing out of the forces of technological and economic structural disarmament, will reduce spending more than anything else.[11] Taking advantage of the new political mood in East–West relations resulting from successful arms control negotiations, countries could utilize economics and technology to proceed with rapid reductions and even disarmament. It may be worthwhile to note the amount of resources spent on the East–West confrontation, by all countries and powers involved, and see its evolution over time. Over $500 billion was spent in military expenditures *on Europe* in 1989. This figure is derived from DOD data on US spending on forces in Europe and SIPRI estimates of: apportioned Soviet expenditure for burden-sharing in the WTO and costs of military assets west of the Urals; European NATO's spending (with small adjustments for excluded out-of-area operations); all spending of non-Soviet WTO (NSWTO) countries; and the defence expenditure of the European non-aligned and neutral countries. The amount represents well over half of world military expenditure. What is more, from 1980 to 1987 it rose by over 30 per cent before beginning a slow descent (see figure 5.1). Although many of these resources will, at least in the short run, be

[10] *Financial Times*, 20 Nov. 1990.

[11] For a discussion, see Deger, S. and Sen, S., SIPRI, *Military Expenditure: The Political Economy of International Security* (Oxford University Press: Oxford, 1990), p. 5.

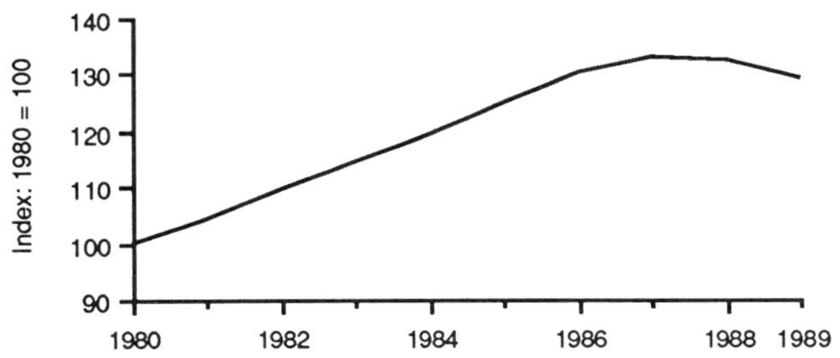

Figure 5.1. Evolution of military expenditure in Europe, 1980–89

retained for military activities, the sum represents the potential available for re-allocation.

The Federal Republic of Germany, France and the United Kingdom

Of all the European NATO countries, the FRG is the most—and most rapidly—affected by the political upheavals of 1989–90. The cost of unification is high; forecasts by economics institutes in the FRG are pessimistic. The budget deficit in 1989 was about 20 billion Deutsche Mark, a negligible share of the gross national product (GNP). In 1991 it will rise to over 140 billion DM—5 per cent of the GNP.[12] However, the long-term prognosis is that the FRG has sufficient economic strength to overcome its immediate difficulties.

Military expenditure in 1991 is set to rise in real terms, although to a level below the combined spending of the FRG and the former GDR, due to the costs of integrating the armed forces and reducing manpower. The terms of unification mean that the FRG has a military manpower ceiling of 370 000, representing 61 per cent of the total strength of the armed forces of the two German states prior to unification. This means that the FRG will have to cut about 40 per cent of its aggregate military forces. Even compared to the pre-unification level (GDR forces excluded), manpower reductions are over one-fourth (26 per cent). Personnel expenditure currently amounts to 50 per cent of total defence spending. Thus a pro-rata calculation would imply a reduction of 13 per cent of military expenditure of the original FRG level. Adding operations and support costs, a cut of approximately 18 per cent can be made after the restructuring is completed.

The procurement figures are more complicated. In the absence of nuclear forces, and as out-of-area operations are banned by the constitution, major weapon procurement accounts for a smaller share of the total in the FRG than in France or the UK, around 19 per cent (less than 12 billion DM) in 1989.

[12] *Financial Times*, 3 Jan. 1990.

However, this figure hides German commitments to specific projects which have a direct impact on West European collaborative weapon programmes. Domestic production accounts for 45 per cent of the procurement budget; co-development (research, design and production in co-operation with other countries) another 25 per cent; co-production (under foreign licence) 20 per cent; and imports the remaining 10 per cent.[13] France and the UK have a domestic production share of 75–80 per cent. Thus, a drastic cut in FRG procurement spending will affect not only domestic but also foreign industry. In particular, if the FRG abandons the European Fighter Aircraft (EFA) co-production venture, there will be economic effects across borders. If production runs are shortened, raising unit costs rapidly, other countries in the consortium may be inclined to back out as well, effectively killing the project.

In France procurement as well as military R&D spending are relatively high, largely because the country, unlike the UK, has borne the full cost of developing and maintaining its independent nuclear force. For the past 20 years about one-third of the military procurement budget has routinely been allocated to the nuclear forces. Considerable economies can be made, but the implications for cuts here would be profound for France.

French defence spending and military capability have also been relatively protected, even sacrosanct, in the changed political environment. However, questions are increasingly being raised, about continuing high levels, and whether fundamental re-orientation is required. In spite of the historical importance of the nuclear triad,[14] discussions have been initiated about the scope and relevance of all its components. Proposals have reportedly been made to scrap the land-based missiles, and even to phase out the strategic bombers. If these are accepted, the current triad, having lost two legs, would be reduced to submarine-launched missiles by the end of the century. Such proposals will be hotly debated since they raise fundamental questions about French security policy that go far beyond simple force restructuring.[15]

A related issue in terms of costs is that the equipment and procurement share of the budget (called Titre V) rose continuously during the 1980s. The corresponding share of personnel and operational costs (Titre II) has fallen.[16] In 1987 the two shares became approximately equal. In 1990 Titre V accounted for around 55 per cent of the total defence budget. Thus, if budgetary reductions are to be applied to procurement expenditure as well, difficult decisions concerning the defence industrial sectors will have to be made. Military procurement is a vital part of overall industrial policy; around 80 per cent of major weapon purchases are made from domestic sources. 'The share of arms in the 1988 defence budget was 41 per cent of the total expenditure. As the largest part of this is produced by French industry, the defence budget has become year after year an industrial budget, inducing some major

[13] Moravcsik, A., 'The European armaments industry at the crossroads', *Survival*, vol. 33, no. 1 (Jan.–Feb. 1990), pp. 65–85.

[14] See chapter 1 in this volume for a description of the French nuclear force structure.

[15] *Defense News*, 22 Oct. 1990, p. 1.

[16] Boucheron J.-M., *1990–1993 Programmation Militaire* (Economica: Paris, 1989).

companies by means of R&D financing support.'[17] Overall, the French national security debate, in the era of arms control, will have to encompass military, political and economic aspects of the problem, all at the same time.

In the UK the share of personnel expenditure has risen steadily since the mid-1980s. At the same time the share of equipment expenditure has fallen. In FY 1989 the two shares were roughly equal, while the estimate for 1990 shows for the first time the personnel share exceeding that of equipment.[18] In short, the trend is the reverse of that of France. This can in part be explained by improvement in service conditions, in part by the 'value for money' policy adopted by the MOD around 1983–84 for its arms procurement. Implemented by Peter Levene, Chief of Defence Procurement, the so called 'Levene reforms' stressed the increased role of competition in tendering, the selection of equipment from a wide range of suppliers, the replacement of cost-plus contracts with fixed-price ones and increased efficiency and better information gathering on the part of the MOD purchasing agency.[19]

The impact of these efficiency measures can be seen from the evolution of the types of contract made through the MOD procurement budget. There are five types—those priced: by competition; by reference to market forces; on estimates at outset with reference to government profit formula; on basis of actual costs with incentives to minimise costs; and on basis of cost-plus formula. The first two give the competitive element of military procurement, the last two provide the purely non-competitive element, while the middle one is mixed in nature. Between FY 1982 and FY 1989, the share of the competitive part rose from 36 to 49 per cent. At the same time, the share of the purely non-competitive element fell from 19 to 9 per cent of the total. Thus, from the demand side at least, the MOD is attempting to open up the restrictive market which has traditionally characterized the military–industrial sector. However, the economic effect on industry itself is yet to be evaluated.

A possible impact of arms control and changed threat perceptions in Europe relates to the future size of the theatre ground forces, which would include the British Army of the Rhine (BOAR, manpower 53 400), the British Berlin forces (2900) and the Home Forces (40 500). The future of the BOAR is uncertain; if not eliminated it will at least be reduced when the issue of foreign troops in the FRG comes up for discussion. These forces cost in aggregate about £3.4 billion, accounting for about 16 per cent of the MOD budget. There is thus substantial scope for cuts in this element of British military spending.

The British military, centred on NATO, has a number of commitments besides defending the UK. These include maintaining army forces in the FRG, naval forces in the North Atlantic, an independent nuclear deterrent, and out-of-area commitments such as in Hong Kong and the Falkland/Malvinas Islands. In FY 1988 the expenditures for these functions were: £2162 million for the 'home base'; £4049 million for forces in the FRG and forward defence;

[17] Schmidt, C, Pilandon, L. and Aben, J., 'Defence spending in France: the price of independence', eds K. Hartley and T. Sandler, *The Economics of Defence Spending* (Routledge: London, 1990).

[18] *Statement on the Defence Estimates, 1990*, vol. 2 (Her Majesty's Stationery Office: London, 1990).

[19] Smith, R., 'Defence spending in the UK', eds Hartley and Sandler (note 17).

£2583 million for the naval forces; £1072 million for the nuclear forces; and £848 million for out-of-area activities. These amounts still leave a large overhead of £8457 million not attributable to any particular function. Nuclear deterrence has been maintained at low costs, principally because of co-operation and collaborative agreements with the USA.

Another interesting feature is that declining home procurement of major weapons, and greater demands for efficiency and unit cost reductions, have gone hand in hand with increasing support for military aid and exports abroad. Military aid, standing at almost £20 million in FY 1990, has grown faster than the aggregate nominal military expenditure in the late 1980s. According to customs data, actual sales of identified defence equipment, which excludes dual-purpose aerospace equipment, have also increased rapidly. In 1989 actual export sales of identified defence products were reported to be over £2.4 billion, as compared to £1.4 billion for 1988. Although recent figures are not available, the export of *additional* aerospace equipment was over £1.8 billion in 1987—an increase of £500 million over three years, reflecting a concern to shield the defence industrial base from too savage cuts by encouraging export.

Weapon procurement and armed forces

Weapon procurement and armed force levels are the areas where the CFE and other arms-limiting processes resulting from the changes now taking place in Europe will have the strongest impact. The trends are clearly indicated by data from the 1980s. Tables 5.9 and 5.10 give information on major weapons procurement for NATO countries, in current and constant prices respectively.

While procurement expenditure on weapons rose steadily for European NATO countries until 1988, the data show a declining trend in 1989 and 1990. The information is clouded by the uncertainty surrounding German data, since it is not clear whether weapon purchases declined after unification, at which the FRG received large stocks of arms from the armed forces of the former GDR. The data for 1990 should therefore be treated with caution. However, the trend reduction is clear enough and is expected to continue, with obvious implications for industrial structure and employment in military industries.

As regards European NATO military personnel, table 5.11 shows a stable level of around 3.6 million men for the period 1981–90. The share of civilian and military employment in the national totals is also stable over the decade: for European NATO the share is about 2.6 per cent, a slight decline from 2.8 per cent in the early 1980s. The situation could change as demands for troop reductions become stronger, as discussed above, particularly for the three major European powers. Demographic trends favour troop reductions—the solution of replacing relatively large draft armies, conscripted on a national scale, with small, professional armies could become more widespread.

Table 5.9. NATO major weapon procurement expenditure, 1981–90

Figures are in local currency, current prices.

		1981	1982	1983	1984	1985	1986	1987	1988	1989	1990
North America											
Canada	m. dollars	1 000	1 332	1 688	1 971	1 941	2 140	2 434	2 486	2 392	2 474
USA	m. dollars	34 487	42 028	50 202	58 328	66 348	72 525	76 362	71 808	76 776	74 054
Europe											
Belgium	m. francs	17 596	17 969	18 853	18 363	18311	19 618	20 360	18 078	15 139	16 088
Denmark	m. kronor	1 803	1 960	2 075	2 048	1 841	1 867	2 182	2 249	2 091	2 115
France	m. francs	29 444	34 637	39 772	42 216	46 492	49 664	55 943	56 564	60 071	58 232
FR Germany	m. DM	10 439	10 847	11 299	11 455	11 730	12 267	12 332	11 896	12 004	10 597
Greece	m. drachmas	29 287	29 966	30 741	41 604	46 687	53 477	67 605	112 141	110 164	130 349
Italy	b. lire	1 707	2 046	2 664	2 843	3 494	3 693	4 900	5 451	56 588	44 463
Luxembourg	m. francs	31	44	36	36	91	74	106	89	114	198
Netherlands	m. guilders	2 135	2 444	2 794	3 012	3 019	2 661	2 359	2 713	2 388	2 474
Norway	m. kronor	1 799	2 147	2 615	2 297	3 846	3 303	3 784	4 018	5 009	5 240
Portugal	m. escudos	3 375	3 318	3 761	4 416	3 675	8 818	16 088	20 356	27 292	18 534
Spain	m. pesetas	70 966	84 291	116 707	170 745	113 380	168 812	210 633	172 918	135 535	119 739
Turkey	b. lira	29	48	56	105	168	334	553	853	1 236	2 719
UK	m. pounds	3 218	3 545	4 122	4 629	4 907	4 762	4 744	4 904	4 731	4 794

Sources: NATO publications; author's calculations. Figures for France are based on national data.

Table 5.10. NATO and EC major weapon procurement expenditure, 1981–90

Figures are in US $m., at constant (1988) prices.

	1981	1982	1983	1984	1985	1986	1987	1988	1989	1990
North America										
Canada	1 169	1 405	1 683	1 883	1 784	1 887	2 058	2 020	1 852	1 838
USA	44 854	51 493	59 581	66 359	72 917	78 219	79 396	71 808	73 244	66 546
Europe										
Belgium	650	611	595	545	518	548	560	492	400	409
Denmark	396	390	386	359	308	301	339	334	296	291
France	7 490	7 878	8 255	8 151	8 492	8 850	9 648	9 496	9 746	9 100
FR Germany	6 843	6 533	6 811	6 743	6 760	7 082	7 100	6 773	6 652	5 703
Greece	689	583	497	569	535	498	541	790	682	663
Italy	2 469	2 540	2 883	2 778	3 128	3 122	3 954	4 188	4 091	3 024
Luxembourg	1.1	1.5	1.1	1.0	2.5	2.0	2.9	2.4	3.0	5.0
Netherlands	1 243	1 344	1 494	1 560	1 523	1 346	1 202	1 373	1 196	1 206
Norway	465	499	560	463	734	588	620	617	735	740
Portugal	74	59	54	49	34	73	123	141	168	101
Spain	1 135	1 180	1 456	1 914	1 168	1 593	1 895	1 484	1 065	904
Turkey	215	271	241	304	3 336	496	559	600	512	748
UK	8 189	8 307	9 240	9 881	9 878	9 270	8 859	8 736	7 830	7 141
European NATO total	**29 217**	**30 197**	**32 473**	**33 317**	**33 417**	**33 769**	**35 403**	**35 026**	**33 376**	**30 035**
NATO total	**75 882**	**83 095**	**93 737**	**101 559**	**108 118**	**113 875**	**116 857**	**108 854**	**108 472**	**98 419**
EC	29 179	29 427	31 672	32 550	32 347	32 685	34 224	33 809	32 170	28 588

Sources: NATO publications; author's calculations. Figures for France are based on national data.

Table 5.11 NATO armed forces, total military personnel, 1981–90

Figures are in thousands.

	1981	1982	1983	1984	1985	1986	1987	1988	1989	1990
North America										
Canada	81	82	81	82	83	85	86	88	88	87
USA	2 168	2 201	2 222	2 222	2 244	2 269	2 279	2 246	2 241	2 189
Europe										
Belgium	109	110	109	107	107	107	109	110	110	108
Denmark	33	30	30	31	29	28	28	30	31	30
France	575	577	578	571	563	558	559	558	554	550
FR Germany	493	490	496	487	493	495	495	495	503	503
Greece	187	188	177	197	201	202	199	199	201	203
Italy	505	517	498	508	531	529	531	533	533	520
Luxembourg	1	1	1	1	1	1	1	1	1	1
Netherlands	108	106	104	103	103	106	106	107	106	104
Norway	39	41	41	39	36	38	38	40	43	51
Portugal	88	89	93	100	102	101	105	104	104	95
Spain	366	372	355	342	314	314	314	304	277	295
Turkey	741	769	824	815	814	860	879	847	780	827
UK	341	334	333	336	334	331	328	324	318	313
European NATO total	**3 586**	**3 624**	**3 639**	**3 638**	**3 630**	**3 669**	**3 693**	**3 651**	**3 560**	**3 603**
NATO total	**5 835**	**5 607**	**5 942**	**5 942**	**5 957**	**6 023**	**6 058**	**5 985**	**5 888**	**5 876**

Sources: NATO publications; author's calculations.

Table 5.12. NATO military and civilian personnel, as share of total labour force, 1981–90

Figures are percentages.

	1981	1982	1983	1984	1985	1986	1987	1988	1989	1990
North America										
Canada	1.0	1.0	1.0	1.0	1.0	1.0	1.0	1.0	0.9	0.9
USA	2.9	2.9	2.9	2.9	2.9	2.8	2.8	2.7	2.7	2.6
Europe										
Belgium	2.8	2.8	2.8	2.8	2.7	2.7	2.7	2.8	2.8	2.8
Denmark	1.7	1.5	1.5	1.5	1.4	1.4	1.4	1.4	1.4	1.4
France	. .	. .	. .	. .	. .	2.9	2.9	2.9	2.8	2.8
FR Germany	2.4	2.4	2.4	2.4	2.4	2.4	2.4	2.4	2.3	2.3
Greece	5.8	5.8	5.2	6.0	6.1	6.0	6.0	6.0	5.7	5.7
Italy	2.5	2.5	2.4	2.4	2.5	2.5	2.4	2.4	2.5	2.4
Luxembourg	0.8	0.8	0.9	0.9	0.9	0.9	0.8	0.8	0.8	0.8
Netherlands	2.6	2.5	2.5	2.2	2.2	2.2	2.2	2.2	2.5	2.4
Norway	2.5	2.6	2.6	2.5	2.3	2.3	2.3	2.3	2.6	2.9
Portugal	2.3	2.4	2.3	2.5	2.6	2.6	2.6	2.6	2.5	2.3
Spain	3.1	3.2	3.0	2.9	2.6	2.5	2.6	2.3	2.2	2.3
Turkey	4.5	4.6	4.8	4.7	4.6	4.9	4.9	4.7	4.5	4.7
UK	2.2	2.1	2.0	2.0	2.0	1.9	1.8	1.8	1.7	1.7
European NATO total	**2.8**	**2.8**	**2.8**	**2.8**	**2.8**	**2.8**	**2.8**	**2.7**	**2.6**	**2.6**
NATO total	**2.8**	**2.8**	**2.8**	**2.8**	**2.7**	**2.7**	**2.7**	**2.7**	**2.6**	**2.6**

Sources: NATO publications; author's calculations.

Table 5.13. Comparative economic and military indicators of the European Community countries, the USA and Japan, 1980

Country	GDP (US $b.)	Population (m.)	Per capita GDP (US $)	1970–80 per capita growth rate (%)	Military expenditure (US $m.)	Armed forces (thou.)	Weapon procurement expenditure (US $m.)
FRG	1 030.5	61.6	16 729	*2.6*	33 807	495.0	5 003
France	810.3	53.9	15 033	*3.0*	32 222	494.7	6 863
Italy	669.9	56.4	11 878	*2.4*	14 174	366.0	2 482
UK	658.3	56.0	11 755	*1.8*	31 100	329.2	8 260
Spain	278.8	37.5	7 435	*3.0*	6 423	342.0	1 265
Netherlands	209.2	14.1	14 837	*2.1*	6 510	115.0	1 178
Belgium	140.2	9.9	14 162	*2.8*	4 614	87.9	663
Denmark	130.3	5.1	20 255	*2.1*	2 235	35.1	360
Greece	50.1	9.6	5 219	*4.0*	2 841	181.5	534
Portugal	33.1	9.8	3 378	*3.3*	1 145	59.5	70
Ireland	27.9	3.4	8 206	*2.4*	525	14.8	56
Luxembourg	5.8	0.4	14 500	*6.8*	60	0.7	1
EC total	**4 017.4**	**317.7**	**12 645**[a]	***3.0***[a]	**135 656**	**2 521.4**	**26 735**
USA	3 851.4	227.8	16 907	*2.7*	206 573	2 050.0	40 281
Japan	2 179.4	116.8	18 659	*4.6*	20 099	241.0	4 174

[a] Average figure.

Sources: SIPRI data base; author's calculations.

Table 5.14. Comparative economic and military indicators of the European Community countries, the USA and Japan, 1989

Figures are in constant (1988) prices.

Country	GDP (US $b.)	Population (m.)	Per capita GDP (US $)	1980–88 per capita growth rate (%)	Military expenditure (US $m.)	Armed forces (thou.)	Weapon procurement expenditure (US $m.)
FRG	1 239.7	61.64	20 112	*1.9*	35 008	503	6 652
France	990.4	56.16	17 635	*1.4*	36 494	554	9 746
Italy	858.9	57.52	14 932	*2.0*	20 559	533	4 091
UK	836.7	57.20	14 628	*2.6*	34 292	318	7 830
Spain	361.5	39.09	9 248	*2.0*	7 583	277	1 065
Netherlands	237.3	14.83	16 001	*1.1*	6 791	106	1 196
Belgium	161.4	9.88	16 336	*1.4*	4 035	110	400
Denmark	108.4	5.13	21 131	*2.2*	2 263	31	296
Greece	46.1	10.03	4 596	*0.9*	3 116	201	682
Portugal	36.2	10.47	3 457	*0.1*	1 415	104	168
Ireland	31.3	3.51	8 917	*1.2*	488	13	41
Luxembourg	8.2	0.38	21 579	. .	79	1	3
EC total	**4 916.1**	**325.84**	**15 220**[a]	***1.8***[a]	**152 123**	**2 750**	**32 170**
USA	4 959.2	249.41	19 884	*2.3*	289 149	2 241	73 244
Japan	2 966.8	123.12	24 097	*3.3*	29 491	249	8 219

[a] Average figure.

Sources: SIPRI data base; author's calculations.

IV. The European Community

In December 1990 the EC Council convened two inter-governmental conferences (IGCs), one on economic and one on political union, of great relevance to future European security. The IGCs are expected to conclude in October 1991, when proposals are to be sent to the national parliaments for ratification. The march towards economic and monetary union is expected to proceed smoothly along the lines laid down in the Delors Report.[20] The progress towards political union is more problematic. There is as yet little agreement among the EC countries on defence matters, and, as the current Gulf crisis has shown, a common security policy is still beyond reach. The Council decided that the union should consider extending its competence in the area of common security and deal with issues such as arms control and disarmament, CSCE matters, economic and technological co-operation in the armament field, co-ordination of arms exports and non-proliferation.[21] However, regardless of central EC policy, the trend towards increasing integration is already set, and military expenditure as well as its components are bound to be affected by the changes.

To clarify the interrelation of security and economic issues, tables 5.13 and 5.14 give data for all EC countries on a number of variables. As a comparison of data for 1980 and 1989 shows, the EC is growing stronger, both in terms of economy and security, highlighting the need for a co-ordinated foreign and defence policy.

As security matters become more important for the EC, a central issue that will have to be resolved is its relationship to NATO. If it is to expand, the EC will likely have to accommodate three neutral countries (Ireland, Austria and Sweden). A coherent defence and security policy will be necessary, even more so if countries of Central Europe, such as Poland and Hungary, are to be allowed to join at some future stage.

Another possible area of dispute is European NATO participation in out-of-area operations, such as in the 1990–91 Gulf War. The UK in particular has shown concern about the costs of its involvement in Operations Desert Shield and Desert Storm. By late January 1991 British operation costs amounted to £3.6 million per day. In addition there were costs accrued through destroyed assets: only to replace the five Tornado strike aircraft lost by 25 January would require an extra expenditure of £105 million. Total British expenditure for the Gulf War was estimated (at an early stage of the hostilities) to be of the order of £2.5–3 billion. There have been increasing demands by the British War Cabinet that the majority of these costs should be met by contributions from EC countries. The FRG has already agreed to defray some of the expenses.[22]

[20] For a discussion of the Delors Report, see Deger, S., 'World military expenditure', SIPRI, *SIPRI Yearbook 1990: World Armaments and Disarmament* (Oxford University Press: Oxford, 1989), p. 158.

[21] *Financial Times*, 17 Dec. 1990, p. 2; *The Economist*, vol. 317, no. 7686/7687 (22 Dec. 1990), p. 27; *Atlantic News*, 19 Dec. 1990, pp. 3–4.

[22] *The Guardian*, 25 Jan. 1991; *Financial Times*, 28 Jan. 1991; *The Guardian*, 30 Jan. 1991.

V. The Soviet Union

After a slow-down in the late 1970s, Soviet military expenditure rose rapidly in the 1980s, exceeding the growth of the national product. The process was accelerated initially by the US spurt in defence spending under the Reagan Administration. It was also sustained by modernization, increased R&D, a new procurement cycle, and military involvement in and aid to the Third World. The military burden rose fast as defence growth exceeded economic growth. The process was halted in 1987, stabilized in 1988 and began to decline in 1989. In 1990 substantial reductions in all categories of forces and expenditures began to be carried out.

Speaking at a defence complex meeting in the Ural industrial city of Nizhniy Tagil in April 1990, President Mikhail Gorbachev said:

> Look at the picture presented by the 11th Five-Year Plan and even the 12th Five-Year Plan. National income was planned to grow by 21–22 per cent, military spending by 45 per cent. Military spending in our country amounted to 18 per cent of national income, more than any other state in the world. It was our foreign policy, which received support and response from all continents and which encouraged positive development throughout the world, which is creating the conditions enabling us to switch these resources to the solution of social problems.[23]

According to preliminary SIPRI estimates, military expenditure grew by over 5.5 per cent per annum in 1980–87, remained approximately constant in 1987–88, fell by 5 per cent in 1989, and is planned to fall by about 10 per cent in 1990. All changes are in real terms, that is, net of inflation. These estimates are based on careful study of Soviet official military data, consistency checks on often contradictory information, as well as numerous statements made by the Government in 1988–90. Considering the trends, the similarity with the USA is striking: an extremely rapid rise followed by stable spending and relatively fast decline. Compared to the USA, the rate of increase of defence spending was faster during the 1980s, but the decline has also been higher.

In 1990 the USSR faced a major economic crisis and strong tendencies towards political disintegration. At the policy-making level the debate was characterized by serious disagreement over how fast to proceed with the marketization of the economy. The radical Shatalin Plan, named after its architect, Soviet economist and former adviser to President Gorbachev, Stanislav Shatalin—to totally change the economic *system* within 500 days rather than proceed with ad hoc *structural* reforms—proved not acceptable to the political leadership, and a compromise programme was adopted.

Official data show that Soviet GNP growth turned negative in 1990, possibly for the first time in 40 years. Soviet economic statistics are suspect; still, it is noteworthy that the State Committee of Statistics (Goskomstat) for the first time stated that the national product had actually *fallen*. While economic growth has been declining over time—a situation often termed growth

[23] *Pravda*, 29 Apr. 1990, pp.1–2, in *Foreign Broadcast Information Service: Soviet Union* (FBIS-SOV-90-083), 30 Apr. 1990, p. 109.

retardation—the output has until recently been rising, albeit at slower rates. The decline was particularly dramatic since *perestroika* was supposed to reverse the economic problems that Gorbachev had inherited from what he had termed the 'pre-crisis' situation of the first half of the 1980s.

Soviet military spending is an explicit function of the economic state of the country, not the least as much hope is placed on the redirection of resources from the military to the civilian sectors. According to the latest available data, from January to September 1990 the GNP fell by 1.5 per cent; the net material product (NMP, GNP minus most service income) by 2.5 per cent; investment by 6 per cent; the inflation rate of the consumer price index was 3.7 per cent; an estimated 2 million were unemployed; strikes occurred in 1700 enterprises; and there were 600 000 internal refugees.[24] In every respect these statistics are worse than for the previous year and reveal a *trend* in economic decline.

As regards the budgetary problems, it was claimed that 'a trend towards reducing the state budget deficit emerged during the first half of the year as a result of the implementation of a package of financial recovery measures, reduction of defence expenditure and increase of certain categories of income'.[25] However, this so-called improvement was due to surpluses in the budgets of the individual republics; the Union budget still showed a deficit of over 25 billion roubles for the period January–July. Oddly, military expenditure was reported to be $41.2 billion for these six months. On a *pro rata* basis this yields over 80 billion roubles in defence spending for the whole of 1990—far in excess of the official budget claim of about 70 billion roubles.

In early November 1990 the Russian Soviet Federated Socialist Republic, the largest and economically and politically most important Soviet republic, launched its own version of the 500-day Shatalin Plan to marketize the economic system.[26] As a model the experiment could have implications for other republics, as well as for the Union. The implementation of the plan—to be carried out in four stages, consisting of 100, 150, 150 and 100 days, respectively—will be watched closely for signs of success and failure. The first stage will entail increasing privatization, including guarantied property rights; partial lifting of price controls on luxury items; introduction of a single commercial exchange-rate for foreign trade; and the control of money supply, coupled with drastic cuts in public expenditure (which includes those on the military). The second stage will entail greater price liberalization, using a supply–demand mechanism for price formation in all but essential goods, and the transformation of state enterprises into joint stock companies, including closure of inefficient plants if necessary (the possibility of defence enterprise bankruptcy remains unlikely). The third stage envisages the abolition of the industrial ministries which now play a management role and a much greater supply of consumer goods. The final stage is the take-off to a capitalist market system, with growing output of light industrial and food products. The last two

[24] 'Results of USSR's socio-economic development during first nine months of 1990', TASS report, 19 Oct. 1990, in FBIS-SOV-90-204, 22 Oct. 1990, pp. 74–75.

[25] *Izvestia*, 21 Oct. 1990, in FBIS-SOV-90-204, 22 Oct. 1990, p. 75.

[26] *International Herald Tribune*, 2 Nov. 1990, p. 2.

stages will be aided by restructuring, with defence enterprises producing increasingly more consumer products, such as appliances and white goods.

Over 70 per cent of the enterprises of the military–industrial complex are based in the Russian Republic, and about one-third of Moscow's industrial output, one-half of its R&D facilities and one-quarter of its industrial work force are military-oriented.[27] Thus, the productivity of resource re-allocation away from the defence sector in Russia is expected to be higher than for the USSR as a whole. The Russian Government has requested the management of defence enterprises to work under its jurisdiction, and to obey its production plans, in return for tax and other economic concessions.[28] However, doubts remain whether Russia, no matter how powerful, will be able to execute the plan on its own within the current framework of the Union. The director of the Kirov plant in the Leningrad region, which produced tanks until January and is considering switching to car manufacturing, has said: 'We are in the absurd situation where Russia has declared sovereignty over all resources on its territory but the Soviet Defence Ministry is still our master'.[29]

The USSR could survive within a federalist structure provided economic prosperity could be given to most of the republics. There are essentially two political models: the USA and the future EC after political union. A loose federation, with a common political structure, defence and foreign policy, but combining the features of an 'economic space' *(ekonomicheskoye prostranstvo)* has also been suggested as an interim solution. Whatever the outcome, cuts in military expenditure to trim the budget deficit, transfers of resources from military to civilian activities, restructuring of the armed forces and transfers in the ownership of the defence industries will all play major roles in the transition. Although developments are uncertain, some possibilities may be mentioned. Given the intractable Union deficit, a reduction of the budget deficit would be welcomed by all republics. A solution to the inefficient and discredited conscription system could be the establishment of a professional army, based on volunteers recruited mainly from Russia and other Slavic republics. The defence industries are mainly located in Russia and the Ukraine; the transfer of ownership to these two republics would not be popular elsewhere. R&D facilities are even more concentrated in regional terms; this national asset would become the property of Russia if decentralization occurs.

Glasnost

Soviet efforts during 1990 in arms control, in fostering a benign foreign policy and in domestic political and economic liberalization are partly motivated by the desire to become an active partner in the international economic community. The USSR wishes to increase trade with Western countries and to participate in the export- and finance-led boom that has typified capitalist

[27] Cooper, J., 'The Soviet defence industry and conversion', *RUSI Journal*, autumn 1990, pp. 51–56.

[28] *Izvestia*, 13 Nov. 1990.

[29] *Financial Times*, 13 Nov. 1990, p. 26.

Table 5.15. Soviet military expenditure and armed forces, official figures, 1989–91

	1989	1990	1991
Military expenditure (b. roubles)	77.3	71	67.3
Armed forces (thou.)	4 258	3 993	3 760

Source: Soviet Government publications; SIPRI data base.

economies in recent years. However, the response of Western governments has been cautious. More important, certain forms of 'political conditionality' have been attached to Western assistance to the USSR. These tend to link economic relations to the domestic political situation as well as to Soviet behaviour in international relations. In this context, Soviet military expenditure has acquired a particularly important role. In 1990 the Group of Seven (G7) leading industrial countries[30] made it very clear in their summit meeting declaration that they expect Soviet defence spending to go down fast if greater economic assistance was to be provided. The Houston Economic Declaration of July 1990 stated: 'We also agreed that further Soviet decisions to introduce more radical steps towards a market-oriented economy, *to shift resources substantially away from the military sector* and to cut support to nations promoting regional conflict will all improve the prospect for meaningful and sustained economic assistance'.[31]

Although the era of negotiated quantitative arms control may be passing, it will remain significant. What is becoming more important are the *perceptions* that influence the behaviour of the state actors in international relations. Nations will be able to live more peacefully if they perceive a more benign environment and a lower level of threat emanating from potential adversaries. Confidence-building measures and the role of defence doctrines will become increasingly crucial. In the specific case of the USSR, greater transparency and faster reduction of military expenditure will enhance perceptions regarding stability, and in the process enhance security. It is therefore particularly important now to understand the detailed mechanics of Soviet defence data and evaluate them independently.

Since 1989 the USSR has been providing relatively detailed accounts of its military expenditure, armed forces and weapon assets. Table 5.15 gives data for military expenditure and armed forces planned for 1989–91. Although the aggregate figure is much higher than earlier, implausible values, controversy continues as to the level of Soviet defence spending. Western intelligence estimates, on which US and NATO perceptions are built, still give much higher levels. However, as discussed below, there is agreement that Soviet military spending is rapidly declining. All major categories (personnel, procurement and R&D) are being cut, to a greater or lesser extent. There is also wide agreement that the size of the armed forces revealed by the

[30] The Group of Seven includes Canada, France, the FRG, Italy, Japan, the UK and the USA.

[31] *Houston Economic Declaration*, Declaration of the 1990 Economic Summit of Industrialized Nations, Houston, Texas, 11 July 1990, para. 44; emphasis added.

Table 5.16. Soviet active forces, January 1990

Category	Number
Strategic nuclear forces	
ICBM launchers	1 398
equipped with MIRVs	760
SLBM launchers	924
equipped with MIRVs	440
Heavy bombers	162
configured to carry ALCMs	97
Total	**2 484**
Total nuclear warheads	10 000
Conventional forces	
Combat aircraft	8 207
Combat helicopters	4 014
Tactical missile launchers	1 723
Tanks (including amphibious crafts)	63 900
Armoured personnel carriers	76 520
Multiple rocket launchers, etc.	66 800
Submarines	260
nuclear powered	113
Large surface ships	157
aircraft-carrying	4
assault landing	41
Armed forces	3 993 000

Source: *Izvestia*, 16 Dec. 1989.

Government is of the correct order of magnitude. Table 5.16 gives information on weapon assets held at the beginning of 1990.

In many countries—particularly, as shown above, the USA—budgetary and financial constraints can be used specifically by the Parliament or Congress to impose discipline on the military sector and thus achieve a form of unilateral arms control. In the USSR the situation has been very different.[32] The MOD is responsible for military personnel and O&M spending only. The 20 billion rouble budget, often quoted prior to 1989, was precisely limited to meet these requirements. Pensions were paid from special funds related to social security.

As regards weapon procurement, the MOD is a customer of enterprises in the military–industrial complex. Since 1989 this has consisted of six ministries for arms production (aviation, defence, electronics, general machine-building, radio and ship-building); two general ministries (for civil aviation and communications); and a State Committee for Computing and Information. In an earlier re-organization in 1988 the Ministry of Machine-Building for Light and Food Industry and Household Appliances was transferred from the civilian to the military–industrial sector. In 1989 the Medium Machine-Building Industry

[32] Alexander, A. A., *Perestroika and Change in Soviet Weapons Acquisition*, Rand report R-3821-USDP (Rand Corp.: Santa Monica, June 1990).

(nuclear weapons and warheads) was moved from the defence sector to the 'fuels and energy industrial complex'. Thus the structure was made similar to that in the USA, where nuclear weapons are the prerogative of the DOE. Despite the fact that the MOD is the only buyer of military products (including exports), the financial allocations are made from the state budget in general. The Ministry of Finance opens credits in the central Bank (Gosbank), from which plants and enterprises draw funding for operating expenditure.

This procedure has two curious consequences. First, it allows for cost inefficiency, since the MOD is primarily concerned with product specification, not that of costing. Second, if expenditure exceeds allocation, subsidies can be provided by the Government by moving funds from other sectors or ministries or even by allocating funds in excess of existing budgets. If enterprises run up deficits these can be met by issuing further credits from the *Gosbank* that are balanced by increasing money supply. Military expenditure on procurements can also be kept artificially low through accounting artefacts.

There is even more financial independence in the conduct of military R&D. The approximately 200 design and research institutes directly involved in military research (excluding basic research, which is done under the auspices of the Academy of Sciences) work in close co-ordination with customers and buyers. However, neither the MOD nor the industrial ministries are directly involved in the financing, which is covered by state funds. As General Vitaliy Shabanov, Deputy Minister of Defence for Armaments, has said: '[R]esearch and development projects for the creation of new weapons are state contracts. Only a small part of such work was done under direct contracts with the Defence Ministry. This system has both positive and negative elements . . . It is negative that the Defence Ministry does not have economic leverage to influence the progress of research and development'.[33]

Again, the contradictory nature of financial inefficiency coexisting with the possibility of (artificially) low expenditure is clear enough. The MOD now wishes to control expenditures through a proper defence budget, and to use its market power as a monopsonist to control pricing. It is possible that by 1991 such changes will be effected, bringing Soviet procurement in line with most other developed countries. This will contribute to greater financial efficiency and 'value for money', although certain problems (such as the creation of natural monopolies) will persist. More financial discipline at a time of budgetary restrictions will incite enterprises to search for measures to cut costs. There is reason to believe that a leaner but stronger military industry might emerge from this restructuring.

SIPRI measures the rouble values of Soviet defence spending through an extensive analysis of financial and economic data pertaining to the four major categories of aggregate military expenditure. Such financial data may well understate the resource costs to the economy. In addition, the opportunity costs of resource diversion from the civilian sector consequent to a defence buildup is not easy to estimate as each rouble has quite different productivity

[33] Lebedev, Y., 'On the Principles of the Defensive Military Doctrine, interview with General of the Army Vitaliy Shabanov', Novosti, 23 Aug. 1990, in FBIS-SOV-90-190, 1 Oct. 1990, p. 62.

in the two sectors. However, such problems are not specific to the military economy, but are intrinsic to the Soviet system as a whole. According to the most recent SIPRI estimates, the USSR spent around 100–105 billion roubles in 1989. This is about one-third more than official data would indicate, but substantially less than estimates made by US intelligence agencies. Based on official *economic* statistics, this would give a defence burden share of 11–12 per cent of GNP and 16–17 per cent of NMP in 1989. These shares represent a downward trend from the peak reached in 1987.

As regards the dollar value of Soviet military expenditure, SIPRI estimates give a figure of $263–$276 billion in 1989 prices. The average of around $270 billion is lower than the estimate for the USA. This contradicts the claim by the Central Intelligence Agency (CIA) that the USSR outspent the USA at least until 1988 (the latest year for which figures are available). In part the discrepancy can be explained by the use of different conversion rates. More importantly, however, the means and items of comparison differ. SIPRI estimates consider existing Soviet forces, with rouble costs expressed in dollars, using purchasing power parities estimated by SIPRI. The CIA measure considers the dollar cost of replicating the Soviet armed forces in the USA. In other words, the maintenance costs of Soviet conscripts are calculated on the basis of the wage costs of the equivalent number of US volunteer forces. If the same method were to be used to cost other conscript forces, such as German or French, the military expenditure estimates for these countries would be higher as well.[34] In the 1989 SIPRI *Yearbook* a similar costing method was in fact used, in the context of NATO burden-sharing, to show that the European NATO countries had a higher defence burden than the USA.

To sum up, until 1989 Soviet military expenditure was very high. In relative terms, measured as military burden, it was at least twice that of the USA, or around 12 per cent of GNP. If Soviet statistics on the national product are flawed, being lower than officially claimed, then this burden could be three times that of the USA, possibly around 18 per cent. If indirect costs—such as state subsidies to defence industries, opportunity costs of priorities allowed to the military sector, and loss (through conscription) of skills and personnel with high productivity in civilian sectors—are added, the share of the defence sector could have exceeded 20 per cent of GNP. There is no example in recent history of a society that has allocated so much to its military during peacetime.

Perestroika

Since around 1988–89 Soviet military expenditure has fallen rapidly. Despite major economic, social and military difficulties, restructuring has taken place.

The breakdown of Soviet military expenditure into its constituent parts (personnel, procurement, O&M, pensions, construction, nuclear weapons and R&D) reveals a number of interesting issues. First, it shows the political perceptions of the leadership as it decides, for example, whether one category of

[34] The CIA method is similar in terms of Soviet dollar costing.

spending should be favoured at the expense of another, or what the internal trade-offs are between various parts of the military budget. Second, the nature of the cuts or increases indicates whether the Government wishes to reduce or protect the importance of the category in question. Third, the amount of non-financial resource transfer—skilled labour, engineers and scientists, and industrial facilities—will be determined by the evolution of each category.

In 1990 the USSR for the first time submitted to the United Nations a detailed categorization of defence spending according to the standardized format of the UN Reduction of Military Budget Programme (UNRMB). This format consists of a matrix giving, within a consistent accounting framework, a detailed, element by element, financial value, as shown in table 5.17.

Analysts agree that the USSR initiated substantial cuts in defence spending in 1988. In their 1990 joint annual Report to Congress, the CIA and the Defense Intelligence Agency (DIA) claim that the decline was on the order of 4–5 per cent in 1988–89.[35] If anything, the process accelerated in 1990.

Konversiya

The best way to construct an analytical framework to analyse resource transfer is to look at the categories of military expenditure and observe the qualitative and quantitative indicators of the 'peace dividend'.

As regards personnel, the force reduction of half a million men is expected be completed by the beginning of 1991 when the Soviet armed forces will have a strength of 3 760 000. This number excludes the approximately 500 000 security forces (300 000 internal troops and 200 000 border guards). The 12 per cent reduction may be only the beginning; the USSR may need to reduce troop strengths even further, due to major problems with manning.

It is becoming increasingly difficult for Moscow to maintain an adequate supply of conscripts. Continuing ethnic violence in the non-Russian republics, along with strident demands for autonomy and independence, has made it problematic to fill up conscription quotas and even to safeguard the cohesion of the armed forces. The use of the Army to preserve law and order cannot help but have a negative effect on morale; indiscipline is rife. The end of the war in Afghanistan has posed the same sort of disciplinary problems that the US armed forces faced at the end of the Viet Nam War. There is no permanent corps of non-commissioned officers (NCOs) that can interact between officers and conscripts. Educational standards are inadequate, particularly for non-Slavic conscripts, making it increasingly difficult for the armed forces to function in the high-technology environment of modern weapon systems.

General Mikhail Moiseyev announced in November 1990 that the armed forces will be reduced to 3 million by the year 2000.[36] In addition, 1300

[35] 'The Soviet economy stumbles badly in 1989', Paper presented by the Central Intelligence Agency and the Defense Intelligence Agency to the Technology and National Security Subcommittee of the Joint Economic Committee, US Congress, 20 Apr. 1990.

[36] *Krasnaya Zvezda*, 18 Nov. 1990; *Financial Times*, 19 Nov. 1990.

generals, 220 000 other officers and 250 000 NCOs will be cut. Substantial pay rises have been offered that will be reflected in the 1991 personnel budget.

Following the precedent set in the USA in the early 1970s, the idea of having an all-volunteer army is vigorously being debated in the USSR. The main constraint is economic—it would be impossible to pay adequate salaries to professional soldiers at current manning levels. The USSR has twice as many soldiers as the USA, yet military personnel expenditure is about 12 per cent of the total; in the USA the corresponding share is about 26 per cent (both figures are for 1989). If a volunteer force is to be created, the size of the armed forces needs to be brought down to the US level of about 2 million. Given the 'long overdue need to improve the material and living conditions of servicemen and members of their families',[37] pay rates have in fact been raised, and the Supreme Soviet has approved an additional 1 billion roubles to improve wages and social facilities. Housing shortages, exacerbated by the withdrawal of forces from Eastern Europe, is a major problem. Construction spending will need to rise after a sharp fall in 1989–90. Despite the need for budgetary austerity, pension fund spending has already risen as well, and this trend is likely to continue as more soldiers are retired. About one-fifth of the forces cut are volunteer officers who will need compensation for early retirement.

Although aggregate expenditure rose rapidly in the first half of the 1980s, procurement spending on major weapons, particularly for conventional forces, did not increase commensurately. According to Soviet sources, a major increase in procurement spending was not envisaged until the 12th Five-Year Plan (FYP, 1986–90), which called for a 5–5.5 per cent per annum rise.[38] The situation has changed radically, however, and since 1987 or 1988 weapon spending has fallen at an increasingly accelerated rate. In part in response to the CFE Negotiation, but mainly as a consequence of changed doctrines and threat perceptions, significant cuts have been made in weapon systems such as tanks and APCs. The 40 per cent reduction in tank production announced by Marshal Sergey Akhromeyev before the US Congress House Armed Services Committee[39] has even been accepted by the DIA, although the latter insists that levels are still too high. According to Marshal Akhromeyev, in 1989 about 1700 tanks were produced; the figure for 1990 could drop to 1000 if the postulated cut is made. These figures include exports, around 300 tanks per annum according to independent estimates.[40] The production of fighters, bombers and fighter/bombers continued to decline steadily, accelerating the

[37] *Krasnaya Zvezda*, 14 Oct. 1989, in FBIS-SOV-89-199, 17 Oct. 1989, p. 55.

[38] Vid, L., 'Guns into butter, Soviet style', *Bulletin of the Atomic Scientists*, Jan.–Feb. 1990, pp. 17–19.

[39] *Soviet Views on National Security Issues in the 1990s*, Hearings before the Committee on Armed Services, US House of Representatives, 100th Congress (US Government Printing Office: Washington, DC, 21 July 1989).

[40] See analysis of Akhromeyev's speech in *New York Times*, 22 July 1988; also *Soviet Views on National Security Issues in the 1990s* (note 39). According to a French study prepared for the National Assembly, *Rapport sur le projet de loi de finances pour 1989, Annexe*, 1600 tanks per year were procured in 1980–88; according to the Defence Intelligence Agency, 3300 tanks were procured in 1988 as 'Production for the Soviet Military', but this estimate seems too high; see *Allocation of Resources in the Soviet Union and China*, Hearings before the Joint Economic Committee, 101st Congress, 14 Apr. and 7 July 1989 (US Government Printing Office: Washington, DC, 1990), p. 157.

Table 5.17. Soviet military expenditure, detailed submission to the United Nations, 1989

Figures are actual outlays in m. roubles, at 1989 prices.

Resource costs	Land forces	Naval forces	Air forces	Other combat forces	Central support, admin. and command		Para-military forces	Military assistance			Un-distrib-uted	Total milex	Civ. def.
					Support	Command		Home territory	Abroad	UN peace-keeping			
Operating costs	9 950	2 737	2 226	2 540	2 506	121	1 113	..	..	..	2 239	23 432	114
Personnel	4 943	1 241	1 296	1 626	..	94	537	..	..	..	2 239	11 976	61
Conscripts	..	..	..	..	..	..	..	..	..	..	..	..	..
Other military, incl. reserves	3 393	953	1 092	1 392	..	82	336	..	..	..	2 239	9 487	..
Civilian	1 550	288	204	234	..	12	201	..	..	..	..	2 489	61
Operations and maintenance	5 007	1 496	930	914	2 506	27	576	..	..	..	..	11 456	53
Materials	2 824	450	144	229	1 632	6	268	..	..	..	..	5 553	14
Maintenance	1 158	931	732	486	..	6	252	..	..	..	..	3 565	..
Purchased services	1 025	115	54	199	874	15	52	..	..	..	..	2 334	..
Rent costs	..	..	..	..	..	..	..	..	..	..	..	..	..
Other	..	..	..	..	..	..	..	..	..	..	..	4	39
Procurement and construction	10 093	7 193	7 360	6 771	5 521	..	740	..	..	..	2 031	39 709	9
Procurement	9 273	6 531	6 941	5 465	4 361	..	637	..	..	..	2 031	35 239	..
Aircraft, engines	..	210	1 845	949	..	..	94	..	..	..	..	3 098	..
Missiles (incl. conventional warheads)	995	695	562	1 621	..	..	..	..	..	..	..	3 873	..
Nuclear warheads and bombs	..	..	..	..	..	..	..	..	..	..	2 031	2 031	..
Ships and boats	..	2 993	..	..	..	..	205	..	..	..	..	3 198	..
Armoured vehicles	2 137	..	..	..	2	..	30	..	..	..	..	2 169	..
Artillery	418	12	..	..	..	..	1	..	..	..	..	431	..
Other ordnance	819	..	..	..	36	..	17	..	..	..	..	872	..

Ammunition	2 256	487	757	..	..	..	30	..	..	..	..	3 530	..
Electronics, communications	779	495	927	767	2 591	..	185	..	..	..	..	5 744	..
Non-armoured vehicles	..	..	..	..	..	..	50	..	..	..	..	50	..
Other	1 869	1 639	2 850	2 128	1 732	..	25	..	..	..	..	10 243	..
Construction	820	662	419	1 306	1 160	..	103	..	..	..	..	4 470	9
Airbases, airfields	..	..	..	..	..	..	..	..	..	..	..	..	..
Missile sites	..	..	..	..	..	..	..	..	..	..	..	..	..
Naval bases and facilities	..	..	..	..	..	..	..	..	..	..	..	..	..
Electronics, etc.	..	..	..	..	..	..	..	..	..	..	..	..	..
Personnel facilities	..	..	..	..	..	..	..	..	..	..	..	..	..
Medical facilities	..	..	..	..	..	..	..	..	..	..	..	..	..
Training facilities	..	..	..	..	..	..	..	..	..	..	..	..	..
Warehouses, depots, etc.	..	..	..	..	..	..	..	..	..	..	..	..	..
Command and administration facilities	..	..	..	..	..	..	..	..	..	..	..	..	..
Fortifications	..	..	..	..	..	..	..	..	..	..	..	..	..
Shelters	..	..	..	..	..	..	..	..	..	..	..	..	..
Land	..	..	..	..	..	..		..	..	..	..	..	..
Other	..	..	..	..	..	..	..	..	..	..	..	..	..
Research and development	996	2 160	2 734	7 484	140	394	9	..	..	..	250	14 137	..
Basic and applied research	..	..	..	..	..	..	..	..	..	..	..	..	..
Development, testing and evaluation	..	..	..	..	..	..	..	..	..	..	..	..	..
Total	21 009	12 090	12 320	16 795	8 167	515	1 862	..	673	..	4 520	77 951	123

Source: United Nations, 'Union of Soviet Socialist Republics, instrument for standardized international reporting of military expenditure, fiscal year 1 January–31 December 1989', document no. A/INF/45/5/add.1 (UN: New York, 12 Oct. 1990), pp. 8–14. The UN format has a column for strategic forces. In the Soviet submission, however, it is claimed that the Soviet strategic forces do not have a clearly designated structure, falling as they do within various branches of the armed forces, and that therefore no figures are available for this entry.

trend present for most of the decade. According to the DOD, the current annual production rate of Soviet fighters and fighter/bombers is half that of 1980. With increasing interest in aircraft exports, domestic procurement is bound to drop significantly. Even though military helicopters are not included in the CFE Treaty's treaty-limited equipment (TLEs), procurement of helicopters has been reduced by 30 per cent compared to the early 1980s.

The picture is mixed in terms of naval assets. The USSR has repeatedly signalled its strong support of naval arms control. Reports of naval shipyards converting production to build merchant marine ships are well documented. The famous Baltiyskiy Zavod shipyard, where the Kirov Class cruisers were built, has now ceased production of naval vessels.[41] All nuclear-armed submarines have been withdrawn from the Baltic,[42] and there is less interest in 'showing the flag' in the Mediterranean and the Far East. The US Navy has begun to withdraw from the former US Cam Rahn Bay naval base in Viet Nam, which in turn, according to reports, has inofficially invited the USA to lease the base again.[43] At the same time, the DOD claims that Soviet production of major surface warships has actually increased in 1989. Due to the ease of verification, DOD information tends to be more reliable on shipbuilding and naval assets than on other categories of weapons such as tanks.[44]

Reduction in procurement allows a direct transfer of resources through utilization of excess capacity in the military–industrial complex (as described above) and in civilian industrial ministries that produce for the military sector. This process, termed industrial conversion, has speeded up dramatically since 1988. It should be noted, however, that even in 1988 the share of civilian products in the total output of the military–industrial sector was around 40 per cent according to Premier Nikolai Ryzhkov,[45] and the share is rising quickly. The precise figures are contradictory and tend to change with time, partly because of the re-organization of the military–industrial complex itself and partly because of unpredictable and differential inflation rates. However, by 1990–91 civilian output is expected to be 50 per cent, and by the end of the 13th FYP in 1995 at least 60 per cent and possibly 65 per cent of the total.

The industrial conversion plan has set up 10 major areas for the 'civilianization' of the defence industry: food processing and agricultural machinery; textile manufacturing machinery; equipment for public catering; consumer durables, particularly white goods; electronics; computers; medical equipment; communications equipment; civilian aircraft and parts; and civilian shipbuilding. The list gives a clear indication of the shortages in the economy.

Three types of industrial conversion are taking place, and all are relevant to the issue of resource transfer.

1. A few enterprises are being totally changed to cater for civilian production alone. Thus the possibility of 'de-conversion', often feared by

[41] Moscow radio service, 30 Sep. 1990, transcript in FBIS-SOV-90-190, 1 Oct. 1990, p. 62.
[42] Moscow television service, 4 Oct. 1990, transcript in FBIS-SOV-90-196, 10 Oct. 1990, p. 88.
[43] *Washington Times*, 10 April 1990.
[44] US Department of Defense, *Soviet Military Power 1990* (DOD: Washington, DC, 1990), p. 38.
[45] See *Allocation of Resources in the Soviet Union and China* (note 40).

Table 5.18. Soviet industrial conversion of three enterprises announced in the 1988 UN speech by President Gorbachev, total conversion by 1990

Figures are percentages.

Enterprise	Military output	Civilian output	Civilian share of total production in 1988	Share of workers employed in civilian production in 1988
Uryuzan engineering works	Cartridges	Refrigerators, chains	81.0	86.0
Ioshkar-Ola engineering works	Fuses	Bicycles, technical equipment	62.2	58.3
Leninska Kuzuiba shipyard, Kiev	ASW ships	Trawlers, dredgers	80.0	82.0

Sources: 'Reply of the USSR to the ILO questionnaire on the conversion of manpower employed in the armaments industry and related activities', Research Working Paper no. 16, World Employment Programme, International Labour Organization, Geneva, 1989; author's estimates.

Western analysts, is to be eliminated in some plants and complexes. President Gorbachev had in his December 1988 UN speech briefly announced that three such plants were to be fully converted. More details are now available; these are summarized in table 5.18. It is important to note that in 1988 the three enterprises had a large volume and share of civilian output, as indicated by the last two columns. This first step in total conversion was thus relatively easy. Six other military enterprises are to go through the same process, including the Baltiyskiy Zavod shipyard and the Kirov tank factory in Leningrad. Full details are not available.

2. As regards partial conversion, in which enterprises would maintain some military production, a large body of literature charting the extent of this venture is now available. The military–industrial complex is now a major producer of light industrial goods as well as intermediate investment products (machineries) for other sectors such as catering, agriculture and food processing. However, the difficulties and problems of implementation have been formidable, and the scenario for the future is not altogether optimistic.

3. Seven civilian industries in the Machine-Building and Metal-Working (MBMW) sector have been producing for the military, including enterprises taking part in the production of armoured vehicles and missile launchers.[46] These are being converted more rapidly. In the civilian sector a total of 34 enterprises are to be totally converted, and all defence production will cease.[47]

[46] *Allocation of Resources in the Soviet Union and China* (note 40), p. 136.

[47] Andreev, V., 'Conversion in the USSR: first steps and results', Paper presented at the Conference on Science and World Affairs, 40th Annual Pugwash Conference, Egham, UK, 15–20 Sep. 1990.

The practical problems of conversion are now well documented. Socialist economies are good at meeting specific targets—and achieving production plans—when large amounts of resources and inputs are utilized. The system is inherently wasteful in a financial sense, although impressive gains can be made in terms of output. When priority areas are identified, resources are typically thrown in 'too much too soon'. All of this is contrary to the logic of civilian production, requiring relatively cheap and rapid production runs without wasting too much money, men and material. This the defence industry has yet to learn. It is possible that the privatization of enterprises and the transformation of the MOD from owner to customer might increase efficiency.

As regards military R&D, the most technologically sophisticated sector of the Soviet economy, transfer of resources is in full swing. In the late 1980s the USSR and the USA were the largest R&D spenders, far in excess of any other country; their combined expenditure was 80 per cent of the world total. The USSR also led the world in terms of the number of scientists and engineers involved in military research.[48] After decades of priority funding, absorbing the best manpower and resources—a one-way street of technological transfer from civilian sectors without much in the way of corresponding spin-offs—the military R&D sector in the USSR was in 1990 fully involved in civilian and/or dual-technology research. The 1989 budget allocation of 15.3 billion roubles represents a real reduction after a decade of growth. However, not all of this money was spent; the actual expenditure was 14.1 billion roubles (see table 5.17). According to preliminary budgetary estimates, Soviet military R&D is expected to fall by 13 per cent between 1989 and 1990—far more than the US Government proposed for military R&D in the USA during the same period.

The thorny question of conversion costs remains. Some Soviet analysts, particularly in the military community, point at numerous social costs, such as unemployment, the loss of privileges, the movement of skilled personnel and possibly even the destruction of the country's scientific and technological base. It is not possible yet to evaluate seriously such concerns and to balance the costs against the obvious advantages and benefits of resource transfer. From a purely financial point of view, some direct costs will initially have to be incurred to protect wages, retool factories and subsidize those enterprises needing more time to adjust. In 1990 expenses for wage protection have initially been put at 350 million roubles. This may have risen to 1 billion roubles by the end of the year. Estimates by *Gosplan* show that around 10 billion roubles will be required for new machineries and equipment until the end of the 13th FYP (1991–95). Although the costs are high, the rewards of resource transfer should under these conditions be even higher. However, it should be noted that according to data emanating from within the military–industrial complex, 63 billion roubles will be required if it is to convert successfully. This sum seems too high and may be politically motivated, but if it is true the costs will cancel any gains in the foreseeable future.

[48] Deger, S. and Sen, S., 'Re-orientation of military R&D towards civilian applications', Paper presented at the Conference on Science and World Affairs, 40th Annual Pugwash Conference, Egham, UK, 15–20 Sep. 1990.

Table 5.19. Disaggregate military expenditure for Poland, 1989

Figures are in b. current zlotys; percentage shares are in brackets.

Category	Value	
Personnel (including civilian pay)	1 246	(*57.8*)
Operations and maintenance	380	(*17.6*)
Procurement (including research and development)	500	(*23.2*)
Construction	28	(*1.3*)
Total	**2 154**	(*100*)

Source: *Polish Army Facts and Figures* (Polish Ministry of National Defence: Warsaw, 1990).

The main difficulty in resource re-allocation is not economic but political. If the Union disintegrates and the republics become more autonomous the role of the military would be questioned. If foreign policy and defence remain with the central authorities, military production and research are concentrated in the RSFSR and a (predominantly Slav) professional armed force is created, the nature of and debate on Soviet military expenditure will need to be altered.

VI. Central and Eastern Europe

With the disintegration of the WTO it is now more appropriate to discuss the NSWTO countries under the headings of Central Europe (Poland, Czechoslovakia and Hungary) and Eastern Europe (Bulgaria and Romania). In 1990 the problems for these countries were rather economic and political than military or security-oriented.[49]

Burden-sharing analyses indicate that the USSR tended to bear the larger part of costs associated with WTO collective security. Since the WTO is not an integrated alliance, and since all Soviet troops are scheduled to leave the three countries of Central Europe, possibly by 1991, the question of who will replace them will have to be faced in the future. The broader region is politically unstable, with the threat of disintegration hanging over the USSR and Yugoslavia, and with minority problems creating friction within and between a number of countries. Economic developmental failures raise the question of the legitimacy of governments and, in a longer perspective, even that of states. The potential for conflict remains high. Until a proper peace order is created the issue of defence spending will remain high on the agenda.

Aside from data made available in the course of the CFE Negotiation, information is still lacking about military spending and capabilities, particularly for Bulgaria and Romania. This is partly due the chaotic state of public finances and the difficulty of adapting to statistical *glasnost*. Transparency has never been a strong feature of socialist countries, while foreign observers have been less keen in their scrutiny of Eastern Europe than of the USSR.

Disaggregate data for Polish and Czechoslovak military expenditures are available for the first time (see tables 5.19 and 5.20). For Poland, the figures

[49] The economic issues are discussed in chapter 6 in this volume.

Table 5.20. Disaggregate military expenditure for Czechoslovakia, 1989

Figures are actual outlays in b. current korunas; percentage shares are in brackets.

Category	Value	
Military personnel	8.3	(*23.6*)
Operations and maintenance (including civilian pay)	11.4	(*32.5*)
Procurement	12.2	(*34.8*)
Construction	1.8	(*5.1*)
Research and development	1.3	(*3.7*)
Total	**35.1**	(*100*)

Source: UN Instrument for standardized international reporting of military expenditures; author's estimates.

for 1989 are based on national definitions; the categories are not strictly comparable to those used by SIPRI. The four categories presented in table 5.19 correspond to the following definitions in the Polish White Paper: personnel upkeep; training, exploitation of armament and installations and force support costs; armament and technical equipment; and investment. The most interesting feature of the disaggregate figures is the high share of personnel expenditure. Even though it includes MOD civilian personnel, the share of almost 58 per cent is remarkable, in particular for a conscript army. Conscripts amount to two-thirds of all manpower. Weapon procurement (including R&D, which is not insignificant for the defence industry) is correspondingly small. This could mean that the Soviet Army was supplying some weapons at subsidized prices. Alternatively, the figures represent the major cuts made in 1989 by the new Government. It would be useful for analysts to receive detailed figures for longer periods of time so that trends could be established.

Military spending in 1990 has been announced to be 10 083.4 billion zlotys. This represents a fivefold rise over the figure for 1989. It is not possible to estimate whether there has been a real decline, since the inflation estimates are very uncertain. It is clear, however, that the change is not as dramatic as it was in 1988–89, when military spending fell by over 10 per cent. It will not be easy to maintain rapid reductions in light of the uncertain developments in the USSR and elsewhere. Polish defence expenditure will be modified not only by military threat perceptions but also by regional stability considerations of a more general nature.

After a considerable rise in the early to mid-1980s, Polish defence spending has since fallen consistently in real terms. As in the USSR, the turning-point seems to have been 1987. As a proportion of the GDP the military burden was in 1986, 3.6 per cent; in 1987, 3.4 per cent; in 1988, 3 per cent; and in 1989, 2.8 per cent. The defence share of total government expenditure was in 1986, 8.3 per cent; in 1987, 8.5 per cent; in 1988, 7.7 per cent; and in 1989, 6.3 per cent. For both measures of defence effort the trend has been downward.

Manpower has been slashed in 1989–90 and is expected to be cut by 25 per cent, with a level of just over 300 000, in 1990. Conscription time has been reduced, and there have been cuts in procurement. Precise figures are not available, but indirect information, such as reduced orders to defence industries and the presence of substantial excess capacity (60–80 per cent), clearly show that weapon purchases have fallen. Some assets have been destroyed: a submarine and 80 aircraft. Thus O&M spending is declining.

Czechoslovak data for 1989 disaggregated military spending are derived from the Government's first ever UNRMB submission in 1990. The level of detail allows a full inspection of defence spending and all its components. Table 5.20 has been constructed from the UNRMB data. Operating costs (personnel and O&M) account for about 56 per cent, with capital costs (procurement, R&D and construction) accounting for the remaining 44 per cent. This is similar to spending figures for some West European armed forces. Major weapon purchases constitute about 20 per cent of aggregate spending—a high figure compared to European NATO. This could reflect the role of the Czechoslovak defence industry, which was one of the largest in the WTO.

In nominal terms the Czechoslovak military budget increased by 7 per cent in 1990. In the past, inflation has been small so that this would constitute a real increase. With the increasing pace of economic reforms, however, inflation is expected to be high in 1990. Hence military expenditure will fall, but, as in Poland, the decline will not be as dramatic as in 1989.

The same type of cuts are occurring in Czechoslovakia as elsewhere in the region: cuts in manpower, conscription, assets, procurement and O&M spending, and conversion of defence industries to civilian output to use excess capacities resulting from lower demands for weapons at home and abroad.

VII. The Asia–Pacific region

Political change and arms control are playing important roles in the Asia–Pacific region as well, although perceptions are moving faster than events. Major political changes, both in terms of domestic policy and foreign relations, were actively discussed in 1990. China has an entente with the USSR; Viet Nam has removed its troops from Cambodia; Japanese–Soviet relations are improving; and President Gorbachev's state visit to Tokyo in 1991 is expected to produce significant results, particularly in terms of the last outstanding dispute of World War II—the return to Japan of the Soviet-occupied Kurile Islands. Unification of the two Korean republics, although distant, is being discussed informally. Suggestions have been made to institute some confidence- and security-building measures, a specific proposal being to form a Conference of Security and Co-operation in Asia (CSCA).[50]

In terms of defence capability and military expenditure the prospects are mixed. Over the last few years there has been some 'spontaneous' arms control, with China and more recently Viet Nam taking the lead in reducing

[50] *Far Eastern Economic Review*, vol. 150, no. 50 (13 Dec. 1990), pp. 25–32. For concrete proposals on the CSCA, see Deger, S., 'Research agenda for defence, disarmament and development—a perspective in 3-D', Paper prepared for the International Conference on Defence and Development Insights from South East Asia, Bangkok, Jan. 1990.

their large, standing armies. However, defence spending *growth* for East Asia (Japan excluded) is one of the regional highest in the developing world. Not counting Chinese demobilization, the number of soldiers has increased in the region. Arms production—aided by greater procurement, activist industrial policies and export promoting industrialization—is increasing, in particularly in countries such as South Korea and Singapore. In 1990 the most significant trends are to be observed in China and Japan, discussed in some detail below.

China

China is one of the poorest countries in the world, with per capita income in 1980 estimated to be $340. However, since then market reforms, initiated in 1979, have led to phenomenal growth rates, with a per annum rate of growth of GDP of over 10 per cent throughout the 1980s. It has in 10 years managed to double the per capita income, and hopes to do so again by the year 2000. However, this rapid expansion has brought with it large-scale inflation and unemployment, both non-existent before 1980. In 1988 a tightening of the economy began, with contractionary policies to control inflation rates exceeding 20–25 per cent. Popular demands for political pluralism, following the liberalization of the economy and the concommitant opening up to the rest of the world, were violently suppressed by the Government in 1989 and 1990.

The military has not been immune to these changes, which resulted in giving it lowest priority. A transfer of resources from the military to the civil economy began, whereby the military would not be able to expand until economic growth was achieved. A fundamental change in military doctrine was the abandonment of the Maoist concept of 'people's war' in favour of that of 'people's war under modern conditions'—with the emphasis on 'modern'. Military expenditure was reduced throughout the 1980s, the share of defence spending in the central budget and in national income went down, procurement was curtailed or postponed, the armed forces were cut and foreign commitments were shelved. The military industries were streamlined, defence enterprises shifted to civilian production to utilize excess capacity, outfits run by the military (such as railways and bases) were turned over to civilian authorities and scientific resources were transferred to the civilian sector.

A considerable amount of information has become available in China on defence matters. In particular defence economics is an expanding science, with many discussions taking place on the cost–benefit aspects of security.[51] However, there is still extreme reluctance on the part of the Government to publish details of the military budget. Official announcements are confined to single-line entries in the state budget, and it is widely believed that reported military spending is an underestimate. SIPRI estimates show that defence spending is almost double the reported value, and could have exceeded 53 billion yuan in 1990. This amounts to a value of over $11 billion in 1988

[51] Yang, R. H. (ed.), *Yearbook on PLA Affairs 1988–89* (Sun Yat-sen Centre for Policy Studies: Kaohsiung, Taiwan, 1989).

Table 5.21. Chinese military expenditure, official figures, 1955–90

Year	Current b. yuan	Constant (1988) b. yuan	Constant (1988) US $b.	As share of central government expenditure (%)	As share of national material product (%)
1955	6.5	12.2	3.3	*24.1*	*8.2*
1956	6.1	11.5	3.1	*20.0*	*6.9*
1957	5.5	10.3	2.8	*18.1*	*6.1*
1958	5.0	9.4	2.5	*12.5*	*4.5*
1959	5.8	11.4	3.0	*10.5*	*4.7*
1960	5.8	10.9	2.9	*8.9*	*4.8*
1961	5.0	8.1	2.2	*13.6*	*5.0*
1962	5.7	9.3	2.5	*18.6*	*6.2*
1963	6.6	11.1	3.0	*19.6*	*6.6*
1964	7.3	12.1	3.3	*18.3*	*6.3*
1965	8.7	14.2	3.8	*18.6*	*6.3*
1966	10.1	16.9	4.5	*18.6*	*6.4*
1967	8.3	13.7	3.7	*18.8*	*5.6*
1968	9.4	15.3	4.1	*26.2*	*6.7*
1969	12.6	21.4	5.7	*24.0*	*7.8*
1970	14.5	25.5	6.9	*22.4*	*7.5*
1971	16.9	29.5	7.9	*23.1*	*8.2*
1972	15.9	27.8	7.5	*20.8*	*7.5*
1973	14.5	25.4	6.8	*18.0*	*6.3*
1974	13.3	23.2	6.2	*16.9*	*5.7*
1975	14.2	27.9	7.5	*17.4*	*5.7*
1976	13.4	25.9	7.0	*16.7*	*5.5*
1977	14.9	28.3	7.6	*17.7*	*5.6*
1978	16.8	31.7	8.5	*15.1*	*5.6*
1979	22.3	41.2	11.1	*17.5*	*6.6*
1980	19.3	33.4	9.0	*16.0*	*5.3*
1981	16.8	28.2	7.6	*15.1*	*4.3*
1982	17.6	29.0	7.8	*15.3*	*4.1*
1983	17.7	28.6	7.7	*13.7*	*3.7*
1984	18.1	28.4	7.6	*11.7*	*3.2*
1985	19.1	26.9	7.2	*10.4*	*2.7*
1986	20.1	26.4	7.1	*8.6*	*2.4*
1987	21.0	25.3	6.8	*8.6*	*2.3*
1988	21.8	21.8	6.9	*8.1*	*1.9*
1989	25.2	21.7	5.8	*8.4*	*1.9*
1990	29.0	23.7	6.4	*8.7*	*2.0*

Source: State Statistical Bureau of the PRC, *China Statistical Yearbook*, various years (China Statistical Information and Consultancy Service Centre: Beijing); author's estimates.

prices and exchange-rates. The official figure for 1990 was 28.9 billion yuan in the initial budget estimate, later increased slightly to 28.97 billion yuan.

Even though the levels are suspect, the official budget figures give an idea of *trends*. It is believed that the change of military spending over time reflects strategic and economic variables, and as such the data are of considerable interest. Table 5.21 shows Chinese defence spending from 1950 to 1990. The first column gives the official figures, the others give SIPRI estimates based on economic data provided by the International Monetary Fund (IMF).

During the 1950s all three economic indicators of military effort—real military expenditure as well as its share in central government expenditure (CGE) and NMP—fell steadily. This was the result of defence co-operation with the USSR and the inflow of Soviet aid, technology and technical manpower. In the 1960s all three indicators of defence effort rose rapidly, for three main reasons: Soviet assistance and technical help were abruptly withdrawn in 1960; China began constructing its independent nuclear force; and relations with the USSR worsened, with border clashes taking place along the Ussuri and Amur rivers. Another factor was the escalating US involvement in Viet Nam. In the 1970s the defence effort slowed down again, partly as a result of economic problems and systemic inefficiency. The short but bloody war with Viet Nam saw a peak in 1979, after which a period of conversion began.[52]

Since about 1982 Chinese defence spending has fallen continuously. The decline speeded up from the mid-1980s, coinciding with a large-scale demobilization. The manpower of the People's Liberation Army (PLA) was reduced by a quarter—1 million men—in the late 1980s and stood at just over 3 million in 1990. Procurement has been cut, postponed and streamlined. Military reforms, such as the introduction of ranks and the reorganization of military districts, have been carried through in an effort to create a more professional army. Defence exports have been encouraged in order to bring in foreign exchange, which can then be used to import needed technology. R&D spending has not been cut; its proportion in the total budget has gone up and is estimated to be over 10 per cent. However, demands for efficiency have been paramount. For example, in 1988 the research budget of the PLA Navy was reportedly slashed by 60 per cent due to there being 'insignificant, substandard, redundant research projects which could not be put into broad applications'.[53]

As noted above, Chinese military expenditure increased in 1990 for the first time since the mid-1980s. The increase is significant, almost 12 per cent. More important, its share in CGE has gone up considerably, from an average 8 per cent in 1986–89 to over 11 per cent. As an indication of the *perceptions* of the leadership, this reflects the increasing importance given to the armed forces. It is not clear where the extra funds will go. Some analysts believe that the bulk will be spent on personnel—on pay rises, better living conditions, pensions and help with economic adjustments after leaving the services, and perhaps as payment for support obtained during the 1989 student revolts. There is considerable ill feeling in the forces regarding pay and conditions, and conscript quotas are not easy to fill. Even though martial law was lifted in January 1990, the leadership still needs the military's support in domestic politics.

[52] Chinese attempts at resource transfer from the military to the civilian sectors have been described elsewhere. See Deger and Sen (note 11), pp. 99–102.

[53] Beijing Radio, 3 Feb. 1988, in *Foreign Broadcast Information Service: China*, 4 Feb. 1988, p. 10.

There is little evidence that procurement could increase significantly. Imported technology, essential to modernization, was initially affected by the sanctions imposed by the West. These have all been lifted. However, arms exports dropped after the end of the Iraq–Iran War, and China has been careful not to break the UN embargo on Iraq. Foreign exchange is scarce, and this could be the main hindrance in producing and procuring new-generation equipment. The 'Peace Pearl' project to modernize the avionics of the J-8 fighters, involving the US firm Grumman, was abandoned in 1990, possibly due to financial difficulties. The project contract was just over $0.5 billion.

Procurement expenditure would also be affected by Sino-Soviet *rapprochement*, begun with President Gorbachev's visit to China in 1989. Premier Li Peng's visit to the USSR cemented relations. The USSR has announced troop cuts in Asia that include forces stationed along the Chinese border. However, China remains wary of modern weapon transfers resulting from Soviet withdrawals from the ATTU zone. Most significant, the two countries have held discussions on defence trade, technology transfer and co-production. If these are successful, Chinese procurement could rise, particularly if the PLA Air Force purchases the Soviet MiG-29 fighters.

The years 1989 and 1990 could constitute a turning-point for China. Either the liberalization trends established over the past 10 years will be reversed, and a more conservative regime giving greater priority to the military will be established, or the attempts made since July 1989 to centralize the economy and repress political dissent will be short-lived.

Japan

Japan is the sixth largest military spender in the world. In 1990 its defence spending exceeded $30 billion (1988 prices). From 1980 to 1989 its military growth rate was 4.3 per cent per annum. Significant weapon stocks have been built up since 1985, with major weapon procurement costs of over $8 billion in 1990, placing the country fifth in the world in terms of military purchases.

The Japanese military, known as the Self Defence Force (SDF), is constitutionally bound to the strict defence of national territory. In 1990, following a debate in the Diet as to whether forces should be sent to the Persian Gulf, it was reiterated that Japanese military personnel cannot be sent abroad. The defence burden is about 1 per cent and will remain at that level. A commonly felt tenet in Japan holds that the country's economic might should be a substitute for military power, and that its economy remains its primary contribution to global security, defined in the broadest possible sense.

Demands have increasingly been made by other countries that Japan take on a greater share of economic burden-sharing. This debate has also been conducted within the country. A 1986 report from the powerful Ministry of International Trade and Industry stated:

> The cost of maintaining the international currency and trading systems as well as the cost of maintaining international security and world politics, and the international

Table 5.22. Japanese budgetary expenditure of the Mid-Term Defense Plan, FYs 1986–90

Budget figures are in current b. yen; growth rate figures are percentages.

Category	1986	1987	1988	1989	1990
Defence budget	3 343.5	3 517.4	3 700.3	3 919.8	4 159.3
Growth rate	*6.6*	*5.2*	*5.2*	*5.9*	*6.1*
Personnel andprovisions	1 508.6	1 543.9	1 578.9	1 613.6	1 668.0
Growth rate	*6.7*	*2.3*	*2.3*	*2.2*	*3.4*
Equipment and material	1 835.0	1 973.6	2 121.5	2 306.3	2 491.3
Growth rate	*6.5*	*7.6*	*7.5*	*8.7*	*8.0*

Source: *Japan Defense Agency, Outline of Japan's Defense Budget for Fiscal Year 1990* (Foreign Press Center: Tokyo, June 1990).

economic order in such areas as (foreign) assistance—that is the burden of the so called international public goods—has been borne chiefly by the United States in the 20th Century. In sharing this burden Japan has been behind West Germany, France and Britain. Japan should play a positive role in maintaining the international order by expanding economic assistance, increasing trade through further market opening, and spending more on R&D, basic R&D in particular, thus raising the share of the yen in external assistance and public foreign currency reserves, and by assuming a larger share of the international public goods burden through raising imports and R&D expenditure to the average international level.[54]

By 1990 many of these concerns had been met and some resolved. Japan's civilian R&D is the second highest in the world; its trade surplus is the highest, helping capital transfers; import restrictions are being reduced; and it has become the world's largest donor of foreign aid.

The level of annual military spending is indirectly guided by the Mid-Term Defense Plan (MTDP). This five-year plan provides approximate targets of military assets that the SDF should have or acquire over the period. The MTDP estimate (*Chugyo*) relates weapon acquisition with expenditure and determines the annual budget, although the relation is not precise. With 1990 ended one MTDP, having run for FYs 1986–90. The total defence spending for this period was set at 18.5 trillion yen in 1985 prices. Nominal expenditure, in TY prices, exceeded 18.6 trillion yen. Given Japanese inflation rates during this period, the total should be about 19 billion yen in TY prices. Spending has thus fallen marginally below target but remains substantial.

The FY 1990 budget shows military expenditure to be 41.6 billion yen—an increase of 6.1 per cent over 1989 in nominal terms and around 3 per cent in real growth. Table 5.22 gives the details for FYs 1986–90. Procurement expenditure has increased faster than personnel and provisions, showing the importance of 'investment' as contrasted to operating expenditures.

[54] Quoted from MITI Report referred to in Niioka, S., 'Japan's defence spending', eds Hartley and Sandler (note 17), pp. 253–75.

As in other areas of government procurement, domestic sources are emphasized in military procurement. In 1987, 91 per cent of total defence procurement was spent within the country. Of the rest, 4.1 per cent was US Foreign Military Sales (FMS), and 5.9 per cent was general commercial imports, also mainly from the USA. FMS purchases in 1987 accounted for less than $0.5 billion (67 billion yen). Between 1950 and 1957, 60 per cent of procurement came from abroad. In the early 1980s the foreign share had fallen below 15 per cent and in the late 1980s that share was less than 10 per cent.

Although the value of direct imports is low in absolute terms, technological co-operation and competition with the USA are bound to increase. Just as is the case in the wider economic and trade-related areas, there is potential for complications also in military procurement and trade. As spending on major weapon procurement has increased rapidly over the 1980s, pressure has increased to 'open' up the defence market. Major weapon spending increased from $4.2 billion in 1980 to over $8 billion in 1990 (in 1980 constant prices).

This pressure to import more from the USA could increase as the domestic market for the US defence industry declines. In an instance from the Viet Nam War era, Japanese indigenous plans to manufacture the PXL anti-submarine warfare (ASW) patrol aircraft were in 1972 dropped in favour of imports and licensed production of the US P-3C Orion. The chequered history of the Fighter Support Experimental (FS-X)—the next-generation aircraft to replace the F-1 support fighter—shows that co-operation will not be easy where economic and industrial policies are involved.[55] The same comments apply to the burden-sharing debate and Japanese payments for US bases.

Japanese military R&D is very small in absolute terms, accounting for 0.8 per cent of Government R&D—one of the lowest shares in the industrial world. After the MTDP emphasized rapid increases in defence-related research, however, military R&D has increased from 1.5 per cent of the defence budget in FY 1984 to 2.5 per cent in FY 1990. In 1990 its value stood at about $750 million. Current military R&D emphasizes four areas: the FS-X; a new communications system for enhanced command, control and intelligence capabilities; short-range surface-to-air missiles; and ASW capability through the short-range G-RX4 torpedo.

VIII. The Third World

Although Third World military expenditure is only a small proportion of the world total—around 15 per cent—it stands out in a number of ways. First, from 1965 to the mid-1980s, Third World military expenditure increased faster than that of the Western countries. Second, as a proportion of GDP the ratio has in aggregate been high—over 4 per cent, in excess of most West European countries—causing grave economic problems for impoverished societies. Third, as a share of central government budgets it has also been high, preventing resource transfer to public goods such as health or education.

[55] Deger and Sen (note 11), pp. 110–13.

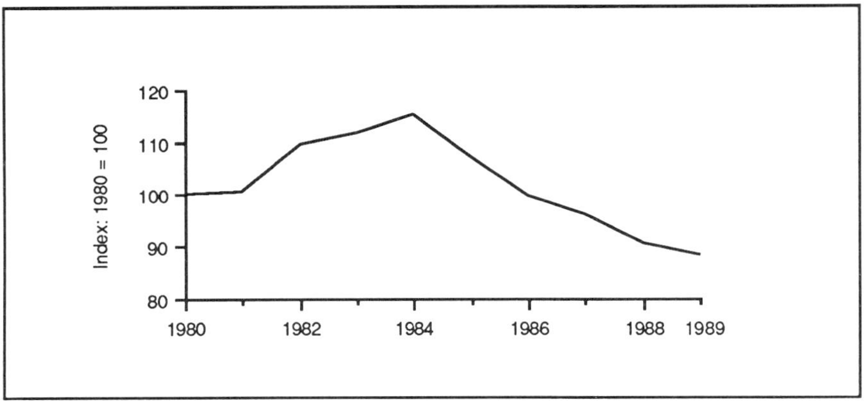

Figure 5.2. Third World military expenditure, 1980–89

In 1990 over 15 per cent of central government expenditure was allocated to the military—possibly higher than shares for health or education.

Since 1984 Third World military expenditure has fallen rapidly, principally as a result of economic crises. Figure 5.2 shows the trend. Preliminary estimates for 1990 show that levels did not fall but may even have risen slightly. This may imply that a plateau has been reached, raising concern that defence spending may again rise if and when economic fortunes allow. It should be noted, however, that the aggregate increase was mainly due to defence spending growth in a few countries and regions, in particular in China, India, Pakistan and in most countries of the Middle East. Propelled by high economic growth, the newly industrializing countries of South-East Asia—Singapore, South Korea and Taiwan—also contributed to the increase.

The Iraqi invasion of Kuwait, and the response of the US-led coalition, prompted increased procurement expenditure and arms imports, possibly signalling a new arms race in the region.The US Government's repudiation of Egypt's military debt and interest arrears may allow Egypt to spend more than economic constraints have hitherto allowed. To achieve a strategic balance the USA will also have to supply more arms to Israel under the FMS programme. Saudi Arabia's military expenditure, which has fallen in recent years, is also expected to rise sharply, partly to offset the costs of the US-led Operation Desert Shield against Iraq. With regards to the latter, although the UN sanctions were effective in bringing down Iraqi procurement spending, personnel cost rose sharply as the armed forces were augmented. Hard data are not available but it is estimated that the *additional* defence spending of the region as a whole will amount to $4 billion in 1990.

IX. Conclusion

Three factors affect the security environment, threat perceptions and technological and economic constraints that contribute to the evolution of military expenditure. The first is long-term and may be termed technological and structural disarmament. Unnecessarily sophisticated technology at extremely high cost makes acquisition programmes more difficult to justify. The second factor is more recent: the change in the political climate in Europe, and the relative success of formal arms control measures. Both these developments have contributed to the fall in military spending witnessed in 1990. However, the build-down is still modest. A third factor, the concern for stability, and the recognition that political and economic insecurity can be as destabilizing as military threats, means that a threshold for further disarmament may soon be reached. Until these structural elements of instability are addressed, a disarmament dividend is unlikely.

Appendix 5A. Tables of world military expenditure, 1981–90

SAADET DEGER and SOMNATH SEN, assisted by Phitsamone Ljungqvist-Souvannavong

Sources and methods are explained in appendix 5B.

Table 5A.1. World military expenditure, in current price figures

Figures are in local currency, current prices.

		1981	1982	1983	1984	1985	1986	1987	1988	1989	1990
NATO											
North America											
Canada	m. dollars	6 289	7 655	8 562	9 519	10 187	10 811	11 529	12 180	12 713	1 3172
USA	m. dollars	169 888	196 390	218 084	238 136	263 900	282 868	289 391	295 841	303 093	298 364
Europe											
Belgium	m. francs	125 689	132 127	136 615	139 113	144 183	152 079	155 422	150 647	152 917	157 723
Denmark	m. kroner	10 301	11 669	12 574	13 045	13 343	13 333	14 647	15 620	15 963	16 143
France	m. francs	129 708	148 021	165 029	176 638	186 715	197 080	209 525	215 073	224 985	232 926
FR Germany	m. D. marks	52 193	54 234	56 496	57 274	58 649	60 130	61 354	61 638	63 178	70 648
Greece	m. drachmas	142 865	176 270	193 340	271 922	321 981	338 465	393 026	471 820	503 032	597 929
Italy	b. lire	9 868	12 294	14 400	16 433	18 584	20 071	23 788	26 590	28 436	29 642
Luxembourg	m. francs	1 715	1 893	2 104	2 234	2 265	2 390	2 730	3 163	2 995	3 349
Netherlands	m. guilders	11 296	11 921	12 149	12 762	12 901	13 110	13 254	13 300	13 571	13 520
Norway	m. kroner	9 468	10 956	12 395	12 688	15 446	16 033	18 551	18 865	20 198	22 392
Portugal	m. escudos	51 917	63 817	76 765	92 009	111 375	139 972	159 288	194 036	229 344	243 869
Spain	m. pesetas	400 940	465 695	540 311	594 932	674 883	715 306	852 767	835 353	920 381	973 490
Turkey	b. lira	313	448	557	803	1 235	1 868	2 477	3 789	6 683	13 136
UK	m. pounds	12 004	14 203	15 605	17 104	18 156	18 581	19 125	19 439	20 749	21 792

WTO											
Bulgaria	m. leva	874	989	965	1 093	1 127	1 404	1 547	1 751	1 605	1 657
Czechoslovakia	m. korunas	21 349	22 220	23 332	24 387	25 512	26 435	27 362	28 374	28 213	25 089
German DR	m. marks	10 705	11 315	11 970	12 830	13 041	14 045	15 141	15 654	14 871	. .
Hungary	m. forints	19 060	20 050	21 900	22 700	37 700	38 800	41 500	49 200	47 760	44 620
Poland	b. zlotys	85	176	191	251	315	466	576	889	2 154	10 083
Romania	m. lei	10 490	11 340	11 662	11 888	12 113	12 208	11 597	11 552	11 753	11 786
USSR	m. roubles	. .	. .	. .	. .	. .	. .	. .	. .	. .	. .
Other Europe											
Albania	m. leks	917	912	888	986	1 700	978	1 055	1 080	1 075	1 030
Austria	m. schillings	12 245	13 334	15 362	15 554	16 786	17 940	16 972	16 597	17 946	18 009
Finland	m. markkaa	4 128	5 182	5 656	6 082	6 555	7 245	7 636	8 419	9 226	9 623
Ireland	m. pounds	207	252	244	275	293	321	311	329	333	350
Sweden	m. kronor	17 467	18 500	19 550	21 164	22 762	24 211	25 662	27 215	29 399	32 362
Switzerland	m. francs	3 349	3 727	3 862	4 009	4 576	4 282	4 203	4 458	4 679	5 090
Yugoslavia	b. new dinars	99	118	155	247	460	968	1 971	5 247	61 125	398 180
Middle East											
Bahrain	m. dinars	80.7	106	62.3	55.6	56.6	60.4	60.3	70.4	70	76.6
Cyprus	m. pounds	17.5	17.9	19.1	19.9	18.5	13.7	16.7	20.4	26.8	30
Egypt	m. pounds	1 238	1 435	1 801	2 173	2 108	2 493	2 742	2 862	3 415	3 640
Iran	b. rials	346	341	340	363	455	486	459	505	483	469
Iraq	m. dinars	1 350	2 400	3 200	4 300	4 000	3 600	4 350	4 000	4 000	4 150
Israel	m. new shekels	53.2	113	309	1 626	4 055	4 936	5 684	6 093	7 373	8 584
Jordan	m. dinars	160	179	196	197	219	243	253	256	252	280
Kuwait	m. dinars	291	370	416	434	469	430	380	408	438	. .
Lebanon	m. pounds	654	1 215	3 554	2 030	2 448	3 740	. .	10 640	. .	97 486
Oman	m. riyals	522	581	670	728	745	665	584	519	510	520
Saudi Arabia	m. riyals	75 723	87 695	84 311	77 817	71 992	62 418	60 726	55 750	55 000	57 090
Syria	m. pounds	9 653	10 703	11 309	12 601	13 000	14 440	14 327	16 638	25 881	. .
United Arab Emirates	m. dirhams	7 672	7 268	7 042	7 093	7 500	6 900	5 800	5 800	5 376	5 824
Yemen Arab Republic	m. rials	2 016	2 933	3 104	2 585	2 616	2 808	3 124	3 660	4 575	. .
Yemen PDR	m. dinars	56.0	57.5	65.8	67.0	65.3	68.8	72	76	80	. .

		1981	1982	1983	1984	1985	1986	1987	1988	1989	1990
South Asia											
Bangladesh	m. taka	3 210	4 190	5 080	5 325	5 790	7 495	9 080	9 931	11 200	11 402
India	m. rupees	45 371	53 193	61 945	70 834	83 651	105 291	124 965	129 878	142 000	154 375
Nepal	m. rupees	273	337	430	493	601	866	1 153	1 304	1 565	1 600
Pakistan	m. rupees	17 731	22 637	26 915	30 689	35 110	38 861	43 995	48 599	54 479	61 458
Sri Lanka	m. rupees	1 051	1 117	1 653	2 194	5 140	7 926	10 103	7 190	7 233	7 500
Far East											
Brunei	m. dollars	416	480	530	534	617	700	568	679	..	..
Hong Kong	m. dollars	1 521	1 478	1 537	1 523	1 639	1 530	1 645	1 676	..	..
Indonesia	b. new rupiahs	2 153	2 613	2 858	3 106	2 856	3 089	3 058	3 164	3 378	3 204
Japan	b. yen	2 388	2 532	2 712	2 911	3 118	3 296	3 473	3 655	3 865	4 099
Korea, North	m. won	3 009	3 242	3 530	3 819	3 935	3 976	3 971	3 886	4 060	4 466
Korea, South	b. won	2 831	3 163	3 406	3 573	3 957	4 372	4 915	5 753	6 226	6 638
Malaysia	m. ringgits	4 693	4 975	4 820	4 370	4 320	4 215	6 142	4 160	4 638	5 232
Mongolia	m. tugriks	630	716	726	764	764	790	837	900	850	800
Myanmar (Burma)	m. kyats	1 712	1 643	1 630	1 760	1 973	1 858	1 875	2 155	2 371	..
Philippines	m. pesos	6 746	7 778	8 530	8 288	7 827	8 662	9 268	10 972	16 447	17 680
Singapore	m. dollars	1 507	1 659	1 640	2 204	2 516	2 403	2 439	2 659	2 845	3 040
Taiwan	b. dollars	117	136	139	138	152	158	164	179	186	204
Thailand	m. baht	37 375	41 250	45 875	49 500	52 275	51 825	53 125	54 655	57 176	64 956
Oceania											
Australia	m. dollars	3 767	4 371	4 992	5 601	6 298	6 932	7 305	7 535	8 079	8 617
Fiji	m. dollars	3.6	4.2	4.7	4.5	4.5	4.8	9.1	10.3	11.6	11
New Zealand	m. dollars	549	628	656	724	825	1 017	1 211	1 340	1 382	1 371
Africa											
Algeria	m. dinars	3 481	3 893	4 477	4 631	4 793	5 459	5 805	6 070	6 756	8 419
Angola	m. kwanzas	15 060	15 060	23 295	31 943	34 306	34 572	..	26 161	23 438	21 094
Benin	m. francs	5 400	7 821	9 500	9 280	10 190	10 610	9 367	11 420	10 405	..
Botswana	m. pulas	28.5	25.2	28.2	34.9	41.7	64.5	124	90.1	93.0	107
Burkina Faso	m. francs	9 216	10 800	11 170	11 780	11 810	17 724	15 241	16 003	16 000	..

Burundi	m. francs	2 700	3 300	3 200	3 900	4 200	4 780	3 910	3 198	4 414	4 671
Cameroon	m. francs	21 415	41 015	63 105	73 658	81 920	86 905	83 150	77 889	50 000	57 120
Cent. Afric Rep.	m. francs	4 029	5 000	6 500	6 500	6 189	5 892	5 610	5 500	5 500	. .
Chad	m. francs	. .	. .	15 000	17 496	17 000	16 850	10 307	20 000	15 517	17 069
Congo	m. francs	11 250	16 500	18 600	21 596	25 000	25 625	26 200	20 440	23 580	25 000
Côte d'Ivoire	m. francs	25 000	28 400	29 658	30 706	31 320	33 547	35 336	36 250	37 193	. .
Ethiopia	m. birr	760	802	845	897	923	972	1 182	1 407	1 687	1 856
Gabon	m. francs	25 600	29 100	33 000	35 100	42 900	47 100	43 407	40 000	40 680	40 000
Ghana	m. cedis	488	587	894	1 605	3 432	4 605	6 659	4 603	8 028	11 334
Kenya	m. shillings	2 182	2 662	2 778	2 523	2 395	3 342	3 909	3 945	4 328	4 774
Liberia	m. dollars	51.6	46.9	25.3	25.2	24.4	23.0	25.8	27.4	28.1	30
Libya	m. dinars	1 310	1 330	1 107	1 096	1 096	819	549	582	524	. .
Madagascar	m. francs	23 500	27 200	29 600	31 730	33 520	39 830	39 200	39 200	40 000	40 000
Malawi	m. kwachas	36.0	29.0	26.1	26.6	28.6	46.1	47.8	61.65	71.5	75
Mali	m. francs	8 600	9 700	10 200	11 100	13 400	13 000	13 300	18 000	20 000	20 000
Mauritania	m. ouguiyas	3 293	2 931	2 639	. .	. .	. .	. .	. .	. .	. .
Mauritius	m. rupees	47.7	30.8	34.4	36.5	36.1	36.3	38.5	64.9	81.8	80
Morocco	m. dirhams	5 047	5 814	4 675	4 960	6 453	6 837	7 190	7 630	8 375	9 216
Mozambique	m. escudos	5 741	6 900	8 300	10 300	10 300	11 214	29 600	50 400	80 000	105 000
Niger	m. francs	4 286	4 232	4 389	4 775	5 075	5 325	5 175	5 365	. .	. .
Nigeria	m. nairas	1 319	1 113	1 179	928	976	957	810	1 270	1 689	2 108
Rwanda	m. francs	2 500	2 622	2 693	2 500	2 760	3 050	2 979	2 800	3 000	. .
Senegal	m. francs	21 565	23 505	25 110	27 046	28 235	28 490	28 784	29 630	28 476	30 000
Sierra Leone	m. leones	17.5	17.9	18.6	23.3	29.4	64.5	101	125	250	. .
Somalia	m. shillings	824	826	1 300	1 786	1 751	2 300	3 800	3 500	7 000	. .
South Africa	m. rands	2 615	2 967	3 314	3 922	4 414	5 412	6 717	7 835	9 873	10 038
Sudan	m. pounds	131	139	212	361	468	562	723	968	1 831	2 187
Swaziland	m. emalangeni	12.0	16.2	16.0	16.1	15.7	15.9	16.8	21.5	24	25
Tanzania	m. shillings	2 122	2 433	2 651	3 201	4 277	7 073	11 025	16 250	21 574	. .
Togo	m. francs	6 202	6 138	6 328	7 007	8 632	9 200	13 047	13 047	13 000	13 000
Tunisia	m. dinars	113	284	364	296	357	413	434	460	460	350
Uganda	m. shillings	54.1	82.3	144	327	782	1 157	4 805	8 500	8 000	8 000
Zaire	m. zaires	316	873	723	1 928	2 013	2 700	5 000	6 500	14 869	. .

		1981	1982	1983	1984	1985	1986	1987	1988	1989	1990
Zambia	m. kwachas	154	148	161	148	167	480	637	717	896	900
Zimbabwe	m. dollars	284	296	353	398	436	554	661	720	804	800
Central America											
Costa Rica	m. colones	317	528	928	1 140	1 202	1 426	1 504	1 586	1 660	2 040
Cuba	m. pesos	1 011	1 109	1 133	1 386	1 335	1 307	1 300	1 350	1 377	1 400
Dominican Rep.	m. pesos	126	128	129	164	191	202	250	298	346	484
El Salvador	m. colones	322	395	442	534	630	964	885	1 002	1 118	. .
Guatemala	m. quetzales	161	208	231	270	371	378	495	645	623	600
Haiti	m. gourdes	105	104	102	110	131	138	150	. .	. .	. .
Honduras	m. lempiras	125	160	240	335	445	450	450	500	823	. .
Jamaica	m. dollars	81.8	98.8	97.8	104	124	125	125	. .	. .	. .
Mexico	b. pesos	37.9	47.4	90.3	181	297	470	894	1 470	1 673	2 024
Nicaragua	m. cordobas	1.3	1.7	3.4	4.9	26.8	91	921	93 827	. .	. .
Panama	m. balboas	46.5	55.0	60.0	88.0	92.0	105	105	113	124	. .
Trinidad and Tobago	m. dollars	371	563	545	490	465	465	. .	. .	. .	. .
South America											
Argentina	m. australes	3.9	8.9	31.2	236	1 387	2 727	5 863	28 224	834 815	. .
Bolivia	t. bolivianos	8.0	19.0	58.0	721	94 677	299 374	327 547	400 300	489 214	. .
Brazil	b. cruzados	0.3	0.8	1.4	4.7	16	45	131	1 023	14 198	158 899
Chile	m. pesos	94 810	117 831	124 901	182 203	194 877	258 675	277 417	385 145	446 768	537 373
Colombia	m. pesos	35 830	44 661	69 531	91 753	105 092	135 712	176 989	265 484	398 226	566 886
Ecuador	m. sucres	5 848	6 870	8 833	12 086	19 743	25 598	35 442	52 595	83 839	. .
Guyana	m. dollars	96	108	142	156	192	276	237	253	403	370
Paraguay	m. guaranies	10 581	11 566	11 676	12 826	15 937	20 097	26 885	32 643	57 978	. .
Peru	b. intis	0.5	1.5	2.5	3.9	11.9	23.9	37	104	2 254	. .
Uruguay	m. new pesos	4 770	5 168	5 877	7 708	12 831	22 828	36 831	59 962	108 275	. .
Venezuela	m. bolivares	8 952	9 905	8 488	9 800	9 457	10 520	15 197	17 585	32 404	45 379

Table 5A.2. World military expenditure, in constant price figures

Figures are in US $m., at 1988 prices and exchange-rates.

	1981	1982	1983	1984	1985	1986	1987	1988	1989	1990
NATO										
North America										
Canada	7 353	8 077	8 534	9 093	9 362	9 535	9 747	9 897	9 843	9 786
USA	220 955	240 616	258 828	270 923	290 026	305 076	300 890	295 841	289 149	268 113
Europe										
Belgium	4 657	4 502	4 323	4 139	4 092	4 261	4 287	4 097	4 035	4 012
Denmark	2 260	2 323	2 342	2 287	2 234	2 153	2 275	2 320	2 263	2 219
France	32 995	33 668	34 252	34 104	34 103	35 118	36 137	36 105	36 494	36 393
FR Germany	34 216	33 786	34 054	33 712	33 796	34 719	35 320	35 097	35 008	38 016
Greece	3 360	3 428	3 128	3 717	3 688	3 152	3 144	3 326	3 116	3 041
Italy	14 269	15 262	15 585	16 057	16 634	16 964	19 199	20 429	20 559	20 160
Luxembourg	62	63	64	64	63	66	75	86	79	85
Netherlands	6 575	6 555	6 497	6 608	6 533	6 633	6 753	6 729	6 791	6 590
Norway	2 447	2 545	2 656	2 558	2 946	2 853	3 037	2 895	2 963	3 161
Portugal	1 142	1 142	1 099	1 021	1 036	1 166	1 212	1 348	1 415	1 323
Spain	6 413	6 518	6 738	6 669	6 952	6 772	7 672	7 171	7 583	7 531
Turkey	2 316	2 528	2 393	2 325	2 467	2 772	2 647	2 664	2 770	3 418
UK	30 549	33 283	34 981	36 511	36 548	36 173	35 713	34 629	34 292	32 470
EC	137 010	141 062	143 529	145 373	146 168	147 693	152 272	151 839	152 123	152 339
WTO										
Bulgaria	718	810	780	877	800	1 071	1 180	1 337	1 122	1 053
Czechoslovakia	3 473	3 454	3 589	3 716	3 838	3 962	4 097	4 241	4 159	3 521
German DR	5 068	5 357	5 667	6 075	6 181	6 656	7 176	7 419	7 048	. .
Hungary	1 597	1 571	1 599	1 531	2 375	2 321	2 285	2 343	1 944	1 434

	1981	1982	1983	1984	1985	1986	1987	1988	1989	1990
Poland	4 117	4 262	3 796	4 332	4 730	5 945	5 863	5 657	3 904	2 864
Romania	1 578	1 458	1 425	1 437	1 470	1 483	1 407	1 402	1 426	1 362
USSR	. .	. .	. .	. .	. .	. .	. .	. .	. .	. .
Other Europe										
Albania	153	152	148	164	283	163	176	180	179	172
Austria	1 238	1 278	1 426	1 366	1 429	1 501	1 401	1 344	1 417	1 371
Finland	1 496	1 714	1 726	1 733	1 765	1 895	1 919	2 013	2 070	2 033
Ireland	512	532	466	484	489	516	485	502	488	499
Sweden	4 539	4 380	4 253	4 263	4 268	4 357	4 431	4 442	4 508	4 492
Switzerland	2 761	2 907	2 926	2 949	3 255	3 022	2 926	3 047	3 100	3 219
Yugoslavia	2 420	2 137	1 994	2 082	2 249	2 491	2 300	2 082	1 810	1 786
Middle East										
Bahrain	226	273	156	139	145	158	161	187	183	200
Cyprus	50	48	49	48	43	31	37	44	55	60
Egypt	5 392	5 442	5 889	6 070	5 252	5 013	4 607	4 089	4 021	3 652
Iran	12 321	10 230	8 523	8 082	9 705	9 339	7 679	7 353	5 747	5 133
Iraq	14 007	21 952	28 596	31 590	23 506	16 531	17 073	12 868	10 720	9 268
Israel	6 887	7 314	8 000	8 420	5 249	4 318	4 134	3 811	3 830	3 807
Jordan	535	557	581	562	607	673	703	689	539	522
Kuwait	1 246	1 470	1 579	1 629	1 733	1 574	1 382	1 463	1 518	. .
Lebanon	59	96	262	107	93	97	. .	26	. .	106
Oman	859	1 016	1 296	1 478	1 517	1 730	1 189	1 350	1 326	1 352
Saudi Arabia	18 557	21 614	20 899	19 513	18 666	16 684	16 384	14 887	14 522	15 213
Syria	3 635	3 526	3 511	3 582	3 152	2 573	1 601	1 482	2 070	. .
United Arab Emirates	2 088	1 955	1 966	2 091	2 211	2 004	1 587	1 580	1 395	1 439
Yemen Arab Republic	322	456	457	339	323	325	340	375	390	. .
Yemen PDR	249	234	241	243	225	224	221	220	232	. .
South Asia										
Bangladesh	203	235	261	247	243	283	313	313	321	303
India	5 819	6 325	6 582	6 955	7 778	9 006	9 822	9 332	9 609	9 550

Nepal	23	26	29	33	37	45	54	56	62	58
Pakistan	1 466	1 767	1 974	2 122	2 299	2 459	2 658	2 700	2 805	2 906
Sri Lanka	65	63	82	93	214	306	362	226	204	173
Far East										
Brunei	245	265	290	283	319	356	287	314	. .	. .
Hong Kong	309	271	256	235	245	223	226	215	. .	. .
Indonesia	2 596	2 505	2 451	2 410	2 116	2 163	1 960	1 877	1 882	1 700
Japan	20 628	21 291	22 400	23 504	24 672	25 924	27 289	28 521	29 491	30 483
Korea, North	1 349	1 454	1 583	1 713	1 765	1 783	1 781	1 743	1 821	2 003
Korea, South	5 103	5 318	5 535	5 675	6 135	6 593	7 195	7 865	8 057	7 827
Malaysia	2 132	2 129	1 990	1 742	1 716	1 664	2 406	1 589	1 723	1 884
Mongolia	210	239	242	255	255	263	279	300	283	266
Myanmar (Burma)	528	481	452	465	488	421	340	337	300	330
Philippines	815	854	851	550	422	463	478	520	705	676
Singapore	816	866	845	1 107	1 258	1 218	1 230	1 321	1 381	1 433
Taiwan	4 432	5 000	5 043	5 007	5 526	5 704	5 891	6 348	6 282	6 562
Thailand	1 808	1 895	2 031	2 174	2 240	2 182	2 181	2 161	2 146	2 392
Oceania										
Australia	5 070	5 309	5 524	5 934	6 272	5 334	6 166	5 910	5 916	5 951
Fiji	4	4	4	4	4	4	7	7	7	6
New Zealand	768	756	735	765	754	822	845	879	859	801
Africa										
Algeria	1 016	1 066	1 138	1 107	1 036	1 050	1 040	1 026	1 053	1 208
Angola	502	502	777	1 065	1 144	1 152	. .	872	781	703
Benin	29	40	44	41	43	43	35	38	32	. .
Botswana	30	24	24	27	30	42	75	50	46	48
Burkina Faso	42	43	41	42	39	60	53	54	54	. .
Burundi	30	34	31	33	34	38	29	23	28	28
Cameroon	133	225	296	311	341	336	303	262	168	191
Central African Rep.	18	20	23	22	19	18	18	19	18	. .
Chad	. .	. .	61	59	54	62	39	67	55	60

	1981	1982	1983	1984	1985	1986	1987	1988	1989	1990
Congo	60	78	81	84	91	91	91	69	76	78
Côte d'Ivoire	117	124	122	121	121	121	127	122	124	. .
Ethiopia	469	475	496	486	420	490	611	680	756	786
Gabon	117	114	117	118	134	139	129	134	128	122
Ghana	23	23	16	20	39	42	44	23	32	32
Kenya	243	247	231	190	160	214	238	222	222	220
Liberia	67	58	30	30	29	26	28	27	20	25
Libya	4 452	4 520	3 762	3 725	3 725	2 784	1 866	1 978	1780	. .
Madagascar	50	44	40	39	37	39	33	28	26	23
Malawi	44	33	26	22	21	30	25	24	25	24
Mali	39	42	42	44	51	47	47	60	67	67
Mauritania	71	56	50	. .	. .	. .	. .	. .	. .	. .
Mauritius	5	3	3	3	3	3	3	5	5	5
Morocco	999	1 042	788	744	898	876	896	929	988	1 032
Mozambique	61	58	55	53	42	36	75	101	107	94
Niger	15	13	14	14	15	16	17	18	. .	. .
Nigeria	914	717	616	347	346	322	248	283	267	310
Rwanda	45	42	40	35	38	43	40	37	39	. .
Senegal	120	111	106	103	95	90	95	100	95	100
Sierra Leone	24	19	12	9	6	8	4	4	5	. .
Somalia	60	49	57	41	29	28	36	21	20	. .
South Africa	3 003	2 970	2 956	3 137	3 036	3 139	3 355	3 468	3 808	3 407
Sudan	194	163	191	242	216	208	239	215	271	216
Swaziland	13	15	13	12	10	9	8	10	10	9
Tanzania	144	127	109	98	97	121	146	164	175	. .
Togo	25	22	21	24	30	31	44	44	44	43
Tunisia	231	509	599	449	502	549	538	536	499	361
Uganda	82	83	116	185	190	104	128	80	40	40
Zaire	34	69	32	57	48	44	46	35	39	. .
Zambia	140	120	109	84	69	130	121	87	49	. .
Zimbabwe	387	364	353	331	334	371	394	400	395	350

Central America										
Costa Rica	21	19	25	27	25	27	24	21	19	20
Cuba	1 303	1 429	1 460	1 786	1 721	1 685	1 676	1 740	1 775	1 804
Dominican Republic	73	69	66	66	56	54	58	49	42	42
El Salvador	219	241	238	258	249	288	212	150	141	. .
Guatemala	135	174	184	208	241	180	209	246	159	140
Haiti	28	26	23	23	25	25	31	. .	. .	. .
Honduras	89	105	145	194	249	241	235	.186	279	. .
Jamaica	38	43	38	32	30	26	25	. .	. .	. .
Mexico	1 296	1 015	959	1 161	1 208	1 027	842	647	613	587
Nicaragua	279	292	445	473	810	352	352	348	350	. .
Panama	51	58	62	90	93	106	105	113	92	. .
Trinidad and Tobago	194	264	222	176	155	144	. .	. .	. .	. .
South America										
Argentina	5 711	4 927	3 897	4 056	3 087	3 194	2 966	3 225	3 000	. .
Bolivia	243	238	202	182	201	169	162	170	181	. .
Brazil	3 362	4 532	3 276	3 703	3 857	4 428	3 908	3 899	3 900	3 000
Chile	1 394	1 574	1 313	1 597	1 307	1 451	1 299	1 572	1 559	1 511
Colombia	484	483	629	715	660	716	758	887	1 058	1 164
Ecuador	142	143	124	129	165	174	186	174	160	. .
Guyana	37	34	40	35	37	50	30	25	21	20
Paraguay	69	71	63	58	57	55	60	59	84	. .
Peru	492	785	671	487	568	641	534	806	500	. .
Uruguay	294	268	205	173	167	169	166	167	167	. .
Venezuela	1 663	1 678	1 354	1 392	1 207	1 204	1 357	1 213	1 200	1 200

Table 5A.3. World military expenditure as a percentage of gross domestic product

	1980	1981	1982	1983	1984	1985	1986	1987	1988	1989
NATO										
North America										
Canada	1.8	1.8	2.0	2.1	2.1	2.1	2.1	2.1	2.0	2.0
USA	5.4	5.7	6.3	6.5	6.4	6.6	6.7	6.4	6.1	5.8
Europe										
Belgium	3.3	3.4	3.3	3.2	3.1	3.0	3.0	2.9	2.7	2.5
Denmark	2.4	2.5	2.5	2.5	2.3	2.2	2.0	2.1	2.2	2.1
France	4.0	4.1	4.1	4.1	4.0	4.0	3.9	4.0	3.8	3.7
FR Germany	3.3	3.4	3.4	3.4	3.3	3.2	3.1	3.1	2.9	2.8
Greece	5.7	7.0	6.8	6.3	7.1	7.0	6.2	6.3	6.4	6.8
Italy	2.1	2.1	2.3	2.3	2.3	2.3	2.2	2.4	2.5	2.4
Luxembourg	1.0	1.1	1.0	1.1	1.0	0.9	0.9	1.1	1.1	1.0
Netherlands	3.1	3.2	3.2	3.2	3.2	3.1	3.1	3.1	3.0	2.9
Norway	2.9	2.9	3.0	3.1	2.8	3.1	3.1	3.3	3.2	3.2
Portugal	3.5	3.5	3.5	3.3	3.3	3.1	3.2	3.1	3.2	3.9
Spain	2.3	2.4	2.4	2.4	2.4	2.4	2.2	2.4	2.1	2.1
Turkey	4.3	4.9	5.2	4.8	4.4	4.5	4.8	4.2	3.8	3.9
UK	4.7	4.7	5.1	5.1	5.3	5.1	4.9	4.6	4.3	4.0
WTO										
Bulgaria	3.0	3.0	3.3	3.1	3.3	3.4	4.0	4.2	4.4	. .
Czechoslovakia	3.1	3.1	3.1	3.2	3.3	3.3	3.4	3.4	3.4	3.7
German DR	4.2	4.4	4.5	4.5	4.7	4.6	4.8	5.0	5.0	. .
Hungary	2.5	2.4	2.4	2.4	2.3	3.6	3.6	3.4	3.5	2.8
Poland	3.0	3.1	3.2	2.8	2.9	3.0	3.6	3.4	3.0	1.8
Romania	1.7	1.6	1.5	1.5	1.4	1.4	1.3	1.2	1.2	1.9
USSR	. .	. .	. .	. .	. .	. .	. .	. .	. .	. .

Other Europe										
Austria	1.2	1.2	1.2	1.3	1.2	1.2	1.3	1.2	1.1	1.1
Finland	1.9	1.9	2.1	2.1	2.0	1.9	2.0	1.9	1.9	1.9
Ireland	1.9	1.8	1.9	1.7	1.7	1.7	1.7	1.6	1.5	1.4
Sweden	3.0	3.0	2.9	2.8	2.7	2.6	2.6	2.5	2.4	2.4
Switzerland	1.9	1.8	1.9	1.9	1.9	2.0	1.8	1.7	1.7	1.6
Yugoslavia	4.9	4.5	4.0	3.8	3.7	3.8	3.9	3.6	3.3	2.9
Middle East										
Bahrain	4.8	5.9	7.5	4.3	3.8	4.2	5.1	5.3	5.0	10.7
Cyprus	1.4	2.0	1.7	1.7	1.5	1.2	0.9	0.9	1.0	1.4
Egypt		6.5	6.3	6.7	6.9	5.8	6.1	6.2	4.8	4.5
Iran	5.4	4.3	3.4	2.6	2.5	3.0	3.0	. .	. .	. .
Iraq	6.3	12.3	18.4	24.3	29.1	26.0	24.2	24.3	23.0	. .
Israel	25.0	23.5	19.0	20.2	21.4	14.4	11.3	10.2	9.1	9.2
Jordan	13.8	13.7	13.5	13.8	13.1	13.6	14.8	15.0	15.0	11.0
Kuwait	3.5	4.4	6.0	6.8	6.8	7.9	8.6	7.0	7.3	6.5
Lebanon	4.1	2.4	4.3	12.0	. .	. .	. .	. .	. .	. .
Oman	19.7	21.0	22.2	24.5	23.9	21.6	23.8	17.6	17.8	15.8
Saudi Arabia	16.6	14.5	21.1	20.3	20.9	22.0	22.4	22.7	19.8	. .
Syria	17.3	14.7	15.6	15.4	16.7	15.6	14.4	11.3	9.2	. .
United Arab Emirates	5.8	6.3	6.5	6.8	7.0	7.6	8.7	6.7	6.7	5.4
Yemen Arab Republic	15.0	12.6	14.7	14.2	10.4	8.4	7.3	7.2	. .	. .
Yemen PDR	17.8	19.7	18.7	19.1	17.7	16.7	22.2	18.4	18.5	. .
South Asia										
Bangladesh	1.4	1.3	1.5	1.6	1.4	1.3	1.5	1.6	1.6	1.6
India	3.0	3.0	3.1	3.1	3.2	3.3	3.7	3.9	3.5	3.3
Nepal	1.0	0.9	1.1	1.2	1.2	1.3	1.6	1.8	2.2	. .
Pakistan	5.7	5.9	6.6	6.9	6.8	6.8	7.1	7.1	6.7	6.7
Sri Lanka	1.5	1.2	1.1	1.4	1.4	3.2	4.4	5.1	3.2	2.9

	1980	1981	1982	1983	1984	1985	1986	1987	1988	1989
Far East										
Brunei	3.9	4.5	5.3	6.5	6.5	7.7	. .	. .	. .	. .
Hong Kong	1.0	0.9	0.8	0.7	0.6	0.6	0.5	0.5	0.4	. .
Indonesia	3.8	3.7	4.2	3.7	3.5	3.0	3.0	2.5	2.3	2.0
Japan	0.9	0.9	0.9	1.0	1.0	1.0	1.0	1.0	1.0	1.0
Korea, North	10.7	11.5	11.8	12.3	12.0	. .	. .	9.5	8.7	8.8
Korea, South	5.9	6.0	5.8	5.3	4.9	4.9	4.7	4.5	4.6	4.4
Malaysia	6.4	8.1	7.9	6.9	5.5	5.6	5.9	6.1	6.3	4.6
Mongolia	. .	. .	. .	. .	. .	11.2	11.0	11.3	11.7	. .
Myanmar (Burma)	3.9	4.1	3.6	3.3	3.3	3.6	3.2	3.0	3.1	. .
Philippines	2.2	2.2	2.3	2.2	1.5	1.3	1.4	1.3	1.3	1.7
Singapore	5.0	5.1	5.1	4.5	5.5	6.5	6.3	5.8	5.5	5.1
Taiwan	6.6	6.7	7.3	6.8	6.1	6.4	5.9	6.3	6.0	6.0
Thailand	5.1	4.8	4.9	5.0	5.0	5.0	4.7	4.3	4.0	3.2
Oceania										
Australia	2.6	2.6	2.7	2.8	2.8	2.8	2.8	2.6	2.2	1.9
Fiji	0.4	0.3	0.4	0.4	0.4	0.3	0.3	0.6	0.7	. .
New Zealand	1.9	2.1	2.1	2.0	1.9	1.9	2.0	2.0	2.1	1.9
Africa										
Algeria	2.1	1.8	1.9	1.9	1.8	1.7	1.7	1.7	1.5	1.9
Angola	12.8	13.8	11.9	16.5	22.0	28.4	28.4	. .	21.5	. .
Benin	1.9	1.8	1.9	2.2	2.0	2.0	1.9	. .	. .	. .
Botswana	3.7	3.7	2.7	2.4	2.4	2.1	2.7	4.2	2.7	1.9
Burkina Faso	2.7	2.8	3.0	2.9	3.0	2.5	3.5	3.0	2.8	. .
Burundi	2.9	3.0	3.5	3.1	3.2	3.0	3.4	2.7	2.2	2.6
Cameroon	1.2	1.1	1.7	2.2	2.1	2.2	2.1	2.1	2.1	. .
Central African Rep.	1.7	2.1	2.0	2.6	2.3	2.0	1.8	1.7	1.8	1.7
Chad	. .	. .	7.0	7.8	5.7	6.0	3.8	. .	. .	. .
Congo	2.8	2.1	2.3	2.3	2.3	2.6	4.0	. .	3.2	3.6
Côte d'Ivoire	1.2	1.1	1.1	1.1	1.1	1.0	1.0	1.2	1.2	1.2

Ethiopia	8.5	8.4	8.4	8.4	9.0	8.9	8.9	10.0	12.2	13.6
Gabon	2.1	2.4	2.4	2.6	2.3	2.6	4.0	4.3	4.5	. .
Ghana	0.4	0.7	0.7	0.5	0.6	1.0	0.9	0.9	0.5	0.6
Kenya	3.8	3.6	3.8	3.6	2.9	2.4	2.9	3.0	2.6	. .
Liberia	2.8	4.8	4.3	2.3	2.4	2.3	2.2	. .	. .	. .
Libya	10.0	14.0	15.0	13.0	14.5	15.2	12.7	. .	8.6	7.4
Madagascar	2.8	3.0	2.7	2.4	2.3	2.2	2.2	1.8	1.4	1.3
Malawi	4.4	3.3	2.4	1.9	1.6	1.5	1.8	1.6	1.7	1.6
Mali	2.3	2.4	2.4	2.4	2.7	2.3	. .	. .	3.2	3.3
Mauritania	9.7	7.6	6.9	5.7	. .	. .	. .	. .	. .	. .
Mauritius	0.5	0.4	0.3	0.3	0.2	0.2	0.2	0.2	0.2	0.3
Morocco	6.3	6.6	6.5	4.9	4.7	5.4	5.1	5.0	4.2	4.3
Mozambique	7.0	8.0	10.7	12.1	11.7	10.4	. .	. .	. .	. .
Niger	0.7	0.7	0.6	0.7	0.7	0.7	0.8	0.8	0.8	. .
Nigeria	2.5	2.3	1.8	1.9	1.3	1.2	1.2	0.7	0.9	1.1
Rwanda	1.9	2.0	2.0	1.9	1.6	1.6	1.9	. .	1.7	1.7
Senegal	3.1	2.8	2.8	2.7	2.7	2.5	2.2	2.0	. .	. .
Sierra Leone	1.0	1.0	0.8	0.7	0.7	0.6	1.1	. .	0.5	. .
Somalia	4.9	4.3	3.4	3.8	2.7	1.8	1.8	1.8	3.0	. .
South Africa	3.9	3.7	3.7	3.7	3.7	3.7	3.7	4.0	3.9	4.2
Sudan	2.3	2.0	1.7	2.1	3.9	2.6	2.1	. .	2.0	. .
Swaziland	2.1	2.2	2.9	2.6	2.3	1.8	1.7	. .	. .	. .
Tanzania	4.0	4.3	4.2	3.9	3.8	3.8	4.7	4.7	5.2	. .
Togo	2.2	2.4	2.3	2.2	2.3	2.6	2.5	2.6	3.2	. .
Tunisia	2.2	2.7	5.9	6.6	4.7	5.2	5.9	5.5	5.3	4.9
Uganda	2.2	3.8	2.7	3.0	5.0	5.9	3.8	3.5	1.7	0.8
Zaire	2.5	1.3	2.8	1.2	1.9	1.4	1.3	1.5	1.1	1.2
Zambia	3.5	4.4	4.1	3.9	3.0	2.4	3.7	3.2	3.2	. .
Zimbabwe	7.1	6.4	5.7	5.7	6.2	5.7	6.2	6.5	7.3	7.9
Central America										
Costa Rica	0.6	0.6	0.5	0.7	0.7	0.6	0.6	0.5	0.4	0.4
Cuba	9.9	8.8	9.1	8.8	10.1	9.6	10.2	10.7	11.3	. .

	1980	1981	1982	1983	1984	1985	1986	1987	1988	1989
Dominican Republic	1.5	1.7	1.6	1.5	1.6	1.4	1.3	1.3	1.1	0.8
El Salvador	2.8	3.7	4.4	4.4	4.6	4.4	4.9	3.8	3.7	3.5
Guatemala	1.8	1.9	2.3	2.6	2.9	3.3	2.4	2.8	3.2	2.6
Haiti	1.4	1.4	1.3	1.2	1.1	1.2	1.4	. .	. .	. .
Honduras	2.4	2.3	2.8	4.0	5.2	6.4	6.0	6.6	5.6	8.4
Jamaica	1.3	1.6	1.7	1.4	1.1	1.1	0.9	0.8	. .	. .
Mexico	0.6	0.6	0.5	0.5	0.6	0.7	0.6	0.5	. .	. .
Nicaragua	4.4	5.3	6.0	10.3	10.9	23.2	20.9	34.2	28.3	. .
Panama	1.2	1.2	1.3	1.4	1.9	1.9	2.0	2.0	2.5	2.7
Trinidad and Tobago	2.0	2.3	2.9	2.9	2.6	2.6	2.7	. .	. .	. .
South America										
Argentina	6.4	7.1	6.0	4.6	4.5	3.5	3.7	3.4	3.0	3.0
Bolivia	4.0	5.3	4.5	3.9	3.4	3.4	2.8	2.9	3.1	3.9
Brazil	1.3	1.3	1.6	1.2	1.2	1.1	1.2	1.1	1.1	1.2
Chile	6.7	7.4	9.5	8.0	9.6	7.6	8.0	6.8	7.8	6.5
Colombia	1.8	1.8	1.8	2.3	2.4	2.1	2.0	2.0	2.3	2.6
Ecuador	1.8	1.7	1.7	1.6	1.5	1.8	1.9	2.0	1.7	1.5
Guyana	6.5	6.0	7.5	9.7	9.2	9.8	12.4	7.0	7.0	. .
Paraguay	1.4	1.5	1.6	1.4	1.2	1.1	1.1	1.1	1.0	1.3
Peru	5.3	6.0	8.5	8.1	5.6	6.4	6.6	5.0	2.5	3.0
Uruguay	2.9	3.9	4.0	3.2	2.6	2.4	2.3	2.1	2.1	2.1
Venezuela	2.7	3.1	3.4	2.9	2.4	2.0	2.1	2.1	1.9	1.4

Table 5A.1: Military expenditure figures are given in local currency at current prices. Figures for recent years are budget estimates.

Table 5A.2: This series is based on the data given in the local currency series, deflated to 1988 price levels and converted into dollars at 1988 period-average exchange-rates. Local consumer price indices (CPI) are taken as far as possible from *International Financial Statistics* (IFS) (International Monetary Fund: Washington, DC). For the most recent year, the CPI is an estimate based on the first 6–10 months of the year. Period-average exchange-rates are taken as far as possible from the IFS. For WTO countries, purchasing power parities (PPP) are used.

Table 5A.3: The share of gross domestic product (GDP) is calculated in local currency. GDP data are taken as far as possible from the IFS. For some socialist economies, gross national product (GNP) or net material product (NMP) is used.

Appendix 5B. Sources and methods

I. Methods and definitions

Since the publication of the first *SIPRI Yearbook* (1968/69), SIPRI has provided annual 10-year time series data on world military expenditure. The main purpose of the data is to provide an easily identifiable measure, over time, of the scale of resources absorbed by the military in various countries. Expenditure data are only indirectly related to military strength, although the change in data over time can be utilized to measure the perception of governments towards military capability.

In recent years, the information available on world military expenditure has increased in quantitative terms while there has been a decline in the quality of information provided. Compared to the past there are now many more sources. At the same time, however, the reliability of the available data has gone down. In addition to the primary sources of national budgets and documents published by international organizations, the military expenditure project also studies over 50 specialist journals, annual reference volumes and newspapers.

In 1990 there were major specific problems with respect to data collection, and all data should be treated as preliminary. The unification of Germany implies that allocations of defence expenditure for the FRG are as yet provisional. The cost of the Persian Gulf conflict cannot be dealt with adequately as information is classified or unknown. Countries in Eastern and Central Europe are in a state of change which affects the information base of public finance in general and of military spending in particular. In addition, very high inflation rates quickly make budgetary data irrelevant.

The NATO definition of military expenditure is utilized as a guideline. Where possible, the following items are included: all current and capital expenditure on the armed forces, in the running of defence departments and other government agencies engaged in defence projects as well as space projects; the cost of paramilitary forces and police when judged to be trained and equipped for military operations; military R&D, tests and evaluation costs; and costs of retirement pensions of service personnel, including pensions of civilian employees. Military aid is included in the expenditure of the donor countries. Excluded are items on civil defence, interest on war debts and veterans' payments. Calendar year figures are calculated from fiscal year data where necessary, on the assumption that expenditure takes place evenly throughout the year.

It should be stressed that even though SIPRI provides military expenditure in constant prices, it does not encourage close comparison between individual countries. Priority is given to the choice of providing a uniform definition *over time for each country* to show a correct time trend, rather than to adjusting the figures for single years according to the common definition. In addition, the recent phenomenon of violently fluctuating exchange-rates (and their lack of correlation to inflationary differentials) makes common dollar figures more difficult to compare. In the absence of explicit military prices, obeying purchasing power parity, the present system must therefore be kept.[1]

[1] For an earlier discussion of methodology, see *World Armaments and Disarmament: SIPRI Yearbook 1984* (Taylor & Francis: London and Philadelphia, 1984), appendix 3B, pp. 132–36.

II. Main sources of military expenditure data

Estimates of military expenditure are made on the basis of national sources, including budgets, White Papers and statistical documents published by the government or the central bank of the country concerned. The reference publications listed below are also used. Journals and newspapers are consulted for the most recent figures.

NATO

Financial and Economic Data Relating to NATO Defence, annual press release (NATO: Brussels).

Non-Soviet WTO

Alton, T. P., Lazaricik, G., Bass, E. M. and Badach, K., 'East European defense expenditures, 1965–1982', *East European Economies: Slow Growth in the 1980s*, vol. 2: *Economic Performance and Policy*, selected papers submitted to the Joint Economic Committee, US Congress (US Government Printing Office: Washington, DC, 1985); UN instrument for standardized international reporting of military expenditure.

Annual reference publications

Europa Yearbook (Europa Publications: London)
Government Finance Statistics Yearbook (International Monetary Fund: Washington, DC)
Military Balance (Brassey's: Oxford)
Statistical Yearbook (United Nations: New York)
Statistical Yearbook for Asia and the Pacific (United Nations: Bangkok)
Statistik des Auslandes (Federal Statistical Office: Wiesbaden)
World Military Expenditure and Arms Trade (US Government Printing Office: Washington, DC).

6. Debt, financial flows and international security

SOMNATH SEN

I. Introduction

In the early post-cold war era, economic security has become as important as traditional security concerns. The demands for stability, in an anarchic but interdependent international system, require that developmental problems be addressed as seriously as those related to military security. These concerns are as applicable to Third World countries as to the East European countries that in 1989–90 began the process of political pluralism and economic liberalization. The debt problem, which continues to affect a large part of the world, is an indicator of economic insecurity; as such it needs to be solved in the interests of global peace.

In 1990 both change and continuity characterized the evolution of debt, development and security. According to International Monetary Fund (IMF) forecasts, the total stock of debt owed by debtor countries exceeded for the first time $1300 billion (see table 6.1). This was more than anticipated, yet this figure does not include data for a number of countries (particularly socialist countries' loans to developing economies). Preliminary SIPRI estimates show that while aggregate debt has since surpassed $1400 billion, the level seems to be stabilizing, and the rapid debt growth of 1982–86 has evened out. However, forecasts show that debt-service payments will jump from approximately $160–170 billion in the late 1980s to almost $200 billion in 1991.

Table 6.1. The Third World external debt, selected regions, 1982, 1988–91

Figures are in US $b., current prices.

Region	1982	1988	1989	1990	1991
Africa	122.4	203.7	208.5	225.1	235.6
Latin America and the Caribbean	331.2	409.3	408.0	414.1	414.5
Total Third World debt	839.2	1 234.8	1 237.0	1 302.6	1 353.8
Total Third World debt-service payments	135.9	165.8	153.0	167.6	190.1

Source: International Monetary Fund, *World Economic Outlook*, Oct. 1990.

Official development assistance (ODA) fell in 1989 below its 1988 peak and the short-term trend is downwards. There was greater competition for financial capital for investment as the US trade deficit remained stubbornly high and as the Federal Republic of Germany reduced its trade surplus to accommodate the structural requirements for unification. The Iraqi invasion of Kuwait produced grave uncertainty, with both interest rates and oil prices initially moving upwards, with a resulting future debt-burden for oil importers.

Institutional changes in 1989–90, such as extra resources for the IMF and the World Bank, debt reduction schemes under the Brady Plan,[1] and debt forgiveness by official creditors to the poorest countries, particularly in Africa, have been helpful. It is believed that the corner may have been turned and that the 1990s will bring a major transformation to the debt problem. However, the situation will remain critical for a number of years.

It is useful at the start to set out the connections between the debt problem and security. The latter is defined in the broadest possible terms to include economic, political and military aspects. There are a number of linkages:

1. By reducing the import of essential commodities, the debt crisis can cause developmental failure. Lack of imported intermediate investment goods and technology reduces investment and productivity; this hampers economic growth. Low growth and developmental problems affect economic security.

2. A reduction in economic security erodes the political legitimacy of governments. In extreme cases it may also erode the legitimacy of states.

3. The ability of governments to spend on social, economic and infrastructural programmes is hampered as increasing amounts of revenue are tied to public and publicly guaranteed debt-servicing. For many Third World countries debt-servicing and military expenditure take up 40–80 per cent of all central government revenue, leaving very little for other essential needs.

4. Conflicts, particularly intra-state conflicts, increase as a result of developmental failures; this has a spill-over effect on inter-state relationships.

5. Arms imports, particularly in the Third World but also on the part of some European countries, have been a major contributory factor towards debt creation. While many of these debts were incurred by military and autocratic regimes, payments have often to made by successive democratic governments.

6. North–South relations are affected. The political influence of the major powers is eroded if they fail to curb the debt problem, reducing their influence when it comes to conflict resolution in the South.

7. The breakdown of the world economic order, in part due to the debt crisis, can lead to major security problems as in the 1930s.

8. It is possible to use the problems inherent in the debt crisis in a positive manner. Transfer of resources from the major powers to other countries can be linked to military expenditure reductions in recipient countries, success in arms control and the implementation of confidence- and security-building measures. However, this requires greater policy co-ordination than is currently present in the international security system.

9. The donor or creditor countries can do much more in transferring resources to the debtors. In particular, reductions in military expenditure consequent to the European peace process can potentially release huge

[1] See Sen, S., 'Debt, financial flows and international security', SIPRI, *SIPRI Yearbook 1990: World Armaments and Disarmament* (Oxford University Press: Oxford, 1990), p. 206.

Table 6.2. Third World annual growth/decline of GDP per capita, selected regions, 1980–89

Figures are percentages.[a]

Region	1980–88	1989	Decline/growth 1980–89
Sub-Saharan Africa	– 2.4	+ 0.3	– 17.4
Latin America and the Caribbean	– 0.7	– 0.6	– 6.0
East Asia	+ 7.0	+ 3.2	+ 77.3
Highly indebted countries	– 0.6	– 0.7	– 5.4

[a] (–) decline in per capita income; (+) increase in per capita income.

Sources: *The World Bank Annual Report 1990* (World Bank: Washington, DC, 1990); *World Development Report 1990* (Oxford University Press: Oxford, 1990); author's calculations.

amounts of resources which could be utilized productively to alleviate the economic burden of poorer economies.

Table 6.2 gives an indication of the Third World's economic insecurity. Annual growth rates for sub-Saharan Africa, Latin America and the highly indebted countries have been persistently negative. The last column shows the total decline in per capita income over the 1980s. Taking the decade of the 1980s as a whole, the decline in sub-Saharan Africa was over 17 per cent. The unevenness of growth performance can be judged by comparing the data with that of East Asia, where real income per head rose by over 77 per cent.

There are a number of indices that can be used to represent the economic costs and weaknesses arising out of external debt. The most often used are various forms of debt-service ratios and debt stock shares in macro-economic variables such as gross national product (GNP) or exports. An increasingly used index is the 'vulnerability measure'. Total foreign exchange *requirements* are calculated by adding up the current account balance (import minus export) and maturities of debt that need to be paid. Total *resources* are then calculated by adding reserves and unused credits. The difference between available resources and needed requirements, due shortages of foreign exchange and a debt overhang, is the 'vulnerability' of that country.

The primary focus of this 'vulnerability measure' is on the creditor countries, since it reflects whether debts will be paid or not in the long run and whether debts will be serviced or not in the short term. An alternative vulnerability index is required which will highlight the problems of the debtors and focus attention on their difficulties. As discussed below, the fundamental problem is not necessarily due to the stock of debt itself but to the abnormally high interest rates and debt-servicing amounts that are to be paid. The most startling feature of the late 1980s is the emergence of *negative net transfers* whereby the debtors are paying more to the creditors than they are receiving in new money (see table 6.3). The sum of principal and interest payments now substantially exceeds disbursements, for the world as a whole as well as for

Table 6.3. Third World long-term debt, financial flows, official development assistance and arms imports, 1985–89

Figures are in US $b.

Year	Total debt	Debt-service[a]	Principal	Interest	Net transfer[b]	ODA	Arms imports
1985	781.3	107.3	53.1	54.2	– 19.7	29.4	32.5
1987	1 001.3	122.5	68.9	53.6	– 34.2	41.6	43.8
1989	988.5	129.8	70.3	59.5	– 42.9	46.7	39.3

[a] Debt-service is expressed in total and as principal and interest.

[b] Net transfer is the remainder of new loans minus debt-service.

Sources: *World Bank Annual Report 1990* (World Bank: Washington, DC, 1990); US Arms Control and Disarmament Agency, *World Military Expenditures and Arms Transfers 1989* (US Government Printing Office: Washington, DC, 1990); Organization for Economic Co-operation and Development, *Development Co-operation* (OECD: Paris, 1990); author's estimates.

the highly indebted countries.[2] This feature is used to construct an alternative 'vulnerability index', discussed in section V below.

II. Official development assistance, military expenditure and arms exports of donor countries

Overall in the 1980s, as total indebtedness rose so did debt-service payments. Many indicators reveal that the economic burden to the Third World was high. While long-term debt stock seemed to have stabilized at around $1 billion, a negative net transfer has emerged, whereby Third World debtors pay back more per year (in principal and interest payments) than they receive in new money. Official development assistance (ODA) can barely compensate for this negative transfer—a financial haemorrhage that takes away foreign exchange and investment opportunities from poor countries. Table 6.3 gives data which reveal this trend. In 1989, net transfer—how much the poor paid to the rich—plus arms imports by the Third World exceeded $80 billion. Comparable data show that ODA barely exceeded $53 billion. Net transfer and arms therefore exceeded foreign aid by about $29 billion. Western assistance alone amounted to $46.7 billion—a real decline of 1.7 per cent from the previous year.

In an increasingly interdependent world, developing economies need finance and foreign exchange for investment and development. However, the lack of ODA and an accumulating debt overhang, whereby new loans are difficult to obtain due to past indebtedness, has meant that developing countries are increasingly short of development finance. Equally important, there exists a world-wide shortage of financial capital, since a few deficit countries absorb most of the world's surplus funds. The low US savings rate, and high government spending prompted by defence expenditure, has meant

[2] World Bank, *World Bank Annual Report 1989* (World Bank: Washington, DC, 1989).

Table 6.4. Current account balance of selected trading countries, 1987–91

Figures are in US $b.

	1987	1988	1989	1990	1991
Japan	87.0	79.6	57.2	47.5	55.8
FR Germany[a]	45.8	50.4	55.4	48.9	38.4
Asian NICs[b]	30.4	27.8	21.3	13.1	11.8
Total of surplus countries	**163.2**	**157.8**	**133.9**	**109.5**	**106.0**
USA	– 162.3	– 128.9	– 110.0	– 97.0	– 99.7
UK	– 6.0	– 26.0	– 31.3	– 26.6	– 21.6
Total of deficit countries	**– 168.3**	**– 154.9**	**– 141.3**	**– 123.6**	**– 121.6**

[a] Figures for 1990 and 1991, after the unification of Germany, do not include data for the eastern part (former GDR).

[b] Newly industrializing countries: Hong Kong, Singapore, South Korea and Taiwan.

Source: International Monetary Fund, *World Economic Outlook*, Oct. 1990

that the USA is absorbing a disproportionate amount of international capital. The unification of the two German states implies that surplus capital in the FRG will increasingly be tied to financial and infrastructural modernization programmes in the former GDR, leaving less to the rest of the world. Table 6.4 shows the external balance of the major surplus and deficit countries. In 1987–91 the USA and the UK soaked up the financial surplus produced by Japan, the FRG and the newly industrializing countries (NICs) of the Far East.

ODA levels in the late 1980s are somewhat higher than at the beginning of the past decade. However, there is genuine concern that they will fall or become more difficult to obtain. Expectations about the international disarmament dividend are still high, but whether defence spending reductions can be translated into aid is a matter of political choice and not economic decision-making. The potential for resource transfers is high, as table 6.5 shows. In the

Table 6.5. Official development assistance as share of GNP and military expenditure; military expenditure as share of GDP, major donor countries, 1989

Figures are percentages.

Country	ODA/GNP	ODA/Milex	Milex/GDP
Canada	0.44	22.0	2.0
France[a]	0.54	15.4	3.7
FR Germany	0.41	14.6	2.8
Italy	0.42	17.5	2.4
Japan	0.32	32.0	1.0
UK	0.31	7.4	4.2
USA	0.15	2.5	5.9
USSR	0.24	2.2	11.0

[a] Excluding overseas territories.

Source: Organization for Economic Co-operation and Development, *Development Co-operation* (OECD: Paris, 1990); SIPRI data base; author's calculations.

Table 6.6. US military assistance to the Third World, loans and grants, FYs 1982, 1985–88

Figures are in US $b., current prices.

Military assistance	1982	1985	1986	1987	1988
Loans	3.1	2.4	2.0	1.0	0.8
Grants	1.1	3.4	3.9	4.1	4.1
Total	**4.2**	**5.8**	**5.8**	**5.1**	**4.8**

Source: Agency for International Development, *Overseas Loans and Grants, July 1, 1945–September 30, 1988* (Office of Planning and Budgeting, USAID: Washington, DC, 1989).

USA and the USSR, foreign aid amounts to about $2–2.5 per $100 of military expenditure. Thus even a transfer of modest amounts could make a substantial difference. A reduction in military expenditure by 1 per cent would permit an increase in US ODA by 40 per cent. For the Soviet Union the potential increase in foreign assistance would be of the order of 45 per cent.

In 1990 the authoritative annual *World Development Report* of the World Bank focused on poverty in the Third World. Among its policy initiatives to better the lot of the 1.1 billion people in the Third World who live below the poverty line, the Report claimed: 'The world is at a turning point. . . . This offers a unique opportunity to cut military spending and increase international assistance. A cut of just 10 percent in military spending by the countries of the North Atlantic Treaty Organization would pay for a doubling of aid. The resources can be made available—although little will be achieved unless they are used effectively.'[3]

However, this optimistic prognosis will become meaningless if the major powers are not responsive to Third World economic needs. A first step could be the trend reduction in military assistance and resource transfer towards economic aid. Unfortunately, the conflict in the Persian Gulf makes this less likely in the near future. Table 6.6 shows that US military assitance remained

Table 6.7. Soviet military exports, aid and defence debt creation in the Third World, 1982–88 (even years)

Figures are in US $b., current prices.

Allocation	1982	1984	1986	1988
Hard currency sales	4.3	3.8	2.8	3.3
Foreign military aid	6.2	6.1	6.0	7.8
Residual (debt creation)	5.5	6.4	6.7	8.0
Total	**16.0**	**16.3**	**15.5**	**19.1**

Source: *Allocations of Resources in the Soviet Union and China*, Hearings before the Subcommittee on National Security Economics, Joint Economic Committee, 14 Apr. and 7 July 1989, 101st Congress (US Government Printing Office: Washington, DC, 1990).

[3] World Bank, *World Development Report 1990* (Oxford University Press: Oxford, 1990), p. 4.

Table 6.8. Third World debt to the Soviet Union, 10 major debtor countries, 1989

Figures are in b. roubles.

Country	Debt stock	Share in total Third World (%)	Principal	Deferred (1986–88)
Cuba	15.5	*19.6*	15.1	2.4
Mongolia	9.5	*12.0*	9.0	2.0
Viet Nam	9.1	*11.5*	8.9	1.6
India	8.9	*11.2*	8.9	1.6
Syria	6.7	*8.5*	6.5	1.0
Iraq	3.8	*4.8*	3.5	1.4
Afghanistan	3.1	*3.9*	2.9	0.6
Ethiopia	2.9	*3.7*	2.8	0.8
Algeria	2.5	*3.2*	2.5	0.6
North Korea	2.2	*2.8*	2.2	0.4
Total 10 major debtor countries	**64.2**	***81.1***	**62.3**	**12.4**
Total Third World	**79.2**	***100***	**76.5**	**14.2**

Source: *Izvestia*, 2 Mar. 1990, in Foreign Broadcast Information Service, *Daily Report—Soviet Union (FBIS-SOV)*, FBIS-SOV-90-043, 3 Mar. 1990, p. 83.

at around $5 billion—over 60 per cent of total aid—in the latter half of the 1980s.

US intelligence estimates of Soviet military exports, aid and debt creation are shown in table 6.7. As the source is not unbiased and as the rouble/dollar exchange rate use may not be appropriate, these figures should be treated with caution. However, the data reveal orders of magnitude. Third World countries have run up sizeable debts to the USSR due to arms imports alone.

In 1990, the USSR released for the first time information on all Third World debt (military and non-military) owed to itself (table 6.8). It is interesting to note that among the top 10 debtors, 8 are *actively* involved in regional arms races, conflicts and wars (the exceptions are Mongolia and Algeria).

III. Security and development in the Third World

There is an implicit contradiction between military security and economic development in the Third World. Small states are particularly vulnerable to regional threats, as the Iraqi invasion of Kuwait demonstrates. While appropriately utilized ODA can alleviate some of the threats to economic security, high defence spending can swamp the beneficial effects of international resource transfers. As table 6.9 shows, for many Third World countries the ratio of ODA to military expenditure is low. For example, Nicaragua spent almost four times more on defence than it received in economic assistance in 1988.

The combination of two basic 'unproductive' expenditures—in the military and on debt servicing—can overwhelm other items of government spending. Thus, the allocations for health, education, social services, infrastructure and economic investment, are crowded out by states struggling to meet external

Table 6.9. Official development assistance as share of GDP and military expenditure; military expenditure as share of GDP, selected recipient countries, 1988

Figures are percentages.

Country	ODA/GDP	ODA/Milex	Milex/GDP
Argentina	0.3	10.0	3.0
Chile	0.05	0.6	7.8
China	0.5	12.8	3.9
Costa Rica	5.7	1 425.0	0.4
Côte d'Ivoire	2.9	241.7	1.2
Egypt	4.7	75.8	6.2
El Salvador	10.2	261.5	3.9
India	1.1	29.7	3.7
Israel	6.2	68.1	9.1
Mozambique	16.7	167.0	10.0
Nicaragua	2.3	7.7	30.0
Nigeria	0.3	30.0	1.0
Pakistan	2.6	37.7	6.9
Paraguay	1.5	150.0	1.0
Peru	1.2	24.0	5.0
Zaire	10.7	713.0	1.5
Zimbabwe	5.1	87.9	5.8

Sources: Organization for Economic Co-operation and Development, *Development Co-operation* (OECD: Paris, 1988); SIPRI data base; author's calculations.

debt obligations and pay for their military security. Table 6.10 shows that for a wide range of countries, an overwhelming share of government revenue, from 38 to more than 100 per cent, is being spent on debt and defence, leaving little for other items essential for economic security.

In 1990 the South Commission—a high-level independent group of politicians and experts from the developing countries alone—presented a report providing a wide-ranging review of developmental problems. Aside from other issues, the report also emphasizes militarization, arms exports and defence spending:

> The international community has the duty to put in place a framework that would guarantee the security of all nations against external threats, including incursions by mercenaries. Nevertheless, it remains an unfinished task of the countries of the South to work out effective mechanisms for settling international and internal conflicts through peaceful means. These mechanisms, together with the strengthening of democratic processes, can play an important role in curbing military expenditure.[4]

IV. The cost of the Gulf conflict to the Third World

1990 will be remembered as the year of the first major conflict of the post-cold war era. The Iraqi invasion of Kuwait on 2 August 1990 could have a

[4] *The Challenge to the South* (Oxford University Press: Oxford, 1990), p. 34.

Table 6.10. External public debt-service and military expenditure as shares of current government revenue, selected Third World countries, 1988

Figures are percentages.

Country	External debt-service	Military expenditure	Debt-service plus military expenditure
Argentina	22.3	15.2	37.5
Colombia	54.3	16.7	71.0
Chile	19.6	24.6	44.2
Egypt	10.8	18.1	28.9
Indonesia	51.6	12.0	63.6
Jordan	67.1	51.4	118.5
Morocco	26.2	20.5	46.7
Pakistan	20.6	41.8	62.4
Philippines	49.0	9.1	58.1
Sri Lanka	24.7	16.8	41.5
Zimbabwe	22.8	18.1	40.9

Sources: *World Development Report* (World Bank: Washington, DC, 1990); SIPRI data base; author's calculations.

significantly adverse impact on the economic security of the Third World. However, it is hard to judge the aggregate impact. Some specific difficulties for developing countries include: the initial rise in the price of oil, leading to a decline in energy use and a fall in gross domestic product (GDP); a corresponding increase in the value of oil imports, leading to balance-of-payments deficits, increase in debt stock and reduction in the import of other essential commodities; loss of exports to relatively rich countries such as Kuwait and Iraq; and loss of migrant workers' remittances from the Gulf. Added to this is the costs of increased military expenditure, in the region as well as elsewhere.

It should be noted that the Iraqi invasion of Kuwait has an economic dimension as well. Iraq was technically bankrupt, burdened by a huge debt it was unable to service. Much of it was accumulated during the war with Iran, and the majority of debt obligations seems to be related to arms imports.[5] Iraq's identified debt to OECD countries, and to multilateral institutions and private banks, amounted to $14.5 billion by 1988 (the latest year for which precise figures are available).[6] In most cases this amount excludes military debt, which cannot be identified from normal debt and trade statistics. Making tentative but modest adjustments for arms-related debt, the total indebtedness to Western countries is about $20 billion. Aggregate debt owned to the USSR is around $6.1 billion. Adding estimated debts to Arab and Gulf countries, Iraqi international indebtedness could have amounted to $76–$86 billion at the beginning of 1990. Thus, faced with a post-war reconstruction burden and low oil prices, Iraq may have resorted to an old-fashioned war of booty.

[5] See chapter 7.

[6] Organization for Economic Co-operation and Development, *Financing and External Debt of Developing Countries: 1989 Survey* (OECD: Paris, 1990).

Table 6.11. Economic costs of the Gulf crisis until mid-January 1991

Figures are in US $b., current prices. Items are explained in text.

Type of cost	Value
Impact on GDP of Third World oil importers	16.7
Impact on balance of payment and debt	13.8
Additional regional military expenditure	3.9
Operation Desert Shield	7.3
Additional expenditure by Saudi Arabia	11.5
Total	**53.2**

Source: Author's estimates.

The embargo and economic sanctions sponsored by the UN were swift and effective. By the end of 1990 the Iraqi economy was hurt and military imports had ceased. The effectiveness of sanctions can be judged by the testimony to Congress by CIA Director William Webster that 97 per cent of Iraq's exports had been eliminated and 90 per cent of Iraq's imports had been stopped by the end of 1990.[7] By closing pipelines, a total embargo on Iraqi oil was achieved.

The military option, vigorously pursued by the USA, led to the mobilization against Iraq of a massive force of over half a million troops from more than 30 nations.[8] Although Operation Desert Shield is predominantly a US operation, allied help in manpower and material was large.[9] Initial cost estimates for FY 1991 for the USA alone were increased from $15 billion to $25 billion.[10]

SIPRI has made independent estimates of the aggregate financial costs that had accumulated by the outbreak of hostilities on 17 January 1991 (table 6.11). The impact on GDP of the factors mentioned above, particularly higher oil prices and increases in interest rates, is calculated and estimated from basic data provided by the World Bank;[11] the same is true for the impact on balance of payments which will give rise to further international indebtedness. Regional military expenditure will rise by almost $4 billion according to SIPRI estimates. Operation Desert Shield is expected to cost as much as $25 billion. In addition, the Saudi Government is expected to make contributions over and above the sums paid for Desert Shield (for transport, accommodation and fuel to the multinational forces) such as compensation for countries which have imposed embargoes and aid to friendly developing countries.

Oil exporters are expected to gain significantly. Press reports claim that Saudi Arabia itself has made a windfall profit of over $13 billion from higher oil prices. However, at an aggregate level these gains will cancel out with other unaccounted losses, such as those made by industrial countries, in particular by East European countries, which stand to lose significantly. Due

[7] *Guardian Weekly*, 23 Dec. 1990, p. 7.

[8] For a discussion of military mobilization in the Persian Gulf conflict, see chapter 19 in this volume.

[9] For force deployment figures for Operation Desert Shield, see chapter 19 in this volume.

[10] 'Cost of Operation Desert Shield', *Financial Times*, 5 Dec. 1990.

[11] *World Debt Tables 1990–91, Vol. 1: Analysis and Summary Tables* (World Bank: Washington, DC, 1990), p. 23.

to problems in verifying the data, specific costs to some directly affected Third World countries are also not accounted for. These are discussed below. Overall, the figures presented in table 6.11 are net costs to the international community, affecting in particular the Third World, and are of the correct order of magnitude.

The total financial costs for the international community, excluding war costs accrued after 16 January 1991, comes to over $53 billion. It is worth noting that this is nearly as much as the ODA that Third World countries are now receiving.

Poorer front-line states, as well as those countries heavily dependent on migrant workers' remittances, tend to have special problems. They also have higher economic losses not fully accounted for in the global calculations given above. Countries such as Bangladesh, Egypt, India, Jordan, Pakistan, Sri Lanka, Turkey and Yemen have encountered major economic problems as a result of the conflict; the specific economic impact may be illustrated by developments in Egypt, Turkey and Bangladesh. Egypt, which has historically shouldered the burden of the Palestinian cause, has suffered intensively in this conflict. While Turkey is prosperous relative to the other countries in this group, it merits special attention for its membership in NATO and the OECD, notwithstanding its Third World status and low per capita income. Bangladesh, one of the 10 poorest countries in the world, has been particularly hard hit due to its dependence on remittances and aid from the Gulf countries.

According to domestic and international official sources, Egypt has incurred economic costs on a number of accounts. The specific amounts of *annual* financial losses break down as follows: deposits kept in Iraqi banks ($0.54 billion); workers' lost remittances ($2.0 billion); lost customs duties from overseas workers bringing goods home ($0.25 billion); revenue lost from the Suez Canal operations due to embargo on Iraqi ships ($0.255 billion); lost tourism ($0.45 billion); additional defence spending ($0.5 billion).[12] The total amounts to almost $4 billion. Offsetting these costs have been gains from the rise in the price of oil amounting to a total of $1.1 billion and the possible cancellation of military debts to the USA, saving a further $0.8 billion in interest payments. The net effect of these expenditures and savings amounts to a cost to Egypt of at least $2.1 billion. This corresponds to over 5.5 per cent of the countrty's GDP. It should also be noted that these figures essentially represent 'first round' reductions in income; additional multiplier effects could well double the amounts stated. In addition, the fate of the staggering $14 billion deposited in Kuwaiti banks is unknown. If these funds are not recovered, Egypt could well lose over $16 billion net due to the Gulf conflict.

Many of the same factors affect Turkey as well. For example, the closure of the pipeline through which Iraqi oil was transported, and the loss of cheap oil, has contributed an additional $1 billion to Turkey's oil bill. The aggregate amount, calculated by the Government, is of the order of $5 billion of lost

[12] Details from official government sources reported in *The Independent*, 3 Sep. 1990; data also from the SIPRI data base and from Organization for Economic Co-operation and Development, *OECD Economic Outlook December 1990* (OECD: Paris, 1990).

incomes and additional costs to the country.[13] This is over 7 per cent of Turkey's GDP. Again, multiplier effects could increase the impact.

Bangladesh has been a major loser in this conflict. According to UN officials,[14] losses due to reductions in remittances, exports and Kuwaiti foreign aid amount to $0.5 billion in 1990. This is equivalent to 2.5 per cent of the Bangladesh GDP. With multiplier effects added, around 5 per cent of GDP is lost principally due to these three items alone. In addition, there will be increasing debt burdens, growing outlays for higher priced oil, increased interest payments for foreign loans and other indirect costs.

V. Eastern Europe

The debt crisis bedevils many of the non-Soviet Warsaw Treaty Organization (NSWTO) countries; its solution is a *sine qua non* of future peace through prosperity. Without its effective resolution, full integration in the international economic order, actively sought by these economies, is not possible. In addition, domestic stability will be reduced if the 'shortage economy', with open or repressed inflation, continues to operate. Imports are essential both for consumption and investment. Maintenance of adequate consumption in the short to medium term requires foreign goods. As for investment, to reverse growth retardation (a characteristic feature of the economies of the previous socialist countries) imported intermediate investment goods and technology are also essential. Yet, the debt overhang makes it impossible to continue with such essential imports.

In 1990 Eastern Europe's national output is expected to decline by about 5.3 per cent. Poland's per capita income in 1990 is reported to be less than what it was in 1979. The price of oil is expected to be a major destabilizing factor, especially from 1991 when all countries are expected to pay in hard currency for their Soviet oil. For example, Hungary may have to incur costs of about 2.2 per cent of its GDP as a result of switching from rouble to dollar trade with the USSR. Harsh measures have been taken to prevent inflation, and these are having an effect on the living conditions of people. Unemployment in Poland is about 5 per cent of the labour force; a few years ago it was zero. Other countries in the region also expect unemployment rates to go up even though they are currently small. In 1990, unemployment in Hungary was 1 per cent of the labour force. Wage increases have remained substantially below inflation levels, contributing to the rapid fall in workers' real standard of living.[15]

International assistance to Eastern Europe, particularly to Poland and Hungary, has been adequate but not overwhelming in 1990. Most aid is now co-ordinated by the European Community (EC), through a programme called PHARE (*Pologne Hongrie: Assistance à la Restructuration Economique*). Set

[13] See *Financial Times*, 10 Jan. 1991.

[14] Reported in *Guardian Weekly* (note 6).

[15] For a perceptive analysis of all these countries, see Economic Commmission for Europe, *Economic Survey of Europe in 1989–1990* (United Nations: New York, 1990).

Table 6.12. WTO gross debt, selected years, 1981–89

Figures are in US $b., current prices.

Country	1981	1985	1987	1988	1989
Bulgaria	3.2	3.7	6.3	7.9	9.5
Czechoslovakia	4.6	3.5	5.3	5.7	6.9
GDR	15.4	14.4	18.9	19.5	21.2
Hungary	8.7	11.7	17.7	17.3	20.6
Poland	25.9	29.8	38.8	39.2	41.0
Romania	10.2	6.6	6.0	2.8	1.0
Total NSWTO	**68.0**	**69.7**	**93.0**	**92.4**	**100.2**
USSR	26.5	28.0	36.5	40.9	48.0
Total WTO	**94.5**	**97.7**	**129.5**	**133.3**	**148.2**

Source: Organization for Economic Co-operation and Development, *Financial Market Trends* (OECD: Paris, Feb. 1990).

up in 1989 specifically to channel aid to Poland and Hungary, as the name implies, its scope was enlarged in early 1990 to include the other countries in the region. Individual association agreements to help trade and financial relations with the EC, not specifying explicit membership commitments, have been signed with Czechoslovakia, Hungary and Poland. As part of another EC effort, the European Bank for Reconstruction and Development (EBRD) was established with headquarters in London. It is the first financial institution to be owned by all governments in Europe, cutting across the East–West divide. Ownership vests in 34 countries.

In spite of favourable developments, conditions remain critical and potentially destabilizing. At least for Poland, Hungary and Bulgaria, many of the economic problems stem from the debt overhang. The Balkan countries also face traumatic economic difficulties, inseparable, in the case of Yugoslavia, from an ongoing political disintegration.

The critical link between economic crisis and political chaos, and its adverse impact on foreign policy and arms control, is the most obvious in the case of the USSR. By early 1991 the Soviet armed forces seemed to be exercising increasing influence on the political process. An authoritative study on the Soviet economy, prepared by the IMF, the World Bank, the OECD and the EBRD, was pessimistic about the prospect of economic reform, claiming that the reforms begun under *perestroika* are not coherent and comprehensive enough to achieve success in the long run.[16] Although the EC has authorized 1.15 billion ECU (European currency units), or $1.6 billion, for food aid and technical assistance, this support is increasingly seen as hinging on the course of Soviet domestic policy. If military crack-downs typified by the intervention in Lithuania in January 1991 continue, large-scale foreign aid is not likely to be forthcoming in the near future.[17]

[16] International Monetary Fund, *The Economy of the USSR* (IMF: Washington, DC, 1990).
[17] *Financial Times*, 17 Dec. 1990.

Table 6.13. Comparison of estimated vulnerability index for Hungary, Poland and all debt-burdened countries, 1986–88

Figures are percentage averages.

Countries	1986	1987	1988
Poland	– 37.8	– 84.3	– 48.5
Hungary	+ 2.3	– 1.0	– 31.0
All debt-burdened countries	– 18.0	– 16.0	– 28.4
SIMIC	– 32.1	– 25.8	– 45.2
SILIC	+ 37.2	+ 50.8	+ 12.0
MIMIC	– 8.8	– 19.2	– 16.7

SIMIC: severely indebted middle-income countries
SILIC: severely indebted low-income countries
MIMIC: moderately indebted middle-income countries

Source: Author's calculations from *World Debt Tables 1989–90*, vols 1 and 2 (World Bank: Washington, DC, 1990).

Data for Warsaw Treaty Organization (WTO) debt are given in table 6.12. The figures for 1989 are preliminary. The debt-service ratio is defined as all interest and amortization on medium- and long-term debt as a percentage of one year's exports. During the late 1980s the six East European economies, with similar structural features, spent almost half of their export earnings on debt servicing. Clearly, this situation is unsustainable.

The debt crisis is tightening the most severe constraint that debtor countries face—that of foreign exchange. This is particularly true for economies, such as in Eastern Europe, which have adequate domestic savings and capacity to absorb (through human capital) productive investment, but which lack foreign resources, particularly in the crucial area of financing technology transfers. The most critical difficulty in this process is negative net transfer, as indicated by the vulnerability index that SIPRI has constructed.

The SIPRI index is the ratio of net transfers over international reserves, the latter showing the stock of foreign resources. A negative value obviously implies that net transfers are negative. As the index declines from positive to negative, the situation clearly becomes worse in terms of resource transfers. When the index reaches 100, a critical level is reached where net transfers away from the economy is enough to exhaust its total international reserves.

Tables 6.13 and and 6.14 give some figures for 1986–88. In table 6.13 data for Poland and Hungary are compared with that of other groups of debtor countries. Table 6.14 is concerned with the six major European debtor nations: Greece, Hungary, Poland, Portugal, Turkey and Yugoslavia. In terms of the SIPRI vulnerability index their situation is critical. In 1987, for example, Poland and Yugoslavia paid back (in net figures) around 100 per cent of their international reserves. In addition, the vulnerability index is declining over

Table 6.14. Estimated vulnerability index for the six major European debtor countries, 1986–88

Figures are percentages.

Country	1986	1987	1988
Greece	+ 10.7	- 12.2	- 22.3
Hungary	+ 2.3	- 1.0	- 31.0
Poland	- 37.8	- 84.3	- 48.5
Portugal	- 14.9	- 18.8	- 12.4
Turkey	+ 6.0	- 7.8	- 12.9
Yugoslavia	- 80.5	- 113.4	- 34.6

Source: Author's calculations.

time for all countries, as a product of higher negative net transfers and lower reserves.

The SIPRI vulnerability index sheds additional light on the debt crisis issue. In 1988 public and publicly guaranteed debt produced a net transfer of $6 billion for the six major European debtor nations. A write-off policy which reduced this amount to zero is a minimalist policy which at least leaves these countries in a sort of neutral equilibrium. By comparison, over $500 billion has been spent every year *in Europe* as military expenditure to preserve security. Consider a hypothetical situation where even half of this massive amount is available as a potential 'disarmament dividend'. It will then require simply 4 per cent of the 'disarmament dividend' to finance the scheme. Yet, the impact in terms of debt alleviation will be very high. The rewards, in terms of economic security and stability in Europe, are immense. Numerous other such calculations can be made. However, the basic issue is clear. Demilitarization will release substantial resources. There is every reason to use these to enhance non-military security and to fight against non-military threats as effectively and as strongly as the previous fight was conducted against military threats.

VI. Conclusion

The euphoria of what has been perceived as the dawning of a post-cold war era was dampened in 1990 with the erruption and escalation of the conflict in the Persian Gulf. However, provided the right lessons are learned, this conflict need not upset the momentum of peaceful structural change set in motion elsewhere. One central lesson must be that military and economic security are interlinked and that it is futile to build a framework for one without considering the other. In similar fashion, domestic and international policies are connected. Developmental needs must thus be met if a lasting security order is to be achieved—otherwise the 1990s will become another 'lost decade'.

7. The trade in major conventional weapons

IAN ANTHONY, AGNÈS COURADES ALLEBECK,
GERD HAGMEYER-GAVERUS, PAOLO MIGGIANO
and HERBERT WULF

I. The arms trade in 1990

The global value of the trade in major conventional weapons in 1990 is estimated to have been $21 726 million in 1985 US dollars. This represents a decrease of 35 per cent from the value for 1989,[1] which was itself a lower figure than those recorded for the years of the mid-1980s. The value of deliveries of major conventional weapons in 1990 is less than 60 per cent of the value recorded for 1987, a peak year.

Within the overall total, the share of deliveries to Third World recipients in 1990 was 55 per cent—the same share as recorded for 1989. The USSR and the USA remained the largest exporters of major conventional weapons in 1990, together accounting for 69 per cent of the total value of deliveries. However, in 1990 the value of US arms deliveries exceeded that of Soviet arms exports, reversing the established rank order. Soviet exports of major conventional weapons fell rapidly in 1990 mainly because of drastically reduced exports to Third World countries—particularly to India, Iraq and Afghanistan—but also as a consequence of political changes in the WTO. In 1989 the USSR and the USA accounted for 37 and 34 per cent, respectively, of the world total; in 1990 their shares were 29 and 40 per cent (see figure 7.1).

The 12 member countries of the European Community accounted for 20 per cent of the total value of major conventional weapons delivered in 1990. This share has been fluctuating throughout the period 1986–90 at around 20 per cent. However, within that total, the share of deliveries accounted for by France, the UK and FR Germany—the three largest arms producers in Western Europe—rose significantly to over 90 per cent. This reflects the decline in the value of exports of major conventional weapons by the Netherlands, Spain and Italy and, to a lesser extent, Belgium and Portugal.

The value of major conventional arms exports by the Third World continued to decline in 1990, representing just over 1 per cent of total exports compared with over 4 per cent in 1987. Exports of major weapons by two of the largest Third World arms producers, Brazil and Israel, declined dramatically in 1990, while exports of major conventional weapons from Singapore and South Korea, which had shown significant growth until the mid-1980s, were virtually eliminated by 1990 (see table 7.1). However, this may exagger-

[1] This reduction should be viewed with caution since the initial estimate for the latest year generally increases as more information becomes available.

Table 7.1. The leading exporters of major conventional weapons, 1986–90

The countries are ranked according to 1986–90 aggregate exports. Figures are in US $m., at constant (1985) prices

Exporters	1986	1987	1988	1989	1990	1986–90
To the Third World						
1 USSR	10 440	10 936	8 658	8 862	4 273	43 169
2 USA	4 981	6 328	3 939	3 465	3 048	21 761
3 France	3 446	2 659	1 413	1 642	1 330	10 490
4 China	1 463	2 553	1 810	817	926	7 569
5 UK	1 091	1 681	1 281	1 187	971	6 210
6 Germany, FR	661	254	367	168	496	1 946
7 Netherlands	132	263	402	661	125	1 583
8 Italy	399	320	362	49	39	1 169
9 Brazil	134	491	338	151	22	1 136
10 Israel	261	267	111	241	31	912
11 Sweden	141	298	240	134	1	813
12 Spain	163	139	193	244	62	802
13 Czechoslovakia	124	198	176	178	58	733
14 Egypt	159	194	216	65	33	668
15 Korea, North	48	103	128	11	11	300
Others	471	543	392	383	415	2 203
Total	**24 114**	**27 228**	**20 025**	**18 256**	**11 841**	**101 464**
To the industrialized world						
1 USA	5 323	6 268	6 564	8 204	5 690	32 050
2 USSR	4 291	3 981	3 901	3 359	2 099	17 631
3 France	650	352	888	936	469	3 293
4 Germany, FR	458	422	903	548	468	2 799
5 Czechoslovakia	373	373	373	259	297	1 674
6 UK	409	135	120	629	249	1 542
7 Sweden	183	191	336	177	114	1 000
8 Canada	278	228	81	51	25	662
9 Poland	92	92	92	92	92	462
10 Italy	58	69	110	119	58	413
11 Switzerland	6	15	19	144	212	394
12 Netherlands	109	2	130	64	27	332
13 Spain	8	0	6	262	12	288
14 Israel	8	73	16	78	8	182
15 Saudi Arabia	39	125	0	0	0	164
Others	54	224	204	332	67	881
Total	**12 338**	**12 549**	**13 741**	**15 253**	**9 885**	**63 767**
To all countries						
1 USSR	14 731	14 916	12 559	12 220	6 373	60 799
2 USA	10 304	12 596	10 503	11 669	8 738	53 811
3 France	4 096	3 011	2 300	2 577	1 799	13 783
4 UK	1 500	1 817	1 401	1 816	1 220	7 752
5 China	1 463	2 553	1 868	874	926	7 684
6 Germany, FR	1 120	676	1 270	716	963	4 745
7 Czechoslovakia	497	570	548	437	355	2 408
8 Netherlands	240	265	532	725	152	1 915
9 Sweden	324	489	575	311	115	1 813
10 Italy	457	389	471	169	96	1 582
11 Brazil	150	507	356	152	24	1 189
12 Israel	269	340	127	318	39	1 094
13 Spain	172	139	199	506	74	1 090
14 Canada	317	265	106	54	60	802
15 Egypt	159	194	216	65	33	668
Others	656	1 047	735	900	760	4 097
Total	**36 453**	**39 777**	**33 767**	**33 509**	**21 726**	**165 232**

Source: SIPRI data base.

Table 7.2. The leading importers of major conventional weapons, 1986–90

The countries are ranked according to 1986–90 aggregate exports. Figures are in US $m., at constant (1985) prices.

Importers	1986	1987	1988	1989	1990	1986–90
Third World						
1 India	3 729	4 582	3 382	3 754	1 541	16 989
2 Saudi Arabia	2 413	2 400	2 046	1 427	2 553	10 838
3 Iraq	2 484	4 440	2 155	1 177	59	10 314
4 Afghanistan	692	768	1 009	2 183	1 091	5 742
5 Korea, North	1 019	631	1 458	1 276	516	4 900
6 Egypt	1 645	2 379	348	139	206	4 717
7 Syria	1 511	1 172	1 172	336	0	4 191
8 Angola	980	1 140	889	74	508	3 592
9 Korea, South	287	604	987	997	249	3 125
10 Iran	738	704	558	336	578	2 913
11 Israel	446	1 629	507	100	21	2 703
12 Pakistan	609	467	467	760	390	2 693
13 Taiwan	825	575	459	391	178	2 427
14 Thailand	94	644	540	489	558	2 325
15 Libya	1 363	294	78	511	0	2 247
Others	5 279	4 797	3 971	4 306	3 393	21 747
Total	**24 114**	**27 228**	**20 025**	**18 256**	**11 841**	**101 464**
Industrialized world						
1 Japan	1 780	1 768	2 176	3 163	2 083	10 971
2 Spain	1 039	1 513	1 580	794	639	5 565
3 Poland	1 057	1 007	1 147	1 179	330	4 719
4 Czechoslovakia	1 077	964	1 054	1 055	422	4 571
5 Turkey	465	1 028	1 219	1 037	623	4 372
6 Greece	156	93	783	1 367	613	3 012
7 Germany, FR	411	301	298	916	1 043	2 970
8 Australia	699	478	579	714	353	2 822
9 Canada	770	702	443	244	289	2 448
10 German DR	515	359	503	502	412	2 292
11 USSR	473	497	483	359	359	2 172
12 Netherlands	702	296	154	761	108	2 021
13 Bulgaria	684	568	187	17	334	1 790
14 Norway	153	395	275	479	348	1 650
15 Yugoslavia	103	234	748	450	14	1 550
Others	2 255	2 345	2 113	2 216	1 914	10 843
Total	**12 338**	**12 549**	**13 742**	**15 253**	**9 885**	**63 768**
All countries						
1 India	3 729	4 582	3 382	3 754	1 541	16 989
2 Japan	1 780	1 768	2 176	3 163	2 083	10 971
3 Saudi Arabia	2 413	2 400	2 046	1 427	2 553	10 838
4 Iraq	2 484	4 440	2 155	1 177	59	10 314
5 Afghanistan	692	768	1 009	2 183	1 091	5 742
6 Spain	1 039	1 513	1 580	794	639	5 565
7 Korea, North	1 019	631	1 458	1 276	516	4 900
8 Poland	1 057	1 007	1 147	1 179	330	4 719
9 Egypt	1 645	2 379	348	139	206	4 717
10 Czechoslovakia	1 077	964	1 054	1 055	422	4 571
11 Turkey	465	1 028	1 219	1 037	623	4 372
12 Syria	1 511	1 172	1 172	336	0	4 191
13 Angola	980	1 140	889	74	508	3 592
14 Korea, South	287	604	987	997	249	3 125
15 Greece	156	93	783	1 367	613	3 012
Others	16 119	15 287	12 361	13 552	10 293	67 612
Total	**36 453**	**39 777**	**33 767**	**33 509**	**21 726**	**165 232**

Source: SIPRI data base.

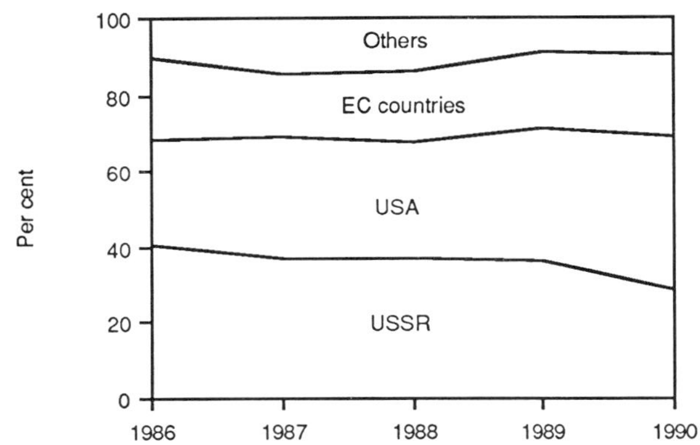

Figure 7.1. Shares of exports of major conventional weapons, 1986–90

ate the decline of arms production in the Third World since the data exclude such items as small arms and artillery with a calibre of less than 100 mm.

In 1990, the major arms-importing country was Saudi Arabia, reflecting the immediate impact of the crisis stemming from the invasion of Kuwait by Iraq. The arms imports of other countries in the Gulf region, notably the United Arab Emirates and Bahrain, also increased significantly in 1990—although in most cases the value of these imports was too low to be represented among the leading 15 importing countries listed in table 7.2. Elsewhere in the Middle East, however, the pattern of reduced arms imports continued. For the period 1986–90 Iraq remained a major importer, although its import of major conventional weapons in 1990 was insignificant. As of December 1990, Israeli arms imports had not returned to the level of the mid-1980s, while there was no evidence that Syria or Libya, previously among the leading arms importers, received any new major conventional weapons.

II. Major arms trade developments in 1990

The crisis around the Persian Gulf, stemming from the invasion of Kuwait by Iraq on 2 August 1990, and the agreements signed in Paris on 20 November 1990 by the 34 heads of government of member states of the Conference on Security and Co-operation in Europe (CSCE) both have implications for the arms trade.

The Gulf crisis notwithstanding, global deliveries of new major weapon systems have continued to fall for several reasons. First, lack of hard currency in many Third World countries that have been major importers in the past is combined with the increased cost of modern weapons. Second, key recipients—notably Afghanistan, Angola, India, Libya, North Korea and Syria—have reduced their imports. Three important arms recipients—Iraq, Jordan and

Kuwait—are now subject to an arms embargo. Third, the transfer of arms now considered surplus as a result of the Treaty on Conventional Armed Forces in Europe (CFE) may reduce demand for new major conventional weapons. Fourth, there is fierce competition for remaining contracts in countries such as Japan, India, Saudi Arabia and South Korea, whose recipient governments will demand more favourable terms, offsets and counter-trade agreements. Examples of all four trends in 1990 are discussed below.

The changed climate in Europe and financial constraints in many countries have reduced the need and the possibilities to buy equipment of the latest generation, and governments may increasingly re-fit and modernize existing equipment. This prospect is discussed in more detail below.

Changes in the global security environment are not confined to Europe. In 1959, the Soviet Union cut off military assistance to China. All Soviet military personnel withdrew from China by the end of 1960, and technology transfer agreements were terminated.[2] Subsequently, the two dominant Asian military powers devoted considerable resources to a military and political confrontation. In 1990, military–industrial co-operation between two of the world's largest arms-producing and arms-exporting countries was resumed.

The impact of the Iraq–Kuwait crisis on the arms trade

One factor which has already influenced and might continue to influence the number and value of arms transfers is the crisis following the Iraqi invasion of Kuwait. At the end of 1990—with the crisis still unresolved—it was difficult to predict its impact. Before full-scale war broke out on 17 January 1991, the crisis in the Persian Gulf had not reversed the overall trend towards reductions in arms transfers.

Iraq's armed forces use arms imported from two dozen countries in addition to a limited number of weapons produced in Iraq. Of the major conventional weapons imported during the period 1980–90, over 80 per cent came from three of the five permanent members of the UN Security Council—whose resolutions the war against Iraq was intended to implement. The USSR accounted for 55 per cent of major conventional weapons imported by Iraq in this period, France 19 per cent and China 8 per cent (see table 7.3). One-third of the major conventional weapons bought by Iraq in this period came from countries whose armed forces joined combat against Iraq with the UN allied forces—notably France, Egypt and Italy.

Iraq has tried to build its own arms industry with technical assistance from foreign countries, including Brazil, Egypt, the USSR and companies in Western Europe.[3]

[2] Bueschel, R. M., *Communist Chinese Air Power* (Praeger: New York, 1968); Gittings, J., *The Role of the Chinese Army* (Oxford University Press: New York, 1967); Sweetman, B., 'Air forces', eds G. Segal and W. T. Tow, *Chinese Defence Policy* (Macmillan: London, 1984).

[3] Middle Eastern arms industries are discussed in Anthony, I. and Wulf, H., 'The trade in major conventional weapons', SIPRI, *SIPRI Yearbook 1990: World Armaments and Disarmament* (Oxford University Press: Oxford, 1990), chapter 7.

Table 7.3. Imports of major weapon systems by Iraq, 1980–90

Figures are SIPRI trend indicator values in 1985 US $m.

Supplier	Aggregate sales in $m.	% of total
USSR	15 002	*55*
France	5 076	*19*
China	2 261	*8*
Egypt	1 108	*4*
Brazil	1 067	*4*
Czechoslovakia	593	*2*
Others	2 262	*8*
Total	**27 369**	***100***

Note: These figures are estimates and do not represent accurate assessments of payments. See appendix 7C for the methodology used.

Source: SIPRI data base.

Following the invasion of Kuwait, all major arms suppliers imposed arms embargoes on Iraq. In some cases, unilateral embargoes preceded actions by the United Nations, which agreed on an arms embargo as part of Security Council Resolution 661 of 6 August 1990. The USA and the USSR announced embargoes in a joint statement on 5 August 1990 and China also announced its decision on 5 August.[4] Resolution 661 is a Security Council decision, rather than a request, and requires member states to prevent 'the sale or supply by their nationals or from their territories or using their flag vessels of any commodities or products, including weapons or any other military equipment, whether or not originating in their territories'.[5]

A multinational naval force, in which the US Navy formed the largest single element, was assembled in the Indian Ocean, the Arabian Sea, the Red Sea and the Mediterranean Sea to enforce Resolution 661. Subsequently, a Polish freighter, carrying a cargo of engines for Iraqi tanks of Soviet origin, diverted to Libya. From there the engines were airlifted to Iraq. Other ships diverted to Yemen from where goods were airfreighted to Iraq. On 25 September 1990, UN Security Council Resolution 670 extended the blockade to cargo aircraft entering or leaving Iraq in an effort to close these loopholes in enforcement. Iraq had continued to receive arms until the eve of the invasion, however. In addition to the Polish shipment noted above, 24 armoured bridgelayers were sent to Iraq from the German Democratic Republic in July.

Iraq—consistently one of the major importers of arms since 1980—has been removed from the global arms market. Arms suppliers will find it difficult to recover monies owed to them under existing contracts while Iraq's global economic reserves are frozen. France is owed 25 billion francs plus 4 billion francs in interest (over $4 billion[6]), of which 4 billion francs is for the

[4] *Beijing Review*, 13–19 Aug. 1990, p. 8.

[5] Resolution 661 (1990) Paragraph 6(c), UN Security Council document S/RES/661(1990), 6 Aug. 1990. For the full text of the resolution see appendix 18A of this volume.

[6] *The Independent*, 6 Aug. 1990.

military, Italy is owed $1.5 billion and the Soviet Union is owed over 3.8 billion roubles by Iraq. The Brazilian company Avibras, which has had a substantial arms transfer relationship with Iraq, is threatened with bankruptcy, apparently as a result of the failure of Iraq to pay for arms delivered. Other countries are in financial difficulties, too: the US State and Defense Departments recommend the cancellation of Egypt's entire $6.7 billion debts for US arms shipments.[7]

Agreements involving Kuwait have also been affected by the Gulf crisis. The US company McDonnell Douglas had contracts worth a total of $1.9 billion to supply 40 F/A-18 Hornet fighters, together with armaments, spare parts and support. Short Brothers, a British subsidiary of the Canadian company Bombardier, had a contract to supply 16 Tucano trainer aircraft, produced in the UK under a licence from the Brazilian company Embraer. These contracts have not been cancelled. Not only the Kuwaiti Government but also significant elements of the Kuwaiti armed forces escaped into Saudi Arabia where they joined the international forces against Iraq. The Saudi Government has agreed to underwrite the costs of ongoing equipment programmes while Kuwait's economic assets are frozen.

In the 1980s, the US Congress restricted US arms sales to Saudi Arabia. The following exchange in a congressional hearing illustrates the basic position of Congress and the Bush Administration prior to 2 August:

> Representative Levine: Do you believe Saudi Arabia could repulse an Iraqi invasion if in fact the Iraqis, with the kind of force it has available, were determined to invade Saudi Arabia?
>
> Mr. Clarke (Assistant Secretary for Politico-Military Affairs, Department of State): I believe if Saudi Arabia has a high quality force, a small force but, nonetheless, a high quality force, it would be able to deter an attack. If the deterrence failed, it would be able to slow that attack down until the United States and other friendly forces were able to do something.[8]

Congress and the Bush Administration support the modernization of Saudi Arabia's armed forces, and neither believes that Saudi Arabia could successfully defend itself against Iraq. However, they disagree on the type and quantities of weapons that are adequate for deterrence and defence.

Several US–Saudi deals were signed before Iraq's invasion of Kuwait and others were in an advanced stage of negotiation. A deal worth $4.4 billion was included in the 'Javits list'—the annual list of proposed arms sales for the following fiscal year—submitted to Congress early in 1990. This deal consisted of over 1000 light armoured vehicles, 27 155-mm calibre howitzers, armoured recovery vehicles and the upgrading of 600 armoured personnel carriers (APCs) and Saudi Arabia's E-3A Sentry Airborne Warning and

[7] *Le Monde*, 20 Feb. 191, p. 1; *Milavnews*, Oct. 1990, p. 7.

[8] *Proposed Tank Sale to Saudi Arabia*, Hearing before the Subcommittee on Arms Control, International Security and Science, and on Europe and the Middle East of the Committee on Foreign Affairs (US Government Printing Office: Washington, DC, 7 Nov. 1989), p. 34.

Control System (AWACS) aircraft.[9] A second deal, worth $3.2 billion, agreed in June, included 315 M-1 Abrams tanks and associated support. A third package, agreed after the Iraqi invasion, included 24 F-15C and D fighter aircraft, 150 M-60-A3 tanks, 200 FIM-92A Stinger portable surface-to-air missiles and large stocks of tank ammunition. This package was worth an estimated $7 billion. A fifth package, introduced in October, included 6 Patriot surface-to-air missile batteries, 12 AH-64 attack helicopters, an additional 150 M-1-A2 Abrams tanks, 10 000 lorries, 9 multiple launch rocket systems (MLRS), 8 C-130 transport aircraft and 7 KC-130 aerial tankers. This deal was estimated to be worth $7.3 billion. Some congressmen continue to find the transfer of F-15 fighter bombers to Saudi Arabia disturbing, and the Administration agreed to defer further requests for Saudi Arabian arms to 1991.[10]

These deals represent a procurement schedule drawn up by the Saudi Government with deliveries stretched over the next 10 years. Major General Donald Kaufman, chief of the US Military Training Mission in Saudi Arabia, described the sales as 'a modernisation rather than a large expansion'.[11] With an annual value of roughly $1.7 billion, agreed sales would significantly raise the value of US–Saudi arms transfers but do not represent a major jump in the value of total Saudi Arabian arms imports.

As reported in the *SIPRI Yearbook 1990*, during the 1980s European suppliers had made a significant inroad into some Middle Eastern countries where the USA had previously been the dominant arms supplier.[12] In Europe, concern has been expressed—particularly by British industry and in the French Parliament—that the net effect of the Gulf crisis will not be an increase in the total number of arms sales, but rather that the crisis will restore to the USA the strong dominance over local arms supplies that it held in the 1970s.[13]

Israel has been granted permission by the United States to use $200 million of its Economic Support Fund for military purchases, and in addition $700 million has been allocated to Israel as a 'drawdown facility'. This means that equipment to the value of $700 million would be made available from US stockpiles to Israel in a crisis.

[9] *World Weapons Review*, 14 Mar. 1990, p. 12; *Aviation Week and Space Technology*, 11 June 1990, p. 31; 'Opponents try to limit arms buy by Saudis', *Defense News*, 25 June 1990, p. 18.

[10] 'Congress faces resolution killing Saudi arms sale', *Defense News*, 15 Oct. 1990; 'Congress clears Saudi arms deal after critics drop opposition', *Defense News*, 29 Oct. 1990, p. 6.

[11] Starr, B., 'Saudi arms buys for update, not expansion', *Jane's Defence Weekly*, 6 Oct. 1990. See also interviews with Lt. General Charles Brown, Director of the Defense Security Assistance Agency, *Jane's Defence Weekly*, 18 Aug. 1990, p. 244, and with Under Secretary of Defense for Policy, Paul Wolfowitz, and Under Secretary of State for International Security Affairs, Reginald Bartholemew, *Jane's Defence Weekly*, 13 Oct. 1990, p. 686.

[12] Anthony and Wulf (note 3), pp. 230–45.

[13] Antoine, P-L., 'The Gulf: a windfall for US industry', *Defence & Armament*, Oct. 1990, pp. 34–38. Branger, J-G., *Avis présenté au nom de la commission de la défense nationale et des forces armées sur le projet de loi de finances pour 1991, Tome V, Défense, recherche et industrie d'armement* (Assemblée Nationale: Paris, 10 Oct. 1990), p. 12; *The Independent*, 6. Aug. 1990, p. 8; Becker, J., 'Baghdad and the blind eyes', *The Guardian*, 2 Nov. 1990, p. 6.

The impact of the CFE Treaty on the arms trade

The CFE Treaty contains no prohibition on arms transfers. In 1989, the Federal German and Soviet foreign ministers, Hans-Dietrich Genscher and Eduard Shevardnadze, suggested that a treaty should be supplemented by a ban on transfers of treaty-limited equipment to the Third World.[14] Neither government insisted on including this suggestion in the CFE Treaty, and it places no restrictions on exports of either new or second-hand arms.

Arms produced for export are exempt from Treaty counting rules under Article III, paragraph 1E, which states that items that 'are awaiting, or are being refurbished for, export or re-export and are temporarily retained within the area of application' are not considered as treaty-limited equipment.

Article VIII of the CFE Treaty states that numerical reductions shall be achieved only in accordance with protocols appended to the Treaty. These protocols stipulate either the destruction of equipment or modification to allow reclassification as non-treaty limited equipment. In practice, however, exporting surplus equipment is the most likely form of Treaty compliance. The CFE Treaty does not enter into force until 10 days after it has been ratified by all states parties. Exports of second-hand treaty-limited equipment from the area of application of the CFE Treaty took place throughout 1990.

In a letter to US Secretary of State James Baker on 13 October 1990, Eduard Shevardnadze explained that between July 1988 and August 1990 the USSR withdrew 10 000 tanks, 25 480 armoured combat vehicles and 24 100 artillery pieces from the CFE Treaty area of application.[15] All of this equipment could legally be exported, and Shevardnadze's letter confirmed that an unspecified number of the tanks were exported.

The USA transferred 700 M-60 tanks to Egypt, 150 M-60 tanks to Saudi Arabia, and 50 M-60 and 300 M-48 tanks to Thailand. Thailand received this equipment at nominal prices, and Egypt paid only the costs of packing and shipping the tanks.

To avoid destroying modern equipment in the CFE Treaty area of application and keeping old equipment elsewhere, the USA and Germany (but not France or the UK) will give or sell equipment to allies—Greece, Portugal, Spain and Turkey—which will then destroy old equipment due for replacement anyway.[16]

Within NATO, the 24 October 1990 meeting of the Conference of National Armaments Directors (CNAD) discussed the preparation of a NATO defence trade agreement.[17] As well as the co-ordinaton of CFE Treaty implementation, the agreement would aim at greater standardization and better value for money in alliance arms procurement. A defence trade task force has been created to

[14] Anthony and Wulf (note 3), pp. 230–31 and 242–43.

[15] See chapter 13 of this volume.

[16] This 'cascading' within NATO is described in Anthony. I., Courades Allebeck, A., Gullikstad, E., Hagmeyer-Gaverus, G. and Wulf, H., 'Arms production', SIPRI, *SIPRI Yearbook 1990: World Armaments and Disarmament* (Oxford University Press: Oxford, 1990), chapter 8, p. 318.

[17] *Atlantic News*, 19 Oct. 1990, p. 2.

study all aspects of improving alliance co-operation on military equipment and report to the first 1991 meeting of the CNAD.[18]

The initiative for the negotiation of a defence trade agreement came from the United States. William Taft, US Ambassador to the North Atlantic Assembly, outlined the idea in a speech to the German Strategic Forum in Bonn on 15 March 1990 and repeated his suggestion at a conference in Brussels in June.[19]

Taft suggested the elimination of direct trade barriers, complete freedom of technology transfer and unlimited foreign ownership of arms producing companies. The chances that these policies—none of which is new—will be adopted are small. In open competition, US companies win most arms contracts. Meanwhile, the US Congress—pointing to the erosion of US industrial competitiveness—insists on supporting the arms industry. Both West European governments and the USA, therefore, lack incentives to remove trade barriers. A new task force report will not solve this problem.

The Sino-Soviet arms transfer relationship

A relaxation of military tensions along the Sino-Soviet border began with unilateral measures announced by President Mikhail Gorbachev on 28 July 1986 in Vladivostok. The resumption of Sino-Soviet military technology transfer in 1990 illustrates how far political relations have improved in four years.

The 1975–78 strategic review in China produced two central conclusions: first, that self-sufficiency came from economic and technological strength, which could only be acquired over the long term;[20] and second, that China's nuclear capability made the chance of war with either the USSR or the USA low. Consequently, defence was allocated the lowest priority of the 'four modernizations' and, during the 1980s, the share of government spending allocated to defence declined consistently.[21] The only area where military expenditure has grown is research and development. Chinese defence industries have sought co-operation with companies in the United States and Europe since the mid-1970s, and contacts expanded in the 1980s.[22]

In 1986 a team of US companies led by Grumman were contracted to install avionics on 55 fighter aircraft designated the F-7M Airguard.[23] The new air-

[18] *Atlantic News*, 27 Oct. 1990, p. 2.

[19] Astor, R. J., 'Arms chaos: maybe a defense GATT?', *International Herald Tribune*, 25 Apr. 1990, p. 8; *Atlantic News*, 30 June 1990, p. 3.

[20] Mukerjee, D., 'China's military choice: slower but surer build-up', *Times of India*, 7 Oct. 1987; 'Gradual improvements for China's equipment', *Jane's Defence Weekly*, 9 Apr. 1988, p. 658; Tai Ming Cheung, 'Modernising China's defence', *Armed Forces*, Sep. 1988, pp. 414–17.

[21] Godwin, P. H. B., 'The Chinese Communist armed forces', *Report for the Political-Military Affairs Division*, Airpower Research Institute (Air University Press: Maxwell AFB, Ala., 1988). However, the modernization of industry, science and technology will obviously have implications for the defence sector.

[22] *China Today: Aviation Industry* (Chinese Social Science Press: Beijing 1989), pp. 106–90.

[23] 'Chinese Air Force developing few new aircraft designs', *Aviation Week & Space Technology*, 7 Dec. 1987, p. 55; 'Grumman in Chinese Fighter deal', *Jane's Defence Weekly*, 19 Nov. 1988, p. 1261;

craft was to have been powered by a General Electric engine and contain a Westinghouse AN/APG-66 radar.[24] The Italian company Aeritalia provided navigation equipment and weapon delivery avionics for a version of the A-5 Fantan fighter.[25] French companies collaborated on a rival version, the A-5K, involving 10 French companies whose work was co-ordinated by Thomson-CSF. The companies involved were Thomson-CSF, Sagem, TRT, Crouzet, Sfena, SFIM, Labinal, ONERA, Jaeger and Auxilec.[26] All three of these programmes, under embargo since June 1989, were cancelled in 1990.[27]

Renewed Sino-Soviet co-operation may compensate for these cancellations. At the 1989 Sino-Soviet summit in Beijing, it was agreed that industrial joint ventures were desirable. In June 1990 a Chinese delegation, including the vice-chairman of the Central Military Commission and the Minister of Aeronautics and Astronautics, opened negotiations in Moscow concerning arms transfers.[28] In September and October 1990, Chinese People's Liberation Army pilots and technicians visited factories in the Soviet Union where the MiG-29 fighter, the Su-24 fighter bomber and military aero-engines are produced. The transfer of 12 Mi-17 helicopters was agreed in 1990, and a deal for the transfer of aerospace technologies was anticipated.

III. The major exporters and importers

Major conventional weapon systems have been imported by 132 countries of the world during the period 1986–90, as table 7.4 shows. While the Soviet Union was the largest arms exporter—in terms of the value of major weapons supplied—it was not the supplier with the highest number of recipients. Thirty-eight countries—mainly from the Third World—imported weapons from the USSR. Within this comparatively small group of recipients Soviet exports were concentrated heavily on nine countries that accounted for 80 per cent of total Soviet exports. These were India, Iraq, Afghanistan, North Korea, Syria and Angola in the Third World, and Poland, Czechoslovakia and the GDR in the WTO. The USA, France and the UK had more clients throughout the period than the USSR. The USA had 77, France 73 and the UK 49 customers for major conventional weapons. This pattern is a traditional one.[29] By far the most important recipient of US weapons was Japan, followed by

'China's F-8II upgrade to include Litton Navigation System', *Jane's Defence Weekly*, 19 Mar. 1988, p. 529.

[24] 'Grumman to upgrade Chinese F-7Ms', *Flight International*, 26 Nov. 1988, p. 7; 'Grumman, China announce joint fighter upgrade', *Defense News*, 28 Nov. 1988, p. 33.

[25] 'Asia Watch: Military—A-5M Fantan', *Asian Aviation*, Nov. 1988, p. 11; 'Updating older combat aircraft: MiG-19 and -21', *International Defense Review*, Dec. 1988, pp. 1590–93.

[26] d'Entremont, X., 'Asiandex, the Peking exhibition', *Defense & Armament Héraclès*, Jan. 1989, pp. 69–73.

[27] *AAS-Milavnews*, June 1990, p. 6; *AAS-Milavnews*, Oct. 1990, p. 6.

[28] 'Will Russia re-arm China?', *Armada International*, May/June 1990, p. 2; 'China seeking Soviet fighters', *Jane's Defence Weekly*, 21 July 1990, p. 70; Tai Ming Cheung, 'A sale is in the air', *Far Eastern Economic Review*, 6 Sep. 1990, pp. 20–21; *Far Eastern Economic Review*, 11 Oct. 1990, p. 8.

[29] For the recipients of the Third World during the period 1971–85, see Brzoska, M. and Ohlson, T., SIPRI, *Arms Transfers to the Third World, 1971–85* (Oxford University Press: Oxford, 1987).

Table 7.4. World trade in major conventional weapon systems, 1986–90

Figures are values of major conventional weapon systems transferred, in US $m., at constant (1985) prices.

	Seller											
Recipient	USSR	USA	France	UK	China	FRG	Czech.	Netherlands	Sweden	Italy	Others	Total
Afghanistan	5 460	140	4	40	53		30	–	–	–	15	5 742
Algeria	783	–	–	28	54	59	–	–	–		6	930
Angola	3 475	11	57	–	–	–	–	–	–		49	3 592
Argentina	–	13	60	–	–	644	–	–	–	37	120	874
Australia	–	2 521	171	50	–	–	–	–	26	6	48	2 822
Austria	–	24	0	–	–	–	–	59	135	24	17	259
Bahamas	–	–	–	14	–	–	–	–	–	–	0	14
Bahrain	–	596	74	10	–	246	–	–	–	–	3	929
Bangladesh	–	10	–	–	298	–	–	–	–	–	75	383
Belgium	–	423	57	–	–	–	–	–	–	–	9	489
Benin	–	–	13	–	–	–	–	–	–	–	0	13
Bolivia	–	53	4	–	–	–	–	–	–	–	10	67
Botswana	–	12	–	18	–	–	–	–	–	–	46	76
Brazil	–	633	504	193	–	90	–	–	43	0	33	1 496
Brunei	–	10	30	–	–	–	–	–	–	3	1	44
Bulgaria	1 750	–	–	–	–	–	37	–	–	–	3	1 790
Burkina Faso	–	–	–	–	–	–	–	–	–	2	1	3
Cameroon	–	–	14	–	–	–	–	–	–	–	3	17
Canada	2	2 053	–	115	–	11	–	–	20	20	227	2 448
Central African Republic	6	–	–	–	–	–	–	–	–	–	0	6
Chad	–	68	30	–	–	–	–	–	–	–	2	100
Chile	–	72	65	330	1	14	–	–	–	–	213	695
China	62	120	187	5	–	–	–	–	–	–	40	414
Colombia	–	144	–	–	–	–	–	–	–	–	189	333
Congo	–	–	2	–	–	–	–	–	–	–	0	2
Costa Rica	–	2	–	–	–	–	–	–	–	–	0	2
Côte d'Ivoire	–	3	8	–	–	–	–	40	–	–	0	51
Cuba	773	–	–	–	–	–	–	–	–	–	36	809
Cyprus	–	–	188	–	–	–	–	–	–	120	66	374
Czechoslovakia	4 571	–	–	–	–	–	–	–	–	–	0	4 571

Denmark	–	163	10	311	–	–	–	–	5	–	22	511
Djibouti	–	–	2	–	–	–	–	–	–	–	6	8
Dominica	–	5	–	–	–	–	–	–	–	–	0	5
Dominican Republic	–	7	–	–	–	–	–	–	–	–	0	7
Ecuador	–	19	3	10	–	–	–	18	–	–	216	266
Egypt	–	3 471	669	24	89	11	–	–	–	161	292	4 717
El Salvador	–	15	–	–	–	–	–	–	–	–	3	18
Ethiopia	378	38	–	–	–	–	147	–	–	16	50	629
Fiji	–	–	3	–	–	–	–	–	–	–	1	4
Finland	346	0	–	55	–	–	–	–	211	2	1	615
France	–	487	–	22	–	–	–	50	–	–	39	598
Gabon	–	5	131	–	–	–	–	–	–	4	3	143
German DR	2 258	–	–	–	–	–	–	–	–	–	34	2 292
Germany, FR	–	2 756	70	78	–	–	–	–	–	8	58	2 970
Ghana	–	1	–	1	–	–	10	–	–	24	1	37
Greece	–	1 601	920	–	–	208	–	22	–	33	228	3 012
Guinea	84	–	1	–	–	–	–	–	–	–	0	85
Honduras	–	91	–	–	–	–	–	–	–	–	5	96
Hungary	1 124	–	–	–	–	–	–	–	–	–	0	1 124
India	12 474	–	1 776	1 147	–	383	–	525	603	–	81	16 989
Indonesia	–	538	60	522	–	112	–	455	–	–	63	1 750
Iran	434	103	–	–	1 488	–	251	–	–	–	637	2 913
Iraq	6 401	237	1 383	–	988	34	132	–	–	11	1 128	10 314
Ireland	–	–	8	29	–	–	–	–	–	–	0	37
Israel	–	2 696	–	–	–	–	–	–	–	–	7	2 703
Italy	–	634	32	–	–	81	–	–	24	–	95	866
Japan	–	10 773	60	138	–	–	–	–	–	–	0	10 971
Jordan	292	144	94	15	–	–	–	–	–	–	312	857
Kampuchea	108	–	–	–	17	–	–	–	–	–	111	236
Kenya	–	–	92	29	–	1	–	–	–	–	63	185
Korea, North	4 406	–	–	–	494	–	–	–	–	–	0	4 900
Korea, South	–	2 887	62	104	–	–	–	–	–	65	7	3 125
Kuwait	126	86	53	77	–	–	–	35	–	–	110	487
Laos	105	–	–	–	2	–	–	–	–	–	5	112
Lebanon	–	–	6	–	–	–	–	–	–	–	82	88
Lesotho	–	0	–	–	–	–	–	–	–	–	5	5
Liberia	–	–	–	–	–	–	–	–	–	–	11	11

	Seller											
Recipient	USSR	USA	France	UK	China	FRG	Czech.	Netherlands	Sweden	Italy	Others	Total
Libya	1 947	55	17	–	–	–	15	–	–	–	213	2 247
Malawi	–	–	3	7	–	14	–	–	–	–	0	23
Malaysia	–	161	9	22	–	10	–	8	–	264	47	521
Mali	26	–	–	–	–	–	–	–	–	–	0	26
Malta	–	–	–	–	–	–	–	–	–	2	0	2
Mauritius	–	–	19	–	–	–	–	–	–	–	3	22
Mexico	–	224	27	14	–	7	–	–	–	–	60	332
Morocco	–	100	35	–	–	–	–	–	–	–	375	510
Mozambique	2	5	–	–	–	–	–	–	–	–	12	19
Myanmar	–	–	–	–	139	–	–	–	–	–	20	159
Nepal	–	–	8	–	2	–	–	–	–	–	0	9
Netherlands	–	1 707	–	–	–	303	–	–	–	–	11	2 021
New Zealand	–	20	–	4	–	–	–	–	–	–	49	73
Nicaragua	488	5	–	–	–	–	–	–	–	–	16	509
Niger	–	–	2	–	–	3	–	–	–	–	0	5
Nigeria	–	1	68	30	–	73	89	–	–	139	32	432
Norway	–	804	15	0	–	264	–	–	564	–	3	1 650
Oman	–	49	56	433	–	–	–	–	–	–	11	549
Pakistan	–	1 070	25	125	1 202	–	–	–	167	16	88	2 693
Panama	–	8	–	–	–	–	–	–	–	–	14	22
Papua New Guinea	12	–	–	–	–	–	–	–	–	–	8	20
Paraguay	–	–	–	–	–	–	–	–	–	–	36	36
Peru	296	261	376	1	–	–	–	–	–	19	67	1 020
Philippines	–	100	–	4	–	4	–	–	–	15	9	132
Poland	4 706	–	–	–	–	–	–	–	–	–	13	4 719
Portugal	–	224	51	10	–	5	–	–	–	–	10	300
Qatar	–	–	324	6	–	–	–	–	–	–	1	331
Romania	1 292	–	119	–	–	–	–	–	–	–	1	1 412
Rwanda	–	–	7	–	–	–	–	–	–	–	0	7
Samoa	–	–	–	–	–	–	–	–	–	–	1	1
Saudi Arabia	–	3 465	2 727	2 811	1 440	–	–	–	–	230	165	10 838
Senegal	–	–	5	–	–	–	–	–	–	–	25	30
Seychelles	8	–	–	–	–	–	–	–	–	–	1	9

Sierra Leone	–	–	–	–	9	–	–	–	–	–	0	9
Singapore	–	808	129	–	–	104	–	–	–	25	1	1 067
Solomon Islands	–	–	–	–	–	–	–	–	–	–	1	1
Somalia	–	–	–	–	–	–	–	–	–	21	22	43
South Africa	0	1	–	–	–	3	–	–	–	–	204	208
Spain	–	4 478	742	11	–	53	–	–	–	136	145	5 565
Sri Lanka	–	40	–	15	85	–	–	–	–	7	87	234
Sudan	–	8	1	0	47	–	–	–	–	5	128	189
Suriname	–	–	–	–	–	–	–	–	–	–	2	2
Sweden	–	232	76	48	–	33	–	–	–	–	11	400
Switzerland	–	15	23	228	–	564	–	–	–	–	0	830
Syria	4 191	–	–	–	–	–	–	–	–	–	0	4 191
Taiwan	–	1 730	–	–	–	–	–	400	–	–	297	2 427
Thailand	–	1 211	87	–	868	34	–	43	–	44	38	2 325
Togo	–	52	16	3	–	–	–	–	–	–	3	74
Tonga	–	–	–	–	–	–	–	–	–	–	2	2
Tunisia	–	46	24	–	–	–	–	–	–	–	0	70
Turkey	–	2 179	20	123	–	1 245	–	183	–	59	563	4 372
Uganda	12	–	–	–	–	–	–	–	–	10	11	33
UK	–	795	496	–	–	–	–	18	–	–	181	1 490
United Arab Emirates	–	46	1 143	102	–	158	–	4	–	32	56	1 541
Uruguay	–	12	58	–	–	–	–	–	–	–	0	70
USA	–	–	6	316	115	14	–	–	3	3	593	1 050
USSR	–	–	–	–	–	–	1 637	–	–	–	535	2 172
Vanuatu	–	–	–	–	–	–	–	–	–	–	1	1
Venezuela	–	130	115	26	–	–	–	27	–	–	102	400
Viet Nam	368	–	–	–	–	–	–	–	–	–	0	368
Yemen Arab Republic	23	–	–	–	35	–	–	–	–	–	5	63
Yemen, People's Democratic Republic	461	–	–	–	–	–	–	–	–	–	0	461
Yugoslavia	1 503	–	41	–	–	–	–	–	–	–	6	1 550
Zaire	–	9	–	–	9	–	–	–	–	–	10	28
Zimbabwe	–	–	4	10	238	–	–	–	–	15	17	284
Total	**60 753**	**53 716**	**13 777**	**7 714**	**7 673**	**4 726**	**2 407**	**1 887**	**1 801**	**1 578**	**9 200**	**165 232**

Notes: Export totals differ from those in table 7.1 because of rounding.
Conventions: – Nil, 0 = < 0.5.
Source: SIPRI data base.

Spain, Egypt, Saudi Arabia, South Korea, the FRG and Israel. The most important customers of France were Saudi Arabia, India, Iraq and the United Arab Emirates. The UK's major importers were Saudi Arabia and India.

The other six major suppliers of weapons for the period 1986–90 recorded in table 7.4 sold major conventional weapons to a much smaller group of countries: China had 22 customers, the FRG 31, Czechoslovakia 10, the Netherlands 15, Sweden 11 and Italy 35.

Table 7.4 also reflects the continuation of an established trend of recipient country policies to diversify the sources of their weapon imports. For economic and political reasons most of the major weapon-importing countries imported from several suppliers, often from both East and West. Exceptions to this rule were Czechoslovakia and Poland, which relied (besides local production) almost entirely on Soviet exports, and Syria, which also imported major conventional weapon systems from the Soviet Union.

The major exporters

The Soviet Union

According to SIPRI data, the USSR remains the largest exporter of major conventional weapons for the period 1986–90, despite a substantial reduction recorded for 1990. With exports totalling $6.4 billion in 1990—approximately one-half of the value for previous years—the USSR is ranked behind the USA for the first time since 1984. In an interview at the end of 1990 Soviet arms export figures were publicly mentioned by I. S. Belousov for the first time.[30] According to his information for the past five years (probably the 1986–90 Five Year Plan) weapons and weapon technology worth 56.7 billion roubles have been exported—of which 9.7 billion in 1990—and 8.5 billion roubles worth were exported free of charge. He confirmed that—aside from WTO countries—the main recipients were Afghanistan, Iraq, Syria, the Peoples' Democratic Republic of Yemen, India, Viet Nam, North Korea, Algeria, Libya, Ethiopia, Mozambique, Angola and Cuba. I. S. Belousov further claimed that weapon exports have been drastically reduced during the current Five Year Plan: missiles by 64 per cent, tanks and armoured personnel carriers 25–30 per cent, artillery 48 per cent, aircraft 53 per cent and ships 56 per cent.The pattern of Soviet arms exports is changing in response to several factors, and the speed and extent of this change are likely to accelerate.

Transfers of advanced aircraft—and in particular the MiG-29 fighter and the Su-24 fighter-bomber—are becoming more important than transfers of armoured vehicles. Apart from the orders for Soviet aircraft from China, noted above, in 1990 India explored the licensed production of the MiG-29 and the purchase of naval versions of the Su-27 fighter. Deliveries of the MiG-29 to Czechoslovakia, Iran, Poland and Romania were confirmed and deliveries to

[30] Published in *Pravitelstvenny vestnik* (Government News) no. 80 (2 Jan. 1991). I. S. Belousov is Deputy Chairman of the Council of Ministers of the USSR and Chairman of the State Committee of the Council of Ministers of the USSR on military industry issues.

Afghanistan were reported but not confirmed. Reports of a sale of MiG-29s to Hungary in 1989 proved to have been false. Sales of the Su-24 to Libya and Iraq in 1989 were confirmed, and sales to Syria were reported.

As noted below, Foreign Minister Shevardnadze called on governments to exercise restraint in supplying arms to areas of conflict, and in 1990 Soviet arms transfer agreements with Ethiopia that expired were apparently not renewed.[31] However, Soviet transfers to Afghanistan and Viet Nam continued.

The 'export option' has entered the debate on the future of Soviet arms production. Managers of arms production plants would like more freedom to export arms to finance capital investment and product development.[32] In Finland, the two Soviet fighter aircraft manufacturers—Mikoyan and Sukhoi—are competing with each other to meet a Finnish Air Force requirement. An export-oriented strategy cannot, however, be an alternative to reducing the existing over-capacities, and furthermore it stands little chance of success.

The USSR has insisted on existing agreements being fulfilled rather than establishing new contracts. For example, both Czechoslovakia and the GDR tried to cancel deliveries of the MiG-29 but were held to contracts by the Soviet Union.[33] In addition, the united Germany will take possession of two Balcom 10 Class corvettes—although not their armament—one Mi-24 helicopter gunship, rocket launchers and 4500 AK-47 rifles together with 50 million rounds of ammunition.[34] However, few new orders are being signed.

The Soviet Ministry of the Aviation Industry anticipates 350–450 million roubles in lost orders since European countries that would previously have ordered Soviet aircraft choose Western aircraft. These aircraft are primarily airliners and air freighters, but anticipated future overseas sales of around 400 MiG-29 fighters are also in jeopardy.[35] Arms-producing companies from the USA and Western Europe are seeking both arms sales and co-operative agreements in the countries of Eastern Europe. These initiatives would depend on political support, but this may be forthcoming. On his visit to Poland in December 1990, US Secretary of Defense Richard Cheney, asked about the possibility of US arms sales to Poland, noted the progressive relaxation of COCOM regulations for members of the Warsaw Treaty Organization (WTO) and replied 'I certainly wouldn't write it off.'[36]

The Soviet decision to conduct all foreign trade in hard currency from 1 January 1991 has major implications for arms exports not only within the former WTO but also with developing countries which have traded on a soft

[31] *Flight International*, 24 Jan. 1990, p. 12; *Air International*, Mar. 1990, p. 106.

[32] Izyumov, A., 'Conversion: an export version', *Moscow News Weekly*, no. 16 (1990).

[33] *Jane's Defence Weekly*, 17 Feb. 1990, p. 284; *Flight International*, 9–15 May 1990, p. 10. The corvettes are likely to be offered for re-sale.

[34] *AAS-Milavnews*, Aug. 1990, p. 10.

[35] *Izvestia*, 10 Mar. 1990; Tai Ming Cheung, 'A sale is in the air', *Far Eastern Economic Review*, 6 Sep. 1990, pp. 20–21.

[36] *Jane's Defence Weekly*, 15 Dec. 1990, p. 1222.

Table 7.5. A comparison of major Third World clients of Soviet arms and major Soviet debtors, 1986–90

Major Soviet debtors	Debt (in m. roubles)	Percentage shares of Soviet arms exports to the Third World 1986-90	Rank as Third World importer of Soviet arms
Cuba	15 490.6	*2*	9
Mongolia	9 542.7	–	–
Viet Nam	9 132.2	*1*	1
India	8 907.5	*29*	14
Syria	6 742.6	*10*	5
Iraq	3 795.6	*15*	2
Afghanistan	3 055.0	*13*	3
Ethiopia	2 860.5	*1*	13
Algeria	2 519.3	*2*	8
North Korea	2 234.1	*10*	4
Leading 10 Third World debtors' percentage of Soviet arms exports to the Third World		*83*	
Leading 10 debtors' percentage of total Third World debt to the USSR		*81*	

Sources: SIPRI data base; *Izvestia*, 1 Mar. 1990 for statistics on debt.

currency basis with the USSR. The dramatic changes of the past five years have undermined one rationale for buying arms from the USSR—obtaining political support from a superpower. If traded on a commercial basis, Soviet goods may no longer retain another comparative advantage—low cost.

Although arms exports have in the past been a major component of total Soviet foreign trade, there is no evidence that arms sales by the USSR bring significant benefits to the economy as a whole. Several of the key recipients of Soviet weapons have not been in a position to pay for the imported weapons, let alone in hard currency. The scattered evidence available on terms of payments suggests that the USSR could not always recover the real cost of production of exported weapons.

Most of the leading Third World debtors to the USSR are also major importers of Soviet weapons. Among the top 10 debtors are 7 of the top 10 weapon importers, as table 7.5 indicates. In the above-mentioned interview I. S. Belousov[37] confirmed the total outstanding debt of approximately 86 billion roubles and said—without giving any details—that some of this debt was due to arms supplies, particularly to Third World countries, although most countries had been on time in their debt service. Although a correlation between debt and arms imports is suggested by the table, and is plausible, neither set of figures should be regarded as truly reliable because information from the USSR remains limited. There is no way to disaggregate the percentage of debt accounted for by arms purchases.

[37] Published in *Pravitelstvenny vestnik* (Government News), no. 80 (2 Jan. 1991).

The USA

US exports of major conventional weapon systems decreased in 1990 by about one-quarter, totalling $8.7 billion. While the politico-military aspects of arms transfers in the United States focus primarily on Middle East policy, questions of industrial competitiveness and the economic benefits of arms transfer relations increasingly focus on East Asia.

Combined civil and military exports by aerospace manufacturers are the single most important foreign sales activity by the USA.[38] In the 1980s, exports by aerospace manufacturers to the Asia–Pacific rim more than doubled and by 1990 this region accounted for almost 30 per cent of US aerospace exports.[39] In 1990 the USA continued to win most contracts in the region. However, South Korea in particular secured favourable financial terms that increased the friction between Congress and the Bush Administration over trade policy *vis-à-vis* Asian countries.

US arms exports continued to be used in support of anti-drug operations, in particular through the Andean initiative. While Congress has questioned the necessity for maintaining military assistance to anti-government forces in Afghanistan, Angola and Kampuchea, there is more support for the Andean initiative, which includes military assistance and arms transfers to Bolivia, Colombia and Peru.[40]

France

France remains the third largest supplier of major conventional weapons, although the $1.8 billion exports in 1990 were less than half of the value for 1986. The French arms industry suffered two serious set-backs. At the beginning of the year Greece—after delivery of 28 Mirage-2000 fighter planes—refused to take delivery of the remaining 12 until renegotiation of the deal was concluded. The Greek refusal came after the disclosure of defects in the radar system of the aircraft.[41]

The French arms industry lost one of its most important customers as a result of imposing the UN embargo on Iraq. At the time of the invasion of Kuwait by Iraq at the beginning of August 1990 the French Government was completing its negotiations to reschedule Iraqi debt. A parliamentary report feared a long-term shift in the regional pattern of arms imports, affecting French industry substantially as the countries of the Middle East region accounted for 47 per cent of French exports of major weapon systems between

[38] *Foreign Assistance Legislation for Fiscal Years 1990–91*, Hearings before the Sub-Committee on Arms Control, Committee on Foreign Relations, House of Representatives (US Government Printing Office: Washington, DC, 8–9 Mar. 1989), pp. 65–69.

[39] According to the Aerospace Industries Association of America, quoted in Flamm, D., 'US aerospace exports to Asia Pacific rim increase substantially', *Asian Defence Journal*, Sep. 1990, pp. 54–60.

[40] 'House votes to put conditions on aid to Angolan rebels', *Congressional Quarterly*, 20 Oct. 1990; 'Use of Cambodia aid questioned', *Congressional Quarterly*, 20 Oct. 1990; 'Update on Andean drug strategy', *US Department of State Dispatch*, 29 Oct. 1990, pp. 219–22.

[41] *Le Monde*, 17 Feb. 1990, p. 3; *World Weapons Review*, 28 Feb. 1990, p. 8; *Jane's Defence Weekly*, 24 Feb. 1990, p. 323.

1986 and 1990. As a solution to the decline in French arms exports the parliamentary report suggests lifting some arms embargoes. Embargoes against South Africa and Taiwan are specifically mentioned; the report qualified these as no longer justified because of positive political developments in these countries. Iraq was in contrast classified as a non-democratic customer capable of using French-made weapons against France.[42] The French Government lifted the temporary embargo on China—imposed as a consequence of the Chinese actions at Tiananman Square—and re-opened the deal concerning air defence missile systems and search radars for the upgrade of two Luda Class destroyers.[43]

The United Kingdom

In 1990, the estimated value of deliveries of major conventional weapons from the UK was $1.2 billion—about 25 per cent below the average annual sales of the previous four years. This figure is significantly lower than the official value of British exports, given for 1988 as £2.08 billion (around $4 billion in 1990 prices). These figures, supplied by the British Customs and Excise, include all military items passing the British customs barrier, whereas the SIPRI estimate is confined to major conventional weapons.[44]

For British arms exports overall, Middle Eastern countries have been the most important customers, accounting for 64 per cent of exports in the period 1986–90. Consequently, there was considerable sensitivity in 1990 to the idea that Saudi Arabia would reconsider its arms import relationship with the UK in the light of the probable increase in US–Saudi arms transfers. In particular, the idea that the 1988 Memorandum of Understanding with Saudi Arabia called 'Al Yamamah II' would lead to fewer contracts than envisaged brought almost frantic denials from the companies concerned.[45]

A series of military-related technology transfers between the UK and Iraq attracted great attention in 1990. These were the interception of nuclear trigger devices at Heathrow airport on 28 March 1990,[46] the sale of components for advanced sea mines,[47] the question of whether eight sections of pipe, heavy forgings and hydraulic shock absorbers seized at Teesport in April were elements of a 1000-mm calibre 'supergun' barrel to be assembled in Iraq,[48] and the allegation that the company Matrix Churchill had sold lathes and

[42] Branger (note 13), p. 13.

[43] *Flight International*, 24 Oct. 1990, p. 11; *International Defense Review*, Aug. 1990, p. 920.

[44] *Statement on the Defence Estimates 1990*, vol. 2, Cmd 1022-II (Her Majesty's Stationary Office: London, Apr. 1990).

[45] *Milavnews*, July 1990, p. 21; Cook, N., 'A piece of the action', *Jane's Defence Weekly*, 6 Oct. 1990, p. 661; *World Weapons Review*, 28 Nov. 1990, p. 6.

[46] *International Herald Tribune*, 3 Mar. 1990, p. 1; *Le Monde*, 30 Mar. 1990, p. 7; *Le Monde*, 31 Mar. 1990, p. 8.

[47] *Financial Times*, 30 Mar. 1990, p. 20.

[48] *Financial Times*, 17 Apr. 1990, p. 1; *Wednesday Report*, 18 Apr. 1990, p. 8; *The Independent*, 19 Apr. 1990, pp. 1–2.

machine tools to Iraq in the knowledge that they would be employed in munitions plants.[49]

These military-related technology transfers raised questions about the effectiveness of existing arms-export licensing procedures and the rigour with which the procedures were implemented.[50] In December 1990 Members of Parliament from both the main political parties called for a parliamentary enquiry into the supply of arms and military-related equipment to Iraq, but there was no call for a wider examination of the other question of arms export regulation.[51]

China

The People's Republic of China remains the fifth largest exporter of major conventional weapon systems for the period 1986–90. In 1990, however, Chinese exports remained at the reduced level of 1989, amounting to $0.9 billion—40 per cent of the 1987 export value. Chinese exports declined mainly as a result of the end of the war between Iraq and Iran, before the approval of the UN embargo.[52] Chinese exports are not likely to rise to the mid-1980s level in the near future. On 22 November 1990 the Chinese Government stated that supplies of arms and military assistance to the Khmer Rouge in Cambodia had been stopped in September.[53] During the five-year period 1986–90 China supplied major conventional weapons to 22 countries: the main customers were Iran, Saudi Arabia, Pakistan, Iraq, Thailand and North Korea.[54] China has intensified co-operation in arms production with Pakistan and is Pakistan's most important supplier. In a deal that has caused concern in neighboring countries, Myanmar (Burma) has ordered Chinese military equipment, including warships and 12 F-6 fighter planes that are already supplied.

The Federal Republic of Germany

For the period 1986–90 the FRG remains the sixth largest exporter of major conventional weapons. With exports increasing to almost $1 billion in 1990, it replaced China as the fifth largest exporter for that year. The FRG was one of the very few arms exporters that increased the sale of major conventional weapons in 1990.

[49] *Financial Times*, 18 Oct. 1990, p. 8; *Financial Times*, 20 Oct. 1990, p. 2; *Jane's Defence Weekly*, 27 Oct. 1990, p. 853.

[50] *The Independent*, 20 Apr. 1990, p. 6; *The Independent*, 21 Apr. 1990.

[51] Pienaar, J., 'Inquiry sought into military trade with Iraq', *The Independent*, 3 Dec. 1990, p. 2.

[52] Observing United Nations sanctions against Iraq has cost China $2 billion, according to Li Junhua, a Foreign Ministry spokeswoman. She refused to say how much of the loss was due to the arms embargo. *Financial Times*, 2 Nov. 1990, p. 4.

[53] *Milavnews*, Dec. 1990, p. 4, See also Whitaker, R., 'Peking ends arms flow to Cambodia', *The Independent*, 23 Nov. 1990, p. 13.

[54] Reports have appeared claiming that Thailand's interest in Chinese weapons is over. The decision to buy Chinese during the 1980s was linked to political considerations as much as for the import of cheap weapons sold by China at 'friendship prices'. Apparently Thai military officers quickly became disillusioned with their quality. *Far Eastern Economic Review*, 4 Oct. 1990, p. 20.

Political attention to arms exports in the FRG did not focus on the trade in major conventional weapons recorded here. Military technology exports to South Africa, Libya and Iraq dominated the public debate.

The parliamentary committee investigating the supply of submarine blueprints to South Africa presented its final report to parliament after four years of work. In the concluding debate on 26 October 1990 the opposition drew the conclusion that the Government, including Chancellor Helmut Kohl, were aware of this violation of the UN embargo against South Africa but did not stop it. Representatives of the coalition Government claimed that the parliamentary investigation could not prove government involvement; instead, the opposition had used the committee as a permanent propaganda instrument to criticize the Government.[55]

Jürgen Hippenstiel, owner of the Imhausen company, was sentenced in June 1990 to five years of imprisonment as the main person accused of the illegal supply of a chemical weapon plant to Libya. Several German companies were involved. Legislative initiatives started in early 1989 as a result of this transfer led to changes in the Foreign Trade Act[56]—imposing more restrictive controls and stronger legal sanctions—and the approval of the law on Improving the Control of Foreign Trade and Payments Transactions and Prohibiting Nuclear, Biological and Chemical Weapons.[57]

In addition the Government tried to close one of the many loopholes in German arms legislation by making the participation in missile development and construction abroad subject to government permission. The move coincided with reports and increasing evidence that numerous German companies had supplied equipment to produce chemical weapons in Iraq, had assisted that country with nuclear technology and had helped to modernize Soviet-made Scud missiles in Iraq. As the Gulf crisis intensified the Government was under international and national pressure to improve legislation and controls. The Weapons of War Act[58] was changed to restrict German participation in arms development and production abroad, an additional 'country list'[59] was added to the Foreign Trade Act to introduce special controls on technology supplies to 53 countries—mainly of the Third World—and the Government guidelines on the control of weapons and arms technology were expanded. According to the new guidelines individuals and companies wanting to export weapons or arms technology will require a certificate of reliability.[60]

[55] Reports on concluding parliamentary discussion in *Das Parlament*, no. 46–47 (9–16 Nov. 1990).

[56] Außenwirtschaftsgesetz, published in *Bundesgesetzblatt*, Teil I, Z 5702 A, no. 36 (27 July 1990), pp. 1457–59 and 1460–61.

[57] Official translation provided by the Ministry of Economics of 'Gesetz zur Verbesserung der Überwachung des Außenwirtschaftsverkehrs und zum Verbot von Atomwaffen, biologischen und chemischen Waffen', published in *Bundesgesetzblatt*, Teil I, Z 5702 A, no. 61 (10 Nov. 1990), pp. 2429–31.

[58] Gesetz über die Kontrolle von Kriegswaffen, published in *Bundesgesetzblatt*, Teil I, Z5702 A, no. 64 (30 Nov. 1990), pp. 2506–19.

[59] Länderliste H, published in *Bundesanzeiger*, herausgegeben vom Bundesminister der Justiz, no. 234 (18 Dec. 1990), p. 6637.

[60] Grundsätze der Bundesregierung zur Prüfung der Zuverlässigkeit von Exporteuren von Kriegswaffen und rüstungsrelevanten Gütern, published in *Bundesanzeiger*, 5 Dec. 1990.

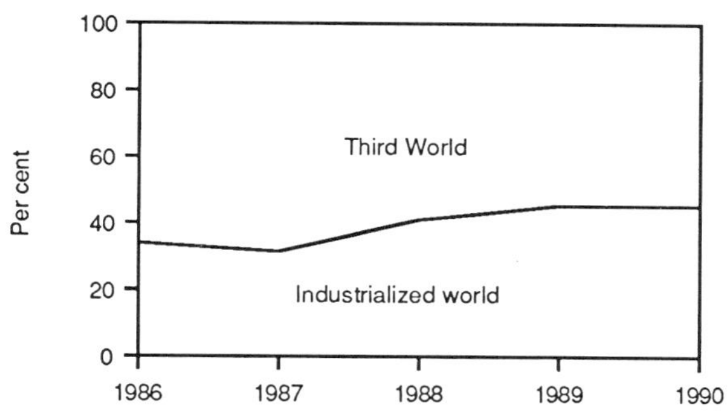

Figure 7.2. Shares of imports of major conventional weapons, 1986–90
Source: SIPRI data base.

The major importers

A small group of countries—India, Saudi Arabia, Iraq, Afghanistan, North Korea and Egypt in the Third World, and Japan, Spain, Poland and Czechoslovakia in the industrialized world—dominated the arms import pattern during the period 1986–90. These 10 countries account for half of all imports of major conventional weapon systems. In 1986–90 the 15 leading importers accounted for almost two-thirds of total imports (see table 7.2).

According to the SIPRI trend indicators there have been changes in the rank order of the leading importers, but virtually the same 15 Third World importers that were reported last year for the five-year period 1985–89 are the leading importers for the period 1986–90 as well. Hungary is no longer represented in the list of leading industrialized importers, as imports of major conventional weapons virtually stopped in 1990. Hungary has been replaced by Norway.

The trend of a shrinking Third World share in the world arms market of major conventional weapons that has been observed for several years was not continued in 1990. The industrialized countries that had increased their share from 34 per cent in 1986 to 45 per cent in 1989 accounted for 45 per cent in 1990, too (see figure 7.2). The decline in imports of weapons in 1990 is a result of similar magnitudes of reductions in both Third World and industrialized countries.

As pointed out in section I several Middle Eastern countries increased their weapon imports as a direct consequence of the Gulf crisis. With large orders still pending at the end of 1990 this tendency is likely to continue in the near future. In contrast to previous years, arms imports in South Asia were substantially reduced in 1990. Although India, Afghanistan and Pakistan remain important weapon importers, the region is no longer the major importing region but has been replaced by the Middle East again. For most other regions SIPRI trend indicators show reduced imports (see appendix 7A).

The overall growth of weapon imports of industrialized countries that SIPRI reported for several years was sharply reversed in 1990. With imports of $11 billion during 1986–90 Japan remains by far the major weapon importer among the industrialized countries. Most industrialized countries reduced their imports in 1990. Exceptions to this rule among the 15 leading importers of 1990 were the FRG, Romania, Bulgaria, the UK and Canada. These five countries increased their imports of major conventional weapons in 1990.

IV. Arms transfer control initiatives

In 1990 several arms transfer control initiatives were launched. In April the UN Department for Disarmament Affairs organized a conference in Florence, Italy on transparency in international arms transfers. At the request of member states and on the basis of a General Assembly resolution from 1989[61] the conference discussed ways and means of providing for more openness and transparency in the world-wide arms trade. Arms transfer control initiatives have also been launched by representatives of the Soviet Union and NATO, between them responsible for the bulk of both arms transfers and military technology development.

In a letter to the UN Secretary-General dated 15 August 1990, Soviet Foreign Minister Eduard Shevardnadze pointed to limiting international sales and supplies of conventional weapons as 'a means of building a new model of security at the time of continuing progress towards mutual understanding and interaction'.[62] Shevardnadze urged the highest legislative bodies of all UN member states to discuss the issue of arms exports and imports and announced that the Soviet Government would 'draft and submit to the Soviet lawmakers an appropriate national law'—suggesting that past Soviet arms exports have not had any legal framework.

In a speech in Istanbul on 18 October 1990, Secretary-General Manfred Wörner suggested that NATO should examine 'new dangers arising from regional conflicts directly affecting the security of our member nations'. In particular, controlling the proliferation of new military technologies required a 'global and enlightened COCOM, based on the cooperative participation of all technologically advanced countries, including the Soviet Union'.[63] One purpose of this organization would be to monitor the arms embargo on Iraq which is expected to remain in place indefinitely.[64]

The Soviet and NATO initiatives illustrate both the changing axis of political confrontation away from its East–West orientation and towards a North–

[61] UN document A/RES/43/75. The resolution also requested the Secretary-General to carry out a study with the assistance of governmental experts on the arms trade. The group of experts met twice in 1990 and will have to submit its report prior to the regular session of the UN General Assembly in 1991. Papers of the Conference are published in United Nations, *Transparency in International Arms Transfers*, UN Disarmament Topical Papers, no. 3 (UN: New York, 1990).

[62] *Izvestia*, 15 Aug. 1990.

[63] *Atlantic News*, 19 Oct. 1990, p. 1.

[64] *The Independent*, 30 Oct. 1990, p. 1.

South orientation and the growing momentum for an arms transfer control process. In particular, the initiatives underline the growing emergence of a US–Soviet discussion of military technology transfer.

There is currently no agreed arms transfer control agenda for many reasons. There are fundamental disagreements about who and what should be controlled, and the UN conference in Florence did not come up with a set of criteria to solve this problem. Responsibility for setting up and operating the infrastructure required to monitor and regulate transfers of military technology, the size and cost of which would be immense, has not been discussed. The 1990 initiatives underline some problems that arms transfer control has yet to overcome as well as suggest the probable future direction of control efforts.

Past initiatives for multilateral discussions of the transfer of military technology have generated resentment and resistance among developing countries, especially where discussions have been extended over dual-use technologies. Controlling arms transfers without controlling arms production has been seen as an infringement of the right to self-defence since it would give arms producers a monopoly over advanced military capabilities.

In anticipation of this objection, Shevardnadze's proposals are qualified, noting that 'the Soviet Union favours a search for new methods of tackling the problem of international sales and supplies of weapons though it is fully aware of its connection with the inalienable right of states to individual and collective self-defence, which is formalised in the UN Charter'.[65] As a further qualification, Shevardnadze notes that 'curbing international supplies of conventional weapons is inextricably linked with greater openness and the elimination of excessive secretiveness'. To complement openness in arms supplies, participation in the standardized reporting of military expenditures to the United Nations would 'help create an objective picture of military potentials of both arms producing and arms importing states'. On 12 October 1990 the Soviet Union provided a breakdown of its military expenditure for the year 1989 according to the Instrument for Standardized International Reporting of Military Expenditure designed by the United Nations. Along with an explanation provided by Deputy Foreign Minister Vladimir Petrovsky, this information represents a new insight into the distribution of the Soviet military effort.[66]

Shevardnadze also suggested precluding 'the possibility of creating offensive weapons potentials' through 'an effective multilateral regime of non-proliferation of some types of missiles and missile technologies'. However, surface-to-surface missiles are not weapons, they are weapon delivery systems. If the focus of arms control is to be delivery systems, rather than warheads, then missiles should logically be considered together with a range of other systems: combat aircraft, artillery, rocket launchers and target-acquisition systems. Moreover, unless Shevardnadze rejects the concept of deterrence, whether or

[65] *Izvestia*, 15 Aug. 1990.

[66] Petrovsky, V., 'Military budget in the light of Glasnost', *Novosti Soviet Press*, no. 45 (Nov. 1990); this is discussed more fully in chapter 5 of this volume.

not ballistic missiles represent an offensive potential depends on the circumstances of their deployment and is not an inherent quality of the systems themselves.

Shevardnadze makes three other proposals in his letter: considering (*a*) 'regional means of restricting international arms traffic, taking into account the states requirements for self-defence and specific conditions of each region'; (*b*) 'restraint on the part of arms suppliers and recipient states in regard to areas of conflicts'; and (*c*) 'regulating the re-export of armaments and combatting illegal arms trafficking'.

These issues were also dealt with at the UN Florence conference on transparency of the arms trade. So far no definitive arms transfer control agenda has been set up. The Gulf crisis has underlined the need for arms transfer control and could thus possibly act as a catalyst for such initiatives.

V. Other forms of military technology transfer

There is a general but gradual movement away from national and towards transnational arms production. An increasing number of co-operative projects involve a flow of technology and components between co-operating companies and countries. Technological, economic and political reasons have contributed to this transnationalization process. A second tendency in arms procurement relates to the increasing importance of retrofitting or upgrading existing weapon systems instead of investing in the development of new major weapon platforms. Both trends contribute to emphasizing technological interdependence between countries and have consequences for arms transfer control and for the reporting of arms transfers.

SIPRI records transfers of major weapon systems but not transfers of components or technology.[67] Significant elements of the arms transfer process are thus not included in the SIPRI statistics. Furthermore, the country of origin of the many suppliers of components and the value of these supplies are omitted.

During their production, many major weapon systems involve international transfers of military technology, either through intra-group sales within a single multinational industrial enterprise or through the award of sub-contracts with overseas suppliers by a prime contractor. This international technology transfer makes it difficult to determine the economic benefits to individual companies or countries of international sales of systems such as the three-nation Tornado multi-role combat aircraft, the Anglo-US Harrier jump-jet or the Franco-German PAH helicopter. However, by looking at some specific programmes it is possible to offer some observations.

Co-operative weapon system development and the arms trade

The speed and range of military-related technology development have meant that no country—not even the United States—can be a leader in all fields. The

[67] See appendix 7C for the methodology used.

technical complexity of designing and building major systems has become too great for any single company to manage alone. Teams are therefore formed between companies, often across national borders. The growth of technological interdependence has been reflected in the trend towards transfers of sub-systems and dual-use military-related equipment, which has multiplied the problems of arms control. Industrial and political considerations led governments in weapon-importing countries to insist on licensed and sub-contractor production, technology transfer, counter-trade and other forms of offsets in military transfers. Financial constraints in the procurement budgets on the one side and rising weapon unit cost on the other have led to strengthening the tendency of pooling resources in joint projects.

There are three different types of project taking the origin of the major sub-systems and armaments of weapons as a criteria: first, genuine national projects that are independent of technology imports or depend only to a small degree on imports; second, projects that are undertaken in one nation, depending to a large extent on imports of technology and sub-systems; and third, projects that are planned and carried out as truly international or multilateral projects. Table 7.6 includes one example of each type.

One of the competing designs to meet the US Air Force requirement for an Advanced Tactical Fighter aircraft is the YF-22A. This is a genuine national programme, reflecting the advanced and diverse nature of the US aircraft industry. US companies occasionally involve foreign companies in programmes. This sometimes reflects a special technical competence—in particular in some West European companies. Sometimes, however, it reflects a political directive to co-operate—for example, through the 1986 amendment to the Defense Authorization Act sponsored by Senators Nunn, Glenn, Roth and Warner, widely known as the Nunn Amendment.

In contrast to the YF-22, the JAS-39 Gripen being built in Sweden depends heavily on imported technology, sub-systems and components. While the Swedish Government controls the size of the production run, from the perspective of technology dependence, this is an international programme.

The programme to develop a European Fighter Aircraft (EFA) was, from the outset, a multilateral programme led by a consortium of companies from four countries—the UK, Germany, Italy and Spain. Its development and production are basically distributed to companies according to the financial contributions of the participating countries.

The involvement of companies from several countries is more typical of technically complex and very expensive systems such as fighter aircraft and warships than for land systems. In main battle tank or artillery production trans-nationalization is not the norm in the major production centres—the USA, Western Europe and the USSR—although it is typical for Third World producers. As a general rule it is possible to conclude that the more expensive and the more complex a weapon system is the more likely is the need for multilateralization. The more advanced the technological base of a country,

Table 7.6. Origin of major components of selected modern fighter aircraft

	Producer[a] (country)		
Component	YF-22A (USA)[b]	JAS-39 (Sweden)	EFA (UK, Germany, Italy, Spain)[c]
Design	Lockheed (USA)	Saab (Sweden)	BAe (UK), MBB (Germany), Aeritalia (Italy), CASA (Spain)
Airframe	Lockheed (USA) forward fuselage; Boeing (USA) wings, aft sections; General Dynamics (USA) mid-fuselage, tail, nose	Saab (Sweden) carbon-fibre wings, on technology by British Aerospace (UK)	BAe (UK) forward fuselage, foreplanes; Aeritalia (Italy), CASA (Spain), BAe (UK) wings; MBB (Germany), Dornier (Germany) mid-fuselage, fin; Aeritalia (Italy), CASA (Spain) rear fuselage
Engine	General Electric (USA) or Pratt & Whitney (USA)	General Electric (USA) modified by Volvo Flygmotor (Sweden)	Rolls Royce (UK), MTU (Germany), FIAT (Italy), SENER (Spain)
Avionics	Lockheed (USA) team leader	Ericsson (Sweden) team leader	BAe (UK) team leader
Digital flight control system	GEC Avionics (UK)	Lear Astronics (UK)	MBB (Germany) team leader
Radar	Westinghouse (USA), Texas Instruments (USA)	Ericsson (Sweden), Ferranti (UK)	Ferranti (UK), Ericsson (Sweden)
Mission and display computer	Texas Instruments (USA)	Ericsson (Sweden)	Smiths Industries (UK)
Head-up display	GEC Avionics (UK)	Hughes Aircraft (USA), Ferranti (UK)	
Air-to-air long-range missile		Hughes (USA) and Raytheon (USA) team leaders	Hughes (USA) and Raytheon (USA) team leaders
Air-to-air short-range missile		BAe (UK), Ford Aerospace/General Electric/ Raytheon (USA)	BAe (UK) team leader; Ford Aerospace/ General Electric/ Raytheon (USA)

[a] The holding company, as the designer and owner of technology, is recorded here rather than the producing subsidiary.

[b] The YF-22A is only one of the two prototypes of an advanced tactical fighter (ATF) competing to be the US fighter of the 1990s, the other one is the Northrop YF-23A.

[c] Four countries participate in the EFA project: the United Kingdom 33%; Germany 33%; Italy 21%; Spain 13%.

Source: SIPRI data base.

independently on a national basis. This long-term tendency towards international collaboration is further intensified by the globally shrinking demand for new arms. Arms procurement in many countries is insufficient to utilize existing arms production capacity. Producers are faced with over-capacities. Collaboration is a means by which both governments and companies try to deal with these problems.

Besides technological criteria, financial and political considerations (primarily intra-alliance standardization) play an important role in forming joint teams and entering into international collaboration. Both governments and companies are interested in joint international projects. Collaborative projects are a means of pooling financial and technological resources to meet research and development problems posed by modern military equipment, to extend production runs and—in some cases—to circumvent restrictive arms export policies.[68]

Retrofit

As the unit cost of new systems continues to increase and procurement budgets fall in real terms, retrofitting existing platforms has become an increasingly attractive option for many governments.

The cost-effectiveness of buying major weapon platforms has been reduced by several factors. First, the performance of weapon systems has superseded the speed and manœuvrability of aircraft, tanks and ships as the central factor in system design as advances in surveillance and guidance systems have reduced the importance of line-of-sight targeting. Second, the increasing use of simulators and computer-aided training devices and the unsupportable costs of training missions have reduced wear and tear on equipment. Third, the use of non-corrosive materials or materials that are resistant to wear and tear, computerized fault diagnosis and greater use of modular design techniques that allow rapid (if expensive) replacement of faulty or worn-out equipment have extended the life of most modern systems.

Through-life improvements have always been a feature of shipbuilding programmes. Since the active life of a ship is likely to be in excess of 30 years, it will typically go through two re-fits—after 10 and 20 years—to upgrade the quality of on-board systems. Increasingly, the capabilities of land and air systems also derive as much from on-board systems as from the airframe or chassis. Looking at the list of aircraft retrofit programmes in table 7.7, it can be seen that many governments now prefer to upgrade existing platforms rather than buy new equipment.

[68] Chapter 8 of this volume analyses the recent phenomenon of internationalization strategies of companies.

Table 7.7. Selected aircraft retrofit programmes in progress in 1990

Recipient	System	Upgrade	Supplier	Quantity	Value ($m.)
Argentina	S-2E Tracker	Engine replacement	USA	6	40
Australia	P-3 Orion	ESM system	Israel	20	70.5
Australia	F-111	Flight controls	USA	22	315
		Inertial-navigation system			
		Terrain-following radar			
		Weapon-aiming systems			
Australia	F/A-18 Hornet	AN/APG-65 radar		73	221
		XN-6 computer			
		AN/AAS-38 Flir	USA		
Belgium	Mirage-5	Navigation system	France	20	
		Head-up display	UK	20	
Belgium, Denmark, Netherlands, Norway	F-16	AN/APG-66 radar	USA	403	1 560
		Head-up displays	UK		
		Computers and data link	USA		
		GPS navigation system	USA		
Canada	CF-18	AN/APG-65 radar	USA		221
Canada	F-5	Head-up display	UK	56	
		Computer systems			
		Inertial-navigation system			
Chile	F-5	EL/M-2023 radar	Israel	14	200
		Inertial-navigation system			
		Head-up display			
		Flight controls			
China	F-7	Skyranger radar	UK		52
Germany, FR	F-4F Phantom	APG-65 radar, computer	USA	110	1 100
Italy	F-104	Radar, ECM and IFF		150	
Japan	F-4	AN/APG-66 radar	USA	100	
		LN-39 inertial navigation system	USA	100	
		ASTAC ESM pods	France	17	
Japan	P-3	AN/ALQ-78 ESM system	USA	60	
Jordan	F-5	Head-up display	UK		
		Inertial-navigation system			
New Zealand	A-4	AN/APG-66 radar	USA	22	
		LN-39 inertial navigation system	USA		
		Type 4510 head up display	UK		
Norway	F-5	Communications system	USA		20
		Electronic countermeasures			
Pakistan	P-3C Orion	Infra-red detection system	USA	3	
		Fire-control radar	USA		
		Sonobuoy reference system	USA		
Qatar	Commando	Navigation system	UK		6.75
		Communications system			
Saudi Arabia	E-3A Sentry	Improved navigation and electronic warfare systems	USA	5	600

Recipient	System	Upgrade	Supplier	Quantity	Value ($m.)
Singapore	A-4	F404-100D engine	USA	24	
		LN-39 inertial navigation system	USA		
		Head-up display	UK		
		Computers and data link	UK		
South Korea	F-4	AN/APG-65 or AN/APG-68	USA	35	
South Korea	F-5	Inertial-navigation system	USA	10	
		Head-up display	UK	10	
Taiwan	S-2T Tracker	Engine replacement	USA	6	40
		AN/APS-509 radar			
		AN/ASN-150 navigation system			
Thailand	F-5	Head-up display	UK	20	38.3
		Inertial-navigation system	USA		
Turkey	F-16	Electronic warfare systems	USA	NA	300
USA	A-7	Head-up display	UK	83	
Venezuela	Mirage-5	Airframe modifications	France	17	296
		Cyrano 4 multi-role radar			
		Head-up display			
		Inertial-navigation system			

Source: SIPRI data base.

The costs of retrofit are considerably less than those of buying new systems. For example, the $1.56 billion to be spent by the governments of Belgium, Denmark, the Netherlands and Norway is a great deal of money. However, it will maintain over 400 modern fighter aircraft in service whereas the same money would only buy between 60 and 90 new aircraft (depending on the designation) of comparable performance.

Appendix 7A. Aggregate tables of the value of the trade in major weapons with the Third World, 1971–90

Table 7A.1. Values of imports of major weapons by the Third World: by region[a]

Figures are SIPRI trend indicator values, as expressed in US $m., at constant (1985) prices. A = yearly figures, B = five-year moving averages.[b]

Region[c]		1971	1972	1973	1974	1975	1976	1977	1978
Middle East	A	6 092	5 837	10 493	6 999	7 014	7 088	10 000	7 709
	B	6 191	6 933	7 287	7 486	8 319	7 762	7 596	7 881
South Asia	A	1 274	1 800	1 047	890	545	1 032	2 092	1 842
	B	1 181	1 174	1 111	1 063	1 121	1 280	1 387	1 751
Far East	A	3 658	7 050	1 844	1 818	1 476	1 570	2 063	3 754
	B	3 360	3 332	3 170	2 752	1 755	2 137	2 970	3 296
Sub-Saharan Africa	A	429	271	466	869	650	1 044	2 523	2 505
	B	335	482	537	660	1 110	1 518	1 526	1 699
South America	A	786	1 019	2 370	1 347	1 600	1 922	2 836	2 350
	B	1 022	1 161	1 424	1 651	2 015	2 011	2 069	2 176
North Africa	A	224	259	293	508	2377	2 399	2 815	2 994
	B	293	342	774	1 186	1 635	2 354	3 386	3 586
Central America	A	136	266	313	300	221	263	583	280
	B	191	238	241	261	320	311	310	306
South Africa	A	104	292	459	533	232	371	171	343
	B	240	333	324	378	353	330	244	219
Total		**12 702**	**16 903**	**17 327**	**13 346**	**14 062**	**15 544**	**22 863**	**22 702**

[a] The values include licensed production of major weapons in Third World countries (see appendix 7B). For the values for the period 1951–70, see Brzoska, M. and Ohlson, T., SIPRI, *Arms Transfers to the Third World, 1971–85* (Oxford University Press: Oxford, 1987).

[b] Five-year moving averages are calculated as a more stable measure of the trend in arms imports than the often erratic year-to-year figures.

[c] The regions are listed in rank order according to the five-year moving average values in the column for 1988. The following countries are included in each region:

Middle East: Bahrain, Egypt, Iran, Iraq, Israel, Jordan, Kuwait, Lebanon, Oman, Qatar, Saudi Arabia, Syria, United Arab Emirates, North Yemen and South Yemen.

South Asia: Afghanistan, Bangladesh, India, Nepal, Pakistan and Sri Lanka.

Far East: Brunei, Fiji, Indonesia, Kampuchea, North Korea, South Korea, Laos, Malaysia, Mongolia, Myanmar (formerly Burma), Papua New Guinea, Philippines, Samoa, Singapore, Solomon Islands, Taiwan, Thailand, Vanuatu and Viet Nam.

1979	1980	1981	1982	1983	1984	1985	1986	1987	1988	1989	1990
6 170	8 436	9 386	11 306	12 211	11 613	10 086	10 401	13 523	8 072	4 436	4 553
8 340	8 601	9 502	10 590	10 920	11 123	11 567	10 739	9 303	8 197	. .	. .
1 423	2 367	2 541	2 690	2 424	2 090	2 809	5 085	5 939	4 916	6 905	3 207
2 053	2 173	2 289	2 422	2 511	3 020	3 669	4 168	5 131	5 210	. .	. .
5 987	3 103	2 981	1 781	2 591	2 916	3 633	3 638	3 254	4 297	3 932	2 075
3 578	3 521	3 289	2 674	2 780	2 912	3 206	3 547	3 751	3 439	. .	. .
908	1 515	2 076	1 734	1 405	2 029	2 007	1 672	1 918	1 408	370	830
1 905	1 748	1 528	1 752	1 850	1 769	1 806	1 807	1 475	1 239	. .	. .
1 635	2 137	3 215	2 509	2 896	2 980	1 219	1 149	1 661	693	1 077	680
2 435	2 369	2 479	2 748	2 564	2 151	1 981	1 540	1 160	1 052	. .	. .
4 612	3 044	3 078	2 893	1 930	1 775	1 794	2 371	1 530	1 093	844	582
3 731	3 819	3 373	2 527	2 086	1 764	1 260	1 000	953	752	. .	. .
299	181	776	1 250	1 122	643	731	661	564	218	287	368
410	536	702	757	853	826	664	486	421	363	. .	. .
102	109	4	4	156	5	4	154	20	28	3	2
146	112	75	56	35	65	68	42	42	42	. .	. .
22 269	**21 188**	**23 964**	**24 271**	**24 483**	**23 727**	**21 551**	**24 114**	**27 228**	**20 025**	**18 256**	**11 841**

Sub-Saharan Africa: Angola, Benin, Botswana, Burkina Faso, Burundi, Cameroon, Cape Verde, Central African Republic, Chad, Comoros, Congo Côte d'Ivoire, Djibouti, Equatorial Guinea, Ethiopia, Gabon, Gambia, Ghana, Guinea, Buinea-Bissau, Kenya, Lesotho, Liberia, Madagascar, Malawi, Mali, Mauritania, Mauritius, Mozambique, Namibia, Niger, Nigeria, Rwanda, Senegal, Seychelles, Sierra Leone, Somalia, Sudan, Swaziland, Tanzania, Togo, Uganda, Zaire, Zambia and Zimbabwe.

South America: Argentina, Bolivia, Brazil, Chile, Colombia, Ecuador, Guyana, Paraguay, Peru, Suriname, Uruguay and Venezuela.

North Africa: Algeria, Libya, Morocco and Tunisia.

Central America: Bahamas, Barbados, Belize, Costa Rica, Cuba, Dominican Republic, Dominica, El Salvador, Guatemala, Haiti, Honduras, Jamaica, Mexico, Nicaragua, Panama, St Vincent and the Grenadines, and Trinidad and Tobago.

[d] Items may not add up to totals due to rounding.

. . Not applicable.

Source: SIPRI data base.

Table 7A.2. Values of exports of major weapons to regions listed in table 7A.1: by supplier[a]

Figures are SIPRI trend indicator values, as expressed in US $m., at constant (1985) prices. A = yearly figures, B = five-year moving averages.[b]

Supplier[c]		1971	1972	1973	1974	1975	1976	1977	1978
USSR	A	6 071	7 931	7 364	5 244	3 610	4 589	7 979	10 052
	B	5 650	6 240	6 044	5 747	5 757	6 295	7 477	8 624
USA	A	3 795	5 821	6 336	4 352	6 866	7 064	9 551	6 862
	B	4 521	4 770	5 434	6 088	6 834	6 939	6 861	6 619
France	A	683	796	1 654	1 279	1 168	1 453	2 284	2 152
	B	827	1 020	1 116	1 270	1 567	1 667	2 025	2 314
China	A	346	409	197	290	251	200	110	362
	B	228	271	298	269	209	242	252	315
UK	A	1 214	1 195	1 309	1 070	1 193	833	1 652	1 215
	B	1 055	1 052	1 196	1 120	1 211	1 193	1 132	1 036
Germany, FR	A	86	108	0	462	269	166	204	258
	B	51	132	185	201	220	272	212	215
Netherlands	A	93	63	169	65	131	357	94	172
	B	78	80	104	157	163	164	190	187
Italy	A	95	141	148	273	144	163	288	327
	B	101	139	160	174	203	239	379	481
Brazil	A	0	0	0	10	25	154	130	120
	B	0	2	7	38	64	88	108	153
Israel	A	1	34	4	67	125	61	59	470
	B	10	22	46	58	63	157	189	205
Other Third World	A	48	129	28	184	146	227	187	95
	B	52	83	107	143	154	168	232	242
Other industrialized, West	A	117	200	64	18	86	151	103	315
	B	122	92	97	104	85	135	156	164
Other industrialized, neutral	A	95	5	10	13	24	63	68	36
	B	25	25	29	23	36	41	136	195
Other industrialized, East	A	60	72	45	19	23	63	155	265
	B	91	68	44	44	61	105	138	162
Total		**12 702**	**16 903**	**17 327**	**13 346**	**14 062**	**15 544**	**22 863**	**22 702**

[a] The values include licensed production of major weapons in Third World countries (see appendix 7B). For the values for the period 1951–70, see Brzoska, M. and Ohlson, T., SIPRI, *Arms Transfers to the Third World, 1971–85* (Oxford University Press: Oxford, 1987).

[b] Five-year moving averages are calculated as a more stable measure of the trend in arms imports than the often erratic year-to-year figures.

[c] The regions are listed in rank order according to the five-year moving average values in the column for 1988.

[d] Other NATO, Australia and Japan.

[e] Austria, New Zealand, Sweden, Switzerland and Yugoslavia.

[f] Other WTO.

– Nil.

. . Not applicable.

Source: SIPRI data base.

1979	1980	1981	1982	1983	1984	1985	1986	1987	1988	1989	1990
11 158	9 341	8 480	7 700	8 038	7 980	9 106	10 440	10 936	8 658	8 862	4 273
9 402	9 346	8 943	8 308	8 261	8 653	9 300	9 424	9 600	8 634	. .	. .
3 961	5 657	6 155	6 989	6 289	5 015	4 114	4 981	6 328	3 939	3 465	3 048
6 437	5 925	5 810	6 021	5 712	5 478	5 346	4 875	4 565	4 352	. .	. .
3 067	2 617	3 508	3 178	3 065	3 217	3 588	3 446	2 659	1 413	1 642	1 330
2 725	2 904	3 087	3 117	3 311	3 299	3 195	2 865	2 550	2 098	. .	. .
339	568	277	682	1 060	1 419	1 217	1 463	2 553	1 810	817	926
331	445	585	801	931	1 168	1 543	1 693	1 572	1 514	. .	. .
767	716	1 077	1 536	676	1 083	1 050	1 091	1 681	1 281	1 187	971
1 085	1 062	954	1 017	1 084	1 087	1 116	1 237	1 258	1 242	. .	. .
162	283	938	323	1 174	1 831	395	661	254	367	168	496
369	393	576	910	932	877	863	702	369	389	. .	. .
196	115	189	154	0	57	38	132	263	402	661	125
153	165	131	103	88	76	98	178	299	317	. .	. .
975	654	1 333	1 350	1 048	831	579	399	320	362	49	39
715	928	1 072	1 043	1 028	841	635	498	342	234	. .	. .
112	250	253	184	298	271	172	134	491	338	151	22
173	184	219	251	236	212	273	281	257	227	. .	. .
227	209	252	365	368	261	155	261	267	111	241	31
244	305	284	291	280	282	262	211	207	182	. .	. .
507	194	485	579	884	631	430	417	564	628	222	189
294	372	530	555	602	588	585	534	452	404	. .	. .
126	124	190	643	997	558	230	235	324	252	249	242
171	279	416	502	524	533	469	320	258	260	. .	. .
491	316	366	202	249	207	263	275	385	282	151	39
255	282	325	268	257	239	276	282	271	226	. .	. .
181	145	460	387	334	367	215	180	202	181	392	109
241	288	302	339	353	297	260	229	234	213	. .	. .
22 269	**21 188**	**23 964**	**24 271**	**24 483**	**23 727**	**21 551**	**24 114**	**27 228**	**20 025**	**18 256**	**11 841**

Appendix 7B. Register of the trade in and licensed production of major conventional weapons in industrialized and Third World countries, 1990

This register lists major weapons on order or under delivery, or for which the licence was bought and production was under way or completed during 1990. 'Year(s) of deliveries' includes aggregates of all deliveries and licensed production since the beginning of the contract. Sources and methods for the data collection, and the conventions, abbreviations and acronyms used, are explained in appendix 7C. Entries are alphabetical, by recipient, supplier and licenser.

Recipient/ supplier (S) or licenser (L)	No. ordered	Weapon designation	Weapon description	Year of order/ licence	Year(s) of deliveries	No. delivered/ produced	Comments
I. Industrialized countries							
Austria							
S: Sweden	500	RBS-56 Bill	Anti-tank missile	1989	1989–90	(300)	Deal worth $80 m
USA	24	M-109-A2 155mm	SPH	1988	1989–90	(12)	Deal worth $36 m; brings total ordered to 109
Australia							
S: France	5	Falcon–900	Transport	1988	1988–90	(5)	For VIP use
Italy	(10)	HSS-1	Surveillance radar	1986	1988–90	(6)	Deal worth $20 m
New Zealand	20	C-130H Hercules	Transport	1990			Bought from Air New Zealand
USA	24	UH-60 Blackhawk	Helicopter	1985	1989–90	24	In addition to previous orders for 30 Blackhawk/ Seahawks
	2	RGM-84A Launch	ShShM launcher	1983			Arming FFG-7 frigates produced under licence; in addition to 4 delivered earlier
	2	RIM-66A Launch	ShAM launcher	1985			Arming FFG-7 frigates produced under licence; in addition to 4 delivered earlier
	. .	AIM-9M	Air-to-air missile	1984	1986–90	(922)	Arming F/A-18 Hornet fighters
	(65)	RIM-67C/SM-2	ShAM/ShShM	(1987)	1989–90	(40)	Deal worth $50 m

L: Germany, FR	10	Meko-200 Class	Frigate	1989			8 for Australia, 2 for New Zealand; option for 2 more
Sweden	6	Type-471	Submarine	1987			Option for 2 more
Switzerland	65	PC-9	Trainer	1986	1987–90	(34)	In addition to 2 delivered directly; 17 for assembly and 48 for production
UK	105	Hamel 105mm	Towed gun	(1982)	1988–90	(52)	Deal worth $112 m
USA	73	F/A-18 Hornet	Fighter	1981	1985–90	(73)	Deal worth $4.8 b incl 2 delivered directly and 18 F/A-18B trainers
	8	SH-60B Seahawk	Helicopter	1986	1990	(2)	In addition to 8 ordered 1985
	2	FFG-7 Class	Frigate	1983			
Belgium							
S: Sweden	28	Helitow	Fire control	1988			To equip A-109 helicopters
USA	. .	AGM-65C	ASM	1989			Arming F-16 fighters
	545	AIM-9M	Air-to-air missile	1988	1989–90	(360)	Arming F-16 fighters; deal worth $49 m
	940	AIM-9M	Air-to-air missile	1989			Deal worth $80 m
	(224)	BGM-71A TOW	Anti-tank missile	(1989)			Arming 28 A-109A Mk-2 helicopters
L: Israel	21	El/M-2310	Battlefield radar	1989	1990	10	Refitted to M-113 APCs to create mobile radars
Italy	46	A-109A Mk-2	Helicopter	1988			Deal worth $317 m incl offsets worth 40%
USA	44	F-16A	Fighter	1983	1988–89	23	Deal worth $625 m incl offsets worth 80%
Bulgaria							
S: Czechoslovakia	18	L-39Z Albatross	Jet trainer	1989	1990	9	
USSR	. .	MiG-29	Fighter	1989	1990	(8)	
	. .	MT-LB	APC	1982	1982–90	(300)	
	1	Koni Class	Frigate	1990	1990	1	
L: USSR	. .	MT-LB	APC	(1980)	1982–90	170	May be 2S1 chassis
Canada							
S: Brazil	. .	EMB-120	Transport	1990			
France	10 000	Eryx	Anti-tank missile	(1987)			Programme suspended

Recipient/ supplier (S) or licenser (L)	No. ordered	Weapon designation	Weapon description	Year of order/ licence	Year(s) of deliveries	No. delivered/ produced	Comments
Italy	10	Skyguard	Air defence radar	1986	1989–90	(5)	Part of ADATS contract
Sweden	12	Giraffe	Fire control radar	(1985)	1988	2	Shipborne version for City Class destroyers
UK	28	EH-101	Helicopter	(1987)			Options for further 17
	7	S-500	Surveillance radar	1987	1988–90	(7)	
	(100)	Javelin	Portable SAM	1990	1990	(100)	For deployment in the Persian Gulf
USA	28	F/A-18 Hornet	Fighter	1989			Attrition replacements
	3	Model 206L	Helicopter	1989	1990	3	
	3	P-3C Update-3	Maritime patrol	1989			
	2	AN/TPS-70	Air defence radar	1990			Deal worth $23 m
	4	Phalanx	CIWS	1987	1989	(1)	Arming Tribal Class frigates
	6	Phalanx	CIWS	1986	1988	(2)	Arming City Class frigates
	6	Phalanx	CIWS	1990			Deal worth $32 m; arming second batch of City Class frigates
	12	RGM-84A Launch	ShShM launcher	(1984)	1989–90	(4)	Arming City Class frigates
	4	Seasparrow VLS	ShAM/PDM launcher	1986			Arming Tribal Class frigates
	12	Seasparrow VLS	ShAM/PDM launcher	1984	1988–89	(2)	Arming City Class frigates; deal worth $75 m incl modifications to missiles
	100	AIM-7M Sparrow	Air-to-air missile	(1987)	1990	50	Arming F/A-18 fighters; deal worth $31 m incl 24 Mk 48 torpedoes
	100	AIM-9M	Air-to-air missile	1988	1989–90	(100)	Deal worth $21 m
	29	RGM-84A Harpoon	ShShM	1988	1989–90	16	Deal worth $47 m incl spares, training and support
	74	RIM-66C/SM-2	ShAM/ShShM	1988	1988–89	(45)	Arming Tribal Class frigates; deal worth $48 m
	128	RIM-67C/SM-2	ShAM/ShShM	1987	1989	(32)	Arming Tribal Class frigates
	168	Seasparrow	ShAM	1984	1989–90	(56)	Arming City Class frigates; deal worth $75 m
L: Germany, FR	. .	Bo-105LS	Helicopter	(1981)	1987–88	(10)	
UK	40	L-119 105mm	Towed gun	1990			Further orders expected
USA	. .	LAV-25	APC	1982	1983–90	(516)	

China							
S: France	4	AS-332	Helicopter	(1987)			For Navy; part of deal worth $183 m incl 4 SA-365Fs
	2	Crotale Naval L	ShAM launcher	1986	1990	(1)	Arming 2 Luda Class destroyers; part of deal worth $91.5 m
	. .	Crotale Naval	ShAM	1987	1990	(36)	Part of deal worth $91.5 m incl Castor 2-C fire-control radar
USA	6	CH-47D Chinook	Helicopter	1989			Deliveries embargoed in June 1989
	4	AN/TPQ-37	Tracking radar	(1987)	1988	2	Deliveries suspended in June 1989 along with deliveries of avionics, 4 Mk 46 torpedoes and 155mm howitzer ammunition
USSR	(24)	Mi-17 Hip-H	Helicopter	1990	1990	(12)	
Cyprus							
S: France	12	AMX-155 Mk-F3	SPH	1990			
	36	AMX-30-B2	Main battle tank	1989	1989–90	24	Deal worth $115 m
	36	VAB	APC	1987	1989	(18)	Armed with HOT anti-tank missiles
	. .	Mistral	Portable SAM	(1988)	1989	(180)	Arming VAB APCs and infantry version
Greece	75	Steyr-4K 7FA	APC	(1990)			Options for 65 more
Italy	30	Skyguard	Air defence radar	1987	1988–90	30	Fire control for new 35mm AAGs
Czechoslovakia							
S: USSR	(34)	MiG-29	Fighter	(1988)	1989–90	(20)	
	. .	SA-13 Launcher	AAV(M)	(1984)	1985–89	(25)	
	. .	AA-10 Alamo	Air-to-air missile	1988	1989–90	(48)	Arming MiG-29 fighters
	. .	AA-11 Archer	Air-to-air missile	1988	1989–90	(48)	Arming MiG-29 fighters
	. .	AA-8 Aphid	Air-to-air missile	1988	1989–90	(96)	Arming MiG-29 fighters
	. .	AT-4 Spigot	Anti-tank missile	1979	1980–89	(2 400)	
	. .	SA-13 Gopher	Landmob SAM	(1984)	1985–89	(330)	
	. .	SA-9 Gaskin	Landmob SAM	1979	1980–89	(1 600)	
L: USSR	. .	BMP-1	MICV	1971	1971–89	(9 100)	70% exported back to USSR

Recipient/ supplier (S) or licenser (L)	No. ordered	Weapon designation	Weapon description	Year of order/ licence	Year(s) of deliveries	No. delivered/ produced	Comments
	. .	BMP-1 Spigot	TD(M)	1979	1980–89	(236)	
	. .	T-72	Main battle tank	1978	1981–90	(735)	
Denmark							
S: France	12	AS-350 Ecureuil	Helicopter	1987	1990	12	Deal worth $67 m incl Helitow sight system and TOW-2 missiles
Germany, FR	. .	RAM	ShAM/PDM	(1985)			Arming 3 Niels Juel Class frigates
Norway	3	Type-207	Submarine	1985	1989–90	2	
Sweden	12	Helitow	Fire control	1987	1990	12	
USA	12	F-16A	Fighter	1988			
	162	AGM-65D	ASM	1989			Arming F-16 fighters; deal worth $24 m
	(196)	BGM-71D TOW-2	Anti-tank missile	1987	1990	(196)	Arming 12 AS-350 Ecureuil helicopters
	840	FIM-92A Stinger	Portable SAM	1988			Deal worth $61 m incl 336 launchers
	. .	Seasparrow	ShAM	1989			From European production
Finland							
S: France	(20)	Crotale SAMS	Mobile SAM system	1990			Deal worth $230 m
	10	TRS-2230/15	3-D radar	1990			Deal worth $200 m
	(360)	Mistral	Portable SAM	1989			Arming Helsinki-2 Class FACs
	(240)	R-440 Crotale	Landmob SAM	1990			
Sweden	. .	Giraffe	Fire control radar	(1987)	1988	(5)	Mounted in Finnish Sisu APCs
	4	RBS-15 Launcher	ShShM launcher	1987			Arming Helsinki-2 Class FACs
	64	RBS-15	ShAM/ShShM	(1987)			Arming Helsinki-2 Class FACs
UK	7	Hawk	Jet trainer	1989			
	. .	Marksman	AAV(G)	1988	1990	1	Deal worth $16 m
	4	Watchman	Surveillance radar	1988	1989–90	(4)	Second order
USSR	. .	MT-LB	APC	(1986)	1986–90	(50)	
	(60)	T-72	Main battle tank	(1986)	1986–90	(60)	

	. .	AT-4 Spigot	Anti-tank missile	(1986)	1986–90	(300)	Part of a $400 m 5-year agreement incl T-72 tanks and MT-LB APCs
France							
S: Brazil	50	EMB-312 Tucano	Trainer	1990			Deal worth $124.5 m, option for 25 more
Nigeria	(14)	SA-330 Puma	Helicopter	1989	1989–90	(6)	Recently upgraded to SA-330L; exchanged for 12 AS-332s
Spain	2	CN-235	Transport	1988	1990	2	Option for 6 more
USA	2	C-130H-30	Transport	1990			Deal worth $58 m
L: USA	80	MLRS 227mm	MRL	1985	1989–90	80	
Germany, FR							
S: France	23	TRS-3050	Surveillance radar	1987	1987–90	10	Improved fire control system for Type 148 FACs
Netherlands	5	Smart	Fire control radar	1989			Fire control radar for Type-123 frigates
USA	3	AN/FPS-117	Air defence radar	1988			
	28	Patriot battery	Mobile SAM system	1984	1989–90	18	
	4	Seasparrow VLS	ShAM/PDM launcher	1986			Arming Type-123 Class frigates
	100	AGM-65A	ASM	1988	1989–90	(24)	
	300	AGM-65D	ASM	(1988)	1989–90	(72)	
	1 200	AGM-65G	ASM	(1988)	1989–90	(300)	
	1 182	AGM-88 Harm	ARM	1987	1988–90	(556)	Arming Tornado fighters
	804	MIM-104 Patriot	SAM	1984	1989–90	450	
L: USA	204	MLRS 227mm	MRL	1985	1989–90	40	
	. .	AIM-120A AMRAAM	Air-to-air missile	1989			Production to begin 1992
	4 500	FIM-92 Stinger	Portable SAM	1983			
	(10 000)	RAM	ShAM/PDM	1985			
German DR							
S: USSR	1	Mi-24 Hind-D	Helicopter	(1988)	1990	1	Part of deal worth $600 m incl anti-tank weapons
	32	MiG-29	Fighter	(1987)	1988–90	24	Deliveries terminated at 24

Recipient/ supplier (S) or licenser (L)	No. ordered	Weapon designation	Weapon description	Year of order/ licence	Year(s) of deliveries	No. delivered/ produced	Comments
	. .	BTR-70	APC	(1982)	1983–90	(1120)	Replacing BTR-60; also designated SPW-70
Greece							
S: France	. .	Stentor	Surveillance radar	(1987)	1988	(2)	Includes agreement for licensed production
	(240)	Magic-2	Air-to-air missile	(1986)	1988–89	(220)	Arming Mirage-2000 fighters
Germany, FR	(96)	NATO Seasparrow	ShAM/ShShM	(1988)			Arming Meko-200 Type frigates
Italy	25	A-109	Helicopter	(1987)			Negotiating
Netherlands	4	Smart	Fire control radar	1989			Fire-control radar for Meko-200 Type frigates
UK	2	S-723 Martello	3-D radar	1990			
USA	20	A-7E Corsair-2	Fighter	1990			
	20	F-4E Phantom	Fighter	1990			
	6	P-3A Orion	Maritime patrol	1990			
	5	SH-2F Seasprite	Helicopter	1990			Option for 3 more
	60	M-48-A5	Main battle tank	1989			Deal worth $26 m; from US stocks
	26	M-88-A1	ARV	1989			Option for 13 more
	4	Phalanx	CIWS	(1987)			Arming Meko-200 Type frigates
	(4)	RGM-84A Launch	ShShM launcher	1989			Arming Meko-200 Type frigates
	(4)	Seasparrow VLS	ShAM/PDM launcher	1988			Arming Meko-200 Type frigates
	1 500	FIM-92A Stinger	Portable SAM	1988	1989–90	(500)	Deal worth $124 m incl 500 launchers
	16	RGM-84A Harpoon	ShShM	1989			Arming first of 4 Meko-200 Type frigates; deal worth $19 m
	(64)	Seasparrow	ShAM	(1988)			Arming Meko-200 Type frigates
	4	Adams Class	Destroyer	1990			
L: Austria	324	Steyr-4K 7FA	APC	1987	1989–90	115	Third order
Denmark	2	PC-55 Class	Patrol craft	1988	1990	1	First of projected 10 to be built in Greek yards
	3	PC-55 Class	Patrol craft	1990			Option for 2 more
Germany, FR	3	Meko-200 Type	Frigate	1988			In addition to 1 frigate delivered directly; deal worth $1.2 b; financial aid from FRG and USA

Italy							
S: Germany, FR	8	Do-228-200	Transport	1990	1990	2	
	..	Kormoran-2	Anti-ship missile	(1986)	1988–90	40	Arming Tornado fighters
USA	2	AV-8B Harrier-2	Fighter	1990	1990	2	Deal worth $111 m; follow-on order expected
	35	M-113-A1	APC	(1988)	1989–90	35	Used as basis for Sidam 25 self-propelled gun
	20	MLRS 227mm	MRL	1985	1989–90	(8)	
	4	AN/FPS-117	Air defence radar	1990			
	20	AN/MPQ-53	Fire control radar	1990			Part of fire-control system for MIM-104 Patriot
	2	HADR	Air defence radar	1985	1988–90	(2)	Part of NADGE system
	2	RIM-67A Launch	ShAM launcher	(1987)			Arming Animoso Class destroyers
	(3 900)	BGM-71D TOW-2	Anti-tank missile	1987	1987–90	(2 500)	Arming A-129 Mangusta helicopters
	(32)	RIM-67C/SM-2	ShAM/ShShM	1987			Arming Animoso Class destroyers
	(16)	UGM-84A Harpoon	SuShM	(1986)	1990	(8)	Arming Sauro Class submarines
L: France	..	Aster	SAM	1988			
	23 000	Milan	Anti-tank missile	1984	1985–90	(7 990)	
	5 000	Mistral	SAM	(1988)			To be built by Italmissile consortium
Switzerland	..	Fledermaus II	Mobile radar	(1970)	1973–90	(180)	
USA	..	AB-206B	Helicopter	1972	1978–90	(625)	Jetranger-3 version available from 1984
	..	AB-212	Helicopter	1970	1971–90	(175)	
	..	AB-212ASW	Helicopter	1975	1975–90	(155)	
	..	AB-412 Griffon	Helicopter	1980	1982–90	(59)	Military version of Bell Model 412; Italy holds marketing rights
	..	CH-47C Chinook	Helicopter	1968	1972–89	(182)	Refit, servicing and maintenance continues
	50	Model 500E	Helicopter	1987	1987–90	(18)	
	..	SH-3D Sea King	Helicopter	1965	1969–89	98	Refit, servicing and maintenance continues
	20	Patriot battery	Mobile SAM system	1988			Part of $2.9 b deal incl 1280 missiles
	(1 100)	AGM-65D	ASM	1988			Italy probable supplier of Spanish and Turkish AGM-65 requirements
	(1 280)	MIM-104 Patriot	SAM	1988			Part of deal worth $2.9 b

Recipient/ supplier (S) or licenser (L)	No. ordered	Weapon designation	Weapon description	Year of order/ licence	Year(s) of deliveries	No. delivered/ produced	Comments
Japan							
S: Italy	3	Sparviero Class	Hydrofoil FAC	(1988)			Deal worth $170 m
UK	1	BAe-125-800	Utility jet	1989	1990	1	
USA	2	C-130H Hercules	Transport	1989	1990	2	
	5	E-2C Hawkeye	AEW	1988	1989–90	5	In addition to 8 previously delivered
	3	E-2C Hawkeye	AEW	1989			Deal worth $214 m incl spares
	2	E-2C Hawkeye	AEW	1990			Deal worth $170 m
	2	King Air C–90	Trainer	1990	1990	2	
	6	MH-53E	Helicopter	(1987)	1989–90	6	
	(28)	Phalanx	CIWS	1985	1987–90	22	Arming Asagiri Class and second batch of Hatsuyuki Class
	(8)	Phalanx	CIWS	1988			Part of Aegis air defence system arming Yukikaze Class
	(4)	RGM-84A Launch	ShShM launcher	1988			Part of Aegis air defence system arming Yukikaze Class
	(8)	Seasparrow VLS	ShAM/PDM launcher	1988			Part of Aegis air defence system arming Yukikaze Class
	55	AGM-84A Harpoon	Anti-ship missile	(1987)	1988–89	(36)	Deal worth $80 m; mix of air-, sea-, and submarine-launched versions
	75	AGM-84A Harpoon	Anti-ship missile	1990			Deal worth $125 m
	. .	FIM-92A Stinger	Portable SAM	(1988)	1990	90	
	(64)	RGM-84A Harpoon	ShShM	1988	1988	(16)	Part of Aegis air-defence system arming Yukikaze Class
	99	RGM-84A Harpoon	ShShM	1989	1990	99	Deal worth $173 m
	(350)	RIM-66C/SM-2	ShAM/ShShM	1988			Part of Aegis air-defence system arming Yukikaze Class
	. .	Seasparrow	ShAM	1980	1981–90	(364)	Arming various Japanese frigates and destroyers
L: UK	(176)	FH-70 155mm	Towed howitzer	1984	1989–90	(66)	Following direct delivery of 197
USA	. .	CH-47D Chinook	Helicopter	(1984)	1989–90	(21)	

	1	EP-3C Orion	Elint	1988			Follow-on orders expected
	55	F-15J Eagle	Fighter/interceptor	1985	1988–90	(33)	MoU signed Dec. 1984; in addition to 100 on order
	(130)	FS-X	Fighter	1988			Based on F-16C; US firms guaranteed 42% of work
	. .	KV-107/2A	Helicopter	(1982)	1984–90	(27)	In addition to 61 produced earlier; improved version
	. .	Model 205 UH-1H	Helicopter	1972	1973–90	(132)	
	(73)	Model 209 AH-1S	Helicopter	1982	1984–90	(55)	
	. .	OH-6D	Helicopter	1977	1982–90	(105)	
	10	P-3C Orion	Maritime patrol	1990	1990	(8)	Deal worth $256 m
	167	SH-3B	Helicopter	1979	1979–90	(167)	Production of spare parts continues
	. .	SH-60J Seahawk	Helicopter	1988	1989–90	(23)	
	40	UH-60J	Helicopter	(1987)	1988–90	(10)	
	25	Patriot battery	Mobile SAM system	(1984)	1988–90	(25)	Part of $2.8 b deal incl 980 licence-produced missiles
	1 330	AIM-7M Sparrow	Air-to-air missile	1990	1990	(150)	Arming F-15 fighters; deal worth $477 m
	. .	AIM-9L	Air-to-air missile	(1982)	1983–90	(4 071)	Arming F-15 fighters
	. .	BGM-71C I-TOW	Anti-tank missile	(1983)	1985–90	(3 209)	Total requirement: up to 10 000
	980	MIM-104 Patriot	Landmob SAM	1984	1989–90	(242)	
	. .	MIM-23B Hawk	Landmob SAM	1978	1978–90	(2 903)	
Netherlands							
S: France	14	Crotale SAMS	Mobile SAM system	1989			Option on further 7; status uncertain
	(168)	R-440 Crotale	Landmob SAM	1989			
Germany, FR	25	Buffel	ARV	1990			
UK	9	Firefly-160	Trainer	1990			
USA	22	MLRS 227mm	MRL	1986	1989–90	(22)	Deal worth $192 m incl 2700 rockets
	4	Patriot battery	Mobile SAM system	(1988)			
	8	RGM-84A Launch	ShShM launcher	1985	1987–90	(4)	Arming 8 Karel Doorman Class frigates
	(40)	AGM-84A Harpoon	Anti-ship missile	1988	1990	(10)	
	290	AIM-9M	Air-to-air missile	1988			Arming F-16 fighters; deal worth $27 m
	256	MIM-104 Patriot	SAM	(1988)			
	(88)	RGM-84A Harpoon	ShShM	1988	1989–90	(40)	
L: USA	53	F-16A	Fighter	1983	1987–90	(20)	Fourth order

Recipient/ supplier (S) or licenser (L)	No. ordered	Weapon designation	Weapon description	Year of order/ licence	Year(s) of deliveries	No. delivered/ produced	Comments
Norway							
S: France	4	AS-332	Helicopter	1990	1990	(2)	
	..	Mistral	Portable SAM	1990			Deal worth $60 m; offsets worth 75%
Sweden	(9)	Giraffe 50	Surveillance radar	1989			Deal worth $90 m
	(360)	RBS-70	Portable SAM	1989			Deal worth $80 m; offsets worth 45%; sixth order
UK	1	SH-3D Sea King	Helicopter	1989			Deal worth $18 m including upgrade of Norwegian Sea King fleet
USA	..	AN/TPQ-36	Tracking radar	1990			Part of a fire control for ground-launched AMRAAM
	..	AIM-120A AMRAAM	Air-to-air missile	1989			Deal worth $12.5 m
	7 612	BGM-71D TOW-2	Anti-tank missile	1985	1987–90	(3 000)	Deal worth $126 m incl 300 launchers and spares
New Zealand							
S: Australia	2	Meko-200 Class	Frigate	1989			Option on 2 more; deal worth $554.7 m
Italy	18	MB-339C	Trainer	1990			Deal worth $206 m, status uncertain
UK	4	Wasp	Helicopter	1989	1990	4	Ex-Royal Navy
Poland							
S: USSR	..	MiG-29	Fighter	(1988)	1989–90	16	
	(6)	SA-N-5 Launcher	ShAM launcher	(1985)	1988–89	(2)	Arming 5 Tarantul Class corvettes; status uncertain
	..	AA-10 Alamo	Air-to-air missile	(1988)	1989–90	(54)	Arming MiG-29 fighters
	..	AA-11 Archer	Air-to-air missile	(1988)	1989–90	(54)	Arming MiG-29 fighters
	..	AA-8 Aphid	Air-to-air missile	(1988)	1989–90	(160)	Arming Mi-24 Hind helicopters
	..	AA-8 Aphid	Air-to-air missile	(1988)	1989–90	(54)	Arming MiG-29 fighters
	(60)	SA-N-5	ShAM	(1985)	1988–89	(16)	Arming 5 Tarantul Class corvettes; status uncertain
	(24)	SSN-2 Styx	ShShM	(1985)	1988–89	(8)	Arming Tarantul Class corvettes; improved Styx
	(3)	Foxtrot Class	Submarine	1986	1987–88	2	
	(6)	Tarantul Class	Corvette	1985	1988–89	(2)	In addition to 2 supplied earlier

L: USSR	. .	An-2	Lightplane	1960	1960–90	(1 550)	In production since 1960; over 11 000 built; most for civilian use
	. .	Mi-2 Hoplite	Helicopter	1965	1965–90	(3 120)	In production since 1965; most for export
	. .	2S1 122mm	SPH	(1980)	1982–90	(480)	Some built for export
	. .	MT-LB	APC	(1980)	1980–90	(195)	
	(1 900)	T-72	Main battle tank	(1978)	1981–90	(750)	
Portugal							
S: France	2	Falcon-50	Transport	1989	1990	(2)	For VIP transport
Germany, FR	10	Alpha Jet	Jet trainer	1990			
	3	Meko-200 Type	Frigate	1986			Deal worth $700 m; 60% from NATO military fund
Italy	24	Aspide	SAM/ShAM	1986			Arming 3 Meko-200 Type frigates
UK	5	Super Lynx	Helicopter	1989			Deal worth $81 m, offsets worth 25%
	2	Watchman	Surveillance radar	1988	1989–90	(2)	Deal worth $9 m incl 2 AN/TPS-44 radars; funded by NATO
USA	3	F-16B	Fighter/trainer	1990			
	. .	Model 205 UH-1A	Helicopter	1989			In exchange for US base in the Azores; ex-USAF; part of a total of 57 helicopters
	. .	Model 209 AH-1G	Helicopter	1989			In exchange for US base in the Azores; ex-USAF; part of a total of 57 helicopters
	5	SH-2F Seasprite	Helicopter	1989			Deal worth $ 69 m; equipping Meko-200 Type frigates; status uncertain
	2	AN/MPQ-54	Guidance radar	1989	1990	1	
	2	AN/TPS-44	Surveillance radar	1988	1989–90	(2)	
	3	HADR	Air defence radar	1985	1988–90	(3)	Part of NADGE air-defence system
	1	Hawk SAMS	Mobile SAM system	1989			In exchange for US base in the Azores; ex USAF
	3	Phalanx	CIWS	1986			Arming 3 Meko-200 Type frigates
	3	RGM-84A Launch	ShShM launcher	1986			Arming 3 Meko-200 Type frigates
	3	Seasparrow VLS	ShAM/PDM launcher	1986			Arming 3 Meko-200 Type frigates
	. .	BGM-71D TOW-2	Anti-tank missile	(1988)			
	24	RGM-84A Harpoon	ShShM	1986			Arming 3 Meko-200 Type frigates
	17	Seasparrow	ShAM	1988			Arming 3 Meko-200 Type frigates

Recipient/ supplier (S) or licenser (L)	No. ordered	Weapon designation	Weapon description	Year of order/ licence	Year(s) of deliveries	No. delivered/ produced	Comments
L: Belgium	100	Jet Squalus	Jet trainer	1989			30 for Portuguese Air Force, 15 for civilian use and 55 for export markets
Romania							
S: USSR	40	MiG-29	Fighter	(1989)	1990	16	
	..	AT-4 Spigot	Anti-tank missile	(1984)	1985–90	(300)	Arming Romanian APCs
L: France	..	SA-330 Puma	Helicopter	1977	1978–90	(150)	
USSR	..	Ka-126	Helicopter	(1987)	1988–90	(11)	
	..	TAB-77	APC	(1970)	1970–90	(2 880)	Romanian version of Soviet BTR-70
Spain							
S: Canada	(8)	CL-215	Amphibian	1989	1989–90	(8)	
France	(2 000)	HOT	Anti-tank missile	1984	1986–90	(2 000)	Incl 150 launchers
	(3 500)	Milan-2	Anti-tank missile	1984	1986–90	(3 500)	Incl 250 launchers
	..	Milan-2	Anti-tank missile	1990			Deal worth $60 m, offsets worth 60%
	3 000	Mistral	Portable SAM	1988	1990	180	
USA	18	AV-8B Harrier-2	Fighter	1990			Deal worth $533 m incl refit of 13 AV-8Bs in service
	72	F/A-18 Hornet	Fighter	1983	1986–90	(72)	60 F/A-18A fighters and 12 F/A-18B trainers
	1	F/A-18 Hornet	Fighter	1990			Attrition replacement
	4	SH-60B Seahawk	Helicopter	(1988)			In addition to 6 previously ordered; equipping FFG-7 Class frigates
	4	RGM-84A Launch	ShShM launcher	1988			Coastal defence version mounted on trucks; for deployment near Gibraltar
	1	RGM-84A Launch	ShShM launcher	1987	1990	1	Arming 1 FFG-7 Class frigate
	250	AGM-65D	ASM	1989			Deal worth $48 m
	250	AGM-65F	Anti-ship missile	1989	1990	(100)	Arming F/A-18 Hornet fighters; mix of F and G versions
	(70)	AGM-84A Harpoon	Anti-ship missile	(1987)			Arming F/A-18 fighters

	32	AGM-88 Harm	ARM	1990			Arming F/A-18 fighters, deal worth $6 m
	200	AIM-120A AMRAAM	Air-to-air missile	1990			Deal worth $132 m
	(400)	BGM-71D TOW-2	Anti-tank missile	1987			
	20	RGM-84A Harpoon	ShShM	1987	1990	20	Arming fourth FFG-7 Class frigate
	16	RGM-84A Harpoon	ShShM	1989			Arming coastal defence bty
	(32)	RIM-67A/SM-1	ShAM/ShShM	(1986)	1990	(32)	Arming fourth FFG-7 Class frigate
L: France	18	AS-332	Helicopter	1986	1988–90	(18)	
Germany, FR	. .	Bo-105CB	Helicopter	(1978)	1981–90	(97)	In addition to 10 purchased directly
UK	5	Sandown Class	Minehunter	(1988)			
USA	1	FFG-7 Class	Frigate	1986	1990	(1)	Based on FFG-7 design; in addition to 3 previously built
	2	FFG-7 Class	Frigate	1990			In addition to 4 previously ordered
Sweden							
S: France	. .			1990			
Spain	1	C-212-200	Transport	1990	1990	1	In addition to 3 previously delivered
L: USA	700	AGM-114A	ASM/ATM	1987	1990	150	Deal worth $65 m; Hellfire coastal defence version
Switzerland							
S: France	12	AS-332	Helicopter	1989			Deal worth $190 m; offsets worth 100%
UK	3	Watchman	Surveillance radar	1990			
USA	54	M-548	APC	1989			Swiss designation RT-68, together with 108 M-109A2 for refurbishment
	. .	AIM-120A AMRAAM	Air-to-air missile	1988			Arming F/A-18 Hornet fighters
	204	AIM-7M Sparrow	Air-to-air missile	1988			Arming F/A-18 Hornet fighters
	(272)	AIM-9L	Air-to-air missile	(1988)			Arming F/A-18 Hornet fighters
	3 500	FIM-92A Stinger	Portable SAM	1988			Licensed production under discussion
L: Germany, FR	345	Leopard-2	Main battle tank	1983	1987–90	218	Deal worth $1400 m incl 35 delivered directly
UK	19	Hawk	Trainer	1987	1990	15	Deal worth $150 m incl training and logistics

Recipient/ supplier (S) or licenser (L)	No. ordered	Weapon designation	Weapon description	Year of order/ licence	Year(s) of deliveries	No. delivered/ produced	Comments
Turkey							
S: France	5	Stentor	Surveillance radar	1987	1988–89	(2)	
	14	TRS-22XX	3-D Radar	1987			
Germany, FR	8	Leopard	ARV	1988	1990	4	
	1	Meko-200 Type	Frigate	1990			Part of deal worth $465 incl 1 to be built in Turkey
Italy	14	SF-260TP	Trainer	1990	1990	6	To be assembled from knock-down kits
	2	Seaguard	CIWS	1989			Arming 2 Meko-200 Type frigates
	2	Seaguard	CIWS	1990			Arming 2 Meko-200 Type frigates
	(48)	Aspide	SAM/ShAM	(1989)			Arming 2 Meko-200 Type frigates
Spain	2	CN-235	Transport	1990			Followed by licensed production of 50
USA	6	Model 209 AH-1S	Helicopter	1990	1990	2	
	5	Model-209 AH-1W	Helicopter	1990			
	6	UH-60 Blackhawk	Helicopter	1990			Deal worth $68 m including support
	6	AN/TPQ-36	Tracking radar	(1986)	1988–89	(2)	
	3	HADR	Air defence radar	1985	1987–90	(3)	Part of NADGE air-defence system
	2	RGM-84A Launch	ShShM launcher	1989			Arming 2 Meko-200 Type frigates
	2	RGM-84A Launch	ShShM launcher	1990			Arming 2 Meko-200 Type frigates
	2	Seasparrow VLS	ShAM/PDM launcher	1990			Arming 2 Meko-200 Type frigates
	(320)	AIM-7F Sparrow	Air-to-air missile	(1983)	1986–90	(320)	
	310	AIM-9F	Air-to-air missile	1990			Deal worth $30 m incl training missiles
	40	RGM-84A Harpoon	ShShM	1990			Arming 2 Meko-200 Type frigates; deal worth $62 m including spares
	(48)	Seasparrow	ShAM	1990			Arming 2 Meko-200 Type frigates
	2	Garcia Class	Frigate	1989	1990	2	Leased from US Navy; armament unclear
L: Germany, FR	4	Meko-200 Type	Frigate	1983	1988-89	2	In addition to 2 built in FRG22
	1	Meko-200 Type	Frigate	1990			Part of deal worth $465 m
	2	Type-209/3	Submarine	1987			Option on 4 more
Italy	26	SF-260TP	Trainer	1990			In addition to 14 delivered directly

Spain	50	CN-235	Transport	1990			Part of deal worth $500 m incl 20 civil versions and 2 delivered directly
USA	152	F-16C	Fighter	1984	1987–90	37	Part of deal worth $4 b with direct delivery of 8 C and D versions
	1 698	AIFV	MICV	1988	1988–90	285	Deal worth $1 b; offsets worth $700 m
	180	MLRS 227mm	MRL	1988			Deal worth $600 m
	. .	FIM-92A Stinger	Portable SAM	1989			Manufacture to begin 1991; part of NATO Stinger programme
UK							
S: USA	8	S-76 Spirit	Helicopter	1989	1990	8	Deal worth $54 m; for Hong Kong
	(11)	RGM-84A Launch	ShShM launcher	1984	1985–90	(11)	Arming Type-22 and Type-23 frigates
	(330)	AIM-120A AMRAAM	Air-to-air missile	(1988)			Status uncertain
	(72)	Trident-2 D-5	SLBM	(1983)			Arming 4 Vanguard submarines
L: Brazil	128	EMB-312 Tucano	Trainer	1985	1987–90	65	Deal worth $145–50 m; option on 15 more
France	. .	Milan	Anti-tank missile	1976	1977–90	73 805	
USA	. .	WS-70	Helicopter	1987	1987	1	
	67	MLRS 227mm	MRL	1985	1989–90	10	
	223	AIM-120A AMRAAM	Air-to-air missile	1988			Licensed production by Euraam (BAe, MBB, AEG and Marconi)
	. .	BGM-71A TOW	Anti-tank missile	1980	1982–90	21 014	
USA							
S: Germany, FR	60	Tpz-1	APC	1989	1989–90	16	Deal worth $31 m
Italy	7	G-222	Transport	1990			Deal worth $80 m
	7	SF-260D	Trainer	1990	1990	7	
	24	Spada	Mobile SAM system	1988			For defence of US air bases in Italy
	16	Skyguard	Air defence radar	1990			For defence of US air bases in Italy
Norway	200	Penguin-3	Anti-ship missile	1990			Deal worth $270 m
Spain	(6)	C-212-300	Transport	1989	1990	1	Test bed for tactical reconnaissance radar

Recipient/ supplier (S) or licenser (L)	No. ordered	Weapon designation	Weapon description	Year of order/ licence	Year(s) of deliveries	No. delivered/ produced	Comments
UK	6	BAe-125-800	Utility jet	1988	1989–90	6	
	10	Sherpa	Transport	1988	1990	10	In addition to 18 previously ordered
L: Israel	86	Have Nap	ASM	1987			For co-production with Martin Marietta; US designation AGM-142
Italy	17	Osprey Class	MCM	1986			US designation Osprey Class
Switzerland	160	ADATS	SAM system	1987	1989	3	Deliveries suspended in 1990
UK	302	T-45 Hawk	Jet trainer	1986	1988	2	
	391	M-119 105mm	Towed gun	1987			Arming US Light Divisions; follows direct delivery of 53
	8	Ramadan Class	Patrol craft	1990			
USSR							
S: Czecho-Slovakia	. .	L-39 Albatross	Jet trainer	(1972)	1974–90	(1 190)	
Poland	. .	Mi-2 Hoplite	Helicopter	1965	1965–90	(2 340)	
Romania	. .	Yak-52	Trainer	(1980)	1981–90	(1 850)	About 200 per year produced for USSR
Yugoslavia							
S: USA	(3)	C-130H Hercules	Transport	(1989)			
USSR	36	MiG-29	Fighter	(1987)	1988–89	(24)	
	. .	AA-11 Archer	Air-to-air missile	1990	1990	(72)	Arming MiG-29 fighters
	(216)	AA-7 Apex	Air-to-air missile	(1987)	1988–89	(144)	Arming MiG-29 fighters
	(216)	AA-8 Aphid	Air-to-air missile	(1987)	1988–89	(144)	Arming MiG-29 fighters
L: USSR	(350)	T-74	Main battle tank	1977	1983–89	(350)	Yugoslavian designation M-84; now produced only for export

II. Third World countries

Afghanistan							
S: China	. .	Type-63 107mm	MRL	(1982)	1982–89	(350)	For Mujahideen; 122mm rockets without launchers supplied from Feb. 1988
	. .	Hong Ying-5	Portable SAM	(1982)	1982–89	(850)	SA-7 copy; for Mujahideen
Egypt	. .	Sakr-18 122mm	MRL	(1988)	1988–90	(30)	For Mujahideen; with large quantities of artillery rockets
	. .	SA-7 Grail	Portable SAM	(1984)	1985–89	(250)	For Mujahideen; unconfirmed
USSR	. .	Mi-24 Hind-D	Helicopter	(1984)	1984–90	(56)	
	. .	MiG-23	Fighter/interceptor	(1988)	1988–90	(53)	
	. .	MiG-29	Fighter	1989			Unconfirmed
	. .	Su-22 Fitter-J	Fighter/grd attack	(1979)	1979–90	(56)	
	. .	Su-25 Frogfoot	Fighter/grd attack	(1986)	1986–90	(60)	
	. .	BM-27 220mm	MRL	1989	1989	(12)	
	. .	BMP-1	MICV	(1979)	1979–90	(266)	May include Czechoslovak-built BMPs
	. .	BTR-70	APC	(1988)	1988–90	(360)	
	. .	D-1 152mm	Towed howitzer	(1987)	1988–90	(147)	
	. .	D-30 122mm	Towed howitzer	(1978)	1978–90	(458)	
	. .	M-46 130mm	Towed gun	(1979)	1979–90	(161)	
	. .	T-55	Main battle tank	(1978)	1978–90	(660)	
	. .	T-62	Main battle tank	(1979)	1979–90	(130)	
	. .	Scud-B Launcher	Mobile SSM system	(1988)	1988	(3)	
	. .	AA-2 Atoll	Air-to-air missile	(1979)	1979–90	(336)	Arming Su-22 fighters
	. .	Scud-B	SSM	(1988)	1988–90	(822)	
Algeria							
S: China	4	Hainan Class	Patrol craft	(1988)	1990	4	
Angola							
S: Spain	(3)	Cormoran Class	FAC	1989			
Switzerland	8	PC-7	Trainer	(1989)	1990	6	

Recipient/ supplier (S) or licenser (L)	No. ordered	Weapon designation	Weapon description	Year of order/ licence	Year(s) of deliveries	No. delivered/ produced	Comments
USSR	. .	Su-22 Fitter-J	Fighter/grd attack	(1985)	1985–90	(14)	
	14	Su-25 Frogfoot	Fighter/grd attack	(1985)	1990	14	
	. .	BRDM-2	Scout car	(1985)	1986–90	(188)	
	. .	D-30 122mm	Towed howitzer	(1985)	1986–90	(200)	D-44 85mm guns also delivered
	. .	M-46 130mm	Towed gun	(1986)	1986–89	(72)	
	. .	T-55	Main battle tank	(1987)	1987–90	(300)	Supplier unconfirmed
	. .	Barlock	Tracking radar	(1985)	1987–88	(7)	
	. .	Flat Face	Tracking radar	(1980)	1981–88	(16)	
	. .	Spoon Rest P-13	Early-warning radar	(1979)	1980–88	(16)	
Argentina							
S: Spain	5	C-212-200	Transport	(1987)	1990	5	
USA	6	SH-2F Seasprite	Helicopter	1990	1990	4	To equip 6 MEKO-140 frigates
L: Brazil	20	CBA-123	Transport	1989			Order for 36; 16 for civilian users
Germany, FR	6	Meko-140 Type	Frigate	1980	1985–90	5	Armed with MM-40 Exocet ShShMs; last 2 will be available for export
	4	Type TR-1700	Submarine	1977			In addition to 2 delivered directly
Italy	. .	A-109 Hirundo	Helicopter	1988			Deal worth $120 m
Bahrain							
S: USA	8	AH-64 Apache	Helicopter	1990			
	8	F-16C	Fighter	1987	1990	8	Partly financed by Saudi Arabia; with electronic countermeasures and laser designator
	4	F-16D	Fighter/trainer	1987	1990	4	
	80	M-113-A2	APC	1989			Deal worth $33 m
	27	M-60-A3	Main battle tank	1990	1990	27	
	. .	MLRS 227mm	MRL	1990			
	450	AGM-114A	ASM/ATM	1990			Arming AH-64 Apache helicopters

	(24)	AGM-65D	ASM	(1987)	1990	24	Arming F-16 fighters
	(48)	AIM-7M Sparrow	Air-to-air missile	(1987)	1990	48	Arming F-16 fighters
	(96)	AIM-9L	Air-to-air missile	1987	1990	(96)	Arming F-16 fighters
Bangladesh							
S: China	(20)	A-5 Fantan-A	Fighter	(1989)	1989–90	(20)	
	..	Fei Lung	ShShM	1988	1989–90	(8)	Arming 2 Jianghu Class frigates
	2	Jianghu Class	Frigate	(1988)	1989–90	2	
Pakistan	50	F-6	Fighter	1989	1990	16	
Botswana							
S: Canada	5	Model 412	Helicopter	1989	1988–90	5	
Switzerland	..	PC-7	Trainer	(1990)	1990	7	
UK	1	BAe-125-800	Utility jet	(1990)	1990	1	To replace BAe 125/400 AC
	2	BN-2A Defender	Lightplane	(1990)	1990	2	
Brazil							
S: France	15	AS-332	Helicopter	1987	1988–89	(10)	
	26	AS-365F	Helicopter	1988	1989–90	26	Part of deal worth $249 m
	..	Magic-2	Air-to-air missile	1988	1989–90	(54)	Arming refurbished Mirage-3 fighters
Indonesia	4	CN-212	Transport	1989			
USA	8	Phalanx	CIWS	1988			Arming 4 Niteroi Class frigates and 4 Inhauma Class corvettes; deal worth $63 m
L: Austria	..	GHN-45 155mm	Towed howitzer	(1985)			Status uncertain
France	16	HB-350M Esquilo	Helicopter	1988	1989–90	16	In addition to 39 previously produced
	10	HB-365F	Helicopter	1988	1990	10	Part of $249 m deal
Germany, FR	..	SNAC-1	SSN	1989			
	(3)	Type-209/3	Submarine	1982			

Recipient/ supplier (S) or licenser (L)	No. ordered	Weapon designation	Weapon description	Year of order/ licence	Year(s) of deliveries	No. delivered/ produced	Comments
Brunei							
S: Germany, FR	(96)	AIM-9L	Air-to-air missile	1989			Arming Hawk-100 fighters
Indonesia	4	CN-235	Transport	1989			
Italy	4	SF-260TP	Trainer	(1989)	1990	4	
UK	8	Hawk-100	Jet trainer	1989			Deal worth $150 m including 8 Hawk-200 fighters
USA	1	UH-60 Blackhawk	Helicopter	(1989)			VIP version
Burma							
S: China	12	F-6	Fighter	1990	1990	(12)	
	12	F-7	Fighter	1990			
	. .	T-63	Light tank	1989	1989–90	(50)	
	(50)	T-69	Main battle tank	1990	1990	50	
	(144)	PL-2A	Air-to-air missile	1990	1990	(50)	Arming F-6 and F-7 fighters
	4	Shanghai Class	Patrol craft	1990			
Poland	(20)	Mi-2 Hoplite	Helicopter	1990			
Yugoslavia	20	G-4 Super Galeb	Jet trainer	1990	1990	4	Option for 10 more; paid for in teak
Cameroon							
S: Canada	3	Model 206L	Helicopter	1989	1990	3	
UK	1	Peacock Class	OPV	1988			
Chad							
S: USA	2	C-130H Hercules	Transport	1989	1990	2	In addition to 4 purchased earlier
Chile							
S: France	2	Falcon-20G	Maritime patrol	1988	1990	2	Part of $210 m deal
	4	AS-365F	Helicopter	1987			To equip County Class frigates; first export of ASW version

	2	MM-40 Launcher	ShShM launcher	(1988)			Modernizing 2 Leander Class frigates
	8	AM-39 Exocet	Anti-ship missile	(1990)	1990	(8)	Arming 4 AS-332 Super Puma helicopters
	(32)	MM-40 Exocet	ShShM/SShM	(1988)	1990	(8)	Modernizing 2 Leander Class frigates
Indonesia	4	AS-332	Helicopter	1988	1988–90	4	Part of deal worth $210 m incl 4 SA-365Fs from France
	6	CN-235	Transport	1988	1989–90	(5)	
Israel	(8)	Barak Launcher	ShAM launcher	1989			For refit into Chilean frigates
	2	Phalcon	AEW&C radar	(1989)			Deal worth $500 m incl 4 Boeing-707s
	(256)	Barak	ShAM/SAM/PDM	1989			
Spain	3	CN-235	Transport	1989	1990	3	Deal worth $65 m
USA	5	Model 530MG	Helicopter	1988	1990	5	
L: South Africa	(400)	G-5 155mm	Towed howitzer	1989	1990	6	
Switzerland	. .	Piranha	APC	1980	1981–90	226	
USA	. .	Model 206	Helicopter	(1988)	1989	1	
	. .	T-35 Pillan	Trainer	1980	1985–90	160	
Colombia							
S: Argentina	3	IA-58B Pucara	COIN	1989	1990	3	
Israel	. .	Barak Launcher	ShAM launcher	1989			Arming F-1500 Type frigate
Spain	3	C-212-300	Transport	1988	1989	1	
Cuba							
S: USSR	(36)	MiG-29	Fighter	(1985)	1989–90	12	
	1	Pauk Class	Corvette	(1989)	1990	1	
Ecuador							
S: Brazil	10	EMB-312 Tucano	Trainer	1988			Deal worth $19 m
Spain	. .	Piranha Class	Patrol craft	1989			Some to be built in Ecuador

Recipient/ supplier (S) or licenser (L)	No. ordered	Weapon designation	Weapon description	Year of order/ licence	Year(s) of deliveries	No. delivered/ produced	Comments
Egypt							
S: Libya	10	L-39 Albatross	Jet trainer	1990	1990	10	
UK	1	Oberon Class	Submarine	1989			May be fitted with UGM-84 Harpoon anti-ship missiles and towed array sonars
	1	Porpoise Class	Submarine	1989			
USA	24	AH-64 Apache	Helicopter	1990			Deal worth $488 m incl Hellfire missiles
	2	E-2C Hawkeye	AEW	1989	1990	1	Deal worth $84 m
	1	F-16D	Fighter/trainer	1988			Deal worth $21 m incl spare parts; attrition replacement
	4	F-16D	Fighter/trainer	1987			
	3	KC-135	Tanker/transport	1989			
	2	S-70C	Helicopter	1990	1990	2	Deal worth $22 m
	2	UH-60 Blackhawk	Helicopter	1988	1989–90	2	
	15	M-1 Abrams	Main battle tank	1988	1990	2	Part of $2 b deal incl 540 to be co-produced
	492	AGM-114A	ASM/ATM	1990			Arming AH-64 Apache helicopters
	7 511	BGM-71D TOW-2	Anti-tank missile	1988	1989–90	(400)	Includes 180 launchers and 504 night-vision sights as well as spare parts
	(170)	MIM-23B Hawk	Landmob SAM	1988	1989–90	(170)	Deal worth $51 m
	29	RGM-84A Harpoon	ShShM	(1988)	1990	29	Deal worth $69 m; submarine-launched versions
L: Brazil	14	EMB-312 Tucano	Trainer	1989			In addition to 120 built previously
France	. .	Sinai 23	Mobile SAM system	1988	1990	4	Integration of Egyptian weapon systems with French fire-control system
UK	. .	Swingfire	Anti-tank missile	1977	1979–90	6 422	
USA	540	M-1 Abrams	Main battle tank	1988			Following delivery of 15; deal worth $2 b
	34	AN/TPS-63	Surveillance radar	1986	1988–90	17	Deal worth $190 m
Ethiopia							
S: Czechoslovakia	. .	T-55	Main battle tank	(1985)	1985–90	(380)	May be Soviet-supplied

Israel	30	T-55	Main battle tank	1990	1990	(30)	
USSR	. .	BM-21 122mm	MRL	(1984)	1984–90	(80)	May be North Korean BM-11
	. .	BRDM-1	Scout car	(1985)	1985–90	(160)	
	. .	BRDM-2	Scout car	1985	1986–90	(80)	
	. .	BTR-60P	APC	(1985)	1985–90	(360)	
	. .	D-30 122mm	Towed howitzer	1985	1985–90	(180)	
	. .	M-46 130mm	Towed gun	1985	1985–90	(80)	
	. .	AT-3 Sagger	Anti-tank missile	1985	1986–90	(400)	
	. .	AT-5 Spandrel	Anti-tank missile	1985			
Fiji							
S: France	2	AS-365	Helicopter	1990	1990	1	
Gabon							
S: France	1	AS-332	Helicopter	1989	1990	1	
	2	P-400 Class	Patrol craft	1985	1988–90	2	
Spain	1	CN-235	Transport	1989	1990	1	
Guatemala							
S: Italy	2	G-222L	Transport	1989			Deal worth $36.3 m
Honduras							
S: USA	8	S-76 Spirit	Helicopter	1989	1990	8	
India							
S: Korea, South	7	Sukanya Class	OPV	1987	1989	1	
UK	11	Sea Harrier	Fighter	1985	1990	3	Deal worth $230 m incl 1 trainer
	10	Sea Harrier	Fighter	1989			
USSR	3	AN-124 Condor	Transport	(1988)	1990	2	
	(8)	Ka-27 Helix	Helicopter	(1985)	1985–90	(8)	
	10	Mi-26 Halo	Helicopter	1988			Second order

Recipient/ supplier (S) or licenser (L)	No. ordered	Weapon designation	Weapon description	Year of order/ licence	Year(s) of deliveries	No. delivered/ produced	Comments
	(8)	SA-N-5 Launcher	ShAM launcher	(1983)	1986–89	(4)	Arming Khukri Class corvettes
	8	SSN-2 Styx L	ShShM launcher	(1985)	1987–90	(5)	Arming Tarantul Class corvettes
	8	SSN-2 Styx L	ShShM launcher	(1983)	1986–89	(4)	Arming Khukri Class corvettes
	. .	SA-N-5	ShAM	(1983)	1986–89	(80)	Arming Khukri Class corvettes
	64	SSN-2 Styx	ShShM	(1985)	1987–90	(20)	Arming Tarantul Class corvettes
	. .	SSN-2 Styx	ShShM	(1983)	1986–89	(16)	Arming Khukri Class corvettes
	8	Kilo Class	Submarine	(1984)	1986–90	7	
	2	Kilo Class	Submarine	1990			In addition to 8 previously ordered
	5	Pauk Class	Corvette	1983	1990	2	
	6	Tarantul Class	Corvette	(1985)	1987–90	(5)	Armed with SSN-2 Styx missiles; to be followed by Indian production
L: France	. .	SA-316B Chetak	Helicopter	(1962)	1964–90	(203)	Also produced for civilian use
	5	TRS-2230	3-D radar	(1983)	1988–90	(5)	In addition to 4 supplied directly
	(42 000)	Milan	Anti-tank missile	1982	1985–90	(26 583)	
Germany, FR	60	Do-228	Transport	1983	1987–90	(28)	Deal worth $440 m for 50 civil and 60 military versions
	2	Type-1500	Submarine	1981	1989	(1)	In addition to 2 delivered directly
Netherlands	212	Flycatcher	Mobile radar	(1987)	1988–90	(26)	In addition to direct deliveries
Poland	4	Polnocny Class	Landing ship	(1985)			Transfer of licence discussed
USSR	(165)	MiG-27	Fighter/grd attack	1983	1987–90	(80)	Follow-on order probable
	. .	BMP-2	APC/ICV	1983	1987–90	(160)	Production under way 1987
	(1 000)	T-72	Main battle tank	(1980)	1987–90	(450)	Production under way 1987; 10% Indian content
	. .	AA-8 Aphid	Air-to-air missile	(1986)			Indian designation Astra
	6	Tarantul Class	Corvette	(1986)			Order may reach 15
Indonesia							
S: Netherlands	. .	F-27 Mk-100	Transport	1990			
UK	14	AR-325	Surveillance radar	1989			

	USA	(48)	AGM-65D	ASM	1987			Arming F-16 fighters; status uncertain
		(96)	AIM-9P	Air-to-air missile	(1986)	1990	(24)	Arming F-16 fighters
L:	France	. .	AS-332	Helicopter	1983	1985–89	(8)	
	Germany, FR	. .	BK-117	Helicopter	1982	1986–90	(12)	Production of 100 planned, most for civil customers
		(80)	NBo-105	Helicopter	1987	1989–90	(12)	Licence to produce up to 100
	Spain	(80)	CN-212	Transport	1976	1978–90	(32)	
Iran								
S:	Brazil	15	EMB-312 Tucano	Trainer	1988	1989–90	(15)	Deal worth $15 m
	China	. .	Type-63 107mm	MRL	(1982)	1983–90	(800)	
		(2)	HQ-2B	SAM system	(1989)			
		(48)	HQ-2B	SAM	1989			For coastal air-defence batteries
	Czechoslovakia	. .	BTR-60P	APC	(1986)	1986–90	(200)	Supplier uncertain
		(90)	T-54	Main battle tank	1989	1989–90	(90)	Includes assistance with anti-tank missile construction
		(90)	T-55	Main battle tank	1989	1989	(90)	
	Korea, North	. .	T-62	Main battle tank	(1983)	1984–90	(180)	
	Romania	. .	Orao	Fighter/grd attack	1989			Unconfirmed; unit cost reported to be $10 m
		(200)	TAB-77	APC	1989	1989	(100)	
		150	T-55	Main battle tank	1989	1989–90	(150)	Ordered with an unspecified number of tank transporters
	USSR	. .	MiG-29	Fighter	1990	1990	14	
		(300)	T-72	Main battle tank	1989			
L:	China	. .	Oghab	SSM	1985	1986–90	900	Chinese Type-83 rocket; local production continues
Iraq								
S:	Brazil	. .	Astros-II SS-30	MRL	(1983)	1984–89	(78)	Delivery suspended 1990
		. .	SS-60	SSM	(1985)	1987–89	(960)	
	Chile	. .	Model 206	Helicopter	(1990)			Status uncertain
	Czechoslovakia	. .	BMP-1	MICV	(1981)	1981–89	(950)	

Recipient/ supplier (S) or licenser (L)	No. ordered	Weapon designation	Weapon description	Year of order/ licence	Year(s) of deliveries	No. delivered/ produced	Comments
Egypt	. .	Sakr Eye	Portable SAM	(1987)			Unspecified number
France	6	AS-332	Helicopter	1988			
	16	Mirage F-1C	Fighter/interceptor	1987			Deliveries embargoed from Aug. 1990 following Iraqi invasion of Kuwait; in addition to 113 previously delivered
	6	AS-365N	Helicopter	1989			Status uncertain
	. .	Tiger	Point defence radar	(1987)	1988–89	(10)	Trailer-mounted versions supplied; some modified as airborne early-warning radar
	36	AM-39 Exocet	Anti-ship missile	1989			Arming AS-332 Super Puma helicopters; status uncertain
	. .	ARMAT	ARM	(1983)	1983–90	(750)	
	(48)	AS-15TT	Anti-ship missile	1989			Arming SA-365 Dauphin helicopters
	240	AS-30L	ASM	(1984)	1986–88	(180)	Arming Mirage F-1s
	350	AS-30L	ASM	1989			Includes limited final assembly; status uncertain
	. .	HOT	Anti-tank missile	(1981)	1981–90	(1 900)	
	. .	Roland-2	Landmob SAM	1981	1982–90	(1 100)	
Germany, FR	6	BK-117	Helicopter	1990	1990	6	VIP transport version
Italy	(10)	Aspide/Albatros	ShAM/ShShM launch.	(1981)			Arming Lupo Class frigates and Wadi Class corvettes
	10	Otomat-2 L	ShShM launcher	(1981)			Arming Lupo Class frigates and Wadi Class corvettes
	(224)	Aspide	SAM/ShAM	(1981)			Arming Lupo Class frigates and Wadi Class corvettes
	(60)	Otomat-2	ShShM	(1981)			Arming Lupo Class frigates and Wadi Class corvettes
	4	Lupo Class	Frigate	1981			
	6	Wadi Class	Corvette	1981			Iraqi designation: Assad Class
USSR	. .	2S3 152mm	SPG	(1986)	1987–89	(100)	Mix of 152mm and 122mm guns unknown
	. .	BM-21 122mm	MRL	(1986)	1986–88	(360)	Production in Iraq continues
	. .	MT-LB	APC	(1982)	1983–88	(800)	Modified in Iraq to carry Egyptian 120mm mortar
	. .	AS-14 Kedge	ASM	(1988)	1988–89	(40)	
L: USSR	. .	Saddam 122mm	Towed howitzer	(1988)	1989–90	(75)	Local content unclear

Israel							
S: USA	18	AH-64 Apache	Helicopter	1989	1990	2	Deal worth $285 m incl support equipment
	. .	Bonanza A-36	Lightplane	1990			
	10	CH-53E	Helicopter	1990			
	3	RGM-84A Launch	ShShM launcher	(1988)			Arming Saar-5 Class corvettes
	539	AGM-114A	ASM/ATM	1990	1990	50	Arming 18 AH-64 Apache helicopters
	. .	FIM-92A Stinger	Portable SAM	1990			
	(48)	RGM-84A Harpoon	ShShM	(1988)			Arming Saar-5 Class corvettes
Jordan							
S: Spain	8	C-101 Aviojet	Jet trainer	1989			In addition to 16 supplied in 1988
USA	(2)	AN/TPQ-37	Tracking radar	(1986)	1989–90	(2)	
L: USA	100	Model 300C	Helicopter	1989			Production for civilian and military customers
Kampuchea							
S: China	24	T-59	Main battle tank	1990	1990	24	For Khmer Rouge
USSR	5	Mi-17 Hip-H	Helicopter	1990	1990	5	
	. .	BTR-60P	APC	(1989)	1990	(40)	
Viet Nam	15	T-55	Main battle tank	1990	1990	15	
Kenya							
S: Canada	2	Dash-8	Transport	1989	1990	2	
France	(1 000)	Mistral	Portable SAM	1990			
Germany, FR	. .	Bo-105	Helicopter	1990	1990	1	
UK	12	EMB-312 Tucano	Trainer	1988	1989–90	6	
Korea, North							
S: USSR	25	MiG-29	Fighter	(1987)	1988–90	25	
	(20)	Su-25 Frogfoot	Fighter/grd attack	1987	1988–90	20	
	. .	BMP-1	MICV	(1984)	1985–90	(102)	Locally modified design

Recipient/ supplier (S) or licenser (L)	No. ordered	Weapon designation	Weapon description	Year of order/ licence	Year(s) of deliveries	No. delivered/ produced	Comments
L: USSR	. .	AT-3 Sagger	Anti-tank missile	1975	1976–90	(1 500)	
	. .	SA-7 Grail	Portable SAM	(1985)	1986–90	(500)	
Korea, South							
S: Germany, FR	3	Type-209/3	Submarine	1987			Deal worth $600 m
	3	Type-209/3	Submarine	1989			In addition to 3 ordered in 1987
UK	20	Hawk	Jet trainer	1990			
	12	Lynx	Helicopter	1988	1990	(6)	Part of deal worth $200 m incl Sea Skua missiles; follow-on order for 20 likely
	1	MBT-3 BL	Bridge layer	1988	1990	1	
	. .	ST-1802	Naval fire control	1989			Fire-control radars for Javelin portable SAMs; part of deal worth $144 m
	(48)	Sea Skua	Anti-ship missile	1988	1990	24	Arming Lynx helicopters
USA	4	C-130H-30	Transport	(1988)	1990	4	
	48	F/A-18 Hornet	Fighter	(1989)			Deal $4.2 b incl 72 licence-produced, status uncertain
	42	Model 209 AH-1S	Helicopter	1986	1988–90	(42)	Deal worth $260 m incl TOW missiles
	8	P-3C Update-3	Maritime patrol	1990			
	9	RF-4C Phantom	Fighter/recce	(1990)	1990	9	Ex-US Air National Guard
	7	UH-60 Blackhawk	Helicopter	1990			Deal worth $44 m incl 2 spare engines and support, prior to licensed production
	80	UH-60 Blackhawk	Helicopter	1990			Deal worth $500 m
	3	AN/FPS-117	Air defence radar	1990			In addition to 5 previously delivered
	(672)	BGM-71D TOW-2	Anti-tank missile	1986	1988–90	(672)	Arming Model-209 helicopters
	704	BGM-71D TOW-2	Anti-tank missile	1987	1990	(250)	
	52	RGM-84A Harpoon	ShShM	1988			Filling reserve stocks
	21	Seasparrow	ShAM	1990			Deal worth $33 m incl training rounds and support
L: France	. .	Crotale	Landmob SAM	(1989)			Based on Crotale missile; developed by a Korean consortium

	Italy	6	Lerici Class	Minehunter	(1986)	1988	1	Class may ultimately be of 10 ships
	Japan	30	BK-117	Helicopter	1990			
	USA	72	F/A-18 Hornet	Fighter	(1989)			In addition to 48 for sale direct from the USA; status uncertain
		(150)	H-76 Eagle	Helicopter	1986			
		. .	Model 500MD	Helicopter	1976	1978–90	205	Over 400 civilian versions produced as well
		272	M-109-A2 155mm	SPH	1983	1985–90	272	
		470	M-109-A2 155mm	SPH	1990			Deal worth $260 m
Kuwait								
S:	Egypt	100	Fahd	APC	1988	1989–90	(60)	Part of $50 m deal incl Amoun air-defence system
		. .	Sakr Eye	Portable SAM	1987	1989–90	(36)	
	UK	16	EMB-312 Tucano	Trainer	1989			Status uncertain
	USA	42	F/A-18 Hornet	Fighter	1988			Deal worth $1.9 b incl Sidewinder, Harpoon, Sparrow and Maverick missiles
		300	AGM-65G	ASM	1988			Anti-ship version; arming F/A-18 Hornet fighters
		200	AIM-7F Sparrow	Air-to-air missile	1988			Arming F/A-18 Hornet fighters
		120	AIM-9L	Air-to-air missile	1988			Arming F/A-18 Hornet fighters
	USSR	245	BMP-2	MICV	1988	1989	(50)	Deal worth $300 m incl anti-tank missiles
		. .	T-72	Main battle tank	1989			Deal worth $700 m, paid partly in oil
		. .	SA-8 SAMS	Mobile SAM system	1988			Deal incl BMP-2 APCs
		(1 220)	AT-5 Spandrel	Anti-tank missile	1988	1989	(240)	Arming BMP-2 APCs
		. .	SA-8 Gecko	Landmob SAM	1988			
	Yugoslavia	200	M-84 155mm	Towed howitzer	1989			May have been delivered before 2 Aug. 1990
		230	T-74	Main battle tank	(1989)			200 tanks, 15 command vehicles and 15 recovery vehicles; part of deal worth $800 m incl 200 152mm howitzers
Laos								
S:	China	2	Y-12	Transport	(1990)	1990	2	

Recipient/ supplier (S) or licenser (L)	No. ordered	Weapon designation	Weapon description	Year of order/ licence	Year(s) of deliveries	No. delivered/ produced	Comments
Liberia							
S: United Arab Emirates	1	DHC-4 Caribou	Transport	1989	1990	1	Refurbished in Malta
Zambia	1	DHC-4 Caribou	Transport	1989	1990	1	Refurbished in Malta
Libya							
S: France	2	Mirage F-1A	Fighter/grd attack	1986	1989	1	
USSR	(15)	Su-24 Fencer	Fighter/bomber	(1988)	1989	(12)	Final batch withheld for non-payment
	. .	SA-5 SAMS	Mobile SAM system	1989			
	. .	AS-14 Kedge	ASM	1989			Arming SU-24D Fencer fighters
Yugoslavia	4	Koncar Class	FAC	1985			Based on Swedish Spica design; contract signed June 1985
Malawi							
S: Germany, FR	1	Do-228-200	Transport	1990	1990	1	
Malaysia							
S: Italy	4	A-109	Helicopter	(1988)	1990	4	For VIP use
Sweden	2	A-19 Class	Submarine	1990			
	2	Draken Class	Submarine	1990			
UK	. .			1988	1989–90	(20)	
	(18)	Hawk-100	Jet trainer	(1990)			Part of a deal worth $740 m incl 10 Hawk-200 aircraft, weapons, training and services
	(10)	Hawk-200	Fighter	(1990)			
	30	FH-70 155mm	Towed howitzer	1988	1989	9	
	20	L119 105mm gun	Towed gun	1988	1989–90	(20)	
	12	DN-181 Rapier	Mobile SAM system	1988			
	2	S-723 Martello	3-D radar	(1988)			Deal worth $190 m
	48	Javelin	Portable SAM	1988			

	576	Improved Rapier	Landmob SAM	1988			
L: UK	..	Harimau	Scout car	1988			Version of Ferret scout car
Mali							
S: USSR	(8)	MiG-21UTI	Jet trainer	(1988)	1989–90	(8)	
Mexico							
S: USA	4	Model 500MD	Helicopter	1990	1990	4	Follow-on order probable
Morocco							
S: France	(28)	AMX-10RC	Scout car	1990			
	..	HOT-2	Anti-tank missile	1987	1988–90	(84)	
Spain	7	CN-235	Transport	1989	1990	7	Deal worth $99 m
USA	200	M-60-A1	Main battle tank	1990			Subject to US approval of terms
Mozambique							
S: Indonesia	..	CN-235	Transport	1988			
Nepal							
S: France	1	AS-332	Helicopter	(1989)	1990	1	
Nicaragua							
S: USSR	6	Mi-17 Hip-H	Helicopter	1988	1990	6	
	4	Mi-17 Hip-H	Helicopter	1989	1990	4	Soviet Government says Mi-17s; US Government says Mi-25s
Nigeria							
S: UK	..	MBT Mk-3	MBT	1990			Deal worth $282 m

Recipient/ supplier (S) or licenser (L)	No. ordered	Weapon designation	Weapon description	Year of order/ licence	Year(s) of deliveries	No. delivered/ produced	Comments
L: USA	. .	Air Beetle	Trainer	1988	1988	1	Version of US RV-6
Oman							
S: Egypt	. .	Fahd	APC	1989	1990	(25)	
UK	16	Hawk-100	Jet trainer	1989			Deal worth $225 m for unspecified mix of Hawk-100 and Hawk-200 versions
	. .	Javelin	Portable SAM	1989	1990	(280)	
USA	(96)	AIM-9L	Air-to-air missile	1990			Arming 16 Hawk-100/200 jet trainers; could be from European production
Pakistan							
S: Australia	50	Mirage-3O	Fighter	1990	1990	32	Deal worth $28 m
China	75	F-7	Fighter	1983	1986–90	75	
	75	F-7	Fighter	1989			
	25	Karakoram 8	Jet trainer	1987			
	. .	T-59	Main battle tank	(1975)	1978–90	(975)	
	. .	Hong Ying-5	Portable SAM	(1988)	1988–90	(300)	Arming M-113 APCs
France	6	Rasit-3190B	Surveillance radar	1988	1989–90	(4)	
UK	24	Shorland S-55	APC	1990			
USA	11	F-16A	Fighter	1988			Second order; deal worth $256 m; attrition replacements
	60	F-16A	Fighter	1989			Deal incl 10 F-100 engines but no air-to-surface armaments; to be funded by Saudi Arabia
	3	P-3C Update-2	Maritime patrol	1988	1990	3	Deal worth $240 m incl spares and support
	6	SH-2F Seasprite	Helicopter	1989	1989	3	Incl 3 SH-2F versions and 3 SH-2G
	(20)	M-198 155mm	Towed howitzer	1988			Deal worth $40 m incl M-109-A2 howitzers and support equipment
	. .	AN/TPQ-36	Tracking radar	(1990)			Congress notified; total cost incl other radar equipment $65 m

	4	AN/TPQ-37	Tracking radar	(1985)	1987–89	(3)	
	44	AGM-84A Harpoon	Anti-ship missile	1990			
	200	AIM-7F Sparrow	Air-to-air missile	1988			Arming F-16 fighters
	360	AIM-9L	Air-to-air missile	1988	1989	(60)	Arming F-16 fighters
	2386	BGM-71D TOW-2	Anti-tank missile	1987			First Pakistani TOW-2 order; with 144 launchers
L: China	. .	T-69	Main battle tank	(1989)	1990	20	Deal worth $1.2 b; Pakistani designation P–90
	. .	Red Arrow-8	Anti-tank missile	1989			
Sweden	(180)	Supporter	Trainer	1974	1977–90	172	Assembly from imported kits began1976; production transferred to Kamra in 1981
	(125)	Anza	Portable SAM	(1985)			Pakistani designation Anza
Papua New Guinea							
S: USSR	4	Mi-17 Hip-H	Helicopter	1990	1990	4	Paid for with coffee beans
Peru							
S: Canada	6	Bo-105L	Helicopter	1989	1989–90	4	
Germany, FR	2	BK-117	Helicopter	1989			Part of deal worth $25–30 m incl 6 Bo-105 helicopters
USSR	18	Mi-17 Hip-H	Helicopter	1989	1990	14	In addition to 15 already delivered
Philippines							
S: Australia	6			1990			Deal worth $200 m
	4	N-24A Nomad	Transport	1989	1990	4	Deal worth $5.3 m
UK	150	FS-100 Simba	Scout car	1990			Deal worth $84 m; offsets worth 110%
USA	15	Bromon BR-2000	Transport	1988			
L: Germany, FR	. .	Bo-105C	Helicopter	1974	1976–89	13	Others built for civil customers
UK	. .	BN-2A Islander	Lightplane	1974	1974–89	30	Others built for civil customers

Recipient/ supplier (S) or licenser (L)	No. ordered	Weapon designation	Weapon description	Year of order/ licence	Year(s) of deliveries	No. delivered/ produced	Comments
Qatar							
S: France	1	Mirage F-1C	Fighter/interceptor	1987	1990	1	Order reduced from 4
	6	TRS-2201	Air defence radar	(1986)	1986–90	(5)	
	. .	Mistral	Portable SAM	1990			
Rwanda							
S: France	1	Falcon-50	Transport	(1989)	1990	1	Second hand
South Africa							
S: Switzerland	1	PC-6	Lightplane	1989	1990	1	For Bophuthatswana
	1	PC-7	Trainer	(1989)	1990	1	For Bophuthatswana
El Salvador							
S: Nicaragua	28	SA-14 Gremlin	Portable SAM	1990	1990	28	For FMLN guerrillas
Saudi Arabia							
S: Canada	1117	LAV-25	APC	1990			
France	12	AS-332	Helicopter	1988	1990	6	Armed with Exocet missiles; deal worth $430 m incl 20 armed speed boats
	(56)	AMX-30 Shahine	AAV(M)	1984	1986–90	(56)	Improved version developed with Saudi financial assistance
	3	Crotale Naval L	ShAM launcher	1990			Arming La Fayette Class frigates
	3	MM-40 Launcher	SShM launcher	1990			Arming La Fayette Class frigates
	48	Shahine-2 L	Mobile SAM system	1984	1986–90	(48)	
	(24)	Crotale Naval	ShAM	1990			Arming La Fayette Class frigates
	(1 000)	HOT-2	Anti-tank missile	1990			
	600	Mistral	Portable SAM	1989			Order may be for up to 1000
	(24)	MM-40 Exocet	ShShM/SShM	1990			Arming La Fayette Class frigates

	4 000	Shahine-2	Landmob SAM	1984	1986–90	(3 400)	Total value of 'Al Thakeb' deal: $4.1 b
	3	La Fayette Cl	Frigate	1989			Deal worth $3.5 b, offsets worth 30%
Italy	(32)	Otomat-2	ShShM	1988	1990	16	Arming 4 F-2000 frigates
Switzerland	300	Piranha	APC	1990			Deal worth $400 m
UK	12	BAe-125-800	Utility jet	1988	1988–89	4	Part of 1988 Tornado deal; for VIP use
	60	Hawk	Jet trainer	1988	1989–90	30	Part of 1988 Tornado deal; some Hawk-200 versions
	20	Hawk-200	Fighter	1988			Part of 1988 Tornado deal
	36	Tornado ADV	MRCA	1988			Al Yamamah II incl 48 Tornadoes, 60 Hawks, 12 BAe-125s, 4 BAe-146s, minehunters, missiles, training and facilities; deal worth $17 b
	12	Tornado IDS	MRCA	1988			Part of 1988 Tornado deal
	. .	WS-70	Helicopter	1988			Part of 1988 Tornado deal; up to 88 expected; order number not finalized; may be reduced due to funding problems
	40	Shorland S-55	APC	1988	1989–90	40	For Gendarmerie
	(60)	Transac GS	APC	(1988)	1989–90	60	Unconfirmed
	(480)	ALARM	ARM	1986			Arming Tornado IDS fighters; status uncertain
	(480)	Sea Eagle	Anti-ship missile	1985			Arming Tornado IDS fighters
	(560)	Sky Flash	Air-to-air missile	(1986)			Arming Tornado ADV fighters
	6	Sandown Class	Minehunter	1988	1990	1	
USA	12	AH-64 Apache	Helicopter	1990			Deal worth $300 including 155 Hellfire missiles
	(8)	C-130H Hercules	Transport	1990			Order may be for 10 aircraft
	12	F-15C Eagle	Fighter	1987	1990	12	
	12	F-15D Eagle	Jet trainer	1990			From US stocks
	7	KC-130H	Tanker/transport	1990			
	8	KC-135	Tanker/transport	1990			Deal worth $600 m incl upgrade of 5 E-3 Sentry AWACS
	15	Model 406CS	Helicopter	1988	1990	15	Deal worth $86 m; armed with TOW missiles
	13	UH-60 Blackhawk	Helicopter	1988	1990	13	Part of deal worth $400 m; 1 for VIP use
	8	UH-60 Blackhawk	Helicopter	1990			Medivac version, deal worth $121 m
	315	M-1 Abrams	Main battle tank	1990			
	150	M-1 Abrams	Main battle tank	1990			Second 1990 order
	27	M-198 155mm	Towed howitzer	1990			

Recipient/ supplier (S) or licenser (L)	No. ordered	Weapon designation	Weapon description	Year of order/ licence	Year(s) of deliveries	No. delivered/ produced	Comments
	220	M-2 Bradley	MICV	1989	1989	2	Deal worth $550 m incl anti-tank missiles and training
	200	M-2 Bradley	MICV	1990			In addition to 220 ordered previously
	150	M-60-A3	Main battle tank	1990	1990	150	
	30	M-88-A1	ARV	1989	1990	17	Part of deal worth $3b; incl heavy trucks spares and support
	12	M-88-A1	ARV	1990			Deal worth $26 m
	9	MLRS 227mm	MRL	1990			
	6	AN/MPQ-54	Guidance radar	1990	1990	6	
	(6)	AN/TPS-43	3-D radar	1985	1987–90	(4)	
	(6)	AN/TPS-70	Air defence radar	1989	1990	(3)	Deal worth $23.5 m
	15	Helitow	Fire control	1988	1990	15	Arming 15 Model-406 helicopters
	6	Patriot battery	Mobile SAM system	1990	1990	6	Deal worth $984 m incl 384 missiles, 6 radars and support
	155	AGM-114A	ASM/ATM	1990			Carried by 12 Apache helicopters
	120	AIM-7F Sparrow	Air-to-air missile	1990	1990	120	Arming F-15C fighters
	671	AIM-9P	Air-to-air missile	1986	1989–90	(400)	
	2 538	BGM-71C I-TOW	Anti-tank missile	1983	1986–90	(2 538)	Deal worth $26 m
	1 750	BGM-71D TOW-2	Anti-tank missile	1990			Deal worth $55 m including 150 launchers
	200	FIM-92A Stinger	Portable SAM	1990	1990	200	
	384	MIM-104 Patriot	SAM	1990	1990	384	
	. .	MIM-23B Hawk	Landmob SAM	1989			
	100	RGM-84A Harpoon	ShShM	1986	1988–90	(60)	
Singapore							
S: France	20	AS-350 Ecureuil	Helicopter	1989			
	(180)	HOT-2	Anti-tank missile	1990			Arming 20 AS-350 helicopters
USA	24	A-4S Skyhawk-2	Fighter/bomber	1989			May be upgrade of Skyhawks in service
	3	AN/TPQ-37	Tracking radar	1989	1990	3	Deal worth $31 m; option on 1 more

	6	RGM-84A Launch	ShShM launcher	(1986)	1990	3	Arming Type 62-001 corvettes
	(6)	RGM-84A Launch	ShShM launcher	(1987)	1988–90	(6)	Arming TNC-45 FACs
	(96)	RGM-84A Harpoon	ShShM	(1986)	1990	(48)	Arming Type 62-001 corvettes
	(48)	RGM-84A Harpoon	ShShM	(1987)	1988–90	(48)	Arming refitted TNC-45 FACs
L: Germany, FR	5	Type 62-001	Corvette	1985	1990	(3)	
Somalia							
S: Libya	32	T-55	Main battle tank	1989	1990	32	
Sri Lanka							
S: Israel	(2)	Dvora Class	FAC	1987			
Sudan							
S: Egypt	. .	Fahd	APC	1989	1989–90	25	
USA	9	V-150 Commando	APC	1988			In addition to about 80 previously ordered
Syria							
S: USSR	(8)	MiG-25 Foxhound	Fighter	(1989)			
	. .	BMP-1	MICV	1977	1977–89	(2 300)	May be from Czechoslovakia
	. .	AT-4 Spigot	Anti-tank missile	(1980)	1981–89	(900)	
	3	Kilo Class	Submarine	(1987)			
Taiwan							
S: Germany, FR	2	Type-209/4	Submarine	1989			Option for four more
USA	12	Commuter-1900	Transport	1989	1989–90	12	
	12	SH-60B Seahawk	Helicopter	1989	1989–90	12	Deal worth $74 m
	7	Phalanx	CIWS	(1989)	1989–90	2	Arming some Gearing Class frigates; deal worth $15 m
	14	RIM-67A Launch	ShAM launcher	1988	1989	(1)	Arming FFG-7 Class frigates to be built under licence

Recipient/ supplier (S) or licenser (L)	No. ordered	Weapon designation	Weapon description	Year of order/ licence	Year(s) of deliveries	No. delivered/ produced	Comments
	. .	AIM-7M Sparrow	Air-to-air missile	1989			Arming upgraded F-104 and Ching Kuo fighter aircraft
	(360)	RIM-67A/SM-1	ShAM/ShShM	1988			Arming FFG-7 Class frigates to be built under licence
	70	RIM-67A/SM-1	ShAM/ShShM	(1989)	1989–90	(20)	Arming 7 ex-US Gearing Class frigates
L: Israel	. .	Gabriel-2	ShShM/SShM	(1978)	1980–90	(465)	Taiwanese designation Hsiung Feng
USA	8	FFG-7 Class	Frigate	1989			Project management by Bath Iron Works
Thailand							
S: Canada	50	Model 412	Helicopter	1990			Deal worth $300 m
China	500	T-69	Main battle tank	1988	1988–90	(300)	Second 1988 order; upgraded with 105mm gun
	360	Type 531	APC	1988	1990	300	Part of deal worth $47 m
	800	Type 531	APC	1988	1990	112	Second 1988 order; mix of Type 531 and Type 85 versions
	55	Type-69 Spaag	AAV(G)	1987	1989–90	(55)	
	. .	C-801	ShShM	1990			Deal worth $40 m; arming 4 Jianghu Class frigates
	2	Jiangdong Class	Frigate	1988			Deal worth $272 m incl 2 Jianghu Class frigates to be refitted before delivery
	2	Jianghu Class	Frigate	1988	1990	2	Part of deal worth $272 m
	2	Jianghu Class	Frigate	1989			In addition to 2 ordered in 1988
	(3)	Romeo Class	Submarine	(1986)			Status uncertain
Germany, FR	3			1990			Follow-on order likely
	(4)	M-40 Type	MSC/PC	1986			In addition to 2 ordered in 1984; order may be for 6
	1	Support Ship	Support ship	1989			Helicopter support ship
Israel	40	Python-3	Air-to-air missile	1989			Status uncertain
Italy	40	AMX	Fighter	(1990)			Deal worth $600 m
Spain	2	C-212-200	Transport	(1990)			In addition to 4 ordered in 1981
Switzerland	20	PC-9	Trainer	1990			Deal worth $90 m

USA	3	C-130H-30	Transport	1988	1989–90	3	Deal worth $66 m
	3	CH-47D Chinook	Helicopter	1990			
	16	F-5E Tiger-2	Fighter	(1990)			
	25	Model 205 UH-1A	Helicopter	1989	1990	10	Deal worth $118 m
	4	Model 209 AH-1G	Helicopter	1988	1989–90	4	
	10	Model 209 AH-1G	Helicopter	1990	1990	4	Deal worth $8 m
	25	Model 212	Helicopter	1990			
	4	S-70C	Helicopter	1989			
	2	SH-2F Seasprite	Helicopter	1989			Equipping last 2 of 6 frigates under construction in China
	20	M-109 155mm	SPH	1988	1990	20	Part of deal worth $63 m
	81	M-125-A1	APC	1990			Deal worth $27 m
	250	M-48-A5	Main battle tank	1990			Deal worth $39 m together with 50 M-60-A1
	11	M-577-A2	CPC	1988			Deal worth $63 m incl 20 M-981s
	50	M-60-A1	Main battle tank	1990			Part of deal worth $459 m incl M-48 tanks, 20 recovery vehicles, small arms and spares
	20	M-88-A1	ARV	1990			Part of a deal worth $459 m
	20	M-981	Support vehicle	1988	1990	20	Deal worth $63 m
	108	Stingray	Light tank	1987	1987–90	(108)	
	2	AN/FPS-117	Air defence radar	1989			Deal worth $43 m
	(16)	AGM-65D	ASM	(1987)			Arming F-16 fighters
	16	AGM-84A Harpoon	Anti-ship missile	1990			Arming 3 P-3 Orion aircraft
	(48)	AIM-9P	Air-to-air missile	(1987)			Arming F-16 fighters
	(48)	BGM-71D TOW-2	Anti-tank missile	(1988)	1989–90	(48)	Arming 4 Model-209 helicopters
L: France	2	PS-700 Class	Landing ship	(1985)	1989	(1)	In addition to 1 ordered 1984; further orders probable
Germany, FR	. .	Fantrainer	Trainer	1983	1986–90	(50)	In addition to 2 delivered directly
UK	3	Province Class	FAC	1987			To be armed with 30mm guns and carry a light helicopter
	1	Province Class	FAC	1989			

Recipient/ supplier (S) or licenser (L)	No. ordered	Weapon designation	Weapon description	Year of order/ licence	Year(s) of deliveries	No. delivered/ produced	Comments
Togo							
S: Norway	1	AS-332	Helicopter	1989	1990	1	Deal worth $7.4 m; for VIP transport
United Arab Emirates							
S: France	1	Mirage-2000	Fighter	1990			Attrition replacement
	2	Crotale Naval L	ShAM launcher	1986	1990	2	
	2	MM-40 Launcher	ShShM launcher	1986	1990	2	Arming Type 62-001 corvettes
	(50)	Crotale Naval	ShAM	1986	1990	(50)	Arming Type 62-001 corvettes
	(208)	Magic-2	Air-to-air missile	1988			Arming Mirage-5 fighters
	(120)	Mistral	Portable SAM	1988	1990	(120)	Arming 2 Type 62-001 corvettes
	(16)	MM-40 Exocet	ShShM/SShM	(1987)	1988	(5)	Arming TNC-45 Class FACs
	(24)	MM-40 Exocet	ShShM/SShM	1986	1990	(24)	Arming 2 Type 62-001 corvettes
	(80)	R-440 Crotale	Landmob SAM	1988			
Germany, FR	2	Type 62-001	Corvette	1986	1990	2	For Abu Dhabi
Netherlands	2	Goalkeeper	CIWS	1986	1990	2	Arming two Type 62-001 corvettes
UK	12	Hawk	Jet trainer	1989			For Abu Dhabi
	12	Hawk-100	Jet trainer	1989			For Abu Dhabi; part of deal worth $340 m incl 12 Hawk trainers
USA	18	AH-64 Apache	Helicopter	1990			
	5	I-Hawk SAMS	Mobile SAM system	1989			Deal worth $168 m
	(108)	AIM-9P	Air-to-air missile	1983	1989–90	108	Arming Mirage-2000 fighters
	(108)	AIM-9P	Air-to-air missile	1989	1990	108	Arming second batch of 18 Mirage-2000 fighters
	(45)	MIM-23B Hawk	Landmob SAM	1989			
Uruguay							
S: France	2	Riviere Class	Frigate	1990	1990	2	Deal worth $17.5 m; armament unclear
USA	4	T-33A	Jet trainer	1990	1990	4	Deal worth $180 000
	6	T-33A	Jet trainer	1990	1990	6	
	1	Cape Class	Patrol craft	1990	1990	1	

Venezuela							
S: Brazil	28	AMX	Fighter	1990			
	100	EE-11 Urutu	APC	1988	1989–90	(60)	
France	8	AS-332	Helicopter	1988	1989–90	8	Deal worth $85 m; originally ordered by Nigeria
	12	Mirage-5OEV	Fighter	1988	1990	1	
	31	AMX-13–90	Light tank	1989	1989–90	(31)	
	(10)	Rassur	Surveillance radar	1988			
	(50)	AM-39 Exocet	Anti-ship missile	(1988)			Arming Mirage-50 fighters
	(100)	Magic-2	Air-to-air missile	1988	1989–90	(40)	Arming Mirage fighters; deal worth approx $30 m
Indonesia	16	Model 412	Helicopter	1988	1989	2	
Netherlands	12	F-5A	Fighter	1990	1990	12	
	. .	Flycatcher	Mobile radar	1988			
Spain	4	Cormoran Class	FAC	1987			
Sweden	70	RBS-70	Portable SAM	1989			
UK	84	Scorpion 90	Light tank	1988	1989–90	(35)	Deal worth $85 m incl support equipment, ammunition and training
USA	18	RGM-84A Harpoon	ShShM	1989			Deal worth $50 m; arming Constitution Class FACs
Zaire							
S: Egypt	12	Fahd	APC	1989	1990	25	
France	13	AMX-13	Light tank	1989			
Zimbabwe							
S: UK	5	Hawk	Jet trainer	1990			

Appendix 7C. Sources and methods

I. The SIPRI sources

The sources of the data presented in the registers are of six general types. Five of these are published sources: newspapers; periodicals and journals; books, monographs and annual reference works; official national documents; and documents issued by international and intergovernmental organizations. The arms trade registers are largely compiled from information contained in around 200 publications searched regularly. The most important sources of arms trade and arms production data are listed below.

Published information cannot provide a comprehensive picture because the arms trade is not fully reported in the open literature. Published reports provide partial information, and substantial disagreement among reports is common. Therefore, the exercise of judgement and the making of estimates are important elements in compiling the SIPRI arms trade data base. Order dates and the delivery dates for arms transactions are continuously revised in the light of new information, but where they are not disclosed the dates are estimated. Exact numbers of weapons ordered and delivered may not always be known and are sometimes estimated—particularly with respect to missiles. It is common for reports of arms deals involving large platforms—ships, aircraft and armoured vehicles—to ignore missile armaments classified as major weapons by SIPRI. Unless there is explicit evidence that platforms were disarmed or altered before delivery, it is assumed that a weapons fit specified in one of the major reference works such as the *Jane's* or *Interavia* series is carried.

II. Selection criteria

The SIPRI arms trade data cover five categories of major weapons: aircraft, armour and artillery, guidance and radar systems, missiles, and warships. Statistics presented refer to the value of the trade in these five categories only. The registers and statistics do not include the trade in small arms, artillery under 100-mm calibre, ammunition, support items, services and components or component technology, except for specific items. Publicly available information is inadequate to track these items satisfactorily.

There are two criteria for the selection of major weapon transfers for the registers. The first is that of military application. The aircraft category excludes aerobatic aeroplanes and gliders. Transport aircraft and VIP transports are included only if they bear military insignia or are otherwise confirmed as military registered. Micro-light aircraft, remotely piloted vehicles and drones are not included although these systems are increasingly finding military applications.

The armour and artillery category includes all types of tanks, tank destroyers, armoured cars, armoured personnel carriers, armoured support vehicles, infantry combat vehicles as well as multiple rocket launchers, self-propelled and towed guns and howitzers with a calibre equal to or above 100 mm. Military lorries, jeeps and other unarmoured support vehicles are not included.

The category of guidance and radar systems is a residual category for electronic-tracking, target-acquisition, fire-control, launch and guidance systems that are either (*a*) deployed independently of a weapon system listed under another weapon category (e.g., certain ground-based SAM launch systems) or (*b*) shipborne missile-launch or point-defence (CIWS) systems. The values of acquisition, fire-control, launch and guidance systems on aircraft and armoured vehicles are included in the value of the respective aircraft or armoured vehicle. The reason for treating shipborne systems separately is that a given type of ship is often equipped with numerous combinations of different surveillance, acquisition, launch and guidance systems.

The missile category includes only guided missiles. Unguided artillery rockets and man-portable anti-armour rockets are excluded. Free-fall aerial munitions (such as 'iron bombs') are also excluded. In the naval sphere, anti-submarine rockets and torpedoes are excluded.

The ship category excludes small patrol craft (with a displacement of less than 100 t) unless they carry cannon with a calibre equal to or above 100 mm, missiles or torpedoes, research vessels, tugs and ice-breakers. Combat support vessels such as fleet replenishment ships are included.

The second criterion for selection of items is the identity of the buyer. Items must be destined for the armed forces, paramilitary forces, intelligence agencies or police of another country. Arms supplied to guerrilla forces pose a problem. For example, if weapons are delivered to the Contra rebels they are listed as imports to Nicaragua with a comment in the arms trade register indicating the local recipient. The entry of any arms transfer is made corresponding to the five weapon categories listed above. This means that missiles and their guidance/launch vehicles are often entered separately under their respective category in the arms trade register.

III. The value of the arms trade

The SIPRI system for evaluating the arms trade was designed as a *trend-measuring device*, to enable the measurement of changes in the total flow of major weapons and its geographic pattern. Expressing the evaluation in monetary terms reflects both the quantity and the quality of the weapons transferred. Aggregate values and shares are based only on *actual deliveries* during the year or years covered in the relevant tables and figures.

The SIPRI valuation system is not comparable to official economic statistics such as gross domestic product, public expenditure and export/import figures. The monetary values chosen do not correspond to the actual prices paid, which vary considerably depending on different pricing methods, the length of production runs and the terms involved in individual transactions. For instance, a deal may or may not cover spare parts, training, support equipment, compensation, offset arrangements for the local industries in the buying country, and so on. Furthermore, to use only actual sales prices—even assuming that the information were available for all deals, which it is not—military aid and grants would be excluded, and the total flow of arms would therefore not be measured.

Production under licence is included in the arms trade statistics in such a way that it should reflect the import share embodied in the weapon. In reality, this share is normally high in the beginning, gradually decreasing over time. However, a single estimate of the import share for each weapon produced under licence is made by SIPRI, and therefore the value of arms produced under licence agreements may be slightly overstated.

IV. Priority sources

Journals and periodicals

AAS Milavnews Air Letter (Romford, UK)
AAS Milavnews News Letter (Romford, UK)
Afrique Défense (Paris)
Air & Cosmos (Paris)
Air Force (Washington DC)
Air International (Bromley, UK)
Armed Forces Journal International (Washington, DC)
Asia-Pacific Defence Reporter (Kunyung, Vic.)
Asian Defence Journal (Kuala Lampur)
Atlantic News (Brussels)
Aviation Week & Space Technology (New York)
Defence & Armament Héràcles (Paris)
Defence (Redhill, UK)
Defence Electronics (Palo Alto, California)
Defence Industry Digest (London)
Defence Intelligence Bulletin (Gutenswil, Switzerland)
Defence Today (Rome)
Defensa (Madrid)
Defense & Economy World Report (Washington, DC)
Defense & Foreign Affairs (Alexandria, Virginia)
Defense & Foreign Affairs Weekly (Alexandria, Virginia)
Defense News (Springfield, Virginia)
Far Eastern Economic Review (Hong Kong)
Flight International (Sutton, UK)
Interavia (Geneva)
Interavia Air Letter (Geneva)
International Defense Intelligence (Greenwich, UK)
International Defense Review (Geneva)
Jane's Defence Weekly (Coulsdon, UK)
Jane's Soviet Intelligence Review (Coulsdon, UK)
JP4 (Florence)
Latin American Regional Report (London)
Latin American Weekly Report (London)
Marine Rundschau (Stuttgart)
Med News (Maisons Laffitte)
Military Technology (Bonn)
Nato's Sixteen Nations (Brussels)
Naval Forces (Farnborough, UK)
Navy International (Haslemere, UK)
Österreichische Militärische Zeitung (Vienna)
Panorama Difesa (Florence)
Proceedings (USNI) (Annapolis, Maryland)
Soldat und Technik (Frankfurt)
Technologia Militar (Bonn)
Wehrtechnik (Bonn)
World Missile Forecast (Newtown, Connecticut)
Worlds Weapons Review (Newton, Connecticut)

Newspapers

Dagens Nyheter (Stockholm)
El País (Madrid)
Financial Times (Europe) (Frankfurt)
Foreign Broadcast Information Service (Washington, DC)
Frankfurter Allgemeine Zeitung (Frankfurt)
Handelsblatt (Düsseldorf)
International Herald Tribune (Paris)
Le Monde (Paris)
Neue Zürcher Zeitung (Zurich)
Süddeutsche Zeitung (Munich)
Svenska Dagbladet (Stockholm)
The Independent (London)
Tribune de l'Expansion (Paris)

Annual reference publications

Asian Recorder (Recorder Press: New Delhi)
Combat Fleets of the World (Naval Institute Press: Annapolis, Maryland)
Defense & Foreign Affairs Handbook (Perth Corporation: Washington, DC)
Jane's All the World's Aircraft (Jane's Publishing Co.: Coulsdon, UK)
Jane's Armour and Artillery (Jane's Publishing Co.: Coulsdon, UK)
Jane's Fighting Ships (Jane's Publishing Co.: Coulsdon, UK)
Jane's Infantry Weapons (Jane's Publishing Co.: Coulsdon, UK)
Jane's Military Vehicles & Support Equipment (Jane's Publishing Co.: Coulsdon, UK)
'Military Aircraft of the World', and 'Missile Forces of the World', annually in *Flight International* (Sutton, UK)
The Middle East Military Balance (Tel Aviv)
The Military Balance (Brassey's: Oxford)
Trends in Conventional Arms Transfers to the Third World (Washington, DC)
World Fighting Vehicles & Ordnance Forecast (Newtown, Connecticut)
World Military Expenditures and Arms Transfers (USGPO: Washington, DC)

V. Conventions

The following conventions are used in appendices 7A and 7B:

. .	Data not available or not applicable
–	Negligible figure (<0.5) or none
( )	Uncertain data or SIPRI estimate

Abbreviations and acronyms

AA	Anti-aircraft
AAG	Anti-aircraft gun
AAM	Air-to-air missile
AAV	Anti-aircraft vehicle
AAV(G)	Anti-aircraft vehicle (gun-armed)
AAV(M)	Anti-aircraft vehicle (missile-armed)

AC	Armoured car
Acc to	According to
ADATS	Air defence and anti-tank system
ADV	Air defence version
Adv	Advanced
AEV	Armoured engineering vehicle
AEW	Airborne early-warning (system)
AEW&C	Airborne early warning and control
AF	Air Force
AFSV	Armoured fire support vehicle
Amph	Amphibious/amphibian
APC	Armoured personnel carrier
Approx	Approximately
ARM	Anti-radar missile
ARV	Armoured recovery vehicle
AShM	Air-to-ship missile
ASM	Air-to-surface missile
ASV	Anti-surface vessel
ASW	Anti-submarine warfare
ATGM	Anti-tank guided missile
ATM	Anti-tank missile
AV	Armoured vehicle
AWACS	Airborne early warning and control system
BL	Bridge-layer
Bty	Battery
CIWS	Close-in weapon system
CG	Coastal gun
COIN	Counter-insurgency
CP	Coastal patrol
CPC	Command post carrier
CS	Coastal surveillance
DOD	Department of Defense (USA)
ECM	Electronic countermeasure
Elint	Electronic intelligence
EW	Early warning
Excl	Excluding/excludes
FAC	Fast attack craft (missile/torpedo-armed)
FMS	Foreign Military Sales (USA)
FY	Fiscal year
Grd	Ground
ICV	Infantry combat vehicle
IDS	Interdictor/strike version
Incl	Including/includes
IRBM	Intermediate-range ballistic missile
Landmob	Land-mobile (missile)
LC	Landing craft (<600 t displacement)
LS	Landing ship (>600 t displacement)
LT	Light tank
LOA	Letter of Offer and Acceptance (USA)

LoO	Letter of Offer (USA)
MAP	Military Assistance Program (USA)
Mar patrol	Maritime patrol aircraft
MBT	Main battle tank
MCM	Mine countermeasure (ship)
MICV	Mechanized infantry combat vehicle
Mk	Mark
MOU	Memorandum of Understanding
MR	Maritime reconnaissance
MRCA	Multi-role combat aircraft
MRL	Multiple rocket launcher
MRS	Multiple rocket system
MSC	Minesweeper, coastal
MSO	Minesweeper, ocean
MT	Medium tank
OPV	Offshore patrol vessel
PAR	Precision approach radar
PC	Patrol craft (gun-armed/unarmed)
PDM	Point defence missile
Port	Portable
RAAF	Royal Australian Air Force
Recce	Reconnaissance (aircraft/vehicle)
RN	Royal Navy (UK)
SAM	Surface-to-air missile
SAR	Search and rescue
SC	Scout car
ShAM	Ship-to-air missile
ShShM	Ship-to-ship missile
ShSuM	Ship-to-submarine missile
SLBM	Submarine-launched ballistic missile
SPAAG	Self-propelled anti-aircraft gun
SPG	Self-propelled gun
SPH	Self-propelled howitzer
SPM	Self-propelled mortar
SShM	Surface-to-ship missile
SSM	Surface-to-surface missile
SSN	Nuclear-powered submarine
SuShM	Submarine-to-ship missile
SY	Shipyard
TD	Tank destroyer (gun-armed)
TD(M)	Tank destroyer (missile-armed)
TG	Towed gun
TH	Towed howitzer
Trpt	Transport
UNITA	National Union for the Total Independence of Angola
VIP	Very important person
VLS	Vertical launch system
3-D	Three-dimensional

8. Arms production

HERBERT WULF

I. Introduction: reductions in arms production

The global trend of a gradual reduction of arms production continued in 1990, with the remarkable exceptions of a few countries. According to information available at the end of 1990, the most noticeable decline in arms production during the year occurred in the three major centres of arms production—the United States, the Soviet Union and Western Europe—while Japan and several Third World countries continued to invest a growing amount in arms procurement. The promising signs that the 1990s would be a decade in which arms production could be substantially reduced and resources shifted from military to civil purposes are still visible, although they suffered a set-back in 1990 from both the Persian Gulf crisis and the severe economic difficulties in the USSR.

A comparison of company arms sales in 1988 and 1989 reveals that the 100 largest arms-producing companies[1] lost 4 per cent of their arms business in 1989. Companies with an interest in arms production are reacting to this situation with a variety of different strategies—including mergers and takeovers at the national and international level, diversification into civil production and dismissal of employees. However, some sectors of the industry have begun to profit from the Gulf conflict.

In 1990 Soviet officials began to acknowledge difficulties with the conversion plan and its implementation. Conversion of factories in the military–industrial sector to civilian production proved to be more difficult than originally predicted by the planners. Conversion was caught in the economic turmoil of changing from a planned to a market economy.

The industrialized countries

In the *United States,* reductions in arms production and arms sales have resulted from the tight US financial situation, which required budgetary cuts, rather than from a conscious disarmament policy. The vast US arms industry, for decades nurtured like the Soviet and the West European industry by the cold war, is rapidly approaching a situation that will lead to fundamental structural changes and a further reduction in size. Conversion of the arms industry to civilian production is not legislated in industrial policy for the United States, an idea that is anathema to free marketers; rather, the contrac-

[1] Arms industries are defined as those companies with production in the seven core business sectors identified by the US Department of Defense; see figure 8.2.

tion and consolidation of military industry are ruled by free-market forces.[2] The financial constraints of the US Government are reflected in a reduction of both prime contract awards[3] since 1986 and company arms sales (see section III). The need to reduce the US budget deficit combined with the perceived disappearance of the Soviet threat to the West made it almost certain that the military budget would be cut drastically and that procurement of military equipment would not be spared from cuts. Since the August 1990 invasion of Kuwait by Iraq, however, the deployment of US forces in the Persian Gulf has been used, although not entirely successfully, to try to reverse this course.

Cuts in planned procurement have been decided in most other NATO countries as well. This has already affected arms sales—although for most countries in *Western Europe* with a time lag of about one year compared to the United States. The arms industry in Western Europe is faced with a long-term situation which is similar to that of its competitors in the United States. In addition the 1990 Conventional Armed Forces in Europe (CFE) Treaty will have long-term indirect effects on industry that will require a reduction of production capacities (see section IV). The compensatory strategy of increasing arms exports—which companies have often adopted when, as a result of cyclical shrinkages in domestic orders, arms sales have dropped—is not a viable alternative today. Global arms exports have declined since 1987, particularly sharply in 1990 (see also chapter 7).

An exception to the general trend of reductions in arms production in the industrialized countries is *Japan*. The Japanese military budget has grown continuously over the past few years and is intended to grow during the next three years as well, although at a somewhat reduced rate. Within the growing budget, the share of arms procurement increased each year in the period 1986–90. This procurement policy, with high investments in several new major weapon systems, which required an expansion of domestic arms production facilities and import of (mainly US) technology, has been reflected in stable company arms sales (see section II).

In the *Soviet Union*, the slow-down of arms production was the result of a declared official policy of both reductions in procurement and research and development (R&D) expenditure and conversion of arms industries to civilian production, as well as a consequence of intensified and chaotic developments in the economy. The conversion process, initiated in 1988, has been far from smooth and efficient. Resistance has been raised at both a theoretical and a practical level. Economists who lost confidence in the capability of the traditional planning apparatus criticized the centralized mechanisms of the top–down approach of ordering managers of military complexes to produce non-military products. On the practical level in the factories, opposition arose

[2] Conversion legislation has been regularly introduced in the USA, although without success. See section III for details.

[3] Prime contract awards are Department of Defense or armed forces procurement orders placed with companies in the USA or abroad. See also section III.

as managers, engineers and workers feared losing traditional privileges and being pushed into less interesting fields of technology (see section V).[4]

Compared to the United States, Western Europe and the Soviet Union, *China* is a second-tier arms-producing country, measured by both the value of total output and the level of technological sophistication in particular. According to official sources, China has continued to convert arms production facilities to production of non-military goods, and the arms industry has become an indispensable part of its state-run civilian economy.[5] This process of conversion is understood in China as the 'transition of defence science, technology and industry from serving only the defence modernization drive to serving the modernization of industry, agriculture, national science and technology, as well as defence'.[6] On the other hand, the two major Chinese weapon-exporting corporations, Poly Technologies (Baoli) and New Era (Xinshidai), both formed in the early 1980s, export weapons and other military equipment in an effort to improve China's balance-of-payments situation. Even though the restrictive Chinese information policy makes a thorough investigation difficult, it seems that the reduction in Chinese arms exports is primarily caused by economic constraints in the importing countries and the loss of Iraq as an important customer rather than restraint imposed by the Chinese Government.

The Third World

The overall positive trend of arms control in Europe has not spread to the Third World. Unlike Europe, the Third World has no forum for an orderly and rational discussion of disarmament or arms control such as the CFE process or of conversion of military industries to civilian production. Third World trends in arms production in 1990 were not uniform. Most of the countries (and companies) which in the past felt compelled to base their arms industry strategy on entering and competing in the world arms market, primarily for economic reasons, suffered from drastic reductions in exports. New producers emerged in the 1970s, generating increasing numbers of competitors. When the import of weapons declined at the end of the 1980s competition became stiffer, and exports by the major Third World arms exporters decreased as well (see figure 8.1). Several companies (in Israel,[7] South Africa, Argentina and

[4] It should be emphasized that, again in 1990, despite *glasnost* and *perestroika*, the information on Soviet production of arms is still far from satisfactory. Although more information has been published than previously, the data are often inconsistent, even contradictory. Thus the description of general trends is based on an unsatisfactory data base.

[5] For the first time, on 15 Dec. 1990 civilian products of the arms sector were shown in a public exhibition in Beijing, China. The Fair '90 on High-Tech & Export Products Transferred from Defence Industry was organized by the China Association for Peaceful Use of Military Industry Technology.

[6] Zhude, J. and Muliang, Ch., 'China's experience: a case study', paper presented at the UN Conference on Conversion: Economic Adjustments in an Era of Arms Reduction, Moscow, 13–17 Aug. 1990, p. 3.

[7] The largest Israeli arms-producing company, Israel Aircraft Industries, was able to recover three years after cancellation of the Lavi fighter programme, partly due to an increase in its non-military production and partly as a consequence of preferential treatment as a supplier to NATO countries. See

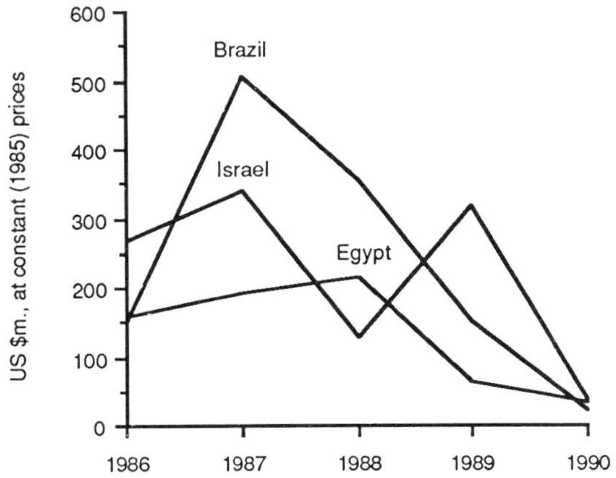

Figure 8.1. The value of major weapon exports from the three largest Third World exporters, 1986–90

Source: SIPRI arms trade data base.

Brazil) faced cancellations of planned weapon programmes and reduced exports and thus severe economic problems, even bankruptcy.

India, Singapore and South Korea are exceptions to the general trend of slowed-down production of arms in the Third World.[8] In India a force modernization plan has been put into effect by the Government. It includes, for example, a large naval expansion programme, a main battle tank project and an air defence system for the Army, and several major fighter aircraft, helicopter and missile projects for the Air Force.[9] Singapore, on the basis of its economic growth, continues to increase its investment in imports and local production of armaments. However, the Singaporean state-owned arms industry is largely dependent on exports. According to the Minister of Trade and Industry and Second Minister for Defence, Lee Hsien Loong, 'for future growth, the defence industries will have to look beyond the SAF [Singapore Armed Forces] . . . They have to develop overseas markets . . . on a commercial basis'.[10] Similarly, in a growth economy, the Government of the Republic of South Korea consistently increased the military budget during the 1980s, and within the growing military budget the share of arms procurement

Milavnews, May 1990, p. 16. Other Israeli arms producers, especially the subsidiaries of Koor Industries, Soltam and Tadiran, were in serious economic difficulties.

[8] In several Middle Eastern countries investments in arms production facilities have been made as well. For a description see Anthony, I. and Wulf, H., 'The trade in major conventional weapons', SIPRI, *SIPRI Yearbook 1989: World Armaments and Disarmament* (Oxford University Press: Oxford, 1989), chapter 7, pp. 230–45.

[9] See appendix 7B for details. See also *Defense & Foreign Affairs*, Apr. 1990, pp. 15–17 and 42–44; and Government of India, Ministry of Defence, *Annual Report 1990–91* (Government Printing Office: New Delhi, 1990).

[10] *Military Technology*, Feb. 1990, p. 8.

was increased as well. The large modernization programme that is being undertaken on the basis of weapon and technology imports has benefited arms-producing companies (aerospace, shipbuilding and heavy engineering) in South Korea.

II. The SIPRI 100

The compilation of data on the 1989 arms sales[11] of the 100 largest arms producers in the OECD (Organization for Economic Co-operation and Development) countries and in the Third World reveals that the above-described trend of reduced arms production has already been reflected in company sales (see appendix 8A). The arms sales of many companies have dropped. The value of the combined arms sales of the 100 largest companies decreased from over $175 billion in 1988 to $169 billion in 1989, a reduction of 4 per cent.

In 1988, corporations from 15 different countries were represented in the SIPRI list of the 100 largest companies;[12] as table 8.1 illustrates, only 14

Table 8.1. Numbers of companies in the list of 100 largest arms-producing companies, grouped by rank and country, 1989

	Rank					Total no. of companies	
Country	1–20	21–40	41–60	61–80	81–100	1989	1988
USA	14	10	10	6	7	47	44
UK	2	1	2	5	4	14	16
France	2	3	2	–	2	9	8
Germany, FR	1	–	1	3	2	7	8
Japan	1	1	1	2	1	6	6
Sweden	–	–	1	2	1	4	4
Italy	–	3	–	–	–	3	3
Switzerland	–	–	1	1	1	3	2
India	–	1	–	–	1	2	2
Spain	–	1	–	–	–	1	1
Netherlands	–	–	1	–	–	1	1
Israel	–	–	1	–	–	1	2
South Korea	–	–	–	1	–	1	1
South Africa	–	–	–	–	1	1	1
Brazil	–	–	–	–	–	–	1

Source: Appendix 8A.

[11] It should be pointed out that for the trade in major weapon systems SIPRI applies its own methodology and pricing system, values being trend indicators in 1985 constant US dollars (see appendix 7C), while arms sales reported in this chapter are given in current US dollars, based on actual sales as reported by companies, governments, journals, newspapers, etc. For the applied methodology and sources, see appendix 8A.

[12] See SIPRI, *SIPRI Yearbook 1990: World Armaments and Disarmament* (Oxford University Press: Oxford, 1990), table 8.2, pp. 326–29.

Table 8.2. National shares of arms sales, 1989 compared to 1988, for the 100 largest producers

No. of companies, 1989	Country/region	Share of arms sales in total sales, 1989 (%)	Share of arms sales in total sales, 1988 (%)	Arms sales, 1989 ($b.)
47	USA	*63.0*	*62.2*	106.2
42	Total Western Europe[a]	*31.1*	*31.4*	52.5
14	UK	*10.2*	*10.6*	17.2
9	France	*9.7*	*9.3*	16.3
7	Germany, FR	*4.5*	*5.1*	7.6
6	Japan	*3.7*	*3.8*	6.2
5	Third World	*2.2*	*2.6*	3.7

[a] This refers to the West European countries of the OECD in appendix 8A.

Source: Appendix 8A.

countries are represented in the list for 1989. Brazil does not appear in this list because of the reduction in arms sales by Brazilian companies, caused by economic difficulties. Furthermore, fewer British, German and Israeli companies are listed among the 100 largest companies as a result of either acquisitions of companies or in some cases reductions in arms sales. The country distribution has therefore shifted somewhat. For 1989, three additional US companies (47 as against 44 in 1988), one more French company and one more Swiss company are listed among the 100 largest companies.

The fact that more US companies appear on the SIPRI list only slightly increased their share of arms sales. The 47 US companies combined account for 63 per cent of the arms sales of the 100 largest companies.[13] Although the total of 42 European companies which form the second large bloc in the SIPRI 100 list represent only five fewer companies than the number of US companies, they account for less than one-third of total arms sales. German companies lost more during 1989, as the reduction in the share of arms sales shown in table 8.2 suggests. French companies increased their respective shares slightly. Japanese companies, on the basis of the Japanese arms procurement programme, maintained the same position. The procurement contracts with the top 20 Japanese arms-producing companies (only six of which appear in the SIPRI 100 list) increased in 1989 to 1.1 billion yen (over $8 billion), an increase of almost 5 per cent.[14]

As discussed in more detail below, not all sectors of the arms industry have been evenly affected by the changes which are taking place. The past trend of electronics companies receiving a growing share of arms procurement has

[13] The calculation in US dollars somewhat distorts the actual changes from 1988 to 1989. For most countries reported in this list, the US dollar was stronger in 1989 than in 1988 (in contrast to 1990, which is not reported here).

[14] Information from the Japanese Office of Defense Production Committee, KEIDANREN, communicated to SIPRI.

Table 8.3. Companies which lost and won the most in arms sales in 1989

Rank	Company	Country	Arms sales, 1989 ($m.)	Change in arms sales, 1989 compared to 1988 ($m.)
Companies which lost arms sales				
3	Lockheed	USA	7 350	– 1 050
70	Thorn EMI	UK	540	– 660
31	LTV	USA	1 580	– 570
6	General Motors	USA	5 500	– 500
9	Northrop	USA	4 700	– 500
10	Rockwell International	USA	4 500	– 500
30	IBM	USA	1 600	– 500
14	United Technologies	USA	4 100	– 400
Companies which won arms sales				
13	Daimler Benz	Germany, FR	4 260	+ 840
4	British Aerospace	UK	6 300	+ 830
2	General Dynamics	USA	8 400	+ 400
94	Mannesmann	Germany, FR	360	+ 360
8	Boeing	USA	4 800	+ 300

Source: Appendix 8A.

continued and is likely to continue in the future as well. Aircraft producers, especially in the United States, lost heavily in this regard, as did the traditional arms manufacturers that produce artillery, tanks and hulls of fighting ships. As a rule, high-technology companies are better placed.

Table 8.3 lists the companies that both lost and won the most in arms procurement, ranked according to their total dollar value loss or gain in arms sales in 1989 compared to 1988; all companies with changes in arms sales of $300 million or more are included. Table 8.3 is also a confirmation of the general trend of reduced arms sales: more companies lost than gained in arms sales. Five companies increased their arms sales by $300 million or more in 1989. Of these five companies, two increased sales by the acquisition of arms-producing companies: Daimler Benz bought the German aircraft and missile producer MBB, and Mannesmann bought the German tank producer Krauss-Maffei.

Only seven companies either entered the SIPRI 100 list for 1989 or dropped off the list for 1988—either because of reduced arms sales or because they were taken over by other companies (see table 8.4).

Companies with an interest in arms production are reacting to the new situation with a variety of strategies. The two most important company responses are mergers, both national[15] and international, and reduction of capacity by laying off employees.

[15] Mergers and takeovers within national boundaries are taking place in the USA and most West European countries. These national activities are not considered here.

Table 8.4. The seven companies which dropped off the SIPRI 100 list for 1988 and the seven new companies in 1989

Rank	Company	Country
Companies in the SIPRI 100 list for 1988 which do not appear in 1989		
Rank 1988		
29	MBB	Germany, FR
56	Plessey	UK
91	Control Data	USA
96	Vickers	UK
97	Krauss-Maffei	Germany, FR
98	Avibras	Brazil
99	Koor Industries	Israel
Companies which appear for the first time in the SIPRI 100 list for 1989		
Rank 1989		
91	Renault Véhicules Ind.	France
92	Morrison Knudsen	USA
94	Mannesmann	Germany, FR
96	Mitre	USA
97	Dyncorp	USA
99	Oshkosh Truck	USA
100	Ascom Holding	Switzerland

Source: Appendix 8A; and *SIPRI Yearbook 1990*, table 8.2.

Cross-border mergers and international takeovers in the defence sector are a somewhat new phenomenon, as was described in the *SIPRI Yearbook 1990*.[16] Tables 8.5 and 8.6 show that this trend accelerated in 1990: many mergers and international takeovers took place, mainly in Europe.[17] US companies concentrated their acquisition activities more within the United States than internationally. In a number of cases in Europe, such co-operation resulted in the dominance of certain market sectors by individual companies. In Europe the main producers of engines, military electronics, missiles and helicopters are in the process of forming joint companies, and the small arms producers are integrated under the umbrella of GIAT of France.

Not all international co-operation leads to the formation of new companies or takeovers. Short of formally founding new companies, so-called 'strategic alliances' are formed to bolster companies' military business in an increasingly uncertain market. Probably the most important example is the co-operation agreement between British Aerospace and General Dynamics, two of the largest arms-producing companies, to 'promote sales or lay the foundation for other collaborative projects'.[18] Another example is the agreed co-

[16] See *SIPRI Yearbook 1990* (note 12), chapter 8.

[17] Not all mergers and takeovers reported in tables 8.5 and 8.6 were finalized in 1990. In several cases monopolies commissions or cartel offices still have to decide on these mergers.

[18] *Financial Times*, 17 May 1990.

Table 8.5. International takeovers in the arms production sector, 1990

Buyer company or companies	Head office	Seller company	Head office
SABCA	Belgium	Dassault Belgique Aviation	France/Belgium
Bombardier	Canada	Learjet	USA
Thomson-CSF	France	NV Philips MBLE Defence	Netherlands
Thomson-CSF	France	Link-Miles	UK
Thomson-CSF	France	Ferranti Sonar division	UK
GIAT	France	Fabrique Nationale (and parts of Beretta, Italy)	Belgium
Aérospatiale Alcatel Selenia	France France Italy	Parts of Space Systems Loral (previously bought from Ford)	USA
MAN	Germany, FR	Steyr-Daimler-Puch	Austria
Sauer (Susco Acquisition)	Germany, FR	Sundstrand-Sauer Co.	USA
Fincantieri Bremer Vulkan AG	France Germany, FR	Sulzer Diesel AG	Switzerland
Finmeccanica	Italy	Ferranti Italiana	UK
Finmeccanica	Italy	FIAR (Italy, owned by Ericsson, Sweden)	Sweden
FIAT	Italy	Enesa	Spain
Thorn EMI	UK	MEL (British military electronics part of Philips)	Netherlands UK
Astra Holdings	UK	PRB	Belgium
Dowty Group	UK	Resdel Engineering	USA
BEI Electronics	USA	4 divisions of Systron Donner (Inertial, Seaton-Wilson, Edcliff and Duncan Electronics)	UK

Source: SIPRI arms production data base.

operation between Vickers Defence Systems in the United Kingdom and GIAT of France to collaborate to reinforce their export positions. Obviously, companies are acting to adjust to the new situation.

The second company strategy in reaction to shrinking orders is the shut-down of factories and lay-off of personnel. A few dozen companies have disclosed or already practised such cost-cutting programmes. Heavy job losses are being experienced in many arms-producing companies in nearly all

Table 8.6. International mergers and formation of new companies in the arms production sector, 1990

Companies	Countries of origin	Name of merged/ new company
Matra	France	Eurodrone
MBB	Germany, FR	
Aérospatiale	France	Eurocopter
MBB	Germany, FR	
Selenia	Italy	CSEE Defense
CSEE	France	
Aérospatiale	France	. .
OFEMA	France	
AIROD	Malaysia	
Thomson-CSF	France	Defence Electronics of Singapore
Allied Ordnance	Singapore	
Thomson-CSF	France	Ferranti-Thomson Sonar Systems
Ferranti International sonar division	UK	
Thomson-CSF guided weapons division	France	Eurodynamics
British Aerospace Dynamics	UK	
MBB	Germany, FR	DEFTEC
Santa Barbara (INI)	Spain	
BMW	Germany, FR	BMW Rolls Royce
Rolls Royce	UK	
CESELSA	Spain	Aeronautical Systems Designers
SD-Scicon	UK	
Ferranti Technologies	UK	Ferranti Tech Asia
Allied Ordnance Company	Singapore	

Source: SIPRI arms production data base.

countries. In the USA, the companies that have far-reaching plans for redundancies are General Dynamics, which wants to reduce gradually its workforce from 103 000 to 70 000 or 65 000; McDonnell Douglas, which plans reductions of 17 000 jobs in the near future; Hughes, which plans reductions of 9000; and Lockheed and Unisys, each planning reductions of 5000.

In the UK, British Aerospace, which substantially increased its arms sales in 1989, at the end of 1990 disclosed plans for rationalization. In the past five years, BAe, the biggest European arms producer, has closed six sites, involving several thousand job losses. At least four more plants are being earmarked for closure, and 5000 more jobs will be lost within the next two to three years. Other British companies that have already laid off or will lay off employees are Cammell Laird, Devonport Dockyard, Ferranti, GEC, Link

Miles, Rolls Royce, Smith Industries, Thorn EMI, VSEL, Vosper Thornycroft and Westland.

In Germany, union representatives fear that MBB, the subsidiary of Daimler's DASA, will lose 2000 jobs; Rheinmetall has announced 1000 redundancies; and the tank producers Krauss-Maffei and MAK have been reducing their capacities since the mid-1980s. Heckler & Koch had to sell the arms-producing part of their company, but whether workers have to be fired has still not been decided.

Thomson CSF and GIAT in France, FN Herstal in Belgium, Oerlikon Bührle in Switzerland, Bofors in Sweden, Philips in the Netherlands, Armscor in South Africa, and Embraer and Avibras in Brazil are all reducing production capacities by cutting employment, some of them drastically as, for example, Embraer, which laid off 4000 of its 12 000-strong workforce in November 1990. Some of these companies are on the verge of bankruptcy, and closures of plants and cuts in personnel are rather the norm than the exception in the arms industry.

III. The United States

In 1990 the US Federal Government revised its procurement plans for the armed forces in 1990. Data on 1989 prime contracts made available by the US Department of Defense (DOD) in 1990, which include information on all major US arms-producing companies, reveal that prime contract awards have already dropped considerably. DOD procurement totalled $139.3 billion during FY 1989. This was a decrease of $12.0 billion (7.9 per cent) from the $151.4 billion awarded during the previous year. $120.0 billion (86 per cent) was for work inside the United States.

Table 8.7. US Department of Defense prime contract awards, FYs 1980–89

Figures are in US $b., at current prices.

	1980	1981	1982	1983	1984	1985	1986	1987	1988	1989
Total	83.7	105.2	124.7	140.4	146.0	163.7	158.8	156.5	151.4	139.3
In USA	66.7	87.2	102.5	121.1	124.9	139.6	136.5	135.3	130.8	120.0

Source: Department of Defense, Directorate for Information, *Prime Contract Awards, Fiscal Year 1989*, Washington, DC, 1990, pp. 7–8.

Table 8.7 illustrates the trend during the 1980s, with the arms buildup during the first Reagan Administration and an increase of nearly 100 per cent in procurement orders from FY 1980 to FY 1985, reaching a peak in FY 1985 both in procurement activity and in business for work in the United States. From this peak of $163.7 billion, procurement orders decreased first with average annual reductions of 2.5 per cent and then with a sharp drop in 1989. Although detailed figures on company arms sales and prime contracts are not

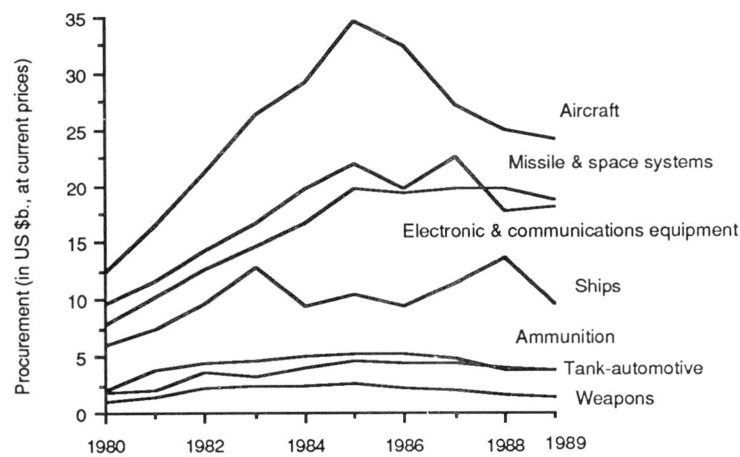

Figure 8.2. US prime contract awards, by business sector, 1980–89

Source: Department of Defense, *Prime Contract Awards, Fiscal Year 1989*, Washington, DC, 1990, pp. 23–27.

yet available for 1990, this trend is bound to continue since procurement expenditures have been reduced.

The various sectors of the US arms-producing industry were differently affected by reduced procurement orders. The seven core sectors of the arms industry, for which data are compiled in figure 8.2, traditionally account for approximately two-thirds of total procurement.[19] The missile and space industry was hardly affected at all. Their prime contracts with the US Government levelled off at approximately the high level that was reached in the mid-1980s. Both the ammunitions industry and those companies producing electronics and communications equipment—although in dollar terms in a different league—lost less than one-sixth of their business compared to the boom year 1985. The value of procurement orders in 1989 was still two-thirds higher than in 1980. Both the missile and the munitions industry have profited from the Gulf crisis. The main burden of the cuts had to be carried by the weapon industry (with a loss of 42 per cent from 1985 to 1989), the aircraft industry (cuts of 30 per cent), the tank-automotive sector (cuts of 27 per cent), and the shipping industry, which is not comparable to the other sectors since their prime contracts fluctuate more owing to large single contracts for major fighting ships such as aircraft-carriers or submarines. The best year for shipping industries was 1988, with prime contracts worth $13.7 billion, followed by reductions of 30 per cent in 1989. In absolute money terms the

[19] The remaining one-third goes to construction, fuels, textiles, services and miscellaneous.

biggest reductions were experienced by aircraft producers, the largest sector in the US arms industry. From a peak point of $34.6 billion in 1985, prime contracts were reduced to $24.2 in 1989, an amount still almost double that of 1980.

Looking at the major projects that were affected by cuts in the original DOD plans, one can assume that business losses which occurred in 1990 were of similar magnitudes to those in 1989. The number of Air Force Stealth B-2 bombers planned for procurement was decreased, as was funding for the C-17 transport plane and for the Advanced Tactical Fighter of the Air Force. Defense Secretary Dick Cheney cancelled the US Navy's A-12 Avenger attack aircraft on 7 January 1991, the largest US weapon programme ever cancelled. According to the two prime contractors, General Dynamics and McDonnell Douglas, 7000 job losses are the likely result.[20]

In one arms industry journal, the situation of the industry was described as follows: 'The American armaments industry has eaten its cake and is now faced with a long series of lean years. The golden age of those policies that favoured a strong growth of military budgets when Ronald Reagan was at the helm is well and truly over . . . No longer can the United States fail to count the cost of its defence spending, for there have been severe blows to its budget'.[21]

The general procurement trend described above is in line with the arms sales figures of the 100 largest companies presented in appendix 8A. Market analysts expect a world-wide growth in military electronics production: 'The percentage of electronics in defense systems has grown from 34% in 1981 to 40% in 1990 and is projected at 43% by the end of the century'.[22] Most companies in missile and/or electronics production (such as Martin Marietta, General Dynamics, Texas Instruments, Raytheon, TRW and Westinghouse Electric) were less affected—some of them even increased their arms sales in 1989—than those producers which specialized in only one or a few types of military product (such as Northrop, Lockheed and Hughes). A realistic assessment of the latter sector of industry can be found in the Northrop company report:

> We are in a difficult and uncertain period for Northrop, and for the defense industry in general. World relations are changing rapidly. The Administration and the Congress are reviewing the size, shape, and nature of our military forces. The availability of funds to modernize the armed services and assure their effectiveness and credibility is also being debated. Unfortunately, the eventual outcome of these deliberations and their ultimate effect on our major programs is far from clear.[23]

Other aircraft producers, particularly Boeing, have prospered because of a boom in commercial aircraft production.

[20] *World Weapons Review*, 9 Jan. 1991, p. 3.

[21] Guérin, M., 'American industry: the lean years', *Defense & Armament Heracles*, no. 91 (Jan. 1990), p. 32.

[22] *Aviation Week & Space Technology*, 19 Mar. 1990, p. 201.

[23] Northrop Corporation, *Annual Report 1989*, p. 2.

US companies tried to weather their slacking business by diversifying their arms production mix (a strategy explicitly tried by Lockheed, Grumman and Martin Marietta),[24] concentrating on the priority procurements in the budget (such as missiles and electronics) or hoping to find long-term growth niches (such as anti-submarine warfare equipment, sensors, countermeasures equipment, and retrofit and upgrade of existing equipment).[25] Companies (such as United Technologies)[26] want to avoid fixed-price contracts, which were introduced by the Pentagon and led to large losses for arms producers. Still other companies (such as General Dynamics tank production facilities)[27] are studying the possibility of shutting down plants and moth-balling them for some time. Others effectively lost their capability as prime contractors and system integrators (such as LTV in military aircraft)[28] or lost interest in arms production and sold or tried to sell their arms production subsidiaries (such as Ford, Chrysler and Honeywell). The long-term effect of the current Gulf conflict on the different sectors of industry is still open to question. One likely outcome is the acceleration of an established trend, that is, more emphasis on high technology, particularly electronics, at the expense of the production of major weapon platforms.

The debate in the United States about the future of the arms industrial base spans over several issues. In academia, the discussion covers issues ranging from a declining economy owing to high military investments,[29] to increasing reliance on foreign technologies which are either imported or purchased from the local subsidiaries of foreign corporations.[30] Worried by a possible erosion of the US arms industrial base, military and political bodies tend to strive for a compromise by which industry is relieved from fixed-price contracts. In contrast to Japan or Western Europe, in the United States strong objections are usually raised against a general industrial policy that would assist industry in adjusting to the new situation or help particularly heavily affected communities or regions.[31] Restructuring and reshaping the arms industry are considered the exclusive responsibility of industry. A report of the Arms Control and Disarmament Agency (ACDA) to Congress, published after

[24] See reports and interviews in *Defence Electronics*, Nov. 1989, p. 17 and Jan. 1990, p. 68; and *Jane's Defence Weekly*, 3 Feb. 1990, p. 210.

[25] *Defense News*, 14 May 1990.

[26] See interview in *Jane's Defence Weekly*, 14 Apr. 1990, p. 718.

[27] *Aviation Week & Space Technology*, 28 May 1990, p. 25.

[28] *Financial Times*, 26 Mar. 1990, p. 18.

[29] For a summary see Weidenbaum M., *Military Spending and the Myth of Global Overstretch*, Significant Issues Series (Center for Strategic Studies, Washington, DC), vol. 11, no. 4 (1989); and Gold, D. and Adams, G., 'Defence spending and the American economy', *Defence Economics*, vol. 1 (1990), pp. 275–93.

[30] A thorough theoretical analysis and a discussion of important studies are presented by Moran, T. H., 'Globalization of America's defense industries', *International Security*, vol. 15, no. 1 (summer 1990), pp. 57–99.

[31] On 24 May 1990 the Senate Banking Committee voted to give the President $250 million to promote private-sector development of key technologies. The committee approved a bill (S 1379) to reauthorize through Sep. 1993 the Defense Production Act, which gives the President discretion to promote the development of strategic materials and technology. Compared to other countries, this amount to promote private industrial activities is, of course, marginal. *Congressional Quarterly*, 26 May 1990, pp. 1641–42.

internal controversies in August 1990, stresses that no special programmes are required because 'companies in the defense industry are well aware that reductions in spending are coming, and their plans for meeting the challenge are well along'.[32]

IV. Western Europe[33]

Arms-manufacturing companies in Western Europe are also faced with a 'shrinking pie'. In 1989, similar to their US competitors, they were affected by cuts in procurement orders and particularly by diminished arms exports. Severe losses for some companies balanced with gains for others. European companies basically held their position on the SIPRI list of the 100 largest companies. The arms industrial base in Western Europe is considerably smaller than that in the United States or the Soviet Union and is concentrated mainly in the United Kingdom, Germany and France. About 70 per cent of the 100 largest arm-producing companies in Western Europe operate from these three countries and account for almost 80 per cent of the total sales (domestic and export) of these 100 West European companies.[34]

Budgetary constraints in many West European countries have grown as a result of competing domestic economic priorities. This pattern is apparent in all West European countries with the exception of the NATO countries Norway, Portugal, Spain and Turkey as well as the neutral countries Sweden and Finland, which planned moderate procurement budget increases in 1990 since they were engaged in the process of modernizing equipment. It is not unrealistic to expect annual cuts of the order of at least 2–3 per cent, in real terms, in the procurement budgets of most West European countries. While these cuts, which have already begun to affect arms sales, were the norm in 1990, this was not the case for military R&D. R&D budgets in the countries of Western Europe are growing much faster than those in the United States. This is the result of a double-track strategy. On the one hand, governments are going ahead with negotiated and unilateral cuts in manpower and equipment and, on the other hand, are keeping options open by continuing the process of developing new and sophisticated weapons. Growing R&D budgets cannot compensate companies for major reductions in production and procurement.

The arms industry in Western Europe has already been affected and will be further influenced by expected cuts in equipment and reductions in procurement budgets. Most immediately affected is Germany. As a result of unification with the GDR and the takeover of the former GDR armed forces (Nationale Volksarmee) and its equipment by the Bundeswehr, cuts in man-

[32] US Arms Control and Disarmament Agency, *Report to Congress on Defense Industry Conversion*, Aug. 1990, p. VI.

[33] This section draws on Anthony, I., Courades Allebeck, A. and Wulf, H., SIPRI, *West European Arms Production: Structural Changes in the New Political Environment*, SIPRI Research Report, Oct. 1990.

[34] These results are based on 1988 data compiled in Anthony, Courades Allebeck and Wulf (note 33), appendix A.

power and equipment in Germany are larger than in any other NATO country. Most of the equipment of the Nationale Volksarmee will be scrapped or exported.[35]

In a 1990 report, SIPRI estimated that in the medium term arms sales (both domestic sales and exports) by the West European arms industry will have to be reduced by at least 15 per cent and possibly by as much as 30 per cent by the mid-1990s. As many as 100 000 of a total of about 1.5 million jobs were lost during the past three years; 350 000 additional jobs (and as many as 500 000 jobs if the arms control process goes beyond the CFE Treaty) might be lost by 1995.[36] Despite such prognoses, it should be recognized that, overall, the macro-economic impact of reduced military production will be marginal since the arms industry is a comparatively small industrial branch in Western Europe. The local impact might be severe if companies get into economic difficulties in regions that already suffer from high unemployment and other structural economic deficiencies.

Larger corporations operating on a European or global scale have reacted to the changing situation with a variety of strategies similar to their competitors in the United States: international mergers and company acquisitions, which in the past were unusual for the arms industry, lay-offs of personnel, closure of factories, concentration on arms sectors where demand remains strong and diversification into civil production. In addition, a small number of companies try to use the technology acquired in arms production to get into the business of the destruction of arms. While in global economic terms the business of destroying arms in the implementation of arms control and disarmament agreements is marginal, for some companies it will be welcome compensation for the loss of business which rested on the arms race.

West European arms producers will be more affected than US companies by arms control agreements that limit conventional forces and equipment in Europe. Those companies in the business of arms production will, in future, operate in circumstances where the NATO force posture has been significantly altered in response to successful disarmament initiatives. As one official describes the situation: 'limitations on a number of key equipments will lead to fewer acquisition programmes, shrinking national demands, reduced export prospects and therefore probable rises in unit costs. This will put increasing pressures on the NATO defence industries, and this may lead to further structural changes in these industries'.[37]

As was pointed out in the 1990 SIPRI research report on the West European arms industry,[38] however, the direct impact of the CFE Treaty on

[35] Of a total of 390 different types of aircraft, 24 MiG-29s will be integrated into the Bundeswehr. In addition, 40 Mi-8 helicopters and 5 SA-5 anti-aircraft missile launchers, but none of the main battle tanks and armoured personnel carriers, will be taken over. For details, see Rühl, L., 'Militärische Vereinigung—eine staatspolitische Aufgabe', *Der Mittler-Brief,* no. 4 (1990), pp. 7–8.

[36] For detailed calculations and the assumptions of this prognosis, see Anthony, I. and Wulf, H., 'The future of the industry: a prognosis', in Anthony, Courades Allebeck and Wulf (note 33), pp. 59–61.

[37] Legge, M., 'NATO defence planning after CFE', *NATO's Sixteen Nations,* June 1990.

[38] Anthony, I., 'Through the looking glass: conventional arms control and West European arms industry', in Anthony, Courades Allebeck and Wulf (note 33), pp. 17–39.

these industries is limited since cuts are, in contrast to those by the Warsaw Treaty Organization, only marginal. The number of aircraft bought by West European governments, for example, will in future be much more limited by budget constraints than by arms control. Similarly, the agreement allows defence ministries to deploy as many non-attack helicopters as they like, but the cost of buying helicopters may be a more important constraint than the CFE Treaty. However, limits that will affect industry have been agreed upon in the sectors of artillery and armoured combat vehicles. Cuts will be required, and the demand for new equipment will be considerably reduced. In spite of this, major new programmes are being undertaken in this area in several West European countries. It is highly unlikely that all of these tank and artillery programmes have a future. Since there exists no agreement concerning the limits and possibilities offered by technological development, reducing the amount of conventional equipment and at the same time increasing its capability by modernization programmes will most likely be an interesting alternative, especially for high-technology companies.[39]

The indirect effects of the CFE Treaty will be considerable because (*a*) the Treaty will contribute to a more benign threat environment, and (*b*) total alliance procurement must keep TLE (treaty-limited equipment) deployment below Treaty ceilings. Thus, government favouritism for national industry can have no future since the number of arms procured within an alliance is regulated by arms control. A process has been set in motion that requires co-ordinated force and procurement planning. Currently, no existing European institution has been given authority by the various governments to carry out this task.[40] Developments in 1990 make the European Community (EC) look like the most likely candidate to take over the responsibility. The heads of state of the EC decided at their 14 December 1990 conference in Rome to promote an evolutionary and continued process to come to a joint policy encompassing all aspects of foreign and security policy.

> As regards *common security*, the gradual extension of the [Political] Union's role in this area should be considered, in particular with reference, initially, to issues debated in international organizations: arms control, disarmament and related issues; CSCE matters; certain questions debated in the United Nations, including peace-keeping operations; economic and technological co-operation in the armaments field; co-ordination of armaments export policy, and non-proliferation.[41]

It is planned eventually to eliminate Article 223 of the Treaty of Rome, which has blocked the EC from any participation in defence and security

[39] It has also been pointed out that if, for economic reasons, future arms modernization focuses primarily on retrofitting, this will have significant consequences for the companies in France, Germany, the UK and Italy, traditionally regarded as systems integrators. Governments in Europe and abroad may prefer to make direct contact with second-tier suppliers of major sub-assemblies to upgrade and modernize weapon systems with new engines, transmissions, guns, radars and communications equipment. See Anthony, I., 'Through the looking glass: conventional arms control and West European arms industry', in Anthony, Courades Allebeck and Wulf (note 33), p. 33.

[40] Courades Allebeck, A., 'The role of the European organizations in the arms industry', in Anthony, Courades Allebeck and Wulf (note 33), pp. 40–53.

[41] EC European Council document ECSN 424/1/90 (OR. f), Rev. 1, p. 6.

issues. This would allow the EC to become more involved in the restructuring of the arms industry. It remains an open question whether governments are prepared to transfer this part of their sovereignty to the European Community.

V. The Soviet Union

Changes in arms production

To arrive at an overview of the size of the Soviet arms industry[42] Western analysts have tried either to estimate which and how many resources (manpower, industrial capacity, R&D, technology, finances, etc.) are invested in the production of arms or to calculate how many major weapon systems are produced. It is not surprising that both methods were inaccurate and required a high degree of skilful guess-work, since the Soviet military sector operated until recently behind a curtain of almost impenetrable secrecy.[43] This situation has improved somewhat, as the Soviet Government has come forward with figures in the CFE Negotiation and has begun to publish more details of its military expenditure, plans to convert military–industrial complexes and, occasionally, data on the number of weapon systems that were produced or are to be reduced. However, the basic dilemma of a severe lack of information still remains.

The controversial debate within the Soviet Union and contradictory public information about the output of the factory complexes under the umbrella of the ministries of defence are indicators that exact information is often not available even in Soviet Union. One of the outspoken critics of the situation, Soviet economist Aleksey Kireyev, asked whether the official numbers were 'statistics or phantoms':

> [I]t was announced that the munitions industry's output was to be reduced, as against the five-year-plan target figures, by 19.5 per cent; tank supply, by 52 per cent; ammunition, by 20 per cent; combat aircraft, by 12 per cent; and combat helicopters, 60 per cent. That's a fine example of glasnost, indeed! What were the original target figures? What kind of reductions—physical or cost—are these? There are no answers to these questions so far.[44]

[42] For a description of the organization of the industry its size and the initial conversion plans, see *SIPRI Yearbook 1990* (note 12), chapter 8, pp. 344–58.

[43] Numerous articles have been published testing the reliability of these data. For a recent debate, see Holzman, F., 'Politics and guesswork: CIA and DIA estimates of Soviet military spending', *International Security*, no. 14 (fall 1989), pp. 101–31. Chalmers, M., 'Soviet weapons procurement in the 1980s: time for *glasnost* in the Pentagon?', *Defense Analysis*, vol. 6, no. 3 (1990), pp. 255–62.

[44] Kireyev, A., 'Crawling towards disarmament', *New Times*, no. 10 (6–12 Mar. 1990), pp. 30–32. In Apr. 1990 during a visit to the USA, prominent Soviet economists said that Soviet economic output was far below the estimates of the CIA; thus military spending might account for 20–25% of the Soviet GNP. See *New York Times*, 24 Apr. 1990. This statement led to a heated public controversy in the USSR between the economists and representatives of the military complex. Academician V. Avduyevskiy, chairman of the Soviet National Commission to Promote Conversion and critic of the military–industrial complex, referred to the controversial figures and mentioned military expenditures of 17.6% of GNP and approximately 200 billion roubles, in *Izvestia*, 7 Feb. 1990, p. 2.

On the basis of scattered, sometimes contradictory information, a Western expert on the Soviet arms industry makes the following estimations: military production, both within and outside the defence complex, accounts for just under 10 per cent of gross industrial output and employs approximately 4.25 million people, over 11 per cent of the total industrial labour force (6.4 million people are estimated to be employed in the military–industrial complex including civilian production). The nine ministries of the defence complex in 1988 possessed 13 per cent of the industrial capital stock. The output of the factories under the ministries of defence is estimated to have amounted to 118 billion roubles in 1988, of which 72 billion roubles was for military production and 46 billion roubles for production of civilian goods. Nearly 90 per cent of the facilities of the defence complex are concentrated in the Russian Soviet Federated Socialist Republic (RSFSR) and the Ukraine.[45]

Cuts in arms production that were announced in the USSR in 1989 included a 52 per cent reduction in tanks, 23 per cent reduction in military aircraft, 60 per cent reduction in military helicopters and 20 per cent reduction in munitions as compared to the original Five Year plan.[46] In the West, it is generally accepted that at least some of the more important of the announced cuts in Soviet military production are being implemented. The US DOD, in the 1990 edition of *Soviet Military Power,* draws the following conclusion:

Soviet 1989 output of military materiel generally fell from 1988, mirroring Gorbachev's January 1989 announcement that output would be reduced. The most pronounced cuts occurred in ground forces materiel. Output of strategic systems was generally level while the number of naval surface units produced actually rose. The production of submarines remained the same. Some of the declines reflect longer-term downward trends; output of conventional ground force equipment as well as helicopters and fighter aircraft have declined since Gorbachev took office in March 1985. However, since 1985 the manufacture of cruise missiles has accelerated.[47]

Comparative figures for the procurement of major weapon systems have often been used by NATO governments to illustrate the size of the Soviet armed forces and the capability of its arms industry. Even with lower production, the Soviet industry produces far more weapons than the United States. According to official US information, Soviet arms production in 1989 was higher than that of the USA in 20 of 23 weapon categories. (The exceptions are long-range ship-launched cruise missiles, anti-submarine warfare fixed-

[45] Cooper, J., 'The contradictions of Soviet defence industry civilianisation', paper prepared for the Second Biennial RAND–Hoover Symposium, 29–30 Mar. 1990; Cooper, J., 'The Soviet defence industry and conversion', *RUSI Journal*, autumn 1990, pp. 51–56.

[46] Vid, L.,'Guns into butter, Soviet style', *Bulletin of the Atomic Scientists*, Jan./Feb. 1990, pp. 16–19; *Pravda*, 26 Sep. 1989; *Krasnaya Zvezda*, 22 Feb. 1990, quoted in Cooper, RAND (note 45), appendix 5, p. 1.

[47] US Department of Defense, *Soviet Military Power 1990* (US Government Printing Office: Washington, DC, 1990), p. 35. Soviet Marshal Sergey Akhromeyev testified before the US House Armed Services Committee on 21 July 1989 and said that Soviet tank output in 1989 was 1700. Tank production would 'be reduced to the tune of over 40 per cent by the end of 1990'. See Institute for Defense and Disarmament Studies, *Arms Control Reporter* (IDDS: Brookline, Mass.), Chronology 1989, sheet 407.B.204, 21 July 1989.

wing aircraft and cruisers.[48]) However, the numbers compiled in the US Government publications need more explanation,[49] not only because differences exist between different government agencies,[50] but also because there are discrepancies between the numbers of weapons produced and the inventories in the Soviet armed forces.[51]

At different times and from different Soviet officials, various bits and pieces of information on past production and the future of the industry have been been given. The following information, collected entirely from official or semi-official sources, is far from consistent. Hence outside observers are faced with a confusing picture.

Military production in the military sector: Prior to conversion (i.e., during the 12th Five Year Plan, 1986–90), production of armaments and other military equipment increased at an annual rate of 5.0–5.5 per cent.[52] By 1991, arms procurement will be reduced by 19.5 per cent and R&D by 13.5 per cent.[53] It has also been announced by the Military–Industrial Commission that reductions would amount to 4.5 per cent in 1989 and 4.7 per cent in 1990.[54] It is not clear whether this means that the remaining part of the announced reductions will occur in 1991. The share of military production in the military industrial complex of 60 per cent in 1988 will be reduced to 40 per cent in 1995.[55] In a press release for the Soviet Conversion-90 exhibition, this share was reported to be 50/50 for 1990 and 35/65 in 1995.[56]

Civilian production in the military sector: Civilian production was planned to be increased by 8.9 per cent in 1989 and 13.1 per cent in 1990.[57] The 40 per cent share of civilian production in 1988 will increase to 60 per cent in 1995.[58] However, the head of the conversion plan at Gosplan, V. I. Smyslov, claimed that the volume of civilian output in the military sector would increase from 43 per cent (not 40) in 1988 to 50 per cent in 1990 and to over 60 per cent in 1995.[59] Allocations for civilian research will increase by 41.1 per cent from 1988 to 1991; the share of civilian R&D projects in the defence-related ministries will amount to 45.8 per cent in 1995, compared to 28.5 per cent in

[48] *Soviet Military Power 1990* (note 47), pp. 38–39.

[49] Chalmers (note 43), p. 261, quotes the Chairman of the House Armed Services Committee, Les Aspin, as saying that the US Government needs to 'uncork the intelligence bottle-neck strangling information about changes in the Soviet military'.

[50] Holzman (note 43), pp. 101–31.

[51] This case has been convincingly explained for tanks, artillery and aircraft by Chalmers (note 43), pp. 255–62.

[52] Vid (note 46), pp. 16–19.

[53] Official figure announced in 1989 and repeated in 1990 by Smyslov, V. I., 'On the basic directions of the draft state programme for the conversion of defence industry in the USSR', paper presented at the UN Conference on Conversion (see note 6), p. 2. The 19.5% figure is interpreted differently, including reductions of output from original plans to reductions of allocations for procurement.

[54] Belousov, I. S. , in *Pravda,* 28 Aug. 1989, quoted in Cooper, RAND (note 45), appendix 3, p. 1.

[55] General Moiseyev, in *Pravda*, 11 June 1989.

[56] Press release for the Conversion-90 exhibition, Munich, Apr. 1990. This different civilian/military distribution might be the result of the military complex having acquired additional civilian factories in 1989.

[57] Belousov, I. S., in *Pravda,* 28 Aug. 1989, quoted in Cooper, RAND (note 45), appendix 3, p. 1.

[58] General Moiseyev, in *Pravda*, 11 June 1989.

[59] Smyslov (note 53), p. 7.

1988.[60] Total output of the defence complex is planned to grow by 45 per cent from 1989 to 1995, and of the civilian component by 82 per cent.[61] If these percentage growth figures, announced by Premier Nikolai I. Ryzhkov, are applied to the present output, then the civil/military relation would be different from that foreseen in the plan. The head of the conversion plan presented the following figures: the total volume of civil output in the defence industry will double, and in 1995 it will exceed 110 billion roubles;[62] non-food consumer goods in the defence complex will increase during the 13th Five Year Plan (1991–95) 1.8 times and equal 71 billion roubles in 1995.[63] According to a report to the Presidential Council of the USSR, the volume of civilian production is to reach 270 billion roubles in 1991–95, double the value of the previous five-year period.[64] It is not clear whether this figure represents total or additional civil production. Adding consumer goods, total production of the defence sector plants for the period 1991 (approximately 70 billion roubles) to 1995 (planned 110 billion roubles) would amount to almost double the 1991–95 figure of 270 billion roubles mentioned above. Other sources claim that consumer goods production in the military industrial–complex rose by nearly 25 per cent over the three years 1987–89 and will increase by almost 23 per cent in 1990.[65]

Employment: According to the conversion plan more than 500 000 people are planned to be made redundant in 1990. However, they do not necessarily have to leave the military–industry complex but are transferred to civilian production.[66] At present 300 000 skilled workers are retrained every year, and over 300 000 specialists have taken up jobs with co-operatives and joint ventures.[67] The Soviet Ministry of the Defence Industry, according to Minister Boris M. Belousov, lost more than 70 000 of its production and industrial personnel during the past two years.[68]

Enterprises: More than 400 Defence Ministry enterprises and over 100 enterprises of other branches of industry are covered by the conversion plan. Half of the defence complex enterprises will reduce their output of military equipment by more than 20 per cent; 34 enterprises of non-defence ministries

[60] Smyslov (note 53), p. 9.

[61] Ryzhkov, N., in *Pravda*, 14 Dec. 1989, quoted in Cooper, RAND (note 45), appendix 3, p. 2.

[62] Smyslov, V. I., in *Izvestia*, 1 Mar. 1990, p. 2; this figure was repeated in *Moscow News*, no. 22 (10–17 June 1990), p. 10.

[63] Smyslov (note 53), p. 5.

[64] Report on a meeting of the Presidential Council of the USSR, 28 Sep. 1990 in TASS (Moscow), 28 Sep. 1990, reported in Foreign Broadcast Information Service, *Daily Report–Soviet Union* (hereafter referred to as FBIS-SOV), FBIS-SOV-90-190, 1 Oct. 1990, p. 43; and Baklanov, O. in *Soviet Press Weekly Review*, Novosti Press Agency, no. 41 (Oct. 1990), p. 13 (Oelg Baklanov is Secretary of the Communist Party Central Committee).

[65] Okritov, N., in *Moscow World Service in English*, 26 July 1990, reported in FBIS-SOV-90-145, 27 July 1990, p. 55 (Nikolay Okritov is chief of the press service of Russia's Federation of Independent Trade Unions). The confusion about consumer goods and civilian output in the military industry arises from the fact that consumer goods are valued in retail prices, i.e., the prices include turnover tax.

[66] Smyslov (note 53), p. 8.

[67] Baklanov, O., in *Soviet Press Weekly Review*, Novosti Press Agency, no. 41 (Oct. 1990), p. 14.

[68] Belousov, B. M., at the 4 July session of 28th CPSU Congress, reported in *Pravda*, 6 July 1990, p. 4.

Table 8.8. Changes in the Soviet arms industry (actual and planned), based on Soviet information, 1986–95[a]

Year	Military output, b. roubles	% of total	% change	Civil output, b. roubles	% of total	% change	Total defence sector output, b. roubles	% of total	% change
Actual									
1986	65	*64*	–	37	*36*	–	102	*100*	–
1987	68	*64*	*+ 5*	39	*36*	*+ 5*	107	*100*	*+ 5*
1988	72	*60*	*+ 6*	46	*40*	*+ 18*	118	*100*	*+ 10*
1989	69	*56*	*– 4*	55	*44*	*+ 20*	124	*100*	*+ 5*
Planned									
1990	66	*51*	*– 4*	62	*49*	*+ 13*	128	*100*	*+ 3*
1991	63	*47*	*– 5*	71	*53*	*+ 15*	134	*100*	*+ 5*
1992	65	*45*	*+ 3*	79	*55*	*+ 11*	144	*100*	*+ 7*
1993	66	*43*	*+ 2*	88	*57*	*+ 11*	154	*100*	*+ 7*
1994	68	*41*	*+ 2*	98	*59*	*+ 11*	166	*100*	*+ 8*
1995	70	*39*	*+ 2*	110	*61*	*+ 11*	180	*100*	*+ 8*

[a] Author's percentage calculations. Detailed figures for 1992–94 were not available. It has been assumed that there are linear changes of military and civilian output planned for the period 1992–95. In addition to the sources below, see *Izvestia*, 7 Feb. 1990, p. 2; and Smyslov, V. I., 'On the basic directions of the draft state programme for the conversion of defence industry in the USSR', paper presented at the UN Conference on Conversion: Economic Adjustments in an Era of Arms Reduction, 13–17 Aug. 1990 (Smyslov, who presents partly contradictory evidence, is Vice-Chairman of the Soviet State Planning Committee).

Sources: Based on Cooper, J., 'The contradictions of Soviet defence industry civilianization', paper prepared for the Second Biennial RAND–Hoover Symposium, 29–30 Mar. 1990, appendix 3. The main sources used by Cooper are: BBC, Summary of World Broadcasts, SU/0416 B/B, 23 Mar. 1989 (N. I. Ryzhkov); *Pravda*, 8 June 1989 (N. I. Ryzhkov); L. Vid, 'Guns into butter, Soviet style', *Bulletin of the Atomic Scientists*, Jan./Feb. 1990, p. 17; *Pravda*, 28 Aug. 1989 (I. S. Belousov); *Izvestia*, 28 Feb. 1990.

and 6 of the defence complex will be totally converted;[69] and 14 per cent of the defence complex will be subject to conversion.[70] The Deputy Minister of Defence for Armaments, Vitaliy Shabanov, says that 422 military-oriented plants will be converted to civilian production by 1995, and 56 of them will be fully retooled to produce consumer goods.[71]

On the basis of parts of this official information Julian Cooper, an expert on the Soviet arms industry, has constructed a table that gives a general impression of actual and planned military and civilian production (see table 8.8).

[69] Smyslov (note 53), p. 6.
[70] Smyslov (note 53).
[71] Shabanov in *Trud*, according to FBIS-SOV-90-189, 28 Sep. 1990, p. 58.

Looking at the scattered and inconsistent information offered by Soviet officials, of course, table 8.8 can offer only indications of orders of magnitude of what is officially intended. According to the collected information, arms production will shrink slightly from the 1988 peak. Compared to the initial Five Year Plan, however, reductions are substantial. It can be concluded that after the initial cuts in 1989 and 1990 the output of military equipment will rise at an annual rate of approximately 2.5 per cent, representing a more or less stable output in real terms, if allowance is made for small price increases. With the dramatic price increases in the Soviet economy in 1990, it is questionable whether weapon production can avoid following this general trend. This would be the case only as long as prices for weapons and other military equipment are politically determined and are not intended to be a reflection of costs that have been incurred. However, this seems not to be the case. Drastic price increases for weapons planned for 1991 are one of the main reasons for a military budget increase from 70.3 billion roubles to 96.5 billion roubles.

The main change, it can be concluded, is not the reduction in arms production but the increase in civilian production in the military complex, by both stepping up civilian production in existing military complex facilities and integrating other civilian factories into the defence sector. Whether this information is a true reflection of reality is doubtful and is also doubted by critics in the USSR. It is highly unlikely that these plans will be implemented exactly as they have been drawn up. Conversion initiatives are already falling behind plans, and in the face of the catastrophic economic developments and the fact that the central authorities are less and less in control of state, regional and local economic policy it seems realistic to assume that the envisioned plans will undergo substantial change.

Export dependence

Parallel to the debate on reductions of arms production and converting production facilities to non-military production, it has been suggested that factories could, instead of converting their facilities, try to export more arms. This runs counter to the declared official Soviet foreign policy, clearly intended to reduce arms exports to the Third World.[72] The general development of introducing free market incentives and creating self-financed factories—even in the military sector—led to such considerations of exploiting the export potential. There are several reasons why the export concept is not a realistic option for keeping the existing facilities occupied: arms industries in potential importer countries—including those in Western Europe—are struggling with their own over-capacities, and it is difficult to imagine that in this situation Soviet weapons will be bought on a large scale by Western governments, even if they are offered at lower prices. Furthermore, many of the Soviet customers in the Third World do not have the hard currency funds available which Soviet companies are looking for. In the past, Soviet weapons

[72] For a discussion of this trend, see chapter 7.

Table 8.9. Estimates of selected Soviet major weapon production and exports, 1980–89[a]

	1980–84	1985–89
Number of main battle tanks produced	13 800	15 000
Thereof % exported:		
to the Third World	*10.5*	*5.5*
to WTO countries	*3.5*	*1.0*
to other countries	*0.5*	*0.5*
total exports	*14.0*	*7.0*
Number of fighter aircraft produced	5 500	3 325
Thereof % exported:		
to the Third World	*14.5*	*21.0*
to WTO countries	*6.0*	*9.0*
to other countries	*1.0*	*1.5*
total exports	*21.5*	*31.5*
Number of major surface ships produced	47	46
Thereof % exported:		
to the Third World	*38.5*	*21.5*
to WTO countries	–	*2.0*
to other countries	*4.0*	*4.5*
total exports	*42.5*	*28*

[a] Percentages do not always add up due to rounding. According to US DOD information, the category 'major surface ships produced' includes carriers, cruisers, destroyers, frigates, corvettes and paramilitary ships of the same class. According to SIPRI information, no carriers or cruisers were exported during the 1980s.

Sources: Production figures are from *Soviet Military Power 1984–1990* (US Government Printing Office: Washington, DC, 1984–90); export figures are from the SIPRI arms trade data base.

were sold on soft terms or not paid for at all.[73] Finally, some of the largest importers of Soviet weapons, the former WTO allies of the USSR, are reducing their inventories as a result of the CFE Treaty and the de facto breakdown of the WTO as an alliance, and they are closing or converting at least some of their own arms factories. It can be expected that their imports will be substantially or even totally reduced.

To arrive at an estimate of Soviet export dependence during the decade of the 1980s, SIPRI has compared the number of weapon systems produced (according to US DOD information) and the weapon systems exported to the Third World, the WTO countries and all other countries, summarized in table 8.9.

In interpreting this table, the following methodological remarks should be observed. The figures come from two different data banks: from the US DOD for production, based on intelligence gathering; and from SIPRI for the export

[73] See chapter 7 on the relationship of the debts that several major arms importers owe the USSR with the arms trade.

figures, based on public sources. Whether the three major weapon categories chosen always match in the two data bases is not known. As pointed out above, the figures given for Soviet tank production by the US Administration are disputed in the USSR.[74] If production was actually much lower, as was claimed in the Soviet Union, export dependence would have been proportionately higher. Furthermore, it should be observed that to arrive at a realistic estimate of the Soviet production/export relation not all the major conventional weapon systems that have been exported by the USSR are included. Second-hand equipment has been excluded since this is likely to come from the armed forces or storage, but not from production. For the same reason, exported equipment that was no longer produced has been excluded in the statistics. Finally, licensed production, that is, Soviet MiG-21 and MiG-27 fighters produced in India and the various types of main battle tanks produced in Czechoslovakia, India, North Korea, Poland and Yugoslavia, have been excluded, although production in these countries depended on Soviet technology and component supplies.

All three selected categories are important export products of the Soviet military–industrial complex. In certain areas (such as strategic bombers, ICBMs and SLBMs) no exports have been recorded at all. The two periods chosen are for the five years just before and the first five years of the Gorbachev Government. According to the information available, of the 13 800 tanks produced in the period before Gorbachev took office in 1985, approximately 14 per cent were exported; most of these main battle tanks were imported by Third World countries. This percentage dropped considerably after 1985, although production was kept at about the same level until it fell in 1989. It will be interesting to observe the future export pattern after the drastic reduction in tank production from 3500 in 1988 to 1700 in 1989.

In contrast to exports of main battle tanks, a higher percentage of Soviet fighter aircraft has been exported: 21.5per cent in the pre-Gorbachev years and 31.5 per cent in 1985–89. This percentage change is the result of reduced production of fighter aircraft but constant export figures. Again, most of these fighter planes were exported to Third World countries—about twice as many as to WTO countries.

The Soviet export dependence on major surface ships was particularly high. Exports fell from 42.5 per cent of total production of major surface ships in the first half of the decade to 28 per cent in the second half. Third World countries were the main customers for this category of conventional weapon systems as well.[75]

The result of table 8.9 corresponds to the general pattern of Soviet exports. No abrupt changes from previous Soviet weapon export policy have emerged

[74] Statement by Marshal Sergey Akhromeyev (note 47).

[75] In an interview at the end of 1990, published in *Pravitelstvenny Vestnik* (Government News), no. 80 (2 Jan. 1991), I. S. Belousov for the first time gave figures for Soviet arms exports. Belousov is Deputy Chairman of the Council of Ministers and Chairman of the State Committee of the Council of Ministers on military industry issues. According to his information, exports of major weapons in several categories (including missiles, tanks, aircraft and ships) fell drastically during the present Five Year Plan. For details, see chapter 7, section III.

in the Gorbachev period. Traditionally, the Soviet Union transferred about 70 per cent of its arms exports to the Third World.

Konversiya or *konvulsiya*

The short history of the present conversion policy in the Soviet Union, initiated as a reaction to the 1987 INF Treaty and its requirement to close production lines for SS-20 missiles and announced at the end of 1988, has already gone through several stages that reach from optimism and great hopes for economic reform in the early phase to criticism of the concept at the present stage. *Konversiya* has in the meantime come to also be called *diversiya* and *konvulsiya*[76] to express the disappointment over the lack of change.

At the beginning, the expected economic benefits from cuts in military production were hoped to contribute to economic reform and renewal of the civilian industry.[77] The poor state of consumer goods production and shortcomings in other industrial sectors and agriculture inspired the State Planning Committee, Gosplan, to assign one or more of the following priorities to the different ministries in charge of arms production: consumer durables, farm machinery, equipment for light industry and food-processing, trade and public catering, medical technology, electronics, computers, communications, television sets and radios.[78]

The State Program for Conversion of the Defense Industry for the Period Through 1995, which has still not been made public, was examined by the Soviet Council of Ministers Presidium in February 1990 after submission by Gosplan, the Soviet Council of Ministers State Commission for Military–Industrial Questions and the Soviet Defence Ministry.[79] On 7 March 1990 the Supreme Soviet Defence and State Security Committee discussed the plan.[80] The picture painted then was optimistic:

> Conversion is already in progress. It now covers some 400 enterprises in the defense complex and 100 civil plants that produced military output . . . By 1995 the country will see a 120 per cent increase in the production of equipment for the agro-industrial complex, and more than 3,000 new types of production lines, installations, and robots will be started up . . . Production of medical equipment will increase by 130 per cent . . . Production of computer equipment will increase by 110 per cent, including a 280 per cent increase in the production of personal computers. Deliveries of civil aircraft will increase by 150 per cent compared with the last five-year plan.[81]

[76] The term was used by the Leningrad *Pravda,* 24 Nov. 1989, quoted by Cooper, RAND (note 45), p. 20.

[77] For background, see *SIPRI Yearbook 1990* (note 8), chapter 8, pp. 352–58.

[78] A summary is given in Izyumov, A., 'The national experience of the USSR', paper presented at the United Nations Conference on Conversion (see note 6).

[79] *Izvestia,* 1 Mar. 1990, quoted in FBIS-SOV-90-044, 6 Mar. 1990, p. 79. The plan was discussed again in Dec. 1990, but it is still not clear whether it has been approved for implementation.

[80] TASS (Moscow), 7 Mar. 1990, quoted in FBIS-SOV-90-3jj, 9 Mar. 1990, p. 33.

[81] *Izvestia,* 1 Mar. 1990, quoted in FBIS-SOV-90-044, 6 Mar. 1990, p. 80. Similar optimistic accounts can be found in later Soviet publications. See, for example, a speech by B. M. Belousov, Soviet Minister of the Defence Industry, at the 4 July session of the 28th CPSU Congress, reported in *Pravda,* 6 July

In May 1990 all Soviet military ministries took a joint initiative—with both political and economic intentions and implications—and organized the Conversion-90 exhibition in Munich, Germany. In the Soviet press the exhibition, intended among other things to co-operate with companies from the West in the non-military sector of the Soviet military factories, was considered an 'undoubted success'.[82] However, two different types of criticism were raised, the longer the conversion plan was discussed and the longer the public failed to see the emergence of immediate results.[83]

The first criticism relates to the general direction of conversion and the way in which conversion is planned. Economists suggest that, instead of increasing civilian production in the military–industrial complex and thus increasing its size and possibly its influence, companies should no longer be under the authority of the defence ministries.[84] It is argued by Academician Vsevolod Avduyevskiy, the Chairman of the Soviet National Commission to Promote Conversion and a critic of the centralized decision-making process in the conversion plan, that: 'the ingrained military–industrial complex mentality needs to be changed to achieve [cost-consciousness]. The military–industrial complex always saw consumer goods as imposed, second-rate output. That mentality is not yet changing'.[85] This criticism relates to both the secrecy and departmental barriers that prevented spin-off from the military sector to the benefit of the consumers and to the lacking market mechanism that would give producers an incentive to look for production alternatives. The early success stories of tank factories producing prams and missile complexes producing dairy equipment are now used to criticize the command-economy top–down type of conversion. 'Aircraft manufacturers have been told to make kitchen furniture . . . Defence people are in the habit of getting everything on a platter the first time they ask for it . . . But now . . . factory managers have asked the central authorities for supplies of resources . . . and have come up against a blank wall perhaps for the first time'.[86] Conversion has been caught in the middle, between the dismantling of the traditional planning system and the introduction of market devices that do not yet work.

A second type of criticism is raised in the industry and the military itself. Resistance in the arms-producing factories to having to switch to less attractive technologies, coupled with fears of loss of privileges, is not uncommon.

1990, quoted in FBIS-SOV-90-130-S, 6 July 1990, p. 32; and interview of *Trud* with Deputy Defence Minister Vitaliy Shabanov, quoted in FBIS-SOV-90-189, 28 Sep. 1990, p. 58.

[82] *Izvestia,* 13 May 1990, quoted in FBIS-SOV-90-098, 21 May 1990, p. 45. Even during the exhibition that was serviced by approximately 200 high-ranking Soviet ministers, chief engineers, scientists, technicians and party officials, the success of the exhibition was permanently underlined. Interviews by the author during the exhibition with more than a dozen officials. However, no precise figures were given of how many products were sold or co-operation agreements signed.

[83] 'Conversion is obviously slipping', according to Kireyev (note 44), pp. 30–32.

[84] The economist Alexei Izyumov requested to 'carry out a more thoroughgoing conversion and remove 20 percent of military plant from the defense ministries'. See Izyumov (note 78); and the reports in the Soviet press. The quote is taken from *Izvestia,* 19 Aug. 1990, quoted in FBIS-SOV-90-164, 23 Aug. 1990, p. 6.

[85] Interview in *Izvestia,* 7 Feb. 1990, quoted in FBIS-SOV-90-031, 14 Feb. 1990, p. 120.

[86] Bordenkov, A., 'Uniformed civilian production', *Moscow News,* no. 22 (10–17 June 1990), p. 10.

Colonel General Yuriy Alekseyevich Yashin, Soviet Deputy Minister of Defence, said in a radio interview:

Conversion cannot, of course, be a panacea for all shortcomings . . . I support a reduction in spending on the development of armaments and military hardware, and that applies particularly to series production. But, I wouldn't want theoretical, basic, and applied research and experimental design work to be reduced in any way, nor the financing of them . . . One of the main ways to achieve reasonable defense sufficiency today is to carry out research and development work.[87]

On 6 September Pravda published an open letter from several dozen managers of the defence industry and high-ranking officials of the ministries of the military industry complex, addressed to the Supreme Soviet. The letter expressed not only concern about the loss of privileges of the industry but also open criticism that no guidelines for 1991 had been given to industry.[88]

In addition to these concerns, motivated by fears that the Soviet military might slip technologically, there is general criticism of the chaos created by changed policies. The newspaper *Krasnaya Zvezda* quotes a chief engineer of an unnamed Urals arms production plant: 'We support the idea of conversion, but look what is happening. There are no new orders. Existing ties have been broken. Supply is starting from nothing. We seem to be no better off than we were at the start'. The report then continues:

There has already been a marked increase in the cost of models of equipment produced by the defense sectors of industry, caused primarily by enterprises' transition to economic accountability [khozraschet] and self financing . . . This chain inevitably leads to the emergence of surplus capacities . . . Conversion of production facilities is leading to the point where people with the highest qualifications are leaving . . . And what about advanced technologies and the obvious slowing down of the implementation of the latest achievements of science and technology? . . . We are rushing to transfer very complex manufacturing lines to produce plates or samovars. What is this? A tribute to fashion or chasing after favourable returns?[89]

In the report to the 28th CPSU Congress, Boris M. Belousov, Soviet Minister of the Defence Industry, criticizes the fact that for 'some incomprehensible reason allocations for equipment were reduced and that therefore the USSR Ministry of the Defense Industry had lost more than 70 000 people from its industrial personnel'.[90]

In April 1990 Gorbachev began to acknowledge difficulties with the conversion plan and its implementation. He hinted at a switch in policy, intending to ask arms-production plant managers to produce more high-technology civilian goods, and in November he repeated that conversion was 'extremely important to the country' but deplorably had 'proved not so simple as it

[87] Moscow Domestic Service, 22 Feb. 1990, quoted in FBIS-SOV-90-039, 27 Feb. 1990, p. 78.

[88] *Pravda*, 6 Sep. 1990, p. 2.

[89] Report of Candidate of Technical Sciences Colonel N. Ustenkov, 'Sharp facets of conversion', *Krasnaya Zvezda*, 4 Apr. 1990, quoted in FBIS-SOV-90-065, 4 Apr. 1990, p. 54. For similar criticism, see also *Izvestia*, 8 Feb. 1990, quoted in FBIS-SOV-90-030, 13 Feb. 1990, p. 114.

[90] Reported in *Pravda*, 6 July 1990, quoted in FBIS-SOV-90-130-S, 6 July 1990, p. 32.

originally seemed'.[91] In summary, one result of *glasnost* has been an open debate and criticism of conversion. Based on publicly available information, in 1990 *konversiya* went through a period of *konvulsiya* and is thus not unlike the state of turmoil of the entire Soviet economy.

The present economic reform, involving de-nationalization of property, de-monopolization, de-regulation of centrally organized materiel supplies and reform of the pricing system in the USSR, has important consequences for the arms industry since it changes the traditional pattern of arms procurement, priority supplies of materiel and the pricing of weapons. Conversion is important for the entire economic reform process since many innovative technologies are concentrated in the military industry. To make use of this technology, the resistance of parts of the military–industrial complex to conversion has to be overcome. Conversion is therefore of utmost political importance.[92]

[91] *Pravda*, 28 Apr. 1990 and 3 July 1990, quoted in Cooper, J., 'The Soviet defence industry and conversion', *RUSI Journal*, autumn 1990, p. 54; and Moscow World Service in English, reported in FBIS-SOV-90-230, 29 Nov. 1990, p. 45.

[92] Schröder, H.-H., '"Demilitarisierung" in der Sowjetunion? Ansätze zu einer Neuordnung der Beziehung von Politik und Militär in der UdSSR (1987–1990)', *Berichte des Bundesinstituts für ostwissenschaftliche und internationale Studien*, no. 49-1990 (Bundesinstitut für ostwissenschaftliche und internationale Studien: Cologne, 1990).

Appendix 8A. The 100 largest arms-producing companies, 1989

IAN ANTHONY, AGNÈS COURADES ALLEBECK, GERD HAGMEYER-GAVERUS, PAOLO MIGGIANO AND HERBERT WULF

The table 8A contains information on the 100 largest arms-producing companies in the OECD and Third World countries in 1989.[1] Companies with the designation *S* in the column for rank in 1989 are subsidiaries; their arms sales are included in the figure in column 6 for the holding company. Subsidiaries are listed in the position where they would appear if they were independent companies. In order to facilitate comparison with data for the previous year, the rank order and arms sales figures for 1988 are also given. Where new data for 1988 has become available, this information is included in the table; thus the rank order and the arms sales figures for some companies which appeared in table 8.2 in the *SIPRI Yearbook 1990* have been revised.

Sources and methods

Sources of data. The data in the table are based on the following sources: company reports, a questionnaire sent to over 300 companies, and corporation news published in the business sections of newspapers and military journals. Company archives, marketing reports, government publication of prime contracts and country surveys were also consulted. In many cases exact figures were not available, mainly because companies often do not report their arms sales or lump them together with other activities. Estimates were therefore made.

Arms sales. The criterion for the rank order of companies is their arms sales in 1989 (column 6). The arms sales figures are based on the sources mentioned above and thus not comparable to the SIPRI arms transfer figures given in chapter 7.

Coverage. The data are for 1989; data in columns 2 and 7 are for 1988. The fiscal year for companies is not always the calendar year. No calculations have been made to adjust fiscal to calendar years.

Exchange-rates. Most figures collected were given in local currencies. To convert figures into US dollars, the period-average of market exchange-rates of the International Monetary Fund, *International Financial Statistics,* was used.

Profit. Profit after taxes is shown for the entire company, not for the arms-producing sector alone. For figures taken from journals and periodicals, it was not always clear whether profit was given before or after taxes.

Employment. The figure shown is either a year-end or yearly average number for the entire company, as published in the sources used.

Key to abbreviations in column 5. A = artillery, Ac = aircraft, El= electronics, Eng = engines, Mi = missiles, MV = military vehicles, SA/O = small arms/ordinance, Sh= ships, and Oth = other

[1] The 24 member countries of the Organization for Economic Co-operation and Development are: Australia, Austria, Belgium, Canada, Denmark, Germany, Finland, France, Greece, Iceland, Ireland, Italy, Japan, Luxembourg, the Netherlands, New Zealand, Norway, Portugal, Spain, Sweden, Switzerland, Turkey, the UK and the USA (Yugoslavia participates with special status). For the countries in the Third World, see appendix 7A.

Table 8A. The 100 largest arms-producing companies in the OECD and Third World countries, 1989[a]

Figures in columns 6, 7, 8 and 10 are in US $ million.

1	2	3	4	5	6	7	8	9	10	11
Rank					Arms sales					
1989	1988[b]	Company[c]	Country	Industry	1989	1988[d]	Total sales 1989	Col. 6 as % of col. 8	Profit 1989	Employment 1989
1	1	McDonnell Douglas	USA	Ac El Mi	8 500	8 500	14 581	*58*	219	128 000
2	3	General Dynamics	USA	Ac MV El Mi Sh	8 400	8 000	10 053	*84*	293	103 000
3	2	Lockheed	USA	Ac	7 350	8 400	9 932	*74*	2	82 500
4	7	British Aerospace	UK	Ac El Mi SA/O	6 300	5 470	14 898	*42*	546	125 600
5	4	General Electric	USA	Ac Eng	6 250	6 250	54 574	*11*	3 939	292 000
6	5	General Motors	USA	Ac Eng El Mi	5 500	6 000	126 932	*4*	4 224	775 000
7	6	Raytheon	USA	El Mi	5 330	5 500	8 796	*61*	529	77 600
8	10	Boeing	USA	Ac El Mi	4 800	4 500	20 276	*24*	973	164 500
9	8	Northrop	USA	Ac	4 700	5 200	5 248	*90*	–81	41 000
10	9	Rockwell International	USA	Ac El Mi	4 500	5 000	12 633	*36*	735	109 000
11	13	Martin Marietta	USA	Mi	4 350	4 300	5 796	*75*	307	65 500
12	12	Thomson S.A.	France	El Mi	4 320	4 470	12 027	*36*	78	100 000
13	15	Daimler Benz	FRG	Ac Eng MV Mi El	4 260	3 420	40 634	*10*	3 622	368 226
S	S	Thomson-CSF (Thomson S.A.)	France	El Mi	4 120	4 320	5 282	*78*	412	21 723
14	11	United Technologies	USA	Ac El Mi	4 100	4 500	19 766	*21*	702	201 000
S	S	Hughes Electronics (General Motors)	USA	Ac El	4 000	4 500	11 400	*35*	781	66 000
S	–	DASA (Daimler Benz)	FRG	Ac Eng El Mi	3 930	0	7 484	*53*	–58	62 959
15	14	Direction des Constructions Navales	France	Sh	3 630	3 580	3 626	*100*	. .	28 000
16	16	TRW	USA	MV Oth	3 050	3 200	7 340	*42*	263	74 300
17	19	Litton Industries	USA	El Sh	2 900	2 920	5 023	*58*	178	50 700

1	2	3	4	5	6	7	8	9	10	11
Rank					Arms sales					
1989	1988[b]	Company[c]	Country	Industry	1989	1988[d]	Total sales 1989	Col. 6 as % of col. 8	Profit 1989	Employment 1989
18	18	GEC	UK	El	2 880	2 970	14 408	*20*	1 118	107 435
19	17	Grumman	USA	Ac El	2 850	3 000	3 559	*80*	67	28 900
20	20	Mitsubishi Heavy Industries	Japan	Ac Mi Sh	2 640	2 840	15 180	*17*	400	43 914
21	21	Westinghouse Electric	USA	El	2 500	2 600	12 844	*19*	922	121 963
S	S	Pratt & Whitney (United Technologies)	USA	Eng	2 500	. .	6 900	*36*	. .	. .
22	22	Unisys	USA	El	2 300	2 500	10 097	*23*	– 639	82 300
23	26	IRI	Italy	Ac Eng El Sh	2 230	2 100	41 285	*5*	910	363 449
24	28	Dassault-Breguet	France	Ac	2 200	2 080	3 059	*72*	46	15 572
25	23	Aérospatiale	France	Ac Mi	2 190	2 300	4 969	*44*	32	36 899
26	25	Texas Instruments	USA	El Mi Oth	2 160	2 150	6 521	*33*	319	73 854
27	32	Tenneco	USA	Sh	1 950	1 670	14 083	*14*	584	90 000
S	S	Newport News (Tenneco)	USA	Sh	1 950	1 670	1 950	*100*	200	28 000
S	29	MBB (DASA)	FRG	Ac El Mi	1 840	1 990	3 336	*55*	16	24 194
28	30	Honeywell	USA	El Mi	1 700	1 800	6 059	*28*	604	65 300
29	31	Lucas Industries	UK	Ac	1 640	1 760	3 655	*45*	313	55 957
30	27	IBM	USA	El Oth	1 600	2 100	62 710	*3*	3 758	383 200
31	24	LTV	USA	Ac MV El	1 580	2 150	6 362	*25*	265	38 000
32	39	ITT	USA	El	1 580	1 390	20 054	*8*	922	119 000
33	34	EFIM	Italy	Ac MV El	1 510	1 520	3 608	*42*	– 18	38 000
34	36	Allied Signal	USA	Ac El Oth	1 500	1 500	12 021	*12*	528	107 100
35	37	Textron	USA	Ac Eng MV	1 400	1 500	7 440	*19*	259	58 000
36	33	Ordnance Factories	India	A SA/O Oth	1 330	1 590	1 387	*96*	. .	. .

37	40	INI	Spain	Ac A MV El Sh SA/O	1 290	1 290	16 118	*8*	693	149 910
38	35	FIAT	Italy	Eng	1 280	1 500	36 695	*3*	2 665	286 294
39	45	Kawasaki Heavy Industries	Japan	Ac Eng	1 270	1 170	5 892	*22*	90	16 833
40	41	SNECMA Groupe	France	Eng	1 260	1 270	3 371	*37*	46	25 702
41	47	GTE	USA	El	1 250	1 100	17 424	*7*	1 417	158 000
42	38	Rolls Royce	UK	Eng	1 220	1 410	4 857	*25*	382	46 000
43	43	E-Systems	USA	El	1 200	1 200	1 633	*73*	83	. .
44	44	Ford Motor	USA	Ac MV El Mi	1 100	1 200	96 932	*1*	3 835	366 641
45	51	Loral	USA	El	1 050	1 000	1 500	*70*	88	. .
S	S	Ingals Shipbuilding (Litton Industries)	USA	Sh	1 050	910	1 050	*100*	. .	14 000
46	53	Oerlikon-Bührle	Switzerl.	Ac A El SA/O	1 040	930	2 903	*36*	– 25	27 326
47	60	Israel Aircraft Industries	Israel	Ac El Mi	1 030	800	1 280	*80*	12	16 600
48	46	GIAT	France	A MV SA/O	1 020	1 150	1 056	*97*	– 107	14 200
S	S	Aeritalia (IRI)	Italy	Ac	1 020	880	1 860	*55*	41	14 903
49	54	Nobel Industries	Sweden	El Mi SA/O	950	910	3 556	*27*	214	22 246
50	59	Siemens	FRG	El	900	800	32 515	*3*	839	365 000
51	57	Gencorp	USA	Ac El Mi SA/O	900	880	1 938	*46*	210	15 100
52	52	FMC	USA	MV Sh Oth	900	950	3 414	*26*	136	24 000
53	48	Matra Groupe	France	Mi Oth	870	1 040	3 462	*25*	95	21 242
54	58	VSEL Consortium	UK	MV Sh	870	830	879	*99*	46	16 610
S	S	Bofors (Nobel Industries)	Sweden	A El Mi SA/O	870	870	898	*97*	. .	7 669
55	61	Mitsubishi Electric	Japan	El	810	790	17 308	*5*	407	47 607
56	49	Philips	Netherl.	El	800	1 010	26 984	*3*	648	304 800
57	55	Hercules	USA	Ac Mi SA/O Oth	800	890	3 121	*26*	– 81	23 290
58	50	Harris	USA	El	800	1 000	2 214	*36*	21	35 100
S	S	MTU (DASA)	FRG	Eng	780	970	1 946	*40*	7	17 654
S	S	Aerojet (Gencorp)	USA	Ac El Mi SA/O	750	800	1 034	*73*	92	. .

1	2	3	4	5	6	7	8	9	10	11
Rank					Arms sales					
1989	1988[b]	Company[c]	Country	Industry	1989	1988[d]	Total sales 1989	Col. 6 as % of col. 8	Profit 1989	Employment 1989
S	–	Telefunken Systemtechnik (DASA)	FRG	El	730	0	1 043	*70*	. .	10 779
S	–	Matra Défense (Matra)	France		710	0	710	*100*	. .	. .
59	62	Sequa	USA	Eng El Oth	700	700	1 959	*36*	56	17 700
60	73	Thiokol	USA	Eng Mi SA/O Oth	660	580	1 170	*56*	18	12 100
61	89	Hunting Associated Industries	UK	SA/O	650	440	1 228	*53*	80	. .
62	70	Motorola	USA	El	650	600	9 620	*7*	498	104 000
63	66	Diehl	FRG	A MV El SA/O	620	610	1 374	*45*	. .	15 300
S	S	Agusta (EFIM)	Italy	Ac	610	490	940	*65*	15	8 426
64	65	AT&T	USA	El	600	650	36 112	*2*	2 707	279 000
65	69	Daewoo	S. Korea	El Sh	600	600	19 981	*3*	115	91 056
S	S	Dornier (DASA)	FRG	Ac El Mi	590	570	1 172	*50*	. .	10 247
66	77	Smiths Industries	UK	El	590	530	1 156	*51*	183	13 606
S	S	Oto Melara (EFIM)	Italy	A MV Mi	580	530	580	*100*	8	2 329
67	80	Avondale Industries	USA	Sh	550	450	708	*78*	– 60	. .
68	76	Eidgenössische Rüstungsbetriebe	Switzerl.	Ac Eng A SA/O	550	550	583	*94*	. .	4 828
69	67	Thyssen	FRG	MV Sh	540	600	18 218	*3*	439	136 091
70	42	Thorn EMI	UK	El	540	1 200	6 094	*9*	521	65 444
71	74	SAAB-SCANIA	Sweden	Ac Eng	530	570	7 219	*7*	156	48 708
S	S	SNECMA (SNECMA Groupe)	France	Eng	530	770	2 108	*25*	13	13 727
72	79	FFV	Sweden	A El SA/O Oth	510	500	993	*51*	6	9 903
73	63	Rheinmetall	FRG	A SA/O	510	650	1 402	*36*	65	13 700
74	75	NEC	Japan	El	510	570	20 011	*3*	543	38 013
75	82	Computer Sciences	USA	El	510	480	1 304	*39*	52	18 300

S	S	Electronique Serge Dassault (Dassault-Breguet)	France	El	500	510	644	*78*	21	4 100
76	64	Toshiba	Japan	El Mi	500	650	22 187	*2*	702	69 643
S	S	CFM International (General Electric and SNECMA)	USA	Ac Eng	500	500	. .	. .	. .	. .
77	71	Teledyne	USA	Eng El Mi	500	600	4 664	*11*	259	43 200
78	81	Racal	UK	El	490	480	3 237	*15*	233	37 414
S	S	Selenia (IRI)	Italy	El	490	380	649	*76*	11	6 655
S	S	CASA (INI)	Spain	Ac	480	502	684	*70*	– 51	10 138
79	83	Olin	USA	Ac El SA/O Oth	480	470	2 530	*19*	124	7 129
80	72	Ferranti-International Signal	UK	El	470	580	1 304	*36*	71	12 700
81	68	Ishikawajima-Harima	Japan	Eng Sh	460	600	4 587	*10*	88	26 178
82	92	Emerson Electric	USA	El	450	400	7 071	*6*	588	72 600
S	S	AVCO (Textron)	USA	Ac	450	650	. .	. .	. .	. .
83	78	Krupp	FRG	MV El Sh Oth	440	510	9 406	*5*	– 240	63 557
84	80	Hindustan Aeronautics	India	Ac Mi	440	480	460	*96*	25	43 403
85	84	Sundstrand	USA	Ac	420	470	1 666	*25*	114	. .
86	95	SAGEM Groupe	France	El	410	350	1 628	*25*	36	16 459
87	87	Devonport Management	UK	Sh	410	450	418	*98*	. .	7 500
S	S	FIAT Aviazione (FIAT)	Italy	Ac Eng	410	660	684	*60*	8	4 651
88	90	Dowty Group	UK	Ac El	400	410	1 199	*33*	90	14 680
S	S	Bell-Boeing (Boeing)	USA	Ac	400	250	. .	. .	. .	. .
89	93	Harsco	USA	MV	400	400	1 351	*30*	– 64	. .
90	85	Westland Group	UK	Ac	390	450	708	*55*	28	. .
91	–	Renault Véhicules Industriels	France	Eng MV	390	340	27 457	*1*	1 457	174 573
S	–	Siemens Plessey Electronic Systems (Siemens)	UK	El	390	0	. .	. .	. .	. .
92	–	Morrison Knudsen	USA	MV Oth	380	300	2 170	*18*	32	. .
S	S	FIAT IVECO (FIAT)	Italy	MV	380	270	3 794	*10*	54	22 969
93	94	Ericsson	Sweden	El	360	390	6 250	*6*	327	69 229

1	2	3	4	5	6	7	8	9	10	11
Rank					Arms sales					
1989	1988[b]	Company[c]	Country	Industry	1989	1988[d]	Total sales 1989	Col. 6 as % of col. 8	Profit 1989	Employment 1989
94	–	Mannesmann	FRG	MV	360	. .	11 878	*3*	268	125 800
S	97	Krauss-Maffei (Mannesmann)	FRG	MV	360	380	718	*50*	3	5 320
95	95	Hawker Siddeley	UK	El	350	390	3 519	*10*	203	42 600
S	–	Sextant Avionique (Thomson-CSF/Aérospatiale)	France	El	350	0	943	*37*	27	9 800
96	–	Mitre	USA	El	350	350	. .	. .	. .	. .
97	–	Dyncorp	USA	Ac El	350	350	. .	. .	. .	. .
98	88	Armscor	S. Africa	A Ac MV El SA/O	340	440	688	*49*	. .	19 000
99	–	Oshkosh Truck	USA	MV	330	300	950	*35*	. .	. .
100	–	Ascom Holding	Switzerl.	El	330	330	1 620	*20*	31	15 351
S	S	Fincantieri (IRI)	Italy	Sh	330	310	1 337	*25*	– 185	20 767

. . Data not available.

– No rank designation for 1988.

[a] Both the rank designation and the arms sales figures for 1988 are also given, in columns 2 and 7, respectively, for comparison with the data for 1989 in columns 1 and 6.

[b] The rank designation in this column may not correspond to that given in table 8.2 in the *SIPRI Yearbook 1990*. A dash (–) in this column indicates either that the company did not produce arms in 1988, in which case there is a zero (0) in column 7, or that it did not rank among the 100 largest companies in table 8.2 in the *SIPRI Yearbook 1990*, in which case figures for arms sales in 1988 do appear in column 7.

[c] Company names in parentheses after the name of the ranked company are the names of the holding companies. The parent companies, with data pertaining to them, appear in their rank order for 1989.

[d] A zero (0) in this column indicates that the company did not produce arms in 1988, but began arms production in 1989, or that in 1988 the company did not exist as it was structured in 1989.

Note: The authors acknowledge the assistance of Bernard Harbor (London), Sami Faltas (Eindhoven), Ernst Gülcher (Antwerp), Peter Hug (Bern), Evamaria Loose-Weintraub (Stockholm), Arcadi Olivares i Boadella (Barcelona), Mario Pianta and Giulio Perani (Rome), Paul Rusman (Haarlem) and Werner Voß (Bremen) in the data collection. Three SIPRI interns also assisted in the preparation of the appendix: Lisa Moore, Ivo Sarges and David Wiley.

9. Ballistic missile proliferation

AARON KARP

I. Introduction

It is widely believed that Third World conflicts are replacing confrontation in Europe as the greatest threat to international security. The proliferation of ballistic missiles, along with chemical and nuclear weapons, is a sign that emerging regional threats must be taken seriously. Concern over Iraqi military capabilities is one manifestation of this view. This shift in international priorities facilitated the expansion and strengthening in 1990 of the most prominent instrument for controlling missile proliferation, the 1987 Missile Technology Control Regime (MTCR)—the goal of which is to restrict the export of missile technology.

The clearest sign of these changes came at the Washington summit meeting of 31 May–3 June 1990, where Presidents Mikhail Gorbachev and George Bush agreed to work together to 'oppose the proliferation of nuclear weapons, chemical weapons, missiles capable of carrying such weapons, and certain other missiles and missile technologies'. After years of uncertainty, the Soviet leader pledged his government's support for the provisions of the MTCR, and both agreed to support 'regional initiatives to reduce the threat of missile proliferation'.[1]

The Iraqi invasion of Kuwait on 2 August 1990 reinforced international recognition of the threat posed by weapon proliferation. Iraqi ballistic missiles came to the forefront of public concern when, from 18 January 1991, President Saddam Hussein used Scud missiles to attack Israel and the multinational forces in Saudi Arabia. His actions exposed the greatest problem unanswered by the MTCR, that of how to deal with ballistic missile proliferation after it becomes established fact.

This chapter assesses the state of ballistic missile proliferation in 1990. Starting with a review of Third World missile programmes, it examines the pace of progress in the countries of greatest concern. The chapter also examines the progress made in strengthening national and multilateral controls on missile proliferation and the effect they can have on Iraq and other countries.

Data describing emerging ballistic missile and related programmes as of 31 December 1990 are summarized in table 9.1. It includes only current programmes. Civilian sub-orbital sounding rockets (SRs) and space launch vehicle (SLV) programmes are also included because of their inherent

[1] 'Joint Statement on Nonproliferation', reprinted in *Arms Control Today*, vol. 20, no. 5 (June 1990), p. 26.

similarities to military rockets. Cruise missiles are not included owing to difficulties in finding comparable data.

II. Developments in ballistic missile capabilities

While the dangers of a missile war in the Middle East escalated, events in 1990 showed that ballistic missile proliferation is actually slowing down. It is premature to conclude whether this is a permanent decline or merely a pause due to unexpected difficulties. Ballistic missile programmes of at least six Third World countries made little or no progress in 1990 while three others stopped altogether. The Iraqi programme stands out not only because of the threat it poses, but also because it is one of the few continuing to make rapid progress.

The reasons for this slow-down are numerous. Economic constraints are especially widespread; development of long-range rockets has turned out to be more costly than many governments anticipated; and domestic economic problems have forced other governments to make stark choices regarding financial priorities. In some cases, improving regional political relations is most salient. In others, bilateral diplomacy has had a visible effect. The most important common factor, however, appears to be the MTCR, which has denied many rocket programmes in Third World countries the technology needed for further progress, regardless of the degree of their commitment or state of their finances. It has also contributed to the emergence of a new international normative standard, by which ownership of ballistic missiles is losing its acceptability.

Programmes making swift progress

Two countries made especially rapid strides in the development and deployment of ballistic missiles in 1990: Iraq and North Korea. Their programmes are insulated against external disruption through an established industrial–technical base, strong political support and financing. In addition, these countries are motivated, by exaggerated regional ambitions and fear of attack, to circumvent the MTCR through organized smuggling and purchases of dual-use technologies. They have also been able to find other governments willing to license sales of smaller missiles and related technology. Their most important successes are based on modification of the Soviet Scud B missile. More ambitious projects have apparently been stymied by international controls.

North Korea

Missile manufacture stands out against the technical weakness of the rest of North Korea's arms industry.[2] In 1987–88 North Korea demonstrated that it could manufacture Soviet-designed Scud missiles by selling 90–100 missiles

[2] Jacobs, G., 'North Korea's arms industry', *Asian Defence Journal*, no. 3 (Mar. 1989), pp. 28–35.

to Iran. North Korea did not receive any Scud missiles from the USSR. Essential assistance came from Egypt, which supplied a few Scud B missiles during the late 1970s. North Korean engineers later copied the Scud for local production.[3] The missile reportedly has an improved range and accuracy of some 10–20 per cent, with the assistance in aerodynamics provided by China and and Japanese electronics.[4]

A second ballistic missile, tentatively known as the Improved Scud (I-Scud) or Scud Product Improvement Programme (Scud-PIP), has been under development. Although based on Scud technology, this appears to be a substantially new missile, more sophisticated than the Iraqi Scud modifications. With a maximum range of 500–600 km, it can reach Japanese territory as well as Beijing and Vladivostok. According to preliminary assessments, the engine has been modified and the guidance package replaced. During 1990 it was test-fired from a facility at Nodong, 100 km north of Pyongyang. The I-Scud could be in limited operational service in 1991.[5]

The I-Scud threat is compounded by concomitant progress with chemical and nuclear weapons. North Korea has reportedly produced chemical warheads for its ballistic missiles and has established a nuclear reactor and a suspected plutonium separation plant at Yongbyon.[6]

Since 'dialogue' between North and South Korea began in the late 1980s there has been no actual negotiation, rather a ritualistic reiteration of basic positions. Pyongyang has made a precondition that US forces be evacuated as a confidence-building measure. Seoul insists that the Korean border must be opened first. During 1990 there were growing signs of flexibility, culminating in a meeting of the prime ministers in September.[7] However, greater progress is necessary before proliferation threats cease to matter.

Iraq

Since its existence was disclosed in 1987, the Iraqi ballistic missile programme has served as President Saddam Hussein's most prominent symbol of military power. Unlike Iraq's chemical and nuclear weapon programme, ballistic missiles were displayed and routinely paraded in Baghdad. During 1990 Saddam Hussein made a series of military threats raising the spectre of attack with chemically armed missiles.[8]

[3] Bermudez, J. S. and Carus, W. S., 'The North Korean "Scud B" programme', *Jane's Soviet Intelligence Review*, vol. 2, no. 4 (Apr. 1989), p. 180. The reverse-engineered Scud was test-fired in 1984. With Iranian financing, 8–10 missiles per month were produced by late 1987.

[4] Bermudez and Carus (note 3), pp. 177–81; Bermudez, J. S., 'New developments in North Korean missile programme', *Jane's Soviet Intelligence Review*, vol. 2, no. 8 (Aug. 1990), pp. 343–45.

[5] Bermudez (note 4); Gertz, B., 'North Korea's ballistic missiles have U.S. worried', *Washington Times*, 4 June 1990, p. 3.

[6] Bermudez, J. S., 'CW: North Korea's growing capabilities', *Jane's Defence Weekly*, 14 Jan. 1990, p. 54; 'Nuclear N. Korea?', *Washington Post*, 13 Apr. 1990, p. 20.

[7] Weisman, S. R., 'Seoul to rebuff disarmament call', *International Herald Tribune*, 6 Sep. 1990, p. 4; Weisman, S. R., 'Koreas end talks, hail new tolerance', *International Herald Tribune*, 7 Sep. 1990, p. 1.

[8] Cowell, A., 'Iraq chief, boasting of poison gas, warns of disaster if Israelis strike', *New York Times*, 3 Apr. 1990, p. 1.

Following Iraq's invasion of Kuwait in August 1990 and the multinational forces' counter-attack on 17 January 1991, these missiles became a direct threat as Saddam Hussein used them to escalate the war. Ballistic missiles are Iraq's most valued striking force, as was seen during the 1988 War of the Cities in the war with Iran. The weapons used then continue to form the backbone of Iraq's missile force. According to Israeli sources, in 1990 Iraq had 200–500 Scud B missiles and various other Scud versions. Other estimates go as high as 1000.[9] Unlike North Korea, Iraq cannot manufacture Scud-type missiles but has been successful in modifying them for improved performance. Most of the improved Scud versions are believed to be the 600-km range al-Hussein version; this range was achieved by reducing the weight of the warhead. Although Iraq has chemical weapons, the consensus among outside observers is that Iraqi Scud versions are armed exclusively with conventional high explosives. Iraq has manufactured an estimated 36 mobile launchers for its al-Hussein missiles and has built a similar number of fixed launchers facing Israel. These were the missiles fired at Israel and Saudi Arabia starting on 18 January 1991.

A second Scud version, the 900-km range al-Abbas, was announced in 1988. This missile appears to have been more demanding for Iraq, and its operational status remains unclear. On 7 January 1990, 12 of the missiles were displayed in a Baghdad military parade, although to some observers these looked like crude mock-ups. Later, Western intelligence sources identified al-Abbas missiles deployed at three Iraqi launch sites. The missiles test-fired on 2 December 1990 also appear to be the al-Abbas.[10]

Iraq has announced many other ballistic missile projects, but their status is even more obscure. The best known of these is the Condor 2 project, in collaboration with Argentina and Egypt. By the beginning of 1990 it appeared that Iraq had been forced to downgrade the priority of the project. Egypt and Argentina had withdrawn, and the numerous Iraqi facilities were unable to continue full-scale development after Western technology and assistance were cut off.[11]

Less is known about the Tammuz. It is most likely that it is the surface-to-surface version of the al-Abed, a three-stage liquid-fuelled space launch vehicle, probably constructed by clustering five Scud missiles. Nothing has been heard of the al-Abed since its first launch in December 1989, although plans to launch two satellites into orbit were repeatedly announced during 1990.[12] Of makeshift design, it is poorly suited for military applications.

[9] Gertz, B., 'Baghdad deploying missiles', *Washington Times*, 29 Mar. 1990, p. 1; Gertz, B., 'Iraqis deploying special launching system', *Washington Times*, 13 Sep. 1990, p. 9; 'Iraq moves Fulcrums forward', *Flight International*, no. 4231, vol. 138 (29 Aug.–4 Sep. 1990), p. 6. The USSR must know how many missiles it has supplied but has kept this information secret for unknown reasons.

[10] Berger, C., 'Iraq puts its "new" missiles on show', *The Independent*, 8 Jan. 1990, p. 9; Gertz (note 10); *International Herald Tribune*, 22–23 Dec. 1990, p. 1.

[11] Slade, A., 'Condor project "in disarray"', *Jane's Defence Weekly*, vol. 13, no. 7 (17 Feb. 1990), p. 295.

[12] Discussions with Gary Milhollin, Washington, DC, Mar. 1990; Furniss, T., 'Iraq plans to launch two science satellites', *Flight International*, vol. 137, no. 4204 (21–27 Feb. 1990), p. 20; 'Satellite launch planned', *Jane's Defence Weekly*, vol. 14, no. 3 (21 July 1990), p. 74.

The Soviet decision to adhere to the UN embargo on trade with Iraq cut off supplies of additional Scud missiles. In anticipation of this problem, Iraq has been working on domestic replacements for the Scud, the 250- to 500-km range Fahd, and the smaller Soviet FROG-7, called the Latif, with a range of 90 km. After the shooting deaths of 21 Palestinian demonstrators in Jerusalem on 5 November 1990, Saddam Hussein announced the existence of a new missile dedicated to their memory, the al-Hijara. Nothing is known about this missile, which probably is a renaming of the al-Abbas.[13]

In addition to basing ballistic missiles in Kuwait, there has been growing evidence that Iraq has enlisted the assistance of sympathetic Muslim governments further afield. In April 1990 US reconnaissance satellites detected construction of facilities described as fixed missile launchers in Mauritania on the Atlantic coast of Africa. According to US officials, Iraq negotiated the facility to establish an unrestricted test range. The accusations were reiterated by President Abdou Diouf of Senegal. Mauritanian officials denied the allegations.[14] One month after the Iraqi invasion of Kuwait, Egyptian sources reported that Iraq had moved a number of Scud missiles to Sudan. This report was denied by Sudanese President Omar Hassan Ahmed al-Bashir.[15]

Until the late 1980s Iraq was able to acquire virtually all the missile technology it required through legal means.[16] The list of countries contributing to Iraq's ballistic missile programme in one capacity or another is long. However, after the War of the Cities and the gas attack on Halabja, Iraq faced mounting difficulties in acquiring additional ballistic missile technology.

The available evidence suggests that since then Iraq has been able to acquire illegally much of the foreign technology it has sought. The extent of this smuggling operation was indicated by the arrest in Britain of Iraqi diplomats and business people in March 1990 for attempting to ship US-made triggers for nuclear weapons.[17] The arrests presaged reports, such as that by German Economics Minister Helmut Haussmann, showing that the extent of European participation in such projects—of which ballistic missiles were but a part—was much greater than previously acknowledged. Many of the more recent contracts appeared illegal, and several dozen firms are under investigation.[18]

Another aspect of Iraqi smuggling was illustrated through the 'supergun' affair. Plans for the supergun, also known as Project Babylon, emerged from

[13] 'Irakische Raketenprogramme mit deutscher Unterstützung', *Neue Zürcher Zeitung*, 10 May 1990, p. 3; White, D., 'Saddam's "new" missile puzzles experts', *Financial Times*, 10 Oct. 1990, p. 7.

[14] Gordon, M. R., 'U.S. fears Iraq is seeking long-range missile site', *New York Times*, 24 Apr. 1990, p. A13; Gertz, B., 'Photos show Iraqi missile launchers in Mauritania', *Washington Times*, 30 May 1990, p. 3; 'Senegal accuses Mauritania of hosting Iraqi missile test site', *Defense and Foreign Affairs Weekly*, 28 May–2 June 1990.

[15] 'Baghdad Scud-B missiles "in Sudan"', *Financial Times*, 4 Sep. 1990, p. 3; 'Sudan says it has no Scuds', *Financial Times*, 5 Sep. 1990, p. 2; 'Iraqi missiles in Sudan', *Mednews*, 15 Oct. 1990, p. 5.

[16] Smith, R. J. and Weiser, B., 'Commerce Dept. urged sale to Iraq', *Washington Post*, 13 Sep. 1990, p. 1.

[17] Kirby, T. *et al.*, 'Iraqis caught smuggling nuclear devices from UK', *The Independent*, 29 Mar. 1990, p. 1.

[18] Kempe, F., 'Germans had big role in helping Iraq arm, internal report shows', *Wall Street Journal*, 2 Oct. 1990, p. 1; 'Raketentechnik nach Iraq', *Frankfurter Rundschau*, 27 Sep. 1990, p. 4.

meetings in Iraq between Gerard Bull of the Space Research Corporation (SRC) and Saddam Hussein in early 1988. This envisioned an enormous gun tube to maximize the energy of rockets, much like an infantry mortar. It called for two 150-metre long tubes with a 1-metre bore. The weapons would fire solid-fuelled rockets weighing about 1500 kg to distances of almost 2000 km.[19] Components for the superguns were seized by British customs officials in April, and the SRC ceased operations shortly thereafter.[20]

Programmes making consistent progress

Three regional powers—India, Israel and Saudi Arabia—have been able to maintain the pace of their ballistic missile and space launch endeavours. These countries are unable to acquire foreign rocket technology today except under special circumstances, but their programmes are distinguished from those in weaker countries by four factors: (*a*) they have accumulated substantial amounts of foreign technology in the past, when it was easily available; (*b*) they have the industrial infrastructure to develop this into an independent technical foundation; (*c*) they have powerful and nationally persuasive reasons for their long-range rocket programmes; and (*d*) they have been able to raise the financial resources needed to keep their efforts going. For these countries, long-range rocket programmes are more robust because they are insulated from external disruption in technology, industry, politics and finances.

India

The vicissitudes of domestic politics have not reduced India's commitment to long-range rocket development. During its 11-month tenure, the Government of V. P. Singh did nothing to alter the Integrated Guided Missile Development Programme (IGMDP) initiated in 1983. Politically, the programme saw its strongest advocates gain in power through the appointment in the summer of 1990 of its director, A. P. J. Abdul Kalam, to the position of Scientific Advisor to the Defence Minister. The appointment of Raja Ramanna as Minister of State for Defence and P. K. Iyengar as chairman of the Atomic Energy Commission brought the two chief architects of India's 1974 nuclear explosion to similarly powerful positions.[21] The collapse of the Singh Government in November 1990 seemed unlikely to change missile policy. The

[19] 'In Iraq soll bereits ein Prototyp der Super-Kanone explodiert sein', *Frankfurter Rundshau,* 23 Apr. 1990, p. 1; 'Iraq: heir to HARP project?', *Jane's Defence Weekly*, vol. 13, no. 17 (28 Apr. 1990), p. 770; 'Supergun details', *Milavnews*, no. 344 (June 1990), p. 18.

[20] James, B., 'After executive is slain, Brussels corporation denies illegal arms sales', *International Herald Tribune,* 27 Mar. 1990, p. 2; Reeves, P., 'Belgian-based parts buyer is dissolved', *The Independent,* 20 Apr. 1990.

[21] 'India enters the missile age', *Amrita Bazar Patrika* (Calcutta), 11 Sep. 1990; Malhotra, I., 'The nuclear option', *Sunday* (Calcutta), 18 Feb. 1990; Srinivasan, S., 'Will the Buddha smile again?', *The Telegraph* (Calcutta), 28 Jan. 1990.

new Government, led by Chandra Shekhar, is ideologically committed to technological self-reliance.[22]

India has four major ballistic missile and space launch projects: the Prithvi, the Agni, the Augmented Satellite Launch Vehicle (ASLV) and the Polar Satellite Launch Vehicle (PSLV). The most successful to date is the most modest, the 240-km range Prithvi, comparable to the US Lance missile. After test launches in 1988 and 1989 the Prithvi is expected to enter series production by Bharat Dynamics in 1991 at a unit cost of 17 million rupees ($1.0 million) and become operational with the Indian Army in 1992–93. The role of foreign technology in the Prithvi is unclear. According to one report it is 15 per cent by value and is expected to decline to 5 per cent in series production. Reports that the Prithvi has a circular error probable (CEP) of 250 metres at maximum range and can carry a payload of 1 tonne are questionable because the CEP cannot be assessed after just two test launches.[23]

The Agni, a 2400-km range 'technology demonstrator' first launched in May 1989, experienced undisclosed set-backs that prevented a second test launch repeatedly scheduled during 1990.[24] The missile is an impressive technological accomplishment in which the nation takes great pride, but the Agni has been developed as a single effort without specific long-term goals. In a series of interviews during 1990, Abdul Kalam argued that India could do much more, noting that it 'has the capability to develop an ICBM, if necessary. It is still to be decided whether we need it'. At the moment other elements of the guided missile programme appear to be receiving more emphasis, including anti-tank and anti-aircraft missiles.[25]

India faces less pressure to develop the Agni into a functional ballistic missile because it has a civilian programme applying similar technology. This overlap minimizes the danger that failure to develop the Agni will lead to the disintegration of its technical base. Of the two space launchers under development, the ASLV is most relevant to military applications. Like the Agni it is derived from the earlier SLV-3, with four solid-fuelled stages. Two launch attempts in 1987 and 1988 were failures, and a third is planned for 1991.[26]

Competing with the ASLV is the PSLV, with a liquid-fuelled second stage, based on the Viking engine developed by France for the Ariane rocket. The PSLV is larger than previous Indian rockets, designed to lift payloads into

[22] The new Prime Minister also took the defence portfolio himself. He depends on the support of the Congress (I) Party, which presided over the birth of India's rocket programmes. Hazarika, S., 'India leader keeps key posts in his new cabinet', *International Herald Tribune*, 22 Nov. 1990, p. 3.

[23] Banerjie, I., 'Aiming for the sky', *Sunday* (Calcutta), 22 Oct. 1989; 'India', *Milavnews*, no. 341 (Mar. 1990), pp. 16–17; 'Bharat Dynamics', *Jane's Defence Weekly*, vol. 13, no. 21 (26 May 1990), p. 1039. In Western and Soviet experience, production follows a much more thorough test programme: over 150 test launches for the Lance. The Lance is described in Cochran, T. B. *et al.*, *Nuclear Weapons Databook, Vol. 1: U.S. Nuclear Forces and Capabilities* (Ballinger: New York, 1984), pp. 284–86.

[24] 'Clearance awaited for second Agni trial', *The Hindu* (New Delhi), 17 Feb. 1990; 'India to test fire IRBM again in September', *Defense and Foreign Affairs Weekly*, 2–8 July 1990.

[25] 'India could develop ICBMs', *Defence*, vol. 21, no. 1 (Jan. 1990), p. 12; 'India's ICBM', *The Patriot* (New Delhi), 12 Jan. 1990, p. 4; 'India can make long-range missiles,' *Indian Express* (New Delhi), 27 Aug. 1990.

[26] 'India to attempt 1991 ASLV launch', *Flight International*, vol. 4209, no. 137 (28 Mar.–3 Apr. 1990), p. 25.

distant orbits. According to the chairman of the Indian Space Research Organization (ISRO), U. R. Rao, the teams assembling the two rockets are in 'a race' to launch first in 1991. Plans for an even larger Geostationary SLV have become more coherent, but uncertainty about the ability to develop adequate engines has delayed the final decision.[27]

Basic decisions about the future of the programmes have not been made. Final development of reliable rockets will require concentration of financial and human resources into a single programme. The strategic thinking that could guide such decision-making matured in 1990, led by the Institute for Defence Studies and Analyses (IDSA) in New Delhi. IDSA director Jasjit Singh has argued for the establishment within the Indian Air Force of a 'Strategic Air Command'. Another analyst extended this argument to advocate acquisition of a mixed force of nuclear armed intermediate-range ballistic missiles and cruise missiles, primarily to reduce Chinese bargaining influence in South Asia.[28]

Israel

The Jericho 2, Israel's most advanced ballistic missile, has become even more shrouded in secrecy than the Israeli nuclear weapon programme. The four or five test launches so far are inadequate for confident production and reliable deployment. No further test launches were reported in 1990. Although assertions that 50–100 of the 1450-km range missiles are deployed are commonplace, there has been no confirmation.

More information has been made public about the Shavit space launcher, apparently based on the Jericho 2. The first Shavit was launched in September 1988 to place in orbit the Ofeq-1 satellite. A second Shavit was launched on 3 April 1990 from a mobile launcher near Palmachim beach 20 km south of Tel Aviv, placing in orbit the 160-kg Ofeq-2 satellite. Director of the Israeli Space Agency and cabinet minister Yuval Neeman said two more Ofeq launches are planned. He emphasized that they 'are not designed for military purposes', but other officials interpreted the launch as a warning of Israel's military capabilities, following President Saddam Hussein's threat to 'burn half of Israel'.[29]

On 9 August 1990 Israeli Aircraft Industries launched the first of four Arrow anti-tactical ballistic missile (ATBM) interceptors planned under a development programme which the USA started in 1988. A second Arrow was launched on 21 December. The Arrow is a 12-metre, conventionally armed missile capable of interceptions at ranges up to 90 km, sufficient to stop

27 'Work progressing on PSLV stages', *The Hindu* (New Delhi), 20 Apr. 1990; 'INSAT-II satellite getting ready for launch', *The Hindu*, 21 Aug. 1990.

28 Panwar, S., 'The shape of India's emerging missile shield', *Strategic Analysis,* vol. 13, no. 3 (June 1990); Singh, J., 'IAF needs a Strategic Air Command', *The Hindustan Times* (New Delhi), 15 June 1989; and Subrahmanyam, K., 'India enters missile age', *Times of India* (New Delhi), 13 July 1990.

29 Labour party leader Shimon Peres said that the launch showed the Iraqi President 'that if he wants to deal with Israel he should look to other means than military': Brinkley, J., 'Israel puts a satellite in orbit a day after threat by Iraqis', 4 Apr. 1990, p. 6; 'Two more Offeq launches planned', *Flight International*, vol. 4212, no. 137 (18–24 Apr. 1990), p. 22.

missiles coming from distances as great as 1000 km.[30] US Secretary of Defense Richard Cheney announced that Washington will offer another $200 million for the Arrow through 1993 (see also chapter 3). The total costs of development and small-scale deployment in 1995 have been estimated at $2 billion, but support for the programme has grown along with fear of Iraq.[31] In the meanwhile, Israel is acquiring Patriot missile batteries with limited ATBM capabilities from the USA as a stopgap.[32]

Saudi Arabia

Saudi Arabia's only ballistic missiles are Chinese-made DF-3 intermediate-range ballistic missiles (IRBMs) received in 1987–88. The 2700 km-range missiles are the most powerful ever transferred from one country to another, with the exception of the US Trident missiles to be sold to the UK in the early 1990s. Saudi Arabia reportedly encountered serious problems bringing the missiles to operational status. According to Israeli disclosures, the missiles did not become operational until May–June 1990. Contrary to earlier reports of 50–60 missiles, it is now understood that Saudi Arabia possesses approximately 120, twice the number deployed by China itself. Two launch sites have been built, at al-Joffer 100 km south of Riyad and at al-Sulaiyil 600 km south. Each site is said to be equipped with four to six unprotected launch pads.[33]

Saudi Arabia has signed the Non-Proliferation Treaty (NPT) and also declared that it will not arm the missiles with chemical or biological weapons. Yet so many large rockets could be militarily significant. Armed exclusively with conventional warheads with 2000 kg of explosive, the force is capable of delivering 240 tonnes, six times as much explosive as Iraqi missiles dropped on Teheran during the 1988 War of the Cities.[34]

Sporadic reports suggest that Saudi Arabia is starting an indigenous rocket development programme. A report in February 1990 revealed that German engineers had been recruited for this undertaking. After extensive debate, the USA agreed to permit Saudi Arabia to purchase a Cray-2 supercomputer. Intended for the Aramco oil company, concerns have been expressed that the computer could be used to facilitate missile development. It appears that the Saudi Arabian missile programme is working primarily on small tactical missiles, especially anti-tank weapons. However, improved relations with China,

[30] 'Arrow missile makes maiden flight', *Armed Forces Journal*, Sep. 1990; 'Chetz data revealed', *Jane's Defence Weekly*, vol. 14, no. 8 (25 Aug. 1990), p. 255; 'Israel successfully launches first anti-ballistic missile', *Aviation Week & Space Technology*, 13 Aug. 1990, p. 23.

[31] Amouyal, B., 'Cheney backs advanced development for Arrow interceptor', *Defense News*, 23 July 1990, p. 8; 'Chetz costs to reach $2b', *Jane's Defence Weekly*, vol. 14, no. 12 (22 Sep. 1990), p. 509.

[32] 'Patriots added to Israeli forces', *Flight International*, vol. 138, no. 4238 (17–23 Oct. 1990), p. 13.

[33] 'Saudi CSS-2 missiles now operational', *Flight International*, vol. 137, no. 4219 (6–12 June 1990), pp. 12–13; 'Saudi CSS-2s are "wild card" in Gulf', *Flight International*, vol. 138, no. 4230 (22–28 Aug. 1990), p. 8; 'Saudi missile setup reported complete', *Washington Times*, 7 June 1990, p. 2.

[34] On the War of the Cities, see McNaugher, T. L., 'Ballistic missiles and chemical weapons: the legacy of the Iran–Iraq War', *International Security*, vol. 15, no. 2 (fall 1990), p. 22. The capabilities of the DF-3 are given in Wade, M., 'The Chinese ballistic missile programme', *International Defence Review*, no. 8 (Aug. 1980), p. 1191.

culminating in an agreement in July 1990 to establish formal diplomatic relations, could provide access to technologies for larger missiles.[35]

Programmes in abeyance

No technological endeavour can be said to have stopped permanently after just one or two years. Such programmes should with greater fairness be said to be moribund. The countries examined here only recently stopped or significantly slowed their ballistic missile programmes. In some other countries—notably Indonesia, Iran, Libya, South Korea, Syria and Taiwan—ballistic missile and long-range rocket development has been in abeyance for a longer period, stymied by international controls and their own technological weakness.[36]

Afghanistan

The Soviet Union began to transfer Scud B ballistic missiles to the Kabul Government in November 1988. By the end of 1989 over 1000 Scud missiles had been fired by the Afghan Army, although reports suggested that the rate of missile attacks had slowed significantly in the last three months of that year.[37] The last major Scud attacks announced by the Mujahideen came in March 1990 during a campaign to quell a revolt in the city of Khost.[38]

The declining use of surface-to-surface missiles corresponds to various factors in Soviet and Afghan government policy. Most important among these were greater efforts in 1990 to reach a settlement with the Mujahideen and Soviet initiatives to improve relations with Pakistan.[39] In addition, the bombardment used a substantial proportion of Moscow's Scud inventory, previously estimated at a total of 2500 missiles. These have assumed a more important role in Soviet military planning and as a potential negotiating tool after the 1987 INF Treaty eliminated larger tactical missiles. Finally, the meagre results of the missile attacks could hardly justify the continued use of weapons costing approximately $1.5 million each.[40] Whether Soviet acceptance of MTCR regulations will halt exports of the Scud—on the borderline of the agreement's thresholds—remains to be seen.

[35] Carnegy, H., 'Saudi purchase of supercomputer angers Israel', *Financial Times*, 17 May 1990, p. 6; 'Saudis get Cray-2', *Mednews*, 28 May 1990, p. 1; 'Bonn bestätigt Kontakte der Raketen-experten mit Saudi', *Frankfurter Rundshau*, 14 Feb. 1990, p. 4.

[36] In these countries little or nothing is changed from the survey in the *SIPRI Yearbook 1989: World Armaments and Disarmament* (Oxford University Press: Oxford, 1989), chap. 9.

[37] 'Red hot news', *Armed Forces Journal International*, vol. 127, no. 8 (Mar. 1990), p. 32; 'Ein neuer sowjetischer Raketentyp für Kabul?', *Neue Zürcher Zeitung*, 23 Sep. 1989, p. 5.

[38] 'Coup attempt fails', *Milavnews*, no. 342 (Apr. 1990), p. 1.

[39] Krauss, C., 'U.S. hopes rebel drive spurs a pact for Afghans', *International Herald Tribune*, 17 Oct. 1990, p. 2.

[40] The total number of Scud-Bs is estimated in Cochran *et al.* (note 23), *Vol IV: Soviet Nuclear Weapons*, p. 221. The Scud-B cost estimate is the author's.

Argentina

Argentine development of the Condor 2 began in 1984 with massive European assistance. Although initial development work was concentrated in Argentina, the missile was widely believed to be intended for the commercial market, especially for Iraq, Egypt and other prospective clients in the Middle East.[41] On being disclosed in the winter of 1987–88, the Condor 2 programme, more than any other individual project, came to symbolize the problem of missile proliferation and growing international anarchy. A meeting of the governments adhering to the MTCR in October 1988 stressed the need to isolate the project. The government of President Raul Alfonsín defended Air Force claims that it was a civilian space launch vehicle.

His successor Carlos Menem demonstrated greater flexibility on this—and other issues—than many observers had expected.[42] On 22 April 1990, Defence Minister Humberto Romero announced that the Condor 2 had been cancelled and reiterated the claim that the Condor 2 was a satellite launcher for peaceful purposes. According to Foreign Minister Domingo Cavallo, the Government was searching for ways to use the Condor missile team in an unspecified international venture to preserve the country's advanced technological capabilities.[43]

The official statement calls attention to some reasons for the cancellation. Several deserve special attention. First, the project had reached a technological juncture at which further progress would significantly increase costs.[44] Second, foreign sources of funding were drying up. Egypt had withdrawn, and Iraq now had the infrastructure to work on ballistic missiles. Argentina would have had to develop the missile itself, a task requiring billions of dollars, which it did not have. Third, international pressure through the MTCR made it increasingly difficult to acquire the foreign technology and assistance it required. Without new technology suppliers, the project ended. From this perspective, the official declaration in April only acknowledged reality. A fourth reason was the lure of a contract with the US Air Force that could lead to the sale of 500 trainer aircraft worth over $3 billion.[45] Thus, it is alleged that the Condor 2 was dropped to appease Washington.[46]

However, the most important factor may have been the evaporation of security motives. Traditional rivalry with Brazil had yielded to a strong trade rela-

[41] Robert Windrem has prepared an exceptionally thorough history in, 'The Condor II', unpublished manuscript.

[42] Mead, G., 'Peronism to take a back seat to privatization drive', *Financial Times*, 30 Mar. 1990, p. 4; Robinson, E., 'From airlines to zoos, Menem is busy selling off Argentina', *International Herald Tribune*, 25 July 1990, p. 11.

[43] 'Argentinien stoppt Entwicklung der Condor-2-Rakete', *Frankfurter Allgemeine Zeitung*, 27 Apr. 1990, p. 8; 'Argentina formally suspends Condor missile project', *Defense and Foreign Affairs Weekly* (30 Apr.–6 May 1990), p. 6; 'Condor project ended', *Milavnews*, no. 343 (May 1990), p. 1; Graham, R., 'Argentina to seek venture for Condor missile team', *Financial Times*, 10 Apr. 1990, p. 9.

[44] Slade (note 11).

[45] Sweetman, B. and Salvy, R., 'Line-up for new US primary trainer', *International Defence Review*, vol. 23, no. 3 (Mar. 1990), pp. 295–98.

[46] 'Air Force chief comments on missile project', DYN, in JPRS-TND, 26 Oct. 1989, pp. 21–22; 'Argentina boosts chances of selling jet to US', *Financial Times*, 4 Apr. 1990.

tionship, including co-operation on aircraft and nuclear programmes. In 1990 substantial progress was being made towards regional integration. Even the 1982 Falklands/Malvinas conflict between Argentina and the UK is still being defused through mutual concessions.[47] Without clear security motives or a high military priority to justify the Condor 2, the Argentine Government and armed forces were unwilling to bear the rising costs of a project facing growing international isolation and condemnation.

However, Argentine interest in long-range rocketry has endured the vicissitudes of domestic and international politics for almost 30 years. Given new financial resources and access to foreign technology, the Condor 2 could be re-initiated within the next few years.

Brazil

In contrast to Argentina, there is no evidence that Brazil has abandoned its work on long-range rockets. Rather, support for the numerous ongoing projects has eroded, and spending has declined to the point where further progress comes slowly, if at all. The largest is the VLS space launch vehicle, a four-stage, liquid-fuelled rocket under development for over a decade. This has been joined by parallel programmes to develop tactical-range ballistic missiles, of which the most prominent are the EE-150 solid-fuelled missile being developed by the firm Orbita and the liquid-fuelled SS-300 under development by Avibras.[48]

Economic difficulties have slowed all Brazilian rocket projects. Although subsystems have been ground tested and smaller developmental rockets have been launched, no complete rocket has been flight tested. The VLS, for example, was intended to place its first satellite in orbit in 1989. More recently officials of the National Institute of Space Research (INPE) have spoken of a flight in 1992, although this seems optimistic, and they are reported to favour continued reliance on foreign launch services through the foreseeable future.[49]

The civilian programme suffered from budget cuts throughout the 1980s that limited planned VLS appropriations of $600 million to $170 million, and a further budget cut of 55 per cent was made in 1989.[50] The ballistic missile projects received only start-up funding from industry and the armed forces. The projects needed central government funding and loan guarantees or generous foreign clients in order to be completed. The rocket firm Avibras declared bankruptcy on 5 January 1990, partially due to Iraq's refusal to pay for artillery rockets ordered during its war with Iran. Its labour force of 6000 was trimmed to 900. Orbita has suffered from the declining fortunes of its two

[47] Barham, J., 'Argentina and Brazil in further integration step', *Financial Times,* 6 Sep. 1990, p. 6; 'Collor and Menem pragmatic on common market', *Financial Times,* 21 Sep. 1990, p. 3; Crawford, L., 'Chile and Argentina sign framework for trade pact', *Financial Times,* 30 Aug. 1990, p. 5.

[48] An exceptional review of Brazil's civil rocket programme is Lopes, R., 'Avibras, PRC to sell satellite launchers', *Folho de São Paulo,* 19 Apr. 1989, in FBIS-LAT, 25 May 1989, pp. 30–34.

[49] 'Obstacles to VLS development reviewed', *Folha de São Paulo,* 14 July 1989, in JRPS-TND, 14 Aug. 1989, pp. 20–22.

[50] Eustaquio de Freitas, J., 'Rocket-launching base prepared in Alcantara', *O Globo* (Rio), 11 Sep. 1989, in JRPS-TND, 6 Oct. 1989, p. 12.

parent companies, Engesa and Embraer; the former declared bankruptcy, and the latter is undergoing restructuring after sales declined during the late 1980s.[51] Unable to make sufficient progress in Brazil, a group of Brazilian missile experts put their services at the disposal of Iraq. Like the collapse of its prestigious military industries, this mission has caused the Brazilian Government considerable embarrassment.[52]

The chief of VLS development, Jayme Boscov, maintains that his programme has been harmed by the US-led technology boycott.[53] US protests, for example, led France to renegotiate a satellite sale that originally offered Ariane rocket engine technology as an offset. However, after extensive debate, the United States agreed to permit Brazil to purchase supercomputers potentially useful in the design of long-range rockets and nuclear weapons after Brazil offered assurances. In another case, Brazil shipped seven steel casings for its VLS rocket to a Chicago firm for annealing treatments. This 'import' required no US licensing, but exporting the treated casings back to Brazil did. Since the casings were Brazilian Government property, the US State Department argued there was no alternative but to permit their return. Interpreted by some as a retreat from the principles of the MTCR, although too small to seriously damage it, this suggests that Brazil may be able to acquire more assistance on the VLS than had seemed possible.[54]

Like Argentina, Brazil faces no credible threat to its military security. Work on nuclear weapons was stopped by the Government of President Fernando Collor de Mello in a highly publicized gesture in September 1990.[55] Brazil refuses to sign the NPT, but discussions have raised hope that it will ratify the Treaty of Tlatelolco to establish a Latin American nuclear weapon-free zone. This would greatly reduce the likelihood of Brazilian ballistic missiles and their danger if they were deployed, although Brazil could still pursue ballistic missiles for commercial reasons.

Egypt

Egypt's withdrawal from the Condor 2 programme in the summer of 1989—the outcome of domestic politics and Iraq's refusal to pay its arms bills—

[51] 'Brazilian firm files for bankruptcy', *Defense News,* 15 Jan. 1990, p. 26; Brooke, J., 'Peace unhealthy for Brazilian arms industry', *New York Times,* 25 Feb. 1990, p. 19; Foss, C. F., 'Engesa bankruptcy filing', *Jane's Defence Weekly,* vol. 13, no. 13 (31 Mar. 1990), p. 573; 'Brazil plane firm Embraer to cut jobs, delay projects', *Wall Street Journal* (European edn), 2 Nov. 1990.

[52] 'Brazilians probe links with Iraq', *Jane's Defence Weekly,* vol. 13, no. 21 (26 May 1990), p. 989; Fisher, S., 'Brazilia embarrassed by military links', *Financial Times,* 18 Sep. 1990, p. 2.

[53] 'Obstacles to VLS development', *Folha de São Paulo,* p. 20; Monteiro, J., 'U.S. said blocking VLS development', *Correio Braziliense,* 6 Aug. 1989, in FBIS-LAT, 4 Oct. 1989, p. 37.

[54] Friedman, A., 'Iraq may gain from US exports to Brazil', *Financial Times,* 8–9 Sep. 1990, p. 2; Wines, M., 'U.S. approves export of rocket parts to Brazil despite fears to link to Iraq', *New York Times,* 7 Sep. 1990, p. 8; Wines, M., 'Supercomputers backed for Brazil', *New York Times,* 19 Oct. 1990, p. 7.

[55] Brooke, J., 'Brazil roots out secret A-bomb project', *International Herald Tribune,* 10 Oct. 1990, p. 1; Fisher, S., 'Brazil lowers its nuclear sights', *Financial Times,* 20 Sep. 1990, p. 4; 'Brasilianisches Atomtestgelände geschlossen', *Neue Zürcher Zeitung,* 21 Sep. 1990, p. 2.

became even firmer following Iraq's invasion of Kuwait.[56] After joining Iraq in a variety of weapons projects, of which the Condor 2 was the largest, Egypt ended all possibilities of continuing by condemning the Iraqi attack. President Hosni Mubarak's leadership of Arab opposition to Iraq and co-operation in the establishment of forces in Saudi Arabia came at considerable cost for Egypt. Iraqi financing and assistance have enabled Egypt to expand and modernize virtually all of its military industries, which now face an uncertain future.

Egypt retains a network of facilities originally intended to manufacture the Condor 2. Closer reliance on the USA would seem to rule out the revival of such an undertaking. The Egyptian military relies on US assistance worth $1.8 billion annually and on sales of advanced military equipment, including F-16 fighters and co-production of the M-1 tank. Washington's cancellation of the Egyptian military debt of $6.7 billion can only strengthen US influence.[57] Technologies supplied by France for smaller missile projects could be relevant if work on ballistic missiles were re-initiated.

Less is known of the state of Egyptian co-operation with North Korea to improve its Scud B force. In the late 1970s or early 1980s Egypt apparently supplied North Korea with a few of its Scud missiles from the Soviet Union. These served as the basis for the North Korean developments described above.[58] The possibility remains that Egypt could acquire additional Scud missiles or improved Scud versions developed by North Korea. Nor can the possibility be discounted that Egypt has a programme of its own to reverse-engineer Scud missiles or for local production. The prosecution of Abdelkader Helmy in 1988 for conspiracy to smuggle missile components out of the USA referred in court documents to the 'Scud-100', an upgrade programme of which nothing else is known.[59] Although Egypt is party to the NPT, press reports indicate that it is negotiating with Argentina over acquisition of a research reactor which could be used to supply a nuclear weapon programme.[60]

Greater insight into Egyptian policy emerged in a letter sent by Foreign Minister Esmat Abdel Meguid to UN Secretary-General Javier Perez de Cuellar calling for elimination of weapons of destruction in the Middle East: 'all weapons of mass destruction, nuclear, chemical and biological, should be prohibited in the Middle East; all nations of the region should meet equal and reciprocal commitments in this regard; and verification measures should be established to ascertain full compliance of all states'.[61]

[56] Ottaway, D. B., 'Egypt drops out of missile project', *Washington Post*, 20 Sep. 1989, p. 32. Iraq owes Egypt over $2.5 billion for various military contracts. See Darwish, A., 'Iraqis "recruited Britons for missile project"', *The Independent*, 3 Mar. 1990, p. 10.

[57] Deputy Secretary of Defense Lawrence Eagleburger, 'Proposal to forgive Egypt's foreign military sales debt', *US Department of State Dispatch*, vol. 1, no. 4 (24 Sep. 1990), pp. 107–108.

[58] Bermudez and Carus (note 3), pp. 177–81.

[59] United States of America *vs.* Abdelkadar Helmy, U.S. District Court for the Eastern District of California, Government's Sentencing Memorandum, 17 Nov. 1990.

[60] *Programme for Promoting Nuclear Non-Proliferation Newsbrief*, no. 10 (summer 1990), p. 10; 'Nuclear weapons plans?' *Milavnews*, no. 343 (May 1990), p. 7.

[61] Berger, C., 'Egypt presses for high-tech weapons ban', *The Independent*, 19 Apr. 1990, p. 14; Murphy, C., 'Egypt's Foreign Minister urges Mideast arms ban', *Washington Post*, 19 Apr. 1990, p. 40.

Pakistan

More than three years after Pakistan announced the test launch of its Hatf I and II ballistic missiles in April 1988, its long-range missile programme appears unable to make real progress. Contrary to Pakistani announcements that the Hatf missiles were domestically developed, they are identical to French sounding rockets manufactured by Aérospatiale, previously transferred to Pakistan for use in atmospheric research. The Hatf project simply mounted these on military gun carriages and fired them in a surface-to-surface trajectory.[62] While it is possible that Pakistan has a large inventory of these missiles, they are not capable of carrying an early-generation nuclear weapon or even large quantities of high explosive.[63]

The Pakistan Army has said that it has a 600 km-range missile under development, and Indian sources maintain that it is developing space launch vehicles with ranges of as much as 2000 km.[64] Aside from the French sounding rockets, it is not evident that the country has been able to acquire the needed technologies. The $100 million space launch programme announced in 1985 is manifestly inadequate for such a task.[65]

In 1990 details emerged of a Pakistani attempt to acquire missile technology and launch facilities through a 1986 contract with a US corporation, ISC Technologies. The Khyber Pass contract, as it was known in Islambad, was expected to be worth $200–300 million. No more than 10 per cent of this was paid before Pakistani officials began to suspect that ISC could not meet its obligations.[66] There is no evidence that Pakistan received any useful missile technologies. Rather it testifies to Pakistani intentions and the difficulties of acquiring missile technology in the face of international controls.

Unlike the countries surveyed above, Pakistan's failure to acquire ballistic missiles is entirely due to international technology controls. The prestige and security motives guiding its efforts—above all, India's continuing progress in long-range rocket development and air defences—remain unchanged. The limited steps towards regional confidence-building agreements by then Prime Ministers Rajiv Gandhi and Benazir Bhutto stalled soon after the agreements were signed. Nor did they address specific issues relating to regional competi-

[62] Author's discussions with officials of the US Department of Defense and the Lawrence Livermore National Laboratory, Apr. 1990.

[63] Under ideal conditions, the French sounding rockets might be capable of ranges of 300 km, the stated maximum range of the Hatf 2. However, at such ranges they could not carry the declared payload of 500 kg, nor is the system accurate, as noted by Army Chief of Staff General Mirza Aslam Beg. Hussain, M., 'Pakistan "responding to the challenge"', *Jane's Defence Weekly*, vol. 12, no. 15 (14 Oct. 1989), p. 779.

[64] Hussain (note 63).

[65] Pakistan has been able to acquire some foreign technologies, such as the design for its first satellite, the Badr-1 micro-satellite launched by China on 16 July 1990. This appears to be based on a British design: Wilkie, T., 'Pakistani micro-satellite may have used Surrey University designs', *The Independent,* 2 Aug. 1990, p. 2; 'June launch for Pakistan's Badr-A', *Flight International*, vol. 137, no. 4211 (11–17 Apr. 1990), p. 24.

[66] Lamb, C. *et al.*, 'Pakistan launches investigation into arms deals involving ISCT', *Financial Times,* 16 Jan. 1990, p. 1; Black, L., 'Chairman "was warned"', *The Independent*, 25 Jan. 1990, p. 28; Pasztor, A., 'U.S. arms dealer to agree to plead guilty, sources say', *Wall Street Journal* (European edn), 15 Oct. 1990, p. 7.

tion for ballistic missiles. The Pakistani long-range rocket programme stands poised to take advantage of changes in the international market. Its position closely resembles some other countries with stalled programmes, especially Iran, Libya and Syria. Pakistani interest in buying M-9 ballistic missiles (600-km range) from China was reported in 1990.[67] Purchases of Scud versions from North Korea could also be possible.

Pakistan has continued to raise its military expenditure and, of greatest relevance to missile ambitions, there is mounting evidence that it has a nuclear weapon. In October 1990 President Bush presented an annual statement to Congress on Pakistani nuclear developments and, for the first time, could not certify that Pakistan 'does not possess a nuclear explosive device'.[68] Despite the suspension of US military assistance, the nuclear programme is among the country's highest military priorities.

South Africa

The test-firing of a long-range 'booster' rocket from South Africa in July 1989 came as a surprise; South Africa had not previously launched even a small ballistic missile. A US report on 28 October 1989 confirmed that the missile was an Israeli Jericho 2, although this was denied by Israel.[69] Another booster rocket was reportedly fired in November 1990, but it is evident that South Africa does not have a mature long-range rocket programme, nor is it investing in one. Combined with dramatic political change and the Government's intention to sign the NPT, South Africa is a declining source of concern.

Withdrawal from Angola and the international settlement in Namibia have resolved South Africa's most serious military problems. Enough political progress towards dismantling apartheid has been made that majority rule is widely regarded as inevitable.[70] In this scenario there is little or no role for weapons of mass destruction and long-range attack. The white leadership seems cognizant that, although they may manufacture such weapons, they will not always be the ones who control their use.[71]

South Africa's changing policies also were manifested in the decision to close the Valindaba pilot enrichment facility, assumed to be the country's source of weapon-grade nuclear material. During the Fourth Review Conference of the NPT, held in Geneva in August–September 1990, South

[67] 'US sources say PRC to sell ballistic missiles to Pakistan', *Defense and Foreign Affairs Weekly*, 16–22 Apr. 1990, p. 1; 'More Chinese missiles?', *Milavnews* (note 60), p. 19.

[68] Drogin, B., 'U.S. officials reportedly conclude that Pakistan has a nuclear bomb', *Los Angeles Times*, 24 Oct. 1990, p. 6.

[69] NBC Nightly News, 28 Oct. 1989.

[70] Jaster, R. S., *The 1988 Peace Accords and the Future of South-Western Africa*, Adelphi Paper no. 253 (International Institute for Strategic Studies: London, autumn 1990).

[71] 'Kürzungen in Armee und Rüstungindustrie Südafrikas', *Frankfurter Allgemeine Zeitung*, 22 Jan. 1990, p. 7; Helmoed-Römer, H., 'Major cuts will go ahead in S. Africa', *Jane's Defence Weekly*, vol. 13, no. 4 (27 Jan. 1990), p. 139; McWilliams, J. P., 'Fallout from political reforms affecting South Africa's armed forces, industry', *Armed Forces Journal International* vol. 128, no. 1 (Aug. 1990), p. 46.

Africa announced that it would sign the Treaty.[72] Joining the NPT does not end concerns over South African nuclear weapons—covert development remains feasible, or the Treaty could be abrogated. However, Treaty adherence would make development of advanced warheads suitable for missile delivery virtually impossible.

A 1990 Report of the UN Secretary-General noted that experience in developing small, tactical missiles 'has brought to South Africa much of the infrastructure, skills and resources required for the initiation and conduct of a long-range rocket or missile programme. However, there is little evidence that South Africa is capable of fully developing such a rocket or missile without substantial foreign technical assistance'.[73] For South Africa, 'Development of domestically manufactured long-range missiles still could be expected to last about 10 years, but many intermediate steps and false paths could be avoided. With detailed production licences, manufacturing assistance and imports of major components, the process probably could be reduced to five years or less, depending upon the degree of national commitment'.[74]

III. Progress in MTCR international controls

The MTCR was heavily criticized in 1990 for failing to stop missile proliferation. In congressional testimony, the Director of US Naval Intelligence noted that both the NPT and the MTCR 'have been largely ineffective and are likely to remain so'.[75] Although the weaknesses of the Regime are well known—they include suppliers outside its membership, poor enforcement by key adherents, opposition among recipient countries and technology already in the hands of many countries—it remains the most important barrier to ballistic missile proliferation. Evidence of its effectiveness is that the programmes of the countries surveyed above are unable to make expected progress. Of related importance is its growing membership, from 7 formal adherents in 1987 to 12 in 1990 and several more *de facto* adherents.

The MTCR was established with the participation of seven countries: Canada, France, FR Germany, Italy, Japan, the UK and the USA. Spain joined in 1989. Australia became a member in 1990. Belgium, Luxembourg and the Netherlands announced their adherence to MTCR export restrictions in May 1990.[76] Belgian adherence is especially important because of the country's role as an exporter of advanced military technology and components. In addition to

[72] von Lucius, R., 'Pretoria zur Unterzeichnung des Atomwaffensperrvertrags bereit', *Frankfurter Allgemeine Zeitung*, 19 Sep. 1990, p. 6. See also chapter 16 in this volume for a report on the NPT Review Conference.

[73] *Report on South Africa's Nuclear-Tipped Ballistic Missile Capability*, UN document A/45/571 (United Nations: New York, 29 Oct. 1990), Annex I, p. 57.

[74] UN (note 73), para. 88.

[75] Rear Admiral Thomas Brooks, quoted in Starr, B., 'MTCR is "largely ineffective"', *Jane's Defence Weekly*, vol. 13, no. 13 (31 Mar. 1990), p. 583.

[76] On Australia, see 'Missile technology control regime targets developing states', *Defense and Foreign Affairs Weekly*, 23–29 July 1990, p. 1. On the Benelux countries, see Mallet, V., 'Benelux joins missile curb accord', *Financial Times*, 17 May 1990, p. 6; 'Benelux joins MTCR', *Jane's Defence Weekly*, vol. 13, no. 21 (26 May 1990), p. 1009.

these 12 current members, four others have decided to join soon: Denmark, Portugal, New Zealand and Norway. At least four other countries have adopted similar export regulations without joining: Austria, Sweden, Switzerland and the USSR.

The Soviet decision to support the MTCR received the most attention. After years of criticizing the MTCR for being discriminatory and incomplete, the first sign of a Soviet policy reversal came at the 7–9 February 1990 Moscow meeting of Soviet Foreign Minister Eduard Shevardnadze and US Secretary of State James Baker. Their Joint Statement noted: 'The sides discussed the problem of non-proliferation of missiles and missile technology. They noted that they are both observing the guiding principles of exports within the framework of the existing regime regarding missiles, which applies to missiles capable of delivering a payload of at least 500 kg a distance of at least 300 km'.[77] This change of policy was expanded into a longer statement agreed by Presidents Bush and Gorbachev at their 31 May–3 June 1990 summit meeting. There the USA won an explicit Soviet commitment to adhere to the MTCR, linked to progress in the START negotiations. The statement also affirmed the need to explore regional initiatives to reduce the threat of missile proliferation.[78]

Among the original MTCR participants, enforcement was the key issue in 1990. This was most visible in FR Germany, which through unification brought the technology suppliers of the GDR into the Regime and passed legislation to strengthen implementation on 1 June 1990. This amends the country's foreign trade law to outlaw co-operation on nuclear and chemical weapons and ballistic missile projects. The law not only raises penalties for illegal arms sales and technology transfers, but also establishes a German legal precedent by forbidding that individuals work on such projects in other countries. In the Bundestag the opposition abstained from voting on the measure in order to express concern that the reforms are inadequate. In particular, the legislation permits technical co-operation on scientific and commercial R&D which critics argue creates chances for circumvention.[79] The reforms have already led to investigation of 59 firms for illegal co-operation with Iraq, including assisting in development of Iraq's extended-range Scud missiles.[80]

French policy on missile proliferation also became more assertive in 1990. Previously one of the least enthusiastic participants in the MTCR, as recently as 7–12 May 1990 France welcomed a high-level delegation from Iraq's

[77] '"Text" of Joint Statement', in Moscow TASS, *Daily Report–Soviet Union (FBIS-SOV)*, FBIS-SOV-90-029, 12 Feb. 1990, pp. 21–22.

[78] 'Joint statement on nonproliferation' (note 1). The members of the MTCR were scheduled to consider formally admitting the Soviet Union at a Jan. 1991 meeting in Tokyo; see 'Soviet entry date fixed', *Jane's Defence Weekly*, vol. 14, no. 5 (4 Aug. 1990), p. 155. However, the meeting was postponed until Mar. 1991; see *Wall Street Journal* (European edn), 11 Mar. 1991, p. 2.

[79] 'Neue Regeln für Waffenexport', *Frankfurter Rundschau*, 31 Mar. 1990, p. 8; 'Bonner Kompromiss für Exportgesetze', *Frankfurter Allgemeine Zeitung*, 7 May 1990, p. 17; 'Raketenbauer sollen daheim bleiben', *Frankfurter Rundschau*, 28 Aug.1990, p. 1. The final Bundestag debate is reprinted in *Das Parlament*, no. 26 (22 June 1990), pp. 4–6.

[80] *Der Spiegel*, 27 Aug. 1990.

Ministry of Industry and Military Industrialization.[81] As French arms exports declined by over 50 per cent in the late 1980s there was widespread speculation that economic pressure would lead to further relaxation of French export restrictions. This does not appear to be the case. Rather, French policy is becoming more restrictive as economic priorities yield to specific security concerns. In an unprecedented statement before an international assembly of armaments engineers on 27 June 1990, General Marcel Benichou warned of 'a new form of proliferation just as dangerous as nuclear proliferation: the proliferation to the third world of missile guidance systems'. These technologies, he said, 'should be closely watched'. France has already refused to sell additional rocket engine technology and co-operated during the Persian Gulf war in coalition efforts to defeat and destroy the equipment it provided to Iraq.[82] Recent evidence suggests that France is preparing, in a significant reversal, to sign the NPT.[83]

Among the most important problems left unresolved by growing support for the MTCR is dual-use technology. In 1990 the leading issue in the USA was exports of civilian manufacturing and research technologies potentially relevant to ballistic missile production. This issue precipitated a series of bureaucratic disputes in the USA.

These disputes centred on the role of the Commerce Department as the Government's institutional export advocate. Critics in and outside government maintain that the Commerce Department has been overly zealous in this role to the detriment of regional security. The dispute is an old one; for 40 years it was debated in terms of exports to the USSR and Eastern Europe. In 1990 the 17-nation Coordinating Committee for Export Controls (COCOM) met repeatedly to relax restrictions on sales of advanced technology to those countries. For the USA, this coincided with rising pressure to raise export barriers against threatening regional actors,[84] but reorienting COCOM and expanding the role of groups like the MTCR require a new consensus.

The promotion by the Commerce Department of a $10 million sale to Iraq of three high-temperature furnaces, which could be used in the manufacture of nuclear weapons and missile components, illustrates this issue. The Department approved the sale in June 1989. Pressure from the Department of Defense (DOD) culminated in an inter-agency conference on 19 July 1990 where the export licence was revoked.[85]

[81] 'Iraqi delegation to Paris', *Mednews*, 28 May 1990, p. 5.

[82] 'France changes its tune on MTCR', *Mednews*, 9 July 1990, pp. 1–2; 'French terms "unacceptable"', *The Statesman* (New Delhi), 10 Nov. 1990.

[83] Müller, H., 'Falling into Line? France and the NPT', Occasional Paper 6 (Programme for Promoting Nuclear Non-Proliferation, University of Southampon, May 1990); Starr, B., 'France, S. Africa may join NPT', *Jane's Defence Weekly*, vol. 14, no. 3 (21 July 1990), p. 78; Porteous, H., 'France plans entry', *Jane's Defence Weekly*, vol. 14, no. 7 (18 Aug. 1990), p. 219.

[84] Hitchens, T., 'CoCom to relax export controls', *Defense News*, 19 Feb. 1990, p. 3; Montagnon, P. and Dawkins, W., 'A small opening for freer trade', *Financial Times*, 4 June 1990, p. 14; Montagnon, 'Tighter export controls likely for Third World', *Financial Times*, 30 Aug. 1990, p. 3.

[85] Smith, R. J. and Weiser, B., 'U.S. barely halted sensitive Iraq sale', *International Herald Tribune*, 14 Sep. 1990, p. 3; Friedman, A., 'US officials ignored objections to "dual-use" exports to Iraq', *Financial Times*, 19 Sep. 1990, p. 2.

Another case—the re-export of steel casings to Brazil mentioned above—illustrates the diplomatic complexity of issues unforeseen by the MTCR.[86] Contrary to statements in the press, the issue remains unlikely to undermine the MTCR. However, it does call attention to the need to continually update the Regime to deal with new manifestations of the proliferation process.

The US Government remains ambivalent about ballistic missile proliferation. While the USA is a leading force behind the MTCR, Israel is seen as an exception to US non-proliferation policy. The Senate passed a bill after the near-fiasco of the Iraqi furnaces to enhance the role of the DOD in reviews of technology transfers to Iraq, Iran, Libya and Syria, but the USA continues to transfer advanced rocket technology to Israel through the Arrow ATBM programme. On 3 April 1990 the US State Department reacted to the Israeli launch of the Ofeq-2 by noting that Israeli leaders have declared that their satellite programme is for peaceful purposes: 'We applaud and welcome the scientific and technological advancement which may result from this program'.[87]

A more serious shortcoming of the MTCR is the failure of major exporters to join. In 1990 US intelligence revealed that China was preparing to ship a new short-range ballistic missile (SRBM) to the Middle East, probably to Iran. At least 50 M18 SRBMs with a range of 80–120 km were reportedly identified at a Chinese port. The allegations were dismissed by the Chinese Foreign Ministry. Other reports indicate that Libya and Syria are continuing discussions with China regarding purchases of longer-range M-series missiles.[88] In a partial change of policy, China voted in favour of the UN embargo on Iraq, previously a leading arms customer.[89] Yet China remains a leading source of instability in this area. Securing Chinese support is the most important immediate goal for the MTCR.

[86] Wines, 'U.S. approves export . . . ' (note 54); Friedman (note 54), p. 2.

[87] State Department daily press briefing, 3 Apr. 1990, p. 14.

[88] Gordon, M. R., 'Beijing avoids new missile assurances', *New York Times,* 30 Mar. 1990, p. 7; Southerland, D., 'China said to sell missiles', *Washington Post,* 29 Mar. 1990, p. 1; 'Libya trying to buy Chinese SSMs, says Israel', *Flight International,* vol. 137, no. 4217 (23–29 May 1990), p. 18.

[89] de Beer, P., 'Chine: un embargo chasse l'autre', *Le Monde,* 24 Aug. 1990, p. 5.

Table 9.1. Ballistic missile and related programmes in the Third World, 1990

Country/ Designation	Type[a]	No. of stages	Weight (kg)	Range (km)	Year first fired	Current status	No. deployed: Launchers	No. deployed: Missiles	Technology supplier	Technology and assistance supplied
Afghanistan[b]										
Scud B	BM	1	6 370	280	1988	In service	12	>1 000	USSR	Launchers, missiles, training
Algeria										
FROG-4	BM	1	2 000	50	Mid-1970s	Retired	12	32	USSR	Launchers, missiles, training
FROG-7	BM	1	2 500	70	Mid-1970s	In service	12	32	USSR	Launchers, missiles, training
Argentina										
									Egypt	Heat shielding, financing
									France	Inertial guidance
									FRG	Design, integration and simulation, launchers
Condor 1	BM	1	2 500	150	1984	Unknown	. .	. .		
Condor 2	BM	2	4 500	≤1 200	. .	Cancelled	. .	. .	Iraq	Financing
									Italy	Propulsion
									Sweden	Warhead fusing
									Switzerland	Management
Brazil										
Sonda 3	SR	2	1 581	(80)	1976	In service	. .	. .	FRG	Design, propulsion
									USA	Training
Sonda 4	SR	2	7 300	(600)	1984	In service	. .	4	FRG	Design, propulsion
SS-60	AR	1	595	60	1983	In service	>12	>100	. .	. .
X-40	AR	1	654	68	1979	In service	. .	>20	. .	. .
EE-150	BM	1	4 500	150	. .	Unknown	. .	. .	. .	. .
SS-300	BM	1	8 000	300	. .	Unknown	. .	. .	. .	. .
VLS	SLV	4	49 000	(6 000)	. .	Development	. .	. .	. .	. .

Country/ Designation	Type[a]	No. of stages	Weight (kg)	Range (km)	Year first fired	Current status	No. deployed: Launchers	No. deployed: Missiles	Technology supplier	Technology and assistance supplied
Cuba										
FROG-4	BM	1	2 000	50	1961	In service	10	30	USSR	Launchers, missiles, training
FROG-7	BM	1	2 500	70	mid-1980s	In service	12	36	USSR	Launchers, missiles, training
Egypt[c]										
Sakr 80	AR	1	660	80	1987	In service	>12	>100	France	Design assistance
FROG-5	BM	1	2 000	50	1968	In service	. .	. .	USSR	Launchers, missiles
FROG-7	BM	1	2 500	70	1973	In service	12	72	USSR	Launchers, missiles
Scud B	BM	1	6 370	280	1973	In service	12	>100	USSR	Launchers, missiles
									N. Korea	Production assistance
Scud 100	BM	1	(7 000)	(600)	1988	Development	. .	. .	USSR	Missiles
									N. Korea	Design assistance
									Iraq	Financing
Condor 2	BM	2	4 500	≤1 200	. .	Cancelled	. .	. .	FRG	Equipment
Greece										
Honest John	BM	1	2 640	37	1959	In service	8	24	USA	Launchers, missiles, training
India										
Centaure	SR	1	530	50	1968	In service	. .	. .	France	Production licence, assistance
									USA	Training
Rohini	SR	2	1 391	130	1972	In service	. .	. .	France	Propulsion assistance
Prithvi	BM	1	4 000	240	1988	Development	. .	. .	. .	. .
Agni	BM	2	21 000	2 400	1989	Development	. .	. .	France	Propulsion, guidance
									FRG	Propulsion, guidance, heat shielding

ASLV	SLV	4	39 000	(4 000)	1987	Development	..	2	France	Propulsion, guidance,
									FRG	Propulsion, guidance, heat shielding, materials
PSLV	SLV	4	137 000	(8 000)	1991	Development	..	..	..	..
GSLV	SLV	..	333 000	(14 000)	..	Planned	..	..	..	..
Indonesia										
RX-250	SR	2	(1 200)	(100)	1987	Development	..	..	France	Training, assistance
SLV	SLV	..	(17 000)	(1 500)	1993	Planned	..	..	..	..
Iran										
Oghab	AR	1	360	45	1987	In service	..	Hundreds	China	Design, production assistance
									N. Korea	Production assistance
Shahin 2	BM	1	580	60	1988	In service	..	..	..	..
Nazeat	BM	1	950	120	1988	In service	..	Hundreds	..	..
..	BM	1	(1 500)	160	..	Development	..	..	..	..
Scud B[d]	BM	1	6 370	280	1985	In service	4	100	Libya	Missiles, launchers
									N. Korea	Missiles
									Syria	Missiles
Iraq[e]										
Ababil 50	AR	1	400	50	1988	Development	..	..	Yugoslavia	Design, assistance
SS-60	AR	1	595	60	1985	In service	30	..	Brazil	Launchers, missiles, training
Sijeel 60	AR	1	588	60	1987	In service	..	..	Brazil	Design, assistance, training
Ababil 100	AR	1	800	100	1989	Development	..	..	Yugoslavia	Design, assistance
FROG-7	BM	1	2 500	70	..	In service	30	>90	USSR	Launchers, missiles, training
Laith	BM	1	2 500	90	1988	Development	..	..	..	..
Scud B	BM	1	6 370	280	..	In service	10	>360	USSR	Launchers, missiles, training
Fahd	BM	1	..	500	..	Development	..	..	..	..

Country/ Designation	Type[a]	No. of stages	Weight (kg)	Range (km)	Year first fired	Current status	No. deployed: Launchers	No. deployed: Missiles	Technology supplier	Technology and assistance supplied
al-Hussein	BM	1	7 000	600	1987	In service	70	<500	USSR	Launchers, missiles
									Brazil	Training
									Egypt	Personnel, assistance
									Sweden	Transporting
al-Abbas	BM	1	8 000	900	1988	Development	. .	. .	USSR	Launchers, missiles
									Egypt	Personnel, assistance
Condor 2	BM	2	4 500	≤1 200	. .	Development	. .	. .	Argentina	Missiles
									Austria	R&D facilities
									Egypt	Equipment, assistance
									FRG	Equipment, assistance
									USA	Equipment
Tamuz-1	BM	3	(48 000)	2 000	. .	. .	. .	. .	USSR	Missiles
al-Abed	SLV	3	48 000	(2 000)	1989	. .	. .	. .	USSR	Missiles
Israel										
MAR 350	AR	1	835	90	1987	In service	. .	. .	Argentina	Financing
Lance	BM	1	1 527	120	1976	In service	12	160	USA	Launchers, missiles, training
Jericho 1	BM	1	(3 000)	480	1968	In service	. .	(50)	France	Design, production assistance
Jericho 2	BM	2	(8 500)	1 450	1986?	Development	. .	. .	. .	. .
Shavit	SLV	2	(25 000)	(7 500)	1988	In service	. .	. .	. .	. .
Korea, North										
FROG-5	BM	1	2 000	50	1969	In service	9	50	USSR	Launchers, missiles, training
FROG-7	BM	1	2 500	70	1970	In service	18	54	USSR	Launchers, missiles, training
Scud B[f]	BM	1	6 370	280	1976	In service	12	Hundreds	Egypt	Sample missiles

Scud PIP	BM	1	(7 000)	(600)	1988	Development	. .	. .	Egypt	Assistance
									Japan	Electronics
Korea, South										
Honest John	BM	1	2 640	37	1959	In service	7	36	USA	Launchers, missiles, training
Nike-Hercules	BM	2	5 200	240	1978	In service	. .	100	USA	SAM missiles
SLV	SLV	3	(30 000)	(4 000)	. .	Planned	. .	. .	. .	. .
Kuwait										
FROG-7	BM	1	2 500	70	1980	In service	4	12	USSR	Launchers, missiles, training
Libya										
FROG-7	BM	1	2 500	70	Mid-1970s	In service	48	>144	USSR	Launchers, missiles, training
Scud B[g]	BM	1	6 370	280	Mid-1970s	In service	80	>240	USSR	Launchers, missiles, training components
M-9	BM	1	6 200	600	. .	Negotiating	. .	. .	China	. .
Ittisalt	BM	1	(6 000)	700	. .	Development	. .	. .	FRG	Design, assistance, components
Pakistan										
Shahpar	SR	2	1 200	(120)	1970s	In service	. .	. .	France	Missiles, training, assistance
									USA	Training
SUPARCO rocket	SR	2	(3 000)	(280)	1980s	In service	. .	. .	. .	. .
Hatf 1	BM	1	(1 500)	80	1987	In service	. .	. .	France	Missiles, training, assistance
Hatf 2	BM	2	(3 000)	280	1988	In service	. .	. .	France	Missiles, training, assistance
. .	BM	1	. .	600	. .	Development	. .	. .	. .	. .
SLV	SLV	3	(15 000)	(1 200)	. .	Planned	. .	. .	. .	. .
Saudi Arabia[h]										
SS-60	AR	1	595	60	1985	In service	. .	. .	Brazil	Launchers, missiles, training
DF-3 (CSS-2)	BM	2	27 000	2 700	1988	In service	12	120	China	Launchers, missiles, training

Country/ Designation	Type[a]	No. of stages	Weight (kg)	Range (km)	Year first fired	Current status	No. deployed: Launchers	No. deployed: Missiles	Technology supplier	Technology and assistance supplied
South Africa										
Jericho 2B	BM	2	(8 500)	1 450	1989	Development	. .	. .	Israel	Missiles, assistance
SLV	SLV	. .	. .	. .	. .	Cancelled	. .	. .	. .	. .
Syria										
FROG-7[i]	BM	1	2 500	70	1971	In service	24	96	USSR	Launchers, missiles, training
SS-21 Scarab	BM	1	1 500	120	1983	In service	12	36	USSR	Launchers, missiles, training
Scud B	BM	1	6 370	280	1975	In service	18	54	USSR	Launchers, missiles, training
M-9	BM	1	6 200	600	. .	Negotiations	. .	. .	China	. .
Taiwan										
Honest John	BM	1	2 640	37	1961	In service	. .	. .	USA	Launchers, missiles, training
Ching Feng	. .	1	1 500	120	1978	In service	. .	. .	Israel	Lance missile design
. .	SLV	(3)	. .	. .	1996	Development	. .	. .	. .	. .
Thailand										
Thanu Fan	AR	1	. .	. .	. .	Development	. .	. .	. .	. .
Turkey										
MLRS	AR	1	308	40	1990	In service	. .	. .	USA	Production assistance, components, training
Honest John	BM	1	2 640	37	1960	Withdrawn	18	54	USA	Launchers, missiles, training
Yemen										
FROG-7	BM	1	2 500	70	1979	In service	12	36	USSR	Launchers, missiles, training
SS-21 Scarab	BM	1	1 500	120	1988	In service	8	24	USSR	Launchers, missiles, training
Scud B	BM	1	6 370	280	1979	In service	6	18	USSR	Launchers, missiles, training

.. Unknown or not applicable

( ) Estimates

[a] Acronyms in this column: AR: artillery rocket (military); BM: ballistic missile; SLV: space launch vehicle; SR: sounding rocket.

[b] Since Oct. 1988, the Soviet Union has supplied the Kabul Government with over 1000 Scud B missiles. Most of these were fired soon after delivery against suspected Mujahideen targets.

[c] Several dozen FROG missiles and at least one Scud B missile were fired at Israel during the 1973 war.

[d] Iran received at least 100 Scud B missiles from Libya, North Korea and possibly Syria during its war with Iraq. Most of these were fired before the cease-fire began on 20 Aug. 1988. Iranian Scud B inventories may have been replenished since then, although this cannot be confirmed.

[e] During the 1980–88 war with Iran, Iraq fired approximately 67 FROG-7 missiles, over 100 Scud Bs and 190 al-Hussein Scud versions. The number of Brazilian SS-60s and other large artillery rockets fired was in the thousands. Little is known about the state of Iraq's missile inventories after the 20 Aug. 1988 cease-fire. In the 1991 Persian Gulf conflict, the US-led counterattack, starting on 17 Jan. 1991, led to the direct destruction of many of these missiles. Others were expended in Iraqi missile launches against Israel and Saudi Arabia.

[f] In 1985 North Korea agreed to supply 90–100 domestically manufactured Scud B missiles to Iran. Most of these were subsequently fired against Iraq.

[g] Although Libya has not used its ballistic missiles in its fighting with Egypt in 1977 or in Chad in 1978–88, it has sold Scud B missiles to Iran for use against Iraq.

[h] Saudi Arabia may also help finance missile production programmes in Egypt and Iraq.

[i] Syria fired approximately 25 FROG-7 missiles at Israel during the 1973 war. Syrian efforts to purchase longer-range missiles in the 1980s were blocked by Western diplomatic pressure and, in the case of the SS-23, the unwillingness of the USSR to sell a weapon system proscribed under the 1987 INF Treaty.

10. Major armed conflicts in 1990

KARIN LINDGREN, BIRGER HELDT, KJELL-ÅKE NORDQUIST and PETER WALLENSTEEN*

I. The conflicts in 1990

In 1990, major armed conflicts were waged in 31 locations (see table 10.1, which presents accounts of the development of the armed conflicts in these locations up to 31 December 1990, from information available as of 31 January 1991). A major armed conflict is defined as prolonged combat between the military forces of two or more governments or of one government and at least one organized armed group, involving the use of weapons and incurring battle-related deaths of at least 1000 persons. In 1989, according to this definition, major armed conflicts were waged in 33 locations.[1] Thus, armed conflicts continue to plague our planet, although encouraging developments have recently taken place in US–Soviet relations, in Eastern and Central Europe, and in arms reduction and disarmament efforts. Three of the major armed conflicts of 1989 did not continue in 1990. The USA–Panama conflict and that in Romania resulted in the victory of one side; and the conflict over Namibia saw a negotiated solution, with Namibia being granted independence in March 1990. One new major conflict emerged in 1990, namely, the internal conflict in Liberia.

One criterion, the threshold of 1000 battle-related deaths, might appear to be somewhat arbitrarily determined.[2] However, it serves the purpose of capturing the most war-like *major* armed conflicts that occur. In addition, the threshold is high enough for reliable data to be found, but low enough normally to include the politically most significant conflict developments. However, for the conflicts waged in 1990 it resulted in the exclusion of one of the most publicized as well as politically and internationally most significant armed conflicts: the Iraq–Kuwait conflict. Sources report that a few hundred

[1] The *SIPRI Yearbook 1990* listed 32 locations, but subsequent information shows that one more major conflict met the criteria: the internal conflict in Laos. See *SIPRI Yearbook 1990: World Armaments and Disarmament* (Oxford University Press: Oxford, 1990), chapter 10. A revised and updated presentation of all major armed conflicts since 1986 is found in Lindgren, K. (ed.), *States in Armed Conflict 1989* (Department of Peace and Conflict Research, Uppsala University: Uppsala, Sweden, 1991); this publication also lists armed conflicts incurring fewer than 1000 deaths. Note that one conflict location can include more than one major conflict as well as additional conflicts with a lower number of battle-related deaths.

[2] The figure refers to battle-related deaths from the beginning of conflict between the current fighting parties.

* Department of Peace and Conflict Research, Uppsala University, Sweden. Some conflict descriptions were prepared by Ramses Amer and Christer Ahlström. Björn Holmberg, Johan Larsson and Magnus Marklund assisted in the data collection. This chapter constitutes part of a continuous project at the Department of Peace and Conflict Research, Uppsala University.

casualties resulted from the Iraqi invasion of Kuwait on 2 August 1990, so this conflict did not reach the level of 1000 battle-related deaths in 1990. Public concern regarding this conflict during the last months of 1990 focused on three issues: the international sanctions, the military forces brought into the region, and the risks of a major military escalation.[3]

In addition, other conflicts with significant implications are not included in the table owing to this threshold figure (or, in some cases, other criteria as well). The conflicts among ethnic and national groups in Spain (Basque) and the Soviet Union (i.a., Armenia, Azerbaijan and Georgia) also fall in this category, as do the rebel invasion in Rwanda (from Ugandan territory) on 30 September and the conflict between the nomadic Touareges in Mali and Niger and their respective governments.

Finally, it should be noted that the criterion refers to the total number of deaths during the conflict, many of which have been going on for a considerable period of time. The Northern Ireland conflict, for example, is included since total deaths number around 2800, although in 1990 fewer than 100 persons were killed. It remains an active major armed conflict, but it was of comparatively low intensity in 1990.

Two of the conflicts waged in 1989 which did not continue in 1990 testify to the impact of changing superpower relations. The fall of the Ceausescu regime in Romania was part of this process in Europe. In this country a constitutional development is under way, although on the first anniversary of the new government in December 1990, discontent with the democratization process was still expressed. In the Namibian conflict, co-operation between the superpowers seems to have been an important factor in the solution process, and it also appears to have had an impact in promoting dialogue between some of the antagonistic parties in South Africa.

A considerable number of the conflicts in 1990 have a long history; for example, the Tamil conflict in Sri Lanka is in its seventh year, the East Timor (Indonesia) conflict in its 15th year, the Eritrean conflict (Ethiopia) in its 28th year, and some of the conflicts in Myanmar (former Burma) started in 1948. The solutions to these conflicts, in other words, require considerable effort and imagination, probably of a different kind than was mustered to end the cold war.

II. Three types of conflict

All the conflicts in the table are over control of government and/or territory. The conflicts can be divided into three groups according to the key issues and parties involved: inter-state conflicts; internal conflicts over government control; and state-formation conflicts, where demands pertain both to territorial and constitutional issues.

The *inter-state conflicts* are those in which the armed forces of (at least) two internationally recognized states are battling each other—a classical

[3] See also chapters 18 and 19 in this volume.

'war'. The *internal conflicts* over government correspond to the notion of a 'civil war': a government facing armed and organized opposition demanding control over the government. The third category, *state-formation conflicts*, includes a government confronting armed and organized opposition, but in this case the opposition demands a territorial and constitutional reformulation of the state, for instance, regional autonomy or independence, or the creation of a new state. Such armed conflicts are classified as state-formation conflicts, although the government might define them as separatist conflicts, and the opposition might see them as liberation struggles.[4]

There was only one major inter-state conflict in 1990, that between India and Pakistan. Low-level military confrontation between these two regional powers took place along the lines separating the forces in the Kashmir region. Tension was related to the more contentious situation regarding the status of the Indian state of Jammu and Kashmir, where the central government during the year faced an uprising by groups in the Muslim majority community.

As in previous years, slightly more than half of the major armed conflicts in 1990 were internal, relating to the control of government in one country. In Nicaragua a successful process was in motion for a solution. An important step was the internationally supervised (and recognized as fair) elections of 25 February 1990. The Sandinista Government unexpectedly lost the election and was replaced with a broad, right-of-centre coalition led by Violeta Chamorro. The disarming of the US-supported Contras remained a problem, and the new government faced some tension in this regard. This process was supervised by the joint OAS–UN Observer Group in Central America (ONUCA), and some 19 000 former Contra soldiers were disarmed. Another promising development took place in Colombia, where the former M-19 guerrilla movement continued its disarmament policy, participated in the elections for president and to the constituent assembly, and made unexpected gains. Among other internal conflicts in 1990, some were on a fairly low level of intensity (e.g., Laos and the Philippines), but most of them saw no particular changes in intensity as compared to 1989. In Chad an additional party joined the conflict and after a military campaign overthrew the government in December. Some headway was made towards peaceful settlements, for instance in Lebanon, after General Aoun gave up his struggle for power and fled to the French embassy. This made it possible to reunite Beirut under the national government backed by Syria. Direct talks continued in some cases, for instance between the fighting parties in Mozambique, Guatemala and El Salvador, with the help of intermediaries. In the case of South Africa, the conflict over majority rule, in which the African National Congress (ANC) had been the main armed opponent, a breakthrough was achieved following the release of Nelson Mandela in February 1990, the legalization of the ANC, and ANC suspension of the armed struggle in August. Mandela's release was a first stage in the South African

[4] This distinction has been developed as part of the Uppsala project and is found to be fruitful when analysing conflict resolution. See Wallensteen, P. (ed.), *States in Armed Conflict 1988* (Department of Peace and Conflict Research, Uppsala University: Uppsala, Sweden, 1989).

Government's intended gradual repeal of apartheid legislation. However, these developments were accompanied by an escalation of violence between followers of the ANC and Inkatha. This related partly to the question of who was to be represented at negotiations with the government. Ethnic loyalties were also involved. As an effect of this violence, more people were killed in South Africa during 1990 than in any year in the 1980s. A new element was bomb attacks by white groups objecting to the process of reconciliation between the Government and the ANC and in which, for instance, offices of the governing Nationalist Party were targeted.

State-formation conflicts, slightly less than half of all the conflicts in 1990, constitute the third group. Among these, few examples of progress towards solution were seen during the year. In Northern Ireland there were some moves towards further talks between several parties, but the main armed group, the IRA, remained isolated, and some of its attacks may have been aimed at preventing this process. Attempts at creating contacts between the warring parties in the Israeli–Palestinian conflict suffered several setbacks. The dissolution of the Likud–Labor coalition government, the suspension of contacts between the USA and the Palestine Liberation Organization (PLO), and the tension created by the Iraq–Kuwait conflict were the main factors behind this development. The Tamil conflict in Sri Lanka was one of the state-formation conflicts which intensified during the year. Another was the Eritrean conflict, where negotiations were initiated, but stalled, in 1989. The conflict changed qualitatively following the capture by the Eritrea People's Liberation Front (EPLF) of the important harbour of Massawa in February. During the year, the conflict was raised in the summit meeting in May between the US and Soviet Presidents. Talks between the Eritrean and Ethiopian sides took place in Washington in October 1990. In several other state-formation conflicts, fighting resulted in lower numbers of deaths in 1990, for example, in Iran, Iraq, East Timor and Western Sahara (Morocco).

In most of the armed conflicts of 1990, the United Nations or regional governmental organizations did not play a direct and public role. However, some important developments took place during the year. In the case of Cambodia, a plan supported by the five permanent members of the UN Security Council was seriously discussed, involving a proposed temporary role for the UN in administration of the country. The Security Council took decisions in support of this plan during the latter half of 1990. In El Salvador a personal representative of the UN Secretary-General was instrumental in the 1990 negotiations. In the case of Eritrea, the Ethiopian Government officially withdrew its objections to UN observer participation in the talks. The West African regional community became involved in the internal conflict in Liberia, where the Economic Community of West African States (ECOWAS) officially tried to reduce the fighting between the different factions by sending in a peace-keeping force.

III. The victims

The highest number of battle-related deaths during the year occurred in Ethiopia (with two major conflicts, one over Eritrea and one over central government), Liberia, India (also with two major conflicts, one in Punjab and one in Jammu and Kashmir, and in addition many hundreds killed in the Hindu temple/Muslim mosque dispute and over other issues in the country) and Sri Lanka (with two major conflicts). In both the Ethiopian and Liberian conflicts, it is likely that more than 10 000 battle-related deaths occurred in 1990. Other conflict locations with a high number of deaths during the year were Chad, El Salvador, South Africa, Peru and Lebanon.

Intensive fighting not only resulted in many military and civilian casualties but also set in motion flows of refugees, particularly into neighbouring countries. In some cases, such as in Ethiopia, Somalia and Sudan, neighbouring countries were also at war, leaving nowhere for civilians to flee. Famine became an imminent fear, particularly as the armed conflicts also prevented the delivery of outside assistance. This illustrates the general situation of civilian victims far outnumbering military victims. In a recent report it was estimated that, in the major conflicts waged in 1988–89, 9 of 10 victims since the start of the conflicts were civilian.[5]

The refugee situation did not improve during the year. Among the 14.5 million refugees of today a large majority come from conflict areas. The Iraq–Kuwait conflict also shows this relation. Tens of thousands of guest workers, mostly from Third World countries, were displaced following the Iraqi invasion of Kuwait.

IV. The 1980s in perspective

During the 1980s the total number of conflict locations increased during the first half of the decade, to 36 in 1986. Since then a slow decline, to 31 in 1990, has occurred. Fewer new major conflicts have emerged, while the process of solution of old conflicts has been slow. The large number of armed conflicts at the end of the 1980s were also active at the beginning of the decade. In addition, there were conflicts in which national governments intentionally prevented resolution activity by internal groups or by the international community. The change in the international climate, associated with changes inside the USSR, may have reduced interest in providing support to parties in conflicts throughout the world. This may mean that some opportunity exists for local peace initiatives (such as in Central America, Southern Africa and South-East Asia). Many of the initiatives related to the South-East Asian region where several old conflicts were being settled: the China–Viet Nam and Laos–Thailand conflicts, and the communist rebellions in Malaysia and

[5] Ahlström, C. and Nordquist, K.-Å, *Casualties of Conflict* (Department of Peace and Conflict Research, Uppsala University: Uppsala, Sweden, 1991).

Thailand. Peace moves were also made in the Cambodia conflict and in the Philippines.

A new dimension of conflicts in the 1980s as compared to the 1970s was the revival of the role of religion in politics. In the Iran–Iraq conflict, this dimension centred on a conflict between secularism and Muslim revivalism, in different forms and with many additional elements as well. Secularism and religion were also important dimensions in the Afghanistan conflict. These two conflicts were the most devastating, especially from a humanitarian point of view, during the decade. Religious 'fundamentalism', a somewhat unclear concept, was also noteworthy in the Punjab conflict (with a link between Sikh religious leaders and political parties) and in the Sri Lanka conflicts (where Buddhist monks espoused strong Singhalese nationalism). In Israel, Jewish orthodox groups became increasingly influential, affecting the conflict with the Palestinians. In Northern Ireland, Sudan and Lebanon, Christian identity is an important dimension of the respective conflicts.

In some situations, the link between ethnic and religious sentiments is very close, but ethnic aspirations may still be differently rooted (notably in history, language and territory). An increasing ethnic dimension of conflicts surfaced in the 1980s, raising the issue of the possible break-up of some states composed of many nationalities. In several cases this resulted in major, protracted armed conflicts (such as in Ethiopia and India) as well as conflicts on lower levels (e.g., Yugoslavia and the Soviet Union). Thus, the 1980s might, in perspective, be regarded as a decade of global religious–political and ethnic revivalism.

The 1980s witnessed an increase in the demand for democratic institutions and human rights in several parts of the world. The swift changes in Eastern and Central Europe in 1989–90 illustrate this trend. In Africa, multi-party systems began to appear, but their strength and significance are still untested. In South America most military governments were replaced by democratically elected governments with a stronger will to adhere to human rights provisions. However, governments were weakened by lack of support in society *vis-à-vis* strong military establishments and other vested interests. Economic conditions inherited by new governments were often chaotic. Finally, the world economy, which was dominated by three dynamic centres of power (the USA, Japan and the European Community), did not contribute to successful economic development in Third World countries—a fact adding to the complexity of conflict situations in this part of the world.

Table 10.1. Major armed conflicts in the world, 1990

Location	Year formed/ year joined[a]	Warring parties[b]	No. of troops in 1990[c]	Deaths[d] Total (incl. 1990)	During 1990	Change from 1989[e]
Europe						
United Kingdom/ Northern Ireland	1969/1969	British Govt *vs.* IRA	29 000* 200–500	1969–90: 2 800	74	0

Comments: The present conflict originates from the division of Ireland in 1922. In 1969 civil rights issues were raised, and since 1970 the Provisional IRA (Irish Republican Army) has constituted the main armed actor on the republican side, demanding reunification under the Republic of Ireland, but without having support from the Republic. The British Govt sought to uphold existing ties between Northern Ireland and Great Britain, to protect the interests of the majority of unionists/loyalists. 1990 showed a pattern of bomb attacks and shootings, with more attacks in the UK than since 1974. In Jan. the Secretary for N. Ireland (P. Brooke) discussed possible devolution to break deadlock. An IRA offer of talks without preconditions was rejected by the British Govt, which extended an invitation to talks on the condition that the IRA laid down arms. Talks between Brooke and political parties in N. Ireland progressed in Apr. and May but were halted in July because of disagreement over the role of the Republic of Ireland. An attack in Apr. caused the death of 4 soldiers in N. Ireland. In June a bomb exploded in London at the Carlton Club, well known as a Conservative Party meeting-place; a British Army officer was killed in Dortmund, Germany, the same month. A close associate of PM Thatcher (I. Gow) was killed by a car bomb outside his home, and a land-mine killed 3 constables and a nun who happened to pass by (July). In Sep. there was an attack in Britain on the former British Governor of Gibraltar, and a bomb exploded in a military barrack in Derby. In Oct. the British Army claimed to have killed a key IRA man in Armagh, N. Ireland. Two weeks later, 6 British soldiers were killed when bombs exploded at 2 security check-points in N. Ireland. Civilians were forced by the IRA to drive the bombs to the check-points; 1 of the drivers was killed. The IRA announced a cease-fire over Christmas. Ulster Freedom Fighters assumed responsibility for some deaths in N. Ireland during the year. In late Nov. additional British soldiers were moved to the area, bringing the Army strength to 11 000.

* British Army, 11 000, plus Royal Ulster Constabulary (the local police force) and Ulster Defence Regiment, gives this total for late 1990.

Location	Year formed/ year joined[a]	Warring parties[b]	No. of troops in 1990[c]	Deaths[d] Total (incl. 1990)	During 1990	Change from 1989[e]
Middle East						
Iran	1972/1979	Iranian Govt *vs.* KDPI	305 000 10 500	1979–90: >17 000*	<50	0

Location	Year formed/ year joined[a]	Warring parties[b]	No. of troops in 1990[c]	Deaths[d] Total (incl. 1990)	During 1990	Change from 1989[e]

Comments: Kurds (in particular the Kurdish Democratic Party of Iran, KDPI), seeking greater autonomy or independence in the north-west, became very active militarily following the overthrow of the Shah in 1979. The establishment of 'liberated zones' led to the 1983–84 campaign by Iranian forces to regain control. During the Iran–Iraq War, Kurdish groups received aid from the Iraqi Govt. In late 1989 the KDPI reportedly intensified its attacks (claiming to have killed over 170 Iranian troops and Revolutionary Guards). The leader of KDPI was assassinated in Austria (July) and a member of the central command of Komala (the Kurdish section of the Communist Party of Iran) was assassinated in Cyprus (Aug.). KDPI reportedly split into 2 factions in 1989. During 1990 the KDPI claimed responsibility for an attack on security forces, killing 15 (May). The Govt also faces opposition from the National Liberation Army (est. 4500), the armed wing of Mujahideen e-Khalq. In 1990 the brother of the Mujahideen e-Khalq leader was assassinated in Switzerland (Apr.).

* Including the 2000 NLA (Iranian National Liberation Army) deaths (in 1988) in the Iran–Iraq War.

Location	Year formed/ year joined[a]	Warring parties[b]	No. of troops in 1990[c]	Deaths[d] Total (incl. 1990)	During 1990	Change from 1989[e]
Iraq*		Iraqi Govt	955 000	1980–89: 5 000–6 000**	. .	*n.a.*
	1961/1980	*vs.* KDP	15 000			
	1975/1980	*vs.* PUK	4 000			

Comments: An attempt at secession by the Kurdish minority in the north-east of Iraq led to general uprising in 1962. In spite of agreement with the Govt in 1970, sporadic clashes continued until 1974 when general hostilities broke out. In 1975 Iran and Iraq concluded an agreement which stopped support from Iran to the Iraqi Kurds, and the resistance declined. During the Iran–Iraq War the Kurdish parties, mainly the Kurdish Democratic Party (KDP) and the Patriotic Union of Kurdistan (PUK), fought the Iraqi Govt with help from Iran. After the cease-fire in the Iran–Iraq War, Iraq moved against the Kurds. Reports of the use of chemical gas by the Iraqi Govt were denied by Iraq. The Iraqi Govt has proclaimed a 30-km 'security zone' along its borders, moved the Kurdish population to other parts of the country and razed Kurdish villages. The main resistance has apparently ceased, and information about military action in 1990 is scarce. The Govt also faces opposition from the Socialist Party of Kurdistan (c. 1500), the Kurdish Worker's Party (a breakaway group from KDP, strength unknown) and the Shi'ite organization Supreme Assembly of the Islamic Revolution in Iraq, but no clashes were reported in 1990.

* Information on fighting in Iraq in 1990 is very scarce, so these data should be treated with caution.

** Figures are for up to 1989, since none are available for 1990. A major part of this range of figures is connected with the reported use of chemical weapons in 1988.

Location	Year formed/ year joined[a]	Warring parties[b]	No. of troops in 1990[c]	Deaths[d] Total (incl. 1990)	During 1990	Change from 1989[e]
Israel/Palestine		Israeli Govt	141 000	1948–90: >11 000	c. 560	+
	1964/1964	*vs.* PLO*	11 000**	Dec. 1987–Dec. 1990: 900–1 000		

Comments: The current conflicting parties were formed in 1948 and 1964, with the formation of the Israeli state and the PLO (Palestine Liberation Organization), respectively. Arab–Israeli wars were waged in the region in 1948–49, 1956, 1967 and 1973. The 1967 and 1973 wars resulted in Israeli occupation of the West Bank and Gaza Strip (1967) and Golan Heights (1973). Israel invaded Lebanon in 1978 and 1982, the latter invasion forcing the PLO HQ to be moved to Tunisia. PLO-related armed groups are based in Lebanon. Since Dec. 1987 a significant popular uprising, the *intifada*, has taken place in the Occupied Territories. PLO leader Yassir Arafat was elected President of the Palestinian State (proclaimed in Nov. 1988) by the PLO Central Committee in Apr. 1989. The first months of 1990 involved (failing) US–Egyptian efforts towards negotiations between Israel and Palestinians over elections in the Occupied Territories based on the US Secretary of State Baker's 5-point initiative (see *SIPRI Yearbook 1990*). Bomb explosions and violent clashes between various groups on both sides were frequent throughout 1990. On 20 May, 4 Palestinian labourers were killed in Rishon Le Zion. This was regarded by Palestinians as a parallel event to the Dec. 1987 deaths of Gaza labourers, which started the *intifada*. Public reaction was therefore particularly serious: i.a., a Palestinian 3-day general strike, a week-long boycott of Israeli workplaces, and a hunger strike the rest of May 1990 by prominent East Jerusalem Palestinians. After a 12-week Israeli Govt crisis, a Likud-led right-wing coalition Govt was formed on 11 June, replacing the former Labour–Likud coalition. A raid on 30 May against Israeli beaches north and south of Tel Aviv by the PLO-related Palestinian Liberation Front (PLF) led to suspension of the US–PLO dialogue, opened in Dec. 1988. Jewish immigration, mainly from the USSR, increased dramatically in 1990, increasing tension between the USA and Israel concerning settlement policy. By Dec. 160 000 Soviet Jews had arrived. On 8 Oct. Israeli police killed over 20 and wounded more than 150 people at Al-Haram al-Sharif/Temple Mount in Jerusalem. The action was criticized in UN Security Council Resolution 672, calling for a fact-finding mission to Israel and the Territories. The resolution demand was rejected by Israel. In Dec. UNRWA (UN Relief and Works Agency for Palestine Refugees) reported 860 Palestinians killed and 60 000 wounded during the then 3-year-long *intifada*. In addition, 322 Palestinians are reported to have been killed by other Palestinians, and 57 Israelis were killed by Palestinians during the same period.

* The PLO is an umbrella organization. Armed action is carried out by member organizations. During the *intifada* period, groups with unclear PLO relations have emerged.
** This figure refers to the total number of PLO-related troops based in Lebanon.

Lebanon		Lebanese Govt (Hrawi/Hoss)/		1975–90: 150 000	>2 350	+ +
		National Army of Lebanon (Lahoud)	10 000			
	1975/1975	Lebanese Army (Aoun)	11 000 –20 000 (Jan.–Oct.)			
	1985/1985	Lebanese Forces (Gaegea)	10 000			
	1979/1979	Amal	5 000			
	1975/1975	Islamic Resistance/Hezbollah	3 500–5 000			
	1964/1964	PLO	11 000*			
	1959/1965	Al Fatah	4 500–5 000			

Location	Year formed/ year joined[a]	Warring parties[b]	No. of troops in 1990[c]	Deaths[d] Total (incl. 1990)	During 1990	Change from 1989[e]
	1969/1969	DFLP (Hawatmah)	1 000			
	1967/1968	PFLP (Habash)	900			
	1968/1968	PFLP–GC (Jibril)	500			
	1968/1975	Palestine Popular Struggle Front	. .			
	. .	Lebanese Forces (Hobeika)	. .			
	1974/1976	FRC (Nidal)	. .			
	1977/1977	Palestine Liberation Front (Yaqoub)	. .			
	1961/1975	Syrian Socialist Nationalist Party	. .			
	1975/1975	Lebanese National Resistance Front	. .			
	1975/1975	Popular Nasserite Organization	. .			
	1975/1975	Lebanese Baath Party	. .			
	1978/1978	SLA	2 500–3 000			
	1976/1976	Syrian Govt	40 000–42 000			
	1978/1982	Israeli Govt	1 000			

Comments: Civil war among Christian, Palestinian, Muslim and Druze groups since 1975. Muslims are estimated to form the majority of the population. Christians dominate political and economic life. Syrian troops have been present since 1976. Israeli invasions in 1978 and 1982. Israel keeps the Israeli-armed South Lebanese Army (SLA) in the Israeli-proclaimed 'Security Zone'. The UN Interim Force in Lebanon (UNIFIL, 5500–6000 in 1990) has been deployed in Lebanon since 1978. Iranian Revolutionary Guards (2000) in the Syrian-controlled Bekaa Valley since 1982. Gen. Aoun (Maronite Christian, self-declared head of Lebanon) fought a 'Liberation War' against Syria (aiming to force Syria out) in Mar.–Sep. 1989, without success. Elias Hrawi (Christian, Syrian-backed) was elected President and Selim Hoss Prime Minister (Nov.). Gen. Aoun, who did not accept Pres. Hrawi's authority, was dismissed as Commander-in-Chief of the Lebanese Army and succeeded by Admiral Lahoud (Nov.). The Taif peace plan (Oct. 1989), aiming to share power more equally between Christians and Muslims, was also dismissed by Gen. Aoun because it did not ensure Syrian withdrawal. Inter-Christian fighting over control of eastern Beirut during Jan.–Sep. 1990, when c. 1200 persons were killed: during early stage of fighting, because Aoun demanded that Lebanese Forces (under Samir Geagea) surrender their weapons; and later, due to Geagea's acceptance of Hrawi's authority and demand for help from the National Army of Lebanon under Lahoud and Syrian forces. There were several limited cease-fires during this period, e.g., in Feb., after the Progressive Socialist Party (Druze, 5000–10 000), the Lebanese Communist Party (1000–2000) and the Syrian Socialist Nationalist Party mobilized forces around Beirut's Christian enclave; Mar., Apr. and May, after mediation by a Christian team and Iraq, respectively. Peace initiatives were also made by the Vatican and the French Govt (June). On 21 Sep. Pres. Hrawi signed constitutional changes in order to implement the Taif peace plan. After an air strike and Syrian-led assault on Gen. Aoun's HQ (reportedly with 10 000 tank-led troops, incl. 6000 Syrians), Gen. Aoun fled on 13 Oct. to the French embassy, where he subsequently was granted asylum. Up to 800 were reported killed (incl. several hundred Syrian men) during the battle. Dany Chamoun, leader of the National Liberation Party and head of a coalition of political parties supporting Gen. Aoun, was murdered on 21 Oct., along with his family. According to the plan to unify Beirut, the various militias had by late Nov. left the capital and surrounding areas, and the 'Greater Beirut Zone' was under control of the Govt and National Army. The militias withdrew to other parts of the country not under control of the Govt. The 2 Shi'ite Muslim groups Hezbollah (seeking an Iranian-style Islamic republic in Lebanon) and the more secular Amal (pro-Syrian, wanting more say in the country's Christian-dominated political system) have been fighting intermittently since 1987 for dominance over the Shia Muslim population, causing over 1000 deaths in 3 years. Battles in southern Beirut, which started in mid-Dec. 1989, continued into 1990. Heavy fighting in southern Lebanon (Iqlim al-Toffah area) in July, with Amal receiving support from the Palestine Liberation Organization (PLO), the Syrian Socialist Nationalist Party and the Lebanese Baath Party. Over 200 killed in 1990. A peace agreement, supervised by Iran and Syria, was signed by Hezbollah and Amal in early Nov. Israeli troops and the South Lebanese Army (SLA) were engaged in fighting, both air attacks (18 during the year) and ground attacks, with Palestinian and Lebanese groups throughout the year. Palestinian and Lebanese groups took responsibility for attacks against Israeli targets. Reported military engagements in 1990, Israel/SLA *vs.* Palestinian/Lebanese groups: Democratic Front for the Liberation of Palestine (DFLP; Jan., Feb., Dec.), Popular Front for the Liberation of Palestine–General Command (PFLP–GC; Mar., Apr., July, Nov.), Popular Front for the Liberation of Palestine (PFLP; Feb., May, Nov.), Fatah Revolutionary Council (FRC, under Abu Nidal; Jan., Dec.), Palestine Liberation Front (May), Palestine Popular Struggle Front (Nov., Dec.), Hezbollah (Jan., Feb., Apr., July, Nov.), Amal (Feb.), and Syrian Socialist Nationalist Party (Nov.). The Lebanese National Resistance Front, which has repeatedly claimed that it has attacked Israeli troops, took responsibility for the death of a US missionary (Mar.). Two militiamen from the Popular Nasserite Organization were killed in a shooting incident in Saida (June). Increased fighting in southern Lebanon in connection with the withdrawal from Beirut. Intra-Palestinian fighting between FRC and Al-Fatah (troops loyal to PLO Chairman Arafat; Sep., up to 80 killed) has also been reported, as has fighting between the Syrian wing of the Lebanese forces (under Hobeika) and Hezbollah (Oct.), Sunni Muslim militia *vs.* PLO forces (Saida, July), and UN troops *vs.* Amal (Feb.). In Dec. Omar Karami was appointed PM. His broadened cabinet was intended to bring together Christian and Muslim factions. However, by late Dec. the Lebanese Front and Phalangist Party had not accepted the offer to join the new cabinet.

* Total strength of PLO-related groups in Lebanon.

Turkey		Turkish Govt	60 000	1984–90: 2 000–2 500	>360*	+ +
	1974/1984	*vs.* PKK	1 500–2 000			

Comments: The Kurdish Worker's Party (PKK), established in 1974, seeks independence from Turkey and has since 1984 staged warfare against the Turkish Govt, mostly in south-eastern Turkey. The estimated strength of PKK is 1500–2000 armed men within Turkey and c. 3500 men in the total force. The Turkish Army has reportedly deployed an estimated 60 000 soldiers, 30 000 police and 18 000 village guards and specially trained police commandos in the region. An

Location	Year formed/ year joined[a]	Warring parties[b]	No. of troops in 1990[c]	Deaths[d] Total (incl. 1990)	During 1990	Change from 1989[e]

increase of PKK activities in 1989 continued into 1990. During the year the struggle was reportedly transformed into a mass nationalist uprising with anti-Govt demonstrations (Mar.) and the PKK's demands receiving more widespread support. Govt imposed restrictions close to censorship on press reports and possibilities of internal exile. Heavy fighting between security forces and PKK reported during Mar.–Oct., leading to a total death toll during 1990 of over 360. PKK reportedly trained in Syrian-controlled Bekaa Valley in Lebanon. Govt also faces increased urban violence from Islamic and left-wing militants; over 20 political assassinations during the year. The Islamic Revenge Movement, the Turkish Islamic Commandos and the Armed Propaganda Union assumed responsibility for attacks (Mar.). The Dev Sol (Revolutionary Left) claimed responsibility for bomb attacks in Ankara and Istanbul (Apr.), said to be perpetrated in protest at Turkey's treatment of its Kurds, and shooting of former head of the Secret Service (Sep.).

* This figures does not include Nov.–Dec. 1990.

South Asia

Location	Year formed/ year joined	Warring parties	No. of troops in 1990	Total (incl. 1990)	During 1990	Change from 1989
Afghanistan		Afghan Govt	58 000	1978–90: 1 000 000*	. .*	*n.a.*
	1978/1978	*vs.* Mujahideen based in				
		Afghanistan,	. .			
		Iran,	126 000			
		Pakistan	40 000			

Comments: After the Apr. 1978 Govt take-over by PDPA (People's Democratic Party of Afghanistan), a civil war began with armed opposition from Muslim groups: the Mujahideen, or Holy Islamic Warriors. In late Dec. 1979, as the Govt under Pres. Amin (leader of the Khalq ('Masses') faction of the PDPA) controlled only the capital and a few urban centres, the USSR intervened militarily on a large scale and installed Babrak Karmal (leader of the Parcham ('Banner') faction of the PDPA) as President. He was succeeded by Mohammad Najibullah in 1986. After 6 years of UN-mediated talks between Afghanistan and Pakistan, an accord was signed in Apr. 1988 in Geneva. Adherence to the accord was to be monitored by UN observers; it contained pledges by Afghanistan and Pakistan of non-interference in each other's internal affairs, provisions for the voluntary return of refugees and the complete withdrawal of Soviet troops by 15 Feb. 1989. The Mujahideen, who had not taken part in the negotiations, declared that they would continue the war. On 23 Feb. 1989 the Pakistan-based Mujahideen formed a Govt-in-exile, the Islamic Interim Afghan Govt (IIAG). Negotiations in 1989 over participation by the Iran-based Mujahideen in the IIAG stalled on the question of representation. Efforts by Pakistan aiming at broadening the IIAG collapsed in Feb. 1990. The same month the Govt presented a peace proposal which included a 6-month cease-fire, UN-supervised elections and demilitarization of Afghanistan. An unsuccessful coup attempt was launched in early Mar. by the Govt Defence Minister Lt.-Gen. Tanay (head of the Khalq faction), leading to up to 200 deaths. Tanay had reportedly been involved in an alleged coup plot against Pres. Najibullah (leader of the Parcham faction) in Dec. 1989. On 28–29 May the constitution of Afghanistan was amended, providing for 'plurality of political parties'. A peace proposal was also presented, consisting of a nation-wide referendum concerning whether a cease-fire should be implemented, to be followed by an international conference. In June the PDPA changed its name to the Homeland Party (Hezb-e Watan), giving it a more national character. Increased fighting in Sep. In early Oct. the Mujahideen seized the provincial capital Tarin Kot. Assault launched by certain factions on Kabul on 12 Oct. Efforts by Pakistan aiming at rapprochement between the factions of the Mujahideen resulted in the Islamabad Agreement on 27 Oct. between some of the Mujahideen factions, which included military co-operation and elections in 13 Mujahideen-controlled northern provinces by 21 Mar. 1991. The Govt has made local cease-fire deals with many Mujahideen commanders. Various Mujahideen groups control c. 80% of the country. In Nov. 1990, according to the Govt but denied by the Mujahideen, talks were held in Geneva between the Govt, moderate Mujahideen leaders and representatives of the former Afghan King Zahir Shah. The talks concerned the formation of an interim coalition Govt headed by the former King and UN-supervised elections. The Govt said that it would accept the plan if it met the approval of the majority of the Mujahideen. The Govt would also accept the halt of weapon deliveries to both sides. The Mujahideen has previously demanded that Najibullah step down. During 1990, talks between the USA and the USSR aimed at a solution of the conflict. The talks concerned stopping arms supplies to both sides, a cease-fire and elections supervised by the UN and the Islamic Conference. Although the USA and the USSR came closer to each other during the year, no agreement was reached: by Dec. the still unresolved issues were the cut-off of arms supplies and the status of Pres. Najibullah. A decrease of the level of fighting has taken place since 1989.

* The figure is likely to include all deaths in connection with the conflict, that is, not only battle-related deaths. According to Soviet sources, the total number of Soviet troops killed in the period 1979–15 Feb. 1989 was 15 000. Figures for the total number of deaths during 1990 are not available.

Bangladesh		Bangladesh Govt	90 000	1975–90: 1 200–3 000	<100*	+
	1971/1982	*vs.* JSS/SB	5 000–7 000			

Comments: The Parbattya Chattagram Jana Sanghati Samiti (JSS, or the Chittagong Hill Tracts People's Coordination Association) and its military wing, the Shanti Bahini (SB, or Peace Force), were formed in 1971. Guerrilla warfare started in 1974–75 when demands for autonomy for the south-eastern Chittagong Hill Tracts (CHT), previously enjoyed under British rule, met no response from the Govt. Bengali (mainly Muslim) settlers, moving into the area, have been attacked by the SB. Bengali settlers have attacked the tribal people of the CHT (inhabited mainly by the Buddhist Chakma tribe but also by Hindus and Christians), reportedly with the support of the Bangladesh Army on some occasions. SB intensified operations in 1981, and in 1984 the Govt sent the Army into the area. Between Oct. 1985 and Dec. 1988, meetings were held between the Govt and the JSS/SB but with little progress. In 1989 the Parliament passed 4 bills designed to provide limited autonomy to the region, and elections to new district councils were held (June). Violence prior to the elections increased the number of refugees in India to c. 65 000. The new laws have split the tribal groups: some want to give the autonomy scheme a chance; others, who regard the new councils as powerless, want to continue the armed struggle. During 1990 both the SB and the Army have stepped up their activities. One-third of the total armed forces are reportedly deployed in the CHT area. Officials put the number killed by SB during the period 1986 to June 1990 at over 1200; other sources

Location	Year formed/ year joined[a]	Warring parties[b]	No. of troops in 1990[c]	Deaths[d] Total (incl. 1990)	During 1990	Change from 1989[e]

put the total death toll at up to 3000. After violent pro-democracy demonstrations in the capital Dhaka and other parts of the country (reportedly up to 100 killed), Pres. Ershad, who came into power in a bloodless coup in 1982, was in Dec. forced to resign.

* Estimated number killed in the conflict between the Govt and JSS/SB.

Location	Year formed/ year joined	Warring parties	No. of troops in 1990	Total (incl. 1990)	During 1990	Change from 1989
India		Indian Govt	1 262 000*	1983–90: >19 800	>3 800***	+
	1947/1981	*vs.* KCF	>8 000**			
	1947/1982	*vs.* JKLF	. .			
	1990/1990	*vs.* Hizbul Mujahideen	. .			
	. .	*vs.* Student's Liberation Front	. .			
	. .	*vs.* All-Umar	. .			
	1988/1988	*vs.* ABSU/BVF	. .			
	. .	*vs.* BPAC	. .			
	. .	*vs.* ULFA	. .			
	1967/1967	*vs.* Naxalites, People's War Group	. .			
	. .	*vs.* NSCN	. .			

Comments: Several Sikh groups, e.g., the Khalistan Commando Force (KCF), the Khalistan Liberation Force, the Council of Khalistan, the Bhindranwale Tiger Force and the Babbar Khalsa, have waged an armed struggle against the central Govt to create an independent Khalistan in the province of Punjab. In June 1984 the Indian Army stormed the Golden Temple, the main Sikh shrine. Sikhs assassinated PM Indira Gandhi in Oct. 1984. In 1988 at least 2500 persons died in the Punjab conflict, the death-toll easing off somewhat after the Govt 'Operation Black Thunder' (May 1988), ending occupation of the Golden Temple by Sikh groups, many of whom surrendered. In 1989, 1044 were killed. PM V. P. Singh offered concessions seeking to stop secessionist struggle in Jan. 1990. He offered all Sikh deserters from the Indian Army new Govt employment and review of the detention of jailed Sikhs. A Sikh leader, Harminder Singh Sandhu, was assassinated in Amritsar in Jan. In Apr. the Indian Parliament extended direct rule of the state for 6 months, a decision which was renewed in Oct. for an additional 6 months. The struggle in Punjab caused the death of some 3800 people in 1990. Another major armed conflict has emerged in Jammu and Kashmir, where several groups (mainly the Jammu and Kashmir Liberation Front (JKLF), Hizbul Mujahideen, Student's Liberation Front and All-Umar) demand that Kashmir is to become an independent state or part of Pakistan. Violence increased significantly following a Govt crack-down on Muslim separatists on 20 Jan. 1990. The locally elected Govt was dismissed in Jan., and a Governor responsible directly to New Delhi was installed. Muslim separatists stepped up their campaign for an independent Kashmir, and in Mar. the Govt acknowledged that the conflict had reached an 'unprecedented point'. Curfews were imposed, causing severe conditions for the civilian population. A large part of the Hindu population had fled Jammu and Kashmir by Apr. A large number of Muslims were being detained; although no accurate figures are available, it is estimated that the numbers reached thousands. The situation in Jammu and Kashmir led to a sharp increase in tension between the Indian Govt and the Pakistani Govt (see below). In end-June the separatists stepped up their military activity against the Indian security forces. Presidential rule was imposed on Jammu and Kashmir on 19 July. In Aug. Indian Security Forces arrested the Commander-in-Chief of the JKLF and inflicted a severe blow to the organization. An alliance of 6 militant groups, Hizbul Mujahideen, moved to the forefront of the separatist struggle. The 1990 death toll in Jammu and Kashmir is estimated at 2000 people. In Assam, the Bodo people, mainly militia drawn from the All Bodo Student's Union (ABSU) and its armed wing, Bodo Voulantiers Force (BVF), with a few hundred fighters and the Bodo People's Action Committee (BPAC), continued their campaign for a separate state, which had started in Mar. 1988. The Govt is also faced with armed opposition from the United Liberation Front of Assam (ULFA). In Nov. 1990 PM Chandra Sekhar imposed direct rule from Delhi over Assam, and the army initiated pursuit of separatists. The New Delhi Govt is also faced with armed opposition in Andra Pradesh (Naxalites, People's War Group, formed in 1967), resulting in 180 deaths in 1990, and Nagaland (National Socialist Council of Nagaland, NSCN). Religious cleavages in Uttar Pradesh over a religious site in Ayodhya led to serious Hindu–Muslim clashes throughout India in Oct. Hindu militants and holy men converged in Ayodhya, at the site of a Muslim mosque (the Babri Masjid mosque), with the intention to tear the mosque down in order to build a Hindu temple at the same site. The president of the Bharatiya Janata Party (BJP), Lal Kishen Advani, was arrested, and the BJP withdrew its support for V. P. Singh's Govt. In a following no-confidence vote in Nov., PM Singh's Govt was unseated. The religious dispute claimed c. 600 deaths (Oct.–Dec.). The new PM, Chandra Shekhar, pleaded that the separatists in Punjab and in Jammu and Kashmir would end their military struggle. Talks between Punjab separatists and PM Shekhar commenced in late Dec. A similar offer to reach a negotiated settlement with separatists in Jammu and Kashmir was rejected by JKLF.

* Total armed forces. Not all of these are necessarily engaged in actual combat.
** 1989 figure. Figures for 1990 are not available.
*** Only in Punjab.

India–Pakistan	1947/1982	Indian Govt *vs.* Pakistani Govt	1 262 000* 550 000*	1971: 11 000 1982–90: <600 (mil.)	<100	+

Comments: Since independence in 1947 there have been 3 armed conflicts between India and Pakistan, first over partition (1947), then over Kashmir (1965) and East Pakistan/Bangladesh (1971). The Simla Agreement was signed in 1972 to observe a line of control (LOC) along the former cease-fire line between Baltistan (Pakistan) and Jammu and Kashmir (India). Since 1982 renewed sporadic fighting has occurred on the Karakoram Range (Siachen Glacier) over a 72-km-long LOC determination, culminating in 1987. In 1989 a military experts' meeting of Indians and Pakistanis took place in Islamabad to discuss the situation in the Siachen Glacier area. Long-standing mistrust between the 2 countries has been increased by trouble in the Provinces of Punjab and Jammu and

Location	Year formed/ year joined[a]	Warring parties[b]	No. of troops in 1990[c]	Deaths[d] Total (incl. 1990)	During 1990	Change from 1989[e]

Kashmir in India. Tension increased between the states as violence stepped up in Kashmir in Jan. 1990. A 'War of Words' commenced over the Jammu and Kashmir issue: India claiming that Pakistan was fanning militant Muslim separatists in Jammu and Kashmir; Pakistan claiming that it was not giving material support to the 'freedom fighters' inside Jammu and Kashmir, and making diplomatic efforts directed to other Muslim countries in order to convene a special session of the UN General Assembly to discuss the situation in Jammu and Kashmir. Furthermore, Pakistani PM Bhutto urged that a plebiscite called by the UN in 1949 be held and welcomed third-party mediation in order to defuse border tensions. Border skirmishes involving small-arms fire across the LOC were reported throughout 1990. Pakistani demonstrators attempting to cross into Indian-controlled Jammu and Kashmir have been shot by Indian forces on several occasions. Tension between the 2 countries reached unprecedented heights, raising concern over a fourth war between the 2 countries as troops and military *matériel* were sent to the border areas. Attempts in July to defuse tension by bilateral talks were mainly unsuccessful, due to a lack of will on the part of India to discuss the status of Jammu and Kashmir. Discussions were also hampered because of strong domestic political pressure in both countries not to make concessions. Concern was also raised of a possible use of nuclear weapons in an armed conflict between the 2 states. Exchanges of artillery fire along the LOC took place in Aug., Sep. and Oct. Pakistani attempts to raise the Jammu and Kashmir issue in the UN in Oct. were rejected by India. Tension between the 2 countries was reduced in the final months of the year.

* The Indian and Pakistani Govts had c. 350 000 and 100 000 military and para-military troops, respectively, in the Kashmir region at the height of the tension (spring/summer).

Location	Year formed/ year joined	Warring parties	No. of troops in 1990	Total (incl. 1990)	During 1990	Change from 1989
Myanmar		Myanmar Govt	200 000–250 000	1948–51: 8 000	. .*	*n.a.*
(formerly Burma)	1948/1949	*vs.* KNU	4 000–8 000	1950: 5 000		
	1948/1948	*vs.* KIA	8 000	1981–84: 400–600 yearly		
	1949/1949	*vs.* Mon State Party	3 000	1985–87: >1 000 yearly		
	1965/1965	*vs.* SSA	. .	1988: 500–3 000		
	1989/1989	*vs.* BNUP	8 000–12 000**			
	1989/1989	*vs.* Noom Suk Harn	. .			
	1989/1989	*vs.* National Democratic Army	. .			

Comments: More than 20 anti-Govt organizations have fought against the central Govt since 1948. The Burma Socialist Programme Party (BSPP) came to power after a military coup in 1962. BSPP changed its name to the National Unity Party in 1988 after pro-democracy demonstrations. These were quelled in Aug.–Sep. 1988 with up to 3000 reported killed (500 acc. to official sources). A military take-over followed, and the State Law and Order Restoration Council

(SLORC) took power. The Govt receives support from China (military supplies and training), and the Army has expanded since 1988. In 1989 the country's name was changed to the Union of Myanmar. In 1975 the National Democratic Front was formed, consisting of 11 non-communist anti-Govt resistance groups, with a total of 20 000–23 000 soldiers in 1990. In 1988 the Democratic Alliance of Burma (DAB) was established, composed of 23 ethnic resistance armies, underground student groups and other anti-Govt organizations. In late 1990 DAB made an appeal to foreign govts for weapon support. In addition, the All-Burma Student Democratic Front opposes the Govt and consists of people who took part in the 1988 uprising and subsequently fled to rebel camps on the Thai border. The Burma Communist Party (BCP), after a mutiny, in 1989 split into at least 4 groups organized along ethnic lines: the Burma National United Party (BNUP, and its armed wing, the BNU Army, primarily Wa hill tribesmen), Nom Suk Harn (the Young Brave Warriors, Shan-dominated), the National Democratic Army (former BCP's Mekong River Division, led by a Chinese volunteer) and the Burma National Democratic Front (BNDF). Communist ideology has been discarded. Other BCP troops of Kachin origin joined the Kachin Independence Army (KIA). During 1990, no reports of activity from the 200–300 men who after the split reportedly still regarded themselves as the BCP. The Govt also faces opposition from other ethnic groups seeking autonomy, such as the Karen National Union (KNU), the Mon State Party and the Karenni National Progressive Party. Since the military take-over the Govt has intensified its offensive against the minority groups along the Thai border, primarily against KNU and the Mon State Party. KNU has reportedly lost all but 2 of its military camps since 1989. Reports that Govt troops entered Thai territory and airspace (in 1990 they used aircraft for the first time against the rebel groups) during its offensives. The Govt objective is reportedly to gain control of the teak forest in the area (timber concessions sold to Thai companies). The number of refugees in Thailand rose in 1990 to c. 45 000. The main political opposition movement, the National League for Democracy (NLD), won 80% of the seats in elections on 27 May to the new parliament, despite many of its leaders being under house arrest and major repression during preceding months (incl. the eviction of people from neighbourhoods known to support the NLD to 'satellite towns'). However, there are no signs of a hand-over of power from SLORC to a new Govt. Demonstrations by monks and students (mainly in the city of Mandalay, the centre of rebellion since the pro-democracy uprising in 1988) Aug.–Oct.; 4–7 people reportedly killed (Aug.) by Govt troops. Monks refused to conduct religious ceremonies for soldiers and their families, which led to Govt-ordered military raids on monasteries in the Mandalay Area (Oct.). In Dec. opponents of the Govt formed a parallel Govt, the National Coalition Govt of the Union of Burma. In late Dec. the NLD was outlawed.

* Since the military take-over, information about battle-related deaths has been scarce.
** Reported number of troops during the second half of 1989. Figures for 1990 are not available.

Sri Lanka		Sri Lankan Govt,	65 000	1983–90: 17 500–20 000	3 500–4 000	–
	1987/1987	Indian Govt (IPKF)	40 000 (Jan.) – (Apr.)			
	1976/1983	*vs.* Tamil Tigers (LTTE)	2 000–3 000			
	1969/1987	*vs.* JVP	<1 200			

				Deaths[d]		
Location	Year formed/ year joined[a]	Warring parties[b]	No. of troops in 1990[c]	Total (incl. 1990)	During 1990	Change from 1989[e]

Comments: Tamil groups, principally the Tamil Tigers (Liberation Tigers of Tamil Eelam, LTTE), demand autonomy or secession for the northern parts of Sri Lanka. An agreement (1987) between India and Sri Lanka placed Indian troops (Indian Peace-Keeping Forces, IPKF) on the island, ostensibly for the Tamils' own safety. During Oct. 1987–Mar. 1990 the IPKF and LTTE forces fought each other: IPKF losing 1155 soldiers while, acc. to Tamil sources, LTTE lost 683. The Sinhalese People's Liberation Front (JVP), which opposes the partial Tamil autonomy agreement between Sri Lanka and India, became a new party to the conflict in Nov. 1987. In Jan. 1990 the Emergency Laws were again extended. The Indian-backed North-Eastern Provincial Council (consisting mainly of EPRLF members with Muslim SLMC (Sri Lanka Muslim Congress) support) defiantly declared 'an independent state of Eelam' as the last 2000 IPKF left Sri Lanka on 24 Mar. Approximately 90 000 EPRLF (Eelam People's Revolutionary Liberation Front) supporters and their leaders left for India during the following weeks. At the same time, LTTE members began to return from jungle hide-outs; among them being their previously declared dead leader, Prabakaran. LTTE cadres took administrative control over Tamil-populated areas in the north and east. In accordance with the previous (June 1989) peace agreement and cease-fire with the Govt, the LTTE had set up a political body, the People's Front of Liberation Tigers (PFLT), to compete in elections for the North-East Council, and they began to establish PFLT offices in the main Tamil towns. Directly after the IPKF withdrawal, peace talks between the Govt and the LTTE were resumed in order to define conditions for an election and the respective subsequent roles of the political and military wings of the LTTE. Simultaneously, reports indicated that the LTTE were constructing new bunkers in the bush. Only 5 days after the Govt had made significant political concessions (6 June), the LTTE started attacking police stations and Army convoys. After abortive truce talks the LTTE launched a major military offensive on 11 June into the Eastern Province and went on to capture the regional capital Jaffna, where heavy fighting ensued for control of the fort. In Aug. the LTTE besieged the garrison at Jaffna, which later fell to the LTTE, only to be recaptured (12 Sep.) by the Govt after heavy air attacks. In a heavy counter-offensive on 13 Sep., Govt forces broke the 3-month-old siege of the Jaffna fort, relieving over 200 troops and police. Repeated Govt air attacks on the town, which LTTE still occupied, alienated the Tamil civilians. On 26 Sep. the Govt decided to abandon the fort, denying that it had any strategic value. The Govt had deployed 1600 troops in the Jaffna area. During 11 June–24 Sep. the LTTE were, acc. to the Govt, estimated to have lost about 2000 while the Govt had lost 700 soldiers. Other sources put the total battle deaths (June–Sep.) at 4000. On 18 Oct. the Govt launched a new general offensive against the LTTE, and 150 people (incl. civilians) were killed in 5 days of fighting. In a LTTE offensive, a large Army camp at Mankulam was retaken from the Govt. The Govt claimed to have killed a further 200 LTTE soldiers, losing 100 of its own soldiers (Nov.) The LTTE declared a unilateral cease-fire from 31 Dec., but the Govt decided not to respond until it had seen practical manifestations of Tamil restraint during the following week. The Govt has decided to expand the Army, which had only 12 000 troops in 1984, to 100 000 troops within the next two years. Militarization of the Muslim population commenced in 1988 when, in response to Tamil attacks, the 'Muslim Jihad' was established to defend Muslims in particular in the Eastern Province. Subsequently, the Govt also trained and armed Muslim youths in order to defend their villages after the SLMC (a party advocating a separate Muslim provincial Govt) had called for security for the

Muslims. LTTE attacks on Muslims (who had participated in the Indian-backed Tamil Regional Govt, dissolved by Pres. Premadasa on 7 July) increased in early Aug. 1990. Over 500 Muslims were believed to have been killed since hostilities recommenced in June. The LTTE claimed that the massacre of 130 people at prayer was carried out by Govt forces who were trying to drive a wedge between the Muslim and Tamil people. In southern and central Sri Lanka, the Govt counter-offensive against the JVP Sinhalese group continued in 1990, but with greatly diminished ferocity. A police raid (29 Dec. 1989) against the HQ, in which the leader and 6 staff members died, marked the end of effective JVP opposition. The Govt claimed to be detaining 6700 JVP members in Feb. 1990. The pro-Govt vigilante group 'Eagles of the Central Hills' continued harassing JVP supporters, killing several of them in an attack on 24 Mar.

Pacific Asia

Cambodia		Cambodian Govt	50 000–70 000	1979–89: >25 300*	..*	–
	1975/1979	*vs.* DK (KR)	30 000–45 000			
	1979/1979	*vs.* KPNLF	10 000–15 000			
	1979/1979	*vs.* FUNCINPEC/ANS	15 000–20 000			

Comments: Border clashes between Kampuchea and Viet Nam during 1977–78 ended with a Vietnamese invasion (Dec. 1978) which ousted Democratic Kampuchea (DK), i.e. the Khmer Rouge (KR), from power (Jan. 1979). Viet Nam announced that it had completed the final withdrawal of its troops from Cambodia in Sep. 1989, although this was not internationally verified. Armed opposition to the Govt is made up of a coalition of DK, Khmer People's National Liberation Front (KPNLF) and Front Uni pour un Cambodge Indépendant, Neutre, Pacific et Coopératif/Armée Nationale Sihanoukiste (FUNCINPEC/ANS), forming the Coalition Govt of Democratic Kampuchea (CGDK) in 1982, which changed its name to the National Govt of Cambodia (NGC) in Feb. 1990. Fighting in 1990 was on a lower scale than during the 1980s, a contributing factor being the withdrawal of Vietnamese troops in 1988–89. In Jan. and early Feb. 1990 the NGC maintained the military initiative from late 1989. From mid-Feb. through Apr. 1990 the Cambodian Govt took over the initiative. From May the NGC, in particular DK, have reportedly regained the initiative. During 1990 several meetings were held between China and Viet Nam to discuss the situation in Cambodia. An informal meeting was held in Jakarta with participation of the Cambodian parties, ASEAN, Laos and Viet Nam as well as representatives from Australia and France (Feb.–Mar.). Tokyo meeting (June) between the Govt, FUNCINPEC and KPNLF led to agreement on the formation of a Supreme National Council (SNC) and on a cease-fire. However, it was never implemented due to DK boycott of the meeting. The USA announced that it intended to withdraw its support for the NGC at the UN and open a dialogue with the Vietnamese Govt on the situation in Cambodia (July) A dialogue was also opened with the Cambodian Govt (Sep.). At their sixth meeting during 1990 regarding Cambodia, the 5 permanent members of the UN Security Council agreed on a peace plan for Cambodia (Aug.). A Sino-Soviet agreement to stop supplying arms to the Cambodian parties (Sep.). Reports indicate that China has continued to send weapons to the NGC parties. Agreement in Jakarta between the 4 Cambodian parties to accept the peace plan of the 5 permanent members and on the creation of a SNC (Sep.). The UN Security Council unanimously adopted resolution 668 on Cambodia, in which the peace plan was endorsed (Sep.). The UN General Assembly unanimously adopted a resolution on 'The situation in Cambodia' (Oct.) which for the first time was made

Location	Year formed/ year joined[a]	Warring parties[b]	No. of troops in 1990[c]	Deaths[d] Total (incl. 1990)	During 1990	Change from 1989[e]

possible because no Cambodian delegation claimed to represent the Cambodian state at the UN after failure between the 4 parties to agree on who should lead a common delegation to the UN. A new UN peace plan was presented in Nov. Two inconclusive meetings were held by the SNC in Thailand (Sep.) and France (Dec.).

* For figures for battle-related deaths in this conflict before 1979, see *SIPRI Yearbook 1990*, page 405, and note *p*, page 418. Regarding battle-related deaths during 1979–89: the only figure available is from official Vietnamese sources, indicating that 25 300 Vietnamese soldiers died in Cambodia. An estimated figure for the period 1979–89, based on various sources, is >50 000, and for 1989, >1000. Figures for 1990 are not available.

Location	Year formed/ year joined	Warring parties	No. of troops in 1990	Total (incl. 1990)	During 1990	Change from 1989
Indonesia		Indonesian Govt	283 000	1975–90: 15 000–16 000 (mil.)*	. .*	*n.a.*
	1975/1975	*vs.* Fretilin	300–1 400			
	. . /1989	*vs.* Aceh Merdeka	. .			
	. . /1989	*vs.* National Liberation Front of Aceh	. .			
	. .	*vs.* Free Papua Movement	500–600			

Comments: The Revolutionary Front for an Independent East Timor (Fretilin) proclaimed the independent state of the Democratic Republic of East Timor (a former Portuguese colony) in Nov. 1975. Indonesia invaded in Dec. 1975, and in July 1977 East Timor became Indonesia's 27th province. By late 1978 most of Fretilin's resistance was quelled. Negotiations between Fretilin and the Govt broken off in 1983, and since the mid-1980s low-level warfare has prevailed. Human rights groups report up to 200 000 deaths in 1975–89, most caused during the first years after the invasion. Demonstrations by youths in support of Fretilin, in connection with the Pope's visit in Oct. 1989, continued into 1990. Fretilin (with 1400 men, acc. to Fretilin and 300–400 acc. to other sources) remains in the eastern part of East Timor, reportedly split into 3 groups during the year. Few clashes reported. 14 000 Indonesian troops deployed, acc. to official 1989 figures (others put the figure higher) in East Timor. In early Nov. 1990 Fretilin stated it wanted to have unconditional peace talks with the Govt. The Govt also faces opposition from the Free Papua Movement (c. 100 armed men of estimated total strength of 500–600) in Irian Jaya and from Aceh Merdeka (Free Aceh) and the National Liberation Front of Aceh (who have reportedly killed more than 70 members of the 8000 deployed security forces during the year) in Sumatra (Aceh Province, predominantly Muslim); the groups are seeking independence from Indonesia. Fighting in the Aceh Province escalated during the year, and some reports estimate the total number of deaths (soldiers, guerrillas and civilians) at around 1000.

* The 15 000–16 000 military deaths refers to the East Timor conflict. No reports are available for the number of deaths in East Timor during 1990.

Laos		Laotian Govt	55 000	1975–90: >1 000**	. .	*n.a.*
	1975/1975	*vs.* opposition groups*	2 000			

Comments: Since the Pathet Lao gained power on its own in 1975 several opposition groups have been active militarily. However, after the Hmong tribesmen resistance suffered a major set-back at Phu Bia mountain in northern Vientiane province in 1978, the security threat to the Laotian Govt has been minor. Four opposition groups representing right-wing forces, neutralists and tribal peoples formed the United National Front for the Liberation of the Lao people (also known as United Lao National Liberation Front, ULNLF) at a conference held in the southern province of Champassak in Sep. 1980. This Front formed a Provisional Revolutionary Govt-in-exile in Dec. 1989. This was followed by an increase in military activity in Dec. 1989 and spring 1990.

* The largest being ULNLF, formed in 1980.

** This currently low-level conflict has since 1975 claimed over 1000 battle-related deaths. However, figures higher than this level vary significantly.

Philippines		Philippine Govt	108 500	1972–90: >37 500*	<400*	–
	1968/1986	*vs.* NPA	16 000–20 000			
	1982/1986	*vs.* RAM	3 000			
	1972/1986	*vs.* MNLF	15 000			
	1990/1990	*vs.* military faction	200–600			

Comments: The main conflict is between the Govt and NPA (New People's Army), affiliated with the Communist Party of the Philippines (CPP). In spite of cease-fire and contacts after the formation of the Aquino Govt (Feb. 1986) the conflict has continued, but the strength of NPA is believed to have declined, partly due to internal friction. In Mar. 1990 fighting between Govt forces and NPA flared up, leading to the death of nearly 90 people. Several assassinations in the capital Manila were attributed to the NPA (e.g., death of 2 US airmen in May and a leading member of the security forces in June). In July NPA unilaterally declared a cease-fire after the Cabanatuan City earthquake disaster. Under a cease-fire arrangement NPA released a US hostage in early Aug. In late Aug. the Govt offered the NPA and other armed opposition groups a cease-fire, and in early Sep. limited cease-fires with NPA and with RAM (Reform Army Movement)-affiliated soldiers were announced, pertaining to earthquake-affected areas. However, on 24 Sep. NPA called off its 2-month cease-fire claiming that the Govt had pursued attacks against NPA and criticized the Govt for entering into a secret deal with the USA over US bases in the country. In Oct. increased NPA attacks were reported in different parts of the country, killing at least 39 people. In Nov. an NPA ambush in Mindanao led to the death of 27 soldiers and guerrillas. A second major conflict concerns the status of the island of Mindanao, with a large Muslim population. New autonomy laws were objected to by the largest Muslim armed force, MNLF (Mindanao National Liberation Front, and its armed wing, the Bangsa Moro Army), and several skirmishes occurred in Jan. Altogether 50 000 people were killed in the Mindanao conflict during 1972–86. In Oct. local powers were transferred to 4 Muslim provinces that voted for autonomy. The third serious issue was the aftermath of the 1 Dec. 1989 coup attempt by RAM and other elements in the armed forces. It included, for instance, a local governor linked to this opposition who was defeated in a shooting incident by armed forces loyal to Pres. Aquino (Mar.). A wave of bomb explosions occurred in Manila in July and Aug., attributed to RAM. In Oct. 3 military camps and 2 cities in Mindanao were seized by officers

Location	Year formed/ year joined[a]	Warring parties[b]	No. of troops in 1990[c]	Deaths[d] Total (incl. 1990)	During 1990	Change from 1989[e]

and soldiers, declaring an independent republic. They gave up after air force attacks. Bombs in Sep. against US-owned factories were attributed to military opposition.

* Not including the Mindanao conflict.

Africa

Angola		Angolan Govt	100 000	1975–89: >25 600*	. .*	*n.a.*
	1975/1975	*vs.* UNITA	65 000			
	1975/1975	*vs.* FLEC	. .			
	1975/1975	*vs.* FNLA	. .			

Comments: The Govt faces armed opposition by UNITA (National Union for the Total Independence of Angola), which is militarily supported by the USA and Zaire (supply route), as well as FLEC (Front for the Liberation of the Enclave of Cabinda) and FNLA (Angolan National Liberation Front). The Angolan (MPLA–PT, Popular Liberation Movement of Angola–Worker's Party) Govt is militarily supported by the USSR. The conflict was formed when the power-sharing agreement at independence between the groups (except FLEC) failed. Stalled peace efforts between the Govt and UNITA in 1989. In late Dec. 1989 the second military offensive of the year against UNITA ('Final Assault') was launched by the Govt, reportedly the largest since 1975 and resulting in intensified fighting in Jan. 1990. In early Feb. UNITA lost control over the southern town Mawinga. The Govt amnesty law, which came into effect 1 year earlier, was extended on 4 Feb. A meeting in early Feb., aiming at getting the peace process moving, between the heads of state of Gabon, Congo, Cameroon and Zaire, resulted in peace proposals. In early Mar. UNITA proposed cease-fire talks under the mediation of Zaire Pres. Mobuto, conditioned on the withdrawal by the Govt from the past 2 months' territorial gains. After a meeting with the heads of state of Gabon, Congo and Sao Tome and Principe, held in early Apr., the Govt announced willingness to hold direct talks with UNITA. A few days later, UNITA withdrew previously made conditions for talks. Direct talks in late Apr. in Portugal resulted in a time-table for further talks. A cease-fire was not discussed. The Govt pulled back its forces from Mawinga in early May. The withdrawal was made in the context of logistical problems, hampering effects of the rainy season and increased presence of UNITA in the north where they reportedly were building up a new HQ. UNITA launched an offensive in May. A UNITA proposal in early June for a 3-month cease-fire from 22 June was dismissed by the Govt. A second round of talks in mid-June in Portugal was terminated when the UNITA delegation was called home because of 'communication difficulties' with their Angolan HQ. No concrete results were reported. In early July the MPLA central committee announced that the country would evolve towards a multi-party system (one of UNITA's main demands) conditioned on the end of fighting. A third and fourth round of talks (Aug. and

Sep., respectively) in Portugal, with the USA and the USSR participating (Sep.), led to a principle agreement on monitoring procedures for an eventual cease-fire. The Govt launched an offensive in the south on 22 Oct. A few days later the MPLA–PT central committee endorsed reforms envisaging a multi-party system and said that this would take place despite the fighting. A revision of the constitution is to follow and elections are to be held after a cease-fire agreement is signed. A fifth round of talks in Portugal mid-Nov. In early Dec. the multi-party system reforms were formally approved by the MPLA–PT party congress. The USA and the USSR drew up a peace plan in mid-Dec. which included a cease-fire, internationally observed elections and halt of arms supplies to both sides once a cease-fire is agreed upon by the warring parties. Cuba continued its UN-monitored military withdrawal during the year, to be completed in mid-1991. It was temporarily suspended in Jan. after Cuban troops (who do not participate in the fighting) were attacked by UNITA. A reported faction of FLEC attacked an oil installation in Apr. FNLA took responsibility for a bomb blast at Luanda's international airport in late Dec. According to the Govt, 1003 former members of the resistance movement UPA/FNLA (Union of the Angolan People/Angolan National Liberation Front) surrendered in Jan.

* Figures are for up to 1989. During Jan.–Oct. 1990 the Govt and UNITA claimed to have killed over 1800 of each other's soldiers, but each only admitted to having incurred small losses. According to the Govt, UNITA in addition killed 1720 civilians during the same period.

Chad		Chad (Habré) Govt,	17 000	1965–90: 33 800	5 800	+ +
	1989/1989	French Govt,	1 800 (Dec.)			
	1982/1987	*vs.* Islamic Legion	2 000–5 000			
	1989/1989	*vs.* MPS	2 500–5 000			
	1973/1979	(*vs.* Libyan Govt (Aozou Strip))	(2 000)			

Comments: Different Chadian factions have been fighting each other since 1965. In 1982 Hissène Habré seized *de facto* power. In 1987 the conflict became a struggle between the combined forces of the Govt and previously Libyan-backed Oueddai against Libya. The Habré Govt has received support from France (troops), the USA (military equipment and diplomatic support during the Reagan Administration, and diplomatic support during the Bush Administration) and Iraq (military equipment). A cease-fire was agreed between Chad and Libya in 1987 and a peace accord (Algiers Treaty) signed in 1989. The Strip has been occupied by Libya (still with 2000 troops in 1990) since 1973 and it has been fought over since 1979. During 1990 the main challenge to the Habré Govt consisted of the Mouvement Patriotique du Salut (MPS) and the Islamic Legion (IL). The MPS is led by former Chadian Commander-in-Chief Idriss Déby, who fled to Sudan after a failed coup in Apr. 1989. The IL arose from the remnants of the anti-Habré forces of Muslim (northern) Chad after their 1987 defeat. In an attempt to further his policy of 'national reconciliation' Pres. Habré arranged for a referendum on his single (UNIR) party rule and a constitution limiting the national assembly to a 5-year tenure. Elections were held on 8 July 1990. Most of the fighting in 1990 took place in the Ourouba region on the eastern border with Sudan, the tribal homeland of the Zaghawa people, among whom Déby has his base. On 25 Mar. 1990 the MPS, with IL support and using MPS bases in Sudan and camps in south-eastern Chad, took the Chadian border garrisons at Bahai and Tine, which were recaptured by the Govt on 27 Mar. Govt troops crossed into Sudan and killed 7 injured insurgents in hospital at Kouttoum. Both Libya and Sudan had repeatedly denied assisting anti-Habré forces. On 30 Mar. French troops from the 1300-strong garrison in the capital N'djaména were sent to the regional capital Abéché to replenish the French garrison from

Location	Year formed/ year joined[a]	Warring parties[b]	No. of troops in 1990[c]	Deaths[d] Total (incl. 1990)	During 1990	Change from 1989[e]

100 (as stipulated in the Algiers Treaty) back to a force of 800 men. On 31 Mar., 170 paratroops were sent from France to N'djaména to protect the Govt. The heaviest fighting since 1987 took place around Iriba (Apr.). Déby claimed the deaths of 763 Govt soldiers, while Govt sources claimed to have killed 730 'pro-Libyan fighters'. On 10 Apr. French forces reportedly counter-attacked Déby's fighters, driving them back to border positions. Several opposition supply convoys were attacked (Apr./May) by Govt forces operating up to 200 km inside Sudan: e.g., on 19 Apr., killing 517 men, all claimed to be part of a consignment of 3000 men sent from Libya to reinforce the 2500 IL soldiers already in Dafur province (Sudan). At an extraordinary meeting of the cabinet and the UNIR Executive Bureau on 31 July, a communiqué was issued on the evolution of the Chad–Libya conflict. The statement claimed that, despite the negotiation efforts of the OAU and the heavy defeats suffered by the Libya–Sudan alliance in Oct. 1989 and Mar./Apr. 1990 in the border area, it was thought that preparations were being made to launch a total war against Chad. It also said that, to that end, a special Libyan–Sudanese brigade was known to be escorting opposition convoys from Kufra in Libya to Darfur province in Sudan. However, at a meeting in Morocco on 22–23 Aug. 1990 (after other meetings during the year) the Libyan and Chad Govts agreed to submit their Aozou Strip dispute to the adjudication of the International Court of Justice at The Hague. A new MPS offensive was launched on 10 Nov. The MPS denied the Govt claim that the offensive was an IL invasion and the Libyan claim that it was merely a tribal conflict. After 17 days of fighting, during which Habré lost 1700 soldiers (two-thirds of his best troops), the garrison at Abéche was withdrawn to N'djaména. 1300 civilians and 800 MPS soldiers were also reported killed. The French 'Epervier' (Sparrowhawk) force abstained from fighting. A further 150 Legionnairs were sent to Chad on 17 Nov. The vanguard of the MPS entered N'djaména on 1 Dec. as Habré sought refuge in Cameroon and was granted asylum in Senegal. Déby proclaimed himself President on 5 Dec., committed to a policy of 'political pluralism, non-alignment and anti-colonialism'.

Ethiopia		Ethiopian Govt	438 000	1962–90: 500 000*	>10 000*	0
	1970/1971	*vs.* EPLF	40 000–50 000			
	1976/1976	*vs.* TPLF	30 000–40 000			
	1975/1980	*vs.* EPDM	. .			
	1977/1977	*vs.* OLF	7 000			
	1974/1975	*vs.* EPRP	. .			
	1975/1975	*vs.* ALF	. .			

Comments: The Ethiopian Govt is involved in 2 major conflicts and a host of minor conflicts. (1) The conflict with Eritreans has persisted since 1961, following the incorporation of Eritrea into the Ethiopian Empire. Main guerrilla movement fighting for Eritrea's independence is today the Eritrean People's Liberation Front (EPLF). It has established *de facto* territorial control over some parts of Eritrea. In 1990 a mobilization of forces on both sides. In early Jan. 1990

EPLF used gunboats on the Red Sea. On 8 Feb. EPLF initiated a military offensive, capturing the important port of Massawa after intensive fighting. Massawa has since then been exposed to Ethiopian bombings. The port had been a main point of entry for delivery of aid to famine-stricken areas. Following this, EPLF surrounded Asmara, the capital of Eritrea, which since then has had only air links to the outside world. Ethiopian losses in Mar.–May estimated at 30 000 soldiers: killed, wounded and captured. Heavy battles continued in May and June around Asmara. In Sep. there was again heavy fighting around Asmara as well as Govt air raids on Massawa. Peace talks between Govt and EPLF that were initiated during the previous year stalled in Nov. 1989 (EPLF demanded UN observers but the UN did not agree to participate). As a result of the summit meeting in May between Presidents Bush and Gorbachev a joint statement was issued calling for a UN conference to mediate an end to the conflict. The Govt announced it would no longer object to UN observers in negotiations with EPLF. However, EPLF declared that it wanted a UN referendum on the future of Eritrea, rather than continued talks. In Oct. EPLF and Govt representatives met in Washington with the US Assistant Secretary of State for African Affairs. In Dec., after separate negotiations with the World Food Program, the Govt and EPLF agreed to re-open Massawa for food relief shipments to prevent famine in the area. (2) There has been a conflict since 1974 with the Tigray People's Liberation Front (TPLF) demanding a change of Govt in all of Ethiopia, but with a right for Eritrea to choose whether to remain part of Ethiopia or become independent. In the early 1980s EPDM (Ethiopian People's Democratic Movement), with similar demands, also began an armed struggle. In 1990 the forces involved were larger than in previous years. Govt claimed to have recaptured, by mid-Jan. and early Feb., towns in Gondar and Shoa provinces previously lost to TPLF. In Mar. some TPLF forces were again reported in Shoa province. In June severe battles took place around Dessie, in the Wollow province. TPLF forces remained c. 100 km from the capital Addis Ababa. Peace talks took place in Rome in Mar. but were not renewed. Other conflicts in Ethiopia are the following. There is an armed conflict with the Oromo Liberation Front (OLF), who demand an independent Oromia, which would require that the Ethiopian borders are radically redrawn. OLF captured towns in the Asosa region close to the Sudan border in early Jan. Attacks were also reported in Aug. and Sep. In addition, sporadic attacks by the Afar Liberation Front (ALF) took place in 1990. The Ethiopian People's Revolutionary Party (EPRP) was active during the year, claiming to have occupied towns in the Lake Tana area in July and Aug. In Mar. the Govt announced it would change to a market economy. In May 1990, 12 generals who were involved in a coup attempt in May 1989 were executed. In late June a call was issued by Ethiopian authorities for 'non-stop recruitment', i.e. general mobilization. The number of Soviet military advisers was reported to have declined from 1500 to 600.

* The 500 000 deaths refer to the Eritrean conflict and include both military and civilian deaths. It is unclear whether the figure includes all deaths in connection with the conflict, i.e., not only battle-related deaths. The >10 000 deaths during 1990 is an estimate of the number killed in all ongoing conflicts in Ethiopia, i.e., not only the Eritrean conflict. Exact figures are not available.

Liberia		Liberian (Doe) Govt	5 000–7 800	1989–90: 10 000–13 000	. .	*n.a.*
	1989/1989	NPLF	200–14 000*			
	1990/1990	INPLF	400–4 000*			
	1990/1990	ECOMOG	2 500–6 000			
	1990/1990	Burkina Faso Govt	. .			

				Deaths[d]		
Location	Year formed/ year joined[a]	Warring parties[b]	No. of troops in 1990[c]	Total (incl. 1990)	During 1990	Change from 1989[e]

Comments: Pres. Samuel Doe, who came to power in 1980 after a coup, has survived numerous plots and attempts against his regime. In late Dec. 1989, 60–200 men (some of whom had been involved in a coup attempt in 1985) crossed into Liberia from neighbouring Ivory Coast, with the goal of overthrowing Doe. The group (reportedly financed by Libya and trained in Burkina Faso) called itself the National Patriotic Forces of Liberia (NPLF) and was led by Charles Taylor, a former high-ranking civil servant who fled Liberia in 1983, charged with embezzlement and who described himself as a 'die-hard capitalist'. In Mar. NPLF was split when a former Army captain and close associate of Taylor, Prince Johnson (who had parted with Taylor in Feb.), formed a rival splinter group, INPLF (Independent NPLF), and claimed that Taylor was 'a Libya-trained socialist and criminal'. The incursion, which initially seemed to crumble, gained momentum as the conflict assumed an increasingly ethnic character. In late May the rebels were said to control two-thirds of the country. Attempts by the USA to get the parties to talk failed in May. In early June the Govt and NPLF announced readiness for talks based on proposals by religious leaders. Talks were held mid-June in Sierra Leone under the mediation of the Liberian Council of Churches: the Govt demanded that NPLF drop its armed struggle and contest in the 1991 elections while NPLF demanded Doe's resignation as a precondition for cease-fire and elections by the end of 1990. The talks resulted in a mutual promise of self-restraint in using military force and free movement of humanitarian aid. A few days later the Govt issued an amnesty to all members of opposition groups, incl. NPLF, and allowed previously forbidden political parties. The talks were agreed to continue late June, thereby allowing time for the Govt delegation to consult with Doe, but did not take place since NPLF had set Doe's resignation as a precondition for further talks. Heavy fighting reached the capital Monrovia in early July. A Govt offer on 7 July of unilateral cease-fire was rejected by NPLF. At talks between NPLF and the Govt in Sierra Leone in mid-July, mediated by ECOWAS (the Economic Community of West African States), NPLF rejected the ECOWAS peace formula (sending a peace-keeping force to establish a cease-fire, an interim Govt and free elections), pointing out that Doe must resign and leave the country. NPLF announced a few days later that they would not attend any more talks. At the end of July Taylor proclaimed himself President and promised to hold elections within 6 months. Amid threats to their citizens and the burden of Liberian refugees, on 9 Aug. ECOWAS decided to send a peace-keeping force, ECOMOG (ECOWAS Monitoring Group), to Liberia. The decision was welcomed by INPLF and Doe while the NPLF threatened to attack any intervention in Liberia's 'internal matters' and described it as a 'plot'. NPLF, e.g., regarded Guinea and Nigeria (members of ECOMOG) to be pro-Doe. On 18 Aug. a cease-fire was agreed between INPLF and the Govt. On 24 Aug., after cease-fire talks (21–22 Aug.) between ECOWAS and NPLF had failed, ECOMOG arrived in Monrovia and fighting against NPLF broke out almost immediately. Burkina Faso reportedly started to send troops and arms to NPLF in Aug. since it opposed the ECOMOG intervention. At an ECOWAS meeting in Gambia on 30 Aug. Amos Sawyer, an exiled opposition leader, was declared head of a broad-based interim Govt until elections are held, due to take place in Oct. 1991. On 9 Sep. Doe was captured by the INPLF at the ECOMOG HQ in Monrovia, where he had been invited by INPLF for talks concerning the creation of a common force against NPLF. The same day Johnson proclaimed himself President, while forces loyal to Doe appointed Gen. David Nimley as constitutional successor of Doe and interim president. The INPLF killed Doe the following day. A peace proposal by

Sawyer in mid-Sep. consisting of an interim legislative assembly was rejected by NPLF since they failed to receive the presidency. Efforts by the USA led to a temporary cease-fire from 22 Sep. and lasted for a few days. On 1 Oct. ECOMOG launched an offensive alongside INPLF (who were promised 4 seats in the Sawyer interim legislative assembly) and the forces of Nimley, forcing NPLF to retreat. An ECOWAS-sponsored meeting held in Gambia in late Oct., aiming at a cease-fire, failed since NPLF refused an agreement. Amid setbacks in the fighting and reportedly the end of support by Libya and Burkina Faso, NPLF agreed to a cease-fire in late Nov. Fighting in early Dec. between INPLF and forces under Nimley. On 21 Dec. the parties agreed to form an interim Govt. By the end of the year most of the country was still controlled by NPLF.

* Different assessments during the year.

Morocco/ Western Sahara	1975/1976	Moroccan Govt *vs.* Polisario	80 000–150 000 5 000–20 000	1975–89: 10 000–13 000*	. .	*n.a.*

Comments: The former Spanish colony of Western Sahara was divided between Morocco and Mauritania in 1975. Morocco annexed the Mauritanian half in 1979, following Mauritanian withdrawal and agreement with the Popular Front for the Liberation of Saguia el Hamra and Rio de Oro (Polisario). Polisario (and its military wing, Sahrawi People's Liberation Army) is fighting for independence for the Saharan Arab Democratic Republic and is based mainly in Algeria. Morocco has completed construction of a wall to keep Polisario out. In Aug. 1988 Morocco and Polisario accepted a UN peace plan, incl. a referendum (to be supervised jointly by the UN and the Organization for African Unity, OAU) to decide upon the status of the territory (part of Morocco or an independent state). Meeting between King Hassan of Morocco and leaders of Polisario in Jan. 1989 was followed by a truce. Fighting renewed in late Sep. when no progress was made to achieve agreements on details in the peace plan. After a visit by UN Secretary-General de Cuellar in Mar. 1990 (second visit in a year) the parties agreed on a new truce. Tribal leaders met in Switzerland under UN auspices (June) to discuss who will be allowed to vote in a possible referendum. In June the UN Security Council approved the peace plan and preliminary plans to establish a UN Mission for the Referendum in Western Sahara (MINURSO) and UN-supervised talks held in Switzerland (July).

* Figures for up to 1989. Military activity during 1990 was low.

Mozambique	1975/1976	Mozambican Govt, Zimbabwe Govt, *vs.* MNR	60 000 12 000 10 000–20 000	1985–89: 7 000–9 000 (mil.)* 100 000 (civ.)*	. .	*n.a.*

Comments: The MNR (National Resistance Movement or RENAMO), which has been fighting the Mozambican (Frelimo) Govt since 1976, receives weapons and supplies from non-Govt organizations (e.g., right-wing and/or religious organizations) and individuals, and support with base facilities, supply routes and training from S. African and Kenyan territories. The Govt receives different forms of military aid, mainly from the USSR, the UK and France. In addition, military co-operation takes place between the Govt and Zimbabwe. MNR also makes raids into neighbouring countries. The Govt amnesty, announced in 1987, has been without major success. Mediation efforts, initially started by the Mozambican Christian Council in Aug. 1988, were taken over by the Presi-

Location	Year formed/ year joined[a]	Warring parties[b]	No. of troops in 1990[c]	Deaths[d] Total (incl. 1990)	During 1990	Change from 1989[e]

dents of Zimbabwe and Kenya in Oct. 1989. In July 1989 Pres. Chissano presented a 12-point peace plan which demanded acceptance of the 1-party system constitution as a precondition for talks. It was rejected by MNR which in Oct. presented a 16-point peace plan demanding a multi-party system and general elections. Indirect peace talks, opened in Aug. 1989, bogged down in early Oct. A 7-point proposal for peace talks, similar to the previously made 12-point Govt plan saying that changes of the political system should be made by peaceful means, was put forward by the USA in early Dec. The following mediation efforts centred on a document urging both sides to negotiate without preconditions. This was agreed by the Govt in late Dec. Pres. Chissano presented a draft constitution in Jan. 1990, proposing changes in a liberal direction inside the 1-party system. The Govt launched its first offensive of 1990 in Feb., followed by a second push in May, when also MNR launched an offensive. After having solved the differences on date and place, the first direct talks ever were held in Rome on 8–10 July. After these talks the Govt announced the end of the mediation efforts by Zimbabwe and Kenya which had not been free of friction: Zimbabwe is considered by MNR as pro-Mozambican because of its military support of the Govt, while Kenya is seen by the Govt as pro-MNR because of MNR presence on Kenyan territory. The Frelimo Political Bureau decided on 31 July in favour of abandoning the 1-party system. This, together with the announcement in July of general elections to be held in 1991, met the 2 foremost demands of the MNR. A second round of talks was held in Rome in mid-Aug. in which MNR insisted that Kenya be reinstated as mediator. A third round of talks was held in Rome in Nov., after being postponed by the MNR in protest at a Govt military offensive launched in Sep. On 30 Nov. a new constitution came into effect, providing for political pluralism. It was rejected by the MNR as invalid, claiming it had been decided in an undemocratic manner, now arguing for power-sharing and indicating the need for changes of the new constitution. On 1 Dec. a partial cease-fire to be implemented from 15 Dec. and limited to 2 transport corridors was agreed. It was to be monitored by an 8-nation Joint Verification Commission (JVC) and confined the Zimbabwean troops (whose complete withdrawal is an MNR demand for a total cease-fire) to the corridors. A fourth round of talks began on 18 Dec., at which the JVC was constituted. The USA, UK, S. Africa and Portugal have also been involved in peace efforts during the year. According to UNICEF, at least 494 000 children and infants are estimated to have died directly or indirectly in 1980–88 as a result of the conflict.

* Figures are for up to 1989, since figures are not available for 1990.

Location	Year formed/ year joined	Warring parties	No. of troops in 1990	Total (incl. 1990)	During 1990	Change from 1989
Somalia		Somalia Govt	65 000	1981–90: 50 000– 60 000	>1 000 (est. mil.)	+ +
	1981/1981	*vs.* SNM	10 000			
	1989/1989	*vs.* SPM	1 000–3 000			
	. . /1990	*vs.* USC	1 000			

Comments: The Somali National Movement (SNM), composed largely of the Isaaq clan in the north of the country, has waged armed struggle against the Barre Govt since 1981. In 1988 fighting escalated, following SNM's establishment of bases inside Somalia. Territorial control has been kept since then. In 1989 armed opposition, the Somali Patriotic Movement (SPM), also emerged in the south of the country, continuing its armed struggle in 1990. Efforts were made to create a joint front of the 2 movements. In Aug. 1990 a third movement, the United Somali Congress (USC), was reported to initiate renewed armed action after a time of little activity, killing an Army general. In Aug. a joint command structure of the 3 movements was set up. In May a manifesto was issued by elders (local clan leaders) on a peaceful solution of the conflicts in Somalia. In July a Govt committee for peace talks was created. Italy was reported to have contacted the SNM, but talks did not take place. Fighting intensified in Sep. and Oct. In Oct. the Govt announced a new constitution, allowing only the propagation of Islam. It also provided for a multi-party system. In late Nov. the 3 opposition groups refused to participate in talks with the Govt. In Nov. and Dec. these groups reported significant military advances in different parts of the country. The unpopularity of the Govt was underscored in a massacre following peaceful protests against the Govt during a sporting event in Mogadishu (July). At various times the conflicts in Somalia spilled over into other countries: attacks were made on a village on the Kenyan border (Apr.), shooting took place in Djibouti involving Somali forces (May), and a Soviet ship was seized by SNM in the Gulf of Aden (July). The total number of deaths in the armed conflicts was in Sep. 1990 estimated at more than 50 000 since 1988, mostly involving civilians.

South Africa		South African Govt	77 400*	1984–90: >7 750**	>3 400**	+ +
	1950/1984	*vs.* ANC	6 000–10 000			
	1979/1983	*vs.* Inkatha *vs.* ANC***				
	1990/1990	*vs.* white rightist groups	80–550			

Comments: The conflict over the apartheid politics of the Nationalist Party (NP), which has formed the Govt since 1948, has continued since 1950. Since the early 1960s the African National Congress (ANC, and its military wing Umkhonto we Sizwe, Spear of the Nation) has been main armed opponent. The armed struggle intensified after 1984 and consisted mostly of sabotage and bombs, but in 1989 a shift was made to non-armed action. On 2 Feb. 1990 Pres. F. W. de Klerk announced the impending release of ANC leader Nelson Mandela and the lifting of the ban on ANC and other organizations, saying that 'the time for negotiation has arrived'. Mandela was released on 11 Feb. On 2–4 May the Govt and ANC representatives met in Cape Town, and measures to improve relations were agreed. On 7 June de Klerk announced that the state of emergency would not be renewed. On 6 Aug. an agreement was signed by de Klerk and Mandela, suspending the ANC armed struggle and setting a stage for peaceful settlement. On 31 Aug. the NP announced that it would be open to non-white members. Violence has increasingly shifted to becoming less structured battles between supporters of ANC (and affiliated organizations) *vs.* Inkatha, the Zulu-based organization led by KwaZulu homeland leader Chief Buthelezi. Inkatha was established in 1975 with ANC blessing, but disputes began in 1979 and fights emerged in 1983 between the 2 groups. The political significance related to rivalry with ANC on who is to negotiate with the Govt. In Natal rival groups fighting for control over the townships of Pietermaritzburg and Durban in Feb. led to more than 100 deaths. Shootings in Sebokeng, Transvaal, in Mar. caused deaths of at least 11 people. ANC–Inkatha-violence deaths for the first 6 months were estimated at 1600, 75% of whom died in Natal. Intensification of tribal

Location	Year formed/ year joined[a]	Warring parties[b]	No. of troops in 1990[c]	Deaths[d] Total (incl. 1990)	During 1990	Change from 1989[e]

deaths (Zulu *vs.* Xhosa) followed in Aug. and Sep. in the Johannesburg area, with 750 deaths in 5 weeks, incl. a massacre of train commuters in Johannesburg (Sep.). In Sep. the Govt announced operation 'Iron Fist' to quell the fighting, and some opportunity was opened for talks between ANC and Inkatha. However, by Dec. more than 1000 deaths were estimated in the Johannesburg area alone, and no meeting had taken place between the 2 African leaders. In Oct. emergency laws were lifted in Natal, to promote further talks between ANC and the Govt. The political changes in S. Africa also affected the so-called independent states and homelands. A coup in the homeland of Ciskei on 4 Mar. ousted Pres. Lennox Sebe by Brig. Gqozo. In riots that followed, 27 people were killed. S. African troops were sent in. In Boputhatswana 14 people were killed on 7–8 Mar. when demanding reintegration of the homeland into S. Africa. In Venda a coup in Apr. brought to power a regime in favour of reintegration. In Gazankulu up to 30 people were killed in demonstrations, demanding resignation of the homeland chief minister. In late Aug. a referendum was announced in Transkei on possible reincorporation into S. Africa. A coup attempt in Transkei in Nov. was severely repressed by the ruler, Holomisa. Another related development concerned the emergence of white right-wing armed groups. In June and July a series of bomb attacks were linked to different ultra-rightist groups. In Sep. a right-wing leader was arrested in connection with bombings and arms theft.

* Total armed forces, including National Service, excluding reserves and homelands. In addition, the S. African police number 60 000.

** Including deaths connected with the struggle between ANC and Inkatha supporters.

*** In the conflict between supporters of Inkatha and ANC, the S. African Govt has a history of favouring one party over the other, thus creating a triangular conflict.

Location	Year formed/ year joined	Warring parties	No. of troops in 1990	Total (incl. 1990)	During 1990	Change from 1989
Sudan		Sudanese Govt	75 700	1983–90: >33 000 (mil.)	1 000 (est. mil.)	0
	1980/1983	*vs.* SPLA/SPLM	55 000			
	1990/1990	*vs.* military factions	. .			

Comments: Since 1983 the Sudanese People's Liberation Army/Movement (SPLA/SPLM) has been fighting the central Govt to increase autonomy of the southern region and to repeal the Islamic Law (Sharia) introduced for the entire country—in contradiction to the peace agreement of 1972, as held by SPLA. Since June 1989 the Armed Forces of Sudan have held power under Brig.-Gen. el-Bashir. Direct talks chaired by former US Pres. Carter on peace occurred in Dec. 1989, without result. SPLA, with increased size of its armed force, made military advances around the cities of Juba and Yei in early Jan. 1990. A local cease-fire was agreed to allow rescue workers to leave Juba (Jan.). In June Govt forces made air raids on Torit, a town held by SPLA for more than 1 year. Heavy fighting was reported in the Upper Nile province in June and July. In the Equatoria province, the Govt continued to hold the major towns (Yei and Juba). In Mar. US Assistant Secretary of State for African Affairs presented in Khartoum a settlement proposal involving 4 phases. The Sudanese Govt finally rejected the proposal in June, as it was seen to internationalize the conflict. Subsequently, several other countries offered to mediate. However, fighting continued, particularly severely in the Equatoria and Upper Niles provinces (Aug.–Nov.). In Oct. both sides proposed 'a period of tranquillity' to allow a

UNICEF child vaccination programme, but no agreement was reached. A proposal for a temporary cease-fire to allow refugees in Kharom to return home also went unheeded (Oct.). Acc. to the Govt, 2 coup attempts were defeated (Mar. and Apr.), and 31 officers were executed. SPLA was accused of involvement, but this was denied by the SPLA. In Mar. Govt human rights violations were criticized by Africa Watch. In Nov. a further coup attempt failed, and mass arrests were carried out.

Uganda		Uganda (NRM) Govt	75 000	1986–90: >11 000 (mil.)	. .	*n.a.*
	1986/1986	*vs.* UPDM/UPDA	5 000			
	1987/1987	*vs.* UPA	2 000			
	1987/1987	*vs.* UDCM	200–800			

Comments: The NRM (National Resistance Movement) Govt seized power in Jan. 1986 after having defeated the Govt under Gen. Tito Okello. The UPDM (Uganda People's Democratic Movement), formed in 1986, is fighting the Govt in the northern part of the country with its armed wing, the UPDA (UPD Army). In June 1988 the UPDA signed a peace agreement with the Govt which provided for their integration in the NRA (NR Army). The agreement, signed by a local commander (Lt.-Col. John Angelo Okello), was rejected by the political wing, and fighting by the remaining UPDA faction continued, led by Odong Latek. In July 1990 a second peace agreement between the Govt and UPDM (reportedly no longer in effective political existence) was signed, providing for their absorption into the NRA, but was rejected by the UPDA military commanders. The Govt also faces armed resistance from the HSM (Holy Spirit Resistance Movement), initially led by Alice Lakwena. The movement, fighting in the north, launched a large-scale offensive against the Govt in early 1987, leading to its near extinction by the end of the year, having reportedly suffered 7000–10 000 deaths. In Apr. 1987 the movement was reorganized under Joseph Kony, who incorporated men from the UPDA, but has today only 200–800 men left due to war losses and surrenders to the Govt. In 1990 the movement changed its name to UDCM (United Democratic Christian Movement). UPA (Uganda People's Army), formed in 1987, is fighting the Govt in the east. A peace agreement in Apr. 1988 between UPA and the Govt was not adhered to since a faction of the UPA decided to continue fighting. Ugandan armed opposition has been crippled by war losses and large-scale surrenders and is reportedly no longer a serious threat to the Govt. In 1990 the fighting seemed to continue on a relatively small scale.

Central and South America

Colombia		Colombian Govt	130 000*	1980–90: >8 500**	1 000**	–
	1949/1978	*vs.* FARC	5 000–6 000			
	1965/1978	*vs.* ELN	1 500–3 000			
	1968/1977	*vs.* EPL	800–1 500			

Location	Year formed/ year joined[a]	Warring parties[b]	No. of troops in 1990[c]	Deaths[d] Total (incl. 1990)	During 1990	Change from 1989[e]

Comments: Since the 1970s, bombings, kidnappings and armed attacks have been staged by several revolutionary groups. The Simon Bolivar Guerilla Co-ordinating Committee was formed in 1987, then consisting of the Fuerzas Armadas Revolucionarias de Colombia (FARC), the April 19 Movement (M-19), the Partido Revolucionario de los Trabajadores (PRT), the Ejército Popular de Liberación (EPL), Quintin Lame and Camilista Union (representing the Ejército de Liberación Nacional (ELN)) and thus constituting the bulk of an est. total of 30 000 guerrilla combatants (1987/88) in the country, active on 60 fronts. Peace talks in 1987 between the Govt and the Committee were in 1988 followed by a Govt peace plan. In 1989 the then largest guerrilla group, M-19, declared a unilateral cease-fire as part of its transition to civil political life. Presidential elections on 27 May 1990 gave the M-19 candidate unexpectedly strong support. Elections to National Assembly on 9 Dec. confirmed M-19 support, giving it 19 of 72 seats, threatening more than a century of political dominance by the Liberal and Conservative Parties. The National Council for Normalization was set up by the Govt in Feb., aiming at duplication of the M-19 process for the other armed opposition groups. On 14 June the EPL started concentrating troops in a 'neutral zone' as a first step towards demobilization. A Govt air attack on 9 Dec. against the FARC HQ claimed the death of 50 soldiers and rebels, but on 22 Dec. ELN, together with FARC, announced their positive attitude towards peace talks with the Govt, although guerrilla activities continued throughout the year. A small guerrilla group, the PRT, will return to civil political life in Jan. 1991, following an announcement on 29 Dec., if pardoned by the Govt. From Jan. to Sep. 1990, 1000 persons died and 182 disappeared in Colombia for 'political reasons'. The Center for Criminology Investigation, Bogotá, reported the violent deaths of 22 500 persons in 1990, incl. guerrilla, narcotics and common crime violence.

* Active forces in the Army, Navy and Air Force.
** Politically related deaths (i.e., excluding deaths resulting from fighting between Govt and cocaine cartels). The figures does not include Oct.–Dec. 1990.

El Salvador		Salvadorean Govt	44 000	1979–90: 76 000	1 500–2 000	--
	1976/1979	*vs.* FMLN	6 000–8 000			

Comments: The FMLN (Farabundo Martí Front for National Liberation) is a coalition of 5 armed opposition groups (People's Revolutionary Army, ERP; Popular Liberation Forces, FPL; Armed Forces of National Resistance, FARN; Revolutionary Party of Central American Workers, PRTC; and Armed Forces of Liberation, FAL) fighting a guerrilla war against the Salvadorean Govt, which throughout the 1980s was supported militarily by the USA. On 20 Nov. 1990 the FMLN launched an offensive against Govt troops, claiming more than 1000 casualties (incl. civilians). The offensive resembled that in Nov. 1989. In 1990 the FMLN obtained surface-to-air missiles, causing a major shift in the military balance between Govt and FMLN. These offensives mark an escalation of violence that has occurred in the past 3 years. The Nov. 1990 offensive was terminated on 31 Dec. In early 1990 UN Secretary-General de Cuellar agreed to mediate between the Govt and FMLN, sending his envoy, Alvaro de Soto. Several talks between the FMLN and Govt under UN auspices were held during

1990. On 24 July, the Govt and FMLN agreed, i.a., to set up an international commission verifying compliance with human rights principles in El Salvador. Further talks between the Govt and FMLN were blocked in Oct. on the issue of the future of the Govt Army. On 27 June the US House of Representatives approved a congressional decision on a conditional blocking of 50% (i.e., $48 million) of 1990 US military aid, depending on Govt and FMLN negotiation postures and the Govt's procedures against murderers of 6 academic Jesuits in Nov. 1989. For Jan.–June 1990, the non-governmental Commission for Human Rights in El Salvador claimed that 695 were murdered, 481 detained and 92 disappeared. Most of these were attributed to the Army, guerrillas and right-wing death-squads. Attacks and clashes between the FMLN and Govt were frequent throughout the year, causing the total death figure since 1979 to reach 76 000, incl. c. 25 000 Govt and FMLN soldiers.

Guatemala		Guatemalan Govt	43 000	1962–90: 20 000–60 000	<500	*n.a.*
	1967/1968	*vs.* URNG	1 000–2 000			

Comments: The armed opposition against right-wing military Govts dates back to the 1960s. In 1982 the Guatemalan National Revolutionary Unity (URNG) was formed to co-ordinate the forces of 4 groups (Ejercito Guerrillero de los Pobres, EGP; Partido Guatemalteco del Trabajo, PGT; Fuerzas Armadas Rebeldes, FAR; and Organizacion del Pueblo en Armas, ORPA). In 1982–83 a massive counter-insurgency campaign by Govt forces cut the strength of the armed opposition. In 1985, military rule was ended by the election of Christian Democrat Vinicio Cerezo as President. A reduction of political violence followed, but rose again in 1988 and 1989. 1990 was politically dominated by the presidential elections (held in Nov.; a second round is scheduled for Jan. 1991) and talks between the URNG and representatives of civil political life. In Feb. URNG staged raids against the Panamerican highway, followed by Govt air raids against suspected guerrilla positions. In mid-Dec. URNG destroyed the Moca bridge, an important communication link between Mexico and Guatemala. In Mar. the US ambassador returned to Washington for consultations because of increased human rights violations. The National Commission for Reconciliation, set up under the 1987 Esquipulas II process, held talks in Mar. in Oslo with the URNG, where the latter agreed on meeting, i.a., representatives of political parties. A Govt body, the Fiscalía de los Derechos Humanos, reported 304 death squad-type murders and 233 'disappearances' in 1990. The non-Govt Human Rights Commission of Guatemala announced 585 people killed by state security or paramilitary death-squads in the first 8 months of 1990. The figures confirm human rights organizations' claim about increased death-squad activity in 1990.

Nicaragua		Nicaraguan Govt	74 000 (Jan.)* 28 000 (Dec.)	1981–90: >30 000 (mil.)	<100 (mil.)	– –
	1981/1981	*vs.* Contras	(12 000)**			

Comments: The armed conflict between the Nicaraguan Govt and Contras (counter-revolutionaries) ended in 1990. In the wake of the Sandinista revolution against the Somoza regime in 1979, the US-supported Contras was founded in 1981. It staged major offensives against Govt troops in 1983–84, with continuing attacks throughout 1985–86 but diminishing in 1987. After the Esquipulas II Agreement of Aug. 1987, Pres. Ortega proposed direct talks with the Contras. In Feb. 1989 the 5 Central American presidents met in La Paz, El Salvador, as part of the Esquipulas II process, agreeing that the Contras be disbanded in 90

Location	Year formed/ year joined[a]	Warring parties[b]	No. of troops in 1990[c]	Deaths[d] Total (incl. 1990)	During 1990	Change from 1989[e]

days. Following a summit meeting in 1989 in Tela, Honduras, where the 5 presidents agreed to set up a joint Organization of American States–UN Support and Verification Commission for the disbanding of Contra forces, the UN Security Council decided in Nov. 1989 to set up ONUCA (UN Observer Group in Central America). In sharp contrast to several opinion polls, the outcome of the 25 Feb. 1990 elections was a 55% victory for the UNO (Unión Nacional Opositora) opposition and its presidential candidate Violeta Chamorro. Despite election results Contra groups clashed with Govt troops in Mar. and Apr., resulting in the death of at least 10 soldiers. A demobilization agreement on 19 Apr. between the Govt and Contras projected 10 June as the final date. By 26 June at least 19 000 Contra soldiers had handed over their weapons and other equipment to ONUCA. Following demobilization, former Contra members staged internal disturbances in Oct., Nov. and Dec., i.a., over conditions for returning soldiers, claiming at least 11 lives.

* Throughout 1990 a continuing reduction of the Army took place. Military conscription was ended by 7 Dec.

** The Contras were disarmed in 1990. Agreements between the Govt and the Contras to disarm the RN (Nicaraguan Resistance, the largest Contra group) estimated at 15 000, put the final deadline at 29 June; Yatama (Atlantic coast Indians), estimated at 2000, 20 June; and Frente Sur (Southern Front, operating in Nicaragua, originally from Costa Rican bases), estimated at 2500, 25 June.

Location	Year formed/ year joined	Warring parties	No. of troops in 1990	Total (incl. 1990)	During 1990	Change from 1989
Peru		Peruvian Govt	120 000	1981–90: 11 500–20 000	3 400	+
	1980/1981	*vs.* Sendero Luminoso	5 000			
	1984/1986	*vs.* MRTA	500			

Comments: A group which splintered in 1970 from the Communist Party, the Sendero Luminoso (Communist Party of Peru, for the Shining Path of José Carlos Mariátegui) describes itself as 'Maoist', with the goal of bringing Indian governance back to Peru. Increased guerrilla activity in Dec. 1988 (after a period of low-intensity conflict) continued in 1989–90. In 1989 Sendero Luminoso extended its territorial occupation. Following a 3-month surveillance, the movement's Lima stronghold was in spring 1990 subsequently detected by the police. A major goal, to halt the 10 June presidential elections, was unsuccessful on the whole; turnout was 'relatively high'. In areas of strong Sendero Luminoso support, such as the Ayacucho area, increased civil resistance has been noticed. On 28 Sep. some 70 guerrilla soldiers were killed by Ashaninka Indians in a jungle fight. A large number of attacks, such as murders of Govt officials, massacres of Indians and bomb attacks, have continuously taken place throughout the country, staged by the Sendero Luminoso and the MRTA (Movimiento Revolucionario Tupac Amaru). However, on 3 Oct. the MRTA leader indicated a respite in violent actions after the presidential elections. Govt security forces were increasingly criticized during 1990 for human rights violations in alleged guerrilla counter-insurgency. A Peruvian Senate Commission reported 3384 deaths from political violence in 1990. The figure includes 'civilians, soldiers, policemen and Marxist guerrillas'. By 1990, the Sendero Luminoso struggle had claimed up to 20 000 lives.

[a] 'Year formed' is the year in which the two or more warring parties last formed their conflicting policies or the year in which a new party, state or alliance involved in the conflict came into being. 'Year joined' is the year in which the armed fighting last began or the year(s) in which armed fighting recommenced after a period for which no armed combat was recorded. For conflicts with very sporadic armed combat over a long period, the 'year joined' may also refer to the beginning of a period of sustained and/or exceptionally heavy combat.

[b] In the list of warring parties for each conflict, note that one side is always a government. The non-governmental warring parties are listed by the name of the organization conducting armed operations. Only those parties which were active during 1990 are listed in this column.

[c] The figures for 'No. of troops in 1990' are for total armed forces (rather than for army forces, as in the *SIPRI Yearbooks 1988–90*), unless otherwise indicated by a note (*). Where a range of figures is given, these are the highest and the lowest figures that were given in the sources used.

[d] The figures for deaths refer to total battle-related deaths during the conflict. The figures exclude, as far as data allow, civilian deaths owing to famine and disease. *'Mil.'* and *'civ.'* refer to estimates, where available, of *military* and *civilian* deaths; where there is no such indication, the figure refers to total military and civilian battle-related deaths in the period or year given. Information about the conflicts which covers a calendar year is by necessity more tentative for the last months of the year. Experience has also shown that the reliability of figures is improved over time; they are therefore revised each year. Where a range of figures is given, these are the highest and the lowest figures that were given in the sources used.

[e] The 'change from 1989' is measured as the increase or decrease in battle-related deaths in 1990 compared with deaths in 1989. Although based on data that cannot be considered totally reliable, the symbols represent the following changes:

+ +	increase in battle deaths of more than 100%
+	increase in battle deaths of less than 100%
0	stable rate of battle deaths (+ or – 10%)
–	decrease in battle deaths of less than 50%
– –	decrease in battle deaths of more than 50%
n.a.	not applicable, since conflict not recorded for 1989.

Sources: For additional information on these conflicts, see chapters in previous editions of the *SIPRI Yearbook*—Lindgren, K., Wilson, G. K., Wallensteen, P. and Nordquist, K.-Å., 'Major armed conflicts in 1989', *SIPRI Yearbook 1990: World Armaments and Disarmament* (Oxford University Press: Oxford, 1990), chapter 10; Lindgren, K., Wilson, G. K. and Wallensteen, P., 'Major armed conflicts in 1988', *SIPRI Yearbook 1989: World Armaments and Disarmament* (Oxford University Press: Oxford, 1989), chapter 9; Wilson, G. K. and Wallensteen, P., 'Major armed conflicts in 1987', *SIPRI Yearbook 1988: World Armaments and Disarmament* (Oxford University Press: Oxford, 1988), chapter 9; and Goose, S., 'Armed conflicts in 1986, and the Iraq–Iran War', *SIPRI Yearbook 1987: World Armaments and Disarmament* (Oxford University Press: Oxford, 1987), chapter 8.

The following other reference books were used: *Amnesty International Årsrapport 1989 [Annual report 1989]* (Amnesty International: Stockholm, Sweden); Brogan, P., *World Conflicts* (Bloomsbury: London, 1989); *Defense and Foreign Affairs Handbook* (Copley: Washington, DC, 1976); Gantzel, K.-J. and Meyer-Stamer, J. (eds), *Die Kriege nach dem Zweiten Weltkrieg bis 1984* (Weltforum: Munich, 1986); Gunson, P., Thompson, A. and Chamberlain, G., *The Dictionary of Contemporary Politics of South America* (Routledge: London, 1989); International Institute for Strategic Studies, *The Military Balance 1990–1991* (Brassey's: London, 1990); Janke, P., *Guerrilla and Terrorist Organisations: A World Directory and Bibliography* (Harvester

Press: Hemel Hempstead, 1983); Jongman, B., *War, Armed Conflict and Political Violence* (Polemological Institute, National University: Groningen, the Netherlands, 1982); Kaye, G. D., Grant, D. A. and Emond, E. J., *Major Armed Conflict, A Compendium of Interstate and Intrastate Conflict 1720 to 1985*, report R95 (Operational Research and Analysis Establishment [ORAE], Canadian Department of National Defence: Ottawa, 1985); Keesing's, *Political Dissent* (Longman: Harlow, Essex, 1983); Laffin, J., *The World in Conflict 1989* (Brassey's: London, 1989); Lindgren, K. (ed.), *States in Armed Conflict 1989* (Department of Peace and Conflict Research, Uppsala University: Uppsala, 1991); Minority Rights Group, *World Directory of Minorities* (Longman: Harlow, Essex, 1989); Munro, D. and Day, A. J., *A World Record of Major Conflict Areas* (Edward Arnold: London, 1990); Sivard, R., *World Military and Social Expenditures* (World Priorities Inc.: Washington, DC, annual); *The Statesman's Yearbook* (Macmillan: London, annual); Small, M. and Singer, J. D., *Resort to Arms, International and Civil Wars 1816–1980* (Sage: Beverly Hills, Calif., 1982); Wallensteen, P. (ed.) *States in Armed Conflict 1988* (Department of Peace and Conflict Research: Uppsala, Sweden, 1989); research reports on particular conflicts; the SIPRI Arms Trade Project data base; and information available at the Department of Peace and Conflict Research, Uppsala University, in the continuous research project on armed conflicts.

The following journals, newspapers and news agencies were consulted: *Africa Confidential* (London); *Africa Events* (London); *Africa News* (Durham); *Africa Research Bulletin* (Oxford); *Africa Reporter* (New York); *African Defense* (Paris); *Armed Forces* (London); *Amnesty Press* (Stockholm); *Asian Defence Journal* (Kuala Lumpur); *Boston Globe* (Boston, Mass.); BBC World Service News (London); *Central America Report* (Guatemala City); *Christian Science Monitor* (Boston, Mass.); *Dagens Nyheter* (Stockholm); *Defense and Foreign Affairs* (Alexandria, Va.); Dialog Information Services Inc. (Palo Alto); *The Economist* (London); *Far Eastern Economic Review* (Hong Kong); *Financial Times* (London and Frankfurt); *The Guardian* (London); *Horn of Africa Bulletin* (Uppsala); *India Today* (New Delhi); *Jane's Defence Weekly* (Coulsdon, Surrey); *IDSA Journal* (New Delhi); *Indian Express* (New Delhi); *The Independent* (London); *International Defence Review* (Geneva); *International Herald Tribune* (Paris); *Kayhan International* (Teheran); *Keesing's Contemporary Archives* (Harlow, Essex); *Latin America Weekly Report* (London); *Mexico and Central America Report* (London); *The Middle East* (London); *Nyheter från Latinamerika* (Stockholm); *New Statesman & Society* (London); *Newsweek* (New York); *Selections from Regional Press* (Institute of Regional Studies: Islamabad); *New York Times* (New York); *Pacific Defence Reporter* (Kunyung); *Der Spiegel* (Hamburg); *The Statesman* (Calcutta); *Svenska Dagbladet* (Stockholm); *Teheran Times* (Teheran); *Time* (New York); *The Times* (London); *Upsala Nya Tidning* (Uppsala;) *US News & World Report* (Washington, DC); *Washington Post* (Washington, DC); *Washington Times* (Washington, DC); and *World Reporter* (Datasolve: London).

Part III. Developments in arms control

11. US–Soviet nuclear arms control

REGINA COWEN KARP

I. Introduction

The year 1990 began with considerable hope that a START (Strategic Arms Reduction Talks) treaty could be ready for signature by the end of the year. At the December 1989 summit meeting in Malta, US and Soviet leaders felt sufficiently confident about progress towards the conclusion of a START treaty to instruct their negotiating teams to resolve all major outstanding disputes in advance of a summit meeting in the spring of 1990 (held in Washington, DC, on 31 May–3 June 1990). This would pave the way for the signing of a START treaty during a second summit meeting to be held in Moscow in December 1990, a meeting which never took place.

Indeed, much progress was made in the first half of 1990 especially concerning an agreement on counting rules for long-range nuclear-armed air-launched cruise missiles (ALCMs) and in what form to limit sea-launched cruise missiles (SLCMs). Yet, other problems, mainly to do with questions of verification and compliance (as is often the case with negotiations in their final stages), emerged in the second half of the year. The meeting between Secretary of State James Baker and Soviet Foreign Minister Eduard Shevardnadze on 10–11 December 1990 in Houston, Texas, succeeded in resolving some issues, but others have yet to be resolved. Thus not enough progress was achieved to conclude the treaty at a December summit meeting.

Hopes of signing a START treaty at a summit meeting in February 1991 were dashed at the end of January with the US announcement that the summit meeting would be postponed until late spring. While the officially stated reason for the postponement of the meeting was the war in the Persian Gulf, the recent use of force by the USSR in the Baltic republics and the continuing problems with concluding the START treaty undoubtedly influenced President George Bush's decision not to attend a planned summit meeting in Moscow in February.[1] Depending upon the impact of developments, both in the Gulf and in the USSR, on US–Soviet relations, the question of when a START treaty might be signed must remain open.

This chapter surveys the major issues—both resolved and unresolved—that were addressed in the START negotiations during 1990 and reviews the status of the Defence and Space Talks. The chapter concludes with a review of major treaty provisions and an overall assessment of the emerging START treaty.

[1] 'The superpowers, citing Gulf war, postpone summit', *International Herald Tribune*, 29 Jan. 1991, pp. 1 and 8; 'Bush and Gorbachev postpone summit', *Financial Times*, 29 Jan. 1991, pp. 1 and 18.

II. Resolved issues

The issues that were resolved in 1990 fall into two categories: (*a*) those that were resolved in time for and at the June 1990 summit meeting in Washington between Presidents George Bush and Mikhail Gorbachev; and (*b*) those that were resolved at the December meeting between Secretary of State Baker and Foreign Minister Shevardnadze in Houston, Texas. This section discusses the negotiating positions of both sides and the solutions reached at the two meetings.

Issues resolved for and at the June 1990 summit meeting

Air-launched cruise missiles

A thorny issue carried over into 1990 was that of ALCMs. The problem these systems posed was twofold: (*a*) the range at which these missiles should be constrained under START; and (*b*) how these missiles should be counted against the agreed warhead ceiling of 6000.

The US position on ALCMs was based on the view that strategic bombers, because they are slow-flying and therefore more suited to retaliatory rather than first-strike missions, should be less constrained under START than ballistic missiles. Therefore, bomber loadings should not count as heavily against the START warhead ceiling as ballistic missiles with multiple warheads. Regarding ALCM-capable bombers, the US negotiating position was to discount the actual number of ALCM warheads carried by these bombers in favour of an agreed number of 10. The USA was also interested in imposing START limits on ALCMs of a range of 1500 km and above.

In sharp contrast to the US approach, the Soviet approach to counting ALCM warheads and constraining the range of the missiles was influenced more by the larger size of the US strategic bomber force than by considerations of first- and second-strike missions. The intention was to make as many warheads as possible count against the overall warhead ceiling of 6000, forcing the USA to choose on which strategic carriers to deploy what number of warheads. Thus the USSR proposed that warhead numbers attributed to ALCM-capable bombers should be based on the maximum number for which each bomber type is equipped. The Soviet approach also aimed to achieve a sub-limit of 1100 warheads for all ALCMs carried on heavy bombers and a constraint on ALCM range of 600 km and above.[2]

At the meeting of the US and Soviet foreign ministers in Moscow on 8–9 February 1990, ALCM counting rules were agreed under which warhead numbers of 10 and 8 are attributed to US and Soviet ALCM-equipped bombers, respectively.[3] In order to reflect the different capacities of US and

[2] Interview with Ambassador Richard Burt, *Arms Control Today*, vol. 20, no. 2 (Feb. 1990), p. 4; *Wireless File*, EUR-103 (United States Information Service, US Embassy: Stockholm, 22 Jan. 1990), pp. 2–3 (hereafter *Wireless File*).

[3] *Wireless File*, EUR-502, 9 Feb. 1990, p. 2.

Soviet bombers, it was agreed that the first 150 US ALCM-capable bombers should be counted as carrying 10 ALCMs each, although they are permitted to carry up to 20. It was also agreed that the first 210 Soviet ALCM-capable bombers would be counted as holding 8 ALCMs each, although they are permitted to carry up to 12. For both sides, each additional bomber equipped for long-range nuclear-armed ALCMs will be attributed with the number of missiles for which it is actually equipped.[4]

This counting rule and attribution arrangement favours the US negotiating position. While the USA would have preferred a 10-warhead counting rule for its entire ALCM-capable bomber force, the ceiling of 150 bombers still permits the USA to exploit the capability of these aircraft to carry as many as 20 ALCMs. Thus the USA could, for the first 150 bombers, deploy twice the number of warheads accounted for under START. The total Soviet bomber force is below the 210 ALCM-capable bomber ceiling, with ALCM-capable Bear and Blackjack bombers numbering 75 and 15 respectively. Thus even if the USSR exploited the ALCM counting rule of 8 by loading 12 ALCMs on existing bombers, the difference in capability compared with that of the USA would be marginal.[5]

The February meeting between Baker and Shevardnadze did not settle the issue of ALCM range. In addition, the question of how a START agreement should deal with conventionally armed ALCMs re-emerged. It had come up in 1989 but was largely overshadowed by the more prominent issue of ALCM counting rules.

At a meeting between the US and Soviet foreign ministers in Washington on 4–6 April 1990,[6] the USA offered to reduce its preferred limit on ALCM range from 1500 km to 1000 km, but the USSR, still preferring a range limit of 600 km, rejected the offer.[7] The USSR further insisted on the inclusion of a provision concerning conventionally armed ALCMs in an ALCM agreement. The USA, however, maintained that conventionally armed ALCMs could not be dealt with in the START framework.

During intense negotiations in Moscow on 16–19 May 1990, the foreign ministers resolved outstanding ALCM issues.[8] The USA agreed to the Soviet demand for counting all nuclear-armed ALCMs with a range of 600 km and above as accountable under START. Regarding conventionally armed

[4] *Wireless File*, EUR-310, 21 Feb. 1990, p. 8; *Jane's Defence Weekly*, 17 Feb. 1990, p. 28; Reifenberg, J., 'Washington strebt ein zweites Abkommen über strategische Waffen an', *Frankfurter Allgemeine Zeitung*, 17 Feb. 1990, p. 6; *Wireless File*, EUR-409, 22 Feb. 1990, p. 37; *Wireless File*, EUR-502, 9 Feb. 1990, p. 2; Starr, B., 'START: the USA's dilemma', *Jane's Defence Weekly*, 10 Mar. 1990, p. 437; 'US–Soviet Joint Statement', *Wireless File*, SFF-501, 1 June 1990, p. 66.

[5] International Institute of Strategic Studies, *The Military Balance 1990–1991* (Brassey's: Oxford, 1990), pp. 212–13.

[6] *Wireless File*, EUR-111, 9 Apr. 1990, pp. 15–18.

[7] Mann, P., 'Soviets ready to resolve START pact despite clash over cruise missiles', *Aviation Week & Space Technology*, vol. 132, no. 16 (23 Apr. 1990), p. 66; 'US–Soviet Joint Statement' (note 4), p. 66.

[8] *Wireless File*, no. 101, 24 May 1990, pp. 9–10.

ALCMs, the two sides agreed that they would not be included in a START agreement.[9]

Soviet concerns about the agreed ALCM counting rule, however, emerged towards the end of 1990.[10] Soviet concerns reportedly focused on the 50 per cent discount for the first 150 ALCM-capable bombers the ALCM counting rule would grant to the USA.[11] In an effort to secure the same discount for its ALCM-capable bombers, the USSR proposed a new counting rule under which the first 180 Soviet ALCM-capable bombers would be counted as carrying 8 but could be equipped with 16 ALCMs. Bombers beyond the 180 ALCM limit would be counted as equipped. The USA agreed to this change in ALCM counting rules for Soviet ALCM-capable bombers.[12] While it may be politically important for the Soviet leadership to be able to point to the same 50 per cent discount as granted to the USA, it remains questionable if the USSR will have the necessary number of ALCM-capable bombers to exploit the new counting rule.

Sea-launched cruise missiles

The US position on limiting nuclear-armed SLCMs hinged upon agreement on effective verification measures. For the past several years, the USA has argued that an effective SLCM verification system had not been found.[13] In the absence of a satisfactory system, the USA instead proposed that both sides make non-binding declarations of the number of SLCMs which they intend to deploy. The USA further proposed that the number of SLCMs should not be accountable under the START treaty ceiling of 6000 warheads.[14]

In contrast, the USSR was aiming for much tighter controls of SLCMs, proposing levels of 400 and 600 for nuclear- and conventionally armed SLCMs, respectively, to be codified in a legally binding and fully verifiable agreement.[15] At the meeting between Secretary of State Baker and Foreign Minister Shevardnadze in Wyoming in September 1989, the USSR offered to deal with the SLCM issue outside the START framework but as part of an agreement on naval arms control.[16] The USSR insisted that a resolution of the SLCM issue had to be found in order to make a START agreement possible. The USA, which so far has been adamantly opposed to naval arms control, rejected the idea of including SLCMs in a naval arms control agreement.[17]

[9] Mann, P., 'Cruise missile accord advances START treaty', *Aviation Week & Space Technology*, vol. 132, no. 22 (28 May 1990), p. 18; *Congressional Quarterly*, 26 May 1990, pp. 1665–66; Lockwood, D., 'Bush, Gorbachev concur: START to finish by year's end', *Arms Control Today*, vol. 20, no. 6 (June 1990), p. 28.

[10] Lockwood, D., 'February START summit uncertain, negotiations inch toward finish', *Arms Control Today,* vol. 21, no. 11 (Jan./Feb. 1991), pp. 23–24.

[11] See Mann (note 7).

[12] See Mann (note 7).

[13] See Cowen Karp, R., 'US–Soviet nuclear arms control', SIPRI, *SIPRI Yearbook 1990: World Armaments and Disarmament* (Oxford University Press: Oxford, 1990), pp. 423–42.

[14] *Wireless File*, EUR-103, 22 Jan. 1990, p. 3.

[15] See Cowen Karp (note 13).

[16] See Cowen Karp (note 13).

[17] *Wireless File*, EUR-103, 22 Jan. 1990, p. 3.

An unexpected breakthrough came at the February meeting between Baker and Shevardnadze when the USA secured Soviet agreement in principle to adopt a declaratory approach in which each side would state the number of SLCMs it planned to deploy.[18] Actual deployments would not be limited, nor would deployments count against START limits. This declaration would be only politically but not legally binding and would include no provisions for verification.[19] The Soviet acceptance of the US approach to dealing with the SLCM problem was regarded as a substantial breakthrough by US negotiators, marred only by the lack of agreement concerning the range of the missiles and whether conventionally armed SLCMs should be included. The USA maintained that SLCMs with a range of 300 km and above should be included. The USSR held the position that both nuclear- and conventionally armed SLCMs of a range in excess of 600 km should be included.[20]

When the two foreign ministers met again in April in Washington, the USSR withdrew its previous agreement to the declaratory, politically binding settlement and instead revived its original negotiating position, demanding strict limits for nuclear and conventional SLCMs as part of a legally binding agreement.[21] The confusion among negotiators in both camps as to what had actually been agreed at the February meeting was great. While the precise circumstances that seem to have caused this misunderstanding remain unclear, observers at the time felt that the tense atmosphere surrounding the US–Soviet relationship over the Lithuanian independence issue or resistance in the Soviet armed forces to the concessions might have led Soviet leaders to retract their SLCM concessions.[22]

Uncertainty about the Soviet position persisted until the preparatory talks in May for the June summit meeting during which agreement on SLCMs was finally achieved. The SLCM agreement, which will not be part of the START treaty, consists of a political obligation by both sides not to deploy more than 880 nuclear-armed SLCMs with a range in excess of 600 km. For the duration of the START treaty (15 years), the two sides will announce planned SLCM deployments, beginning with announcements to cover the first five years of the treaty.[23] The USA and the USSR also agreed on a data exchange to cover SLCMs with ranges between 300 and 600 km. There will be no verification of actual SLCM deployments.[24]

[18] *Wireless File*, EUR-310, 21 Feb. 1990, p. 8; *Wireless File*, EUR-502, 9 Feb. 1990, p. 2. With reference to an SLCM agreement, the latter source quotes Secretary Baker as calling it a 'pretty much irrevocable agreement'.

[19] See note 18.

[20] *Wireless File*, EUR-406, Apr. 1990, p. 3.

[21] Lockwood, D., 'START talks falter, early summit scheduled', *Arms Control Today*, vol. 20, no. 5 (May 1990), p. 24; Mann (note 7), p. 66.

[22] Walker, M., 'Washington drafts START II declaration', *The Guardian*, 1 May 1990, p. 5; Steele, J., 'Way clear for arms cut treaty', *The Guardian*, 19 May 1990, p. 1; Fieldhouse, R., 'Cruise missile compromise surfacing', *Bulletin of the Atomic Scientists*, vol. 46, no. 5 (June 1990), p. 21; Keeny, Jr, S. M., 'Lithuania and arms control', *Arms Control Today*, vol. 20, no. 3. (Apr. 1990), p. 2.

[23] Mann (note 9), p. 18; 'Ambassador Burt assesses status of START', *Wireless File*, no. 110, 7 June 1990, p. 13 (transcript of WorldNet Interview of 7 June 1990); 'US–Soviet Joint Statement' (note 4), p. 67.

[24] See Mann (note 23).

In its essential points, the SLCM agreement constitutes an abandonment on the part of the USSR of its negotiating position. While the USSR can claim that the agreement imposes an upper limit on US SLCM deployments, this ceiling is actually higher than planned US deployments of 758 nuclear-armed SLCMs; hence the numerical constraints are unconfining.[25] Indeed, due to budgetary pressures in the USA, ultimate SLCM deployment is expected to be below 758 systems.[26] The USSR did achieve a thorough discussion of the SLCM issue, but it failed to achieve the ultimate objective to have a legally binding agreement with full verification provisions and low numerical levels.[27]

Deployed mobile ICBMs

While there is still disagreement on a warhead sub-limit for ICBMs (see below), the issue of a sub-limit on the number of warheads deployed on mobile ICBMs was settled in February.

The USA sought to limit the number of warheads on mobile launchers to 800, whereas the USSR aimed for a limit of 1200 warheads. In a compromise between the two positions, the agreed warhead number on mobile ICBMs was set at 1100.[28] In light of currently deployed mobile Soviet launchers and planned mobile US launchers, the sub-ceiling of 1100 gives both sides the opportunity to expand current and planned deployments. The USSR has 825 warheads deployed on 285 mobile launchers (10 warheads each on 60 SS-24s and 1 warhead each on 225 SS-25s).[29] The USA is also not expected to reach the 1100-warhead ceiling. It has currently deployed 50 MX ICBMs (each armed with 10 warheads) with the intention of making the MX missile rail-mobile. However, the MX programme has faced increasing criticism within the Bush Administration and from Congress and is unlikely to be deployed in a mobile mode.[30]

[25] See Cowen Karp (note 13).

[26] See Lockwood, D., 'START talks stalled, 1990 finish in jeopardy', *Arms Control Today*, vol. 20, no. 9 (Sep. 1990), p. 17; 'Potential strategic forces under START', *Arms Control Today*, vol. 20, no. 7 (Sep. 1990), pp. 24–25; Fieldhouse (note 22), p. 21.

[27] *Congressional Quarterly*, 26 May 1990, pp. 1667–68; Lockwood, D., 'Bush, Gorbachev concur: START to finish by year's end', *Arms Control Today*, vol. 20, no. 5 (June 1990), p. 28; Krepon, M., 'Put a ceiling on nuclear-armed SLCMs', *Defense News*, 18 June 1990, pp. 29–30; Leopold, G., 'Naval arms control sails through rough waters', *Defense News*, 9 July 1990, pp. 4 and 33.

[28] Burt, R., 'The Strategic Arms Reduction Talks—a look at the endgame and beyond', *NATO Review*, vol. 38, no. 4 (Aug. 1990), p. 25.

[29] See IISS (note 5), pp. 212–13.

[30] On 17 Oct. 1990, the the House–Senate joint conference committee agreed to authorize a funding pool of $680 million for the MX and the single-warhead Midgetman missiles. The conference also adopted a House provision stating that no more than one mobile ICBM should be deployed. See *Wireless File*, no. 201, 17 Oct. 1990, p. 3. The Bush Administration had requested $2.4 billion for ICBM modernization; the Senate Armed Services Committee bill had called for $750 million and the House Armed Services Committee bill had proposed $610 million. For a review of the debate on the MX missile, see 'Administration budget for strategic weapons faces heavy opposition', *Defense News*, 11 June 1990, p. 10; 'Panel cancels MILSTAR, MX funds', *Defense News*, 16 July 1990, p. 3; 'Both chambers ready plans for long-term reductions', *Congressional Quarterly*, 28 July 1990, pp. 2426–31; 'Senate votes to save Stealth by narrow margin', *Congressional Quarterly*, 4 Aug. 1990, pp. 2528–32; 'Senate approves its version of defense spending bill', *Congressional Quarterly*, 20 Oct. 1990, pp. 3526–31; 'Washington roundup', *Aviation Week & Space Technology*, vol. 133, no. 24 (10 Dec. 1990), p. 19.

Non-deployed missiles

The principal problem with non-deployed missiles is that their exact number is extremely difficult to ascertain. These missiles are not deployed in silos or on launchers but are either in production, kept for training purposes, or kept in storage. Depending on the category of missile, the fact that a large number of missiles could be stored and therefore escape arms limitation is a potentially worrisome issue. The US view was that non-deployed ballistic missiles pose a greater threat to stability than non-deployed cruise missiles.[31] Therefore, the USA sought to limit only the number of non-deployed ballistic missiles in the belief that, left uncontrolled, these missiles could provide the USSR with a treaty break-out option. The USSR held the view that non-deployed cruise missiles should also be limited.[32]

At the February meeting between Baker and Shevardnadze in Moscow, the two sides agreed that numerical limits should be imposed only on non-deployed mobile ballistic missiles. In addition, they agreed to an accord limiting the location and movement of all non-deployed ballistic missiles.[33] The agreement represents a compromise between the US and Soviet positions. The USA did not achieve numerical limitations on Soviet non-deployed silo-based systems but managed to avoid numerical limitations on non-deployed cruise missiles. Nevertheless, the agreement favours the USA in that it caps the total Soviet mobile missile force, an important issue for the USA because it, unlike the USSR, has not deployed a mobile ICBM system.[34]

Issues resolved at the December 1990 foreign ministers' meeting[35]

After the June summit meeting, three major issues dominated the negotiation agenda: (*a*) US–British nuclear collaboration and its potential for circumventing treaty limits; (*b*) the possible upgrading of the medium-range nuclear-capable Soviet Backfire bomber to intercontinental range; and (*c*) the modernization potential of the Soviet SS-18 ICBM.

US–British nuclear collaboration and treaty circumvention

The issue of treaty circumvention was raised by Soviet negotiators in early 1990. It concerned the present and future role of the USA in helping the United Kingdom, through the transfer of technology and weapon systems, to maintain its independent nuclear deterrent. Under an ongoing agreement, the USA will sell 64 Trident II SLBMs to the UK.[36] The USSR expressed concern

[31] *Wireless File*, no. 110, 7 June 1990, pp. 15–16.
[32] See note 25.
[33] *Wireless File*, EUR-406, 5 Apr. 1990, p. 3.
[34] See note 33.
[35] *Wireless File*, no. 235, 6 Dec. 1990, p. 7; *Wireless File*, no. 237, 10 Dec. 1990, pp. 1 and 5–6; *Wireless File*, no. 238, 11 Dec. 1990, pp. 5–7; *Wireless File*, no. 239, 12 Dec. 1990, pp. 7–10; *Wireless File*, no. 240, 13 Dec. 1990, p. 9.
[36] *Wireless File*, no. 110, 7 June 1990, p. 11. See also 'Ambassador Burt assesses status of START' (note 23), p. 14; and Lockwood (note 26), p. 17.

that current and future US–British nuclear co-operation could lead to a circumvention of treaty provisions.[37] On numerous occasions, however, the USA made it clear that its long-standing co-operation with Britain would take precedence over an arms control treaty with the USSR.[38]

While this issue had the potential to become a major obstacle to the conclusion of a START treaty, it has been resolved.[39] Reportedly, the USA will submit a unilateral statement to the effect that 'existing patterns of cooperation' would not be affected by the START treaty, and the USSR would submit a unilateral statement emphasizing that a shift in the strategic balance caused by US nuclear transfers could give the USSR grounds for withdrawal from the treaty.[40] These expected unilateral statements would not reflect a harmonization of viewpoints, but they make it possible for the issue to be set aside for the time being.

The Backfire bomber

A concern for the USA was the potential range of the Backfire bomber. Through in-flight refuelling, the range of this bomber could be extended, enabling it to execute strategic missions. The issue was previously raised by the USA during the SALT II negotiations, and in 1979 a satisfactory solution was found. The USSR submitted a unilateral statement not to extend the range of the Backfire.[41] At the June 1990 summit meeting, the USSR offered the same commitment.[42] Reportedly, the USA accepted the Soviet offer during a meeting of the US and Soviet foreign ministers in New York on 5 October 1990.[43] The USSR agreed to make a politically binding statement to limit its deployment of Backfire bombers to 500 (300 non-naval and 200 naval) and not to upgrade this bomber to intercontinental range.[44]

SS-18 modernization

The USSR has already agreed to a halving of its SS-18 ICBM force, not to make these missiles mobile, and not to replace them with a new type of heavy missile.[45] However, the USSR refused to agree to a US proposal to limit the

[37] See note 36.
[38] See note 36.
[39] *Wireless File*, no. 237, 10 Dec. 1990, p. 6.
[40] See Lockwood (note 10). See also note 35.
[41] The Soviet Backfire Statement was submitted by President Brezhnev to President Carter on 16 June 1979. A Soviet commitment not to upgrade the Backfire bomber can be expected, in its relevant passages, to be similar to the SALT II Backfire Statement: 'The Soviet side informs the US side that the Soviet "Tu-22M" airplane, called "Backfire" in the USA, is a medium range bomber, and that it does not intend to give this airplane the capability of operating at intercontinental distances. In this connection, the Soviet side states that it will not increase the radius of action of this airplane in such a way as to enable it to strike targets on the territory of the USA. Nor does it intend to give it such a capability in any other manner, including by in-flight refuelling'. See US Arms Control and Disarmament Agency, *Arms Control and Disarmament Agreements: Text and Histories of the Negotiations,* 1990 edn (US Government Printing Office: Washington, DC, 1990), p. 300.
[42] *Wireless File*, SFF-603, 2 June 1990, p. 55.
[43] *Wireless File*, no. 194, 5–7 Oct. 1990, pp. 11–12.
[44] See note 35.
[45] 'US–Soviet Joint Statement' (note 4), p. 67.

number of SS-18 test flights to two per annum. Such a limitation on the number of test flights would curb the ability of the USSR to modernize its SS-18s.[46] The USSR is in the process of deploying the SS-18 Mod. 5, an upgraded version of the SS-18, and the SS-18 Mod. 6, a single-warhead missile.[47] Critics of a START treaty in the USA have focused on the SS-18 modernization issue; to them the imposition of strict limits on present and future capabilities of the SS-18 ICBM is a key pre-condition for successful passage of the treaty through the US Senate.[48] At the December meeting between the US and Soviet foreign ministers in Houston, a compromise solution was reached. The USA would no longer insist on test flight limits in exchange for a Soviet commitment not to increase the throw-weight and launch-weight of the SS-18 Mod. 5.[49]

III. Unresolved issues

At the close of 1990, four issues were known to await settlement: (*a*) a sub-ceiling for ICBM warheads; (*b*) access to ballistic missile telemetry data; (*c*) continuous monitoring of mobile-ICBM production sites; and (*d*) verification of strategic bombers.

Sub-ceiling for ICBM warheads

Both sides are agreed that neither side should have more than 6000 accountable warheads on 1600 accountable delivery vehicles. They have also agreed on a ballistic missile warhead sub-ceiling of 4900 and pledged not to deploy more than 1540 warheads on so-called heavy ICBMs.[50]

The USA, aiming to cap the size of the Soviet ICBM force, has proposed to limit the number of warheads to be carried by ICBMs to 3000–3300. The Soviet position has been not to accept limitations on the number of ICBM warheads unless the USA also agrees to limit the number of SLBM warheads,[51] a weapon category in which the USA has a substantially larger force than the USSR. The objective on each side has been to obtain a sub-ceiling that would favour its existing force strengths in a particular weapon category while constraining the freedom of the other side to choose the ratio between ICBM and SLBM warheads under the aggregate START limits. This issue is not as important as it was a few years ago, and an agreement should be

[46] *Congressional Quarterly*, 26 May 1990, p. 1668; Bunn, M., 'SS-18 modernization: the Satan and START', *Arms Control Today*, vol. 20, no. 6 (July/Aug. 1990), pp. 13–17; *Wireless File*, SSF-603, 2 June 1990, pp. 54–55; 'Ambassador Burt assesses status of START' (note 23), pp. 12–18.

[47] US Department of Defense, *Soviet Military Power 1990* (US Government Printing Office: Washington, DC, 1990), pp. 51–52.

[48] See Bunn (note 46); Skantze, L. (Ret. Gen., US Air Force), 'Defense debate ignores Soviet ICBM', *Defense News*, 16 Apr. 1990, p. 19; Mann (note 9), p. 19; Gaffney, Jr, F., 'Heralded cut in Soviet ICBMs illusory', *Defense News*, 11 June 1990, pp. 19–20.

[49] See note 35.

[50] These basic treaty provisions were agreed at the Washington summit meeting of 7–10 Dec. 1987. See *Wireless File*, EUR-203, 6 Feb. 1990, p. 21.

[51] *Congressional Quarterly*, 26 May 1990, p. 1666.

possible. The US negotiating team has already achieved deep cuts in Soviet ICBMs, particularly with regard to the number of Soviet SS-18s which will be halved under the START provisions. START also provides for a 10-warhead limit on heavy ICBMs. These agreements plus the 1100-warhead ceiling on mobile ICBMs and the Soviet pledge to halt production of the SS-24 will effectively serve to keep the number of warheads on Soviet ICBMs at significantly lower levels than current holdings.[52]

Telemetry encryption

For purposes of effective verification, both sides require access to data emitted by ICBMs and SLBMs during test flights. At the February meeting of the US and Soviet foreign ministers, it was agreed that neither side would encrypt missile test flight data.[53] While this issue appeared to have been resolved, US negotiators in late 1990 became concerned over specific technical aspects of telemetry encryption.

The USA is concerned that a possible adoption by the USSR of US telemetry transmission techniques may prevent the USA from gathering vital Soviet missile telemetry data. If the USSR adopted the US technique of broadcasting telemetric data at low power and high frequency for its ICBM test flights, the USA would be unable to obtain telemetric data. The USSR does not have a comparable problem with the existing US practice of telemetry transmission because US ICBM test flights take place over the Pacific Ocean, where Soviet surface ships can receive the signals. To forestall a change in current Soviet practice, the USA has proposed a three-year moratorium on changes in telemetry transmission techniques, after which time the issue would be subject to discussion.[54]

Monitoring production sites of mobile ICBMs

The USA unlike the USSR, does not have final missile assembly plants but assembles all of its mobile missiles at their deployment sites. Because of the disparity in assembly procedures, it has been difficult for the two sides to agree on the locations at which the missiles will be inspected.

The USSR has reportedly offered to permit US inspections of its mobile SS-24 and SS-25 final assembly plants located at Pavlograd and Votkinsk, respectively.[55] In return, the USSR is interested in gaining access to the production facilities for the first stages of the MX and Midgetman missiles located in Magna and Brigham City, Utah, respectively. It is conceivable that, despite the different purposes of the US and Soviet plants, a deal could be

[52] 'US–Soviet Joint Statement' (note 4), p. 65–68.
[53] *Jane's Defence Weekly*, 10 Mar. 1990, p. 437; *Wireless File*, EUR-406 (5 Apr. 1990), p. 5.
[54] See note 35. Secretary Baker reportedly identified three problem areas: telemetry encryption, portal monitoring of missile plants and inspection of bomber production facilities. See 'The superpowers, citing Gulf war, postpone summit' (note 1), pp. 1 and 8.
[55] See Lockwood (note 10).

struck on inspecting these plants, especially since under the 1987 INF Treaty the USA has already accepted the principle of final assembly plant inspection.[56]

ALCM convertibility of non-ALCM-capable bombers

Under the START treaty a bomber equipped with gravity bombs and SRAMs is counted as carrying only one warhead regardless of how many weapons it actually carries, whereas specified numbers of bombers armed with ALCMs will, as noted above, count as carrying 10 warheads for the USA and 8 for the USSR. Therefore, it is important to both sides to have the ability to verify that bombers declared as non-ALCM-capable are not clandestinely equipped with ALCMs. Such action would circumvent the warhead discount rule that applies to ALCM-capable bombers. Under START, each side will verify through on-site inspection that the other side applies the ALCM counting rule to which it has agreed. Reportedly, the USSR has raised concerns about the possibility that the USA might secretly convert B-1B bombers from the role of penetrator to that of ALCM-carrier. The B-1B has in the past been tested with long-range nuclear-armed ALCMs, but the USA does not plan to equip it with these missiles. Nevertheless, the USSR has proposed design changes to the interior of the B-1B weapon bays in order to foreclose the option of deploying this bomber with ALCMs.[57]

A second ALCM-related issue raised by the USSR concerns on-site inspection of the US B-2 bomber force which is to be deployed without ALCMs. The USSR has reportedly suggested that although the B-2 has not been tested with long-range nuclear-armed ALCMs, if it is tested with a *conventionally armed* ALCM, Soviet inspectors should be granted access to the bomber in order to ascertain that it could not also be equipped with nuclear-armed ALCMs.[58] While the USSR can make a credible case for B-1B on-site inspections (this bomber has been tested with long-range nuclear-armed ALCMs), the case for on-site inspection of the B-2 bomber appears to be strained.

IV. The Defence and Space Talks

Despite the Soviet decision to de-link the START negotiations from resolution of the strategic defence issue, the negotiating position of both sides at the Defence and Space Talks remains unchanged. The USSR maintains its opposition to the US Strategic Defense Initiative (SDI) and its endorsement of the traditional interpretation of the 1972 ABM Treaty. The USA continues its

[56] See note 35. See also Article 11, paragraph 6(b) of the 1987 INF Treaty, reproduced in SIPRI, *SIPRI Yearbook 1988: World Armaments and Disarmament* (Oxford University Press: Oxford, 1988), appendix 13A, pp. 395–406.

[57] See note 35. See also 'The superpowers, citing Gulf war, postpone summit' (note 1), pp. 1 and 8.

[58] See Lockwood (note 10); note 35; 'The superpowers, citing Gulf war, postpone summit' (note 1), pp. 1 and 8.

efforts to persuade the USSR to engage in steps leading towards 'a cooperative transition to a stabilizing balance of offensive and defensive forces',[59] but the two sides have not moved closer to an agreement.

The USA has submitted a number of proposals aimed at facilitating discussions of how a transition from offensive to a mix of offensive and defensive forces could be brought about. The USSR, however, has rejected these proposals because they would necessarily involve a reinterpretation of the ABM Treaty. The USSR has not been opposed to US suggestions of data exchanges, visits to laboratories and observation of tests. But the USSR regards these steps as confidence-building measures,[60] making the existing ABM Treaty more effective, rather than as first steps towards an eventual move away from Treaty provisions.

In the Joint Statement issued at the June summit meeting, Presidents Bush and Gorbachev called upon negotiators to continue the Defence and Space Talks without delay after a START treaty has been achieved.[61] The statement further elaborated that the negotiations should focus on how to achieve an appropriate relationship between strategic offence and defence.[62] This wording leaves ample room for the position each side holds on the relevance of strategic defences to strategic stability. Unless the two sides agree on a mutually acceptable definition of the provisions in the ABM Treaty and the role of strategic defences in their security relationship, the Defence and Space Talks cannot be expected to yield results.

The US SDI programme and with it the Bush Administration's commitment to keep the strategic defence deployment option open have experienced considerable set-backs in the congressional debates on the 1991 defence appropriations bill. The Administration's SDI funding request of $4.7 billion was reduced to $2.3 billion and $3.7 billion by the House of Representatives and Senate, respectively. The subsequent House–Senate budget conference settled SDI appropriations at $2.9 billion.[63] Thus the downward trend in SDI funding is continuing, and a deployment decision on space-based interceptors has been delayed even further than anticipated. The SDI programme seems to be well on the way to becoming a long-term research programme rather than a deployment option.[64]

Beyond the financial problem, with the end of the cold war and the onset of a much improved relationship with the USSR, the SDI programme has lost much of its erstwhile sense of urgency. In order to maintain the credibility of

[59] *Wireless File*, EUR-103, 22 Jan. 1990, pp. 4.

[60] Smith, D. J., 'The Defence and Space Talks: moving towards non-nuclear defences', *Nato Review*, vol. 38, no. 5 (Oct. 1990), pp. 17–21.

[61] 'Joint Statement on Future Negotiations on Nuclear and Space Arms and Further Enhancing Strategic Stability', Statement reproduced in *Arms Control Today*, vol. 20, no. 5 (June 1990), p. 23.

[62] See note 61.

[63] *Wireless File*, EUR-205, 23 Oct. 1990, p. 9.

[64] Finnegan, P., 'Cheney presses committees on SDI, threatens to recommend Bush veto', *Defense News*, 6 Aug. 1990, pp. 13 and 23; Finnegan, P., 'House SDI critics plan to bolster Senate's spending limits', *Defense News*, 13 Aug. 1990, pp. 3 and 44; Finnegan, P., 'House ignores veto, slashes SDI', *Defense News*, 24 Sep. 1990, p. 38; 'Scaled-down SDI makes sense', *Defense News*, 15 Oct. 1990, p. 46; Zimmerman, P., 'Senate nudges SDI back to its roots', *Defense News*, 3 Sep. 1990, pp. 43–44.

the SDI effort in spite of these positive political developments, SDI supporters in Congress and the SDI Organization began to highlight the threat posed to US and allied security by the increasing spread of missile technology in the Third World. The onset of the war in the Persian Gulf and Iraq's ballistic missile capability have been used to make the case for defences against attacks by third countries and terrorists. Congressional opponents of the SDI programme have rejected this redirection of the SDI effort, arguing that rather than spending billions of dollars on an SDI system that would only give limited protection in any case, the spread of ballistic missile technology should be curbed through multilateral diplomatic efforts.[65]

Proponents of the SDI programme received unexpected support during 1990. In an unprecedented series of articles in Soviet newspapers and journals, Soviet military and political analysts questioned the merit of the official Soviet position on the ABM Treaty and the US SDI programme.[66] The authors carefully avoided endorsing the SDI programme but show an increased sensitivity to US proposals for discussions of measures that would lead to an incorporation of strategic defences into the security strategies of both countries. Furthermore, the articles show an increasing awareness of the potential threat posed to Soviet territory by countries that have the capability to launch ballistic missiles and argue for a reconsideration of Soviet official policy on ballistic missile defence.[67]

V. The emerging START treaty

The START treaty stipulates that neither side should have more than 6000 accountable nuclear warheads on no more than 1600 accountable nuclear delivery vehicles. Within the warhead ceiling, START provides for sub-limits of 4900 warheads on ballistic missiles, of which a maximum of 1540 and 1100 warheads may be deployed on heavy ICBMs and mobile ICBMs, respectively. These provisions require both sides to undertake significant reductions in their currently held strategic nuclear arsenals.

In order to comply with START limits and sub-limits, the USA will have to cut 1006 ICBM warheads or 41 per cent of its total ICBM warheads and 1760 SLBM warheads or 34 per cent of its total SLBM warheads. Thus in order not

[65] Finnegan, P., 'Cooper: Mideast crisis underlines SDI's importance', *Defense News*, 3 Sep. 1990, p. 6; *Wireless File*, no. 168, 29 Aug. 1990, p. 7; *Wireless File*, no. 193, 3 Oct. 1990, p. 16.

[66] *Wireless File*, no. 216, 7 Nov. 1990, pp. 11–12; *Wireless File*, no. 110, 7 June 1990, p. 10; Aleksandrov, M., 'Defense domination versus nuclear containment', *Soviet Military Review*, no. 12 (1989), pp. 50–51; Zhirnov, Y., 'Laser technology, laboratory inspections noted', *Komsomolskaya Pravda*, 24 Oct. 1989, p. 1, in Foreign Broadcast Information Service, *Daily Report–Soviet Union (FBIS-SOV)*, FBIS-SOV-89-211, 2 Nov. 1990, p. 4; Belous, V. (Maj. Gen., USSR), 'Reasonable compromises possible on SDI', *Sovetskaya Rossiya*, 23 Mar. 1990, p. 5, in Foreign Broadcast Information Service, *Daily Report–Soviet Union (FBIS-SOV)*, FBIS-SOV-90-058, 26 Mar. 1990, pp. 1–3); Dokuchayev, A. (Lt Col., USSR), 'ABM system role in deterrence is viewed', *Krasnaya Zvezda*, 5 Oct. 1990, p. 2, in Foreign Broadcast Information Service, *Daily Report–Soviet Union (FBIS-SOV)*, FBIS-SOV-90-201, 17 Oct. 1990, pp. 41–43; Smith, D., 'Soviets view SDI in new light', *Defense News*, 5 Nov. 1990, p. 24.

[67] See note 66.

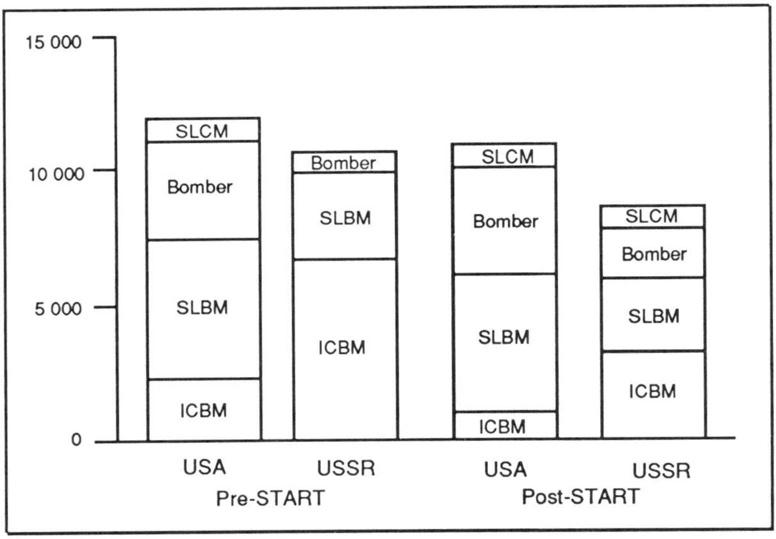

Figure 11.1. Pre- and potential post-START treaty nuclear warhead holdings
Source: See table 11.1 and the notes.

to exceed the 4900 warhead sub-limit on ballistic missile warheads, the US will have to cut 36 per cent of its total ballistic missile warheads.

The USSR will have to cut 3080 ICBM warheads (49 per cent) from its present ICBM arsenal and 1930 warheads (53 per cent) from its SLBM warhead holdings. Overall, the number of Soviet ballistic missile warheads will be cut by 50 per cent. In the absence of an agreed sub-limit on ICBMs and SLBM warheads, both sides are free to mix ICBMs and SLBMs under the 4900 sub-limit on ballistic missiles, and the 1540 and 1100 sub-limits on heavy ICBM and mobile ICBM warheads, respectively. Thus projected cuts on ICBMs and SLBMs, both with regard to warheads and delivery systems, assume particular force trade-offs (see table 11.1). In theory, however, either side could cut its ICBM/SLBM forces differently than suggested here.

As a consequence of mandatory cuts in ballistic missile warheads, the US ICBM force is expected to decline from 2450 warheads to 1444, and its SLBM force from 5216 to 3456 warheads. The Soviet ICBM force is expected to decline from 6280 warheads to 3200 warheads; and its SLBM force from 3626 to 1696 warheads (see table 11.1). Again, in the absence of an ICBM/SLBM warhead sub-limit, these projections could vary.

START counting rules favour warheads on strategic bombers. The treaty provides for heavy discounts of gravity bombs and SRAMS and grants a 50 per cent discount for specified numbers of ALCM-capable bombers. The intention of the treaty is to reward a shift away from warheads on ballistic missiles to warheads on strategic bombers. As shown in table 11.2, START permits an increase in the currently existing bomber warhead number of 280 for the USA and 1286 for the USSR.

Table 11.1. Pre- and post-START treaty nuclear force levels: warheads (estimates)

	USA		USSR	
	Pre-START[a]	Post-START[b]	Pre-START[a]	Post-START[b]
ICBM	2 450	1 444[c]	6 280	3 200[c]
SLBM	5 216	3 456[c]	3 626	1 696[c]
Bomber	4 300	4 580[d]	974	2 260[d]
SLCM	350	880[e]	. .	880[e]
Total	**12 316**	**10 936**[f]	**10 880**	**8 564**[f]

[a] Estimates of existing strategic nuclear force levels vary. Estimates used here are taken from chapter 1 of this volume, pp. 16–21. For different baseline estimates, see *The Military Balance 1990–91* (Brassey's: Oxford, 1990), pp. 212–13; and *Arms Control Today*, vol. 20, no. 7 (Sep. 1990), pp. 24–25 (hereafter *ACT*). The figure for SLCMs in the present US inventory is taken from *ACT*. The Soviet SS-N-21 SLCM is deployed, but the number is not available. See US DOD, *Soviet Military Power 1990* (US Government Printing Office: Washington, DC, 1990), p. 53. Estimates for strategic force potentials are taken from the relevant sections in *ACT*.

[b] Estimates of post-START strategic force levels rest on assumptions about procurement decisions, rates of procurement, ballistic missile force mixes (especially regarding ICBM/SLBM deployment trade-offs in the absence of an agreed ICBM/SLBM sub-limit), operational loadings of bombers and the desire/ability of either side to exploit liberal bomber counting rules. Depending on decisions and developments in these areas, actual warhead numbers are likely to vary. None of the estimates in this table violates START provisions.

[c] ICBM/SLBM force projections are based on the START ballistic missile sub-limit of 4900 warheads. The US ICBM figure assumes a force of 50 MX missiles with 10 warheads each, and a mix of 944 single-warhead Midgetman, 'down-loaded' Minuteman III and some Minuteman II missiles. If the Midgetman missile is not procured, 'down-loading' of the Minuteman III may not be an option if the USA wants to maintain overall ICBM numbers. For the USSR, the ICBM figure assumes 360 SS-24 warheads, 740 SS-25 warheads (in order to comply with the START mobile missile sub-limit of 1100 warheads), 1540 SS-18 warheads and 560 SS-24 (silo-based) warheads. The US SLBM figure assumes 18 Trident submarines each carrying 24 missiles with 8 warheads each. Accordingly, the USA would phase out its remaining Poseidon force. For the USSR, the SLBM figure assumes 14 Delta IV submarines carrying 16 missiles with 4 warheads each (896 warheads) and 4 Typhoon submarines carrying 20 missiles with 10 warheads each (800 warheads). The USSR currently has 7 Delta IV and 6 Typhoon submarines. The SLBM figures used here assume a doubling of the Delta IV force, phasing out of 2 Typhoons and the decommissioning of the entire Yankee Class submarines and the older Delta force.

[d] Strategic bomber force projections assume likely operational loadings. Thus for the USA a bomber force consisting of 75 B-2s carrying 16 bombs and SRAMs (1200 warheads), 95 B-1Bs carrying 16 bombs and SRAMs (1520 warheads) and 93 B-52Hs carrying 20 ALCMs each (1860 total warheads) is assumed. Because START counts bombers equipped with bombs and SRAMS as 1 warhead and counts each ALCM-carrying bomber as equipped with only 10 ALCMs, the USA is allowed to equip its strategic bomber force with a considerably larger number than is accountable under START. Thus only 1100 warheads are counted while actual warheads number 4580. Since the number of B-2 bombers to be deployed remains unsettled and is likely to be well below 75, potential warhead totals could vary significantly. Alternatively, if fewer B-2s are procured, the procurement of the B-1B could be stepped up to fill the gap. For the future Soviet strategic bomber force, 130 Bear-H bombers carrying 10 ALCMs each (1300 warheads) and 60 Blackjack bombers carrying 16 bombs and SRAMs each (960 total warheads) are assumed. With a counting rule of 8 ALCMs per ALCM-capable bomber, the Soviet ALCM bomber force will count as only 1040 warheads, rising to a total of 1100 if non-ALCM-capable bombers are added. Soviet production of Blackjack bombers reportedly has encountered severe difficulties. Should these persist, the target number of 60 used here may not be reached; see chapter 1 of this volume, pp. 16–21.

Table 11.1. *contd.*

[e] As agreed by the USA and the USSR, each side will periodically declare planned SLCM deployments. The figure of 880 used here reflects the agreed deployment ceiling.
[f] This total includes 576 US and 528 Soviet warheads which would not be accountable under START if the US-proposed SLBM launcher exemption (for submarines in overhaul) of 72 is agreed upon.

With regard to strategic force modernization, the START treaty bans the following categories of strategic weapons, basing modes and activities: (*a*) new types of heavy ICBMs; (*b*) heavy SLBMs and launchers of heavy SLBMs; (*c*) mobile launchers of heavy ICBMs; (*d*) new types of ICBMs and SLBMs with more then 10 re-entry vehicles; (*e*) test flights and deployment of existing types of ICBMs or SLBMs with a number of re-entry vehicles greater than the number specified in the Joint Statement from the Washington summit meeting of December 1987;[68] (f) rapid reload of ICBM launchers; and (*g*) long-range ALCMs equipped with multiple independently targetable nuclear warheads.[69]

Finally, as stated in the Washington summit meeting communiqué, 'the treaty will be accompanied by the most thorough and innovative verification provisions ever negotiated'.[70]

START verification provisions are indeed ambitious. They include on-site inspections and continuous monitoring of mobile ICBM production sites. Each side is allowed to conduct short-notice inspections of START-accountable systems especially in order to verify the number of warheads on re-entry vehicles on deployed ballistic missiles. Inspections to verify missile eliminations and inspections of suspect sites have also been agreed. In order to complement on-site inspections, the two sides have agreed on the use of national technical means of verification. These provisions are enhanced through a ban on the denial of telemetric information.

Both sides have also undertaken to exchange data on the number, location and technical characteristics of START-accountable nuclear weapons before the START treaty is signed. These data will be updated regularly.

There has also been an agreement concerning the verification of mobile ICBMs, one of the most difficult verification issues. The agreement foresees strict limitations on the movement of deployed mobile ICBM launchers and missiles. Because the number of non-deployed ICBMs for mobile launchers has been limited, their location and number will also be verified. All ICBMs for mobile launchers will be identifiable by tags. Finally, the two sides have agreed to establish a Joint Compliance and Inspection Commission.[71]

[68] 'START Chronology 1981–1990' (US Arms Control and Disarmament Agency: Washington, DC, 6 Feb. 1990), p. 8.
[69] 'US–Soviet Joint Statement' (note 4).
[70] 'US–Soviet Joint Statement' (note 4).
[71] 'US–Soviet Joint Statement' (note 4), p. 68.

Table 11.2. Estimated warhead cuts/gains under the START treaty

	USA		USSR	
Category	Cuts/gains	%	Cuts/gains	%
ICBM	– 1 006	*– 41*	– 3 080	*– 49*
SLBM	– 1 760	*– 34*	– 1 930	*– 53*
Ballistic missile[a]	– 2 766	*– 36*	– 5 010	*– 50*
Bomber	+ 280	*+ 6*	+ 1 286	*+ 32*

[a] ICBM and SLBM warheads.

Source: Calculated from table 11.1.

An assessment

The emerging START treaty has several major shortcomings. Despite imposing cuts on Soviet ICBMs, these cuts are not sufficient to reduce the vulnerability of the US ICBM force. Treaty provisions address accountable warheads and delivery systems, not actually existing nuclear arsenals. Thus the potential size of post-START strategic forces is obscured.

With regard to strategic nuclear force modernizations, the treaty permits both sides to continue with modernizations currently under way. As stated in the US–Soviet Joint Statement from the Washington summit meeting, 'except as specifically prohibited modernization and replacement of strategic offensive arms may be carried out'.[72] In effect, both sides are permitted to replace ageing forces with more modern and more lethal systems, while implementing START-mandated force cuts by retiring older, less capable forces.

On the Soviet side, large numbers of older ICBMs, SLBMs and SSBNs (nuclear-powered ballistic missile submarines) are being retired in favour of more capable modern missiles and platforms. Although the post-START treaty Soviet ICBM force will be smaller than that of 1990, it will be more accurate and, with about one-third of the ICBM force consisting of mobile missiles, less vulnerable.

The USA is equally entitled to proceed with its strategic modernization programmes such as the mobile MX and Midgetman ICBMs and the B-2 bomber. The problem for the USA is not, in the first instance, with START provisions (many of which the USA itself initially proposed) but with the fact that the USSR appears to be more able to exploit these provisions than the USA. At present, none of the three major US strategic modernization programmes has sufficient congressional backing to assure programme completion.

The US ICBM force will be faced with a modern, highly accurate Soviet ICBM force that even under START is more than double the size of the US ICBM force (see table 11.1). Under these conditions, the USA will have failed to secure one of its primary START objectives, which has been to reduce the

[72] 'US–Soviet Joint Statement' (note 4), p. 67.

Table 11.3. Total estimated cuts in warheads

	USA	USSR
Present force levels	12 316	10 880
Potential force levels	10 936	8 564
Force cuts[a]	1 380	2 316
Change (%)	*– 11*	*– 21*

[a] Total force cuts are calculated by subtracting force gains (bomber warheads, SLCMs and SLBM warheads on submarines in overhaul) from total ballistic missile warhead cuts.

Source: Calculated from table 11.1.

Soviet ICBM threat to US land-based systems. For the USA, existing START provisions for ballistic missiles strengthen strategic stability only if ICBM mobility can be achieved. Otherwise, a situation might arise in which the current silo-based US ICBM force of 1000 missiles could be faced with 3200 Soviet ICBM warheads capable of killing hard targets. This warhead-to-target ratio (3.2:1) is unlikely to be overlooked by those US Senators to whom this ratio is a key yardstick for assessing strategic stability.

The warhead ceilings and sub-ceilings envisaged by the START treaty will leave both sides with numbers of nuclear warheads significantly larger than the originally postulated cuts of 50 per cent. For the USA, START-mandated cuts need amount only to 11 per cent of present warhead numbers, while the USSR would have to implement cuts of 21 per cent (see table 11.3). Total force reductions are smaller than the cuts mandated for ballistic missiles would indicate because the impact of mandated cuts is offset by the very generous counting rule for warheads on strategic bombers, the exclusion of SLCMs from the treaty and the exemption of ballistic missile launchers (and their warheads) on submarines in overhaul. Thus the number of post-START US and Soviet strategic nuclear warheads could exceed the START accountable warheads of 6000 by 4936 and 2564 warheads, respectively, without violating START provisions (see table 11.4). Whether both sides will actually reach these projected numbers in the post-START treaty environment is secondary to the fact that the emerging treaty permits either side to deploy warheads in numbers greatly in excess of the accountable 6000; hence this number obscures rather than illuminates what might be actual numbers of warheads deployed.

Taken together, the cuts mandated by START will not bring about a significant reduction in the total number of nuclear warheads. If permissible warhead levels are not reached, it will not be as a result of START-imposed restrictions but as a result of unilateral decisions not to exploit START counting rules or the failure to procure a particular weapon system. Although the START treaty encourages a shift towards placing warheads on bombers rather than ballistic missiles, with advances in stealth technology likely to decrease warning time, this kind of force restructuring may not bring about the desired increase in strategic stability. While START does cut the Soviet SS-18

Table 11.4. Warhead limits and potential levels under the START treaty

	USA	USSR
Warhead limit under START	6 000	6 000
Warhead number permitted under START	10 936	8 564
Warhead number above START limit	4 936	2 564

Source: Calculated from table 11.1.

force by half, the remaining half has a hard-target kill capability against the US silo-based ICBM force. Together with other highly capable Soviet ICBMs, the SS-18 poses a formidable threat to the US ICBM force, and START exposes rather than alleviates this problem. The START treaty cuts total ballistic missile forces but allows the cuts to be made among older systems in favour of more modern ones. Thus the START treaty contributions to strategic stability, with respect to cuts in individual systems and total warheads, and to force restructurings, are disappointing.

VI. Conclusion

In light of the prospective START treaty's shortcomings, it is not surprising that the debate among policy-makers and analysts about the future of strategic nuclear arms control has already moved beyond the START treaty which remains unsigned. Indicative of this development is the Joint Statement from the Washington summit meeting which calls for future negotiations to reduce 'the concentration of warheads on strategic delivery vehicles as a whole including measures related to the question of heavy missiles and MIRVed ICBMs'.[73] Prior to the summit meeting, members of the US Congress and the Bush Administration had already proposed the elimination of MIRVed mobile ICBMs.[74] These proposals would in the first instance lead to the elimination of the US MX and the Soviet SS-24 programmes but are likely to cover the whole issue of land-based ICBMs.

The commitment to continue the process of arms reductions may serve to enhance support for the START treaty during the ratification debates in the USA and the USSR because it makes the weakness of some provisions appear relative rather than fixed and leaves room to redress their shortcomings in future negotiations. However, in light of the above assessment, it is difficult to foresee a smooth passage for START through the US Senate and the Supreme Soviet.

[73] 'Documents from the US–Soviet summit', *Arms Control Today*, vol. 20, no. 5 (June 1990), p. 23.

[74] Gordon, M. R., 'Arms control skids into a curve', *International Herald Tribune*, 10 Apr. 1990, p. 8; Walker, M., 'Washington drafts START-2 declaration', *The Guardian*, 1 May 1990, p. 5; Barber, L., 'White House split on missiles clouds arms control proposal', *Financial Times*, 16 Jan. 1990; Leopold, G., 'Dugan reaffirms AF missile stance', *Defense News*, 3 Sep. 1990, pp. 4 and 76; Adams, P., 'Pentagon considers arms control plan: Soviet mobile SS-24s for US mobile MX', *Defense News*, 26 Mar. 1990, p. 33.

While the USSR appears to be more able to exploit START provisions than does the USA, the START treaty may nevertheless encounter problems during the ratification process in the USSR. Soviet legislators may well react negatively to the treaty when they reflect upon the failure of the Soviet negotiating team to secure concessions from the USA in return for de-linking the defence and space issue from START, abandoning the demand for a legally binding and fully verifiable accord on nuclear-armed SLCMs, acceding to a strategic bomber counting rule which favours the already larger US strategic bomber fleet, and the deep cuts START imposes on Soviet ICBM and SLBM forces while leaving the USA with a significantly higher number of total warheads.

US senators are likely to be concerned about the fact that the negotiating policy of successive administrations presumed a corresponding relationship between arms control objectives and the procurement of strategic nuclear systems, many of which are not yet in place and are unlikely to be deployed in the desired basing mode or in the number originally planned. While Congress itself has largely been responsible for procurement cuts and delays, the fact that the START treaty assumes a different kind of US force structure than is likely to be in place when the treaty comes up for ratification will be seen as a failure on the part of the Bush Administration to harmonize domestic constraints on force procurement with its stance at the negotiating table.

If a second START treaty is to yield more far-reaching results than the first, it will need to be based on a reassessment of the mission of strategic nuclear forces in a political environment that has fundamentally changed since the inception of the START negotiations in 1982. The question for both sides is: Does the political environment permit a more modest deterrence strategy based on fewer nuclear weapons? A thorough re-evaluation of present nuclear targeting and procurement plans is necessary to answer this question. Deep reductions in nuclear forces can be achieved only be if this re-evaluation is undertaken.[75] In its absence, nuclear arms control is likely to continue to serve existing targeting plans and make only a limited contribution to US–Soviet relations.

[75] Congressional leaders have begun to ask questions about strategic targeting requirements as contained in the existing Single Integrated Operational Plan (SIOP). See Leopold, G. and Munro, N., 'Hill seeks voice in nuclear war plan', *Defense News*, 5 Nov. 1990, pp. 1 and 36. For a discussion of post-START treaty strategic arms control possibilities, see Clyne, S., 'From START to finish: beyond the Strategic Arms Reduction Talks', *Pacific Research*, vol. 3, no. 3 (Aug. 1990), pp. 3–6; Chernoff, F., 'START or finish? The future of strategic arms control and profound force reductions', *Defense Analysis*, vol. 6, no. 3 (Aug. 1990), pp. 235–54; Slocombe, W. B., 'Strategic stability in a restructured world', *Survival*, vol. 32, no. 4 (July/Aug. 1990), pp. 299–312; Kortunov, A. and Fedorenko, A., 'After the treaty, what's in store? Reflections after the Shevardnadze–Baker meeting in Washington', *Moscow New Times*, no. 16 (17–23 Apr. 1990), pp. 10–11, in Foreign Broadcast Information Service, *Daily Report–Soviet Union (FBIS-SOV)*, FBIS-SOV-90-084, 1 May 1990, pp. 2–4; Vinogradov, M. (Lt Gen., USSR) and Belous, V. (Maj. Gen., USSR), 'Flaws, implications of START draft viewed', *Sovtskaya Rossiya*, 23 Aug. 1990, p. 3, in Foreign Broadcast Information Service, *Daily Report–Soviet Union (FBIS-SOV)*, FBIS-SOV-90-165, 24 Aug. 1990, p. 1–3; Feiveson, H. A. and von Hippel, F. N., 'Beyond START: how to make much deeper cuts', *International Security*, vol. 15, no. 1 (summer 1990), pp. 154–80; Speed, R. D., *Strategic Forces: Future Requirements and Options* (Center for Technical Studies on Security, Energy and Arms Control, Lawrence Livermore National Laboratory: Livermore, Calif., Nov. 1990).

12. The implementation of the INF Treaty

STEPHEN IWAN GRIFFITHS

I. Introduction

This chapter briefly reviews the record of the implementation of the Treaty between the United States of America and the Union of Soviet Socialist Republics on the elimination of their intermediate-range and shorter-range missiles (the INF Treaty), as of 31 December 1990. It presents up-dated information on inspections and eliminations, and includes a brief account of the problems that arose during implementation of the INF Treaty in 1990.[1]

II. Inspections, eliminations and INF institutions

Inspections

As of 31 December 1990, the United States and the Soviet Union had carried out 388 and 171 cumulative inspections, respectively. A break-down of the inspections by category is presented in table 12.1.

Table 12.1. US and Soviet missile inspections as of 31 December 1990

Country	Baseline	Short-notice	Elimination	Close-out
USA	117	50	115	106
USSR	31	51	79	10

Sources: On-Site Inspection Agency, 'OSIA Fact Sheet' (Office of Public Affairs, OSIA: Washington, DC, 30 Nov. 1990), p. 2; *Wireless File,* no. 235 (United States Information Service, US Embassy: Stockholm, 6 Dec. 1990), p. 7; telephone interview with an official at the OSIA Office of Public Affairs, Jan. 1991.

Eliminations

By the end of 1990, both parties were approaching the final elimination of all missiles and launchers. The Soviet Union has only 66 of one system (the SS-20 missile), and the United States 180 of two systems (Pershing II and ground-launched cruise missiles, GLCMs), to eliminate before the 1 June

[1] For a more comprehensive assessment of the contents of the Treaty, see Griffiths, S. I., 'The implementation of the INF Treaty', SIPRI, *SIPRI Yearbook 1990: World Armaments and Disarmament* (Oxford University Press: Oxford, 1990), pp. 443–58. For an account of the negotiations and the text of the INF Treaty, see Dean, J., 'The INF Treaty negotiations', SIPRI, *SIPRI Yearbook 1988: World Armaments and Disarmament* (Oxford University Press: Oxford, 1988), pp. 375–489. See also Carter, A., SIPRI, *Success and Failure in Arms Control Negotiations* (Oxford University Press: Oxford, 1989), especially chs 7–8, pp. 172–229.

Table 12.2. US and Soviet missile elimination data as of 31 December 1990

Country	Type	Total	Deployed	Non-deployed	Eliminated	To be eliminated[a]
USA	Pershing II	234	120	114	181	53
	GLCM	443	322	121	315	128
	Pershing 1A	169	0	169	169	0
Total		**846**	**442**	**404**	**665**	**181**
USSR	SS-20	654	405	249	588	66
	SS-23	239	127	112	239	0
	SS-4	149	60	89	149	0
	SS-5	6	0	6	6	0
	SS-12	718	85	633	718	0
	SSC-X-4	80	0	80	80	0
Total		**1 846**	**677**	**1 169**	**1 780**	**66**

[a] This column includes both deployed and non-deployed missiles to be eliminated.

Sources: SIPRI correspondence with the US Government; *The First Anniversary of the INF Treaty* (Novosti Press Agency Publishing House: Moscow, 1989), pp. 13–14; *Wireless File*, EUR–413 (US Information Service, US Embassy: Stockholm, 4 Jan. 1990); telephone interview by the author with an official at the OSIA Office of Public Affairs, Jan. 1991.

1991 deadline for the final elimination of all missiles, launchers, support structures and equipment. The elimination target for 1990 was that both sides by 1 November could only possess deployed launchers capable of carrying no more than 171 warheads. Both sides fulfilled this obligation ahead of schedule.[2] In April 1990, the Federal Republic of Germany also announced that it had begun preparatory work for the destruction of its Pershing 1A missiles.[3]

INF institutions

Apart from reporting the satisfactory functioning of the On-Site Inspection Agency (OSIA), the Nuclear Risk Reduction Centre (NRRC) and the Special Verification Commission (SVC), very little can be said about the activities of these institutions in 1990.[4]

The SVC, which exists to improve the viability and effectiveness of the Treaty, held three sessions in 1990: Session Seven (5 June–19 July); Session Eight (17–25 September); and Session Nine (18 October–21 November).[5]

The workings of the SVC are secret, and only brief statements reporting the satisfactory conclusion of meetings were issued by either the USA or the

[2] For a complete chronology of INF Treaty implementation, see table 12.1 in Griffiths (note 1), p. 444.

[3] See Institute for Defense and Disarmament Studies, *Arms Control Reporter* (IDDS: Brookline, Mass.), sheet 403.B.748, Apr. 1990. See also 'First INF GLCM's removed from West Germany', *Wireless File*, EUR-423 (United States Information Service, US Embassy: Stockholm, 19 Apr. 1990), p. 11.

[4] For a description of these institutions, see Griffiths (note 1), pp. 450–540.

[5] Six sessions were held in 1989. See Griffiths (note 1), pp. 451–52.

USSR. The current heads of the SVC delegations are US Ambassador Steven E. Steiner and Soviet Ambassador Mikhail N. Strel'Tsov.[6]

III. Problems and disputes over INF Treaty implementation

Despite the fact that the overall implementation of the Treaty proceeded very smoothly in 1990, a small number of minor problems and disputes arose.

Since October 1988, US X-ray equipment at the portal monitoring facility at Votkinsk has been a source of controversy between the United States and the Soviet Union.[7] Despite the fact that OSIA Lieutenant Colonel Joe Wagovich on 3 January 1990 reported that the Cargoscan X-ray scanning device would 'become operational in the Soviet Union in the near future', problems continued until early summer.[8] In fact, Cargoscan did become operational on 9 February, despite continued protests from the Soviet Union.

At the beginning of March, the United States requested permission to use Cargoscan to inspect a railcar, but the Soviet Union refused. In an effort to resolve the problem, the United States dispatched a team of technicians to Moscow. However, before the group arrived, the Soviet Union moved the canister in question and two others out of Votkinsk.[9] The United States was not permitted to X-ray them.[10] On 11 March, Secretary of State James Baker lodged what is considered to be the highest-level protest of the entire period of implementation of the Treaty.[11] In a related matter, there were unsubstantiated allegations that Soviet officers had brandished pistols to get the canisters past US inspectors. V. Shchukin, a deputy minister of the Soviet defence industry, denied this report.[12] The Soviet Union responded to the protest of the United States by demanding $2 million in 'reparations' for 'blockading' the Votkinsk plant. The incident ended on 21 March 1990, when the United States agreed to narrow the technical operational scope of Cargoscan.[13]

The other major dispute of 1990 concerned the transfer of previously unreported SS-23 missiles to the German Democratic Republic, Bulgaria and Czechoslovakia by the Soviet Union. A US inquiry into the transfer was apparently prompted by an article in a GDR newspaper.[14]

[6] See *Wireless File*, no. 203 (United States Information Service, US Embassy: Stockholm, 19 Oct. 1990), p. 12.

[7] For a discussion of the development of this problem, see Griffiths (note 1), pp. 455–56.

[8] *Aviation Week & Space Technology*, vol. 130, no. 21 (22 May 1989) p. 95. See also *Wireless File*, EUR-413 (US Information Service: US Embassy, Stockholm, 4 Jan. 1990).

[9] See 'Votkinsk incident called serious', *Wireless File*, EUR-501 (US Information Service: US Embassy, Stockholm, 16 Mar. 1990).

[10] See Dybvik, R., 'Soviets refuse X-raying of missiles leaving Votkinsk', *Wireless File*, EUR-409 (United States Information Service, US Embassy: Stockholm, 15 Mar. 1990), pp. 15–16.

[11] See 'Baker letter helps prompt INF resolution', *Wireless File*, EUR-411 (United States Information Service, US Embassy: Stockholm, 22 Mar. 1990).

[12] See Institute for Defense and Disarmament Studies, *Arms Control Reporter* (IDDS: Brookline, Mass.), sheet 403.B.743, Mar. 1990; and Institute for Defense and Disarmament Studies, *Arms Control Reporter* (IDDS: Brookline, Mass.), sheets 403.B.745 and 403.B.746, Apr. 1990.

[13] See note 12. See also 'Controversies mar Soviet INF compliance', *Arms Control Today*, vol. 20, no. 3 (Apr. 1990), pp. 29 and 32.

[14] See *Arms Control Today* (note 13), p. 29.

In early February 1990, GDR Defence Ministry spokesman Colonel Uwe Hempel stated that the GDR had begun to destroy 24 conventionally armed SS-23 missiles, four launch ramps, four transporters, and technical and repair infrastructure.[15] The major concern of the United States was whether these missiles had been transferred before or after the signing of the INF Treaty. If they had been transferred after the signing, then this would have constituted a major violation of the Treaty.

On 6 March 1990, US Undersecretary of State Reginald Bartholemew called in a Soviet embassy official to answer questions about the missiles. The United States made it clear that a formal complaint would be made unless the Soviet Union could adequately explain the situation.[16] At the end of March, the Soviet Union stated that it had transferred SS-23s to Bulgaria, as well as the GDR and Czechoslovakia.[17] The Soviet Union emphasized however, that the missiles had been transferred before the signing of the INF Treaty. Soviet Foreign Minister Eduard Shevardnadze reportedly told Secretary of State James Baker that neither he, President Gorbachev nor the treaty negotiators had known about the transfers.[18] It was reported that the Governments of Czechoslovakia and the GDR had decided to eliminate their SS-23s, but the Bulgarian Government has apparently not agreed to any such undertaking.[19] By the end of 1990, the whole issue appeared to be unresolved, at least as far as the US Government was concerned. This will probably remain the case until all of the SS-23s—whether or not they constitute a violation of the Treaty—have been eliminated.[20]

IV. Conclusion

The implementation of the INF Treaty proceeded according to schedule in 1990. Apart from a few minor controversies which caught the media's attention, the task of implementing the INF Treaty has gone smoothly. Unless unforeseen difficulties arise, the United States and the Soviet Union should complete their missile elimination obligations before the deadline of 1 June 1991. Notwithstanding the relative lack of media attention overall, the implementation of the INF Treaty to date represents a considerable achievement for both the United States and the Soviet Union.

[15] See IDDS (note 12), sheet 403.B.744, Mar. 1990.

[16] See IDDS (note 12), sheet 403.B.744, Mar. 1990.

[17] See *Arms Control Today* (note 12), p. 29. See also 'US still seeking information on transfer of SS-23s', *Wireless File*, EUR-402 (United States Information Service, US Embassy: Stockholm, 29 March 1990); and 'Soviet SS-23 missiles confirmed in East Germany', *Wireless File*, EUR-202 (United States Information Service, US Embassy: Stockholm, 27 Mar. 1990).

[18] See *Arms Control Today* (note 13), p. 29. See also Smith, R. J., 'Soviets shifted missiles', *International Herald Tribune*, 7/8 Apr. 1990, p. 3.

[19] See *International Herald Tribune* (note 18), p. 3; IDDS (note 12), sheet 403.B.747, Apr. 1990.

[20] *The President's Unclassified Report on Soviet Noncompliance with Arms Control Agreements* (White House, Office of the Press Secretary: Washington, DC, 15 Feb. 1991).

13. Conventional arms control in Europe

JANE M. O. SHARP

I. Introduction

This chapter reviews negotiated and unilateral measures of conventional arms control in Europe during 1990. Section II highlights the main provisions of the Treaty on Conventional Armed Forces in Europe (CFE), signed at the Paris summit meeting of the Conference on Security and Co-operation in Europe (CSCE) on 19 November 1990.[1] Section III analyses the CFE Negotiation (CFE I), focusing on how issues were resolved during 1990.[2] Section IV explores the impact of the Treaty on NATO and WTO equipment holdings in the Atlantic-to-the-Urals (ATTU) zone defined by the Treaty. Section V reviews the efforts by the Joint Consultative Group to resolve discrepancies in the data exchanged at Treaty signature. Sections VI and VII outline the unilateral reduction measures undertaken by both alliances parallel to the CFE Negotiation. Section VIII describes the Vienna Document 1990 on Confidence- and Security-Building Measures (CSBMs) adopted on 17 November 1990 by the 34 CSCE states.[3] Section IX evaluates CFE I and the Vienna CSBMs and assesses the prospects for follow-on talks, the CFE IA negotiations, which formally opened on 26 November 1990.

II. Provisions of the CFE Treaty

The CFE Treaty limits five categories of equipment deployed by the 16 member states of the North Atlantic Treaty Organization (NATO) and the six member states of the Warsaw Treaty Organization (WTO) in an area that stretches from the Atlantic to the Urals. Eight separate protocols provide instructions on how states must dispose of equipment in excess of CFE limits, a schedule for the exchange of data and information, a schedule for inspections, and a mechanism designed to resolve discrepancies in data exchanges and ambiguities in Treaty interpretation and compliance.

[1] For the text of the Treaty on Conventional Armed Forces in Europe see appendix 13A. Integral to the CFE Treaty (but not included in the appendix) are eight protocols, with annexes, and three declarations with respect to naval-based aircraft, general personnel strength and the personnel strength of the German armed forces in particular.

[2] For more detail on the first five rounds of the CFE Negotiation, see Sharp, J. M. O., 'Conventional arms control in Europe', SIPRI, *SIPRI Yearbook 1990: World Armaments and Disarmament* (Oxford University Press: Oxford, 1990), pp. 459–507.

[3] For the text of the Vienna Document 1990 of the Negotiations on CSBMs convened in accordance with the relevant provisions of the Concluding Document of the Vienna Meeting of the CSCE, see appendix 13B.

Table 13.1. Alliance limits on stored and active TLE

	TLE						
Zone[a]	Battle tanks	Artillery	ACVs	AIFVs	HACVs	Combat aircraft	Attack helicopters
Active TLE							
Sub-zone IV.4	7 500	5 000	11 250				
Sub-zone IV.3	10 300	9 100	19 260				
Sub-zone IV.2	11 800	11 000	21 400				
Flank zone	4 700	6 000	5 900				
Total in ATTU zone	**16 500**	**17 000**	**27 300**				
Stored TLE	3 500	3 000	2 700				
Odessa MD	400	500					
S.Leningrad MD	600	400	800				
Sub-zone IV.2	2 500	2 100	1 900				
Active and stored TLE in ATTU zone	**20 000**	**20 000**	**30 000**	**18 000**	**1 500**	**6 800**	**2 000**
Single-country limits	13 300	13 700	20 000	16 800	1 000	5 150	1 500
Kiev MD	2 250	1 500	2 500				

[a] Sub-zones IV.4 and IV.3 nest inside sub-zone IV.2; the flank zone is outside sub-zone IV.2. Ceilings for the entire ATTU zone (IV.1) equal the sum of IV.2 plus the flank states.

Note: ACV: armoured combat vehicle; AIFV: armed infantry fighting vehicle; HACV: heavy armoured combat vehicle; MD: Military District.

Sub-zone IV.4	NATO:	Belgium, Germany, Luxembourg, the Netherlands
	WTO:	Czechoslovakia, Hungary, Poland
Sub-zone IV.3	Sub-zone IV.4, plus	
	NATO:	Denmark, France, Italy, the UK
	WTO:	USSR (Baltic, Byelorussian, Carpathian, Kiev MDs)
Sub-zone IV.2	Sub-zone IV.3, plus	
	NATO:	Portugal, Spain
	WTO:	USSR (Moscow and Volga-Ural MDs)
Flank zone	NATO:	Greece, Iceland, Norway, Turkey
	WTO:	Bulgaria, Romania, USSR (Leningrad, North Caucasus, Odessa, Transcaucasus MDs)

Definitions, counting rules and zonal limits

Article II defines the area of application and five categories of treaty-limited equipment (TLE). For each category there are ceilings for each alliance group of states and each single country in the ATTU zone. There are sub-zonal limits within the alliance-wide ceilings for tanks, artillery and ACVs, but none for combat aircraft or specialized attack helicopters.

Articles IV and V define the sub-zones (as in figure 13.1) and set out the zonal limits shown in table 13.1. Article VI sets out the sufficiency rule—the limits for a single state party may not exceed one-third of the TLE in the area of application.

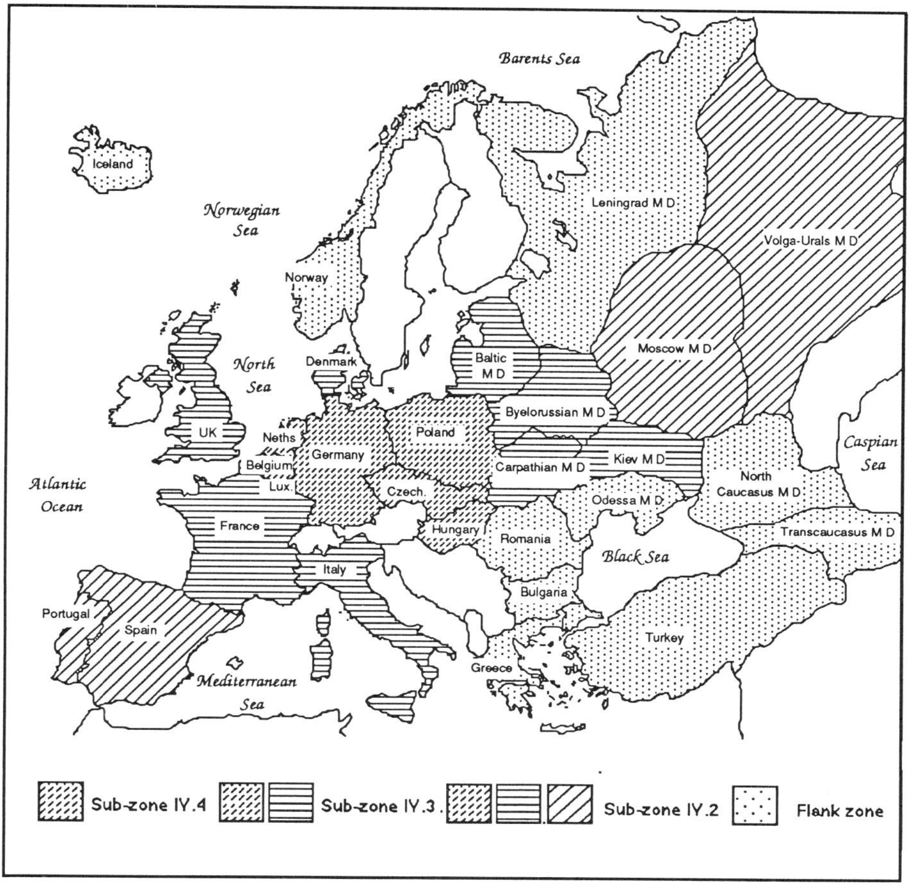

Figure 13.1. Sub-zones in the area of application of the CFE Treaty

Article XII sets limits on AIFVs deployed with paramilitary forces assigned to peacetime internal security functions. The Treaty limits a single country's paramilitary forces to 1000 AIFVs. Those in excess of 1000 must be counted against the CFE limits. Paramilitary forces are not included in the Soviet declared inspection sites or objects of verification (OOVs) but are subject to challenge inspections.

Article II.1 (S) excludes 'look-alikes' from CFE limits. Field ambulances and similar vehicles often look like ACVs and AIFVs from the outside and are subject to inspection to ensure that parties are not evading limits on these items. Article III lists circumstances under which TLE is exempt from limits.

The reduction provisions

Article VIII stipulates that TLE in excess of CFE limits in the ATTU zone must be inactivated in one of eight ways: destruction, conversion to non-military purposes, placement on static display, use as ground targets, modification, ground instructional purpose, reclassification or recategorization.

Section III of the Protocol on Reduction specifies five methods of destruction: severing, explosive demolition, deformation, smashing and the use of target drones. However, it appears that little material will be destroyed as states have been rather ingenious in relocating, converting and recategorizing equipment.

Conversion to non-military purposes

Section VIII of the Protocol on Reduction sets out rules and procedures for the conversion of battle tanks and ACVs to non-military purposes.

Battle tanks (T-54s, T-55s, T-62s, T-64s and Leopard 1s) and ACVs (BMP-1s and BTR-60s) can be converted into 14 different kinds of listed non-military vehicle: general-purpose prime movers, bulldozers, fire-fighting vehicles, cranes, power unit vehicles, mineral fine-crushing vehicles, quarry vehicles, rescue vehicles, casualty evacuation vehicles, transport vehicles, oil rig vehicles, clean-up vehicles for oil spills, ice breakers and environmental vehicles. The Treaty allows each state party to convert 15 per cent of its ACVs (not to exceed 3000) and 5.7 per cent of its tanks (not to exceed 750) or 150 items, whichever is the higher (section VIII, para. 2).

Reclassification of combat-capable aircraft

A separate Protocol on Reclassification sets out the rules for reclassifying combat aircraft. Su-15U, Su-17U, MiG-15U, MiG-21U, MiG-23U, UIL-28 and MiG-25U aircraft can be converted into unarmed trainer aircraft within 40 months of entry into force of the Treaty, but no more than 550 aircraft can be removed from Treaty limits in this way (section I, para. 2).

Recategorization of helicopters

The Protocol on Helicopter Recategorisation sets out the rules by which multi-purpose attack helicopters may be recategorized as combat support helicopters, not subject to Treaty limits, by removing attachments for weapons and all integrated fire control and targeting systems. Such recategorization must be notified and certified by inspection.

Information exchange

The Protocol on Notification and Exchange of Information (section VII) requires exchange of data: (*a*) upon signature of the Treaty, after which the parties had 90 days in which to make corrections (i.e., until 18 February 1991); (*b*) 30 days after entry into force (i.e., after all states parties have ratified the Treaty); (*c*) on 15 December each year; and (*d*) on completion of the 40-month reduction period.

Information is to be exchanged on: (*a*) the structure of land, air and air defence forces, especially the command organization that connects land and

Table 13.2. Passive quotas for inspections of declared and undeclared sites

	I. Baseline validation	II. Reduction phase	III. Residual validation phase	IV. Residual levels
Declared sites	20% of OOVs	10% OOVs p.a.	20% of OOVs	15% OOVs p.a.
Challenge inspection	15% of declared sites quota	15% of declared sites quota	15% of declared sites quota	23% of declared sites quota

Source: Based on Verification Technology Information Centre, *The VERTIC Guide to the CFE Treaty* (VERTIC: London, 1990), p. 14.

air forces; (*b*) overall holdings in each TLE category; (*c*) location, numbers and types of TLE; (*d*) OOVs and declared sites; (*e*) sites from which TLE has been withdrawn; (*f*) any changes in organizational structure or force levels; (*g*) entry into service of TLE; (*h*) entry and exit of TLE to and from the ATTU zone; and (*i*) TLE in transit in the ATTU zone for more than seven days.

Verification

The verification provisions include national and multinational technical means (NTM and MTM) of verification of declared and undeclared sites, and of reduction and certification facilities. States are unequally endowed with the technical and analytical capabilities to conduct efficient inspections; while all are capable of on-site inspections, not all have access to satellite reconnaissance data.[4] The Treaty makes no provision for pooling of information from NTM.

For inspection purposes the Treaty is divided into four phases: (*a*) the baseline validation phase I lasts 120 days after ratification and covers intensive inspection of baseline data; (*b*) phase II is the three-year 'reduction period' following the 120-day validation period; (*c*) the residual validation phase III lasts 120 days and entails intensive inspection of the new baseline data following reductions; and (*d*) the residual level phase IV lasts for the duration of the Treaty (unlimited).

The Protocol on Inspection provides for four types of inspection: inspections of declared sites; challenge inspections; inspections to witness reductions; and inspections to witness certification (sections VII, VIII, IX and X).

Each state has a quota of passive inspections it must accept in each Treaty phase. No more than 50 per cent of a passive quota can be taken up by a single state party, but states can transfer their inspection rights to other parties. Passive quotas are expressed as a percentage of each party's OOVs in the ATTU zone. OOVs are units in the military force structure which hold TLE and are located at declared sites (section I (J)).

[4] Kokoski, R. and Koulik, S. (eds), SIPRI, *Verification of Conventional Arms Control in Europe: Technological Constraints and Opportunities* (Westview Press: Boulder, Colo., 1990).

Joint Consultative Group

The Joint Consultative Group (JCG) was established to resolve any ambiguities about compliance, to consider measures to enhance viability and effectiveness of the Treaty, to update lists of equipment, to resolve technical questions, to work out rules of procedure, and to consider conference proposals and matters of dispute. The JCG is to meet twice a year for four-week sessions. Decisions will be by consensus and, unless otherwise agreed, proceedings will be confidential (Protocol on the Joint Consultative Group).

Follow-on negotiations

Article XVIII of the Treaty commits the states parties to follow-on negotiations (CFE IA), with the same mandate and same states, to be completed before the 1992 CSCE follow-up meeting in Helsinki. The main objective of the CFE IA negotiations is to limit the personnel strength of conventional armed forces in the ATTU zone. The CFE IA negotiations were scheduled to begin immediately after signature of the CFE Treaty, but no serious discussion took place in the first round because of the dispute over data presented in November 1990.[5]

Declarations associated with the CFE Treaty

Annexe II of the CFE Treaty contains three declarations: (*a*) an agreement of the 22 CFE states not to exceed 430 land-based naval aircraft for each alliance group of states, with a single-country limit of 400; (*b*) the unilateral declaration of the German Government to limit German armed forces to 370 000 troops, of which no more than 345 000 can be air and ground forces; and (*c*) the declaration by the 22 states that they would not increase their military manpower parallel to the CFE IA negotiations.[6]

III. How issues were resolved at the CFE Negotiation in 1990[7]

The counting rules

Stored versus active units

Initially NATO countries proposed limits on TLE in active units only (to exclude Prepositioned Organizational Material Configured to Unit Sets—POMCUS—earmarked for US reinforcements to Europe in a crisis). WTO proposals initially set limits on both active and stored TLE, precisely in order to limit US POMCUS. In October 1990 a compromise was reached that set limits on both categories. Poland was troubled during 1989 that US protection

[5] 'CFE: no plenary sessions', *Atlantic News*, no. 2301 (27 Feb. 1991), p. 2.
[6] *Atlantic News*, no. 2273 (Nov. 1990), p. 23.
[7] See also note 2.

of its POMCUS stocks might lead to Soviet insistence on its own prepositioned equipment in non-Soviet WTO (NSWTO) countries. This fear was alleviated in 1990, however, as Hungary and Czechoslovakia negotiated Soviet withdrawals from their territories outside the CFE context. Norway and Turkey insisted on limits on stored equipment in the flank zone to prevent a buildup of Soviet equipment from Central Europe on their borders.

Stationed forces

NATO's initial proposals in March 1989 and the WTO response in May 1989 both included limits on those forces stationed outside national territory on the territory of an ally. NATO's purpose was to limit Soviet forces stationed in Europe, while not prohibiting the stationing of forces outside national territory in Western Europe. US forces are stationed in several allied countries, and several allied countries station forces in Germany (see section VII).

NATO lost interest in stationed force limits during 1990, however, as political events forced the pace of Soviet withdrawals from the NSWTO states. In February and March Czechoslovakia and Hungary negotiated the total withdrawal of Soviet forces by mid-1991; and in July Chancellor Helmut Kohl and President Mikhail Gorbachev agreed that all Soviet forces would leave German territory before the end of 1994.[8] From the NATO perspective, limits on stationed forces then became redundant and did not survive into the Treaty text. NATO thus avoided limits that would constrain the further development of multinational units as contemplated in the London Declaration issued at the NATO summit meeting in early July 1990.[9]

Single-country sufficiency rules

In order to further limit Soviet deployments in the NSWTO states, NATO initially proposed a sufficiency rule that limited each country to no more than 30 per cent of the cuts for the entire ATTU zone and 60 per cent of the ceiling for each alliance group of states. The WTO countries agreed to this in their 1989 proposals but in 1990, after the GDR joined NATO and the other WTO states insisted on Soviet withdrawals, the USSR argued for a higher sufficiency rule (40 per cent) to compensate for its lost allies. The NSWTO states considered 40 per cent to be too high, so new sufficiency rules were proposed by Secretary of State James Baker and Foreign Minister Eduard Shevardnadze on 3 October in New York; these are shown in table 13.3.

[8] Chancellor Kohl and President Gorbachev agreed on the combined German troop levels and the schedule for Soviet withdrawals from Germany on 16 July 1990 in Zheleznovodsk, USSR.

[9] London Declaration on a transformed North Atlantic Alliance, issued by the Heads of State and Government participating in the Meeting of the North Atlantic Council, London, 5–6 July 1990, reprinted in Rotfeld, A. D. and Stützle, W. (eds), SIPRI, *Germany and Europe in Transition* (Oxford University Press: Oxford, 1991), p. 151, para. 14; see also proposals for multinational forces by van Eekelen, W., 'Security: for a strong West European pillar', *International Herald Tribune*, 8 Mar. 1990; Gordon, M. R., 'NATO weighing new look with combined allied units', *International Herald Tribune*, 23 May 1990; *Le Monde*, 25 May 1990; Hershberg, J., 'One army for a new Europe', 19 June 1990; Mather, I. and Claveloux, D., 'Joint East–West European army on NATO agenda', *The European*, 29 June–1 July 1990.

Table 13.3. Baker–Shevardnadze sufficiency rules, 3 October 1990

TLE	ATTU zone ceiling	Alliance ceiling	Single-country ceiling	% of ATTU ceiling
Tanks	40 000	20 000	13 300	*33.25*
Artillery	40 000	20 000	13 700	*34.30*
ACVs	60 000	30 000	20 000	*33.33*
Combat aircraft[a]	13 600	6 800	5 150	*37.50*
Attack helicopters	4 000	2 000	1 500	*37.50*

[a] In Annexe II each alliance is allowed 430 additional land-based naval aircraft, for which the single-country limit is 400. See *Atlantic News*, no. 2273 (21 Nov. 1990), p. 3.

Intra-WTO ceilings

At a number of meetings during October 1990 the NSWTO states complained that the single-country sufficiency rules agreed by Baker and Shevardnadze were still too high. Poland especially objected to the allocation of 13 300 tanks to the USSR. Eventually, on 3 November 1990 the WTO agreed on allocations (shown in table 13.6 below) that cut the Soviet ceiling for tanks from 13 300 to 13 150 and for artillery from 13 700 to 13 175.

As well as cutting the Soviet ceilings, the WTO changed the balance among the NSWTO states. Czechoslovakia was the most willing to give up equipment, with approximately 30 per cent of the NSWTO TLE before, and only 19 per cent after, the CFE Treaty. Poland, by contrast, was more anxious to build up its national forces. Poland argued for the highest allocation of tanks among the NSWTO states and increased its TLE from 19 per cent to 24 per cent. In an uncharacteristically assertive demonstration of independence from Moscow even Bulgaria insisted on a higher allocation of tanks than it was initially assigned.[10] Romania and Bulgaria were each allocated 21 per cent of CFE TLE and Hungary 14 per cent.[11]

Intra-NATO ceilings

NATO made public its holdings of TLE in November 1990 at Treaty signature, but did not release data on allocations among the allies until early December. Apparently the delay was caused by disputes about modernization of Greek and Turkish holdings and precisely which of the TLE should 'cascade' from the better-endowed allies in central Europe (see below). NATO cuts were allocated to give alliance ceilings below Treaty limits for tanks, artillery, ACVs and combat aircraft. Allocations for combat helicopters total 2000 which gives most allies headroom to build up to their national ceiling. Only France, Germany, Italy and the Netherlands are required to reduce or

[10] Interviews by the author with CFE delegates.

[11] Forsberg , R. *et al.*, 'WTO pushes post-CFE forces in Eastern Europe closer to parity', *Vienna Fax*, no. 27 (5 Nov. 1990).

recategorize helicopters; Belgium, Canada, Denmark, Portugal, Spain, Turkey, the UK and the USA were all assigned more than their current holdings. Table 13.7 (below) shows intra-NATO allocations of CFE ceilings.

Manpower

Initially the NATO states wanted the CFE Treaty to limit only three categories of ground-force equipment: tanks, artillery and ACVs. The WTO states wanted a much broader agenda that would include manpower as well as naval and nuclear forces. The 1989 mandate excluded naval and nuclear forces. The Bush Administration persuaded NATO to include manpower and air power in round II of the Negotiation in the summer of 1989, but it was dropped in August 1990 when it became clear that agreements outside the context of CFE would impose limits on Soviet and German manpower.

Parallel to round V (15 January–22 February 1990), the NSWTO states pressed the USSR for deeper cuts and faster withdrawals from their territory. In Washington critics accused the Bush Administration of a CFE policy that encouraged higher levels of Soviet troops in Europe than would be the case if politics took its natural course. To counter these complaints, in his State of the Union address on 31 January 1990, President George Bush proposed lower levels for stationed Soviet and US forces in Europe (195 000 each in Central Europe) with a separate allowance of 30 000 for the USA in the European periphery to compensate for Soviet proximity to Europe. The Bush proposal was presented as a formal NATO proposal in Vienna on 8 February, and accepted by all NATO and WTO foreign ministers, meeting at the Open Skies Conference in Ottawa later in the month. Nevertheless, the proposal drew strong criticism from Pentagon officials and congressional Democrats in Washington, who claimed it would undermine US flexibility and ability to reinforce Europe in a crisis.[12]

On the last day of round V, the USSR and the GDR informally floated a proposal to limit manpower for each alliance group of states in Central Europe (the Benelux countries, Czechoslovakia, Denmark, the FRG, the GDR and Poland) to 700 000–750 000.[13] In round VI (15 March–26 April 1990), the USSR began to back away from the manpower agreement which it had accepted in February. More generally, the CFE Negotiation began to lose steam as the USSR grew more anxious about the prospect of losing the GDR to NATO through German unification, and in May Shevardnadze suggested that the Two-plus-Four talks were the optimum forum in which to limit German forces.[14]

[12] Smith, R. J., 'Array of forces oppose Ottawa troop accord', *International Herald Tribune*, 2 Mar. 1990; Pringle, P., 'Democrats say Europe cuts too hasty', *The Independent*, 28 Feb. 1990.

[13] Grinevsky on troop cuts, TASS, 6 Mar. 1990, Foreign Broadcast Information Service, *Daily Report–Soviet Union (FBIS-SOV)*, 7 Mar. 1990; Chernyshev, V., 'NATO's perfidious proposals curtailed', TASS, 5 Mar. 1990, FBIS-SOV-90-044, 6 Mar. 1990, pp. 6–7.

[14] *Isvestia*, 30 May 1990; Shevardnadze, E., 'Europe: a generation's mission', in Rotfeld and Stützle (note 9), pp. 106–109.

In the event, parallel to round VII (17 May–19 November 1990) the FRG announced cuts in combined German forces to 370 000, and Kohl and Gorbachev agreed that all Soviet forces would leave German territory by 31 December 1994. Federal Foreign Minister Hans-Dietrich Genscher announced the decision to cut German forces to a CFE plenary session on 30 August 1990, and the new ceilings are acknowledged in Article 3 of the Treaty on the final settlement with respect to Germany signed on 12 September 1990.[15] With Soviet and German manpower effectively reduced outside the CFE process, there was no need, from the US perspective, for reciprocal cuts in US forces, especially as the US Defense Department had always argued that the proposed limits would have undermined US flexibility to reinforce Europe.

Thus, in early August in Irkutsk, James Baker and Eduard Shevardnadze agreed to drop manpower limits from the CFE Negotiation.[16] Nevertheless, in Vienna in late August, Oleg Grinevsky, the Soviet Ambassador to the CFE Negotiation, again raised the possibility of a cut in US stationed manpower as part of a CFE agreement. He suggested a ceiling of 70 000–80 000 US troops in the ATTU zone, the same number that Senator Sam Nunn and others had earlier suggested would be an optimum residual force in Europe.[17] In New York in early October, however, Baker and Shevardnadze confirmed the Irkutsk understanding not to seek manpower limits until the CFE IA negotiations in 1990–92.[18]

Ground force equipment

Tanks and ACVs proved difficult to define and limit because of the many different varieties of equipment in the ATTU zone. Soviet delegates reported concern that NATO could easily convert its HACVs into battle tanks.[19] For many months France and the UK resisted the inclusion of light tanks but they gave in when armoured vehicles were redefined (as battle tanks, ACVs, AIFVs and HACVs) and ceilings were raised. Final East–West differences over tank and ACV numbers were resolved in mid-June by a Franco-Polish compromise.[20]

Artillery was the first TLE category to be defined, on 17 October 1989 in round III. NATO delegates were well satisfied with CFE limits on tanks and

[15] Treaty on the final settlement with respect to Germany, Art. 4; see appendix 17C in this volume.

[16] The decision in Irkutsk in Aug. was referred to by US Secretary of State James Baker in Sep. 1990. See Friedman, T. L., 'NATO members to weigh adding troops to Gulf force', *New York Times*, 11 Sep. 1990, and 'Pact on European armies may skip troop limits to speed accord', 12 Sep. 1990.

[17] Smith, R. J., 'Soviet proposes deep force cuts by West', *Washington Post*, 9 Sep. 1990.

[18] Mortimer, E., 'Moscow gives ground in conventional weapons talks', *Financial Times*, 8 Oct. 1990.

[19] For discussion of Soviet concern about NATO conversion of HACVs into tanks see Institute for Defense and Disarmament Studies, *Arms Control Reporter* (IDDS: Brookline, Mass.), sheets 407.B.361–63, 368–69 and 376–78, Apr., May and June 1990.

[20] French Ambassador François Plaisant and Polish Ambassador Wlodzimierz Konarski presented the compromise agreement in parallel statements to the CFE plenary session in Vienna, 14 June 1990.

Table 13.4. Spring 1990 negotiating positions on combat aircraft

Aircraft type	NATO proposal	Hungarian proposal	Soviet proposal	CFE Treaty alliance limits
Total combat aircraft	5 200	6 400	7 700	6 800[a]
Strategic interceptors	500	800	1 500	–
Combat trainers	–	–	1 500	–
Tactical aircraft[b]	4 700	–	4 700	–
Land-based naval aircraft	–	–	unlimited	430[c]
Strategic bombers	–	–	unlimited	–

[a] Single-country limit 5150.
[b] Including medium-range bombers, ground-attack and reconnaissance aircraft.
[c] Single-country limit 400.

ACVs, but disappointed that more Soviet artillery holdings were not cut, particularly as artillery transferred from the ATTU zone seems even less likely to be destroyed than tanks and ACVs.

Aircraft

The CFE Treaty limits each group of alliance states to 6800 combat aircraft and 2000 attack helicopters, with single-country limits of 5150 and 1500, respectively. Aircraft proved difficult to define because Soviet and NATO doctrines differ about the use of air power.

The USSR first proposed inclusion of air power, but Soviet delegates soon realized that it had far more equipment in this category than NATO and proposed the exclusion of several categories of air power from CFE limits: those designated for training purposes, as defensive interceptors and for naval missions. The USSR was particularly reluctant to accept limits on its land-based naval aircraft, given the fact that US carrier-based air power, even in European waters, was outside the CFE mandate. For their part, NATO countries were particularly anxious to limit Soviet Backfire bombers since these were not yet limited by a bilateral US–Soviet START agreement.

During the spring of 1990 three proposals emerged to limit combat aircraft: from NATO, from Hungary and from the USSR, as shown in table 13.4.

Throughout the summer this problem looked so intractable that some delegates felt that aircraft would be left to the next stage of the CFE Negotiation,[21] but in early October, Shevardnadze told Baker that the minimum acceptable level for Soviet combat aircraft, including strategic Backfire bombers, was 5150. Once this was settled it was relatively easy to set the alliance ceilings at 6800 and to agree on a separate annexe (II) setting alliance ceilings on land-based naval aircraft at 430, with single-country limits of 400. Later in October, negotiations within the WTO allocated aircraft among the NSWTO states as shown in table 13.6 (below).

[21] Moxon, J., 'CFE could exclude combat aircraft cuts', *Flight International*, 20–26 June 1990, p. 4.

Attack helicopters

The WTO states included helicopters in their 23 May 1989 proposals during round II.[22] President Bush persuaded NATO to include aircraft and helicopters in its proposal of 13 July 1989. NATO was embarrassed to discover many more helicopters in its inventories than initially estimated and worked around this problem by raising limits and redefining helicopters in several different categories. Definitions were agreed on 17 October 1990 when several categories of helicopter were distinguished. Combat helicopters include both attack helicopters and combat support helicopters. Only attack helicopters are subject to CFE limits. They comprise two sub-groups: specialized attack helicopters and multi-purpose attack helicopters. A separate protocol lays out provisions for disarming and recategorizing helicopters as combat support helicopters and thus outside Treaty limits.

During the negotiation helicopter limits were raised to 2000 for each alliance, with a single-country limit of 1500.

A point of dispute throughout the summer of 1990 was the Soviet claim that the Mi-24, which NATO considered to be a dedicated attack helicopter, was largely used for reconnaissance and fire control missions. Eventually NATO agreed to permit the USSR a special allowance of 100 Mi-24 helicopters equipped for reconnaissance. Excess Mi-24s will be limited and counted as specialized attack helicopters regardless of how they are equipped.[23]

Sub-zonal limits

Anxiety on the NATO flanks

Establishing sub-zonal limits on TLE was a matter of special concern for Norway and Turkey, both fearing that Soviet TLE moved from Central Europe would end up in Soviet MDs on their borders. Initial proposals set no specific limits for the flank zones. During 1990, however, Norway complained that Soviet combat aircraft withdrawn from Hungary had turned up on the Kola Peninsula, where the USSR has many underused airfields. Turkish officials complained that not only troops returning from Afghanistan, but also those being withdrawn from Central Europe, were redeployed in Soviet MDs on Turkey's northern border. In October 1990 Norway and Turkey succeeded in imposing special limits on both active and stored TLE in the outer zone. Total storage in the flank zone must not exceed 1000 tanks, 900 artillery pieces and 800 ACVs (see table 13.1). All other stored TLE must be in sub-zone IV.2. Norway insisted not only on limits for active and stored TLE in the flank zone but also on what may temporarily pass through the flank zone (Articles V.1 (B) and V.1 (C)).

[22] Sharp (note 2), pp. 484–85.

[23] Remme, K. D., Forsberg, R. and Leavitt, R., 'Update on helicopters', *Vienna Fax*, no. 22 (10 Sep. 1990).

Greece was unhappy that south-eastern Turkey was excluded from the ATTU zone because Turkey launched its invasion of Cyprus in July 1974 from the port of Mersin. Because of Greek sensitivity on this point the definition of the western border of south-eastern Turkey was left deliberately vague: 'and thence to the sea' (Article II (B)).

Soviet concerns

No sooner had the 22 delegates in Vienna agreed on the new flank-zone ceilings than the USSR complained (on 31 October) that it could not meet these requirements, unless the Kiev MD was transferred to sub-zone IV.3. Kiev was the MD to which troops returned from Afghanistan and is also the area in which the USSR claims it needs to maintain capabilities to control ethnic unrest that may re-emerge between Armenia and Azerbaijan.[24] In meetings with the Soviet military in Moscow on 7–8 November, Secretary of State Baker agreed to the transfer of the Kiev MD to sub-zone IV.3. This was a reversal of the Soviet Foreign Ministry position agreed to in early October and was later cited by the legal representative of the Soviet General Staff as an example of the military's vigilance in safeguarding the state's security interest in the negotiations.[25] The other 20 delegates agreed in Vienna immediately afterwards, but maintained the same zonal ceilings overall with specific ceilings for TLE in the Kiev MD, namely, that active plus stored TLE must not exceed 2250 tanks, 2500 ACVs and 1500 artillery pieces (see table 13.1).

Hungary moves to sub-zone IV.4

While Norway, Turkey and the USSR wanted to adjust zonal limits, and the zones, to meet traditional military security interests, Hungary insisted on being moved from sub-zone IV.3 to IV.4 for political reasons. Hungary had always objected to being left out of the Mutual (and Balanced) Force Reduction—M(B)FR—Talks and, especially in the new post-bloc Europe, wanted to be considered a Central European not a Balkan power. Hungarian military officers questioned the wisdom of this until it was clear that Hungarian equipment would not be more severely limited in sub-zone IV.4 than in IV.3, and in particular not more limited than Romanian equipment.[26] This problem could arise again in the CFE IA talks, because Hungarian officials will want to avoid an unfavourable manpower ratio with Romania.

Verification

Compared to his predecessors, Gorbachev has been more willing to open up Soviet territory to inspection and more open with data on Soviet budgets and

[24] Interviews by the author with CFE delegates.

[25] Interview with Lt General F. I. Ladygin, Chief of the Soviet Armed Forces General Staff Legal Directorate, by V. Izgarshev: 'On a partnership basis: the Paris meetings repercussions', *Pravda*, 7 Dec. 1990, FBIS-SOV-90-237, 10 Dec. 1990, pp. 3–6.

[26] Interviews by the author with CFE delegates.

force postures. Nevertheless, in the last weeks of negotiation differences emerged between the USSR and the other participants on the issue of inspections, with the USSR insisting on fewer annual inspections than the other delegates would have preferred. In general the Europeans, both NATO and NSWTO states, resented US acceptance of Soviet proposals to reduce inspection requirements and of Soviet definitions of what constitutes an OOV.[27]

NATO initially proposed that the number of inspections a state must accept should be a function of the total TLE and its geographic size. In New York in early October, however, the USA accepted the Soviet approach that based the number of inspections on the structure of a nation's armed forces. The CFE Treaty defines an OOV as any formation or unit holding TLE, any permanent storage site, and any TLE reduction site (Protocol on Inspection, section I.1(J)). When asked in October how many OOVs the USSR had in the ATTU zone, Ambassador Oleg Grinevsky told Western delegates in Vienna that they would have 1600 at the time of CFE signature in mid-November and 1500 at the end of the 40-month reduction phase.[28] In fact when information was exchanged just prior to signature of the Treaty, the USSR claimed only 895 OOVs in the ATTU zone; this number was later adjusted to 910.[29]

The inspection regime

During 1990, it became clear that the CFE Treaty would be concluded between a relatively cohesive NATO and six increasingly individualistic states, several of whom wanted to participate in a compliance regime not as members of the WTO but as members of a multilateral 22-state body. In October 1990, for example, Hungary warned that it would not be able to ratify a CFE agreement that did not permit intra-alliance inspections, that is, that did not allow Hungarian inspections of Soviet territory. While expressions of NSWTO independence have generally been welcomed by NATO delegates in Vienna, this Hungarian enthusiasm to inspect the USSR was a mixed blessing for the West. To the extent NSWTO states insist on inspecting Soviet sites, this will be at the expense of inspections of the West. Each state party can inspect all other parties in the ATTU zone, but no party can conduct more than five inspections per annum on the territory of a party in its own alliance group of states (Protocol on Inspection, section II, para. 14). Table 13.5 shows the number of OOVs declared at Treaty signature and the number of inspections each country must accept in each reduction period.

The other 21 parties to the CFE Treaty were concerned about the low Soviet OOV number because it cuts down the number of inspections they had expected to be able to make of Soviet territory, a concern that grew more acute when it was clear that the USSR had transferred almost 80 000 pieces of TLE

27 Interviews by the author with CFE delegates.

28 Grinevsky did not make a formal statement to this effect in a CFE plenary session but informally to a number of Western delegates.

29 Soviet statement to the JCG, 14 Feb. 1991.

Table 13.5. Passive and challenge inspections in the reduction period

State	Declared OOVs	Inspections each state must accept[a] Phase I	Phase II	Phase III	Phase IV
Belgium	53	10 (1)	5 (1)	10 (1)	7 (1)
Bulgaria	84	16 (2)	8 (1)	16 (1)	12 (2)
Canada	12	2 (1)	1 (1)	2 (1)	1 (1)
Czechoslovakia	250	50 (1)	25 (1)	50 (1)	37 (1)
Denmark	61	12 (1)	6 (1)	12 (1)	9 (1)
France	286	57 (8)	28 (4)	57 (8)	42 (9)
Germany	450	90 (13)	45 (6)	90 (13)	67 (15)
Greece	109	21 (3)	10 (1)	21 (3)	16 (3)
Hungary	55	11 (1)	5 (1)	11 (1)	8 (1)
Iceland	0	0 (1)	0 (1)	0 (1)	0 (1)
Italy	158	31 (4)	15 (2)	31 (1)	23 (1)
Luxembourg	0	0 (1)	0 (1)	0 (1)	0 (1)
Netherlands	99	19 (2)	9 (1)	19 (2)	14 (3)
Norway	56	11 (1)	5 (1)	11 (1)	8 (1)
Poland	118	23 (3)	11 (1)	23 (3)	17 (4)
Portugal	61	12 (1)	6 (1)	12 (1)	9 (2)
Romania	167	33 (5)	6 (1)	12 (1)	9 (1)
Spain	90	18 (2)	9 (1)	18 (2)	13 (3)
Turkey	120	24 (3)	12 (1)	24 (3)	18 (4)
UK	228	45 (6)	22 (3)	45 (6)	34 (8)
USA	164	32 (4)	16 (2)	32 (4)	24 (5)
USSR	895	179 (26)	89 (13)	179 (26)	134 (30)

[a] Maximum allowed challenge inspections are in parentheses.

Source: Lewis, P. M., 'Verifying the CFE Agreement' (Verification Technology Information Centre: London, Feb. 1991).

from the zone. While the number of inspections that the USSR must accept still falls within the statistically acceptable range for confidence in the verification regime (i.e., 134 passive and 30 challenge inspections annually for the duration after the reduction period ends) the other states parties will need to share information with each other to make up for their reduced rights.[30]

Production monitoring

Ideally, a CFE Treaty regime would monitor TLE production but this proved impossible to negotiate. The USA, for example, proposed monitoring in the ATTU zone, but France and Britain considered this discriminatory since it would involve intrusive inspections of their defence industries, while excluding those in the USA, Canada and Soviet territory east of the Urals. Eventually the USA withdrew its proposals.

[30] On statistical sampling see Lewis, P. M., 'Implementation of verification methods', Kokoski and Koulik (note 4), pp. 168–88.

IV. Impact of the CFE Treaty on NATO and WTO TLE holdings[31]

Much less equipment will be destroyed as a result of the CFE Treaty than had been predicted because the USSR transferred most of the excess Soviet TLE from the ATTU zone before it was signed. On 19 November 1990, the day the Treaty was signed, a BBC world service newscaster announced that the Treaty would result in the destruction of 250 000 pieces of equipment.[32] Several Soviet spokesmen, including Chief of General Staff Mikhail Moiseyev, said that the Treaty required the destruction of 120 000 pieces of TLE.[33] Colonel General Fyodor Ladygin was closer to the mark when he stated that the Treaty would require NATO cuts of 16 000 TLE items and WTO cuts of 34 000, of which Soviet cuts would be 1300 combat aircraft, 600 tanks, 9600 ACVs and 760 artillery pieces.[34]

According to data released at its signature, the Treaty requires 44 829 TLE items to be removed from the ATTU zone and fewer than 27 000 and 7000 items to be destroyed for the WTO and NATO, respectively. Table 13.6 shows the required WTO cuts to achieve CFE-imposed ceilings by 1 January 1995. Table 13.7 shows NATO holdings and required cuts to observe CFE limits, and table 13.8 summarizes the impact of the Treaty on both alliances.

WTO tanks and ACVs

One effect of the CFE Treaty will be to encourage European states to deploy lighter, more mobile forces, leading to some concern about leaner and meaner, rather than more manifestly defensive forces.[35] When the CFE talks began it was assumed that the USSR would have to destroy many more tanks than NATO to comply with equal alliance ceilings. According to WTO data submitted in January 1989 WTO tank holdings were 59 470, of which 41 580 were Soviet tanks and 17 890 were NSWTO tanks. With a CFE alliance tank ceiling of 20 000 and a single-country limit for the USSR of 13 150, the WTO would have had to cut 39 470 tanks, of which 28 430 would be Soviet and 11 040 NSWTO. Between July 1988 and Treaty signature, however, the USSR moved over 20 000 tanks from the ATTU zone and out of CFE jurisdiction, thereby avoiding the obligation to destroy many of its TLE pieces in excess of CFE ceilings.

[31] The analysis in this section is based on the initial data exchanged in Nov. 1990 and is subject to adjustment by the 22 states parties to the CFE Treaty.

[32] The need to reduce 250 000 pieces is also cited by Reuters, 'Soviet Union to cut defence spending', *Financial Times*, 23 Nov. 1990, p. 2.

[33] Interview with Mikhail Moiseyev, *Isvestia*, 22 Nov. 1990, FBIS-SOV-90-226, 23 Nov. 1990, p. 4; Chernyshev, V., 'About the USSR's conventional arms in Europe', *Krasnaya Zvezda*, 27 Nov. 1990, FBIS-SOV-90-229, 28 Nov. 1990, pp. 2–3.

[34] *Pravda*, 7 Dec. 1990.

[35] Hall, R., 'Where next for the Soviet division?', *Jane's Soviet Intelligence Review*, vol. 2, no. 12 (Dec. 1990), pp. 538–40.

Table 13.6. WTO holdings in November 1990, cuts required and 1995 CFE ceilings

Country	Battle tanks	Artillery	ACVs	Combat aircraft	Attack helicopters
Bulgaria	2 416	2 474	2 010	387	44
Cuts	941	724	10	152	23+
CFE	1 475	1 750	2 000	235	67
Czechoslovakia	3 035	3 485	4 359	369	56
Cuts	1 600	2 335	2 309	24	19+
CFE	1 435	1 150	2 050	345	75
Hungary	1 345	1 047	1 720	110	39
Cuts	510	207	20	70+	69+
CFE	835	840	1 700	180	108
Poland	2 850	2 300	2 377	654	128
Cuts	1 120	690	227	194	2+
CFE	1 730	1 610	2 150	460	130
Romania	2 851	3 819	3 135	407	104
Cuts	1 476	2 344	1 035	–	23+
CFE	1 375	1 475	2 100	430	120
Total NSWTO	**12 497**	**13 125**	**13 601**	**1 927**	**371**
Cuts	**5 647**	**6 300**	**3 601**	**277**	**129+**
CFE	**6 850**	**6 825**	**10 000**	**1 650**	**500**
USSR	20 694	13 828	29 348	6 445	1 330
Cuts	7 544	653	9 348	1 295	170+
CFE	13 150	13 175	20 000	5 150	1 500
Total WTO	**33 191**	**26 953**	**42 949**	**8 372**	**1 701**
Cuts	**13 191**	**6 953**	**12 949**	**1 572**	**299+**
CFE	**20 000**	**20 000**	**30 000**	**6 800**	**2 000**

Source: 'The distribution of the WTO conventional ceilings', *Rzeczpospolita* (official gazette of the Polish Government), 16 Nov. 1990; data from Márton Krasznai, Ministry of Foreign Affairs, Budapest, 3 Dec. 1990.

When data were exchanged in late November, the USSR revealed that it had only 20 694 tanks in the ATTU zone, requiring cuts of 7544 by 1 July 1995. Thus it cut its ATTU zone holdings by 50 per cent unilaterally and is required to cut only 18 per cent under the Treaty (see table 13.11 below). In addition to the reduction of Soviet tanks, the CFE Treaty also requires tank cuts in the NSWTO states as shown in table 13.6.

According to the Protocol on Reduction, of the 7544 tanks the USSR must reduce, 750 may be converted and 6794 destroyed. Each of the five NSWTO states may convert at least 150 tanks. Romania, Poland and Czechoslovakia can convert 162, 162 and 183 tanks, respectively, so that the total of NSWTO tanks that may be converted is 807, leaving 4840 that must be destroyed (table 13.8 below).

The USSR may convert 3000 ACVs and must destroy 6348. The NSWTO must cut 3601 ACVs, of which they can convert 2028 and destroy 1573. The WTO must destroy 11 634 tanks, 7921 ACVs and 6953 artillery pieces: that is, of the 33 093 items of ground force TLE the WTO must reduce, 26 528 items must be destroyed and 6585 can be converted (table 13.8).

NATO tanks and ACVs

NATO data exchanged at Treaty signature indicate that alliance deployments will remain below Treaty limits (table 13.7). Tank holdings at Treaty signature were 25 091, which would require a cut of 5091 to comply with the Treaty limit for each alliance group of states of 20 000, but NATO decided to reduce to 19 142. This implied reductions of 5949, of which 2107 can be converted according to the Protocol on Reduction, leaving only 3842 to be destroyed. NATO holdings of ACVs at Treaty signature were 34 453 and allocations of CFE residuals came to 29 822. Of the 4631 to be reduced, all can be converted so none need to be destroyed. Thus, out of the 12 914 items of NATO TLE to be reduced, 6738 can be converted and 6176 must be destroyed: 3842 tanks and 2334 pieces of artillery.

Artillery holdings

When the CFE Negotiation began WTO artillery holdings in the ATTU zone were considered vastly superior in number to those of NATO: 42 400 pieces on 1 July 1988 (see table 13.11). Parallel to the CFE Negotiation, however, more than 20 000 Soviet artillery pieces were transferred east of the Urals, so that when Soviet data were presented in Vienna in November 1990, they showed only 13 828 Soviet artillery pieces, compared to 20 620 for NATO. (NSWTO artillery was almost the same as Soviet holdings at 13 125.) The Soviet allocation of artillery in the WTO ceiling is 13 175 pieces so the USSR is required to cut almost precisely the same number as NATO: 653 pieces for the USSR, 620 pieces for NATO. Unilateral Soviet withdrawal of artillery from the ATTU zone represents 74 per cent of 1988 holdings, Treaty-imposed cuts represent only 1 per cent (table 13.11).

Combat aircraft

The Treaty does not require any cuts in NATO air power. According to the data exchanged at CFE Treaty signature, NATO had 5939 combat aircraft in the ATTU zone and the WTO had 8372, of which 6445 were Soviet and 1927 NSWTO. NATO can add 723 combat aircraft and 264 attack helicopters to its holdings in the ATTU zone—probably not the result anticipated by the Soviet policy makers who argued for inclusion of air power in the negotiation. The Treaty requires cuts in WTO combat aircraft, but each state party can disarm and reclassify up to 550 aircraft (of which no more than 130 can be MiG-25Us—Protocol on Procedures Governing Reclassification of Aircraft) so of the 1295 aircraft the USSR must reduce, only 745 need be destroyed, and of the 277 the NSWTO states must reduce all can be converted.

Table 13.7. NATO holdings in November 1990, cuts required and 1995 CFE ceilings

Country	Battle tanks	Artillery	ACVs	Combat aircraft	Attack helicopters
Belgium	359	376	1 282	191	0
Cuts	25	56	183	41+	46+
CFE	334	320	1 099	232	46
Canada	77	38	277	45	12
Cuts	0	0	0	45+	1+
CFE	77	38	277	90	13
Denmark	419	553	316	106	3
Cuts	66	0	0	0	9+
CFE	353	553	316	106	12
France	1 358	1 330	4 125	700	429
Cuts	52	38	305	100+	77
CFE	1 306	1 292	3 820	800	352
Germany	7 133	4 644	9 598	1 064	357
Cuts	2 967	1 939	6 152	164	51
CFE	4 166	2 705	3 446	900	306
Greece	1 725	1 941	1 639	480	0
Cuts	10+	63	895+	170+	18+
CFE	1 735	1 878	2 534	650	18
Italy	1 912	2 222	3 591	584	169
Cuts	564	267	252	66+	27
CFE	1 348	1 955	3 339	650	142
Netherlands	913	838	1467	196	91
Cuts	170	231	387	34+	22
CFE	743	607	1 080	230	69
Norway	205	532	146	90	0
Cuts	35	5	79+	10+	0
CFE	170	527	225	100	0
Portugal	146	334	259	96	0
Cuts	154+	116+	171+	64+	26+
CFE	300	450	430	160	26
Spain	854	1 373	1 259	252	28
Cuts	60	63	329+	58+	43+
CFE	794	1 310	1 588	310	71
Turkey	2 888	3 202	1 554	589	0
Cuts	93	321+	1 566+	161+	43+
CFE	2 795	3 523	3 120	750	43
UK	1 198	636	3 193	842	368
Cuts	183	000	17	584	16+
CFE	1 015	636	3 176	900	384
USA	5 904	2 601	5 747	704	279
Cuts	1 898	109	375	80+	239+
CFE	4 006	2 492	5 372	784	518
Total NATO	**25 091**	**20 620**	**34 453**	**5 939**	**1 736**
Cuts	**5 949**	**2 334**	**4 631**	**723+**	**264+**
CFE	**19 142**	**18 286**	**29 822**	**6 662**	**2 000**

Sources: CFE Declarations and Residual Ceilings, British Ministry of Defence, Nov. 1990; *Hansard*, 19 Nov. 1990, pp. 15–16.

Table 13.8. Summary of the impact of the CFE Treaty on NATO and the WTO

	Battle tanks	Artillery	ACVs	Combat aircraft	Attack helicopters
CFE limits					
ATTU	40 000	40 000	60 000	13 600	4 000
Alliance	20 000	20 000	30 000	6 800	2 000
Country[a]	13 150	13 175	20 000	5 150	1 500
November 1990 holdings					
ATTU	58 282	47 573	77 402	14 311	3 437
NATO	25 091	20 620	34 453	5 939	1 736
WTO	33 191	26 953	42 949	8 372	1 701
USSR	20 694	13 828	29 348	6 445	1 330
NSWTO	12 497	13 125	13 601	1 927	371
ATTU zone					
CFE cuts	18 282	7 573	17 402	711[b]	(563+)
Agreed cuts[c]					
NATO cuts	5 949	2 334	4 631	(723+)	(264+)
To convert	2 107	0	4 631		
To destroy	4 388	2 334	0		
WTO cuts	13 191	6 953	12 949	1 572	(299+)
To convert	1 557	0	5 028	827	
To destroy	11 634	6 953	7 921	745	
Soviet cuts	7 554	653	9 348	1 295	(170+)
To convert	750	0	3 000	550	
To destroy	6 794	653	6 348	745	
NSWTO cuts	5 647	6 300	3 601	277	(129+)
To convert	807	0	2 028	277	
To destroy	4 840	6 300	1 573	0	

[a] The WTO agreed on lower single-country limits for the USSR.

[b] 711 is the net cut in ATTU zone combat aircraft; the WTO must cut 1572, but NATO has headroom to add 723.

[c] Each alliance allocated CFE cuts below Treaty limits.

Cascading

To avoid destruction of their most modern equipment, NATO defence ministers agreed in principle during 1989 to 'cascade' modern TLE in excess of CFE limits down to their less well endowed allies, who in turn agreed to destroy older equipment. On 7 December 1990, Secretary General of NATO Manfred Wörner told reporters that defence ministers had agreed to transfer over 3000 tanks, over 1000 ACVs and 176 artillery pieces from Germany, the Netherlands and the USA to Denmark, Greece, Norway, Portugal, Spain and Turkey.[36] Czechoslovakia, Poland and Hungary have also expressed the

[36] *Aviation Week & Space Technology*, 17 Dec. 1990, pp. 70–71.

Table 13.9. NATO cascading plans, December 1990

	Recipient					
Item/donor	Denmark	Greece	Norway	Portugal	Spain	Turkey
Tanks	110	700	100	800	530	1 050
From USA	–	30	–	800	530	200
Germany	110	500	100	–	–	850
Netherlands	–	170	–	–	–	–
ACVs	–	150	125	100	100	600
From USA	–	150	125		100	
Germany	–	–	–	–	–	600
Netherlands	–	–	–	100	–	–
Artillery	36	70	–	–	–	70
From USA	36	70	–	–	–	70
Total TLE items cascaded	**146**	**920**	**225**	**900**	**630**	**1 720**

Source: Hitchens, T., 'NATO to move weapons from central Europe', *Defence News*, 17 Dec. 1990.

Table 13.10. Impact of cascading on TLE destruction by donor

Donor/item	CFE required cuts by donor	TLE cascaded to allies	Conversions permitted	TLE still to destroy
Germany				
Tanks	2 967	1 560	406	1 000
ACVs	6 152	600	1 439	4 113
Artillery	1 939	–	–	1 939
USA				
Tanks	1 898	1 560	336	2
ACVs	375	375	847	0
Artillery	109	176	–	0
Netherlands				
Tanks	170	170	150	0
ACVs	387	100	120	157

desire to benefit from a similar cascading programme, especially of equipment formerly held by the Nationale Volks Armee (NVA) in the GDR.[37] Table 13.9 shows the initial cascading distribution among the NATO allies. Table 13.10 indicates how cascading affects the number of TLE items that donor countries will have to destroy.

[37] AP, 'Hungary bids for Berlin arsenal', *The Times*, 4 Jan. 1991.

V. Resolving discrepancies in the initial data exchange

The Treaty established a Joint Consultative Group in which discrepancies in data and ambiguities in Treaty compliance could be resolved. The Treaty also provided a 90-day period from Treaty signature in which to correct any mistakes in the initial data exchange; that is, from 18 November 1990 until 18 February 1991. The first meeting of the JCG was called on 29 November 1990, ended on 13 December and reconvened on 19 January 1991. It continued to meet every second week beyond 18 February because Soviet delegates were unable to clarify their data in the 90 days allocated for that purpose. In addition, several Western countries sought to resolve differences with the USSR bilaterally. US Ambassador James Woolsey led a US delegation to Moscow on 6–7 December. A few days later in Houston, Texas, Woolsey and James Baker met with Eduard Shevardnadze, Dmitri Yazov and Mikhail Moiseyev, apparently with little success because the following week, at the NATO ministerial meeting in Brussels, Baker told reporters that serious problems remained.[38]

Two major and two relatively minor problems arose with the Soviet data. First, in late 1990, US intelligence sources claimed that the USSR had under-reported its holdings in the ATTU zone by some 20 000–40 000 TLE items. Second, the other 21 states parties objected to Soviet claims that TLE associated with three motor rifle divisions, resubordinated to coastal defence, were exempt from CFE limits because naval forces were not included in the CFE mandate. Less serious but still troubling, the USSR also claimed exemption for TLE beyond those already provided for in Article III of the Treaty, namely, for TLE with the strategic rocket and DOSAAF forces.[39] Finally, the USSR declared only 895 OOVs, that is, half the number it had acknowledged previously, but was able to provide plausible explanations of the discrepancy at the JCG.[40]

The resolution of these issues was complicated by two factors. The first was the difference in attitude towards the CFE Treaty between the Soviet Foreign Ministry which, under Shevardnadze, focused on its diplomatic advantages, and the Soviet General Staff, which feared that it would undermine Soviet security.[41] These differences were apparent even before Shevardnadze's resignation in December, but emerged full-blown in early 1991 as senior military spokesmen stepped into the vacuum left by his departure and went on the offensive in the JCG on CFE data issues. Some Western observers detected an

[38] Reuters, 'NATO tells Soviets to amend flaws in its arms reporting', *International Herald Tribune*, 19 Dec. 1990.

[39] DOSAAF stands for 'Voluntary service for cooperation with the Army, Aviation and Navy' and is a youth training scheme analagous to the ROTC (Reserve Officer Training Corps) in the USA.

[40] Moscow World Service, 'General Chervov assures CFE treaty compliance', 15 Dec. 1990, FBIS-SOV-90-242, 17 Dec. 1990, p. 1; see also Mokovskiy, O., 'Chervov clarifies CFE inspection site numbers', TASS, 7 Dec.1990, FBIS-SOV-90-236, 7 Dec. 1990, p. 2; OOV data set reprinted in 'High level discussions focus on Soviet force data', *Basic Reports from Vienna*, no. 12 (17 Dec. 1990), p. 2.

[41] See, for example, Soviet military officers reactions to Shevardnadze's resignation and the CFE Treaty in 'Officers react to Shevardnadze resignation', *Krasnaya Zvezda*, 22 Dec. 1990, FBIS-SOV-90-247, 24 Dec. 1990, pp. 81–82.

attempt to sabotage the Treaty altogether, while others believed the USSR was merely squeezing the last ounce of exemptions and concessions out of the West before finally giving in. The second, and related, problem in data resolution was that, parallel to the CFE Negotiation, the USSR was completing a programme of unilateral withdrawals from Eastern Europe and of restructuring its slimmed-down forces. To the extent that the transfers were made before Treaty signature, they were a breach of the spirit but not of the letter of the agreement. Nevertheless, Western countries demanded that the USSR continue to inform the other 21 CFE states parties about the future disposition of its TLE east of the Urals and asked for assurances that the Soviet military would not be allowed to generate a new strategic reserve from the transferred equipment.

Did the USSR under-report TLE in the ATTU zone?

Even though the Soviet TLE transfers were not illegal, their scale, especially the acceleration of shipments in late October and early November 1990, complicated resolution of the large discrepancy between Soviet data presented at the CFE Negotiation on 18 November 1990 and US intelligence estimates of Soviet TLE in the ATTU zone delivered on 20 November to the White House.[42]

Gorbachev announced the Soviet force reductions in December 1988, and the Soviet military provided regular bulletins both of withdrawals of units from its WTO allies (as well as from Mongolia) and of the restructuring of its tank and motor rifle divisions in the ATTU zone (see below).[43]

US satellites monitored the movement of large amounts of military equipment east of the Urals during 1989–90, and in early October 1990 there were reports of 7000 transferred tanks with another 10 000 apparently awaiting shipment in the Kiev MD.[44] When Baker and Shevardnadze met in Irkutsk in August 1990, and again in New York in early October (for the CSCE meeting to prepare the Paris summit meeting), Baker asked for more detail about CFE treaty-limited equipment leaving the ATTU zone. The response came in a press conference that Soviet Chief of Staff Mikhail Moiseyev gave at the Pentagon on 5 October and in a letter from Eduard Shevardnadze to James Baker on 13 October 1990. Moiseyev said that TLE had been transferred as part of the unilateral withdrawals from Czechoslovakia and Hungary, the switch to a structure of 'defensive sufficiency', and the implementation of the unilateral reduction of Soviet armed forces by 500 000.[45] Shevardnadze's letter gave more details of the tanks, ACVs and artillery that had been moved from the ATTU zone between July 1988 and August 1990. He suggested that

[42] Walker, M., 'Arms deal back on track as US admits error', *The Guardian,* 2 Feb. 1991.
[43] These are documented in Sharp (note 2), pp. 459–74.
[44] International Institute for Strategic Studies, *Military Balance 1990–1991* (Brassey's: Oxford, 1990), p. 31.
[45] 'Moiseyev continues US visit', *Izvestia,* 8 Oct. 1990, p. 3.

Table 13.11. ATTU zone holdings, transfers of Soviet TLE and required CFE cuts

	1 July 1988 (A)	1 Aug. 1990 (B)	19 Nov. 1990 (C)	Transfers (A→B)	(B→C)	(A→C)	CFE ceiling	CFE cut by Jan. 95	% cut unilat.	% cut by CFE
Tanks	41 580	24 898	20 694	16 682	4 204	20 886	13 150	7 544	*50*	*18*
Artillery	42 400	18 300	13 828	24 100	4 472	28 572	13 175	653	*74*	*1*
ACVs	57 800	32 320	29 348	24 480	2 972	27 452	20 000	9 348	*49*	*16*
Total	**141 780**	**75 518**	**63 870**	**65 262**	**11 648**	**76 910**	**46 325**	**17 545**	*55*	*12*

Note: Data in columns A, B and C are from Shevardnadze's letter of 13 October 1990.

the ATTU holdings would not change much before CFE Treaty signature except that Soviet tanks in the ATTU zone would be reduced to about 21 000. In October, James Baker and General Colin Powell (Chairman of the Joint Chiefs of Staff) expressed satisfaction with these explanations. But the data presented by the Soviet military at Treaty signature also showed further cuts in artillery and ACVs. As Foreign Ministry officials later acknowledged, the Soviet General Staff had not kept the Foreign Minister informed of plans to move as much TLE out of the ATTU zone as possible before Treaty signature.[46]

Table 13.11 summarizes the transfers of Soviet TLE out of the ATTU zone east of the Urals between July 1988 and November 1990.

Shevardnadze's October letter explained that of the more than 20 000 tanks that would have been withdrawn from service in the ATTU zone by the end of the year, 10 000 came from units withdrawn from the groups of Soviet forces stationed on the territory of their European allies and 10 000 from restructuring Soviet tank and motor rifle divisions throughout the ATTU zone. Of these 20 000 tanks 4000 had already been destroyed, exported or converted into 'prime movers, fire engines or trainers', 8000 would be redeployed with divisions in the Asian MDs of the USSR, and 8000 would either be destroyed or 'converted in accordance with the State Conversion Program'. TASS reported in late November that of the artillery pieces removed from the ATTU zone 500 were eliminated, 1100 were cascaded to Soviet forces in Asia, and 16 400 were stored in depots in Siberia and Central Asia for gradual replacement of older systems as they 'exhaust their service life'. Of the ACVs transferred from the ATTU zone, 11 200 were transferred to Soviet forces in Asia and 4700 were stored.[47]

After the official data exchange in November US intelligence sources claimed that the Soviet military had under-reported by 20 000–40 000 TLE items. One source cited differences of between 6000–11 000 tanks, 12 000 ACVs, 12 000 artillery pieces and 3000 combat aircraft.[48]

[46] Interview with an unnamed 'senior official of the USSR Ministry of Foreign Affairs', who participated in the Vienna talks, by Dmitri Yakashkin, 'The end of the Vienna agreements', *Moscow News*, no. 3 (20–27 Jan. 1991), p. 13.

[47] Chernyshev, V., 'Chernyshev on CFE Treaty, Soviet implementation', TASS, 22 Nov. 1990, FBIS-SOV-90-228, 27 Nov. 1990, pp. 1–2.

[48] *Arms Control Reporter*, sheet 407-B-413, Nov. 1990.

On 22 January 1991 a British delegate to the JCG asked the USSR for information that appeared to be missing in data reports on the Western Group of Forces stationed in the former GDR; specifically, he wanted details of the 7th Tank Division, the 12th Tank Division, the 281st Artillery Brigade and the 214th Artillery Training Brigade.[49] As early as January 1989, however, General Secretary Erich Honecker had announced that the 7th and 12th tank divisions, as well as an aerial storm brigade, three training regiments and three independent battalions, would be withdrawn from the GDR in 1990 as part of the Gorbachev cuts announced in December 1988.[50] In mid-December 1990, General Nikolai Chervov, speaking on Moscow radio, noted that the 7th Tank Division had been disbanded during the past year, its personnel demobilized and its tanks transferred to Siberia.[51] Other Soviet military spokesmen have explained in various forums which military units had been disbanded in the unilateral reduction programme. For example, on 14 June 1990 the Soviet Defence Ministry issued a statement on disbanding the 13th Guards (from the Southern Group of Forces in Hungary), and the 32nd Guards and 25th Tank Division (from the Western Group of Forces in the GDR).[52] It would have been easier for Western delegates to clarify Soviet CFE data if the Soviet General Staff had been more specific not only about military units remaining in the ATTU zone on 18 November 1990, but also about all those disbanded since 1988.

Meanwhile, in Washington in early February 1991, State Department spokesmen admitted that intelligence estimates about Soviet holdings in the ATTU zone delivered to the White House on 20 November were six weeks old.[53] Intelligence estimates in late January 1991 showed a discrepancy of less than 10 000, of which 2000–3000 were genuine 'smoking guns', that is, specific sightings of equipment that had not appeared in Soviet data on 18 November 1990.[54] These new estimates implied that the Soviet military could have transferred nearly as much TLE out of the ATTU zone as claimed in their 18 November data exchange. Nevertheless, in the wake of US reaction to Soviet action in the Baltic republics, and several weeks into the war against Iraq, the Bush Administration put off a scheduled summit meeting with Mikhail Gorbachev. In addition, James Baker told the House Armed Services Committee that he recommended further delay in submitting the Treaty

49 Cable from US CFE delegation to State Department, 22 Jan. 1991, reprinted in *Arms Control Reporter*, sheet 407.D.77–79, 1991.

50 Erich Honecker at a dinner for Swedish Prime Minister Ingvar Carlsson in East Berlin, 23 Jan. 1989; see McCartney, R., *Washington Post*, 24 Jan. 1989.

51 Nikolai Chervov on Moscow World Service, 15 Dec. 1990, 'General Chervov assures CFE Treaty compliance', FBIS-SOV-90-242, 17 Dec. 1990, p. 1.

52 'Defence Ministry statement on troop reductions', TASS, 14 June 1990, FBIS-SOV-90-116, 15 June 1990, p. 87.

53 Walker, M., 'Arms deal back on track as US admits error', *The Guardian*, 2 Feb. 1991.

54 Jack Mendelsohn, remarks at the Arms Control Association press conference, Washington, DC, 6 Feb. 1991; *Vienna Fax*, vol. 2, no. 1 (6 Feb. 1991).

Table 13.12. TLE in Soviet units resubordinated to naval infantry

	Tanks	ACVs	Artillery
77th Guards Archangelsk (Leningrad MD)	265 (T-80)	30	160
3rd Guards Klaipeda (Baltic MD)	271 (T-72B)	433	384
126th Guards Simferopol (Odessa MD)	271 (T-64A, T-64B)	433	294
Sub-totals	807	896	838
Total TLE items		**2 541**	

Source: Starr, B., 'Soviet move to evade CFE', *Jane's Defence Weekly*, 22 Dec. 1990, p. 1257.

to the Senate for ratification because although one data problem looked close to resolution other serious ones remained.[55]

Resubordination of the motor rifle divisions

The most serious data problem was the resubordination of three Soviet motor rifle divisions (MRDs) to naval infantry—a measure that assigned more tanks to the Soviet Navy than were held by British ground forces. The Soviet military announced this measure to James Baker in Houston in December, when Eduard Shevardnadze apparently also heard it for the first time; again revealing the lack of communication between the foreign and defence ministries in Moscow.[56] Yazov and Moiseyev claimed that the NATO countries, which had insisted on excluding naval forces from the CFE mandate, could hardly impose limits on Soviet naval TLE. The Western response was that equipment was treaty-limited if it was land-based in the ATTU zone, unless specifically excluded by the Treaty. Moreover, Poland, Romania and Spain had each counted naval infantry TLE in their ATTU holdings.[57] The resubordinated divisions and their TLE holdings are listed in table 13.12.

At the JCG in Vienna, the Soviet delegates offered what they called a compromise on the resubordinated MRDs, namely, that they would make a no-increase commitment in coastal defence forces. The response of the 21 other delegations was to repeat their earlier insistence that land-based TLE, whether nominally assigned to naval infantry or conventional ground force units, must be subject to CFE limitations. Meanwhile, in Washington a bipartisan group of five senators who served on the foreign affairs, intelligence and armed services committees (Joseph Biden, David L. Boren, Frank H. Murkowski, Sam Nunn and John W. Warren) told James Baker that they would consider ratify-

[55] AP, 'Baker seeks delay in Europe forces pact', *International Herald Tribune*, 7 Feb. 1991; Barber, L. and Boulton, L., 'US signals change in policy over Soviet contacts', *Financial Times*, 7 Feb. 1991.

[56] Interviews by the author with CFE delegates in London and Washington, DC.

[57] *Vienna Fax*, 6 Feb. 1991, p. 1.

ing the CFE Treaty only if there was a legally binding provision that counts the TLE of the Soviet naval infantry.[58]

VI. Unilateral WTO reductions

Soviet reductions and withdrawals

As indicated above, deeper equipment cuts occurred in the ATTU zone by unilateral measures during 1989–90 than will be imposed by the CFE process. Soviet military manpower cuts are also likely to dwarf cuts imposed by a CFE IA agreement. In late December 1990 Soviet military spokesmen announced that 500 000 troops had been demobilized as promised by Gorbachev in December 1988.[59] More cuts are anticipated in 1991: on 14 June 1990 the Soviet Defence Ministry issued a statement noting that the agreements to withdraw completely from Hungary and Czechoslovakia considerably extended the scale of unilateral reductions from those announced in December 1988. The statement implied that three tank divisions were disbanded in Eastern Europe and three in the western MDs rather than six from Eastern Europe as initially planned. Colonel-General German Burutin reported that 200 000 men had been moved out of Czechoslovakia, Germany, Hungary, Mongolia and Poland; of these, 44 000 had left Hungary and 57 000 had left Czechoslovakia.[60] Other reports indicated that some 42 000 troops had left the GDR and another 15 500 were scheduled to withdraw from Poland. Thus, 158 500 Soviet troops (compared to the 50 000 proposed in 1988) were withdrawn from the NSWTO countries in the two-year period since Gorbachev's UN speech: 28 000 in 1989 and 130 500 in 1990 (see table 13.13).

Soviet reports of troop withdrawals are not always consistent with each other, but table 13.13 shows how the 500 000 troop cuts appear to have been distributed over the two-year period since December 1988: 200 500 foreign-based and 299 500 inside the USSR. During 1989 approximately 23 000 troops were demobilized from Mongolia, 54 000 from the Western MDs (of which 20 000 came from the Leningrad MD), 60 000 from the Southern theatre of military operations (TVD) and 100 000 from the Far East.[61] Burutin's report indicated another 19 000 troops were cut from Mongolia in 1990. During 1990, Soviet authorities repeated earlier plans to cut 200 000 in the Far East by the end of the year, noting that 120 000 would be from forces in Eastern Siberia.[62]

[58] Smith, R. J., 'Senators balk at arms treaty: panel wants Soviets to back down on European forces', *International Herald Tribune*, 9–10 Feb. 1991.

[59] Interview with Dmitri Yazov, Zalchanchenk, M., 'Yazov reviews past year and future tasks', TASS, 2 Jan. 1991, FBIS-SOV-91-002, 3 Jan. 1991, pp. 48–51.

[60] Moskovoskiy, O., 'East Europe troop withdrawal second stage complete', TASS in English, 27 Dec. 1990, FBIS-SOV-90-249, 27 Dec. 1990, p. 1.

[61] Sharp (note 2), pp. 469–71.

[62] Khartov, G., 'Admiral views US superiority in Far East', Moscow World Service, 11 Aug. 1990, FBIS-SOV-90-156, 13 Aug. 1990, p. 67; Gorbachev interview with Asahi Shimbum, 'Gorbachev meets journalists', *Pravda*, 30 Dec. 1990, FBIS-SOV-90-215, 31 Dec. 1990, pp. 5–7.

Table 13.13. Soviet forces and cuts, December 1989–December 1990

Deployment (divisions)	December 1988	1988–89 cut	December 1989	1989–90 cut	December 1990	1988–90 cut
Foreign-based						
Western (19) (GDR)	380 000	(12 000)	368 000	(30 000)	338 000	42 000
Northern (2) (Poland)	62 000	(3 500)	58 500	(12 000)	46 500	15 500
Central (5) (Czechoslovakia)	75 000	(1500)	73 500	(55 500)	18 000	57 000
Southern (4) (Hungary)	61 000	(11 000)	50 000	(33 000)	17 000	44 000
Total NSWTO (30)	**578 000**	**(28 000)**	**550 000**	**(130 500)**	**419 500**	**158 500**
Mongolia	75 500	(23 000)	52 500	(19 000)	33 500	42 000
Other foreign	25 000	–	25 000	–	25 000	–
Total foreign-deployed	**678 500**	**(51 000)**	**627 500**	**(149 500)**	**478 000**	**200 500**
In USSR						
Far East (56)	1 120 000	(100 000)	1 020 000	(20 000)	1 000 000	120 000
South (30)	600 000	(60 000)	540 000	–	540 000	60 000
Central (26)	520 000	–	520 000	–	520 000	–
West (ATTU zone) (63)	1 250 000	(54 000)	1 196 000	(65 500)	1 130 500	119 500
Strategic air forces	90 000	–	90 000	–	90 000	–
Total in USSR	**3 580 000**	**(214 000)**	**3 366 000**	**(85 500)**	**3 280 500**	**299 500**
Total Soviet forces	**4 258 000**	**(265 000)**	**3 993 500**	**(235 000)**	**3 758 500**	**500 000**

Sources: Mishin, Y., *Agumenty i Fakty*, no. 6 (11–17 Feb. 1989); Yazov, D., *Krasnaya Zvezda*, 13 Apr. 1989; TASS report, 'Soviet troop numbers abroad cited for UN session', *Krasnaya Zvezda*, FBIS-SOV-89-244, 21 Dec. 1989, p. 1; International Institute for Strategic Studies, *Military Balance 1988–1989* (IISS: London, 1988); Ministers of Defence of the Warsaw Treaty Member States, 'On the relative strength of the armed forces and armaments of the WTO and NATO in Europe and adjacent waters', *Pravda*, 30 Jan. 1989; NATO, *Conventional Forces in Europe: The Facts*, Brussels, 28 Nov. 1988; 'Gorbachev meets journalists', *Pravda*, 30 Dec. 1990, FBIS-SOV-90-215, 31 Dec. 1990, pp. 5–7; 'Yazov reviews past year and future tasks', TASS, 2 Jan. 1990, FBIS-SOV-91-002, 3 Jan. 1991, pp. 48–51; Moskovoskiy, O., 'East Europe troop withdrawal second stage complete', TASS in English, 27 Dec. 1990, FBIS-SOV-90-249, 27 Dec. 1990, p. 1.

In early 1989, NATO countries welcomed Gorbachev's 1988 initiative to cut forces unilaterally, and a report issued on 9 July 1990 by the Armed Services Committee of the US House of Representatives noted that NATO's warning time of a Soviet attack had increased by a factor of five during the previous year and that 'NATO will soon possess superior conventional forces' to those of the USSR shorn of its allies.[63] By late 1990, however, most

[63] Aspin, L., *et al.*, *The Fading Threat: Soviet Conventional Power in Decline* (US Government Printing Office: Washington, DC, 9 July 1990).

Western leaders viewed the shipment of Soviet equipment to east of the Urals as an effort to evade CFE limits.

If Western attitudes towards Soviet withdrawals went sour, there was even more of a backlash in Moscow where conservatives complained that a combination of Gorbachev's ill-advised unilateral cuts and Shevardnadze's over-eager diplomacy severely undermined Soviet security. Critics claimed that the USSR lost not only its former numerical superiority over NATO in conventional forces and its protective East European glacis, but also the economic, political and military co-operation of its WTO allies.[64] In the Congress of People's Deputies, members of the right-wing Soyuz group, especially two young Air Force colonels, Viktor Alksnis and Nikolai Petrushenko, condemned Shevardnadze's arms control diplomacy and foreign policy initiatives; not least for acquiescing in the US military buildup in the Persian Gulf against a former Soviet ally, Iraq.[65] Even commentators normally supportive of Gorbachev's reforms and Shevardnadze's diplomacy complained that events were moving too quickly for the country to adjust.[66]

Soviet withdrawals from Eastern Europe continued nevertheless, since even conservatives in Moscow knew that the era of Soviet occupation and control of Central European allies was over. On 25 February 1991, the military wing of the WTO was formally abolished at a meeting in Budapest attended by Dmitri Yazov, Alexander Bessmertnykh and all the other WTO defence and foreign ministers.[67]

Soviet withdrawals from Germany

In November 1989, *Pravda* reported that 11 620 Soviet troops and 247 artillery pieces had been withdrawn from the GDR. During 1990 another 30 000 Soviet troops returned to the USSR, but withdrawals slowed down in the middle of the year as the Soviet leadership grew increasingly anxious about the unification of the two German states and the potential loss of the GDR to NATO.

In February and March the Soviet Government insisted that the stay of foreign troops in the two German states was a matter for the four powers, which still retained 'obligations arising from the second world war'.[68] In late April, GDR Defence and Disarmament Minister Rainer Eppelman met Soviet Defence Minister Dmitri Yazov in Moscow and later announced that the USSR was willing to halve its 380 000 troops in the GDR.[69] No withdrawal

[64] Interview with Lt. General Igor Sergeyev, *Moscow News*, vol. 8, no. 9 (1990), p. 11; see also interview with V. Litov in *Sovietskaya Rossiya*, 1 Jan. 1991, FBIS-SOV-91-008, 11 Jan. 1991, pp. 1–4.

[65] Peel, Q., 'Gorbachev told confidence in him is exhausted', *Financial Times*, 19 Nov. 1990; Remnick, D., 'Soviet military draws the line: the time for words is over', *International Herald Tribune*, 22–23 Dec. 1990; Nelan, B. W., 'Broadside from the right', *Time*, 31 Dec. 1990, p. 14.

[66] Kondrashov, S., 'Coachmen, do not drive the horses too fast', *Izvestia*, 3 Jan. 1991, FBIS-SOV-90-002, pp. 1–3.

[67] Denton, N., 'Comecon row marks end of Pact', *Financial Times*, 26 Feb. 1991.

[68] O. Grinevsky at the CFE Negotiation, reported by *Arms Control Reporter*, 16 Feb. and 15 Mar. 1990, sheets 407.B.312, 341–42.

[69] Foreign Broadcast Information Service, *Daily Report–Eastern Europe* (*FBIS-EE*), FBIS-EE-90, 30 Apr. 1990.

dates were set, however, and in May Soviet spokesmen announced that Soviet withdrawals from the GDR had ceased as a result of the acute housing shortage in the USSR as well as apprehension about German unification.[70] In September, the Bonn Government tried to ease the housing problem with a promise of DM 12 billion to support Soviet troops in Germany temporarily, as well as to transport, retrain and rehouse them in the USSR. The German Government also offered Moscow credits worth DM 3 billion.[71] In mid-July a Commander of the Western Group of Soviet Forces in the GDR, Vasiliy Kasachenko, said that withdrawals had halted, not only because of the housing shortage but also because of accelerated withdrawals from Hungary and Czechoslovakia.[72]

A new round of Soviet–German talks on troop withdrawals began in early February 1991 between German Defence Minister Gerhard Stoltenberg and General Matvei Burkalov. Burkalov said that withdrawals were on schedule and manpower in the Western Group of Soviet Forces was down to 338 000. Of the remaining Soviet troops 30 per cent were scheduled to leave in 1991, 30 per cent in 1992, 30 per cent in 1993 and the remaining 10 per cent in 1994. Because of difficulties with the Polish and Czechoslovak authorities, Soviet troops were being shipped by sea from the East German ports of Rostock and Wismar.[73]

Soviet withdrawal from Poland

Two divisions of Soviet Army troops (approximately 40 000 men) remained in Poland after World War II, primarily to protect the lines of communication between the USSR and the Northern Group of Soviet Forces in Germany. These forces were augmented in 1984 when the USSR moved the headquarters of the Western TVD to Legnica in Silesia. The Legnica HQ controlled Soviet military operations in the NSWTO countries as well as in the Baltic, Byelorussian and Carpathian MDs.[74] The HQ was withdrawn in July 1990 because the unification of Germany made Legnica too forward a base to maintain a credibly defensive posture, as outlined by the WTO in 1987.[75]

In late 1988, when Gorbachev announced his programme of unilateral cuts, the Northern Group of Forces comprised approximately 62 000 troops: 40 000 Army, 17 000 Air Force and 5000 naval infantry troops.[76] Poland was not initially included as one of the WTO states from which Soviet troops were to be withdrawn. Polish officials insisted on inclusion, however, and during 1989

[70] Clarke, D. L., 'Soviets halt troop pullout from East Germany', *Report on Eastern Europe*, vol. 1, no. 22 (1 June 1990).

[71] Gennrich, C., *Frankfurter Allgemeine Zeitung*, 12 Sep. 1990; Fisher, M., 'Germany to pay $8 billion for departing Soviet army', *Boston Globe*, 11 Sep. 1990.

[72] Interview in TASS, FBIS-SOV-90, 20 July 1990.

[73] Marsh, D., 'Bonn talks on Soviet pullout open today', *Financial Times*, 6 Feb. 1991.

[74] MccGwire, M., *Military Objectives in Soviet Foreign Policy* (Brookings: Washington, DC, 1987), pp. 128–29.

[75] Clarke, D. L., 'Soviets withdraw headquarters from Poland', *Report on Eastern Europe*, 17 Aug. 1991, p. 9.

[76] Ostrowski, Z, 'Major concentrations of Soviet troops in Poland', *Warsaw Voice*, 17 Feb. 1991, p. 9.

some 3500 Soviet troops, 87 tanks, 16 artillery pieces and 88 aircraft were withdrawn.

While the majority of the Polish population objected as strongly as those in Hungary and Czechoslovakia to Soviet occupation forces, some Polish officials argued that the presence of Soviet troops and WTO membership were important guarantees of post-1945 borders with Germany. In early February 1990, for example, in a speech to the Annual World Economic Forum in Davos, Switzerland, General Wojciech Jaruzelski ruled out a Soviet withdrawal from Poland until a united Germany guaranteed not to seek to regain territory lost to Poland at the end of World War II.[77] After Czechoslovakia and Hungary had negotiated total Soviet withdrawal agreements in February and March, domestic pressure forced the Polish Government to follow suit.

New negotiations aimed at total withdrawal rather than mere reductions of Soviet forces began in April 1990.[78] In May an air assault brigade was withdrawn.[79] Reports in early June suggested an arrangement whereby 10 000 Soviet troops would remain in Poland as long as Soviet troops remained in Germany,[80] but reports in August suggested that between 10 000 and 12 000 Soviet troops would leave Poland during 1990, leaving some 45 000 in place.[81]

In September Chancellor Kohl agreed to recognize the Oder-Neisse line as a permanent border, after which the Polish Foreign Minister, Krzysztof Skubiszewski, called for Soviet withdrawals on the same schedule as those from Hungary and Czechoslovakia.[82] Skubiszewski suggested cuts of 8000 in 1990 and on 24 September Deputy Defence Minister of Poland Janusz Onyszkiewicz said that Soviet forces would leave Poland in 'about two years'.[83] On 14 November 1990, Genscher and Skubiszewski signed a treaty confirming the existing Polish–German border.[84] Thereafter, Poland grew more assertive in negotiating with the USSR. In December, Poland insisted that Soviet troops from Germany could transit through Poland only if ammunition was transported by sea not land, if they were escorted through Poland by Polish officials and if the Northern Group of Soviet Forces also withdrew. Negotiations grew more tense in January after Soviet action in the Baltic republics. General Viktor Dubinin, Commander in Chief of the Northern Group, said there was no question that Soviet troops would leave Poland eventually but not until Soviet troops left Germany. Dubinin also complained that Poland was trying to humiliate Soviet troops by treating them like occupation forces or prisoners of war, for example, wanting to ship them home in sealed cars, disarmed and with no military equipment.[85] In mid-February 1991, Lev

[77] Jaruzelski's speech reported in *Washington Post,* 5 Feb. 1990.
[78] Bobinski, C., 'Poland seeks Soviet pullout', *Financial Times,* 27 Apr. 1990.
[79] IISS (note 45), p. 30.
[80] *Arms Control Reporter,* sheet 407-E-1.16, July 1990.
[81] *Warsaw Gazeta Wyborcza,* 3 Aug. 1990, FBIS-EE, 7 Aug. 1990.
[82] Gorski, M., 'Poland presses for early pullout', *The Guardian,* 10 Sep. 1990.
[83] Reuters, *Current News, Arms Control Reporter,* sheet 407-E-1.21, Dec. 1990.
[84] *Rzeczpospolita,* 15 Nov. 1990.
[85] Dubinin's remarks were reprinted in the official Polish daily *Rzeczpospolitica,* no. 14 (17 Jan. 1991), p. 7 and in *Gazeta Wyborcza,* 16 Jan. 1991. See also Reuters, 'Doubt on Moscow's pullout of troops', *The Times,* 17 Jan. 1991; UPI, 'Moscow is stalling on troop pullout, Poland says', *International*

Klepacki, the Soviet chargé d'affaires in Warsaw,[86] echoed General Dubinin's earlier statement that Soviet troops would remain in Poland as long as they remained in Germany, but Grzegarz Kostrzewa Zorbas, the chief Polish negotiator, insisted that all Soviet troops must leave in 1991.[87]

USSR out of Czechoslovakia

During 1989, 1500 Soviet troops, 200 tanks and 20 combat aircraft were withdrawn from Czechoslovakia. Another 3800 troops, 500 tanks and 180 aircraft were scheduled for withdrawal in 1990. Once the Communist Government had been toppled, however, the new government wanted a more rapid withdrawal. To ease their return home, President Vaclav Havel promised to provide Soviet troops with prefabricated housing in the USSR. On 26 February 1990, Czechoslovakia and the USSR signed an agreement for the complete withdrawal of all Soviet troops and equipment by 1 July 1991.[88]

Withdrawals appeared to go relatively smoothly, despite the problems of different railway gauges in Central Europe and the USSR, but were marked by Soviet demands for compensation for the fixed assets they left behind, and Czechoslovak outrage at the ecological damage perpetrated by the Soviet forces.[89] On 27 December 1990 Colonel General German Burutin, deputy head of the Soviet General Staff, told TASS that 57 000 troops, 1260 tanks, 1060 artillery pieces and 80 aircraft had left Czechoslovakia. The remaining 16 000 troops and equipment would leave by 1 July 1991.[90]

USSR out of Hungary

Over 10 800 Soviet troops, 447 tanks, 176 artillery pieces and 76 combat aircraft were withdrawn from Hungary in 1989.[91] Negotiations aimed at complete withdrawal of Soviet forces began in January 1990. An agreement setting 30 June 1991 as the deadline for complete withdrawal was signed in Moscow by Eduard Shevardnadze and Foreign Minister Gyula Horn on 9 March 1990.[92] In March 1990 the Southern Group of Soviet Forces to be moved out of Hungary comprised 50 000 troops, 50 000 family members, 860 tanks, 1500 ACVs, 600 self-propelled vehicles, and 560 000 t of fuel, ammunition and other supplies.

Herald Tribune 18 Jan. 1991; Battiata, M., 'Soviet army and Poland stalled over withdrawal', *International Herald Tribune*, 22 Jan. 1991.

[86] Lloyd, J. and Dempsey, J., 'Soviet troops continue to march home', *Financial Times*, 13 Feb. 1991.

[87] Bobinski, C., 'Poland stands firm on troop withdrawal', *Financial Times*, 14 Feb. 1991.

[88] *Arms Control Reporter*, sheet 407.E.1.7, Mar. 1990.

[89] '"CSFR" makes counter demands', *Krasnaya Zvezda*, 22 Nov. 1990, FBIS-SOV-90-227, 26 Nov. 1990, p. 21.

[90] Moskovskiy, O., 'East Europe troop withdrawal second stage complete', TASS, 27 Dec. 1990, FBIS-SOV-90-249, 27 Dec. 1990, p. 1.

[91] TASS, 'Military withdrawal from Hungary continues', 29 Dec. 1988, FBIS-SOV-90-001, 2 Jan. 1990.

[92] Text of the agreement from Nepszabadsag, 11 Mar. 1990, reprinted in English, FBIS-SOV-90-049, 13 Mar. 1990, p. 29.

As in Czechoslovakia, withdrawals from Hungary were marked by Soviet demands for compensation for fixed assets left behind, and Hungarian demands for reparations from the USSR for damage to the Hungarian environment by the Soviet military.[93] By late December the Soviet General Staff reported that approximately 75 per cent of the Southern Group of Forces had left Hungary: 44 000 men, 1078 tanks, 680 artillery pieces and 160 aircraft. The remaining units are scheduled to leave by 1 July 1991.[94]

Non-Soviet WTO unilateral cuts

All the NSWTO governments, except Romania, followed Gorbachev's 1988 initiative with announcements of cuts in their own national forces.[95] During 1990 the NSWTO states adopted increasingly independent positions at the CFE Negotiation in Vienna, and gradually withdrew their national forces from Soviet control and from Soviet joint exercises and training schemes. In late 1990, however, as the WTO countries met to allocate forces within their assigned CFE alliance ceilings, it was clear that the problems of developing national defence forces, to deal with possible new threats, were beginning to engage the attention of the newly emerging democratic governments.[96] Freed from the Soviet straight-jacket, they were unsure how to plan for their own national defence, whether to maintain independent defence industries, what kind of air defence system they could afford, and whether to co-operate with each other or to seek security assistance and guarantees from the West. For Hungary it seemed of primary importance to leave the WTO structure as soon as possible and to maintain a balance of forces with Romania. Polish officials appeared more anxious about a security vacuum and the need for some kind of guarantee from the Western allies. Initially, Czechoslovakia seemed to put the most faith in the CSCE to provide a security umbrella. By February 1991, however, President Vaclav Havel was calling for some form of 'associate membership' in NATO for the Central European powers.[97] In 1990 Poland, Hungary and Czechoslovakia were also beginning to explore co-operation among themselves as a means to join the European Community and to gain security guarantees from the West.[98] On 27 February, however, the defence ministers of Poland and Czechoslovakia signed a military co-operation agreement without Hungary.[99]

Bulgaria and Romania were more isolated than the three Central European powers. Bulgaria still looked to the USSR for protection and was anxious to maintain the strongest possible forces on its border with Turkey. Romania was

[93] Barany, Z. D., 'Not a smooth ride: Soviet troop withdrawals from Hungary', *Report on Eastern Europe*, Radio Free Europe, vol. 1, no. 24 (15 June 1990), pp. 20–28.

[94] Note 60.

[95] For details of cuts in NSWTO national forces, in 1989, see Sharp (note 2).

[96] WTO CFE allocation meetings were held 10–11 Sep. in Bratislava, 18 Sep. in Sofia, 22–23 Sep. in Prague, 2 Oct. in New York, 4 Oct. in Warsaw, 9 Oct. in Prague, 29 Oct. in Prague, and 3–4 Nov. in Budapest.

[97] *Financial Times*, 6 Feb. 1991.

[98] 'Post-Pact co-operation', *Jane's Defence Weekly*, 3 Mar. 1991, p. 300; *Atlantic News*, 20 Feb. 1991, p. 4.

[99] *Atlantic News*, no. 2303 (6 Mar. 1991), p. 4.

the last of the NSWTO states to call for an end to the military structure of the WTO.[100]

Poland

In February 1990 Poland produced details of its armed forces and its new defensive doctrine.[101] This document was primarily a statement asserting Polish national sovereignty and independence from the USSR in military matters.

As of 1 January 1990 Polish armed forces totalled 314 000 men, 206 000 ground forces, 38 000 air forces, 21 200 seamen, and 48 200 air defence forces.[102] Rear Admiral Piotr Kolodziejczyk said in August that 100 000 men would be cut as part of the restructuring of Polish forces, but defence officials stressed the need to modernize the armed forces since unrest in the USSR could spill over its borders.[103]

At the end of 1990 total armed forces numbered 305 000.[104] In October Poland cut conscription from two years to 18 months, negotiated a military training scheme with the Bundeswehr and transferred 10 per cent of its armed forces from its western border with Germany to its eastern border with the USSR.[105] Poland also began to diversify its arms supplies, seeking MiG-29s from Germany (that had belonged to the NVA) as Soviet aircraft priced in hard currency were no longer cost effective.[106]

Hungary

During 1990, Hungary was the most anxious non-Soviet ally to leave the WTO.[107] Throughout 1990, however, the Hungarian Government was persuaded by Poland and Czechoslovakia to maintain the political skeleton of the alliance during the next stage of disarmament negotiations. In any event, for the foreseeable future the Hungarian budget is unlikely to sustain a credibly independent national defence posture. In order to generate some foreign earnings from its arms reductions, before a CFE Treaty required destruction of excess TLE, Hungary sold off MiG-21 aircraft, as well as T-52, T-54 and T-55 tanks. Some tanks were also exported to Sweden, as scrap metal. Hungary also

[100] Reuters, 'Romania wants to end Warsaw Pact', *International Herald Tribune*, 8 Feb. 1991.

[101] 'Doctryna obronna Rzeczypospolitej Polskiej' [The Defence Doctrine of the Republic of Poland], in the official journal *Monitor Polski*, Feb. 1990. See also Sadykiewicz, M. and Clarke, D. L. (eds), 'The new Polish defense doctrine: a further step towards sovereignty', *Report on Eastern Europe*, vol.1 , no. 18 (4 May 1990), pp. 20–23.

[102] Polish Ministry of National Defence, *Polish Army Facts and Figures*, Warsaw, 1990, pp. 15–20.

[103] Reuters, cited by *Arms Control Reporter*, sheet 407-E-1.20, Sep. 1990.

[104] Cielemecki, M., 'In Poland', *Warsaw Voice*, 17 Feb. 1991, p. 9.

[105] Bellamy, C., 'Polish forces look west as they shift to east', *The Independent*, 12 Dec. 1990.

[106] 'Poland seeks German arms as goodwill gesture', *Krasnaya Zvezda*, 6 Oct. 1990, FBIS-SOV-90, 10 Oct. 1990, p. 34.

[107] On 26 June, the Hungarian Parliament voted to leave the WTO by a vote of 232 : 0 with 4 abstentions; Budapest radio, FBIS-EE-90, 27 June 1990.

commercialized its military airfields in the hope of becoming an important air link on the continent.[108]

In August Foreign Minister Geza Jerzensky announced at the CFE plenary the reduction of military service from two years to 18 months as of 1 February 1991.[109] In November, a Defence Ministry spokesman, Colonel Gyorgy Keleti, announced that Hungary would dismantle its 107 FROG-7 missiles and 18 launch pads, as well as 24 Scud B missiles and 9 launch pads. Keleti said Hungary had no use for such offensive systems.[110]

In November 1990 Defence Minister Lajos Fur said that troops associated with the missiles would be demobilized (some 800 men) and the armed forces would be reduced from 107 000 to 80 000 during 1991.[111] The military budget, however, would increase by 70–75 billion florints. Colonel Keleti said that Hungary wanted extra funds to purchase former NVA equipment from Germany; including 360 T-72 tanks, 350 BPM infantry fighting vehicles, over 1000 trucks, 72 artillery pieces, 50 000 AK-74 rifles, 50 million rounds of ammunition, 100 000 anti-tank guided missiles and 200 000 grenades.[112]

Czechoslovakia

During 1990, like Poland but unlike Hungary, Czechoslovakia emphasized transforming rather than abandoning the WTO. Foreign Minister Jiri Dienstbier believed the WTO should be an important partner of NATO in negotiating further cuts in military forces and in creating a new all-European security structure.

Relative to those of other NSWTO states Czechoslovak troops were well armed and equipped and could afford to save funds by forgoing the purchase of new equipment.[113] In January 1990, the new Czechoslovak Government declared cuts in its armed forces and a new defensive military doctrine. Major General Anton Slimak, Chief of the General Staff, announced that Czechoslovakia would take out of service and dismantle 850 tanks, 51 combat aircraft and 165 ACVs. All bridging equipment would also be dismantled and Czechoslovakia would also gradually stop producing tanks. As of late April 256 tanks had been destroyed, the engines saved for other vehicles and the metal sold as scrap. Forces would be restructured to emphasize defensive missions. Conscription would be cut from 12 to 9 months and fewer reserves would be called up.[114] The armed forces would be reduced from 200 000 to 140 000 by the end of 1993; 20 000 troops would be cut by the end of 1990, another 20 000 by end of 1991 and a third cut of 20 000 during 1992. Czechoslovak forces would no longer participate in WTO exercises, and

[108] Donovan, P., 'Hungary sells off Soviet air base', *The Guardian*, 18 Sep., 1990; Betts, P., 'Hungary to convert military airfields', *Financial Times*, 18 Sep. 1990.

[109] *Die Presse* (Vienna), 10 Aug. 1990.

[110] Budapest domestic radio, 16 Nov. 1990, transcription in FBIS-EE-19, Nov. 1990.

[111] *Arms Control Reporter*, sheet 407.E.1.23, Jan. 1991.

[112] *Washington Times*, 3 Jan. 1991.

[113] See interview of Lt General Josef Vincenz, First Deputy Chief of the General Staff with Jan Oberman in *Report on Eastern Europe*, vol. 1, no. 29 (20 July 1990), pp. 14–17.

[114] *Bratislava Verejnost*, 18 June 1990, FBIS-EE-90, 22 June 1990.

troops were being redeployed away from the western border with Germany to the eastern border with the USSR. Czechoslovakia has also made arrangements for regular bilateral military contacts, and training schemes in France and the UK.

In addition to planned cuts in national forces, on 25 January 1990 Dienstbier declared that Czechoslovakia would no longer export arms, 'without taking into account what the pragmatists say or whether it will be a blow to the state coffers'.[115] This proved difficult to put into practice as much military equipment was already in the export pipeline. Later statements indicated that existing contracts would be honoured but production would be cut by 25 per cent by 1993. But the conversion programme was controversial when workers were laid off, and in January 1991, Slovak Prime Minister, Vladimir Meciar announced that Slovakia would defy the Prague-imposed ban on arms production and export, and would resume the production and export of heavy weaponry.[116] On 13 June, the Ministry of Defence announced a cut in the defence budget from 35 062 billion korunas ($2.19 billion) in 1989 to $31 180 billion korunas (1.9 billion) in 1990.[117]

Bulgaria and Romania

In late 1989 Major General Ivan Petrov announced that a five-year plan had been drawn up to increase the share of civilian production by the Bulgarian defence industries.[118] In February 1990, Chief of Staff Colonel General Christo Dobrev announced that forces were being restructured and tanks being reduced by one-third and ACVs by 10–30 per cent. Defence spending was cut from 1713 to 1605 billion leva.[119] In August the National Assembly announced a cut in the length of conscription from 24 to 18 months.[120] By the end of 1990 Bulgaria still held 72 SS-23 missiles, but suggested in November that the missile complex would soon be demolished.[121]

Romania was the only WTO country not to announce unilateral cuts after Mikhail Gorbachev's UN speech of December 1988, largely because the armed forces were so poorly equipped. In March 1990, Romania was denied a request to train its military personnel in Britain.[122]

VII. Unilateral NATO reductions in the ATTU zone

NATO also made unilateral cuts parallel to the CFE Negotiation but not to the same degree as the WTO countries. NATO countries reduced their forces in the ATTU zone during 1990 for two main reasons. In the latter half of the year

[115] As reported in the *New York Times*, 25 Jan. 1990.

[116] Colitt, L., 'Slovakia will defy Prague's arms export ban to protect defence jobs', *Financial Times*, 10 Jan. 1991; and 'Arms and the man in Slovakia', *Financial Times*, 22 Jan. 1991.

[117] *Arms Control Reporter*, sheet 240.B.1.13, July 1990.

[118] Khristo Marinchev, in *Sofia Rabotnichesko Delo*, 27 Dec. 1989, FBIS-EE-18, Jan. 1990.

[119] Bulgarian Telegraph Agency (BTA), 14 Feb. 1990, FBIS-EE-15, Feb. 1990.

[120] Staevski, I., *Otechestven Vestnik* (Sofia), FBIS-EE-20, Aug. 1990.

[121] *Sofia Mladezh*, 30 Nov. 1990, FBIS-EE-90, 4 Dec. 1990.

[122] *The Times*, 16 Mar. 1990, cited by *Arms Control Reporter*, sheet 407.E.1.11, June 1990.

the Iraqi invasion of Kuwait generated a US-led military buildup in the Gulf region.[123] This involved the temporary transfer of over 150 000 troops and over 5000 TLE items from the ATTU zone to the Gulf (table 13.14). The main NATO contributors to the buildup were the USA, Britain and France, but Italy, the Netherlands, Belgium and the FRG also sent token forces to south-east Turkey and to the Gulf. The second reason for cuts was a combination of fiscal constraints in many NATO countries and a lessening of the threat from the USSR. These conditions stimulated a rethinking of NATO strategy during 1990 that generated plans to cut military spending, to reduce manpower, to close bases in the ATTU zone, and to restructure slimmed-down forces. Some of these cuts occurred in 1990, others were scheduled over the next few years, independent of, but parallel to, the CFE IA negotiations.

NATO transfers from the ATTU zone before the invasion of Kuwait

Unlike Soviet TLE transferred out of the ATTU zone, most of the NATO TLE transferred to the Gulf were included in NATO data on ATTU holdings in November 1990. The exceptions were US transfers out of the ATTU zone in early 1990. Between 13 February and 31 May 1990, the USA withdrew 900 tanks from the ATTU zone. Of these 110 were returned to the continental USA (CONUS), 60 were sold to Morocco and 700 to Egypt. These were primarily M-60A1 tanks, vintage early 1960s, that some Western analysts consider were made obsolete when the Soviet T 64 appeared in 1965.[124].In August, an additional 300 surplus US M-60A1 and 350 even older M-48A5 tanks were sold to Thailand from stocks in the ATTU zone.[125] These 1550 old tanks would have been subject to CFE destruction requirements had they not been withdrawn from the ATTU zone; in this sense they are analogous to the TLE that the USSR transferred behind the Urals. Similarly, 177 155-mm howitzers were withdrawn from the ATTU zone earlier in the year to modernize the equipment of CONUS-based reserve forces.[126]

NATO TLE transfers to the Persian Gulf after the invasion of Kuwait

Since August 1990, a number of NATO countries have transferred TLE from the ATTU zone to the Gulf region. By January 1991 these amounted to 5507 TLE items (see table 13.15). Those TLE that survived the Gulf war will be liable to CFE-imposed destruction to the extent that returning them to Europe put NATO TLE above limits for the ATTU zone.

[123] See also chapter 19 in this volume.

[124] Zaloga, S., 'Soviet tank development revealed', *Armed Forces Journal International*, Dec. 1990, p. 24.

[125] Lewis Young, P., 'Thailand set for surplus equipment upgrades', *Armed Forces Journal International*, Dec. 1990, p. 46.

[126] Information from the Office of the Secretary of Defense, Washington, DC.

Table 13.14. NATO manpower and TLE from the ATTU zone to the Gulf, August 1990–January 1991

Country	Air and ground manpower	Combat aircraft[a]	Attack helicopter	Battle tanks	ACV	Artillery
USA[b]	100 000	148	126	1 777	1 200	657
UK[c]	37 000	121	50	230	600	100
France[d]	11 660	40	120	40	190	38
Italy	500	16[e]				
Canada	500	18				
Belgium	500	18				
Netherlands	150					
Germany	1 700	18	0	0	0	0
Total manpower	152 010					
Sub-totals		379	296	2 047	1 990	795
Total TLE items				**5 507**		

[a] Aircraft data obtained from Donald Kerr at the International Institute of Strategic Studies.
[b] Information from the Office of the Secretary of Defense, Washington, DC.
[c] Information from the British Ministry of Defence, London.
[d] Information from WEU, *The Gulf Crisis: Chronology of Events from 4th December to 15th January 1991*, A/WEU/DEF (91)2, 21 Jan. 1991; and *Le Figaro*, 5 Feb. 1991.
[e] Six in the Allied Mobile Force in south-east Turkey and 10 Tornados in the Gulf.

Transfer of US ground force equipment

The first transfer of US forces to the Gulf was from CONUS rather than Europe.[127] One reason that US forces were not transferred from Europe to the Persian Gulf initially was that NATO leaders objected to US troops based in Europe being rotated for duty in the Gulf, much as their predecessors had objected to US troops being shipped from Europe to South-East Asia in the late 1960s.[128] US personnel in USAF medical and other support units were sent from Europe early to the Gulf but no ground combat troops were sent until November. During September and October 836 tanks left the ATTU zone: 220 were M-60s, of which 150 went to Saudi Arabia, 37 to Bahrain, and 33 to Oman; 616 were M-1A1s and all these went to Saudi Arabia to upgrade the equipment of the 24th Infantry Division (transferred from Georgia to Saudi Arabia in August) and the 1st Cavalry Division already deployed there.

In November, however, the USA decided to send 250 000 more troops to the Gulf, of which approximately 100 000 were from US bases in Europe. The US Army also froze the release and retirement of all personnel for the duration of the crisis, so as to maintain a reinforcement capability for the Gulf.[129]

[127] See also chapter 19 in this volume.
[128] Moore, M., 'A plan to rotate US troops: forces in Europe may be sent to Gulf', *International Herald Tribune*, 20–21 Oct. 1990.
[129] AP, 'GIs in Europe: half go to Gulf', *International Herald Tribune*, 10–11 Nov. 1990; Wilson, G. C. and Smith, R. J., 'US to delay retirement of tens and thousands of troops', *International Herald Tribune*, 23 Nov. 1990.

Table 13.15. US ground force TLE to the Gulf from the ATTU zone in 1990

	Tanks	ACVs	Artillery Howitzers/MLRS/mortars			Attack helicopters
1st Armored Division (Ansbach)	348	600	72	18		36
2d Armored Brigade (Garlestedt)	116		24	18		
3d Armored Division (Frankfurt)	348	600	72	18		36
2d Armored Cavalry Regiment (Nurnberg)	129		24	18		18
Corps artillery			108			
Six mortars per battalion					285	
12th Aviation Brigade						36
TLE to Gulf	**1777**	**1200**		**657**		**126**
In November	941	1 200		657		126
Pre-November	836					

Sources: Office of the Secretary of Defense, Washington, DC; Ministry of Defence, London; Western European Union, *The Gulf Crisis: Chronology of Events from 4th December 1990 to 15th January 1991*, A/WEU/DEF (91)2, 21 Jan. 1991, pp. 38–40.

Table 13.15 shows the 3760 US ground force TLE transferred from European bases in the ATTU zone to the Gulf. In January 1991, Soviet military action in the Baltic republics heightened concern that Western Europe was more vulnerable than at any time since World War II with most US forces previously earmarked for European contingencies in the Gulf.[130]

Transfer of US combat aircraft

In September 1990, 24 F-15 aircraft were delivered on a priority basis to the Saudi Air Force. Twelve of these came from the 32d Tactical Fighter Squadron (TFS) at Soesterberg, in the Netherlands, and 12 from the 36th TFS at Bitburg, Germany.[131] These squadrons were later made up to full strength with aircraft from the USA and do not represent any change in the TLE holdings in the ATTU zone.

Later in 1990, 148 combat aircraft were transferred from USAF bases in the ATTU zone to the Gulf: 24 F-16 Cs from the 614th TFS in Spain; 20 F-111Es from the 77th TFS in Upper Heyford, UK; 60 F-111Fs from the 492d, 493d, 494th and 495th TFS from Lakenheath, UK; 20 F-4Cs from the 23d, 81st and 480th TFS from Spangdahlem, FRG; and 24 A-10As from the 509th and 511th TFS in Brentwater/Woodbridge, UK. These aircraft represent TLE included in the US ATTU holdings exchanged at CFE signature.

[130] Walker, M., 'New Moscow nightmare for the Pentagon', *The Guardian*, 20 Jan. 1991.
[131] *Atlantic News*, no. 2281 (12 Dec. 1990).

British transfers

In contrast to most of NATO, the UK decided to send forces to the Gulf soon after the invasion of Kuwait.[132] A substantial British air and naval presence was already in the region when Operation Granby deployed ground forces to the Gulf in October. These included the 7th Armoured Brigade (Desert Rats) from Germany.[133] In November, the 4th Armoured Brigade joined the 7th to form the 1st Armoured Division, bringing the British contingent in the Gulf to 42 000 men: 31 000 Army, 6000 RAF and 5000 naval.[134]

French transfers

Of the European NATO countries, France sent the second largest contingent to the Gulf. After Iraqi troops ransacked the French embassy in Kuwait City in September, President François Mitterrand ordered Operation Daguet which by mid-January 1991 had transferred 11 660 men, bringing the French strength in the region to 15 400 air and ground force manpower. Approximately 3750 men are permanently stationed in Djibouti.[135] Sent from Europe were one regiment of artillery, two regiments of light armour, one heavy tank regiment and one helicopter regiment. CFE TLE with these regiments include 38 artillery pieces, 190 ACVs, 40 main battle tanks, 120 attack helicopters and 40 combat aircraft. There were also 20 French ships in the region.[136]

Other NATO transfers

Britain and France each sent several squadrons of fighter aircraft to the Gulf. In addition, in response to a request from Turkey, the Supreme Allied Commander, Europe (SACEUR) deployed NATO's multinational Allied Mobile Force (AMF) to south-eastern Turkey in early January 1991.[137] The AMF contingent included a battery of Patriot anti-aircraft missiles from the Netherlands,[138] 18 Belgian Mirage fighter aircraft from Biersset, 8 Italian F-104 Starfighters from Gioia del Colle, and 18 German Alpha jets from

[132] Helm, S., 'Britain rethinks military strategy', *The Independent*, 4 Sep. 1990; Bellamy, C. and Brown, C., 'UK will send 6000 men and 120 tanks to Gulf', *The Independent*, 13 Sep. 1991.

[133] Lemoyne, J., 'Elite Britsh troops join buildup in the Gulf', *International Herald Tribune*, 18 Oct. 1990.

[134] Malet, V., 'Britain to double troop presence in Gulf', *Financial Times*, 23 Nov. 1990.

[135] Western European Union, *The Gulf Crisis: Chronology of Events from 4th December 1990 to 15th January 1991*, A/WEU/DEF (91)2, 21 Jan. 1991, pp. 38–40 cites 3800 men in Djibouti; IISS (note 45), p. 66 cites 3650.

[136] Information from the French Embassy, Stockholm; see also 'Les regiments present', *Le Figaro*, 5 Feb. 1991; WEU (note 136), pp. 38-40.

[137] The AMF was established in 1960 and is made up of land and air forces from 8 NATO countries to reinforce the NATO flanks (mostly Norway and Turkey) in a crisis. Land forces designated for a northern flank contingency are from Canada, Italy, Luxembourg and the UK, and for a southern contingency are from Belgium, Germany, and the USA. Air squadrons for a northern contingency come from Canada, the Netherlands, the UK and the USA; air squadrons for northern flank contingencies come from Belgium, Germany and Italy; see George, B. (ed.), *Jane's NATO Handbook 1989–1990* (Jane's: London, 1989), pp. 145–47.

[138] Patriot missiles are not CFE-limited.

Oldenberg.[139] The Belgian aircraft were stationed at Diyarbakir and the German and Italian units at Malatya. Both bases are outside the ATTU zone which excludes south-eastern Turkey.[140] Italy also sent 10 Tornados to the Gulf. All these combat aircraft are CFE TLE included in NATO ATTU zone holdings.[141]

NATO budget cuts, base closures and restructuring in the ATTU zone

At the NATO summit meeting in early July 1990, the allies agreed to 'prepare a new allied strategy moving away from "forward defence", where appropriate, towards a reduced forward presence and modifying "flexible response" to reflect a reduced reliance on nuclear weapons'.[142] Several studies were under way during 1990 aiming for a new strategy document in 1991. These included the Joint Strategy Group (JSG) in NATO's Planning and Policy Division which was preparing a military strategy concept to replace MC 14/3 and, at SHAPE, General Galvin and his military staff were preparing various ways of implementing the new strategy.[143]

The assumption of these studies was that the WTO was no longer a military threat, even though the military power of the USSR, or even of the Russian Republic, still casts a long shadow over Western Europe. Regardless of the results of CFE IA, NATO will have to manage with fewer forces, especially fewer stationed forces in a united Germany. On 25 September, the ambassadors of the Western allies that station troops in Germany exchanged letters with the Foreign Ministry in Bonn to adapt the previous force stationing agreements to the new situation, in particular to recognize the end of four-power authority over Germany.[144]

The goal was to cut the number of stationed NATO forces in Germany from 407 000 to approximately 150 000 during the 1990s, of which approximately 75 000 would be US and 75 000 non-US.[145] NATO forces would be restructured into multinational units spread throughout NATO Europe.[146] General John Galvin, NATO's SACEUR, endorsed the concept of multinational forces, as did many continental Europeans including the Secretary General of the Western European Union,[147] but many German and US military officers

[139] Davidson, I., 'European capitals resolve their levels of commitment to the conflict', *Financial Times*, 17 Jan. 1991; Kinzer, S., 'German military gets ready to fight', *International Herald Tribune*, 4 Jan. 1991.

[140] Article II of the CFE Treaty (see appendix 13A).

[141] CFE Protocol on Existing Types, section 1.4 on combat aircraft.

[142] London Declaration (note 9), para. 20.

[143] Frinking, T. and Bereuter, D., *Interim Report of the Special Committee on Alliance Strategy and Arms Control*, North Atlantic Assembly, Nov. 1990, p. 4.

[144] *Atlantic News*, no. 2257 (28 Sep. 1990), p. 4.

[145] Stoltenberg, G., *Die Welt*, 30 Aug. 1990.

[146] 25 Sep. Bonn communiqué; and communiqué after NATO ministerial session, 18 Dec. 1990.

[147] van Eekelen, W., *Future European Defence Cooperation and the Role of the WEU*, European Strategy Group Occasional Paper, Sep. 1989; Interview with General Galvin at the Pentagon, 17 Sep. 1990, *ACE Output* (SHAPE public information office), vol. 9, no. 4 (Nov. 1990). See also Jacchia, E., 'Urgently begin the restructuring job', *International Herald Tribune*, 5 July 1990; White, D., 'NATO wonders how to re-order the ranks', *Financial Times*, 21 June 1990.

envisage a restructured NATO force in which the US presence is separate from a European multinational force. This would allow faster US reaction to out-of-area contingencies and be more acceptable to the US Congress which will not want US forces hampered by European vetoes. Lt General Henning von Ondaerza, German Army Chief of Staff, proposed bi-national units, with one partner clearly dominant.[148]

Non-US stationed forces in Europe currently number approximately 160 000. Parallel to the CFE IA talks the UK will halve its forces in Germany, from 55 000 to 27 000 in the British Army on the Rhine (BAOR) and from 11 000 to 5500 in the RAF; France will withdraw about 20 000 of its 55 000; Belgium will cut its 25 000 troops to 3500 by 1995. In 1991, Canada will withdraw 1400 of its 7000 and the Netherlands will cut 750 of its 8000 forces stationed in Germany. If Canada and the Netherlands were each to contribute 3500 (the same number of troops as Belgium plans to contribute to the new NATO multinational force), and if Britain and France were each to maintain 32 750, this would meet the goal of 75 000 non-US stationed forces in Germany.

US base closures and budget cuts

Base closures in the ATTU zone. In late January 1990 US Secretary of Defense Richard Cheney announced the closure of nine bases in Western Europe; the latter included seven USAF bases: three in the UK, one in Italy, one in the FRG, one in Turkey and one in Greece. The other two were a naval base at Nea Makri in Greece and a munitions storage site at Eskisehir in Turkey. Cheney said the budget cuts and base closures represented the first steps in responding to the reduced threat from the East as well as tighter budget constraints.[149]

In September Cheney announced the closure or trimming back of activity at 150 additional foreign bases including 95 to be closed and 14 to be trimmed back in Germany.[150] Reports from Washington suggested that the USA intended to withdraw 60 000 troops from Europe by 1997 and, on 26 September, Cheney said 40 000 US troops would leave over the next 12 months: 30 000 Army and 10 000 USAF.[151]

In October the Defence Authorization Bill cut the number of troops that could be deployed in Europe (the European Troop Strength Ceiling) from the previous level of 326 414 to 261 855; reflecting the proposed cuts of 60 000.[152] As noted, however, the 100 000 US troops transferred from the ATTU zone to the Gulf may not return to Europe.

[148] Schulte, H., 'Finding a new role for NATO in a new Germany', *Jane's Defence Weekly*, 26 Jan. 1991, p. 113.

[149] Moore, M., 'Pentagon plans to close US and overseas bases', *International Herald Tribune*, 30 Jan. 1990.

[150] Horvitz, P. F., 'US military to end or trim operations at 150 foreign sites', *International Herald Tribune*, 19 Sep. 1990.

[151] Barber, L., 'US will withdraw 40,000 troops from Europe', *Financial Times*, 27 Sep. 1990; Helm, S., 'Pentagon pull-back gives a clue to cuts', *The Independent*, 20 Sep. 1990.

[152] *US Congressional Record*, 23 Oct. 1990, p. H. 11955.

In January 1991, at least one of the US bases in the UK that had been scheduled for closure (Fairford in Gloucestershire) was reactivated to accommodate B-52 bombers for Gulf missions.[153]

Budget and manpower cuts. In April Cheney announced budget cuts of $2.4 billion, a 50 per cent cut in the B-52 bomber procurement, delays in advanced fighter aircraft purchases and 25 per cent manpower cuts by 1997. The Army would be cut from 764 000 to 580 000 in active duty forces, that is, from 18 to 14 divisions, and from 776 000 to 645 000 in the National Guard reserve force, that is, from 10 to 8 divisions.[154] The Air Force would be cut from 545 000 to 476 000 by 1997 and the Navy would shrink from 549 to 488 ships. There followed several months of speculation about how defence industries would adapt to cuts in the budget and who would benefit from the peace dividend.[155]

Through March 1991, US participation in the Gulf War was largely funded by other members of the anti-Iraq coalition (Japan, Saudi Arabia and Germany) and did not affect the US defence budget. In February 1991, however, Cheney warned the Congress that Soviet behaviour in the Baltic republics and reluctance to clarify its CFE data could justify a budget rise and a slow-down in the planned 25 per cent cut in US forces over the next five years.[156]

British budget cuts and base closures

In June, Minister of Defence Tom King proposed spending cuts of £600 million, from a budget of £21 billion, and the cancellation of 33 Tornado aircraft. In July, King presented the government's proposed 'Options for Change' to the House of Commons.[157] These included cuts in manpower over the next five years: the Army from 160 000 to 120 000, the RAF from 90 000 to 75 000 (representing the loss of 7 of 16 fighter squadrons), cuts in the BAOR from 55 000 to 20 000–25 000 and the Royal Navy from 63 000 to 60 000. No cuts were proposed in planned nuclear forces, though purchase of US Trident missiles could be delayed.[158]

The Ministry of Defence announced in November that the UK would close two RAF bases in Germany in 1991 (Wildenruth near the Netherlands border

[153] Whitney, C. R., 'US will fly B-52 raids from bases in Britain', *International Herald Tribune*, 1 Feb. 1991.

[154] Tyler, P. E., 'US Army outlines plans to cut quarter of troops by 1997', *International Herald Tribune*, 6 Apr. 1990.

[155] See, for example, Rifkind, J., 'Put the Green Dividend to work on tomorrow', *International Herald Tribune*, 9 May 1990; and four articles in the *New York Times* on 'Shrinking the military': Gordon, M. and Eckholm, E., 'Global change and budget cuts test Pentagon', 20 May 1990; Gordon, M. R., 'Army and Air Forces fix sights on the changing face of war', 21 May 1990; Ekholm, E., 'In detente and cutbacks, Navy has powerful foes', 22 May 1990; Gordon, M. R., 'Stocking the atomic arsenal: how much deterrence to buy?', 23 May 1990.

[156] Ridell, P., 'Friendship at a crossroads', *Financial Times*, 8 Feb. 1991.

[157] *Hansard*, 26 July 1990, columns 470–73.

[158] Riddell, P., 'Defence budget squeeze delays Trident purchase', *Financial Times*, 1 Oct. 1990.

and Gutersloh near Essen),[159] and in December that Army manpower would be cut even further than was proposed in July, down to 100 000 overall.[160]

French budget cuts and restructuring

Bucking the trend towards reduced defence spending elsewhere in Europe, in September 1990 Defence Minister Jean-Pierre Chevènement asked for a 1991 budget 3.3 per cent higher than spending in 1990; barely an increase when 2.8 per cent inflation is taken into account.[161] In early July Finance Minister Pierre Beregovoy asked Chevènement to cut defence spending by 7 per cent from the projected 1991 budget of 202 billion francs. Chevènement refused, saying that those who proposed budget cuts should first propose which weapons systems should be cut to show where the money could be saved.[162] Beregovoy persuaded President François Mitterrand of the need for a cut, however, and the budget was trimmed back to 195.4 billion francs in December; an increase of 2.7 per cent over 1990, but short of 7.4 billion francs compared to Chevènement's earlier request.[163]

Immediately after the NATO summit, in July 1990, President Mitterrand cut the length of conscription from 12 to 10 months and announced that French forces would leave Germany once four-power responsibility for Germany was over; presumably when the Two-plus-Four agreement on German unification had been ratified by France, the UK, the USA and the USSR.[164] German officials asked Mitterrand to reconsider and, in mid-September, after two days of talks with Chancellor Kohl, Mitterrand announced that French forces might not be totally withdrawn and certainly not before 1995. Fifty per cent would be withdrawn in the first phase. No immediate cuts were anticipated in the Franco-German brigade or the French garrison in Berlin.[165]

In August Chevènement announced that the headquarters of French forces in Germany, currently in Baden-Baden, would be moved to Strasbourg, in preparation for eventual French withdrawal from Germany.[166]

In October, Chevènement said that reductions in the threat to Europe would allow France to field lighter and more mobile forces.[167] The army would take the brunt of initial French manpower cuts and in November the Ministry of

[159] Dickson, T., 'Two RAF bases to be closed in Germany', *Financial Times*, 14 Nov. 1990.

[160] Bellamy, C., 'Gulf deployment cripples Rhine Army', *The Independent*, 30 Dec. 1990.

[161] *Arms Control Reporter*, sheet 240.B.1.17, Oct. 1990.

[162] *Jane's Defence Weekly*, 21 July 1990.

[163] Isnard, J., 'Le project de budget pour 1991 entre deux lois de programmation: un temps mort dans le debat strategique', *Le Monde*, 7 Nov. 1990, p. 13; 'French 1991 defense budget: the army takes a beating', *International Defense Review*, Oct. 1990, p. 1085; Michaud, P., 'French defence budget awaits new four year plan', *Defence*, Dec. 1990, p. 760.

[164] *Le Monde*, 8–9 July 1990.

[165] Lewis, J. A .C., 'French to begin force pullout', *Jane's Defence Wekly*, 1 Sep. 1990; Gow, D., 'France to pull out entire Rhine Army', *The Guardian*, 19 Sep. 1990.

[166] *Jane's Defence Weekly*, 1 Sep. 1990, p. 301.

[167] Chevènement statement, Ministry of Defence, 19 Oct. 1990, cited by de Briganti, G., 'French will improve military ties', *Defense News*, 3 Dec. 1990, pp. 8–26.

Defence announced that the 3rd Armoured Division would be dissolved, involving the withdrawal of 9000 troops from Germany.[168]

German budget cuts and restructuring

In June 1990, Defence Minister Gerhard Stoltenberg said that he would cut FRG defence spending by 3 per cent (to DM 52.6 billion) to compensate for the costs of unification.[169] In January 1991, after the first elections in a united Germany, the new government confirmed the 1991 defence budget at DM 52.4 billion, down from DM 61 billion in 1990. Fighter aircraft would be cut by 500, from current levels of 620 in the Bundeswehr and 400 in the NVA. Germany would not cut development funds for the European Fighter Aircraft (EFA) project but neither would it commit additional funds for production.[170]

As noted above, the FRG Government pre-empted the imposition of CFE cuts with a unilateral decision to cut the manpower of combined German forces to 370 000, as part of the package it offered to Gorbachev to gain Soviet approval of German unification and membership of a unified Germany in NATO.[171] The reduction will take place over the same time frame as Soviet withdrawals from Germany—before 31 December 1994.

VIII. The CSBM Negotiations[172]

Since March 1989, parallel to the CFE talks, the 35 (later 34) states that participate in the Conference on Security and Co-operation in Europe (CSCE) conducted negotiations in Vienna on Confidence- and Security-Building Measures (CSBMs). This CSBM forum produced a Document adopted by the 34 CSCE Heads of State on 21 November 1990, in Paris.[173] The Vienna Document both widened the parameters of the first- and second-generation CSBMs negotiated in Helsinki (1975) and in Stockholm (1986), and generated new measures appropriate to the changing political situation in Europe.[174]

[168] Reuters, '9,000 French troops to be withdrawn', *Financial Times*, 13 Nov. 1990.

[169] *Die Welt* cited in *Arms Control Reporter*, sheet 240-B-1.13, July 1990.

[170] *Atlantic News*, no. 2289 (18 Jan. 1991), p. 3.

[171] Note 15, p 185.

[172] Background to the earlier CSBM negotiations and the Helsinki and Stockholm documents can be found in the *SIPRI Yearbooks*: The Document on Confidence Building Measures and Certain Aspects of Security and Disarmament, an integral part of Chapter I of the CSCE Final Act (Helsinki, 1 August 1975) devoted to questions relating to security in Europe, is reprinted in SIPRI, *World Armaments and Disarmament: SIPRI Yearbook 1976* (Almqvist & Wiksell: Stockholm, 1976), pp. 359–62; the 1986 Stockholm Document is reprinted in SIPRI, *SIPRI Yearbook 1987: World Armaments and Disarmament* (Oxford University Press: Oxford, 1987), pp. 355–69; for a comparison of the 1975 CBMs and the 1986 CSBMs, see Darilek, R., 'The future of conventional arms control in Europe, A tale of two cities: Stockholm, Vienna', *SIPRI Yearbook 1987*, pp. 339–54.

[173] Vienna Document 1990 of the Negotiations on Confidence- and Security-Building Measures convened in accordance with the relevant provisions of the Concluding Document of the Vienna Meeting of the Conference on Security and Cooperation in Europe, reproduced here as appendix 13B.

[174] See appendix 13B and Ghebali, V.-Y., 'Confidence-building measures: paragraph-by-paragraph analysis of the Helsinki and Stockholm regimes', United Nations Institute for Disarmament Research (UNIDIR), Research Paper No. 3, Geneva, Mar. 1989; Rotfeld, A. D., 'CSBMs in Europe: A future oriented concept', eds R. D. Blackwill and S. Larrabee, *Conventional Arms Control and East–West*

From an early focus on transparency of military forces in an East–West context, the CSBM talks in 1990 sought a regime that would inhibit the use of military force by any European state against any other.[175]

The Vienna CSBMs

In January 1989 the Concluding Document of the third CSCE follow-up meeting mandated new talks to build upon and expand the Stockholm Document with a view to generating a third generation of CSBMs that would further reduce the risk of war in Europe.[176]

The CSBM proposals in Vienna came from the three main groups of states: the 16 NATO states, the 7 (later 6) WTO states and the 12 neutral and non-aligned (NNA) states. As negotiations developed, the cohesion within each group varied. The USA (with its reluctance to engage seriously in the CSBM process) and Turkey (preoccupied with the vulnerability of its south-eastern region) were most often at odds with the other NATO states. The 6 WTO states separated into the Central European powers (Czechoslovakia, Hungary and Poland), whose delegates were usually anxious to co-operate with the West, the two Balkan states (Bulgaria and Romania), and the USSR. Among the 12 NNA states, the four neutral states—Austria, Finland, Sweden and Switzerland—were relatively cohesive with Switzerland and most often at odds with the other three. Within the sub-group of the 8 non-aligned states Cyprus and Malta were the most active.[177]

1990 CSBM Negotiations in Vienna: rounds V,VI and VII

In preparation for round V (15 January–23 February) NATO's High Level Task Force gathered together seven previous CSBM proposals that were judged ripe for consensus. Delegates reported French enthusiasm for the package, but negative responses from the USA to proposals 'not made in Washington, DC'. Anticipation of a CSCE summit meeting spurred activity, however, and on 16 February 1990 four drafting groups were set up based on the working groups established in 1989: A 1 dealt with verification and information, A 2 with communication, consultation and military contacts, B 1 covered notification and observation, and B 2 covered constraints and the annual calendar of military activities.

Round V of the CSBM Negotiations was devoted primarily to the seminar on military doctrine (16 January–5 February), remarkable for the independent views exhibited by the NSWTO states and the discomfort of senior members of the Soviet delegation who, despite having initiated the idea of a doctrine

Security (Duke University Press: Durham, N.C., 1989); Holst, J. J., 'Confidence and security building in Europe: achievements and lessons', paper presented in Seoul, Korea 10–11 Oct. 1990.

175 See for example the address by James Baker, 'From revolution to democracy: Central and Eastern Europe in the new Europe', Charles University, Prague, 7 Feb. 1990.

176 1989 mandate for the CSBM Negotiations, reprinted in SIPRI, *SIPRI Yearbook 1989: World Armaments and Disarmament* (Oxford University Press: Oxford, 1989), pp. 419–20.

177 Interviews by the author with delegates at the CSBM Negotiations.

seminar in 1987, were manifestly upstaged by the greater openness with which North Americans and Europeans were willing to discuss military issues.[178]

In the time remaining from the seminar, round V made little progress. The USA and the USSR focused primarily on the CFE Negotiation. The European states in both alliances were the most enthusiastic about producing a summit-worthy document by November. The NNA countries seemed inclined to spin the talks out through 1991, fearing loss of a negotiating role between the November 1990 Paris summit meeting and the March 1992 Helsinki follow-up meeting, unless CFE and CSBM forums could be merged after the Paris summit.[179] An encouraging sign of pan-European co-operation emerged on the last day of round V, 23 February, when Austria, France, the GDR, Hungary and Sweden elaborated an earlier proposal to improve communication among the CSCE states.[180] The idea was to create a computerized communication network to supplement normal diplomatic channels as part of a risk reduction regime. On the same day the NATO states also reiterated earlier proposals to exchange information on military budgets and to establish an annual implementation assessment meeting—both of which had been proposed by the WTO states in December 1989 and March 1990 respectively.[181]

Between rounds V and VI the NATO states discussed a number of measures. The UK proposed a new measure, regular peacetime visits to combat air bases. Other measures included notification of infrastructure improvements, greater coverage of air and amphibious activity and new constraints on military activities.

In round VI (15 March–26 April 1990) progress was slowed by Soviet anxieties about German unification. At the CFE Negotiation Soviet delegates stalled by questioning previously agreed understandings on manpower cuts, and at the CSBM talks they resurrected old proposals for constraints on air and naval forces.

At the beginning of round VII (17 May–19 November) NATO presented six proposals, two of which reiterated old themes from earlier proposals (submitted initially in March 1989 and amplified in June 1989) and four of which were largely new ideas. The revisions of earlier proposals were Measures 10 and 11 from the March 1989 NATO proposals, namely: improved access for accredited personnel in foreign embassies and on special assignments dealing with military matters, and new means of communication by co-ordinated computer networks. Of the four new ideas, the first (Measure 15) was based on the human dimension procedures agreed in the concluding document of the Vienna CSCE follow-up meeting in January 1989, and sought a mechanism whereby states could ask for an explanation of unusual military activity. This stemmed from the crash of a Soviet MiG-23 in the Netherlands

[178] See appendix 13D for more on the seminar on military doctrine.

[179] *Arms Control Reporter*, sheet 402.B.259 and interviews with delegates, Mar. 1990.

[180] CSCE/WV/WGA.4.

[181] Borowski, J., 'The Vienna Negotiations on Confidence and Security Building Measures', *Royal United Services Institute (RUSI) Journal*, vol. 135, no. 3 (autumn 1990), pp. 40–44.

in July 1989.[182] The second (Measure 16) was designed to facilitate reporting of hazardous incidents that could cause accidents. This proposal stemmed from two incidents in which US inspectors were threatened at gunpoint by Soviet soldiers during routine inspections; one in a railyard near Schonhausen, GDR, in May 1989, and another in Votinsk, USSR, in March 1990.[183] The third (Measure 17) would permit observation and inspection of air bases; and the fourth (Measure 18) would require reporting of the upgrading of military infrastructure.[184]

There were many proposals in June, albeit mainly elaborations of earlier themes. On 1 June Sweden endorsed earlier proposals on information exchange and the reporting of planned new weapons.[185] Poland and Czechoslovakia reiterated proposals on reporting actual and planned defence budgets.[186] Poland had produced extensive budget information at the CSCE seminar on military doctrine earlier in the year,[187] and Czechoslovakia did so on 13 June, noting it had not done so in a public forum since 1948.[188] On 15 June, the USSR raised the issue of naval restraints again.[189] Czechoslovakia (on 7 June) and a group of nine neutral and non-aligned states (18 June) endorsed the inspection of air bases and air forces.[190]

France circulated a draft CSBM document on 8 June and was particularly anxious that a comprehensive document be ready in time for the Paris summit meeting, not least to undermine the view that serious security issues could only be negotiated between alliances, as at the CFE Negotiation. Nevertheless, despite all the diplomatic activity, delegates reported in the summer of 1990 that progress was hampered by US reluctance to focus as seriously on the CSBM as on the CFE Negotiations, and by persistent disagreement between the USSR, which wanted to include information and constraints on naval forces, and the USA, which adamantly refused to consider naval issues of any kind. The Europeans were neither as enthusiastic as the USSR nor as negative as the USA on naval questions. While the USA insisted that the Madrid mandate precluded any discussion of naval issues, most of the allies and all the NNA states argued that the mandate certainly did not exclude the exchange of static information on naval forces.[191] In most cases the Western and NNA country information is in the public domain in any event, so requiring information on naval forces of the 34 would primarily increase transparency of Soviet naval forces.

[182] Interviews by the author with CSBM delegates.

[183] US Department of State File on CDE inspection 187206 reports the incident on 20 May 1989 at a railyard near Schonhausen, GDR. The Votinsk incident is reported by Walker, M., 'Soviet troops pulled guns on US monitors', *The Guardian,* 17 Mar. 1990.

[184] For the 18 May NATO proposals, see CSCE document CSCE/WV.8, Vienna, May 1990; see also *Arms Control Reporter*, sheets 402.B.267–68, May 1990.

[185] *Arms Control Reporter*, sheet 402.B.269, July 1990.

[186] CSCE document CSCE/WV.10, Vienna, 1 June 1990.

[187] Ministry of National Defence, *Polish Army Facts and Figures*, Warsaw, 1990, 40pp.

[188] Prague domestic radio, FBIS-EE-90, 13 June 1990.

[189] CSCE/WV. 13, Vienna, 15 June 1990.

[190] CSCE/WV/WGB.5, Vienna, 7 June 1990; CSCE/WV/WGA.3/Add.1, Vienna, 18 June 1990.

[191] For interpretations that the Madrid mandate allows exchange of naval information, see *Arms Control Reporter*, sheets 402.B.227, July 1989 and 402.B.265, May 1990.

In August, Soviet delegate Oleg Grinevsky admitted in an interview that naval issues could be set aside until the next phase of the CSBM Negotiations,[192] but at formal plenary sessions the Soviet delegates continued to press for the inclusion of naval data in information exchanges right up to the week before the summit meeting. During September and October reports from Vienna suggested that several other issues would not be resolved before November. In the event, however, a CSBM document, incorporating several important new elements, was adopted by the 34 heads of state in Paris on 21 November 1990.

New elements in the Vienna Document

1. The parties agreed to exchange not only static but also dynamic military data, including current and projected military budgets and plans for new weapon systems. Information will be exchanged no later than 15 December each year (Article I, paras 10–16) and budget data will be submitted according to the UN 'Instrument for standardised international reporting of military expenditures', adopted on 12 December 1980.

2. The parties agreed to establish a Conflict Prevention Centre (CPC) as an element of a risk reduction mechanism designed to reduce the risk of conflict (Article II, para. 17), and to facilitate the reporting of unusual military activities and 'hazardous incidents of a military nature' (Article II, para. 18). The CPC will also serve as the forum for an annual implementation assessment of the 1990 Vienna Document (Article X, paras 151–54).

3. The parties agreed to set up a computerized communications network (Packet Switched Data Network, PSDN), co-ordinated from a centre in the Netherlands, that will complement existing diplomatic channels (Article IX, paras 143–50).

4. The parties agreed to increase military contacts, including visits to air bases at their peacetime locations, previously considered off-limits (Article III, paras 19–34), exchanges between senior military officials and institutions, as well as academics and experts in military studies. The 34 states also scheduled a second seminar on military doctrine for the spring of 1991 (Preamble, para. 4).[193] Annexe I emphasizes that the seminar will not deal with naval issues by reiterating that the zone of the application for the CSBMs and the doctrine seminar covers land territories only.

5. Unlike the Stockholm regime which precluded states from inspecting others in the same alliance, the Vienna CSBMs allow each state to inspect any other.[194] The right to inspect states in the same alliance group is an integral part of the CFE Protocol on Inspection (section II, para. 24) but is only

[192] O. Grinevsky interview by TASS correspondent V. Smelov, FBIS-SOV-90-58, 10 Aug. 1990 p. 3.

[193] For a discussion of the 1990 seminar on military doctrine see appendix 13D.

[194] Annexe IV, 1986 Stockholm Document (note 172), and interpretative statements by Italy on behalf of NATO and by Hungary on behalf of the WTO set out the state parties understandings that they will not inspect other states in the same alliance grouping. See Stockholm Conference, *CSCE Journal*, no. 379, 178th plenary meeting, 19 Sep. 1986.

implicit in the CSBM Document. To ensure there would be no ambiguity about inspection rights in the Vienna CSBM regime, on 17 November 1990 at the 66th CSBM Plenary meeting, the delegation of Hungary, also on behalf of the delegations of Poland and Czechoslovakia, asserted the right to carry out inspections and evaluation visits on the territory of any other participating state.[195] As a measure of interest among the 22 NATO and WTO states in inspecting Soviet territory, it is worth noting that at the Budapest Open Skies Conference in April–May 1990 a ballot, asking the 22 states how they would allocate aerial overflights among them, showed that of a total of 150 flights, 110 would be over the USSR, 7 would be over Bulgaria and the other 33 would be distributed more or less evenly among the other 20 states.[196]

Expanded Stockholm CSBMs

More than half the provisions in the Vienna Document 1990 repeat those of the Stockholm Documents; those on refraining from the threat or use of force, on prior notification, on observation of certain military activities, on annual calendars, on constraining provisions and on verification. However, the Vienna Document also expands several of the Stockholm provisions:

1. The verification regime is strengthened by an evaluation mechanism for the new data required in the annual information exchange (Article VIII, paras 112–42);

2. The Vienna Document 1990 requires a greater amount of information to be notified in advance than the 1986 Stockholm Document. In the section on Prior Notification of certain Military Activities (Article IV) information on notifiable military activities now includes the designation, subordination, number and type of formation and units down to and including brigade/ regiment or equivalent level (para. 42.1.2) In the Stockholm Document the requirement was only to division level (para. 35.1.3).

3. The Vienna Document improves the conditions for observers at military exercises. In the section on Observation of Certain Military Activities (Article V) several new provisions ensure non-interference with inspectors, permit more aerial surveillance and grant more equipment to the observers, as well as equal access to the media of all participating states.

4. The Vienna Document requires lower thresholds, from 75 000 to 40 000 troops for exercises that require two years' advance notice (Article VII, para. 72).

IX. Evaluation and conclusions

The contrast between the rapid progress and upbeat atmosphere at the Vienna negotiations during most of 1989 and 1990 and the deep pessimism surround-

[195] *CSCE Journal*, no. 241/Rev 1 (Nov. 1990), p. 2.

[196] Interviews by the author with delegates to the Open Skies Conferences in Budapest.

ing the deliberations at the JCG in late 1990 and early 1991, amply demonstrates that arms control negotiations reflect rather than affect international relations. Thus the CFE Negotiation made impressive progress during 1989 and 1990 parallel to Gorbachev's renunciation of the Brezhnev doctrine, his willingness to withdraw forces unilaterally from Eastern Europe and his acceptance of a united Germany. But after the Treaty was signed in November 1990, relations between the USSR and the rest of Europe began to deteriorate. In December 1990, conservative elements forced the resignation of Foreign Minister Shevardnadze, the most important Soviet interlocutor with the West. In January 1991 Gorbachev ordered, or sanctioned, a military crackdown in the Baltic republics, the Soviet military was increasingly critical about the damaging effects of Shevarnadze's diplomacy, the loss of the allies and the success of Western forces in the Gulf war. In Vienna the Soviet General Staff were unco-operative in the Joint Consultative Group. Far from building trust and confidence, in early 1991 the Soviet General Staff eroded much of the goodwill that Shevardnadze and the Soviet Foreign Ministry had built up during the previous two years, and as this *Yearbook* goes to press the prospects for ratification of the CFE Treaty look remote. If the Treaty is concluded, however, it could make a substantial contribution to East–West stability in Europe by codifying lower levels of residual forces in the ATTU zone, and cementing the improved political relations between NATO and the Central European and Balkan countries.

The Treaty measures up well to the CFE mandate of January 1989, which sought to establish a secure and stable balance of forces at lower levels; to eliminate the capability of any state or group of states to launch a surprise attack; and to constrain the ability to initiate and sustain large-scale offensive actions in Europe.[197] With respect to the new balance of forces, the numerical superiority of the WTO has been eliminated, although what have been cut are certainly the least-capable elements of each category, and the CFE verification regime will ensure timely detection of any militarily significant Treaty violations. Whether states view the new balance as stable, however, depends on their perception of other's intentions. The NATO–WTO balance codified by the Treaty became meaningless once the military structures of the WTO were abolished. The Treaty thus leaves Europe with a new set of asymmetries. The most significant of these is that NATO forces are now both quantitatively as well as qualitatively superior to those of the USSR—an imbalance that probably reassures most Europeans but certainly not the military and political leadership in the USSR. The Treaty leaves the USSR militarily superior to the collective strength of its former WTO allies. Such an imbalance seemed irrelevant in the *détente* atmosphere of late 1989 and most of 1990, but was unsettling to Poland, Hungary and Czechoslovakia in the aftermath of Soviet

[197] The CFE mandate is reprinted in *SIPRI Yearbook 1989: World Armaments and Disarmament* (Oxord University Press: Oxford, 1989), pp. 420–22.

action in the Baltic republics in January 1991 and could become more so if the Soviet military establishment grows more assertive.[198]

With respect to surprise attack, serious analysts in the West never believed the Soviet military had such a capability,[199] but after the CFE Negotiation and Soviet withdrawals from Eastern Europe even the most conservative NATO planners were confident that any standing-start Soviet surprise-attack capability had been eliminated. The Treaty does little, however, to constrain Soviet capability to initiate and sustain offensive military action. CFE limits for the USSR are only one-third of the Soviet forces facing NATO in 1988, but Soviet force levels remain high, and there are no constraints yet on manpower, logistics, or readiness levels. The USSR remains the strongest state in Europe, for example, with more than four times as many tanks allowed in the ATTU zone (13 150 : 3000) than Nazi Germany had when it launched World War II in 1939.[200]

Few European states worry any longer about German aggression, but the states of Central Europe are nevertheless in the shadow of a united Germany with higher CFE ceilings in every category (except battle tanks) than the FRG had before unification. Although Germany plans to cut manpower unilaterally in the early 1990s, the resulting forces will be equipped with highly sophisticated weaponry that could look provocative if political relations in the region deteriorate. Most of Germany's forces will remain integrated in NATO's multinational structure, but those stationed in the five eastern Länder (Bundeswehr Ost) are national forces outside alliance control.

The Nordic countries complain that the CFE Treaty does nothing to ease the heavy concentration of Soviet and US naval power in the north Atlantic.[201] Norway and Turkey have been consistently downbeat about the impact of CFE on their security, noting that much of the modern Soviet equipment moved from Central Europe has moved closer to their borders. Resubordination of the 77th Guards Division in Archangel upset Norway and that of the 126th Guards Division at Simferpol upset Turkey.

The NSWTO countries had mixed feelings about the Treaty. The euphoria of getting rid of oppressive communist regimes in late 1989 soon gave way to anxiety in 1990, adrift as they were between NATO and the USSR with no security guarantees from either. This anxiety was manifest in the intra-WTO meetings to allocate CFE ceilings as newly democratic countries sought extra

[198] Reisch, A., 'Hungary condemns Soviet moves against Lithuania'; Obrman, J., 'Czechoslovakia reacts to crackdown in the Baltic Republics', and Sabbat-Swielicka, A., 'Polish reactions to the Lithuanian crisis'; *Report on Eastern Europe*, vol. 2, no. 6 (8 Feb. 1991), pp. 25–34.

[199] Mearsheimer, J. J., 'Why the Soviets can't win quickly in central Europe', eds S. Miller and S. Lynn-Jones, *Conventional Forces and American Defence Policy*, 2nd edn (MIT Press: Cambridge, Mass., 1989).

[200] The highest estimate of Nazi tanks in 1939–40 is 3000. See Horne, A., *To Lose a Battle* (Penguin Books: Harmondsworth, 1966), p. 217.

[201] On Nordic anxieties about CFE see Miller, S., 'The superpowers and Nordic security in post cold war Europe', eds B. Huldt, G. Herolf and M. Dufwa, *Towards a New European Order of Peace and Security* (Utrikespolitiska Institutet: Stockholm, forthcoming); Harjula, J. and Jarvenpaa, P. (eds), *The Treaty on Conventional Forces in Europe: A Finnish View*, Pugwash Workshop, 9–11 Nov. 1990; *Report on Finnish Security Policy to the Foreign Affairs Committee of the Finnish Parliament*, 24 Oct. 1990.

equipment to build up national defence forces for the first time in many decades, and in the requests of Germany from Central European defence ministries for surplus NVA equipment.

The reductions imposed by the Treaty would probably have happened unilaterally in any event. Ratification is necessary, however, to activate the inspection provisions that will maintain the stability and predictability of the new lower level of military forces in Europe. Here, the Treaty breaks new ground by including challenge inspections to undeclared sites. The verification regime is more intrusive than that of any other arms control agreement yet concluded and could still prove unacceptable to the Soviet General Staff. The inspection provisions are nevertheless far from perfect. It proved impossible, for example, to negotiate production monitoring, identification of TLE by tagging or aerial inspection.[202]

Some attempt will be made to rectify these omissions in the CFE IA negotiations for which two primary goals have been identified: manpower limits and aerial inspection. Manpower ceilings are important to avoid further singularization of Germany, which made a unilateral commitment to cut personnel in 1990. Aerial inspection is important on two counts: it is the only means by which non-Soviet countries can be confident about the state of military forces east of the Urals, and it can also give the Soviet military more confidence that they can monitor arms production in the continental USA.

The prospects for negotiated deep cuts in manpower are slim. Just as equipment cuts imposed by the CFE Treaty were a small percentage of those achieved by unilateral means during 1989 and 1990, so CFE IA manpower cuts are likely to be a small percentage of cuts achieved unilaterally in the early 1990s. All 16 NATO countries will want ceilings that permit maximum flexibility to restructure forces into either national, binational or multinational forces. At best a CFE IA agreement will set limits close to 1991–92 levels, at worst the CFE IA talks could degenerate into a repetition of the M(B)FR Talks that were also devoted to manpower limits and stymied over data disputes. Prospects for establishing an aerial inspection regime also look slim, unless the NATO countries are prepared to share technology with the other states.

The CSBM Negotiations were overshadowed by the CFE Negotiation during 1989–90 and not all the new proposals offered survived into the Vienna CSBM Document. The USSR failed to stretch the original mandate to include naval CSBMs. The neutral states resisted notification of reservist activity and other measures that might undermine their reliance on rapid mobilization in a crisis, and the NATO states were unable to convince the others to notify improvements to military infrastructure. Nevertheless, some new measures were adopted of which the most important was the establishment of a risk reduction mechanism that includes a computerized communications network, a system of reporting unusual military activity and a Crisis Prevention Centre to head off potential crises before they erupt into violence.

[202] Fetter, S. and Garwin, T., 'Tags', Kokoski and Koulik (note 4), pp. 139–54; and Spitzer, H., 'Aerial inspection and overflights', pp. 89–122 of the same volume.

The Centre for Conflict Prevention (CPC)[203] is, however, not yet adequate to deal with the conflicts most likely to occur in Europe, that is, those within national, albeit disputed, borders. This was brought home starkly to the Baltic states in January 1991 when the Soviet military opened fire on unarmed civilians. At a special meeting of CSCE experts in Vienna, the USSR vetoed a proposal for a special meeting to discuss the Baltic republics. The vote was 33 : 1, reflecting the growing isolation of the USSR from the rest of Europe at the end of 1990.[204] The USSR will have to face its critics, however, at the CSCE Human Rights Conference scheduled for October 1991 in Moscow.

The mandate for the 1991 seminar on military doctrine appears to exclude naval issues, but in a reservation to the Vienna Document submitted on 17 November 1990 the USSR declared its intention to raise naval issues in the next round of CSBM talks.[205] Recent Soviet proposals include the withdrawal of forward naval bases, especially the US and Soviet fleets from the Mediterranean; both the prenotification and the reduction of naval exercises as well as the prohibition of interference with observation of exercises at sea; and limits on combined arms exercises such as those that NATO conducts with air, land and naval forces in north Norway.[206] Both at the CSBM and CFE Negotiations it was clear that, facing a CFE Treaty regime that leaves the USSR inferior to NATO in conventional forces, the Soviet military establishment now seeks an agreement to redress NATO superiority in naval forces, in particular to impose limits on US holdings of sea-launched cruise missiles and carrier-based aircraft.

Given the trend towards smaller, more mobile forces, NATO states will continue to press for advance information about plans to improve infrastructure and measures to keep track of mobilization of reserves. The four neutral states (Austria, Finland, Sweden and Switzerland) seem likely to continue to resist any constraints on their mobilization capability.

The actions of the Soviet 'Black Berets' in the Baltic republics suggest the need for greater control over paramilitary forces.[207] In May 1990, Hungary and Czechoslovakia proposed bilateral CSBMs to curb activity in border areas that might give rise to apprehension and misunderstanding.[208] Clashes on the border between Ukraine and Czechoslovakia in January 1991, and the growing anxiety about mass migrations from the south and east towards the north and west, emphasize the need for procedures governing border patrols.[209]

[203] See also chapter 17 in this volume.

[204] Reuters, 'Latvians accuse Soviet riot police', *Financial Times*, 18 Jan. 1991; Traynor, I., 'Moscow vetoes European conference on Baltics', *The Guardian*, 18 Jan. 1991.

[205] *CSCE Journal*, no. 241/Rev., p. 2, para. 6b.

[206] For more on Soviet proposals for naval CSBMs, see Fürst, A., Heisse, V. and Miller, S. (eds), SIPRI, *Europe and Naval Arms Control* (Oxford University Press, forthcoming).

[207] The Soviet 'Black Berets' are special assignment troops (*Spetsnatz*) and special-purpose militia squads (*Omon*) subordinate to the Soviet interior Ministry in Moscow. See Peel, Q., 'Latvians vote for force to counter wave of terror', *Financial Times*, 22 Jan. 1991.

[208] CSCE document CSCE/WV.9, Vienna, 25 May 1990.

[209] Traynor, I., '14 dead after Czechoslovak guards and Soviets clash on border', *The Guardian*, 25 Jan. 1991.

Appendix 13A. Treaty on Conventional Armed Forces in Europe

Paris, 19 November 1990

The Kingdom of Belgium, the People's Republic of Bulgaria, Canada, the Czech and Slovak Federal Republic, the Kingdom of Denmark, the French Republic, the Federal Republic of Germany, the Hellenic Republic, the Republic of Hungary, the Republic of Iceland, the Italian Republic, the Grand Duchy of Luxembourg, the Kingdom of the Netherlands, the Kingdom of Norway, the Republic of Poland, the Portuguese Republic, Romania, the Kingdom of Spain, the Republic of Turkey, the Union of Soviet Socialist Republics, the United Kingdom of Great Britain and Northern Ireland and the United States of America, hereinafter referred to as the States Parties,

Guided by the Mandate for Negotiation on Conventional Armed Forces in Europe of January 10, 1989, and having conducted this negotiation in Vienna beginning on March 9, 1989,

Guided by the objectives and the purposes of the Conference on Security and Cooperation in Europe, within the framework of which the negotiation of this Treaty was conducted,

Recalling their obligation to refrain in their mutual relations, as well as in their international relations in general, from the threat or use of force against the territorial integrity or political independence of any State, or in any other manner inconsistent with the purposes and principles of the Charter of the United Nations,

Conscious of the need to prevent any military conflict in Europe,

Conscious of the common responsibility which they all have for seeking to achieve greater stability and security in Europe,

Striving to replace military confrontation with a new pattern of security relations among all the States Parties based on peaceful cooperation and thereby to contribute to overcoming the division of Europe,

Committed to the objectives of establishing a secure and stable balance of conventional armed forces in Europe at lower levels than heretofore, of eliminating disparities prejudicial to stability and security and of eliminating, as a matter of high priority, the capability for launching surprise attack and for initiating large-scale offensive action in Europe,

Recalling that they signed or acceded to the Treaty of Brussels of 1948, the Treaty of Washington of 1949 or the Treaty of Warsaw of 1955 and that they have the right to be or not to be a party to treaties of alliance,

Committed to the objective of ensuring that the numbers of conventional armaments and equipment limited by the Treaty within the area of application of this Treaty do not exceed 40 000 battle tanks, 60 000 armoured combat vehicles, 40 000 pieces of artillery, 13 600 combat aircraft and 4 000 attack helicopters,

Affirming that this Treaty is not intended to affect adversely the security interests of any State,

Affirming their commitment to continue the conventional arms control process including negotiations, taking into account future requirements for European stability and security in the light of political developments in Europe,

Have agreed as follows:

Article I

1. Each State Party shall carry out the obligations set forth in this Treaty in accordance with its provisions, including those obligations relating to the following five categories of conventional armed forces: battle tanks, armoured combat vehicles, artillery, combat aircraft and combat helicopters.

2. Each State Party also shall carry out the other measures set forth in this Treaty designed to ensure security and stability both during the period of reduction of conventional armed forces and after the completion of reductions.

3. This Treaty incorporates the Protocol on Existing Types of Conventional Armaments and Equipment, hereinafter referred to as the Protocol on Existing Types, with an Annex thereto; the Protocol on Procedures Governing the Reclassification of Specific Models or Versions of Combat-Capable Trainer Aircraft into Unarmed Trainer Aircraft, hereinafter referred to as the Protocol on Aircraft Reclassification; the Protocol on Procedures Governing the Reduction of Con-

ventional Armaments and Equipment Limited by the Treaty on Conventional Armed Forces in Europe, hereinafter referred to as the Protocol on Reduction; the Protocol on Procedures Governing the Categorisation of Combat Helicopters and the Recategorisation of Multi-Purpose Attack Helicopters, hereinafter referred to as the Protocol on Helicopter Recategorisation; the Protocol on Notification and Exchange of Information, hereinafter referred to as the Protocol on Information Exchange, with an Annex on the Format for the Exchange of Information, hereinafter referred to as the Annex on Format; the Protocol on Inspection; the Protocol on the Joint Consultative Group; and the Protocol on the Provisional Application of Certain Provisions of the Treaty on Conventional Armed Forces in Europe, hereinafter referred to as the Protocol on Provisional Application. Each of these documents constitutes an integral part of this Treaty.

Article II

1. For the purposes of this Treaty:

(A) The term 'group of States Parties' means the group of States Parties that signed the Treaty of Warsaw[1] of 1955 consisting of the People's Republic of Bulgaria, the Czech and Slovak Federal Republic, the Republic of Hungary, the Republic of Poland, Romania and the Union of Soviet Socialist Republics, or the group of States Parties that signed or acceded to the Treaty of Brussels[2] of 1948 or the Treaty of Washington[3] of 1949 consisting of the Kingdom of Belgium, Canada, the Kingdom of Denmark, the French Republic, the Federal Republic of Germany, the Hellenic Republic, the Republic of Iceland, the Italian Republic, the Grand Duchy of Luxembourg, the Kingdom of the Netherlands, the Kingdom of Norway, the Portuguese Republic, the Kingdom of Spain, the Republic of Turkey, the United Kingdom of Great Britain and Northern Ireland and the United States of America.

(B) The term 'area of application' means the entire land territory of the States Parties in Europe from the Atlantic Ocean to the Ural Mountains, which includes all the European island territories of the States Parties, including the Faroe Islands of the Kingdom of Denmark, Svalbard including Bear Island of the Kingdom of Norway, the islands of Azores and Madeira of the Portuguese Republic, the Canary Islands of the Kingdom of Spain and Franz Josef Land and Novaya Zemlya of the Union of Soviet Socialist Republics. In the case of the Union of Soviet Socialist Republics, the area of application includes all territory lying west of the Ural River and the Caspian Sea. In the case of the Republic of Turkey, the area of application includes the territory of the Republic of Turkey north and west of a line extending from the point of intersection of the Turkish border with the 39th parallel to Muradiye, Patnos, Karayazi, Tekman, Kemaliye, Feke, Ceyhan, Dogankent, Gözne and thence to the sea.

(C) The term 'battle tank' means a self-propelled armoured fighting vehicle, capable of heavy fire power, primarily of a high muzzle velocity direct fire main gun necessary to engage armoured and other targets, with high cross-country mobility, with a high level of self-protection, and which is not designed and equipped primarily to transport combat troops. Such armoured vehicles serve as the principal weapon system of ground-force tank and other armoured formations.

Battle tanks are tracked armoured fighting vehicles which weigh at least 16.5 metric tonnes unladen weight and which are armed with a 360-degree traverse gun of at least 75 millimetres calibre. In addition, any wheeled armoured fighting vehicles entering into service which meet all the other criteria stated above shall be deemed battle tanks.

(D) The term 'armoured combat vehicle' means a self-propelled vehicle with armoured protection and cross-country capability. Armoured combat vehicles include armoured personnel carriers, armoured infantry fighting vehicles and heavy armament combat vehicles.

The term 'armoured personnel carrier' means an armoured combat vehicle which is designed and equipped to transport a combat infantry squad and which, as a rule, is armed with an integral or organic weapon of less than 20 millimetres calibre.

The term 'armoured infantry fighting vehicle' means an armoured combat vehicle which is designed and equipped primarily to transport a combat infantry squad, which normally provides the capability for the troops to deliver fire from inside the vehicle under armoured protection, and which is armed with an integral or organic cannon of at least 20 millimetres calibre and sometimes an antitank missile launcher.

Armoured infantry fighting vehicles serve as the principal weapon system of armoured infantry or mechanised infantry or motorised infantry formations and units of ground forces.

The term 'heavy armament combat vehicle' means an armoured combat vehicle with an integral or organic direct fire gun of at least 75 millimetres calibre, weighing at least 6.0 metric tonnes unladen weight, which does not fall within the definitions of an armoured personnel carrier, or an armoured infantry fighting vehicle or a battle tank.

(E) The term 'unladen weight' means the weight of a vehicle excluding the weight of ammunition; fuel, oil and lubricants; removable reactive armour; spare parts, tools and accessories; removable snorkelling equipment; and crew and their personal kit.

(F) The term 'artillery' means large calibre systems capable of engaging ground targets by delivering primarily indirect fire. Such artillery systems provide the essential indirect fire support to combined arms formations.

Large calibre artillery systems are guns, howitzers, artillery pieces combining the characteristics of guns and howitzers, mortars and multiple launch rocket systems with a calibre of 100 millimetres and above. In addition, any future large calibre direct fire system which has a secondary effective indirect fire capability shall be counted against the artillery ceilings.

(G) The term 'stationed conventional armed forces' means conventional armed forces of a State Party that are stationed within the area of application on the territory of another State Party.

(H) The term 'designated permanent storage site' means a place with a clearly defined physical boundary containing conventional armaments and equipment limited by the Treaty, which are counted within overall ceilings but which are not subject to limitations on conventional armaments and equipment limited by the Treaty in active units.

(I) The term 'armoured vehicle launched bridge' means a self-propelled armoured transporter-launcher vehicle capable of carrying and, through built-in mechanisms, of emplacing and retrieving a bridge structure. Such a vehicle with a bridge structure operates as an integrated system.

(J) The term 'conventional armaments and equipment limited by the Treaty' means battle tanks, armoured combat vehicles, artillery, combat aircraft and attack helicopters subject to the numerical limitations set forth in Article IV, V and VI.

(K) The term 'combat aircraft' means a fixed-wing or variable-geometry wing aircraft armed and equipped to engage targets by employing guided missiles, unguided rockets, bombs, guns, cannons, or other weapons of destruction, as well as any model or version of such an aircraft which performs other military functions such as reconnaissance or electronic warfare. The term 'combat aircraft' does not include primary trainer aircraft.

(L) The term 'combat helicopter' means a rotary wing aircraft armed and equipped to engage targets or equipped to perform other military functions. The term 'combat helicopter' comprises attack helicopters and combat support helicopters. The term 'combat helicopter' does not include unarmed transport helicopters.

(M) The term 'attack helicopter' means a combat helicopter equipped to employ anti-armour, air-to-ground, or air-to-air guided weapons and equipped with an integrated fire control and aiming system for these weapons. The term 'attack helicopter' comprises specialised attack helicopters and multi-purpose attack helicopters.

(N) The term 'specialised attack helicopter' means an attack helicopter that is designed primarily to employ guided weapons.

(O) The term 'multi-purpose attack helicopter' means an attack helicopter designed to perform multiple military functions and equipped to employ guided weapons.

(P) The term 'combat support helicopter' means a combat helicopter which does not fulfill the requirements to qualify as an attack helicopter and which may be equipped with a variety of self-defence and area suppression weapons, such as guns, cannons and unguided rockets, bombs or cluster bombs, or which may be equipped to perform other military functions.

(Q) The term 'conventional armaments and equipment subject to the Treaty' means battle tanks, armoured combat vehicles, artillery, combat aircraft, primary trainer aircraft, unarmed trainer aircraft, combat helicopters, unarmed transport helicopters, armoured vehicle launched bridges, armoured personnel carrier look-alikes and armoured infantry fighting vehicle look-alikes subject to information exchange in accordance with the Protocol on Information Exchange.

(R) The term 'in service', as it applies to conventional armed forces and conventional armaments and equipment, means battle tanks, armoured combat vehicles, artillery, combat aircraft, primary trainer aircraft, unarmed trainer aircraft, combat helicopters, unarmed transport helicopters, armoured vehicle launched bridges, armoured personnel carrier look-alikes and armoured infantry fighting vehicle look-alikes that are within the area of application, except for those that are held by organisations designed and structured to perform in peacetime internal security functions or that meet any of the exceptions set forth in Article III.

(S) The terms 'armoured personnel carrier look-alike' and 'armoured infantry fighting vehicle look-alike' mean an armoured vehicle based on the same chassis as, and externally similar to, an armoured personnel carrier or armoured infantry fighting vehicle, respectively, which does not have a cannon or gun of 20 millimetres calibre or greater and which has been constructed or modified in such a way as not to permit the transportation of a combat infantry squad. Taking into account the provisions of the Geneva Convention 'For the Amelioration of the Conditions of the Wounded and Sick in Armed Forces in the Field' of 12 August 1949 that confer a special status on ambulances, armoured personnel carrier ambulances shall not be deemed armoured combat vehicles or armoured personnel carrier look-alikes.

(T) The term 'reduction site' means a clearly designated location where the reduction of conventional armaments and equipment limited by the Treaty in accordance with Article VIII takes place.

(U) The term 'reduction liability' means the number in each category of conventional armaments and equipment limited by the Treaty that a State Party commits itself to reduce during the period of 40 months following the entry into force of this Treaty in order to ensure compliance with Article VII.

2. Existing types of conventional armaments and equipment subject to the Treaty are listed in the Protocol on Existing Types. The lists of existing types shall be periodically updated in accordance with Article XVI, paragraph 2, subparagraph (D) and Section IV of the Protocol on Existing Types. Such updates to the existing types lists shall not be deemed amendments to this Treaty.

3. The existing types of combat helicopters listed in the Protocol on Existing Types shall be categorised in accordance with Section I of the Protocol on Helicopter Recategorisation.

Article III

1. For the purposes of this Treaty, the States Parties shall apply the following counting rules:

All battle tanks, armoured combat vehicles, artillery, combat aircraft and attack helicopters, as defined in Article II, within the area of application shall be subject to the numerical limitations and other provisions set forth in Article IV, V and VI, with the exception of those which in a manner consistent with a State Party's normal practices:

(A) are in the process of manufacture, including manufacturing-related testing;

(B) are used exclusively for the purposes of research and development;

(C) belong to historical collections;

(D) are awaiting disposal, having been decommissioned from service in accordance with the provisions of Article IX;

(E) are awaiting, or are being refurbished for, export or re-export and are temporarily retained within the area of application. Such battle tanks, armoured combat vehicles, artillery, combat aircraft and attack helicopters shall be located elsewhere than at sites declared under the terms of Section V of the Protocol on Information Exchange or at no more than 10 such declared sites which shall have been notified in the previous year's annual information exchange. In the latter case, they shall be separately distinguishable from conventional armaments and equipment limited by the Treaty;

(F) are, in the case of armoured personnel carriers, armoured infantry fighting vehicles, heavy armament combat vehicles or multipurpose attack helicopters, held by organisations designed and structured to perform in peacetime internal security functions; or

(G) are in transit through the area of application from a location outside the area of application to a final destination outside the area of application, and are in the area of application for no longer than a total of seven days.

2. If, in respect of any such battle tanks, armoured combat vehicles, artillery, combat aircraft or attack helicopters, the notification of which is required under Section IV of the Protocol on Information Exchange, a State Party notifies an unusually high number in more than two successive annual information exchanges, it shall explain the reasons in the Joint Consultative Group, if so required.

Article IV

1. Within the area of application, as defined in Article II, each State Party shall limit and, as necessary, reduce its battle tanks, armoured combat vehicles, artillery, combat aircraft and attack helicopters so that, 40 months after entry into force of this Treaty and thereafter, for the group of States Parties to which it belongs, as defined in Article II, the aggregate numbers do not exceed:

(A) 20 000 battle tanks, of which no more than 16 500 shall be in active units;

(B) 30 000 armoured combat vehicles, of which no more than 27 300 shall be in active units. Of the 30 000 armoured combat vehicles, no more than 18 000 shall be armoured infantry fighting vehicles and heavy armament combat vehicles; of armoured infantry fighting vehicles and heavy armament combat vehicles, no more than 1 500 shall be heavy armament combat vehicles;

(C) 20 000 pieces of artillery, of which no more than 17 000 shall be in active units;

(D) 6800 combat aircraft; and

(E) 2000 attack helicopters.

Battle tanks, armoured combat vehicles and artillery not in active units shall be placed in designated permanent storage sites, as defined in Article II, and shall be located only in the area described in paragraph 2 of this Article. Such designated permanent storage sites may also be located in that part of the territory of the Union of Soviet Socialist Republics comprising the Odessa Military District and the southern part of the Leningrad Military District. In the Odessa Military District, no more than 400 battle tanks and no more than 500 pieces of artillery may be thus stored. In the southern part of the Leningrad Military District, no more than 600 battle tanks, no more than 800 armoured combat vehicles, including no more than 300 armoured combat vehicles of any type with the remaining number consisting of armoured personnel carriers, and no more than 400 pieces of artillery may be thus stored. The southern part of the Leningrad Military District is understood to mean the territory within that Military District south of the line East–West 60 degrees 15 minutes northern latitude.

2. Within the area consisting of the entire land territory in Europe, which includes all the European island territories, of the Kingdom of Belgium, the Czech and Slovak Federal Republic, the Kingdom of Denmark including the Faroe Islands, the French Republic, the Federal Republic of Germany, the Republic of Hungary, the Italian Republic, the Grand Duchy of Luxembourg, the Kingdom of the Netherlands, the Republic of Poland, the Portuguese Republic including the islands of Azores and Madeira, the Kingdom of Spain including the Canary Islands, the United Kingdom of Great Britain and Northern Ireland and that part of the territory of the Union of Soviet Socialist Republics west of the Ural Mountains comprising the Baltic, Byelorussian, Carpathian, Kiev, Moscow and Volga–Ural Military Districts, each State Party shall limit and, as necessary, reduce its battle tanks, armoured combat vehicles and artillery so that, 40 months after entry into force of this Treaty and thereafter, for the group of States Parties to which it belongs the aggregate numbers do not exceed:

(A) 15 300 battle tanks, of which no more than 11 800 shall be in active units;

(B) 24 100 armoured combat vehicles, of which no more than 21 400 shall be in active units; and

(C) 14 000 pieces of artillery, of which no more than 11 000 shall be in active units.

3. Within the area consisting of the entire land territory in Europe, which includes all the European island territories, of the Kingdom of Belgium, the Czech and Slovak Federal Republic, the Kingdom of Denmark including the Faroe Islands, the French Republic, the Federal Republic of Germany, the Republic of Hungary, the Italian Republic, the Grand Duchy of Luxembourg, the Kingdom of the Netherlands, the Republic of Poland, the United Kingdom of Great Britain and Northern Ireland and that part of the territory of the Union of Soviet Socialist Republics comprising the Baltic, Byelorussian, Carpathian and Kiev Military Districts, each State Party shall limit and, as necessary, reduce its battle tanks, armoured combat vehicles and artillery so that, 40 months after the entry into force of this Treaty and thereafter, for the group of States Parties to which it belongs the aggregate numbers in active units do not exceed:

(A) 10 300 battle tanks;

(B) 19 260 armoured combat vehicles; and

(C) 9100 pieces of artillery; and

(D) in the Kiev Military District, the aggregate numbers in active units and designated permanent storage sites together shall not exceed:

(1) 2250 battle tanks;

(2) 2500 armoured combat vehicles; and

(3) 1500 pieces of artillery.

4. Within the area consisting of the entire land territory in Europe, which includes all the European island territories, of the Kingdom of Belgium, the Czech and Slovak Federal Republic, the Federal Republic of Germany, the Republic of Hungary, the Grand Duchy of Luxembourg, the Kingdom of the Netherlands and the Republic of Poland, each State Party shall limit and, as necessary, reduce its battle tanks, armoured combat vehicles and artillery so that, 40 months after entry into force of this Treaty and thereafter, for the group of States Parties to which it belongs the aggregate numbers in active units do not exceed:

(A) 7500 battle tanks;

(B) 11 250 armoured combat vehicles; and

(C) 5000 pieces of artillery.

5. States Parties belonging to the same group of States Parties may locate battle tanks, armoured combat vehicles and artillery in active units in each of the areas described in this Article and Article V, paragraph 1, subparagraph (A), up to the numerical limitations applying in that area, consistent with the maximum levels for holdings notified pursuant to Article VII and provided that no State Party stations conventional armed forces on the territory of another State Party without the agreement of that State Party.

6. If a group of States Parties' aggregate numbers of battle tanks, armoured combat vehicles and artillery in active units within the area described in paragraph 4 of this Article are less than the numerical limitations set forth in paragraph 4 of this Article, and provided that no State Party is thereby prevented from reaching its maximum levels for holdings notified in accordance with Article VII, paragraphs 2, 3 and 5, then amounts equal to the difference between the aggregate numbers in each of the categories of battle tanks, armoured combat vehicles and artillery and the specified numerical limitations for that area may be located by States Parties belonging to that group of States Parties in the area described in paragraph 3 of this Article, consistent with the numerical limitations specified in paragraph 3 of this Article.

Article V

1. To ensure that the security of each State Party is not affected adversely at any stage:

(A) within the area consisting of the entire land territory in Europe, which includes all the European island territories, of the People's Republic of Bulgaria, the Hellenic Republic, the Republic of Iceland, the Kingdom of Norway, Romania, the part of the Republic of Turkey within the area of application and that part of the Union of Soviet Socialist Republics comprising the Leningrad, Odessa, Transcaucasus and North Caucasus Military Districts, each State Party shall limit and, as necessary, reduce its battle tanks, armoured combat vehicles and artillery so that, 40 months after the entry into force of this Treaty and thereafter, for the group of States Parties to which it belongs the aggregate numbers in active units do not exceed the difference between the overall numerical limitations set forth in Article IV, paragraph 1 and those in Article IV, paragraph 2, that is:

(1) 4700 battle tanks;

(2) 5900 armoured combat vehicles; and

(3) 6000 pieces of artillery;

(B) notwithstanding the numerical limitations set forth in subparagraph (A) of this paragraph, a State Party or States Parties may on a temporary basis deploy into the territory belonging to the members of the same group of States Parties within the area described in subparagraph (A) of this paragraph additional aggregate numbers in active units for each group of States Parties not to exceed:

(1) 459 battle tanks;

(2) 723 armoured combat vehicles; and

(3) 420 pieces of artillery; and

(C) provided that for each group of States Parties no more than one-third of each of these additional aggregate numbers shall be deployed to any State Party with territory within the area described in subparagraph (A) of this paragraph, that is:

(1) 153 battle tanks;

(2) 241 armoured combat vehicles; and

(3) 140 pieces of artillery.

2. Notification shall be provided to all other States Parties no later than at the start of the deployment by the State Party or States Parties conducting the deployment and by the recipient State Party or States Parties, specifying the total number of each category of battle tanks, armoured combat vehicles and artillery deployed. Notification also shall be provided to all other States Parties by the State Party or States Parties conducting the deployment and by the recipient State Party or States Parties within 30 days of the withdrawal of those battle tanks, armoured combat vehicles and artillery that were temporar-

ily deployed.

Article VI

With the objective of ensuring that no single State Party possesses more than approximately one-third of the conventional armaments and equipment limited by the Treaty within the area of application, each State Party shall limit and, as necessary, reduce its battle tanks, armoured combat vehicles, artillery, combat aircraft and attack helicopters so that, 40 months after entry into force of this Treaty and thereafter, the numbers within the area of application for that State Party do not exceed:

(A) 13 300 battle tanks;
(B) 20 000 armoured combat vehicles
(C) 13 700 pieces of artillery;
(D) 5150 combat aircraft; and
(E) 1500 attack helicopters.

Article VII

1. In order that the limitations set forth in Articles IV, V and VI are not exceeded, no State Party shall exceed, from 40 months after the entry into force of this Treaty, the maximum levels which it has previously agreed upon within its group of States Parties, in accordance with paragraph 7 of this Article, for its holdings of conventional armaments and equipment limited by the Treaty and of which it has provided notification pursuant to the provisions of this Article.

2. Each State Party shall provide at the signature of this Treaty notification to all other States Parties of the maximum levels for its holdings of conventional armaments and equipment limited by the Treaty. The notification of the maximum levels for holdings of conventional armaments and equipment limited by the Treaty provided by each State Party at the signature of this Treaty shall remain valid until the date specified in a subsequent notification pursuant to paragraph 3 of this Article.

3. In accordance with the limitations set forth in Articles IV, V and VI, each State Party shall have the right to change the maximum levels for its holdings of conventional armaments and equipment limited by the Treaty. Any change in the maximum levels for holdings of a State Party shall be notified by that State Party to all other States Parties at least 90 days in advance of the date, specified in the notification, on which such a change takes effect. In order not to exceed any of the limitations set forth in Articles IV and V, any increase in the maximum levels for holdings of a State Party that would otherwise cause those limitations to be exceeded shall be preceded or accompanied by a corresponding reduction in the previously notified maximum levels for holdings of conventional armaments and equipment limited by the Treaty of one or more States Parties belonging to the same group of States Parties. The notification of a change in the maximum levels for holdings shall remain valid from the date specified in the notification until the date specified in a subsequent notification of change pursuant to this paragraph.

4. Each notification required pursuant to paragraph 2 or 3 of this Article for armoured combat vehicles shall also include maximum levels for the holdings of armoured infantry fighting vehicles and heavy armament combat vehicles of the State Party providing the notification.

5. Ninety days before expiration of the 40-month period of reductions set forth in Article VIII and subsequently at the time of any notification of a change pursuant to paragraph 3 of this Article, each State Party shall provide notification of the maximum levels for its holdings of battle tanks, armoured combat vehicles and artillery with respect to each of the areas described in Article IV, paragraphs 2 to 4 and Article V, paragraph 1, subparagraph (A).

6. A decrease in the numbers of conventional armaments and equipment limited by the Treaty held by a State Party and subject to notification pursuant to the Protocol on Information Exchange shall by itself confer no right on any other State Party to increase the maximum levels for its holdings subject to notification pursuant to this Article.

7. It shall be the responsibility solely of each individual State Party to ensure that the maximum levels for its holdings notified pursuant to the provisions of this Article are not exceeded. States Parties belonging to the same group of States Parties shall consult in order to ensure that the maximum levels for holdings notified pursuant to the provisions of this Article, taken together as appropriate, do not exceed the limitations set forth in Articles IV, V and VI.

Article VIII

1. The numerical limitations set forth in Articles IV, V and VI shall be achieved only by means of reduction in accordance with the Protocol on Reduction, the Protocol on Heli-

copter Recategorisation, the Protocol on Aircraft Reclassification, the Footnote to Section I, paragraph 2, subparagraph (A) of the Protocol on Existing Types and the Protocol on Inspection.

2. The categories of conventional armaments and equipment subject to reductions are battle tanks, armoured combat vehicles, artillery, combat aircraft and attack helicopters. The specific types are listed in the Protocol on Existing Types.

(A) Battle tanks and armoured combat vehicles shall be reduced by destruction, conversion for non-military purposes, placement on static display, use as ground targets, or, in the case of armoured personnel carriers, modification in accordance with the Footnote to Section I, paragraph 2, subparagraph (A) of the Protocol on Existing Types.

(B) Artillery shall be reduced by destruction or placement on static display, or, in the case of self-propelled artillery, by use as ground targets.

(C) Combat aircraft shall be reduced by destruction, placement on static display, use for ground instructional purposes, or, in the case of specific models or versions of combat-capable trainer aircraft, reclassification into unarmed trainer aircraft.

(D) Specialised attack helicopters shall be reduced by destruction, placement on static display, or use for ground instructional purposes.

(E) Multi-purpose attack helicopters shall be reduced by destruction, placement on static display, use for ground instructional purposes, or recategorisation.

3. Conventional armaments and equipment limited by the Treaty shall be deemed to be reduced upon execution of the procedures set forth in the Protocols listed in paragraph 1 of this Article and upon notification as required by these Protocols. Armaments and equipment so reduced shall no longer be counted against the numerical limitations set forth in Articles IV, V and VI.

4. Reductions shall be effected in three phases and completed no later than 40 months after entry into force of this Treaty, so that:

(A) by the end of the first reduction phase, that is, no later than 16 months after entry into force of this Treaty, each State Party shall have ensured that at least 25 percent of its total reduction liability in each of the categories of conventional armaments and equipment limited by the Treaty has been reduced;

(B) by the end of the second reduction phase, that is, no later than 28 months after entry into force of this Treaty, each State Party shall have ensured that at least 60 percent of its total reduction liability in each of the categories of conventional armaments and equipment limited by the Treaty has been reduced;

(C) by the end of the third reduction phase, that is, no later than 40 months after entry into force of this Treaty, each State Party shall have reduced its total reduction liability in each of the categories of conventional armaments and equipment limited by the Treaty. States Parties carrying out conversion for non-military purposes shall have ensured that the conversion of all battle tanks in accordance with Section VIII of the Protocol on Reduction shall have been completed by the end of the third reduction phase; and

(D) armoured combat vehicles deemed reduced by reason of having been partially destroyed in accordance with Section VIII, paragraph 6 of the Protocol on Reduction shall have been fully converted for non-military purposes, or destroyed in accordance with Section IV of the Protocol on Reduction, no later than 64 months after entry into force of this Treaty.

5. Conventional armaments and equipment limited by the Treaty to be reduced shall have been declared present within the area of application in the exchange of information at signature of this Treaty.

6. No later than entry into force of this Treaty, each State Party shall provide notification to all other States Parties of its reduction liability.

7. Except as provided for in paragraph 8 of this Article, a State Party's reduction liability in each category shall be no less than the difference between its holdings notified, in accordance with the Protocol on Information Exchange, at signature or effective upon entry into force of this Treaty, whichever is the greater, and the maximum levels for holdings it notified pursuant to Article VII.

8. Any subsequent revision of a State Party's holdings notified pursuant to the Protocol on Information Exchange or of its maximum levels for holdings notified pursuant to Article VII shall be reflected by a notified adjustment to its reduction liability. Any notification of a decrease in a State Party's reduction liability shall be preceded or accompanied by either a notification of a corresponding increase in holdings not exceeding the maximum levels for holdings notified pursuant to Article VII by one or

more States Parties belonging to the same group of States Parties, or a notification of a corresponding increase in the reduction liability of one or more such States Parties.

9. Upon entry into force of this Treaty, each State Party shall notify all other States Parties, in accordance with the Protocol on Information Exchange, of the locations of its reduction sites, including those where the final conversion of battle tanks and armoured combat vehicles for non-military purposes will be carried out.

10. Each State Party shall have the right to designate as many reduction sites as it wishes, to revise without restriction its designation of such sites and to carry out reduction and final conversion simultaneously at a maximum of 20 sites. States Parties shall have the right to share or co-locate reduction sites by mutual agreement.

11. Notwithstanding paragraph 10 of this Article, during the baseline validation period, that is, the interval between entry into force of this Treaty and 120 days after entry into force of this Treaty, reduction shall be carried out simultaneously at no more than two reduction sites for each State Party.

12. Reduction of conventional armaments and equipment limited by the Treaty shall be carried out at reduction sites, unless otherwise specified in Protocols listed in paragraph 1 of this Article, within the area of application.

13. The reduction process, including the results of the conversion of conventional armaments and equipment limited by the Treaty for non-military purposes both during the reduction period and in the 24 months following the reduction period, shall be subject to inspection, without right of refusal, in accordance with the Protocol on Inspection.

Article IX

1. Other than removal from service in accordance with the provisions of Article VIII, battle tanks, armoured combat vehicles, artillery, combat aircraft and attack helicopters within the area of application shall be removed from service only by decommissioning, provided that:

(A) such conventional armaments and equipment limited by the Treaty are decommissioned and awaiting disposal at no more than eight sites which shall be notified as declared sites in accordance with the Protocol on Information Exchange and shall be identified in such notifications as holding areas for decommissioned conventional armaments and equipment limited by the Treaty. If sites containing conventional armaments and equipment limited by the Treaty decommissioned from service also contain any other conventional armaments and equipment subject to the Treaty, the decommissioned conventional armaments and equipment limited by the Treaty shall be separately distinguishable; and

(B) the numbers of such decommissioned conventional armaments and equipment limited by the Treaty do not exceed, in the case of any individual State Party, one percent of its notified holdings of conventional armaments and equipment limited by the Treaty, or a total of 250, whichever is greater, of which no more than 200 shall be battle tanks, armoured combat vehicles and pieces of artillery, and no more than 50 shall be attack helicopters and combat aircraft.

2. Notification of decommissioning shall include the number and type of conventional armaments and equipment limited by the Treaty decommissioned and the location of decommissioning and shall be provided to all other States Parties in accordance with Section IX, paragraph 1, subparagraph (B) of the Protocol on Information Exchange.

Article X

1. Designated permanent storage sites shall be notified in accordance with the Protocol on Information Exchange to all other States Parties by the State Party to which the conventional armaments and equipment limited by the Treaty contained at designated permanent storage sites belong. The notification shall include the designation and location, including geographic coordinates, of designated permanent storage sites and the numbers by type of each category of its conventional armaments and equipment limited by the Treaty at each such storage site.

2. Designated permanent storage sites shall contain only facilities appropriate for the storage and maintenance of armaments and equipment (e.g., warehouses, garages, workshops and associated stores as well as other support accommodation). Designated permanent storage sites shall not contain firing ranges or training areas associated with conventional armaments and equipment limited by the Treaty. Designated permanent storage sites shall contain only armaments and equipment belonging to the conventional armed forces of a State Party.

3. Each designated permanent storage site

shall have a clearly defined physical boundary that shall consist of a continuous perimeter fence at least 1.5 metres in height. The perimeter fence shall have no more than three gates providing the sole means of entrance and exit for armaments and equipment.

4. Conventional armaments and equipment limited by the Treaty located within designated permanent storage sites shall be counted as conventional armaments and equipment limited by the Treaty not in active units, including when they are temporarily removed in accordance with paragraphs 7, 8, 9 and 10 of this Article. Conventional armaments and equipment limited by the Treaty in storage and other than in designated permanent storage sites shall be counted as conventional armaments and equipment limited by the Treaty in active units.

5. Active units or formations shall not be located within designated permanent storage sites, except as provided for in paragraph 6 of this Article.

6. Only personnel associated with the security or operation of designated permanent storage sites, or the maintenance of the armaments and equipment stored therein, shall be located within the designated permanent storage sites.

7. For the purpose of maintenance, repair or modification of conventional armaments and equipment limited by the Treaty located within designated permanent storage sites, each State Party shall have the right, without prior notification, to remove from and retain outside designated permanent storage sites simultaneously up to 10 percent, rounded up to the nearest even whole number, of the notified holdings of each category of conventional armaments and equipment limited by the Treaty in each designated permanent storage site, or 10 items of the conventional armaments and equipment limited by the Treaty in each category in each designated permanent storage site, whichever is less.

8. Except as provided for in paragraph 7 of this Article, no State Party shall remove conventional armaments and equipment limited by the Treaty from designated permanent storage sites unless notification has been provided to all other States Parties at least 42 days in advance of such removal. Notification shall be given by the State Party to which the conventional armaments and equipment limited by the Treaty belong. Such notification shall specify:

(A) the location of the designated permanent storage site from which conventional armaments and equipment limited by the Treaty are to be removed and the numbers by type of conventional armaments and equipment limited by the Treaty of each category to be removed;

(B) the dates of removal and return of conventional armaments and equipment limited by the Treaty; and

(C) the intended location and use of conventional armaments and equipment limited by the Treaty while outside the designated permanent storage site.

9. Except as provided for in paragraph 7 of this Article, the aggregate number of conventional armaments and equipment limited by the Treaty removed from and retained outside designated permanent storage sites by States Parties belonging to the same group of States Parties shall at no time exceed the following levels:

(A) 550 battle tanks;

(B) 1000 armoured combat vehicles; and

(C) 300 pieces of artillery.

10. Conventional armaments and equipment limited by the Treaty removed from designated permanent storage sites pursuant to paragraphs 8 and 9 of this Article shall be returned to designated permanent storage sites no later than 42 days after their removal, except for those items of conventional armaments and equipment limited by the Treaty removed for industrial rebuild. Such items shall be returned to designated permanent storage sites immediately on completion of the rebuild.

11. Each State Party shall have the right to replace conventional armaments and equipment limited by the Treaty located in designated permanent storage sites. Each State Party shall notify all other States Parties, at the beginning of replacement, of the number, location, type and disposition of conventional armaments and equipment limited by the Treaty being replaced.

Article XI

1. Each State Party shall limit its armoured vehicle launched bridges so that, 40 months after entry into force of this Treaty and thereafter, for the group of States Parties to which it belongs the aggregate number of armoured vehicle launched bridges in active units within the area of application does not exceed 740.

2. All armoured vehicle launched bridges within the area of application in excess of the aggregate number specified in paragraph 1 of

this Article for each group of States Parties shall be placed in designated permanent storage sites, as defined in Article II. When armoured vehicle launched bridges are placed in a designated permanent storage site, either on their own or together with conventional armaments and equipment limited by the Treaty, Article X, paragraphs 1 to 6 shall apply to armoured vehicle launched bridges as well as to conventional armaments and equipment limited by the Treaty. Armoured vehicle launched bridges placed in designated permanent storage sites shall not be considered as being in active units.

3. Except as provided for in paragraph 6 of this Article, armoured vehicle launched bridges may be removed, subject to the provisions of paragraphs 4 and 5 of this Article, from designated permanent storage sites only after notification has been provided to all other States Parties at least 42 days prior to such removal. This notification shall specify:

(A) the locations of the designated permanent storage sites from which armoured vehicle launched bridges are to be removed and the numbers of armoured vehicle launched bridges to be removed from each such site;

(B) the dates of removal of armoured vehicle launched bridges from and return to designated permanent storage sites; and

(C) the intended use of armoured vehicle launched bridges during the period of their removal from designated permanent storage sites.

4. Except as provided for in paragraph 6 of this Article, armoured vehicle launched bridges removed from designated permanent storage sites shall be returned to them no later than 42 days after the actual date of removal.

5. The aggregate number of armoured vehicle launched bridges removed from and retained outside of designated permanent storage sites by each group of States Parties shall not exceed 50 at any one time.

6. States Parties shall have the right, for the purpose of maintenance or modification, to remove and have outside the designated permanent storage sites simultaneously up to 10 percent, rounded up to the nearest even whole number, of their notified holdings of armoured vehicle launched bridges in each designated permanent storage site, or 10 armoured vehicle launched bridges from each designated permanent storage site, whichever is less.

7. In the event of natural disasters involving flooding or damage to permanent bridges, States Parties shall have the right to withdraw armoured vehicle launched bridges from designated permanent storage sites. Notification to all other States Parties of such withdrawals shall be given at the time of withdrawal.

Article XII

1. Armoured infantry fighting vehicles held by organisations of a State Party designed and structured to perform in peacetime internal security functions, which are not structured and organised for ground combat against an external enemy, are not limited by this Treaty. The foregoing notwithstanding, in order to enhance the implementation of this Treaty and to provide assurance that the number of such armaments held by such organisations shall not be used to circumvent the provisions of this Treaty, any such armaments in excess of 1000 armoured infantry fighting vehicles assigned by a State Party to organisations designed and structured to perform in peacetime internal security functions shall constitute a portion of the permitted levels specified in Articles IV, V and VI. No more than 600 such armoured infantry fighting vehicles of a State Party, assigned to such organisations, may be located in that part of the area of application described in Article V, paragraph 1, subparagraph (A). Each State Party shall further ensure that such organisations refrain from the acquisition of combat capabilities in excess of those necessary for meeting internal security requirements.

2. A State Party that intends to reassign battle tanks, armoured infantry fighting vehicles, artillery, combat aircraft, attack helicopters and armoured vehicle launched bridges in service with its conventional armed forces to any organisation of that State Party not a part of its conventional armed forces shall notify all other States Parties no later than the date such reassignment takes effect. Such notification shall specify the effective date of the reassignment, the date such equipment is physically transferred, as well as the numbers, by type, of the conventional armaments and equipment limited by the Treaty being reassigned.

Article XIII

1. For the purposes of ensuring verification of compliance with the provisions of this

Treaty, each State Party shall provide notifications and exchange information pertaining to its conventional armaments and equipment in accordance with the Protocol on Information Exchange.

2. Such notifications and exchange of information shall be provided in accordance with Article XVII.

3. Each State Party shall be responsible for its own information; receipt of such information and of notifications shall not imply validation or acceptance of the information provided.

Article XIV

1. For the purpose of ensuring verification of compliance with the provisions of this Treaty, each State Party shall have the right to conduct, and the obligation to accept, within the area of application, inspections in accordance with the provisions of the Protocol on Inspection.

2. The purpose of such inspections shall be:

(A) to verify, on the basis of the information provided pursuant to the Protocol on Information Exchange, the compliance of States Parties with the numerical limitations set forth in Articles IV, V and VI;

(B) to monitor the process of reduction of battle tanks, armoured combat vehicles, artillery, combat aircraft and attack helicopters carried out at reduction sites in accordance with Article VIII and the Protocol on Reduction; and

(C) to monitor the certification of recategorised multi-purpose attack helicopters and reclassified combat-capable trainer aircraft carried out in accordance with the Protocol on Helicopter Recategorisation and the Protocol on Aircraft Reclassification, respectively.

3. No State Party shall exercise the rights set forth in paragraphs 1 and 2 of this Article in respect of States Parties which belong to the group of States Parties to which it belongs in order to elude the objectives of the verification regime.

4. In the case of an inspection conducted jointly by more than one State Party, one of them shall be responsible for the execution of the provisions of this Treaty.

5. The number of inspections pursuant to Sections VII and VIII of the Protocol on Inspection which each State Party shall have the right to conduct and the obligation to accept during each specified time period shall be determined in accordance with the provisions of Section II of that Protocol.

6. Upon completion of the 120-day residual level validation period, each State Party shall have the right to conduct, and each State Party with territory within the area of application shall have the obligation to accept, an agreed number of aerial inspections within the area of application. Such agreed numbers and other applicable provisions shall be developed during negotiations referred to in Article XVIII.

Article XV

1. For the purpose of ensuring verification of compliance with the provisions of this Treaty, a State Party shall have the right to use, in addition to the procedures referred to in Article XIV, national or multinational technical means of verification at its disposal in a manner consistent with generally recognised principles of international law.

2. A State Party shall not interfere with national or multinational technical means of verification of another State Party operating in accordance with paragraph 1 of this Article.

3. A State Party shall not use concealment measures that impede verification of compliance with the provisions of this Treaty by national or multinational technical means of verification of another State Party operating in accordance with paragraph 1 of this Article. This obligation does not apply to cover or concealment practices associated with normal personnel training, maintenance or operations involving conventional armaments and equipment limited by the Treaty.

Article XVI

1. To promote the objectives and implementation of the provisions of this Treaty, the States Parties hereby establish a Joint Consultative Group.

2. Within the framework of the Joint Consultative Group, the States Parties shall:

(A) address questions relating to compliance with or possible circumvention of the provisions of this Treaty;

(B) seek to resolve ambiguities and differences of interpretation that may become apparent in the way this Treaty is implemented;

(C) consider and, if possible, agree on measures to enhance the viability and effectiveness of this Treaty;

(D) update the lists contained in the Protocol on Existing Types, as required by Article II, paragraph 2;

(E) resolve technical questions in order to seek common practices among the States Parties in the way this Treaty is implemented;

(F) work out or revise, as necessary, rules of procedure, working methods, the scale of distribution of expenses of the Joint Consultative Group and of conferences convened under this Treaty and the distribution of costs of inspections between or among States Parties;

(G) consider and work out appropriate measures to ensure that information obtained through exchanges of information among the States Parties or as a result of inspections pursuant to this Treaty is used solely for the purposes of this Treaty, taking into account the particular requirements of each State Party in respect of safeguarding information which that State Party specifies as being sensitive;

(H) consider, upon the request of any State Party, any matter that a State Party wishes to propose for examination by any conference to be convened in accordance with Article XXI; such consideration shall not prejudice the right of any State Party to resort to the procedures set forth in Article XXI; and

(I) consider matters of dispute arising out of the implementation of this Treaty.

3. Each State Party shall have the right to raise before the Joint Consultative Group, and have placed on its agenda, any issue relating to this Treaty.

4. The Joint Consultative Group shall take decisions or make recommendations by consensus. Consensus shall be understood to mean the absence of any objection by any representative of a State Party to the taking of a decision or the making of a recommendation.

5. The Joint Consultative Group may propose amendments to this Treaty for consideration and confirmation in accordance with Article XX. The Joint Consultative Group may also agree on improvements to the viability and effectiveness of this Treaty, consistent with its provisions. Unless such improvements relate only to minor matters of an administrative or technical nature, they shall be subject to consideration and confirmation in accordance with Article XX before they can take effect.

6. Nothing in this Article shall be deemed to prohibit or restrict any State Party from requesting information from or undertaking consultations with other States Parties on matters relating to this Treaty and its implementation in channels or fora other than the Joint Consultative Group.

7. The Joint Consultative Group shall follow the procedures set forth in the Protocol on the Joint Consultative Group.

Article XVII

The States Parties shall transmit information and notifications required by this Treaty in written form. They shall use diplomatic channels or other official channels designated by them, including in particular a communications network to be established by a separate arrangement.

Article XVIII

1. The States Parties, after signature of this Treaty, shall continue the negotiations on conventional armed forces with the same Mandate and with the goal of building on this Treaty.

2. The objective for these negotiations shall be to conclude an agreement on additional measures aimed at further strengthening security and stability in Europe, and pursuant to the Mandate, including measures to limit the personnel strength of their conventional armed forces within the area of application.

3. The States Parties shall seek to conclude these negotiations no later than the follow-up meeting of the Conference on Security and Cooperation in Europe to be held in Helsinki in 1992.

Article XIX

1. This Treaty shall be of unlimited duration. It may be supplemented by a further treaty.

2. Each State Party shall, in exercising its national sovereignty, have the right to withdraw from this Treaty if it decides that extraordinary events related to the subject matter of this Treaty have jeopardised its supreme interests. A State Party intending to withdraw shall give notice of its decision to do so to the Depositary and to all other States Parties. Such notice shall be given at least 150 days prior to the intended withdrawal from this Treaty. It shall include a statement of the extraordinary events the State Party regards as having jeopardised its supreme interests.

3. Each State Party shall, in particular, in exercising its national sovereignty, have the

right to withdraw from this Treaty if another State Party increases its holdings in battle tanks, armoured combat vehicles, artillery, combat aircraft or attack helicopters, as defined in Article II, which are outside the scope of the limitations of this Treaty, in such proportions as to pose an obvious threat to the balance of forces within the area of application.

Article XX

1. Any State Party may propose amendments to this Treaty. The text of a proposed amendment shall be submitted to the Depositary, which shall circulate it to all the States Parties.

2. If an amendment is approved by all States Parties, it shall enter into force in accordance with the procedures set forth in Article XXII governing the entry into force of this Treaty.

Article XXI

1. Forty-six months after entry into force of this Treaty, and at five-year interval thereafter, the Depositary shall convene a conference of the States Parties to conduct a review of the operation of this Treaty.

2. The Depositary shall convene an extraordinary conference of the States Parties, if requested to do so by any State Party which considers that exceptional circumstances relating to this Treaty have arisen, in particular, in the event that a State Party has announced its intention to leave its group of States Parties or to join the other group of States Parties, as defined in Article II, paragraph 1, subparagraph (A). In order to enable the other States Parties to prepare for this conference, the request shall include the reason why that State Party deems an extraordinary conference to be necessary. The conference shall consider the circumstances set forth in the request and their effect on the operation of this Treaty. The conference shall open no longer than 15 days after receipt of the request and, unless it decides otherwise, shall last no longer than three weeks.

3. The Depositary shall convene a conference of the States Parties to consider an amendment proposed pursuant to Article XX, if requested to do so by three or more States Parties. Such a conference shall open no later than 21 days after receipt of the necessary requests.

3. In the event that a State Party gives notice of its decision to withdraw from this Treaty pursuant to Article XIX, the Depositary shall convene a conference of the States Parties which shall open no later than 21 days after receipt of the notice of withdrawal in order to consider questions relating to the withdrawal from this Treaty.

Article XXII

1. This Treaty shall be subject to ratification by each State Party in accordance with its constitutional procedure. Instruments of ratification shall be deposited with the Government of the Kingdom of the Netherlands, hereby designated the Depositary.

2. This Treaty shall enter into force 10 days after instruments of ratification have been deposited by all States Parties listed in the Preamble.

3. The Depositary shall promptly inform all States Parties of:

(A) the deposit of each instrument of ratification;

(B) the entry into force of this Treaty;

(C) any withdrawal in accordance with Article XIX and its effective date;

(D) the text of any amendment proposed in accordance with Article XX;

(E) the entry into force of any amendment to this Treaty;

(F) any request to convene a conference in accordance with Article XXI;

(G) the convening of a conference pursuant to Article XXI; and

(H) any other matter of which the Depositary is required by this Treaty to inform the States Parties.

4. This Treaty shall be registered by the Depositary pursuant to Article 102 of the Charter of the United Nations.

Article XXIII

The original of this Treaty, of which the English, French, German, Italian, Russian and Spanish texts are equally authentic, shall be deposited in the archives of the Depositary. Duly certified copies of this Treaty shall be transmitted by the Depositary to all States Parties.

---

[1] The Treaty of Friendship, Cooperation and Mutual Assistance signed in Warsaw, 14 May 1955.

[2] The Treaty of Economic, Social and Cultural Collaboration and Collective Self-Defence signed in Brussels, 17 March 1948.

[3] The North Atlantic Treaty signed in Washington, 4 April 1949.

Appendix 13B. Vienna Document 1990 of the Negotiations on CSBMs convened in accordance with the relevant provisions of the Concluding Document of the Vienna Meeting of the CSCE

Vienna, 17 November 1990

(1) The representatives of the participating States of the Conference on Security and Co-operation in Europe (CSCE), Austria, Belgium, Bulgaria, Canada, Cyprus, Czech and Slovak Federal Republic, Denmark, Finland, France, Germany, Greece, the Holy See, Hungary, Iceland, Ireland, Italy, Liechtenstein, Luxembourg, Malta, Monaco, the Netherlands, Norway, Poland, Portugal, Romania, San Marino, Spain, Sweden, Switzerland, Turkey, the Union of Soviet Socialist Republics, the United Kingdom, the United States of America and Yugoslavia, met in Vienna from 9 March 1989, in accordance with the provisions relating to the Conference on Confidence- and Security-Building Measures and Disarmament in Europe contained in the Concluding Documents of the Madrid and Vienna Follow-up Meetings of the CSCE.

(2) The participating States recalled that the aim of the Conference on Confidence- and Security-Building Measures and Disarmament in Europe is, as a substantial and integral part of the multilateral process initiated by the Conference on Security and Co-operation in Europe, to undertake, in stages, new, effective and concrete actions designed to make progress in strengthening confidence and security and in achieving disarmament, so as to give effect and expression to the duty of States to refrain from the threat or use of force in their mutual relations as well as in their international relations in general.

(3) Opening statements were made by the Ministers of Foreign Affairs and other Heads of Delegation.

(4) From 16 January to 5 February 1990, the participating States held discussions in a seminar setting on military doctrine in relation to the posture, structure and activities of conventional forces in the zone of application for confidence- and security-building measures.[1] Encouraged by the course of these discussions, the participating States decided to hold a second seminar on military doctrine in the spring of 1991 in Vienna.

(5) The participating States have adopted the present document which integrates a set of new confidence- and security-building measures with measures adopted in the Document of the Stockholm Conference which have been further developed in the light of experience gained.

(6) The participating States recognised that the mutually complementary confidence- and security-building measures which are adopted in the present document and which are in accordance with the mandates of the Madrid and Vienna Follow-up Meetings of the CSCE serve by their scope and nature and by their implementation to strengthen confidence and security in Europe.

(7) The participating States also recognised that the negotiations will continue in accordance with the mandates of the Madrid and Vienna Follow-up Meetings of the CSCE in order to further build upon and expand the results already achieved and that proposals which have been submitted remain subject to further negotiations.

(8) The participating States recalled the declaration on Refraining from the Threat or Use of Force contained in paragraphs (9) to (27) of the Document of the Stockholm Conference and stressed its continuing validity as seen in the light of the Charter of Paris for a New Europe.

(9) The participating States have adopted the following:

I. Annual exchange of military information

Information on military forces

(10) The participating States will exchange annually information on their military forces concerning the military organization, manpower and major weapon

and equipment systems, as specified below, in the zone of application for confidence- and security-building measures (CSBMs).

(11) The information will be provided in an agreed format to all other participating States not later than 15 December of each year. It will be valid as of 1 January of the following year and will include:

(11.1) 1. Information on the command organisation of those military forces referred to under points 2 and 3 specifying the designation and subordination of all formations[2] and units[3] at each level of command down to and including brigade/regiment or equivalent level.

(11.2) 2. For each formation and combat unit[4] of land forces down to and including brigade/regiment or equivalent level the information will indicate:

(11.2.1) the designation and subordination;

(11.2.2) whether it is active or non-active;[5]

(11.2.3) the normal peacetime location of its headquarters indicated by exact geographic terms and/or co-ordinates;

(11.2.4) the peacetime authorised personnel strength;

(11.2.5) the major organic weapon and equipment systems, specifying the numbers of each type of:

(11.2.5.1) battle tanks;

(11.2.5.2) helicopters;

(11.2.5.3) armoured combat vehicles;

(11.2.5.4) anti-tank guided missile launchers permanently/integrally mounted on armoured vehicles;

(11.2.5.5) self-propelled and towed artillery pieces, mortars and multiple rocket launchers (100-mm calibre and above);

(11.2.5.6) armoured vehicle launched bridges.

(11.3) For each amphibious formation and amphibious combat unit[6] permanently located in the zone of application down to and including brigade/regiment or equivalent level, the information will include the items as set out above.

(11.4) For each air formation and air combat unit[7] of the air forces, air defence aviation and of naval aviation permanently based on land down to and including wing/air regiment or equivalent level the information will include:

(11.4.1) the designation and subordination;

(11.4.2) the normal peacetime location of the headquarters indicated by exact geographic terms and/or co-ordinates;

(11.4.3) the normal peacetime location of the unit indicated by the air base or military airfield on which the unit is based, specifying:

(11.4.3.1) the designation or, if applicable, name of the air base or military airfield and

(11.4.3.2) its location indicated by exact geographic terms and/or co-ordinates;

(11.4.4) the peacetime authorised personnel strength;[8]

(11.4.5) the numbers of each type of:

(11.4.5.1) combat aircraft;

(11.4.5.2) helicopters

organic to the formation or unit.

Information on plans for the deployment of major weapon and equipment systems

(12) The participating States will exchange annually information on their plans for the deployment of major weapon and equipment systems as specified in the provisions on Information on Military Forces within the zone of application for CSBMs.

(13) The information will be provided in an agreed format to all other participating States not later than 15 December of each year. It will cover plans for the following year and will include:

(13.1) the type and name of the weapon/equipment systems to be deployed;

(13.2) the total number of each weapon/equipment system;

(13.3) whenever possible, the number of each weapon/equipment system planned to be allocated to each formation or unit;

(13.4) the extent to which the deployment will add to or replace existing weapon/equipment systems.

Information on military budgets

(14) The participating States will exchange annually information on their military budgets for the forthcoming fiscal year, itemising defence expenditures on the basis of the categories set out in the United Nations 'Instrument for Standardised International Reporting on Military Expenditures' adopted on 12 December 1980.

(15) The information will be provided to all other participating States not later than two months after the military budget has been approved by the competent national authorities.

(16) Each participating State may ask for clarification from any other participating State of the budgetary information provided.

Questions should be submitted within a period of two months following the receipt of a participating State's budgetary information. Participating States will make every effort to answer such questions fully and promptly. The questions and replies may be transmitted to all other participating States.

II. Risk reduction

Mechanism for consultation and co-operation as regards unusual military activities

(17) Participating States will, in accordance with the following provisions, consult and co-operate with each other about any unusual and unscheduled activities of their military forces outside their normal peacetime locations which are militarily significant, within the zone of application for CSBMs and about which a participating State expresses its security concern.

(17.1) The participating State which has concerns about such an activity may transmit a request for an explanation to another participating State where the activity is taking place.

(17.1.1) The request will state the cause, or causes, of the concern and, to the extent possible, the type and location, or area, of the activity.

(17.1.2) The reply will be transmitted within not more than 48 hours.

(17.1.3) The reply will give answers to questions raised, as well as any other relevant information which might help to clarify the activity giving rise to concern.

(17.1.4) The request and the reply will be transmitted to all other participating States without delay.

(17.2) The requesting State, after considering the reply provided, may then request a meeting to discuss the matter.

(17.2.1) The requesting State may ask for a meeting with the responding State.

(17.2.1.1) Such a meeting will be convened within not more than 48 hours.

(17.2.1.2) The request for such a meeting will be transmitted to all participating States without delay.

(17.2.1.3) The responding State is entitled to ask other interested participating States, in particular those which might be involved in the activity, to participate in the meeting.

(17.2.1.4) Such a meeting will be held at a venue to be mutually agreed upon by the requesting and the responding States. If there is no agreement, the meeting will be held at the Conflict Prevention Centre.

(17.2.1.5) The requesting and responding States will, jointly or separately, transmit a report of the meeting to all other participating States without delay.

(17.2.2) The requesting State may ask for a meeting of all participating States.

(17.2.2.1) Such a meeting will be convened within not more than 48 hours.

(17.2.2.2) The Conflict Prevention Centre will serve as the forum for such a meeting.

(17.2.2.3) Participating States involved in the matter to be discussed undertake to be represented at such a meeting.

(17.3) The communications between participating States provided for above will be transmitted preferably through the CSBM communications network.

Co-operation as regards hazardous incidents of a military nature

(18) Participating States will co-operate by reporting and clarifying hazardous incidents of a military nature within the zone of application for CSBMs in order to prevent possible misunderstandings and mitigate the effects on another participating State.

(18.1) Each participating State will designate a point to contact in case of such hazardous incidents and will so inform all other participating States. A list of such points will be kept available at the Conflict Prevention Centre.

(18.2) In the event of such a hazardous incident the participating State whose military forces are involved in the incident should provide the information available to other participating States in an expeditious manner. Any participating State affected by such an incident may also request clarification as appropriate. Such requests will receive a prompt response.

(18.3) Communications between participating States will be transmitted preferably through the CSBM communications network.

(18.4) Matters relating to information about such hazardous incidents may be discussed by participating States at the Conflict Prevention Centre, either at the annual implementation assessment meeting at the Centre, or at additional meetings convened there.

(18.5) These provisions will not affect the rights and obligations of participating States under any international agreement concerning hazardous incidents, nor will they preclude additional methods of reporting and clarifying hazardous incidents.

III. Contacts

Visits to air bases

(19) Each participating State with air combat units reported under paragraph 11 will arrange visits for representatives of all other participating States to one of its normal peacetime air bases[9] on which such units are located in order to provide the visitors with the opportunity to view activity at the air base, including preparations to carry out the functions of the air base and to gain an impression of the approximate number of air sorties and type of missions being flown.

(20) No participating State will be obliged to arrange more than one such visit in any five year period.

(21) Prior indications given by participating States of forthcoming schedules for such visits for the subsequent year(s) may be discussed at the annual implementation assessment meetings.

(22) As a rule, up to two visitors from each participating State will be invited.

(23) Invitations will be extended to all participating States 42 days or more in advance of the visit. The invitation will indicate a preliminary programme, including; place, date and time of assembly; planned duration; languages to be used; arrangements for board, lodging and transportation; equipment permitted to be used during the visit; and any other information that may be considered useful.

(24) When the air base to be visited is located on the territory of another participating State, the invitations will be issued by the participating State on whose territory the air base is located. In such cases, the responsibilities as host delegated by this State to the participating State arranging the visit will be specified in the invitation.

(25) Replies to the invitation, indicating the names and ranks of the visitors, will be given not later than 21 days after the issue of the invitation. If the invitation is not accepted in time, it will be assumed that no visitors will be sent.

(26) The visit to the air base will last for a minimum of 24 hours.

(27) In the course of the visit, the visitors will be given a briefing on the purpose and functions of the air base and on current activity at the air base. They will have the opportunity to communicate with commanders and troops, including those of support/logistic units located at the air base.

(28) The visitors will be provided with the opportunity to view all types of aircraft located at the air base.

(29) At the close of the visit, the host State will provide an opportunity for the visitors to meet together and with host State officials and senior air base personnel to discuss the course of the visit.

(30) The host State will determine the programme for the visit and access granted to visitors at the air base.

(31) The visitors will follow the instructions issued by the host State in accordance with the provisions set out in this document.

(32) The visitors will be provided with appropriate accommodation in a location suitable for carrying out the visit.

(33) The invited States will cover the travel expenses of its representatives to and from the place of assembly specified in the invitation.

(34) Participating States should, in due co-operation with the visitors, ensure that no action is taken which could be harmful to the safety of visitors.

Military contacts

(35) To improve further their mutual relations in the interest of strengthening the process of confidence- and security-building, the participating States will, as appropriate, promote and facilitate:

(35.1) exchanges and visits between senior military/defence representatives;

(35.2) contacts between relevant military institutions;

(35.3) attendance by military representatives of other participating States at courses of instruction;

(35.4) exchanges between military commanders and officers of commands down to brigade/regiment or equivalent level;

(35.5) exchanges and contacts between academics and experts in military studies and related areas;

(35.6) sporting and cultural events between members of their armed forces.

IV. Prior notification of certain military activities

(36) The participating States will give notification in writing through diplomatic channels in an agreed form of content, to all other participating States 42 days or more in advance of the start of notifiable[10] military activities in the zone of application for confi-

dence- and security-building measures (CSBMs).

(37) Notification will be given by the participating State on whose territory the activity in question is planned to take place even if the forces of that State are not engaged in the activity or their strength is below the notifiable level. This will not relieve other participating States of their obligation to give notification, if their involvement in the planned military activity reaches the notifiable level.

(38) Each of the following military activities in the field conducted as a single activity in the zone of application for CSBMs at or above the levels defined below, will be notified:

(38.1) The engagement of formations of land forces[11] of the participating States in the same exercise activity conducted under a single operational command independently or in combination with any possible air or naval components.

(38.1.1) This military activity will be subject to notification whenever it involves at any time during the activity:

– at least 13 000 troops, including support troops, or

– at least 300 battle tanks

if organized into a divisional structure or at least two brigades/regiments, not necessarily subordinate to the same division.

(38.1.2) The participation of air forces of the participating States will be included in the notification if it is foreseen that in the course of the activity 200 or more sorties by aircraft, excluding helicopters, will be flown.

(38.2) The engagement of military forces either in an amphibious landing or in a parachute assault by airborne forces in the zone of application of the CSBMs.

(38.2.1) These military activities will be subject to notification whenever the amphibious landing involves at least 3000 troops or whenever the parachute drop involves at least 3000 troops.

(38.3) The engagement of formations of land forces of the participating States in a transfer from outside the zone of application for CSBMs to arrival points in the zone, or from inside the zone of application for CSBMs to points of concentration in the zone, to participate in a notifiable exercise activity or to be concentrated.

(38.3.1) The arrival or concentration of these forces will be subject to notification whenever it involves, at any time during the activity:

– at least 13 000 troops, including support troops, or

– at least 300 battle tanks

if organized into a divisional structure or at least two brigade/regiments, not necessarily subordinate to the same division.

(38.3.2) Forces which have been transferred into the zone will be subject to all provisions of agreed CSBMs when they depart their arrival points to participate in a notifiable exercise activity or to be concentrated within the zone of application for CSBMs.

(39) Notifiable military activities carried out without advance notice to the troops involved, are exceptions to the requirement for prior notification to be made 42 days in advance.

(39.1) Notification of such activities, above the agreed thresholds, will be given at the time the troops involved commence such activities.

(40) Notification will be given in writing of each notifiable military activity in the following agreed form:

(41) A – General information

(41.1) The designation of the military activity;

(41.2) The general purpose of the military activity;

(41.3) The names of the States involved in the military activity;

(41.4) The level of command, organizing and commanding the military activity.

(41.5) The start and end dates of the military activity.

(42) B – Information on different types of notifiable military activities

(42.1) The engagement of formations of land forces of the participating States in the same exercise activity conducted under a single operational command independently or in combination with any possible air or naval components:

(42.1.1) The total number of troops taking part in the military activity (i.e. ground troops, amphibious troops, airmobile and airborne troops) and the number of troops participating for each State involved, if applicable;

(42.1.2) The designation, subordination, number and type of formations and units participating for each State down to and including brigade/regiment or equivalent level;

(42.1.3) The total number of battle tanks for each State and the total number of anti-tank guided missile launchers mounted on armoured vehicles;

(42.1.4) The total number of artillery pieces and multiple rocket launchers (100-mm calibre or above);

(42.1.5) The total number of helicopters, by category;

(42.1.6) Envisaged number of sorties by aircraft, excluding helicopters;

(42.1.7) Purpose of air missions;

(42.1.8) Categories of aircraft involved;

(42.1.9) The level of command, organizing and commanding the air force participation;

(42.1.10) Naval ship-to-shore gunfire;

(42.1.11) Indication of other naval ship-to-shore support;

(42.1.12) The level of command, organizing and commanding the naval force participation.

(42.2) The engagement of military forces either in an amphibious landing or in a parachute assault by airborne forces in the zone of application for CSBMs:

(42.2.1) The total number of amphibious troops involved in notifiable amphibious landings, and/or the total number of airborne troops involved in notifiable parachute assaults;

(42.2.2) In the case of a notifiable amphibious landing, the point or points of embarkation, if in the zone of application for CSBMs.

(42.3) The engagement of formations of land forces of the participating States in a transfer from outside the zone of application for CSBMs to arrival points in the zone, or from inside the zone of application for CSBMs to points of concentration in the zone, to participate in a notifiable exercise activity or to be concentrated:

(42.3.1) The total number of troops transferred;

(42.3.2) Number and type of divisions participating in the transfer;

(42.3.3) The total number of battle tanks participating in a notifiable arrival or concentration;

(42.3.4) Geographical co-ordinates for the points of arrival and for the points of concentration.

(43) C – The envisaged area and timeframe of the activity

(43.1) The area of the military activity delimited by geographic features together with geographic co-ordinates, as appropriate;

(43.2) The start and end dates of each phase (transfers, deployment, concentration of forces, active exercise phase, recovery phase) of activities in the zone of application for CSBMs of participating formations, the tactical purpose and corresponding geographical areas (delimited by geographical co-ordinates) for each phase;

(43.3) Brief description of each phase.

(44) D – Other information

(44.1) Changes, if any, in relation to information provided in the annual calendar regarding the activity;

(44.2) Relationship of the activity to other notifiable activities.

V. Observation of certain military activities

(45) The participating States will invite observers from all other participating States to the following notifiable military activities:

(45.1) – The engagement of formations of land forces[12] of the participating States in the same exercise activity conducted under a single operational command independently or in combination with any possible air or naval components.

(45.2) – The engagement of military forces either in an amphibious landing or in a parachute assault by airborne forces in the zone of application for CSBMs.

(45.3) – In the case of the engagement of formations of land forces of the participating States in a transfer from outside the zone of application for CSBMs to arrival points in the zone, or from inside the zone of application for CSBMs to points of concentration in the zone, to participate in a notifiable exercise activity or to be concentrated, the concentration of these forces. Forces which have been transferred into the zone will be subject to all provisions of agreed confidence- and security-building measures when they depart their arrival points to participate in a notifiable exercise activity or to be concentrated within the zone of application for CSBMs.

(45.4) The above-mentioned activities will be subject to observation whenever the number of troops engaged meets or exceeds 17 000 troops, except in the case of either an amphibious landing or a parachute assault by airborne forces, which will be subject to observation whenever the number of troops engaged meets or exceeds 5000 troops.

(46) The host State will extend the invitations in writing through diplomatic channels to all other participating States at the time of

notification. The host State will be the participating State on whose territory the notified activity will take place.

(47) The host State may delegate some of its responsibilities as host to another participating State engaged in the military activity on the territory of the host State. In such cases, the host State will specify the allocation of responsibilities in its invitation to observe the activity.

(48) Each participating State may send up to two observers to the military activity to be observed.

(49) The invited State may decide whether to send military and/or civilian observers, including members of its personnel accredited to the host State. Military observers will, normally, wear their uniforms and insignia while performing their tasks.

(50) Replies to the invitation will be given in writing not later than 21 days after the issue of the invitation.

(51) The participating States accepting an invitation will provide the names and ranks of their observers in their reply to the invitation. If the invitation is not accepted in time, it will be assumed that no observers will be sent.

(52) Together with the invitation the host State will provide a general observation programme, including the following information:

(52.1) – the date, time and place of assembly of observers;

(52.2) – planned duration of the observation programme;

(52.3) – languages to be used in interpretation and/or translation;

(52.4) – arrangements for board, lodging and transportation of the observers;

(52.5) –arrangements for observation equipment which will be issued to the observers by the host State;

(52.6) – possible authorization by the host State of the use of special equipment that the observers may bring with them;

(52.7) – arrangements for special clothing to be issued to the observers because of weather or environmental factors.

(53) – The observers may make requests with regard to the observation programme. The host State will, if possible, accede to them.

(54) The host State will determine a duration of observation which permits the observers to observe a notifiable military activity from the time that agreed thresholds for observation are met or exceeded until, for the last time during the activity, the thresholds for observation are no longer met.

(55) The host State will provide the observers with transportation to the area of the notified activity and back. This transportation will be provided from either the capital or another suitable location to be announced in the invitation, so that the observers are in position before the start of the observation programme.

(56) The invited State will cover the travel expenses for its observers to the capital, or another suitable location specified in the invitation, of the host State, and back.

(57) The observers will be provided equal treatment and offered equal opportunities to carry out their functions.

(58) The observers will be granted, during their mission, the privileges and immunities accorded to diplomatic agents in the Vienna Convention on Diplomatic Relations.

(59) The participating States will ensure that official personnel and troops taking part in an observed military activity, as well as other armed personnel located in the area of the military activity, are adequately informed regarding the presence, status and functions of observers. Participating States should, in due co-operation with the observers, ensure that no action is taken which could be harmful to the safety of observers.

(60) The host State will not be required to permit observation of restricted locations, installations or defence sites.

(61) In order to allow the observers to confirm that the notified activity is non-threatening in character and that it is carried out in conformity with the appropriate provisions of the notification, the host State will:

(61.1) – at the commencement of the observation programme give a briefing on the purpose, the basic situation, the phases of the activity and possible changes as compared with the notification and provide the observers with an observation programme with a daily schedule;

(61.2) – provide the observers with a map with a scale of 1 to not more than 250 000 depicting the area of the notified military activity and the initial tactical situation in this area. To depict the entire area of the notified military activity, smaller-scale maps may be additionally provided;

(61.3) – provide the observers with appropriate observation equipment; in addition, the observers will be permitted to use their own binoculars, maps, photo and video cameras, dictaphones and hand-held passive night-vision devices. The above-mentioned equipment will be subject to examination and approval by the host State. It is understood that the host State may limit the use of certain equipment in restricted locations, installations or defence sites;

(61.4) – be encouraged, whenever feasible and with due consideration for the security of the observers, to provide an aerial survey, preferably by helicopter, of the area of the military activity. If carried out, such a survey should provide the observers with the opportunity to observe from the air the disposition of forces engaged in the activity in order to help them gain a general impression of its scope and scale. At least one observer from each participating State represented at the observation should be given the opportunity to participate in the survey. Helicopters and/or aircraft may be provided by the host State or by another participating State at the request of and in agreement with the host State;

(61.5) – in the course of the observation programme give the observers daily briefings with the help of maps on the various phases of the military activity and their development and inform the observers about their positions geographically; in the case of a land force activity conducted in combination with air or naval components, briefings will be given by representatives of these forces;

(61.6) – provide opportunities to observe directly forces of the State(s) engaged in the military activity so that the observers get an impression of the flow of the entire activity; to this end, the observers will be given the opportunity to observe combat and support units of all participating formations of a divisional or equivalent level and, whenever possible, to visit units below divisional or equivalent level and communicate with commanders and troops. Commanders and other senior personnel of the participating formations as well as of the visited units will inform the observers of the mission and disposition of their respective units;

(61.7) – guide the observers in the area of the military activity; the observers will follow the instructions issued by the host State in accordance with the provisions set out in this document;

(61.8) – provide the observers with appropriate means of transportation in the area of the military activity;

(61.9) – provide the observers with opportunities for timely communication with their embassies or other official missions and consular posts; the host State is not obligated to cover the communication expense of the observers;

(61.10) – provide the observers with appropriate board and lodging in a location suitable for carrying out the observation programme and, when necessary, medical care;

(61.11) – at the close of each observation, provide an opportunity for the observers to meet together and with host State officials to discuss the course of the observed activity. Where States other than the host State have been engaged in the activity, military representatives of those States will also be invited to take part in this discussion.

(62) The participating States need not invite observers to notifiable military activities which are carried out without advance notice to the troops involved unless these notifiable activities have a duration of more than 72 hours. The continuation of these activities beyond this time will be subject to observation while the agreed thresholds for observation are met or exceeded. The observation programme will follow as closely as practically possible all the provisions for observation set out in this document.

(63) The participating States are encouraged to permit media representatives from all participating States to attend observed military activities in accordance with accreditation procedures set down by the host State. In such instances, media representatives from all participating States will be treated without discrimination and given equal access to those facets of the activity open to media representatives.

(64) The presence of media representatives will not interfere with the observers carrying out their functions nor with the flow of the military activity.

VI. Annual calendars

(65) Each participating State will exchange, with all other participating States, an annual calendar of its military activities subject to prior notification,[13] within the zone of application for CSBMs, forecast for the subsequent calendar year. A participating State which is to host military activities subject to prior notification conducted by any other participating State(s) will include these activities in its annual calendar. It will be transmitted every year, in writing, through diplomatic channels, not later than 15 November for the following year.

(66) If a participating State does not fore-

cast any military activity subject to prior notification it will so inform all other participating States in the same manner as prescribed for the exchange of annual calendars.

(67) Each participating State will list the above-mentioned activities chronologically and will provide information on each activity in accordance with the following model:

(67.1) – type of military activity and its designation;

(67.2) – general characteristics and purpose of the military activity;

(67.3) – States involved in the military activity;

(67.4) – area of the military activity, indicated by geographic features when appropriate and defined by geographic co-ordinates;

(67.5) – planned duration of the military activity, indicated by envisaged start and end dates;

(67.6) – the envisaged total number of troops[14] engaged in the military activity. For activities involving more than one State, the host State will provide such information for each State involved;

(67.7) – the types of armed forces involved in the military activity;

(67.8) – the envisaged level of the military activity and designation of direct operational command, under which this military activity will take place;

(67.9) – the number and type of divisions whose participation in the military activity is envisaged;

(67.10) – any additional information concerning, *inter alia*, components of armed forces, which the participating State planning the military activity considers relevant.

(68) Should changes regarding the military activities in the annual calendar prove necessary, they will be communicated to all other participating States no later than in the appropriate notification.

(69) Should a participating State cancel a military activity included in its annual calendar or reduce it to a level below notification thresholds, that State will inform the other participating States immediately.

(70) Information on military activities subject to prior notification not included in an annual calendar will be communicated to all participating States as soon as possible, in accordance with the model provided in the annual calendar.

VII. Constraining provisions

(71) Each participating State will communicate, in writing, to all other participating States, by 15 November each year, information concerning military activities subject to prior notification involving more than 40 000 troops,[15] which it plans to carry out or host in the second subsequent calendar year. Such communication will include preliminary information on each activity, as to its general purpose, timeframe and duration, area, size and States involved.

(72) Participating States will not carry out military activities subject to prior notification involving more than 40 000 troops, unless they have been the object of communication as defined above.

(73) Participating States will not carry out military activities subject to prior notification involving more than 40 000 troops unless they have been included in the annual calendar, not later than 15 November each year.

(74) If military activities subject to prior notification are carried out in addition to those contained in the annual calendar, they should be as few as possible.

VIII. Compliance and verification

(75) According to the Madrid Mandate, the confidence- and security-building measures to be agreed upon 'will be provided with adequate forms of verification which correspond to their content'.

(76) The participating States recognize that national technical means can play a role in monitoring compliance with agreed confidence- and security-building measures.

Inspection

(77) In accordance with the provisions contained in this document each participating State has the right to conduct inspections on the territory of any other participating State within the zone of application for CSBMs.

(78) Any participating State will be allowed to address a request for inspection to another participating State on whose territory, within the zone of application for CSBMs, compliance with the agreed confidence- and security-building measures is in doubt.

(79) No participating State will be obliged to accept on its territory within the zone of application for CSBMs, more than three inspections per calendar year.

(80) No participating State will be obliged to accept more than one inspection per calendar year from the same participating State.

(81) An inspection will not be counted if,

due to *force majeure*, it cannot be carried out.

(82) The participating State which requests an inspection will state the reasons for such a request.

(83) The participating State which has received such a request will reply in the affirmative to the request within the agreed period of time, subject to the provisions contained in paragraphs (79) and (80).

(84) Any possible dispute as to the validity of the reasons for a request will not prevent or delay the conduct of an inspection.

(85) The participating State which requests an inspection will be permitted to designate for inspection on the territory of another State within the zone of application for CSBMs, a specific area. Such an area will be referred to as the 'specified area'. The specified area will comprise terrain where notifiable military activities are conducted or where another participating State believes a notifiable military activity is taking place. The specified area will be defined and limited by the scope and scale of notifiable military activities but will not exceed that required for an army level military activity.

(86) In the specified area the representatives of the inspecting State accompanied by the representatives of the receiving State will be permitted access, entry and unobstructed survey, except for areas or sensitive points to which access is normally denied or restricted, military and other defence installations, as well as naval vessels, military vehicles and aircraft. The number and extent of the restricted areas should be as limited as possible. Areas where notifiable military activities can take place will not be declared restricted areas, except for certain permanent or temporary military installations which, in territorial terms, should be as small as possible, and consequently those areas will not be used to prevent inspection of notifiable military activities. Restricted areas will not be employed in a way inconsistent with the agreed provisions on inspection.

(87) Within the specified area, the forces of participating States other than the receiving State will also be subject to the inspection conducted by the inspecting State.

(88) Inspection will be permitted on the ground, from the air, or both.

(89) The representatives of the receiving State will accompany the inspection team, including when it is in land vehicles and an aircraft from the time of their first employment until the time they are no longer in use for the purposes of inspection.

(90) In its request, the inspecting State will notify the receiving State of:

(90.1) – the reasons for the request;

(90.2) – the location of the specified area defined by geographical co-ordinates;

(90.3) – the preferred point(s) of entry for the inspection team;

(90.4) – mode of transport to and from the point(s) of entry and, if applicable, to and from the specified area;

(90.5) – where in the specified area the inspection will begin;

(90.6) – whether the inspection will be conducted from the ground, from the air, or both simultaneously;

(90.7) – whether aerial inspection will be conducted using an airplane, a helicopter, or both;

(90.8) – whether the inspection team will use land vehicles provided by the receiving State or, if mutually agreed, its own vehicles;

(90.9) – information for the issuance of diplomatic visas to inspectors entering the receiving State.

(91) The reply to the request will be given in the shortest possible period of time, but within not more than twenty-four hours. Within thirty-six hours after the issuance of the request, the inspection team will be permitted to enter the territory of the receiving State.

(92) Any request for inspection as well as the reply thereto will be communicated to all participating States without delay.

(93) The receiving State should designate the point(s) of entry as close as possible to the specified area. The receiving State will ensure that the inspection team will be able to reach the specified area without delay from the point(s) of entry.

(94) All participating States will facilitate the passage of the inspection teams through their territory.

(95) Within 48 hours after the arrival of the inspection team at the specified area, the inspection will be terminated.

(96) There will be no more than four inspectors in an inspection team. While conducting the inspection the inspection team may divide into two parts.

(97) The inspectors and, if applicable, auxiliary personnel, will be granted during their mission the privileges and immunities in accordance with the Vienna Convention on Diplomatic Relations.

(98) The participating States will ensure that troops, other armed personnel and officials in the specified area are adequately

informed regarding the presence, status and functions of inspectors and, if applicable, auxiliary personnel. The receiving State will ensure that no action is taken by its representatives which could endanger inspectors and, if applicable, auxiliary personnel. In carrying out their duties, inspectors and, if applicable, auxiliary personnel will take into account safety concerns expressed by representatives of the receiving State.

(99) The receiving State will provide the inspection team with appropriate board and lodging in a location suitable for carrying out the inspection, and, when necessary, medical care; however this does not exclude the use by the inspection team of its own tents and rations.

(100) The inspection team will have use of its own maps and charts, photo and video cameras, binoculars, hand-held passive night vision devices and dictaphones. Upon arrival in the specified area the inspection team will show the equipment to the representatives of the receiving State.

(101) The inspection team will have access to appropriate telecommunications equipment of the receiving State for the purpose of communicating with its embassy or other official missions and consular posts accredited to the receiving State.

(102) The receiving State will provide the inspection team with access to appropriate telecommunications equipment for the purpose of continuous communication between the sub-teams.

(103) Inspectors will be entitled to request and to receive briefings at agreed times by military representatives of the receiving State. At the inspectors' request, such briefings will be given by commanders of formations or units in the specified area. Suggestions of the receiving State as to the briefings will be taken into consideration.

(104) The inspecting State will specify whether aerial inspection will be conducted using an airplane, a helicopter or both. Aircraft for inspection will be chosen by mutual agreement between the inspecting and receiving States. Aircraft will be chosen which provide the inspection team a continuous view of the ground during the inspection.

(105) After the flight plan, specifying, *inter alia*, the inspection team's choice of flight path, speed and altitude in the specified area, has been filled with the competent air traffic control authority the inspection aircraft will be permitted to enter the specified area without delay. Within the specified area, the inspection team will, at its request, be permitted to deviate from the approved flight plan to make specific observations provided such deviation is consistent with paragraph (86) as well as flight safety and air traffic requirements. Directions to the crew will be given through a representative of the receiving State on board the aircraft involved in the inspection.

(106) One member of the inspection team will be permitted, if such a request is made, at any time to observe data on navigational equipment of the aircraft and to have access to maps and charts used by the flight crew for the purpose of determining the exact location of the aircraft during the inspection flight.

(107) Aerial and ground inspectors may return to the specified area as often as desired within the 48-hour inspection period.

(108) The receiving State will provide for inspection purposes land vehicles with cross country capability. Whenever mutually agreed taking into account the specified geography relating to the area to be inspected, the inspecting State will be permitted to use its own vehicles.

(109) If land vehicles or aircraft are provided by the inspecting State, there will be one accompanying driver for each land vehicle, or accompanying aircraft crew.

(110) The inspecting State will prepare a report of its inspection and will provide a copy of that report to all participating States without delay.

(111) The inspection expenses will be incurred by the receiving State except when the inspecting State uses its own aircraft and/or land vehicles. The travel expenses to and from the point(s) of entry will be borne by the inspecting State.

Evaluation

(112) Information provided under the provisions on Information on Military Forces and on Information on Plans for the Deployment of Major Weapon and Equipment Systems will be subject to evaluation.

(113) Subject to the provisions below each participating State will provide the opportunity to visit active formations and units in their normal peacetime locations as specified in point 2 and 3 of the provisions on Information on Military Forces to allow the other participating States to evaluate the information provided.

(114) Each participating State will be obliged to accept a quota of one evaluation visit per calendar year for every sixty units,

or portion thereof, reported under paragraph (11). However, no participating State will be obliged to accept more than fifteen visits per calendar year. No participating State will be obliged to accept more than one fifth of its quota of visits from the same participating State; a participating State with a quota of less than five visits will not be obliged to accept more than one visit from the same participating State during a calendar year. No formation or unit may be visited more than twice during a calendar year and more than once by the same participating State during a calendar year.

(115) No participating State will be obliged to accept more than one visit at any given time of its territory.

(116) If a participating State has formations or units stationed on the territory of other participating States (host States) in the zone of application for CSBMs, the maximum number of evaluation visits permitted to its forces in each of the States concerned will be proportional to the number of its units in each State. The application of this provision will not alter the number of visits this participating State (stationing State) will have to accept under paragraph (114).

(117) Requests for such visits will be submitted giving 5 days notice.

(118) The request will specify:

(118.1) – the formation or unit to be visited;

(118.2) – the proposed date of the visit;

(118.3) – the preferred point(s) of entry as well as the date and estimated time of arrival for the evaluation team;

(118.4) – the mode of transport to and from the point(s) of entry and, if applicable, to and from the formation or unit to be visited;

(118.5) – the names and ranks of the members of the team and, if applicable, information for the issue of diplomatic visas.

(119) If a formation or unit of a participating State is stationed on the territory of another participating State, the request will be addressed to the host State and sent simultaneously to the stationing State.

(120) The reply to the request will be given within 48 hours after the receipt of the request.

(121) In the case of formations or units of a participating State stationed on the territory of another participating State, the reply will be given by the host State in consultation with the stationing State. After consultation between the host State and the stationing State, the host State will specify in its reply any of its responsibilities which it agrees to delegate to the stationing State.

(122) The reply will indicate whether the formation or unit will be available for evaluation at the proposed date at its normal peacetime location.

(123) Formations or units may be in their normal peacetime location but be unavailable for evaluation. Each participating State will be entitled in such cases not to accept a visit; the reasons for the non-acceptance and the number of days that the formation or unit will be unavailable for evaluation will be stated in the reply. Each participating State will be entitled to invoke this provision up to a total of five times for an aggregate of no more than 30 days per calendar year.

(124) If the formation or unit is absent from its normal peacetime location, the reply will indicate the reasons for and the duration of its absence. The requested State may offer the possibility of a visit to the formation or unit outside its normal peacetime location. If the requested State does not offer this possibility, the requesting State will be able to visit the normal peacetime location of the formation or unit. The requesting State may however refrain in either case from the visit.

(125) Visits will not be counted against the quotas of receiving States, if they are not carried out. Likewise, if visits are not carried out, due to force majeure, they will not be counted.

(126) The reply will designate the point(s) of entry and indicate, if applicable, the time and place of assembly of the team. The point(s) of entry and, if applicable, the place of assembly will be designated as close as possible to the formation or unit to be visited. The receiving State will ensure that the team will be able to reach the formation or unit without delay.

(127) The request and the reply will be communicated to all participating Sates without delay.

(128) Participating States will facilitate the passage of teams through their territory.

(129) The team will have no more than two members. It may be accompanied by an interpreter as auxiliary personnel.

(130) The members of the team and, if applicable, auxiliary personnel, will be granted during their mission the privileges and immunities in accordance with the Vienna Convention on Diplomatic Relations.

(131) The visit will take place in the course of a single working day and last up to

12 hours.

(132) The visit will begin with a briefing by the officer commanding the formation or unit, or his deputy, in the headquarters of the formation or unit, concerning the personnel as well as the major weapon and equipment systems reported under paragraph (11).

(132.1) In the case of a visit to a formation, the receiving State may provide the possibility to see personnel and major weapon and equipment systems reported under paragraph (11) for that formation, but not for any of its formations or units, in their normal locations.

(132.2) In the case of a visit to a unit, the receiving State will provide the possibility to see the personnel and the major weapon and equipment systems of the unit reported under paragraph (11) in their normal locations.

(133) Access will not have to be granted to sensitive points, facilities and equipment.

(134) The team will be accompanied at all times by representatives of the receiving State.

(135) The receiving State will provide the team with appropriate transportation during the visit to the formation or unit.

(136) Personal binoculars and dictaphones may be used by the team.

(137) The visit will not interfere with activities of the formation or unit.

(138) The participating States will ensure that troops, other armed personnel and officials in the formation or unit are adequately informed regarding the presence, status and functions of members of teams and, if applicable, auxiliary personnel. Participating States will also ensure that no action is taken by their representatives which could endanger the members of teams and, if applicable, auxiliary personnel. In carrying out their duties, members of teams and, if applicable, auxiliary personnel will take into account safety concerns expressed by representatives of the receiving State.

(139) The travel expenses to and from the point(s) of entry will be borne by the visiting State.

(140) The visiting State will prepare a report of its visit which will be communicated to all participating States expeditiously.

(141) Each participating State will be entitled to obtain timely clarification from any other participating State concerning the application of agreed confidence- and security-building measures. Communications in this context will, if appropriate, be transmitted to all other participating States.

(142) The communications concerning compliance and verification will be transmitted preferably through the CSBM communications network.

IX. Communications

(143) The participating States will establish a network of direct communications between their capitals for the transmission of messages relating to agreed measures. The network will complement the existing use of diplomatic channels. Participating States undertake to use the network flexibly, efficiently and in a cost-effective way.

(144) Each participating State will designate a point of contact capable of transmitting and receiving such messages from other participating States on a 24-hour-a-day basis. Each participating State will notify this designation in writing to other participating States not later than 15 April 1991 and will notify in advance any change in this designation.

(145) The technical characteristics of the network are set out in Annex II.

(146) Communications may be in any one of the six working languages of the CSCE.

(147) Details on the use of these six languages are set out in Annex III. The provisions of this annex have been elaborated for the practical purposes of the communication system only. They are not intended to change the existing use of all six working languages of the CSCE according to established rules and practice as set out in the Final Recommendations of the Helsinki Consultations.

(148) Messages will be considered official communications of the sending State. If the content of a message is not related to an agreed measure, the receiving State has the right to reject it by so informing the other participating States.

(149) Participating States may agree among themselves to use the network for other purposes.

(150) All aspects of the implementation of the network may be discussed at the annual implementation assessment meeting.

X. Annual implementation assessment meeting

(151) The participating States will hold each year a meeting to discuss the present and future implementation of agreed CSBMs. Discussion may extend to:

(151.1) – clarification of questions arising

from such implementation;

(151.2) – operation of agreed measures;

(151.3) – implications of all information originating from the implementation of any agreed measures for the process of confidence- and security-building in the framework of the CSCE.

(152) Before the conclusion of each year's meeting the participating States will normally agree upon the agenda and dates for the subsequent year's meeting. Lack of agreement will not constitute sufficient reason to extend a meeting, unless otherwise agreed. Agenda and dates may, if necessary, be agreed between meetings.

(153) The Conflict Prevention Centre will serve as the forum for such meetings.

(154) The first annual implementation assessment meeting will be held in 1991.

* * *

(155) The participating States stress that this new set of mutually complementary confidence- and security-building measures builds upon and expands the results already achieved at the Stockholm Conference and is designed to reduce the risk of military confrontation in Europe and emphasize that its implementation will contribute to these objectives.

(156) Reaffirming the relevant objectives of the Final Act, the participating States are determined to continue building confidence, to lessen military confrontation and to enhance security for all.

(157) The measures adopted in this document are politically binding and will come into force on 1 January 1991.

(158) The Government of Austria is requested to transmit the present document to the Meeting of the Heads of State or Government of the CSCE participating States in Paris and to the Helsinki Follow-up Meeting of the CSCE. The Government of Austria is also requested to transmit the present document to the Secretary-General of the United Nations and to the Governments of the non-participating Mediterranean States.

(159) The text of this document will be published in each participating State, which will disseminate it and make it known as widely as possible.

(160) The representatives of the participating States express their profound gratitude to the Government and people of Austria for the excellent arrangements they are making for the Vienna CSBM Negotiations and the warm hospitality they are extending to the delegations which are participating in the Negotiations.

---

[1] Annex I.

[2] In this context, formations are armies, corps and divisions and their equivalents.

[3] In this context, units are brigades, regiments and their equivalents.

[4] In this context, combat units are infantry, armoured, mechanised, motorised rifle, artillery, combat engineer and army aviation units. Those combat units which are airmobile or airborne will also be included.

[5] In this context, non-active formations or combat units are those manned from zero to fifteen per cent of their authorised combat strength. This term includes low strength formations and units.

[6] Combat unit as defined above.

[7] In this context air combat units are units, the majority of whose organic aircraft are combat aircraft.

[8] As an exception, this information need not be provided on air defence aviation units.

[9] In this context, the term normal peacetime air base is understood to mean the normal peacetime location of the air combat unit indicated by the air base or military airfield on which the unit is based.

[10] In this document, the term notifiable means subject to notification.

[11] In this context, the term land forces includes amphibious, airmobile and airborne forces.

[12] In this context, the term 'land forces' includes amphibious, airmobile and airborne forces.

[13] As defined in the provisions on Prior Notification of Certain Military Activities.

[14] As defined in the provisions on Prior Notification of Certain Military Activities.

[15] As defined in the provisions on Prior Notification of Certain Military Activities.

Appendix 13C. Implementation of the Stockholm Document and calendar of planned notifiable military activities in 1991

AXEL KROHN

This appendix details the notifiable military activities for 1991 planned by the 34 CSCE countries[1] and reviews developments in this field since the implementation of the confidence- and security-building measures (CSBMs) established by the 1986 Stockholm Document.[2] The Stockholm Document entered into force on 1 January 1987 and requires the participating states to exchange annual calendars of military activities, to notify exercises above a certain threshold, to invite all states participating in the Conference on Security and Co-operation in Europe (CSCE) to observe exercises above a certain threshold, and to allow on-site challenge inspections of military exercises as a means of verification.

Participating states must prepare and exchange calendars of notifiable military activities planned for the following year (paragraph 55 of the Stockholm Document) by 15 November each year. Each state must also provide information on military activities involving more than 40 000 troops planned for the second subsequent year (paragraph 59). The annual calendar for 1991 is presented in table 13C.1, which lists information specified in accordance with the requirements of the Stockholm Document (paragraph 56).[3]

The table is a compilation of information from the CSCE states' calendars, and gives an overall picture of all their planned notifiable military activities.[4] The calendars of notifiable military activities are still the best single source of information on the subject. Participating states are required to report such activities occurring on their territory or in which their participation reaches the notifiable level (paragraph 31), to give the planned duration and to state the 14-day period ('start window') within which the military activity is planned to start.[5] Each military activity is listed as one event, regardless of the number of states notifying or participating, or the number of exercises occurring simultaneously. The activities appear in the table in chronological order.

[1] With German unification on 3 Oct. 1990 the CSCE now consists of 34 member states.

[2] The Stockholm Document expanded on the confidence-building measures (CBMs) of the 1975 Helsinki Document and made them mandatory. It is reprinted in SIPRI, *SIPRI Yearbook 1987: World Armaments and Disarmament* (Oxford University Press: Oxford, 1987), appendix 10A, pp. 353–69. The Document covers all of Europe, has lower thresholds for notifications of manœuvres and longer periods of prior notification than the Helsinki Final Act and includes verification provisions. See for a comparison Ghebali, V.-Y., 'Confidence-building measures within the CSCE process: paragraph-by-paragraph analysis of the Helsinki and Stockholm régimes', United Nations Institute for Disarmament Research (UNIDIR), Research Paper No. 3, Geneva, Mar. 1989.

[3] This year there is no advance forecast for 1992, as no military activities above these limits are planned for 1992.

[4] The data in this appendix are based on official information provided to SIPRI by the governments of the CSCE countries.

[5] If states provide the actual exercise dates, these are given in table 13C.1.

Table 13C.1. Calendar of planned notifiable military activities in 1991, as required by the Stockholm Document

States/ Location	Dates/ Start window	Type/Name of activity[a]	Area	Level of command	No. of troops	Type of forces or equipment	No. and type of divisions[a]	Comments
1. UK and Netherlands in Norway	1–2 Mar.	Amphibious exercise 'Adger 91'	Fogsteinene–Lista Lighthouse–Rauna–Bispen–Nagelstad–Grøtteland–Fogsteinene	Norwegian regional command in conjunction with Commander, 3rd Commando Brig., Royal Marines	4 000 (800 Neth.) (3 200 UK)	Amphibious forces	1 brig.	Landing craft and support helicopters; in conjunction with FTX 'Adger 91' (below notifiable level)
2. Sweden in Sweden	11 days, 10–20 Mar.	FTX 'Nordanvind, FMÖ 1991'	Jaekkvik–Oeverkalix–Moudoslompolo–Kebnekajse	General Officer Commanding, Military region Upper Norrland	24 000	Ground and air forces	1 div. (–) + local defence forces	Observers to be invited
3. Italy, USA, UK, Netherlands, Spain and Greece in Italy	13–22 May	Amphibious exercise 'Dragon Hammer 91'	South-west Sardinia	Regional command	3 000	Amphibious forces	. .	. .

4. USA, UK, Belgium, Netherlands, France and FRG in FRG	25 days, 26 Aug.–8 Sep.	FTX/CFX 'Certain Shield 91'	Bad Salzuflen–Burgsteinfurt–Gronau–Bocholt–Wesel–Haltern–Lüdinghausen–Arnsberg–Brilon–Kassel–Münden–Göttingen–Northeim–Einbeck–Eschershausen–Holzminden–Bad Pyrmont	Army group	42 850 (18 000 USA) (12 500 UK) (2 500 Belg.) (7 400 Neth.) (100 France) (2 350 FRG)	Ground and air forces	2 tank divs. 2 tank brig. 1 mech. inf. div. 1 light inf. brig. 1 amb. div. 1 hc. brig. 1 arm. rec. rgt.	Approx. 15 500 US soldiers in conjunction with 'Reforger-91' exercise; includes advance parties which arrive before the exercise starts and stay on after the end of the exercise to support the retransfer of the troops; observers to be invited
5. Denmark, FRG, Netherlands, UK, USA, Belgium, Luxembourg, Italy and Canada in Denmark	1–27 Sep.	FTX 'Action Express 91'	Zealand Group of Islands	Corps	20 000 (9 000 Den.) numbers from FRG, Neth., UK, USA, Belg., Lux., Italy and Canada to be determined	Ground, air and amphibious forces	3 inf. brig. 1 cbt. group, corps units, homeguard, 1 amph. brig. + sub-units	Exercise includes an amphibious landing with approx. 4 000 soldiers; observers to be invited
6. USSR in USSR	6–7 days, 1–14 Sep.	FTX	Leningrad MD–Liinahamari–Zapolyarni–landmark 359–Vuoriyarvi–Kandalksha–landmark 448	Military District	17 000	Ground and air forces	3 mot. inf. div (–)	Observers to be invited
7. France in France	13–20 Sep.	FTX 'Marne'	Sezanne–Bar-Le-Duc–Neufchateau–Troyes	Corps	13 000	Ground forces	1 tank div. 1 log. brig. corps units	. .
8. USSR in USSR	6–7 days, 14–28 Sep.	FTX	Odessa MD–Lesnoye–Berezino–Plakhteevka–Antonovka	Military District	c. 9 000	Ground and air forces	3 mot. inf. div. (–) (300 tanks)	. .

States/ Location	Dates/ Start window	Type/Name of activity[a]	Area	Level of command	No. of troops	Type of forces or equipment	No. and type of divisions[a]	Comments
9. USSR in USSR	6–7 days, 1–14 Oct.	FTX	White Russia MD–Shatsk–Dubrovo–Ushachi–Berezino–Ulla	Military District	c. 10 000	Ground and air forces	2 tank div. (–) (300 tanks)	. .
10. USSR in USSR	6–7 days, 14–28 Oct.	FTX	Carpathian MD–Rava-Russkaya–Yavorov–Gorodok–Nesterov	Military District	17 000	Ground and air forces	4 mot. inf. div. (–)	Observers to be invited

Note: No military activities are notified for the advance forecast for 1992.

[a] See the list of abbreviations below. (–) means that the division is below full strength or not comprised of all its component units.

Abbreviations used in the table:

amb.	air-mobile	div.	division	mech.	mechanized
arm.	armoured	FTX	field training exercise	mot.	motorized
brig.	brigade	hc.	helicopter	rec.	reconnaissance
cbt.	combat	inf.	infantry	rgt.	regiment
CFX	command field exercise	log.	logistic		

States participating in notified military activities in 1991, by activity number:

Belgium: 4, 5	Greece: 3	Sweden: 2
Canada: 5	Italy: 3, 5	UK: 1, 3, 4, 5
Denmark: 5	Netherlands: 1, 3, 4, 5	USA: 3, 4, 5
France: 4, 7	Norway: 1	USSR: 6, 8, 9, 10
FRG: 4, 5	Spain: 3	

States planning no notifiable military activities in 1991: Austria, *Belgium*, Bulgaria, *Canada*, Cyprus, Czechoslovakia, *Denmark*, Finland, *FRG*, *Greece*, Hungary, the Holy See, Iceland, Ireland, *Italy*, Liechtenstein, *Luxembourg*, Malta, Monaco, *the Netherlands*, *Norway*, Poland, Portugal, Romania, San Marino, *Spain*, Turkey and Yugoslavia. (States participating in notifiable activities but not responsible for notification are given in italics.)

Implementation

The information received by SIPRI on implementation of the requirements of the Stockholm Document has become more comprehensive over the years, because more and more governments are willing to provide other documents in addition to the calendars, such as composite calendars and inspection reports, allowing for a comprehensive picture of the actual military activities conducted.[6]

As of January 1991, all 34 participating states have complied with the terms of the Stockholm Document concerning the exchange of annual calendars and forecasts, notifications and observations as well as on-site inspections, and no nation has indicated dissatisfaction with the implementation process.

As the information in the calendars is about planned military activities, changes may occur when the manœuvres actually take place, perhaps affecting the number and type of personnel involved. These changes are communicated to the other participating states within the framework of the formal notifications of each activity.

This was of course also the case in 1990, when political developments in Europe and the Middle East in particular led to the scaling down or even the cancellation of exercises during 1990. Aside from the general process of *détente* in Europe—which to large extent reduced the threat perceptions between NATO and the WTO[7]—exercises were cancelled or scaled down especially as a reflection of events such as the unification of Germany and the withdrawal of Soviet troops from Eastern Europe. Another reason for the scaling down or cancellation of NATO exercises was the emergence of the Gulf crisis during which US troops stationed in Germany were redeployed to the Middle East. The cancellation of the 'Bold Guard' exercises is one example. These manœuvres have been held each year with the participation of West German, Danish, US and British forces, but as they were supposed to take place mainly along the border between the FRG and the GDR, their conduct would have been controversial. US forces withdrew from the exercises because of their involvement in the Gulf.[8] The USA also cancelled its participation in the September air–sea exercise 'Teamwork' off Norway; some of the 6000 US military personnel who were supposed to participate had already left for the Gulf.[9] Although the USSR reinforced one exercise held on GDR territory[10] because of the political events in Eastern Europe most WTO manœuvres were also scaled back or cancelled.[11] Table 13C.2 shows corrections which must be made with regard to military exercises in 1990.

[6] Composite calendars for 1991 and corrections for 1990 were received from the FRG and Sweden. Inspection reports about the 1990 on-site inspections listed in table 13C.5 were received from the FRG, Norway and USSR. However, some of the CSCE countries are still not providing SIPRI with the required material, which is hardly compatible with the ongoing process of increasing transparency and confidence between the European countries. By the end of Jan. 1991 SIPRI had not received any information on planned notifiable military activities from France, Greece, Italy, Liechtenstein, Luxembourg and Malta.

[7] This creates certain 'problems' for the conduct of exercises, as a 'realistic' threat definition becomes much more complicated. See as an example Serra, F. G., 'Exercise "Dragon Hammer" contradictions in threat definition', *International Defense Review*, vol. 23, no. 7 (1990), pp. 745–47.

[8] *Atlantic News*, no. 2247 (24 Aug. 1990), p. 3.

[9] *Atlantic News*, no. 2253 (14 Sep. 1990), p. 4.

[10] From July 1990 the East German National People's Army (NVA) no longer participated in WTO exercises. As the Secretary of State for Disarmament and Defense of the GDR said, such Pact exercises were considered anachronistic by most members of the army during the period of unification with the FRG and the approach to NATO membership. See *Atlantic News*, no. 2244 (20 July 1990), p. 2.

[11] Pitts, D., 'Warsaw Pact's unraveling seen as extensive', *Wireless File*, EUR-103 (United States Information Service, US Embassy: Stockholm, 29 May 1990), p. 7.

Table 13C.2. Military activities at or above the notification/observation threshold which were scaled back in 1990[a]

State(s)/Location	No. of troops reduced from—to	Exercise no. in *SIPRI Yearbook 1990*
USSR and GDR in GDR	13 500—15 900	2 (reinforced)
USSR	17 000—15 500	3
USSR and GDR in GDR	15 700—14 620	6
GDR, USSR, and Poland in GDR	Called off	7
USSR	13 000—9 300	8 (notifiable because of 400 tanks)
UK, USA and FRG in FRG	26 650—9 000	9 (replaced by FTX 'Fall Tide')
FRG, USA, France, and Belgium in FRG	Called off	10
Netherlands, Norway, UK and USA in Norway	18 250—11 000	11
Netherlands, Norway, UK and USA in Norway	10 450—4 000	12
Hungary, Czechoslovakia and USSR in Hungary	8 000	13 (notified voluntarily)
Denmark, FRG, Netherlands, UK and USA in Denmark and FRG	Cancelled	14 and 15
UK and USA in FRG	26 880—8 000	16
Switzerland	15 000—below 13 000	17 (notified voluntarily)
USSR	Cancelled	19
Austria	15 000—13 000	20
Switzerland	15 000—below notification threshold	21

[a] Information from the corrected compository calendar of the FRG for 1990 and from the Swedish Foreign Ministry.

In 1990 changes in the size and conduct of manœuvres occurred in 17 of the 21 military exercises published in the calendar for 1990.[12] Aside from the above-mentioned political reasons, the decreasing number of exercises finally conducted above the notification threshold is also a result of a general trend towards smaller and fewer exercises.

Calendars

The notifications for 1991 show a further remarkable decrease in the total number of planned exercises above the notification threshold. Only 10 military activities are notified for 1991, and no military activity involving more than 40 000 troops is forecast for 1992. Planned military activities for 1991 are as follows: five by NATO countries, one by a neutral and non-aligned (NNA) country and four by the USSR.

[12] The Yugoslavian FTX 'Pester 90', which was cancelled, did not appear in the 1990 calendar because of early notification of its cancellation by the Yugoslav authorities.

No WTO state plans to conduct or participate in any military activity above the notification threshold, a clear sign that the military role of the WTO is coming to an end.[13]

Table 13C.3 shows the annual number of military exercises notified in the annual calendars for 1987–91 and indicates the trend towards fewer military activities above the notification threshold since the implementation of the Stockholm Document in 1987. However, the overall decrease in the number of exercises for 1991 is an indicator of changes in the security pattern in Europe. The basic features are that NATO figures have been reduced by 50 per cent and that the non-Soviet member countries of the WTO will hold no notifiable military activities above the notification threshold at all.

Table 13C.3. Annual numbers of military exercises notified in the annual calendars by NATO, WTO and NNA countries

Bloc	1987	1988	1989	1990	1991	Total
NATO	17	13	11	10	5	56
WTO	25	22	17	7	4[a]	75
NNA	5	3	3	4	1	16

[a] The four exercises are conducted by the USSR in the USSR.

Source: Compiled from *SIPRI Yearbooks 1987/88/89/90*—1987: appendix 10B; 1988: appendix 11A; 1989: appendix 11A; 1990: appendix 13A and the forecast of notifiable military activities for 1991.

Table 13C.4 shows the corrected figures according to the exercises above the notification threshold that were actually held in 1987–90.

Table 13C.4. Annual numbers of military exercises above the notification threshold actually conducted by NATO, WTO and NNA countries

Bloc	1987	1988	1989	1990	Total
NATO	15(9)	15(8)	10(6)	4(2)	44(25)
WTO	25(8)	21(7)	13(5)	5(1)	64(21)
NNA	2(–)	3(3)	3	3(2)	11(5)

Source: Compiled from *SIPRI Yearbooks 1987/88/89/90*—1987: appendix 10B; 1988: appendix 11A; 1989: appendix 11A; 1990: appendix 13A; and information from Streitkräfteamt -KVAE-Gruppe, Bonn.

Compared to the exercises notified in the annual calendars on average the number of exercises above the notification threshold actually conducted was reduced by approximately 1 to 2 military activities per year. This was the case within all three

[13] In a common declaration from the three foreign ministers of Poland, Czechoslovakia and Hungary at a meeting in Budapest in Jan. 1991 it was announced that, aside from the agreed dismantling of the military organization of the alliance by July 1991, the total disbandment of the WTO should be achieved by the end of 1991 or at the latest in Mar. 1992, before the beginning of the CSCE follow-up meeting. *Neue Zürcher Zeitung*, 24 Jan. 1991, p. 3.

groups. 1990 showed a somewhat greater decrease in NATO exercises. NATO conducted only 4 exercises above the notification threshold, instead of the planned 10, the WTO conducted 5 instead of 7 and the NNA states conducted 3 instead of 4. Figures in brackets show the number of exercises to which, because of the number of troops involved, observers had to be invited.

A comparison of tables 13C.3 and 13C.4 shows that from 1987 until 1990 NATO had planned to conduct a total of 56 exercises above the notification threshold, compared to the 44 actually held. For the WTO the ratio is 75:64 and for the neutral and non-aligned countries it is 16:11. This means that NATO countries conducted approximately 79 per cent, the WTO approximately 85 per cent and the NNA countries approximately 68 per cent of their military activities as planned. It is important to note that NATO conducted larger-scale exercies than those of the WTO.

The new trend towards smaller exercises involving fewer troops and less *matériel* but more computerized equipment continued during 1990. 'Centurion Shield', for example, was primarily an exercise in computer simulation rather than strategic deployment of US Army troops. The shift went from tactical and technical training of troops to computer simulations for training commanders and their staffs at battalion through corps levels. In 1990 there were no tanks, 57 per cent fewer troops than in 1989 and approximately 72 per cent fewer tracked vehicles.[14] There were similar developments in the training of Soviet troops. Colonel General A. Demidov, Chief of Ground Forces Combat Training Directorate, for example, called for the development of new computers and training simulators to create a complex 'battlefield environment' and has stated that the Soviet Army is adopting US and NATO training standards, procedures, methodologies and simulation devices.[15]

With respect to the non-Soviet WTO states, the conduct of large-scale coalition manœuvres among the East European countries seems rather unlikely in the future.

Notifications

Notification is to be given by the participating state on whose territory the activity in question is planned (paragraph 30), in writing through diplomatic channels in an agreed form on content, to all other participating states, 42 days or more in advance of the start of notifiable military activities (paragraph 29). The notifications represent the final information sent to all CSCE countries concerning the forthcoming military activity. The information is more detailed than that in the calendars. As far as can be determined from the information provided to SIPRI, all 34 CSCE countries have fulfilled their obligations in providing notification of their military activities.

Observations

The fundamental merit of the Stockholm observation regime is that it makes observation both multilateral and compulsory, creates an interdependence between observable military activities, and lays down the obligations of the inviting state with a cer-

[14] The concept was first tested during 'Caravan Guard 89' by headquarters V Corps; see Hyde, J. C., 'REFORGER 90 had fewer troops and tracks, more computers', *Armed Forces Journal International*, Mar. 1990 p. 23–24; see also appendix 13A in *SIPRI Yearbook 1990*, p. 516.

[15] This plan might not be easily implemented, because of the miserable state of Soviet economy, shortages of resources and a slowdown in R&D cycles in the defence industry currently occupied with conversion related problems. *Jane's Soviet Intelligence Review*, vol. 2, no. 8 (Aug. 1990), p. 384.

tain degree of accuracy.[16] The observation provisions have proven to serve well, both in terms of compliance as well as in confidence building between the participating parties; over the years the yearly observations have become an institutionalized system of regular contacts on the military level.

The Stockholm Document provides for mandatory invitation of observers to notifiable military activities at or above a certain threshold—17 000 regular troops or 5000 amphibious or airborne troops (paragraph 38.4). These observations are compulsory and available to all CSCE states. The observation programme is determined by the host country and the obligations are laid down in the Document.

Observation requirements have been met in 1990. For 1991, five military activities will be above the observation level. NATO countries will conduct two military activities, and the USSR and Sweden will hold two and one, respectively.

Inspections

Means for verifying compliance with the requirements of the Stockholm Document are a most vital part of this arms control agreement. The provision for inspection constituted a breakthrough in conceptual and political terms. It makes the CSBMs an integral part of the approach to arms control and reflects the positive evolution of the Soviet position on mandatory on-site inspection, which before political changes took place under President Mikhail Gorbachev was regarded as a threat to national sovereignty and a certain kind of espionage.[17] This inspection regime, therefore, represents a substantial step towards greater transparency by allowing foreign military personnel to inspect exercising troops.

In accordance with the Document each participating state has the right to conduct on-site challenge inspections with no right of denial (paragraphs 65–66).[18] During 1990, 10 on-site inspections were conducted (see table 13C.5), compared to 16 in 1989, 13 in 1988 and 5 in 1987.[19]

The ratio of inspections conducted was even in 1990. NATO countries held five and the USSR also conducted five inspections. This is in line with the overall ratio for the period 1987–90, when NATO countries conducted 23 inspections and the WTO countries held 21. In 1990 inspections on the territory of the WTO took place in GDR and USSR. Inspections on NATO territory took place in Belgium, the FRG, Italy, the UK and Norway. In 1990 Norway conducted its first on-site inspection since the implementation of the Stockholm Document in 1987.

Summing up the basic contents of the inspection reports received by SIPRI,[20] all the terms and conditions of the Stockholm Document have been met and inspections have taken place in an atmosphere of good will and mutual understanding.

[16] Ghebali (note 1), p. 54.

[17] Ghebali (note 1), p. 67.

[18] After arrival in the host country the inspection teams (four officers) received two helicopters, two cars of high practicability and the necessary means of communication with their embassies.

[19] Even though the actual number of on-site challenge inspections was lower in 1990 than in previous years, it still represents an increase because of the smaller number of military exercises conducted.

[20] By 25 Jan. SIPRI had received inspection reports from the FRG, Norway and the USSR.

Table 13C.5. On-site inspections of military activities conducted in 1990

Date	Inspecting state	Host state	Exercise/area
9–11 Feb	USA	GDR	Haldesleben–Brandenburg–Sewekow
20–22 Mar	FRG	USSR	Kiev MD
22–24 Apr	UK	GDR	. .
13–15 Sep	UK	USSR	Molodechno–Bobruysk–Ushachi
13–15 Sep	Norway	USSR	Molodechno–Bobruysk–Ushachi
6–8 Jan	USSR	Belgium	'Reforger 90'
14–16 Mar	USSR	Norway	'Cold Winter 90'
8–10 May	USSR	Italy	'Dragon Hammer'
26–28 Sep	USSR	UK	. .
10–12 Oct	USSR	FRG	'Quarter Final'

Compliance

No inspection reported a failure to meet the provisions of the Stockholm Document in terms of notification of numbers of personnel and equipment involved in the military activities. However, minor cases of dissatisfaction also occurred during the 1990 inspections, showing that there is still room for greater transparency. The inspection report of the Norwegian team inspecting a Soviet exercise stated that the team's requests for access were granted, except in one case involving overflight of areas containing Northern Fleet installations.[21] An inspection report by the FRG, also inspecting a Soviet exercise, indicates that there were uncertainties about the designations of participating units.[22] Even though the inspectors were told at the pre-inspection briefing that the real designations of troops could be revealed by subtracting the figure 10 from the exercise designations of the divisions and by subtracting the figure 5 from the regimental designations, the inspection team met several soldiers and officers who gave exercise designations which did not reveal the true designations. The same inspection report also stated that during the aerial inspection the pilots repeatedly left the agreed flight route without prior consultation with the inspectors. The FRG Government therefore expressed its hope that in future inspections of Soviet forces the Soviet escort will be more supportive to the inspection team according to the spirit and provisions of the Stockholm Document.[23] However, there was no evidence of non-compliance with the provisions in 1990.

[21] However, the Soviet denial was in line with provisions of the Stockholm Document which exclude inspections of specific areas and sensitive points or naval vessels (paragraph 74). See Report of the Norwegian inspection in the Pechenga area of the Leningrad Military District of the Soviet Union, 11–13 Sep. 1990.

[22] Inspection report by the FRG on an inspection held 20–22 Mar. 1990 for the military activity conducted by the USSR.

[23] Dissatisfaction with the way Soviet escorts sometimes dealt with the inspecting team have been raised several times since the implementation of the inspection regime. However, dangerous incidents

Assessment and outlook

Because of the political events which took place, 1990 was an unusual year. How the future rate and numbers of military exercises will develop once conditions in Eastern Europe have become more settled and the WTO as a military organization is disbanded, remains to be seen. However, in future the conduct of large-scale military activities does not seem very likely in the East European countries. The role of the non-Soviet WTO countries in providing exercise ground for allied manœuvres is at an end. In future military exercises in Eastern Europe and the USSR will probably be held on national scale and territory. The notified exercises of the USSR for 1991 and the non-participation of any non-Soviet WTO state in any exercise above the notification threshold indicate a move in this direction.

1990 also marked the beginning of a new step in the process of enhancing confidence and security building. The Document adopted at the Paris summit meeting at the Conference on Security and Co-operation in Europe on 21 November 1990 strengthened the provisions of the Stockholm Document and also established new measures aiming to enhance future co-operation and transparency in European security.[24] The measures adopted in the document are politically binding and came into force on 1 January 1991.[25]

Since the implementation of the Stockholm Document in 1987 the adopted CSBMs have proven their importance and value by enhancing transparency and openness in the field of military security between the CSCE countries. They established a framework of regular information exchange and personal contacts on the military level. During recent years the obligation of the Stockholm Document has often been exceeded by giving additional information on voluntary basis. As the past experience on the implementation of the CSBMs according to the Stockholm Document was rather positive, expectations for the next generation of CSBMs were rather high. However, the Document of the Paris summit meeting did not reach these expectations in every respect. As far as the deepening and extension of the provisions of the Stockholm Document are concerned, most of the contents remained unchanged.[26] In particular the hope that naval issues would be included in the CSBM framework of prior notifications and constraints failed. The limitations on military exercises are also rather modest as negotiators failed to reach full agreement on limit-

like the one involving a US inspection team inspecting Soviet forces in the GDR in May 1989, where a Soviet guard pointed his weapon at the party and ordered them to halt, did not occur during 1990. See Department of State, *United States Inspection Report on Military Activity in GDR*, File 5362 EUR7171, UNCLAS STATE 187206 (US Department of State: Washington, DC, 19–21 May 1989). The report received from the USA on inspections in 1990 takes note of the increased efforts on the part of escort officers to ensure the safety of the inspection team; United States inspection of a military activity in the GDR, 9–11 Feb. 1990.

[24] Vienna Document 1990 of the negotiations on Confidence- and Security-Building Measures convened in accordance with the relevant provisions of the Concluding Document of the Vienna Meeting of the Conference on Security and Co-operation in Europe.

[25] For a description of the CSBM negotiations and their results, see section VIII of chapter 13 of this volume.

[26] The numerical thresholds for prior notification of certain military activities remained unchanged. Military activities will still be subject to prior notification whenever they involve at least 13 000 troops including support troops or at least 300 tanks; exercises involving at least 200 sorties by aircraft, excluding helicopters; amphibious landing forces involving at least 3000 troops; or whenever a parachute drop involves at least 3000 troops (Article IV, paras 38.1.1; 38.1.2; 38.2.1 of the Vienna Document 1990).

ing them more strictly. Such limits will be taken up in the continuing CSBM Negotiations in 1991.[27]

The only new constraining provision refers to the two-year advance notification requirement for the conduct of large-scale military exercises. Participating states are not to carry out military activities subject to prior notification involving more than 40 000 troops unless they have been the object of communications as defined in the Stockholm Document (Article VIII, paragraph 72).[28]

Improvement was also reached with the specifications given in the notification of the military exercise. They will be more comprehensive and include new information on the designation, subordination and the number and type of units participating down to and including brigade/regiment or equivalent level (Article IV, paragraph 42.1.2).[29]

Conditions for the observation of military activities were also improved. As well as having the right to conduct aerial survey (Article V, paragraph 61.4), the observers are now allowed to use their own binoculars, maps, photographic and video cameras, dictaphones and hand-held passive night-vision devices (Article V, paragraph 61.3).[30] The participating states are also encouraged to invite media representatives from all participating states to attend observed military activities (Article V, paragraph 63).

With respect to the inspection of military activities, any participating state is allowed to address a request for inspection to another participating state on whose territory compliance with the agreed CSBMs is in doubt (Article VIII, paragraph 78). As a result inspections will also be carried out within the alliances. This issue was of special interest to WTO states, which expressed their interest in also inspecting Soviet military exercises.[31]

The relatively few improvements regarding the conduct and inspection of military exercises represent no big step forward in the process of enhancing confidence- and security-building in Europe but rather imply a continuation of established procedures. The next round of CSBM Negotiations will therefore have to establish whether stricter constraining measures can be implemented on military activities, aside from measures to further enhance transparency. So far the more important improvements with respect to CSBMs approved at the Paris summit meeting are the provisions for annual exchange of detailed military information on military budgets and planned weapon systems; the mechanism for consultation and co-operation as regards unusual military activities; and the establishment of a Risk Reduction Centre. These measures of transparency and risk reduction signal promising new directions for a new security system.

[27] See *Vienna Fax*, no. 33 (26 Dec. 1990), p. 1.

[28] This threshold sets a prohibition because all notifiable military activities involving more than 40 000 troops are forbidden if they were not the object of the two-year advance notice. In the Stockholm Document the number of troops was 75 000. See also Ghebali (note 1), p. 66.

[29] The Stockholm Document required only information down to division level.

[30] Previously they were only allowed to use their own binoculars.

[31] The contents of Article VIII of the Vienna Document (note 24) are in accordance with the Stockholm Document. In Annex IV of the Stockholm Document, however, NATO and WTO statements explained that they would not inspect military activities of their own alliance. See Sharp (note 26). At the seminar, the delegation of Hungary made a statement on behalf of the delegations of Czechoslovakia, Hungary and Poland, in which they stated that each of them has the right to carry out inspections and evaluation visits on the territory of the other participating states and is ready to accept such inspections and evaluation visits on its territory under the Vienna Document 1990. CSCE, 66th Plenary Meeting, Journal 241/Rev. 1, 17 Nov. 1990, p. 2.

Appendix 13D. The Vienna Military Doctrine Seminar

AXEL KROHN

As part of the Confidence- and Security-Building Measure (CSBM) Negotiations of the Conference on Security and Co-operation in Europe (CSCE) a seminar held in Vienna, 16 January–5 February 1990, brought together military leaders of 31 of the 35 CSCE countries to discuss military doctrines and their implementation.[1]

Military doctrine is not only important for national armaments dynamics, perhaps leading to international arms competition, but, if it is or is perceived as potentially offensive, it is also a central risk factor which can lead to political and military instabilities. The evaluation of military doctrines and the effort to derive intentions from them would be greatly facilitated by transparency and confidence. The CSBM Negotiations were the obvious forum for the discussion of this issue, but the seminar was not intended to produce an agreement or specific document.

Before the official seminar in Vienna, an informal dialogue on military doctrine between military officers from NATO and WTO countries took place at forums organized by SIPRI, Pugwash and the Institute for East–West Security Studies (New York). A semi-official meeting, held at the Stiftung Wissenschaft und Politik in Ebenhausen, FRG, on 16 June 1989, convened at the initiative of the FRG and Polish foreign ministers became the forerunner to the Vienna seminar. It was the first 'communication of those who cannot communicate'—a thorough exchange of views among military experts, researchers and NATO, neutral and non-aligned (NNA) and WTO diplomats on this issue.[2]

However, the concrete invitation to talk about military doctrines, equipment and threat perceptions was issued by the WTO countries in May 1987 at the meeting of the Warsaw Pact Political Consultative Committee in East Berlin, where the WTO officially proclaimed a defensive military doctrine.[3] The subject was raised again by this body in July 1988 in the communiqué of the meeting of the Political Consultative Committee of the WTO member states in Warsaw. At the opening of the CFE and CSBM negotiations in March 1989 NATO countries accepted the WTO invitation, and from mid-1989 the process of formulating contents and aims was under way.[4]

[1] The Holy See, Liechtenstein, San Marino and Monaco do not maintain armed forces.

[2] The invitation to the CSCE participants was launched by FRG Foreign Minister Hans-Dietrich Genscher and Polish Foreign Minister Tadeusz Olechowski at the opening of the CSBM Negotiations in Vienna, March 1989; Proceedings of the Seminar on Security Concepts / Military Doctrine / Military Strategies, 21–24 June 1989, Stiftung Wissenschaft und Politik, Ebenhausen, 1989; and Sharp, J. M. O., 'Conventional arms control in Europe', SIPRI, *SIPRI Yearbook 1990: World Armaments and Disarmament* (Oxford University Press: Oxford, 1990), p. 503.

[3] Supplement to the WTO Communiqué, *On Military Doctrine of Warsaw Treaty Member States*, Berlin, 29 May 1987.

[4] *Österreichische Militärische Zeitschrift*, vol. 2 (1990), p. 153.

Setting the scene

The Plenary of the CSBM Negotiations adopted the NATO proposal for an *ad hoc* preparatory body to define an agenda, a timetable and organizational modalities for a seminar on military doctrine.[5] The mandate was based on two proposals submitted by the delegations of Poland and Spain (20 June 1989) and a Swedish 'non-paper' (27 June 1989). A decision was taken on 20 October 1989 to hold the seminar as an integral part of the CSBM Negotiations. The purpose of the seminar was defined as follows: 'to contribute to confidence-building through a free and frank exchange of ideas designed to increase understanding of military capabilities and of their use. . . . to allow for a discussion on military doctrine in relation to the posture, structure and activities of conventional forces in the zone. Similarly, other connected issues of military doctrine bearing on military capabilities might be discussed. No final report or document will be elaborated at the seminar.'[6]

The seminar was structured as follows:

1. Opening session;

2. Presentation and discussion by participating states of their current or planned military doctrines or similar concepts against the background of their security policy in accordance with the purpose of the seminar and with attention to the definition of terminology used;

3. Discussion of military doctrine or similar concepts in relation to:

(*a*) Posture and structures of armed forces (including organization, command structures, deployment, support systems, personnel, armament, equipment, state of preparedness, procurement plans);

(*b*) Military activities and training (including exercises, personnel training, and use of relevant manuals);

(*c*) Military budgeting and planning (including preparation and place in national budgets, allocation of resources);

4. Closing remarks;

5. Closure of the seminar.[7]

The mandate made clear that the intention was not to come up with substantially new findings for future CSBMs but rather to contribute to the development of confidence and understanding through a free and open exchange,[8] to permit presentations of the official perceptions of individual countries on agreed issues and to provide an opportunity for military leaders of the participating countries to meet each other personally.[9] Most participants found the seminar valuable and stressed that the productive atmosphere enabled highly sensitive and controversial issues to be presented

[5] CSCE document CSCE/WV.Y, Vienna, 5 May 1989.

[6] For the full text see Institute for Military Security Policy (hereafter IMS), *Seminar on Military Doctrine: Plenary—Statements 1* (IMS: Vienna, Mar. 1990), Journal no. 61 (20 Oct. 1989), p. 2.

[7] The opening statement was given by Prime Minister of Austria Alois Mock. For the agenda see IMS (note 6), Journal no. 61 (20 Oct. 1989), Annex, p. 2.

[8] See 'CSBM Military Doctrine Seminar a unique opportunity', Transcript of Ambassador Maresca on WorldNet, *Wireless File*, EUR-204 (United States Information Service, US Embassy: Stockholm, 9 Jan. 1990), pp. 1–15.

[9] See Ambassador Massimiliano Bandini, Head of the Italian Delegation to the CSBM Negotiations, 'The CSBM negotiations in Vienna: a commitment to build a new European military security system,' *NATO Review*, vol. 38, no. 5 (Oct. 1990), p. 12.

constructively.[10] As Erik Pierre, deputy head of the Swedish CSBM delegation in Vienna stated: 'The fact that the field marshals of Europe are meeting for the first time since the Congress of Vienna in 1815 is in itself a good sign.'[11]

However, the presentations did not reflect the political changes already under way in Europe to the extent merited by their importance. Many papers were still largely based on the security thinking of the era before the political changes in Eastern Europe. While acknowledging the Eastern approach to new security structures, NATO countries were not prepared to present new strategic thinking which reflected political changes in Eastern Europe and the USSR. Nonetheless, the USA, for example, recognized that the changes in Europe are not only military and that the changes in the USSR towards democratization are without precedent, as is the move towards political pluralism in Eastern Europe.[12]

NNA states stressed their traditional commitment to defend their neutrality by purely defensive military force structures and a clear renunciation of any threat to neighboring countries.[13] The Swedish representative stated that the policy of neutrality means non-participation in alliances in peacetime, aiming at neutrality in war. NNA countries prepare to defend themselves from all directions, which rules out any planning or preparation in operational matters with other states.[14] Their armed forces are rather modest in peacetime, seeking to establish a defensive threshold high enough to repel an attack by ready forces; defence planning includes strong augmentation by mobilization.[15] Owing to the structure and size of the NNA countries' armed forces, their defensive intentions and capabilities were not a matter of concern to the other participants, and a critical reflection on the possible future role of neutrality within the changing political and military environment of Europe was not provided.

Because of the former policy of strict secrecy in all military aspects, the presentations of the East European participants were the more interesting ones. Although they were still basically in line with the official general security perceptions of the Soviet Union, they showed first steps in the direction of developing more independent thinking on national security.[16] A Polish diplomat commented that the seminar was too early for some of the East European countries because they were just beginning to make real changes in their doctrine; but all of those nations tried hard to demonstrate the independent nature of their defence plans.[17]

Reasonable defensive sufficiency

The meeting sought to allow clarification of uncertainties and expression of concerns. As the FRG representative said: 'The cause of international tensions does not arise

[10] *Focus on Vienna*, no 17 (Mar. 1990), p. 8.

[11] See Leavitt, R. and Miller, C., 'Doctrine Seminar looks beyond the cold war', *Defense & Disarmament Alternatives*, vol. 3, no. 1 (Jan. 1990), p. 1.

[12] IMS (note 6) *Plenary—Statements 1*, Presentation by General Colin L. Powell, USA, pp. 7–10.

[13] IMS (note 6) *Plenary—Statements 1*, Presentation by the Austrian representative, p. 2.

[14] IMS (note 6) *Plenary—Statements 1*, Presentation by Bengt Gustafsson, Supreme Commander Swedish Armed Forces, p. 1.

[15] The strong dependence on mobilization is not only common in NNA countries. It is increasingly important for other European countries, as the numerical reductions and restructuring of armed forces create the need for 'cadre units' instead of fully manned units.

[16] Basically they reflected the general line of the 1987 WTO declaration where they had already officially adopted a defensive military doctrine for their alliance.

[17] *Vienna Fax*, no. 12 (5 Feb. 1990), p. 1.

from military potentials alone but also from antagonistic political and ideological aims. Nevertheless, armed forces are generally an expression of power and represent a possible threat in the eye of the opposite side, especially if there are differences between declaratory statements and the visible number and structure of forces.'[18]

The USA and other NATO countries requested further explanation of the Soviet term 'defensive sufficiency' and its doctrinal implications. Chief of Soviet General Staff Mikhail A. Moiseyev explained that the USSR understands 'reasonable defensive sufficiency' to be a new concept of the composition and structure of armed forces and military development in general, and that with regard to conventional forces, reasonable defensive sufficiency implies forces able to withstand any military attack, but not able to conduct any large-scale military offensives, which makes the forces 'non-offensive'.[19] He defined Soviet military doctrine as comprising interrelated political and military–technical parts. Whereas the political part is considered to govern the military–technical aspects, it is through the political aspect of doctrine that the USSR expresses its attitude towards wars and the military–political tasks of averting them and strengthening national defence and security.

Moiseyev explained that the essence of the political aspects of the Soviet doctrine can be described by some key provisions. Basic among these is the rejection of war as a means of settling international disputes; the declared intention not to open hostilities against another country under any circumstances; the refusal of territorial claims against any other state; and the belief that rough parity between the alliances remains decisive for averting war. The Soviet military doctrine does not link the future of Soviet security to the military solutions of international problems—these should be settled by peaceful political means.[20] This official definition shows the changes that Soviet military doctrine has recently undergone. Before 1986, prevention of war was not part of the military doctrine, which was basically a guide for the preparation for and waging of war. Reassurance of national security was sought through military preparedness.[21] This seemed to indicate a decrease in power for the military in dealing with national security, especially, as Moiseyev stated, as the Soviet doctrine now recognizes that no weapons can completely guarantee protection for a state by military and technical means alone.[22]

Follow-on forces

Despite NATO's declared defensive intentions, follow-on forces capabilities have been a permanent concern to WTO states because they include military strikes against second-echelon forces and command and control facilities on WTO territory.

WTO participants raised concerns about the NATO forward-defence and follow-on forces attack strategies. Moiseyev, for example, asked for an explanation of those

[18] IMS (note 6) *Plenary—Statements 1*, Statement by the Chief of Staff of the Federal Armed Forces Germany, Admiral Dieter Wellershoff, p. 3.

[19] IMS (note 6) *Plenary—Statements 1*, Statement by the Chief of Staff of the Soviet Forces Major General Moiseyev, pp. 13–14.

[20] IMS (note 6) *Plenary—Statements 1*, Statement by the Chief of Staff of the Soviet Forces Major General Moiseyev, pp. 8–9.

[21] See also Almquist, P., 'The Vienna Military Doctrine Seminar: flexible response vs. defensive sufficiency', *Arms Control Today*, vol. 20, no. 3 (Apr. 1990), p. 22.

[22] Developments in the USSR during the last months of 1990 and early 1991 indicate an increasing importance for the military. President Gorbachev's shift towards seeking stronger support from the conservative military by acknowledging their military and political importance, especially for domestic politics, might strengthen the military factor in Soviet politics again.

provisions of the 'flexible-response' doctrine which indicate an excessively offensive role for NATO armed forces, with massive ground and air strikes throughout the entire depth of the opponent's positions. He asked whether such NATO military doctrine provisions still relate to the problem of strengthening European security in the context of the new realities in European policy.[23]

The US representative stated that 'the follow-on forces attack' concept, which some nations find threatening, is an inherently defensive concept that responds to the geographic realities of NATO's defensive positions and to our perception of the operational tactics and size of potential opponents' armed forces.'[24] He explained further that, because of their smaller size, NATO forces must be able to engage follow-on attacking forces prior to their arrival in the forward area, using interdiction assets which can neither seize nor hold ground. The follow-on forces attack concept thus remedies a lack of strategic depth by means designed to preclude the creation of unfavourable force ratios at NATO's forward line of defence.[25]

However, several WTO delegates complained that Western officials seemed unwilling to concede that aspects of NATO strategy could be perceived as offensive and threatening. Polish Ambassador Wlodzimierz Konarski stated at a press conference that future dialogue must include more examination of NATO's military strategy for deep strikes into Eastern Europe and consider that, even if the West's overall strategy is defensive, its follow-on forces attack capabilities are clearly offensive.[26]

Nuclear deterrence

Redefinition of NATO's strategy of nuclear deterrence, under way later in the year, did not show up at the meeting.[27] Chairman of the US Joint Chiefs of Staff General Colin Powell explained that 'America remains committed to the strategy of Flexible Response. The very nature of Flexible Response is to provide a credible deterrent, and, failing that, to be able to halt hostilities quickly and restore the status quo. This requires a wide range of capabilities from conventional through nuclear . . .'.[28]

The British representative said that at both strategic and sub-strategic level, the UK also places 'highest priority on the . . . provision of nuclear forces. We maintain what we define as a minimum strategic deterrent both nationally and as a contribution to NATO's strategic forces.'[29]

The French representative explained that nuclear deterrence is a dissuasion 'from the feeble to the the strong' (*du faible au fort*), based on a scale of means capable of convincing the adversary that the nuclear risk to his own territory would be greater than the gain he could anticipate from aggression against France. French forces, which all serve the same mission, are structured in accordance with this concept.[30] He also said that nuclear weapons can only be understood as a 'non-use weapon', a

[23] IMS (note 6), *Plenary—Statements 1*, Statement by the Chief of Staff of the Soviet Forces Major General Moiseyev, p. 19.

[24] IMS (note 6), *Plenary—Statements 2*, Statement by J. D. Robinson, USA, p. 6.

[25] Note 24.

[26] Note 17.

[27] NATO is ready for further 'substantial reductions. The tendency is to make a substantial reduction in land-based systems and put increasing emphasis on weapons delivered by dual-capable aircraft'; see 'Nato ponders new strategy', *Jane's Defence Weekly*, vol. 14, no. 24 (15 Dec. 1990), p. 1203.

[28] IMS (note 6), *Plenary—Statements 1*, Presentation by General Colin L. Powell, USA, pp. 7–10.

[29] IMS (note 6), *Plenary—Statements 1*, Presentation by Air Vice Marshal John Willis, UK, p. 2.

[30] IMS (note 6), *Plenary—Statements 1*, Presentation of France, pp. 3 and 4.

diplomatic weapon of equilibrium and resistance to blackmail, wherever it comes from. They make war irrational and victory impossible.[31]

The nuclear element was therefore seen as of vital importance, including the role of and support by theatre nuclear weapons which, beside their function as a means of deterrence and escalation, especially provide the visible link that underscores the commitment of US strategic forces to the defence of NATO.[32]

The USSR took the opportunity to express its negative attitude to US and NATO nuclear concepts, which it found obsolete, contradictory and dangerous, and put forward its old idea of a common 'no-first-use policy'.[33] The Soviet representative defined 'reasonable sufficiency' in the context of strategic nuclear weapons as 'a rough parity between the Soviet Union and the US. Structures may differ, but the potential combat possibilities at any stage of reductions must be comparable'.[34]

The naval question

A concern expressed by the Eastern side is the US refusal to include naval issues in the arms control process, perceived by the USSR as preserving naval superiority at a time when ground forces are reduced to parity. US naval power, especially the capability for sea-launched land attack, is perceived by the USSR as a potential threat and this issue is a longstanding bone of contention between the superpowers. Also, US bases world-wide are perceived as a means of encircling the Soviet Union.[35]

One US representative made it clear that the USA sees the naval role as essential to the military tasks its armed forces must be able to fulfil. Explaining the defensive tasks of US military forces, General Colin Powell laid out four guiding principles:

> First, we have structured our defense posture not for war, but to deter aggression. Deterrence, the cornerstone of our military strategy, is geared toward preventing an attack on the U.S. and on its allies. Second, when we are challenged, we will respond—with force if necessary—to defend our citizens, our allies, our interests and our values. . . . Third, we will honour our commitments to collective security, to partnership with nations who share our fundamental values. . . . Fourth, we will continue to adhere to the distinction as laid down in the Constitution between raising armies and maintaining navies. . . . naval forces can provide security to ourselves and to our allies. But there are two things navies cannot do: they cannot seize territory and they cannot, by themselves, win wars.[36]

Another US representative pointed out that the USA is a maritime nation, dependent on the sea for its economic and political livelihood and linked to its allies by the sea lanes. 'The deployment of naval units is not intended to threaten anyone . . . If the United States is to effectively participate in mutual defense, it is imperative to have forces deployed near these regions. Deployed naval forces are therefore only a threat to those who would threaten US allies or US commitments. U.S. maritime policy has always been to defend these interests, as it maintains open lines of communication

[31] IMS (note 6), *Plenary—Statements 1*, statement of the French representative, quoting from the speech by the Minister of Defense at the Academy Vorochilov in Moscow, 5 Apr. 1989, p. 3.

[32] IMS (note 6), *Plenary—Statements 2*, Presentation by J. D. Robinson, USA , p. 10.

[33] IMS (note 6), *Plenary—Statements 1*, Statement by the Chief of Staff of the Soviet Forces Major General Moiseyev, p. 12.

[34] Note 33, p. 13.

[35] IMS (note 6), *Plenary—Statements 1*, Statement by the Chief of Staff of the Soviet Forces Major General Moiseyev, pp. 10–11.

[36] IMS (note 6), *Plenary—Statements 1*, Presentation by General Colin L. Powell, Chairman, Joint Chiefs of Staff, pp. 5–6.

and trade on the high seas.'[37] These statements make clear that so far this position leaves little room for further movement on this issue.[38]

Defensive objectives and force structure

Common to the East European presentations was the rejection of counter-attack strategies that would shift combat action onto the territory of another state even if this state started a military aggression. The Chief of the General Staff of Czechoslovakia explained the political priorities embodied in this doctrine: no military solutions of international conflicts and no territorial claims against any other country.[39]

All WTO presentations of military doctrine were underlain by the new emphasis on the national character of military security with a general development towards commitment to defend national borders against any aggressor. However, these statements were still declaratory, as restructuring and redeployment of forces within WTO countries has just started. The abandoning of total control by the Communist Party has brought an end to their domination of national security interests and perceptions—the respective parliaments decide on whether or not to use force.

Western analysts pointed out a number of indicators of the WTO ability to launch offensive strikes and indicated that until the newly formulated defensive policies have an impact on WTO force structure, training, exercises and military budgeting, such proclamations would not be convincing. East European representatives took up these Western concerns and explained the structural changes in and reductions of their military forces. Their presentations on force structures ranged from information on force numbers and dispositions to length of service and indicated the changes the armed forces are undergoing in most of the countries. WTO countries attempted to demonstrate that their doctrinal changes towards border defence and denial of offensive operations is also expressed in force structures.

The Polish representative acknowledged that Poland had previously proceeded from the point of view that, in case of aggression against the country, the armed forces, after repelling the attack and inflicting losses on the enemy, would rapidly move hostilities into the aggressor's territory. This is why the Polish armed forces have a comparatively large number of attack aircraft, tanks, artillery and warships.[40] Now, the Polish armed forces are adapting to the changes in the military–political situation and aiming to provide a 'level of sufficiency for defence requirements', which means a capacity to repel aggression while at the same time being unable to conduct large-scale offensive operations.

The Hungarian presentation indicated the general development towards a defence *tous azimuts*, stating that along with unilateral reduction of the Hungarian People's Army they are changing the level of readiness of their troops and abandoning the practice of deploying most—some 60 per cent—combat-ready troops in the Western

[37] IMS (note 6), *Plenary—Statements 2,* Presentation by J. D. Robinson, USA, p. 8.

[38] This was also reflected at the CSBM Negotiations and in the concluding document of the Paris summit meeting. Also in this forum the USSR was unable to implement the question of naval forces into the new CSBMs. Future rounds of negotiation will show whether the USA might be willing to make concessions on this issue. See appendix 13A on the implementation of the Stockholm Document.

[39] IMS (note 6), *Plenary—Statements 1*, Statement by the Chief of Staff of the armed forces of Czechoslovakia Gen. Slimàk, pp. 5 and 9.

[40] IMS (note 6), *Plenary—Statements 1*, Statement by Major-General Henryk Szumski, First Deputy Chief of General Staff of the Polish Armed Forces, p. 1.

part of their country, close to the Austrian border. This is the first step towards the organization of the defence of all borders within the present force posture.[41]

The US presentation made clear that forward-deployed forces are still the tangible evidence of US commitment to the alliance and demonstrate the will to defend shared interests. These forces are relatively small in number and positioned not to gain territory but to deter aggression. There is an inherent guarantee of stability in that the forces are insufficient to conduct a surprise offensive. They would be used to repel aggression—but to restore the pre-attack status quo rather than seize territory.[42]

Training

As the whole emphasis of Soviet and East European presentations was to show a clear development towards non-offensive defence postures, an important issue was the operational and combat training of the armed forces. Moiseyev stated that the USSR has reduced the number of large-scale exercises at division and regiment level by half—NATO countries were invited to follow suite.[43] The basic concern of both sides was that the opponent might use a military exercise as a cover for surprise attack. Large-scale exercises create a potential for such offensive planning.[44]

The trend towards defensive doctrines is also reflected by training in other WTO countries—the Czechoslovak representative stated that because Czechoslovakia is also changing the training and organization of the armed forces, no large-scale exercises will be conducted.[45] The Polish representative stated that defence problems account for approximately 60 per cent of the training curriculum of mechanized units and exercises are so organized that neither their scope, the composition of ground-force units, armed services and arms involved, nor the way in which the training scenario is played out, will constitute a threat to any state.[46]

NATO has generally conducted fewer but larger exercises than the WTO because of the need to train for reinforcement of NATO troops by units from the USA. However, large-scale exercises are decreasing in both NATO and the WTO.[47] The number of military activities which need to be notified to the CSCE countries in accordance with the Stockholm Document shows this trend.[48]

[41] Poland meanwhile also developed a more balanced deployment of forces all over the country. The creation of two new military districts along the Soviet border will lead to the redeployment of forces from the West to provide an 'equal defence' of all borders. See Ripley, T., 'Poland shakes up forces structure', *Jane's Defence Weekly*, vol. 14, no. 23 (8 Dec. 1990), p. 1131.

[42] It was also stressed that, especially in the case of NATO where multinational forces are spread along an extensive front, US forces need strategic mobility. NATO forward defence in Europe cannot be conducted by trading space for time—there is neither strategic depth for defence nor the force structure or logistics base from which to launch significant counter-offensive operations. IMS (note 6), *Plenary—Statements 2*, J. D. Robinson, pp. 3 and 6.

[43] IMS (note 6), *Plenary—Statements 1*, Statement by the Chief of Staff of the Soviet Forces, Major General Moiseyev, p. 18.

[44] Fears about a surprise attack emerging from a large-scale exercise do not come out of the blue; for example, this was the case with the 1968 invasion of Czechoslovakia and the pressure and intimidation connected with the Solidarity movement in Poland in December 1980.

[45] IMS (note 6), *Plenary—Statements 1*, Statement by the Chief of Staff of the armed forces of Czechoslovakia General Slimàk, p. 9.

[46] IMS (note 6), *Plenary—Statements 3*, Statement by H. Szumski (Major-General), Poland, p. 7. However, this raises the question of the kind of training conducted in the remaining 40 per cent.

[47] See appendix 13C, table 13AC.4 in this volume.

[48] The CSBMs established by the 1986 Stockholm Document require participating states to exchange annual calendars of military activities, to notify exercises above a certain threshold, to invite all CSCE states to observe exercises above a certain threshold, and to allow on-site challenge inspections of military exercises as a means of verification. The Stockholm Document is reprinted in SIPRI, *SIPRI*

Another general scheme was the reliance on reserve training in countries with conscript armies. The presentations indicated a reduction of service time. This differs from country to country, for example, 12 months in FR Germany while Sweden is already experimenting with a 5-month term. Other countries still have longer service terms, for example 17 months in the Netherlands or 18 months in Czechoslovakia (where compulsory service has been reduced from 24 months). Some armed forces have different lengths of service for army, navy and air force; for example, Denmark, where the conscripts serve 4–12 months.

Defence budgets

Unlike in Western countries, figures for the defence expenditures of East European countries and the USSR have hardly been available. This seminar provided the first opportunity for Western participants to get an official picture of defence budgeting in WTO countries and made clear to the East European participants that an open defence budget is a precondition for democratic control and transparency. The importance of government control within an open legislative process of budget planning and approval, especially the opportunity to make government figures and decisions available to public scrutiny, was an important aspect of Western presentations. Besides giving information about the composition and development of their defence budgets, Western presentations therefore emphasized the legal framework and parliamentary process that budgeting must undergo in different countries.

With regard to former WTO secrecy the FRG representative said it would be desirable, and certainly a confidence-building measure, if in future the defence expenditures of WTO states were publicly available in a complete, detailed, comparable and transparent manner, if the defence budgets in Europe became verifiable; and if it were thus possible to reconstruct the relevant calculations step-by-step.[49] Implementing this procedure in all European countries will take some time. The first Soviet budget information given to the UN in October 1990 is a step in this direction.[50]

WTO presentations generally showed decreasing expenditures for defence as an outcome of the changing political situation.[51] The Hungarian Chief of Staff stated that Hungary's defence spending had been reduced by over 9 per cent and will be further cut by 11 per cent in 1990.[52] Reductions in the Soviet military budget are leading to a decrease in military production. General Moiseyev pointed out that in

Yearbook 1987: World Armaments and Disarmament (Oxford University Press: Oxford, 1987), appendix 10A, pp. 353–69.

[49] IMS (note 6), *Plenary—Statements 4,* Statement by Director Wolfgang Ruppelt, Ministry of Defence of the Federal Republic of Germany, p. 7.

[50] UN document A/INF/45/5/Add.1, 12 Oct. 1990. Another step forward is the agreement on annual exchange of information on current and projected budgets as well as planned weapon procurement in the Vienna Document 1990 of the negotiations on Confidence- and Security-Building Measures convened in accordance with the relevant provisions of the Concluding Document of the Vienna Meeting of the Conference on Security and Co-operation in Europe, Article I, paras 14–16.

[51] The Polish speaker explained that in addition to the nominal fall in defence expenditures the decrease within the defence budget during past years, in real terms, that is in actual prices, the expenditure for the 1989 Polish defence budget was actually much lower, that is, 20 per cent lower than defence expenditures for 1986. Inflation obviously plays an important role and probably creates problems not only for Poland but also for other countries. IMS (note 6), *Plenary—Statements 4,* Statement by M. Daniluk (Brigadier General), Poland, p. 1.

[52] IMS (note 6), *Plenary—Statements 1,* Statement by Lt. Gen. Lászlo Borsits, Chief of General Staff of the Republic of Hungary on Agenda Item 2, 19 Jan. 1990, p. 8.

1989 the military budget was reduced by 8.9 per cent (from 77.3 to 71 billion roubles). For the next two years a further reduction by over 14 per cent (10 billion roubles) is envisioned. The volume of military production is supposed to go down by 19.5 per cent: tank production—50 per cent; ammunition—30 per cent; helicopters—25 per cent; and armoured personnel carriers and infantry combat vehicles—20 per cent.[53]

Conclusion

The seminar provided an important opportunity officially to present and discuss security issues on very specific levels. It went beyond the pure presentation of military doctrine, with presentations on force structures, training, exercises and military budgeting, giving a picture of the respective national military security thinking, planning and behaviour. While openness on these issues is more common within NATO, it was a major step towards greater transparency and confidence for the WTO participants, many of whom provided such a picture of their country's defence for the first time. Such public revelations would have been unthinkable a few years ago.

The information presented indicated that the WTO was no longer a militarily homogeneous alliance. Despite differences in specifics, developments in individual member countries indicate the establishment of national homeland defence against any possible aggressor. WTO countries no longer deploy their forces against an 'enemy by definition' but are trying to achieve a defence structure for their country with the tasks of a defence *tous azimuts*. The former military doctrine, directed solely against NATO, is gone and obviously together with the diminishing of ideological security objectives the binding force of the military alliance is also diminishing. Trends visible at the Military Doctrine Seminar came to fruition later in the year. At their meeting in Budapest in early November 1990, WTO states spoke in favour of disbanding the alliance's military structure by July 1991, and in February 1991 President Gorbachev suggested that the WTO foreign and defence ministers gather to decide about the liquidation of the WTO military structure by 1 April 1991.[54]

Obviously it will take considerable efforts to reshape forces according to new defensive objectives, because the weapon systems so far deployed in the armed forces hardly match the new tasks. To restructure present forces in accordance with new defensive tasks will require the phasing out and destruction of weapon systems such as tanks and artillery and the acquisition of equipment with a stronger defensive orientation, such as anti-tank weapons, light combat vehicles, and so on. This process will take much time, effort and funds. Even though the on-going process of reducing defence expenditures as well as conversion of defence industries frees resources desperately needed for transfer to the civilian sectors of the crumbling economies of most of the East European countries, money will still be needed for defence procurement. It became clear during the seminar that economic problems are of great importance. Unilateral initiatives by WTO states, such as those in force reductions and defence expenditures, indicate not only the growth of *détente* but also the pressing economic burden of military costs for many countries.

Alongside the valuable aspects of the seminar, it should be noted that Western countries did not give adequate indication of the future development of their military

[53] IMS (note 6), *Plenary—Statements 1*, Statement by Chief of the General Staff of the USSR Armed Forces, Major General Moiseyev, on Agenda Item 2, 16 Jan. 1990, p. 16.

[54] This proposal was made in a letter from Gorbachev to the Presidents of Czechoslovakia, Hungary and Poland; see *International Herald Tribune*, 12 Feb. 1991, p. 1.

doctrine in the light of the changing overall European situation. This discussion is now under way within NATO. The 'London Declaration' of 6 July 1990, and the NATO–WTO 'Joint Declaration' signed in November 1990 declaring that they are no longer adversaries and will establish new friendly relations, are an outcome of this process.[55] The joint meeting of NATO's Defence Planning Committee and Nuclear Planning Group in Brussels ended on 7 December 1990 with the prospect that the Atlantic Alliance's strategy review can be completed before mid-1991.[56] The shape of the future security structure in Europe is not yet defined but the discussion is obviously under way.[57] It is not yet clear how long it will be before the military capabilities in East European countries are restructured according to their defensive intentions.[58]

However, the Military Doctrine Seminar provided an opportunity for NATO and NNA participants to examine the first changes in the military doctrines of WTO countries resulting from their changing political structure and environment, developments which are changing security perceptions and the tasks of their armed forces.[59]

It is clear that an institutionalized mechanism for transparency in security issues will be one of the basic pillars of the future European security system. The exchange of information which took place during the seminar marked in this respect an important innovative step in the process of enhancing confidence- and security-building within the CSCE process. The practical results of the seminar were the above-mentioned exchange of information on military budgets and the agreement to hold on a rotating basis an annual implementation Assessment Meeting to discuss present and future implementation of CSBMs.[60] The seminar was a starting point for an exchange of information on military subjects which will be continued during the next meeting on military doctrine in 1991, as decided at the Paris summit meeting.[61]

[55] For the London Declaration on a Transformed North Atlantic Alliance, see Rotfeld, A. D. and Stützle, W. (eds), SIPRI, *Germany and Europe in Transition* (Oxford University Press: Oxford, 1991), pp. 150–52; for the Joint Declaration of Twenty-Two states, see appendix 17B of this volume.

[56] As far as force posture was concerned, NATO reaffirmed again the need to have conventional and nuclear forces stationed in Europe. The forces will be small (multinational) and will be capable of being rapidly reconstituted during crisis, underlining the need of reassessing the capability to reinforce Allied forces. See *NA AN nouvelles atlantiques atlantic news*, no. 2280 (8 Dec. 1990), p. 1.

[57] See, e.g., 'Grössere Flexibilität der Nato-Streitkräfte, Überlegungen von Sandrats zur Verteidsgungsstruktur Europas,' *Neue Zürcher Zeitung*, 14 Nov. 1990.

[58] See Hall, R., 'Where next for the Soviet Division?', *Jane's Soviet Intelligence Review*, vol. 2, no. 12 (Dec. 1990), pp. 538–40.

[59] Poland, for example, adopted in February 1990 a new national military doctrine which, structured into nine chapters, outlines the new defence concept of Poland which now shows strong similarities to the doctrine of Western countries, as the military doctrine is de-ideologized and in accordance with the basics of international law aims at the defence of the territorial integrity of the Polish state. Interventionist activities like the 1968 invasion of the former CSSR are excluded by the new doctrine. At the seminar Poland also provided the first copies of the Polish White Book—the first time for Poland that compiled defence figures and planning were presented to the public. See 'Eine neue Verteidigungsdoktrin für Polen, Anpassung an die Praxis in demokratisch regierten Ländern', *Neue Zürcher Zeitung*, 26 Feb. 1990; and Sadykiewicz, M. and Clarke, D. L., 'The new Polish defense doctrine: a further step toward sovereignty ,' *Report on Eastern Europe*, vol. 1, no. 18 (4 May 1990).

[60] These two new elements were presented in the aftermath of the Vienna Seminar on Military Doctrines by the NATO countries in a proposal on 23 February 1990. They became part of the Vienna Document 1990 (note 50), Article I, paragraphs 14–16 and Article X, paragraphs 151–54.

[61] Vienna Document 1990 (note 50), p. 8.

14. Multilateral and bilateral talks on chemical and biological weapons

S. J. LUNDIN and THOMAS STOCK*

I. Introduction

Efforts to finalize the Chemical Weapons Convention (CWC) and to strengthen the Biological Weapons Convention (BWC) continued in 1990. The following were the most important events of the year:

1. The US–Soviet bilateral negotiations on chemical weapons resulted in an agreement to destroy the majority of US and Soviet chemical weapons by the year 2002 and to stop chemical weapon (CW) production. Both countries also expressed support of a comprehensive, global CWC.
2. For the first time, it was possible for the Geneva Conference on Disarmament (CD) to mandate that the negotiations begin work on the final text of the CWC. Although there was not sufficient political support to achieve a Convention, much technical work was performed and is reflected in the 1990 draft text of the CWC (the so-called 'rolling text').
3. The crisis in the Persian Gulf did not directly influence the negotiations, but some Gulf countries continued to express a desire to link nuclear and chemical disarmament.
4. The realization that only a comprehensive CWC can possibly stop the spread of chemical weapons was increasingly voiced.
5. Suggestions for chemical weapon-free zones (CWFZ) were made by some Gulf and Far Eastern countries, but no formal negotiations were held.
6. In preparation for the 1991 BWC Review Conference, a number of international meetings and publications focused on strengthening the BWC but did not result in any new measures. New risks to the BWC were emphasized.
7. Environmental agreements were reached which aim to diminish environmental damage and risk to mankind, and which are also relevant to the CWC and the BWC. They include agreements on the release of carbon dioxide and fluorocarbons into the atmosphere and numerous national regulations concerning toxic waste, air pollution, and health and safety.

* Fredrik Wetterqvist of the SIPRI Chemical and Biological Warfare (CBW) Programme assisted in preparing references and data for this chapter. The references were gathered from the SIPRI CBW Programme Data Base and were also kindly provided by J. P. Perry Robinson, Science Policy Research Unit, University of Sussex, UK, from the Sussex–Harvard Information Bank.

II. Bilateral US–Soviet negotiations

The signing of the Memorandum of Understanding at the September 1989 Wyoming summit meeting[1] showed that the major CW possessor states, the USA and the USSR, were genuinely interested in achieving a bilateral disarmament agreement to reduce their CW arsenals. On 29 December 1989, they for the first time exchanged data on their CW stockpiles and facilities in accordance with Phase I of the Memorandum.[2] The head of the Soviet CD delegation declared in January 1990, that in order to achieve a bilateral agreement between the USA and the USSR, three elements would be needed: (*a*) radical reduction of CW stocks on a bilateral basis; (*b*) non-production of chemical weapons, including binary weapons; and (*c*) total renunciation by the USSR and the USA of the use of chemical weapons.[3] Both countries are also bound by the 1925 Geneva Protocol but, like many other parties, both reserve the right to use chemical weapons in retaliation if chemical weapons are used against them or their allies.

During a meeting in Moscow on 7–9 February, Soviet Foreign Minister Eduard Shevardnadze and US Secretary of State James Baker discussed the CWC. An official joint statement on chemical weapons was released stating that both countries had agreed to sign a bilateral agreement at a June 1990 summit meeting.[4] The statement also indicated that the Soviet Union was in the process of giving up its reservations about the 2 per cent proposal made by President George Bush.[5] The following formulation was agreed upon: 'When the Chemical Weapons Convention enters into force, the sides [the USA and the USSR] will further reduce their CW stocks to equal levels at a very small fraction of their present holdings over the first eight years of operation of the Convention. All remaining CW stocks should be eliminated over the subsequent two years'. There was common understanding that for the last two sensitive years 'all CW-capable states must adhere to the Convention'.[6]

[1] See *Chemical Weapons Convention Bulletin*, no. 8 (June 1990), pp. 19–22.

[2] '29 December', *Chemical Weapons Convention Bulletin*, no. 7 (Feb. 1990), p. 17.

[3] 'Batsanov interviewed on chemical weapon talks', LD1601120190, Moscow Domestic Service, 0830 GMT, 16 Jan. 1990 (in Russian) in Foreign Broadcast Information Service, *Daily Report–Soviet Union* (*FBIS-SOV*), FBIS-SOV-012, 18 Jan. 1990, pp. 6–7.

[4] 'U.S–Soviet statement on chemical weapons', LD1002144490, Moscow, TASS International Service, 1405 GMT, 10 Feb. 1990 (in Russian) in FBIS-SOV-90-029, 12 Feb. 1990, p. 1; 'Text of joint statement', LD 1002141490, Moscow, TASS International Service 1230 GMT, 10 Feb. 1990 (in Russian) in FBIS-SOV-029, 12 Feb. 1990, pp. 20–24; Permanent Mission of the Soviet Union (Geneva), 'Joint Soviet–American statement on chemical weapons', *Press Bulletin*, no. 30 (2013), 13 Feb. 1990, pp. 3–4; Permanent Mission of the Soviet Union (Geneva), 'Joint Soviet–American statement on chemical weapons', *Press Bulletin*, 11 Feb. 1990, p. 79; Häggman, L., 'Avtal vid nytt toppmöte', *Svenska Dagbladet* (Stockholm), 11 Feb. 1990, p. 5; 'CW, START pacts move closer' *Jane's Defence Weekly*, vol. 13, no. 7 (17 Feb. 1990), p. 284.

[5] Ember, L., 'U.S.–Soviet pact on chemical arms likely', *Chemical Engineering News*, vol. 68, no. 8 (19 Feb. 1990), p. 5.

[6] See SIPRI, *SIPRI Yearbook 1990: World Armaments and Disarmament* (Oxford University Press: Oxford, 1990), p. 532.

The 14th round of US–Soviet bilateral consultations on the banning of chemical weapons resumed on 20 February in Geneva,[7] and the results of this discussion round, which ended on 8 March,[8] were presented to the CD.[9] In accordance with Phase I of the Memorandum of Understanding, both sides agreed to an exchange of seven visits to three categories of facility: CW storage facilities, production facilities and industrial chemical enterprises. It was also agreed that the first exchange of visits would be made to CW storage facilities in June 1990. The USA and the USSR also began discussing possible co-operation in technology and procedures for safe, expeditious, and economically and environmentally sound CW destruction. A preliminary paper was drafted addressing the 'order of destruction' and was presented to the socialist and Western groups of states at the CD for further discussion.[10]

In an April meeting in Washington, Baker and Shevardnadze reviewed the discussions and provided further guidance for preparation of a bilateral agreement for the summit meeting which was held on 30 May–3 June.[11] At another bilateral meeting in April, an agreement was reached that CW stocks would be reduced by each side to a level of 5000 metric tonnes.[12] The exchange of visits would start with CW storage and production facilities and chemical industry plants in June and August; additional visits were scheduled for early 1991.[13]

Just prior to the summit meeting, President Bush had offered to end US production of chemical weapons if the USSR agreed to the proposed reduction and the schedule for CW destruction.[14] At a meeting between Baker and Shevardnadze in Moscow, the USSR had accepted the US proposal to destroy

[7] 'Soviet–U.S. consultations open in Geneva', LD2002211190, Moscow, TASS, 1935 GMT, 20 Feb. 1990 (in English) in FBIS-SOV-90-035, 21 Feb. 1990, p. 2.

[8] 'Further reportage on CW talks', LD0903104490, Moscow, TASS International Service, 1822 GMT, 8 Mar. 1990 (in Russian) in FBIS-SOV-90-047, 9 Mar. 1990, pp. 1–2; Permanent Mission of the Soviet Union (Geneva), '2. Soviet–US consultations on chemical arms', *Press Bulletin*, no. 49 (2032), 14 Mar. 1990, pp. 3-4.

[9] Conference on Disarmament document CD/PV.541, 8 Mar. 1990, pp. 14–16.

[10] See 'Further reportage on CW talks' (note 8); see also Institute for Defense and Disarmament Studies, *Arms Control Reporter* (IDDS: Brookline, Mass.), sheet 704.B.430, May 1990.

[11] Hoffman, D., 'Summit is set for late May', *International Herald Tribune*, 6 Apr. 1990, pp. 1, 7; 'Start: Vertrag nicht unterschriftsreif', *Süddeutsche Zeitung*, 7–8 Apr. 1990, p. 9.

[12] 'Geneva chemical weapons ban talk resume', LD1004135890, Moscow, TASS, 1311 GMT, 10 Apr. 1990, (in English) in FBIS-SOV-90-070, 11 Apr. 1990, p. 1; 'TASS comment on chemical weapons ban talks', LD1204191590, Moscow, TASS, 1837 GMT, 12 Apr. 1990 (in English) in FBIS-SOV-90-072, 13 Apr. 1990, pp. 1–2; 'Superpowers discuss chemical weapons', *Financial Times*, 11 Apr. 1990, p. 2; 'Neue Verhandlungen über C-Waffen', *Neue Zürcher Zeitung*, 12 Apr. 1990, p. 4.

[13] 'Chemical weapons talks', LD2604193690, Moscow, TASS, 1851 GMT, 26 Apr. 1990 (in English) in FBIS-SOV-90-082, 27 Apr. 1990, pp. 3–4; 'Chemical arms pact hits snag in Geneva', *Defense News*, vol. 5, no. 18 (30 Apr. 1990), p. 30; Permanent Mission of the Soviet Union (Geneva), '2. Spokesman on Soviet–US consultations on chemical weapons', Moscow, Apr. 26, TASS, *Press Bulletin*, no. 80 (2062), 30 Apr. 1990, pp. 3–4; 'Round of chemical weapons talks reported', PM0105092390, Moscow, *Krasnaya Zvezda*, 28 Apr. 1990, 1st edn, p. 5 (in Russian) in FBIS-SOV-90-086, 3 May 1990, pp. 1–2; 'Efforts to reduce chemical weapons stocks noted', LD1105203490, Moscow, TASS, 2004 GMT, 11 May 1990 (in English) in FBIS-SOV-90-093, 14 May 1990, pp. 1–2; 'U.S., USSR agree to reduce chemical weapons stocks', *Wireless File*, EUR-305 (United States Information Service, US Embassy: Stockholm, 25 Apr. 1990), pp. 11–13.

[14] Gordon, M. R., 'Bush offers a cutoff of toxic arms', *International Herald Tribune*, 10 May 1990, pp. 1, 6; Walker, M., 'US offers deal to Russia on chemical arms', *The Guardian*, 10 May 1990, p. 4; Christiansson, L., 'Bush redo stoppa kemiska vapen', *Svenska Dagbladet* (Stockholm), 10 May 1990, p. 4; 'Bush offers deal', *Jane's Defence Weekly*, vol. 13, no. 20 (19 May 1990), p. 947.

up to 5000 metric tonnes of both countries' stockpiles and then to destroy the remaining stockpiles in the first eight years after entry into force of the CWC, to a level of 500 tonnes (i.e., approximately 2 per cent of the US stockpile).[15] The remaining stocks would then be destroyed 'at such time as all chemical weapons capable states have indicated a willingness to come on board and, in effect, accomplish a global ban'.[16] The USA also committed itself to immediately cease CW production; thus all major obstacles were removed shortly before the summit meeting.

On 1 June 1990, President Bush and President Mikhail Gorbachev signed a bilateral agreement to diminish US and Soviet CW stockpiles (see appendix 14A).[17] The key provisions of the agreement are: (*a*) to begin destruction of the vast bulk of declared stockpiles by the end of 1992; (*b*) to destroy at least 50 per cent of them by the end of 1999; (*c*) to reduce them to 5000 agent tonnes by 2002; (*d*) to permit on-site inspection during and after the destruction process to confirm that destruction had taken place; (*e*) to annually exchange data on stockpile levels in order to facilitate monitoring of declared stockpiles; (*f*) to work out the details of the inspection procedure by 31 December 1990; (*g*) to co-operate in developing and using safe, environmentally sound methods of destruction; (*h*) to cease CW production upon entry into force of the agreement rather than waiting for entry into force of the CWC; and (*i*) to take steps to encourage all CW-capable states to become parties to the future multilateral CWC.[18] For the CWC both the USA and the USSR agreed: (*a*) to accelerate their CW destruction under a global CWC so that, by the eighth year after its entry into force, they will have reduced their declared stocks to no more than 500 agent tonnes; and (*b*) to propose the convening of a special conference at the end of the eighth year to determine whether participation in the CWC is at that time sufficient to allow the elimination of CW stocks during the following two years.

A US–Soviet Joint Statement on Proliferation was signed by Presidents Bush and Gorbachev declaring 'their commitment to preventing the proliferation of nuclear weapons, chemical weapons, and missiles capable of

[15] 'Supermächte über Abrüstung einig', *Frankfurter Rundschau*, 21 May 1990, pp. 1, 2; 'Grundsatzeinigung über ein Start-Abkommen', *Neue Zürcher Zeitung*, 22 May 1990, pp. 1, 2; 'U.S–Soviet accord near on arms cuts', *International Herald Tribune*, 19–20 May 1990, pp. 1, 4; 'Superpowers progress in talks on arms reduction', *Financial Times*, 19–20 May 1990, p. 1; Oberdorfer, D., 'Superpowers clear hurdle to arms pact', *International Herald Tribune*, 21 May 1990, pp. 1, 4; 'Supermächte über Abrüstung einig', *Frankfurter Rundschau*, 21 May 1990, pp. 1, 2; Hermam, K. J., 'Abrüstungsschritte zum Gipfel', *Neues Deutschland*, 21 May 1990, p. 1.

[16] Felton, J., 'In the bag: chemical weapons pact', *Congressional Quarterly*, vol. 48, no. 21 (26 May 1990), p. 1664.

[17] Wieland, L. 'Vertrag über Vernichtung chemischer Waffen', *Frankfurter Allgemeine Zeitung*, 2 June 1990, p. 2; Pringle, P., 'Pact on nuclear missiles', *The Independent*, 2 June 1990, p. 1; 'Reduktion von C-Waffen: Ein gutes Beispiel gegen schlechte Sitten', *Europäische Wehrkunde*, vol. 39, no. 6 (June 1990), p. 323; 'Beschränkte Fortschritte im Abrüstungsbereich', *Neue Zürcher Zeitung*, 5 June 1990, p. 3; Simonitsch, P., 'Neue Perspektiven eröffnet', *Frankfurter Rundschau*, 6 June 1990, p. 2; 'Arms talks "completed successfully"', LD0206094390, Moscow, TASS International Service, 0715 GMT, 2 May 1990 (in Russian) in FBIS-SOV-90-108, 5 June 1990, pp. 10–11.

[18] 'U.S.–USSR Chemical Weapons Destruction Agreement', *Wireless File*, SFF-504 (United States Information Service, US Embassy: Stockholm, 1 June 1990), pp. 71–72.

carrying such weapons and certain other missiles and missile technologies'.[19] The two countries agreed *inter alia*: (*a*) to take steps to strengthen the 1925 Geneva Protocol by encouraging states that are not parties to accede to it; (*b*) to improve the effectiveness of their respective export controls to stop the spread of chemical weapons; (*c*) to join with other nations in multilateral efforts to co-ordinate export controls, exchange information and broaden international co-operation to stem CW proliferation; (*d*) to confirm their intent to pursue political and diplomatic actions where specific cases give rise to concerns about the production, use or spread of chemical weapons; (*e*) to affirm their intention to consider the imposition of sanctions against violators of the protocol, including those under Chapter VII of the UN Charter; and (*f*) to confirm their intention to provide active support to the UN Secretary-General in investigating reported violations of the Geneva Protocol.

The bilateral agreement increased political pressure on those still reluctant to support a global convention and refuted the argument that CW capability is needed to deter CW use. In the joint statement, the USA and the USSR declared 'that a multilateral, effectively verifiable chemical weapons convention . . . is the best long-term solution to the threat to international security posed by the use and spread of chemical weapons'.[20] The bilateral agreement also implies that both states have assumed the obligation never again to produce chemical weapons and that the treaty will remain in effect for an unlimited time until it is superseded by a multilateral treaty. In June the USA and the USSR introduced a joint working paper to the CD describing their proposal for the convening of a special conference eight years after entry into force of the CWC for the purpose of determining whether states parties should be required to destroy all of their remaining CW stocks.[21] The head of the Soviet delegation explained that this joint proposal was a compromise taking into account the US 2 per cent proposal and the criticism which that proposal had received initially from the USSR and others at the CD.[22]

Analysis of the joint statement suggests that the USSR accepted the 2 per cent proposal because the USA agreed to stop CW production. The most controversial aspect as it relates to the future CWC is the US–Soviet proposal to hold a 'special conference' eight years after entry into force of the CWC and that the conference should determine whether 'participation in the Convention is sufficient'. Nations which acquired chemical weapons would be required to destroy all of their chemical stockpiles only if the special conference succeeded. Additionally, states attending the special conference would be required to fulfil preconditions by: (*a*) declaring CW possession to the CD before 31 December 1991; (*b*) signing the CWC within 30 days after its opening for signature; and (*c*) becoming a party to the CWC not later than

[19] Conference on Disarmament document CD/1001, 12 June 1990; 'U.S.–Soviet non-proliferation joint statement', *Wireless File*, EUR-113 (United States Information Service, US Embassy: Stockholm, 4 June 1990), pp. 23–24; 'Two powers sign pledge to curb weapons spread', *Wireless File*, EUR-112 (United States Information Service, US Embassy: Stockholm, 4 June 1990), pp. 18–22.

[20] See 'Two powers sign pledge to curb weapons spread' (note 19).

[21] Conference on Disarmament document CD/CW/WP.303, 28 June 1990.

[22] Conference on Disarmament document CD/PV.560, 28 June 1990, pp. 2–13.

one year after its entry into force. The USA and the USSR will both attempt to persuade countries to declare possession of chemical weapons by the end of 1991 whether or not a convention exists.[23] Both the USA and the USSR might feel the need to keep 500 tonnes of chemical weapons for defence purposes if a majority of states do not participate in the special conference. Paradoxically, the proposed formula might have the effect on chemical proliferation that states would be induced to acquire chemical weapons prior to signature of the treaty in order to obtain the same status as the USA and the USSR (especially given the 31 December 1991 time-limit).[24] There has been much opposition to the proposal, especially from the non-aligned and the Group of 21 countries.[25] They argue that the proposal would also create a situation similar to that of the Non-Proliferation Treaty (NPT) in that some countries would be entitled to have chemical weapons and others not. Many countries find this to be discriminatory on the same grounds as they find the NPT discriminatory. Although some aspects of the new US–Soviet co-operation have not benefited the CWC, others most certainly have. It has been possible to work out one part of the provisions for the order of CW destruction, and a working paper to that effect was introduced in 1990.[26]

The inspections called for under the Memorandum of Understanding began in June with the visit of US inspectors to a Soviet CW storage site.[27] In June Soviet experts also visited the Tooele Army Depot in Utah;[28] in July they inspected three other US chemical facilities;[29] and in August they checked a chemical storage facility.[30] In August a team of US specialists also visited Chapayevsk, a production facility in Dzerzhinsk (which had been mothballed since the mid-1940s), and a storage facility for lewisite at Kambarka.[31] At the 16th round of bilateral talks held in Geneva in August,[32] both delegations

[23] Goldblat, J. and Bernauer, T., 'The US–Soviet chemical weapons agreement of June 1990: its advantages and shortcomings', *Bulletin of Peace Proposals*, vol. 21, no. 4 (Dec. 1990), pp. 355–62.

[24] Feinstein, L., 'Chemical talks slowed by U.S. insistence on "security stockpile"', *Arms Control Today*, vol. 20, no. 8 (Oct. 1990), pp. 27–28.

[25] The Group of 21 is an informal organization of some neutral and non-aligned states at the Conference on Disarmament. See Conference on Disarmament document CD/PV.567, 24 July 1990, pp. 9–10.

[26] See note 21.

[27] '6–8 June', *Chemical Weapons Convention Bulletin*, no. 9 (Sep. 1990), p. 11.

[28] 'Chemical weapons talks to be held in U.S', LD1906222990, FBIS-SOV-90-119, 20 June 1990, pp. 1–2; 'U.S. chemical weapons experts, counterparts meet', LD2206133490, Moscow, TASS, 1231 GMT, 22 June 1990 (in English) in FBIS-SOV-90-122, 25 June 1990, p. 19.

[29] '28 July–3 August', *Chemical Weapons Convention Bulletin*, no. 9 (Sep. 1990), p. 16.

[30] Institute for Defense and Disarmament Studies, *Arms Control Reporter* (IDDS: Brookline, Mass.), sheet 704.B.456, Aug. 1990.

[31] 'Soviet–U.S. chemical arms talks end in Geneva', LD1708192890, Moscow, TASS, 1711 GMT, 17 Aug. 1990 (in English) in FBIS-SOV-90-162, 21 Aug. 1990, p. 1; 'Americans tour chemical arms destruction center', LD2208184390, Moscow, TASS, 1804 GMT, 22 Aug 1990 (in English) in FBIS-SOV-90-164, 23 Aug. 1990, p. 1; 'U.S. experts visit chemical weapons facility', LD2408093290 Moscow, TASS, 0927 GMT, 24 Aug. 1990 (in English) in FBIS-SOV-90-165, 24 Aug. 1990, p. 1; 'U.S. delegation visits chemical arms plant', LD2708211590, Moscow Television Service, 1430 GMT, 27 Aug. 1990 (in Russian) in FBIS-SOV-90-167, 28 Aug. 1990, p. 1; 'Chemical weapon destruction "example" to all', PM2908153990, Moscow, *Izvestia*, 25 Aug. 1990, morning edn, p. 2, (in Russian) in FBIS-SOV-90-169, 30 Aug. 1990, pp. 3–4; 'U.S. chemical arms experts visit plants', LD3008094390, Moscow, TASS, 0933 GMT, 30 Aug. 1990 (in English) in FBIS-SOV-90-169, 30 Aug. 1990, p. 1.

[32] See 'Soviet–U.S. chemical arms talks end in Geneva' (note 31).

began work on a document dealing with inspection procedures.[33] In follow-up rounds in October and November, the USA and the USSR worked intensively on the inspection protocol for the bilateral agreement, which was to be completed by 31 December 1990.[34]

III. CW negotiations at the Conference on Disarmament

Despite a new mandate for the negotiations allowing the final elaboration of a CWC, no final drafting of the convention was in fact achieved owing to a number of political difficulties. The most important of these were: (*a*) the problems related to on-site inspections on challenge; (*b*) the US–Soviet proposal that 2 per cent of CW stockpiles could be withheld until eight years after entry into force of the CWC, when a conference between parties would be held to evaluate the convention's effectiveness; (*c*) the political review process in the USA which focused on the challenge inspection; and (*d*) the focus of US–Soviet effort on the bilateral agreement. It should also be mentioned here that the bilateral accord signed at the June summit meeting did not allow very much time for the CD to reconsider the situation. Nevertheless, much useful technical work was conducted.

Negotiation issues

The work of the 1989 *Ad Hoc* Committee on Chemical Weapons of the CD was concluded by 1 February 1990, under the Western Group chairmanship of Ambassador Pierre Morel of France, and a final report was presented to the CD.[35] (A comprehensive overview of the status of the negotiations during the past two years was published in 1990.[36]) When the 1990 session of the *Ad Hoc* Committee began, a new mandate was adopted which requested that the committee 'continue the full and complete process of negotiations, developing and working out the convention'.[37] The mandate no longer restricted the negotiations from achieving a final draft of the CWC, which was a political step forward.

Thirty-eight nations were admitted to the *Ad Hoc* Committee as observers including, for the first time, Israel.[38] Ambassador Carl-Magnus Hyltenius of

[33] 'U.S.–Soviet conclude productive CW talks in Geneva', *News Backgrounder* (United States Information Service, US Embassy: Stockholm, 24 Aug. 1990), pp. 1–2.

[34] 'Talks with U.S. on chemical weapons important', LD1710083190, Moscow, TASS, 0724 GMT, 17 Oct. 1990 (in English) in FBIS-SOV-90-201, 17 Oct. 1990, p. 1.

[35] Conference on Disarmament document CD/961, 1 Feb. 1990.

[36] Molander, J., 'Negotiating chemical disarmament', reprint from *Kungliga Krigsvetenskaps-akademiens Handlingar och Tidskrift*, Jan. 1990, pp. 19–34; Molander, J., 'The chemical weapons negotiations at a critical juncture: is there a way ahead?', *Chemical Weapons Convention Bulletin*, no. 10 (Dec. 1990), pp. 13–19.

[37] Conference on Disarmament document CD/968, 15 Feb. 1990.

[38] Conference on Disarmament document CD/1039, 30 Aug. 1990, p. 57: Austria, Bahrain, Bangladesh, Cameroon, Chile, Denmark, Democratic People's Republic of Korea, Finland, Ghana, Greece, Holy See, Honduras, Iraq, Ireland, Israel, Jordan, Kuwait, Libyan Arab Jamahiriya, Malaysia, New Zealand, Norway, Oman, Portugal, Qatar, Republic of Korea, Senegal, Spain, Sudan, Syrian Arab Republic, Switzerland, Tunisia, Turkey, United Arab Emirates, Uruguay, Viet Nam and Zimbabwe.

Sweden was appointed the new Chairman of the Committee, and three working groups were set up to deal with verification, technical, and legal and institutional issues.[39] Further consultations were held by the Chairman and Friends of the Chairman on specific questions.[40] The *Ad Hoc* Committee re-established the Technical Group on Instrumentation to deal separately with verification by instruments and other technical means.[41] A meeting was again held at the CD with representatives of the international chemical industry on issues relevant to the negotiations.[42] The following discussion illustrates some of the issues which have been dealt with and some which must still be resolved.

Verification by challenge and related legal problems

Challenge inspections based on the principle of 'any time, anywhere' date back to the draft treaty proposed by the USA in 1984. Although the negotiating parties acknowledged the need for challenge inspections to ensure compliance, some countries including India[43] and China[44] disagreed about the scope of such inspections. Since 1987 the 'any time, anywhere' concept has been favoured by countries which were members of the North Atlantic Treaty Organization (NATO) or the Warsaw Treaty Organization (WTO), but it has not been acceptable to most other countries. China, for example, fears political misuse of the challenge inspection and has suggested the 'right to launch appeals concerning the abuse of the right to challenge inspection, and to provide relevant evidence'. The Chinese proposal to determine the 'scope' of the challenge inspection as 'any facility, location or installation relevant to compliance, and the implementation of the convention'[45] can be seen as an attempt to limit the universality of the challenge concept. The debate focuses on the multilateral character of the challenge process and particularly on the question of whether or not a representative of the challenging state should be

[39] The three working groups dealt with the following issues: Working Group A: the protocol on inspection procedures, *ad hoc* verification measures and verification of alleged use of chemical weapons; Working Group B: Articles IV (chemical weapons) and V (CW production facilities) in particular the order of destruction, schedules, guidelines for Schedule 1, definitions, toxicity, thresholds and production capacity; and Working Group C: amendments, other final clauses, including the settlement of disputes, sanctions and the Organization; Conference on Disarmament document CD/1039, 30 Aug. 1990.

[40] The problems discussed in the open-ended consultations were: Article IX (consultations, co-operation and fact-finding); undiminished security and universal adherence to the Convention; function, composition and decision-making process of the Executive Council; Article XI (economic and technological development); Article X (assistance and protection against chemical weapons); old chemical weapons; jurisdiction and control; see also Conference on Disarmament document CD/CW/WP.316, 6 Aug. 1990.

[41] For the report of the group see Conference on Disarmament document CD/CW/WP.306, 17 July 1990.

[42] During this meeting a discussion of the following aspects took place: protection of confidential information; technical aspects of the Convention, in particular the contents of the schedules of chemicals together with their verification regimes; *ad hoc* verification; and possible conclusions to be derived from the national trial inspections carried out thus far.

[43] Conference on Disarmament document CD/PV.575, 21 Aug. 1990, pp. 11–16.

[44] Conference on Disarmament documents CD/PV.551, 12 Apr. 1990, pp. 2–9; CD/PV.576, 24 Aug. 1990, pp. 4–6.

[45] Conference on Disarmament document CD/1031, 10 Aug. 1990.

present. The USA now also seems uncomfortable with the 'any time, anywhere' concept, and in an August speech at the CD, the US ambassador gave no clear indication of the current US position on challenge inspections.[46] In an attempt to soften the phrase 'verification by challenge', Sweden proposed substituting 'suspicion of violation'.[47] There is now broad support for another phrase: 'inspections on request'. As a result of open-ended consultations on Article IX of the draft CWC, the formulation of the scope of the challenge inspection was changed to: 'Each State Party has the right to request an on-site inspection in any other State Party in order to clarify (and resolve) any matter which causes doubts about compliance with the provisions of the Convention'.[48] Despite special consultations held by the Chairman of the Committee[49] there was no resolution of questions about the role of observers, alternative arrangements, managed access and the role of the Executive Council in the challenge process. There were two comprehensive studies of the problems related to challenge inspections.[50]

The question of 'jurisdiction and control' is also unresolved. There are many complicated legal matters related to the extra-territorial rights of countries[51] such as questions about military bases, the rights of companies abroad, the matter of old chemical munitions left in another country by a state party, and so on.[52] An addition was made to the draft CWC[53] to reflect the progress made on this issue and to relate it to other convention provisions. Further elaboration is needed with respect to Article VII (national implementation measures) in order also to address private activities on the territory of a state party or extra-territorially. For example, obtaining entry to a privately owned facility may need to be dealt with by the judicial system of a state party in order to fulfil the provisions of Article VII.[54] This is particularly so for the United States but applies also to other countries.

[46] Conference on Disarmament document CD/PV.574, 16. Aug. 1990, pp. 18–21.

[47] Institute for Defense and Disarmament Studies, *Arms Control Reporter* (IDDS: Brookline, Mass.), sheets 704.B.434–35, Apr. 1990; see also Conference on Disarmament document CD/CW/WP.316, 6 Aug. 1990, p. 4.

[48] Conference on Disarmament document CD/1033, 10 Aug. 1990, pp. 216–19.

[49] Conference on Disarmament document CD/CW/WP. 316, 6 Aug. 1990.

[50] Lüdeking, R., 'Verifying a chemical weapons convention', *Chemical Weapons Convention Bulletin*, no. 9 (Sep. 1990), pp. 1–8; Cooper, G., 'Inspections on request: coming to terms with their scope', *Chemical Weapons Convention Bulletin*, no. 10 (Dec. 1990), pp. 1–3.

[51] Koplow, D. A., 'Long arms and chemical arms: extraterritoriality and the draft chemical weapons convention', *Yale Journal of International Law*, vol. 15, no. 1 (winter 1990), pp. 1–83.

[52] An example relating to bases is the discussion about whether or not the USA can place chemical weapons at its bases in the Philippines despite prohibitions on weapons of mass destruction in the Philippine Constitution; see 'Accord accused of violating chemical weapons ban', HK0702062990, Manila, *Philippines Newsday*, 7 Feb. 1990, p. 9 (in English) in Foreign Broadcast Information Service, *Daily Report–East Asia* (*FBIS-EAS*), FBIS-EAS-90-026, 7 Feb. 1990, p. 54; 'Study says bases can store chemical weapons', HK2401092590, Manila, *Manila Times*, 23 Jan. 1990, p. 2 (in English) in FBIS-EAS-90-018, 26 Jan. 1990, pp. 51–52.

[53] See CD/1033, pp. 187–88.

[54] One factor which may complicate the US policy decision is the Fourth Amendment of the US Constitution, under which admittance to a privately owned facility would not be allowed unless a court order had been obtained. This implies that an inspection, if granted, might be delayed longer than the 48 hours set as a limit for a challenge inspection as early as in the original US Draft for a CWC in 1984 (see Conference on Disarmament document CD/500, 18 Apr. 1984; Institute for Defense and Disarmament Studies, *Arms Control Reporter* (IDDS: Brookline, Mass.), sheet 704.C.1–2, Oct. 1990) and that, in fact,

Prohibition on use and order of destruction

The CWC must encompass the prohibition on use of the weapons covered by the 1925 Geneva Protocol, the BWC and the Environmental Modification (Enmod) Convention. The Group of 21 reaffirmed its position that the CWC should prohibit the use of such weapons under any circumstances.[55] When Iraq invaded Kuwait, the USA again expressed a desire 'to retain the right to use CW in retaliation',[56] which is also reflected in its reservations to the Geneva Protocol.

In 1990 the CD agreed on an order of destruction for chemical weapons and CW production facilities in the annexes to Articles IV and V of the draft CWC.[57] Chemical weapons are to be destroyed annually in equal quantities during the first eight years. By the end of the eighth year the remaining quantity of chemical weapons may not exceed 500 tonnes or 20 per cent of the quantity of chemical weapons declared by a state party, whichever is less. This was the first time that the delegations agreed on an exact number. Specific provisions for binary chemical weapons were also added, including comparative counting rules for the destruction of binary components so that when a particular key component is destroyed a corresponding quantity of the other component is also destroyed on a weight ratio calculation basis. The order of destruction of CW production facilities is based on the principle of levelling out. For example, for a state which is a party to the convention at the time of its entry into force, the overall destruction plan for facilities which produce Schedule 1 chemicals is separated into three destruction periods (years 2–5, years 6–8 and years 9–10).

The Annex on Chemicals

The working out of the Annex on Chemicals[58] is an example of what still remains to be done on other aspects of the CWC. The Annex on Chemicals contains all of the technical provisions which were previously contained in different articles and annexes. They have been grouped together in order to obtain a coherent overview of the definitions, schedules and verification regimes. This tidying up has long been needed to remove the contradictions created in the past by treating the same issues under different articles. The current verification approach for non-prohibited activities under Schedule 2 provides no resolution of the question of how to link existing quantitative

the request for admittance might even be turned down (see Tanzman, E. A., 'Constitutionality of warrantless on-site arms control inspections in the United States', *Yale Journal of International Law*, vol. 13, no. 1 (winter 1988), pp. 21–68). Similar situations may also exist in other countries.

[55] Conference on Disarmament documents CD/PV.567, 24 July 1990, pp. 9–10; CD/PV.532, 6 Feb. 1990, pp. 27–28; CD/PV.543, 15 Mar. 1990, pp. 21–22.

[56] Conference on Disarmament documents CD/PV.535, 15 Feb. 1990, pp. 12–13; CD/PV. 574, 16 Aug. 1990, pp. 18–21.

[57] See CD/1033, pp. 94–121.

[58] See CD/1033, pp. 58–76.

threshold criteria with applicable qualitative criteria for chosen chemical production facilities.[59]

Ad hoc verification

The non-production verification regime in the current draft CWC provides only for the declaration and verification of those facilities whose production exceeds certain thresholds for substances listed in the CWC schedules. This means that only a very limited number of facilities would be covered. Theoretically any chemical industry facility can be used or converted to produce chemical warfare agents or their precursors. The *ad hoc* check is a verification tool which was originally designed as a routine measure that would be at the disposal of the International Organization envisaged under the CWC. It is a complement to the non-production verification regime and provides another element of deterrence against the use of facilities for prohibited activities. In 1990 this approach was further elaborated upon by the Western Group[60] after the FRG introduced the idea of a national register,[61] in which a state party would declare all relevant facilities in its chemical industry as the basis for selecting facilities to be inspected under the *ad hoc* check approach. Checks would be conducted according to a quota system. The proposed national register would serve as a confidence-building measure (CBM). It has been proposed that a limit be placed on the number of *ad hoc* checks which could be requested annually by each state party or by the technical secretariat. In June the USA submitted a working paper on *ad hoc* visits[62] which endorsed the paper submitted by the Western Group.

National trial inspections

By the end of 1990, 20 CD member states had carried out national trial inspections (NTIs) as had 4 non-member states.[63] Most of the countries conducted routine NTIs at chemical industry facilities. After these initial inspections, some non-routine challenge NTIs were carried out at military installations by Czechoslovakia, the Federal Republic of Germany, France, the former German Democratic Republic, the Netherlands, the Soviet Union and the UK.

[59] See Molander, J., 'The chemical weapons negotiations at a critical juncture: is there a way ahead?' (note 36).

[60] Conference on Disarmament document CD/CW/WP.286, 11 Apr. 1990.

[61] Conference on Disarmament document CD/984, 10 Apr. 1990.

[62] Conference on Disarmament document CD/CW/WP.300, 27 June 1990.

[63] *Chemical Weapons Convention Bulletin*, no. 9 (Sep. 1990), pp. 20–22. The following CD member states conducted NTIs: Australia (CD/910), Belgium (CD/917), Brazil (CD/895), Canada (CD/987, CD/1030), Czechoslovakia (CD/900, CD/1021, CD/1022), Egypt (CD/958), the FRG (CD/912, CD/950, CD/975, CD/983), the GDR (CD/899,CD/996,CD/1020), France (CD/913, CD/960, CD/1029), Hungary (CD/969), India (CD/988), Iran (CD/1040), Italy (CD/893), Japan (CD/WP.228), the Netherlands (CD/924, CD/925, CD/1006), the Soviet Union (CD/894, CD/966), Sweden (CD/WP.216), the United Kingdom (CD/WP.249, CD/921, CD/1012), the United States (CD/922, CD/WP.301), Yugoslavia (CD/982); the following non-CD members conducted NTIs: Austria (CD/999), Finland (CD/WP.233), Norway (CD/WP.285), Switzerland (CD/WP.247).

In July the UK presented to the CD the results of a detailed two-year programme of at least six NTIs.[64] Among other conclusions, the UK reported that the concept of 'managed access' may be the key to establishing a balance between the protection of legitimate security interests and the degree of intrusiveness necessary for effective verification,[65] and also that no British site had been deemed sufficiently sensitive from the point of view of national security that access had been denied. The results of the British verification programme, particularly those related to challenge inspections, support the concept of inspection on request 'any time, anywhere' without the right of refusal. The sites chosen were inspected under the concept of 'managed access'. A system of random selective access was used (i.e., only a given percentage of buildings within a site or a part of a site, and/or a given percentage of rooms within a building and/or items within a room were available for inspection) in combination with shrouding of sensitive displays, material and equipment.[66] Multilateral trial inspections (MTIs) also need to be conducted in the future.

During the 1990 session some additional countries provided statistics about the production, consumption and transfer of chemicals under the schedules.[67]

Visit to Munster and the meeting with industry

The FRG[68] again invited the CD members and observer countries to visit its destruction facility at Munster to see new destruction techniques, methods to detect traces of chemical weapons or their breakdown products and the applicability of non-destructive techniques to the examination of munitions.[69] The destruction technology demonstrated in Munster is quite effective and is currently used to destroy old chemical weapons which are occasionally still found in the FRG.

In June another in the series of meetings with representatives of the chemical industry was held at the CD. The views of the industry representatives were reflected in a number of contributions in the following areas: (*a*) protection of confidential information; (*b*) technical aspects of the CWC, in particular the content of the schedules of chemicals and their verification regimes; (*c*) *ad hoc* verification; and (*d*) possible conclusions which can be

[64] Conference on Disarmament document CD/1012, 11 July 1990.

[65] Conference on Disarmament document CD/PV.564, 12 July 1990, pp. 2–6.

[66] See Cooper (note 50).

[67] A number of CD member states—Bulgaria, Hungary, Japan, Poland, Romania, the Soviet Union and Sweden—and participating non-members—Austria, Chile, Denmark and Finland—provided relevant data in 1990; see Conference on Disarmament documents CD/1017, CD/969, CD/CW/WP.281, CD/985, CD/1014, CD/CW/WP.264; CD/CW/WP.280, CD/971, CD/1042, CD/991, CD/CW/WP.297.

[68] Conference on Disarmament document CD/37, 12 July 1979.

[69] Conference on Disarmament document CD/1026, 3 Aug. 1990. One of the methods demonstrated is of particular interest: the SNAL (sample now, analyse later) method which makes it possible to obtain a large number of microsamples from the air or from a production process by absorption on a running tape over a relatively long period (months). The tape can be stored and analysed later in a mass-spectrometer. The miniaturization and relative freedom of maintenance make the method attractive in contexts where minimal access and long sampling periods are preferable, see Conference on Disarmament document CD/CW/WP.204, 19 July 1988.

drawn from the NTIs thus far carried out.[70] The chemical industry expressed the view that it does not favour the use of 'open-ended' categories or general descriptions of potentially dangerous chemicals. The chemical industry also suggested that there should be a review mechanism to decide on the inclusion of new chemicals and the removal of old ones from the schedules, and that even the exemption of chemical warfare agents from the schedules in cases of legitimate peaceful use should be discussed. The industry representatives agreed that occasional routine on-site inspections or challenge inspections were preferable to long-term, on-site monitoring activities.

Protocol on inspection procedures

In 1989 the first draft of a protocol for inspection procedures had been introduced into the rolling text, and in 1990 much additional work was conducted on it.[71] The common position of the USA and the USSR, particularly on challenge inspections conducted pursuant to Article IX (consultations, co-operation and fact finding), helped to accelerate elaboration of this part of the protocol. There was also development of another part of the protocol which deals with inspection procedures in the event of allegation of CW use.

Size and cost of the International Inspectorate

In 1990 the *Ad Hoc* Committee on Chemical Weapons and other bodies focused on the possible organization, tasks, size and cost of a Technical Secretariat for the CWC. Canada presented a study in which the annual operating cost of an International Inspectorate was estimated at $120 million for the first 10 years after entry into force of the CWC.[72] The cost would gradually diminish, as would the number of inspectors, particularly after destruction of all CW stockpiles. On the request of the Office of the Assistant to the Secretary of Defense (Atomic Energy), the US Institute for Defense Analyses made a similar study of the tasks of a Technical Secretariat from the US perspective.[73] The US Congressional Budget Office released a report in which the cost for on-site inspection and compliance with the bilateral CW agreement is estimated at $15–70 million annually.[74] Nevertheless it should be remembered that a number of political decisions still have not been made

[70] See statements from Conseil Europeen des Federations de l'Industrie Chimique (CEFIC), Brussels: 'CEFIC statement on confidential business information'; 'Executive summaries of the CEFIC statements on: (*a*) Schedules, (*b*) Confidential business information, (*c*) *Ad hoc* checks and the chemical industry'; 'CEFIC statement on schedules'; '*Ad hoc* checks and the chemical industry: discussion paper'; see also Chemical Manufacturers Association, 'Draft CMA position papers for consideration in Geneva', June 1990.

[71] See CD/1033, pp. 154–84.

[72] Conference on Disarmament document CD/1037, 23 Aug. 1990; *The Chemical Weapons Convention and the International Inspectorate: A Quantitative Study* (University of Saskatchewan: Saskatoon, Aug. 1990).

[73] Grotte, J. H., Leibbrandt, S. D. and Schultz, D. P., 'Inspection costs for a multilateral chemical weapons convention', IDA Paper P-2383 (Institute for Defense Analyses: Alexandria, Va., June 1990).

[74] 'Verification costs: forecast', *Trust and Verify*, no. 15 (Nov. 1990), p. 3.

which will influence the composition of the Technical Secretariat and the International Inspectorate. The size and cost of the International Inspectorate will be greatly influenced by whether it conducts routine inspections (*ad hoc* checks) or rare challenge inspections. The Technical Secretariat will also require modification after entry into force of the CWC in order to adequately accomplish the tasks assigned to it.

Technical contributions

A number of countries, including Canada, Finland and Norway (the latter two observers at the CD), provided substantial technical input to the negotiations in 1990. Finland consulted experts from several laboratories about an international test in 1989, and as a result of these discussions an inter-laboratory comparison test, a so-called 'round robin', was co-ordinated by the Finnish Research Group in Helsinki. The outcome was presented in a CD working paper[75] and published in the Finnish report series, Methodology and Instrumentation for Sampling and Analysis in the Verification of Chemical Disarmament.[76] Current procedures for sample preparation and analysis were tested to determine whether standardized operating procedures would be required for the CWC. Specially prepared samples were distributed to laboratories in different countries;[77] the laboratories were free to choose the method of preparation and analysis. The main objective of the test was to qualitatively identify the agents and their degradation products. In 1990 the Finnish Government began two courses as part of a training programme in analytical methods and the use of the instrumentation, knowledge of which will be needed to meet the verification tasks of the CWC.[78] Finland also chaired the Technical Group on Instrumentation which discussed necessary technical equipment for monitoring and verifying compliance.[79]

Norway presented another in its series of reports to the CD on the use of different sampling and analytical techniques for chemical warfare agents.[80]

Canada continued to make a substantial contribution to the work of the CD by submitting two technical reports on the use of new agents in cases of

[75] Conference on Disarmament document CD/CW/WP.288, 11 Apr. 1990.

[76] *International Inter-Laboratory Comparison (Round-Robin) Test for the Verification of Chemical Disarmament: F.1. Testing of Existing Procedures* (Ministry for Foreign Affairs of Finland: Helsinki, 1990).

[77] Laboratories in the following countries were involved: Australia, Canada, the FRG, Finland, France, the Netherlands, Norway, Sweden, Switzerland and the UK.

[78] Conference on Disarmament documents CD/PV.516, 11 July 1989; CD/932, 12 July 1989.

[79] Conference on Disarmament documents CD/CW/WP.272, 22 Jan. 1990; CD/CW/WP.306, 17 July 1990.

[80] Conference on Disarmament document CD/1019, 23 July 1990; see also Royal Norwegian Ministry of Foreign Affairs, *Research Report on Verification of a Chemical Weapons Convention: Use of Sorbent Extraction in Verification of Alleged Use of Chemical Warfare Agents, Part IX* (Royal Norwegian Ministry of Foreign Affairs: Oslo, July 1990).

alleged use and for toxicity determination.[81] Efforts continued in 1990 to study the industrial applicability of a chosen key-precursor (thiodiglycol).[82]

National implementation measures

The national implementation measures (Article VII of the CWC) which must be taken in order to implement the provisions of the CWC and to show compliance with it are of particular importance in the pre-convention phase.[83] Several countries, such as Argentina and Hungary, have expressed awareness of the need to begin efforts to prepare for the establishment of a future National Authority under the CWC.[84] SIPRI recently presented a study focusing on these problems.[85]

Offer to host the International Organization

Several offers to host the International Organization of the CWC have been made.[86] During the opening session of the CD in February 1990, the Austrian Foreign Minister reaffirmed an earlier offer by his country to provide financing and a site in Vienna for the International Organization. He noted that Vienna already hosts a number of international organizations, foremost among them the International Atomic Energy Agency (IAEA).[87] In August, the Netherlands repeated its offer to host the International Organization at The Hague and to provide financial support, training help and other necessary assistance.[88]

[81] Conference on Disarmament documents CD/993, 26 Apr. 1990; CD/1038, 23 Aug. 1990.

[82] Conference on disarmament document CD/CW/WP.279, 16 Mar. 1990; an earlier working paper was submitted by Canada in 1989 entitled 'Pinacolyl alcohol', Conference on Disarmament document CD/CW/WP.259, 14 Aug. 1989.

[83] Conference on Disarmament document CD/994, 30 Apr. 1990; *Role and Function of a National Authority in the Implementation of a Chemical Weapons Convention* (University of Saskatchewan: Saskatoon, Aug. 1990).

[84] At the Sep. 1989 Government–Industry Conference against Chemical Weapons in Canberra, Australia presented information about establishing a National Authority. In 1990 Hungary was among those countries to inform the CD (Conference on Disarmament document CD/PV.537, 22 Feb. 1990, pp. 2–9) of its efforts to to establish a national body. Argentina reported that it had set up a Commission for the Study of the Draft Convention to serve *inter alia* as a preliminary body until the establishment of an Argentine National Authority as envisaged under Article VII (see Conference on Disarmament document CD/PV.570, 2 Aug. 1990, pp. 18–21).

[85] Stock, T. and Sutherland, R. (eds), *National Implementation of the Future Chemical Weapons Convention*, SIPRI Chemical & Biological Warfare Studies, no. 11 (Oxford University Press: Oxford, 1990).

[86] In 1989 Belgium offered to host the International Organization; see Conference on Disarmament documents CD/PV.424, 23 July 1987, p. 15; CD/PV.506, 25 April 1990, pp. 12–13.

[87] Conference on Disarmament documents CD/PV.532, 6 Feb. 1990, p. 15; CD/972, 21 Feb. 1990.

[88] Conference on Disarmament documents CD/PV.575, 21 Aug. 1990, pp. 5–7; CD/PV.532, 6 Feb. 1990, p. 12; 'Holländisches Angebot für die C-Waffen-Agentur', *Neue Zürcher Zeitung*, 25 Aug. 1990, p. 5.

Proposal for a 1991 ministerial level meeting

During 1990 many delegations, including that of the USSR,[89] supported a French suggestion to organize a meeting in early 1991 'at the ministerial level to examine the status of the negotiations at that time, identify approaches needed to complete them, set a relatively short time-limit and give the corresponding instructions to delegations'.[90] One issue which might be addressed at such a meeting is the opposition to the 2 per cent proposal. However, no final decision was taken about such a meeting, perhaps owing in part to the outbreak in August of the conflict in the Persian Gulf.

The CD session: new organizational features

Summing up the results of the 1990 CD session, the Chairman of the *Ad Hoc* Committee listed the following as its major achievements: (*a*) elaboration of the rolling text on verification of alleged use; (*b*) establishment of an order of destruction; (*c*) development of the text of the amendments; and (*d*) introduction of methods for the settling of disputes.[91] New material related to jurisdiction and control, old chemical weapons and 'measures to redress a situation' is included in the current rolling text.

At the end of the summer session, two new organizational features were agreed upon. For the first time in the history of the CD, the new chairman of the *Ad Hoc* Committee on Chemical Weapons will be a Soviet citizen—Ambassador Sergei Batsanov, head of the Soviet Delegation.[92] Neither the USA nor the USSR has taken an active organizational role in the disarmament negotiations since the days of the co-chairmanship of the two countries during the time of the Eighteen-Nation Disarmament Committee (ENDC) and the Conference of the Committee on Disarmament (CCD). If the USA is given the chairmanship in 1992, this would imply that the two major players at the negotiations will assume greater responsibility for finalizing the drafting process. It has also now been agreed that from 1991 the work of the CD will consist of three annual sessions.[93] This will facilitate co-ordination of work at the CD with that conducted on the national level and allow for more effective utilization of the sessions.

For the inter-sessional working period in November and December 1990 and January 1991, the Chairman proposed setting up three different working groups to address various specific provisions of the rolling text.[94]

[89] Conference on Disarmament document CD/PV.574, 16 Aug. 1990, p. 10.

[90] Conference on Disarmament document CD/PV.570, 2 Aug. 1990, pp. 6–17.

[91] Conference on Disarmament document CD/PV.575, 21 Aug. 1990, pp. 2–5.

[92] See note 91.

[93] Conference on Disarmament document CD/8/Rev. 3, 21 Aug. 1990, pp. 1–7.

[94] The inter-sessional work was conducted on 26 Nov.–21 Dec. 1990 and 8–18 Jan. 1991; the working groups dealt with the following subjects: Working Group A: review of verification parts of the annexes to Articles IV, V and VI; review of the Protocol on Inspection Procedures, Parts I and II; Working Group B: the Annex on Chemicals and related issues; Article III; Working Group C: measures to redress a situation, settlement of disputes; amendments (improvement of the text); financial aspects of the Organization.

At the United Nations, the 45th General Assembly adopted by consensus three resolutions on CBW.[95] The first urged the CD to intensify its efforts during 1991 to conclude the CWC; the second called on the Secretary-General to provide the necessary service for the Third BWC Review Conference; and the third once more strongly reaffirmed the Geneva Protocol. It also supported the use of the power of the Secretary-General to investigate future cases of violation of the Geneva Protocol. The First Committee also adopted a study commissioned by the Secretary-General on the role of the UN in the field of verification and empowered him to *inter alia* start work on the creation of a data bank covering all aspects of verification and compliance related to the BWC and the future CWC. This data bank would be administered by the UN Department for Disarmament Affairs.[96] No decision was taken on the matter by the General Assembly.

IV. Chemical weapon-free zones and other measures to reduce the CW threat

During 1990 a number of suggestions were made for the creation of chemical weapon-free zones (CWFZ). Iraq proposed a zone encompassing nuclear, chemical and biological weapons, arguing that all weapons of mass destruction should be treated alike.[97] President Hosni Mubarak of Egypt and the Egyptian delegation to the CD also made a similar suggestion.[98] In December Soviet Foreign Minister Shevardnadze suggested at a meeting with US Secretary of State Baker that a nuclear and chemical weapon-free zone be discussed for the Middle East after resolution of the Persian Gulf crisis. Iraqi Foreign Minister Tariq Aziz and Israeli Prime Minister Yitzhak Shamir agreed, in principle, to this proposal.[99]

In 1990 a regional conference was held in Jordan under the auspices of the University of Jordan and the Quaker UN Office in Geneva.[100] Specific problems related to CW disarmament and the Middle East were discussed, but these suggestions were no longer given serious consideration when the Persian Gulf crisis broke out. A seminar on the implications of chemical weapons and

[95] United Nations General Assembly document A/45/777, 21 Nov. 1990; 'UN Committee discusses chemical weapons ban', LD1511085390, Moscow, TASS, 0724 GMT, 15 Nov 1990 (in English) in FBIS-SOV-90-222, 16 Nov. 1990, p. 1.

[96] '15 November', *Chemical Weapons Convention Bulletin*, no. 10 (Dec. 1990), p. 12.

[97] Conference on Disarmament document CD/PV.548, 3 Apr. 1990, pp. 19–22; 'Envoy addresses UN on mass destruction weapons', JN2504193390, Baghdad, INA, 1850 GMT, 25 Apr. 1990 (in Arabic) in Foreign Broadcast Information Service, *Daily Report–Near East & South Asia* (*FBIS-NES*), FBIS-NES-90-081, 26 Apr. 1990, p. 15.

[98] Conference on Disarmament document CD/989, 20 Apr. 1990; 'Mubarak addresses opening', NC2805170090, Cairo Domestic Service, 1318 GMT, 28 May 1990 (in Arabic) in FBIS-NES-90-103, 29 May 1990, pp. 7–10.

[99] 901211 and 901212, SHIB Rolling Chronology, Sussex–Harvard Information Bank, Science Policy Research Unit, University of Sussex, Brighton, UK, printout of 24 Jan. 1991.

[100] The Conference 'Towards a universal Chemical Weapons Convention' took place in Ma'in, Jordan, on 7–9 May 1990; see Institute for Defense and Disarmament Studies, *Arms Control Reporter* (IDDS: Brookline, Mass.), sheet 704.B.441, May 1990.

missile proliferation in Asia and the Pacific was held in Canberra on 8–9 March 1990.[101]

V. The Biological Weapons Convention

States parties to the BWC continued to provide confidence-building information about their convention-related activities to the UN Department for Disarmament Affairs as agreed at the Second Review Conference in 1986.[102] In 1990 five of the states parties provided the requested information for the first time.[103] There is thus still not general participation in the information exchange by all parties to the BWC. This is cause for concern and measures should be taken to remedy the situation at the Third Review Conference in September 1991. There is a need to clarify more precisely what data should be provided in the information exchanges. States parties should be encouraged to participate in the exchange or to indicate why they have refrained from doing so. Serious reconsideration of the information exchange process could lead to more substantial, relevant results. Overloading the process with too much information which is of dubious value creates compliance difficulties. This issue and others relevant to the BWC are discussed in two recent SIPRI publications.[104]

Discussion and studies have been carried out in preparation for the 1991 Review Conference by governments and other interested parties. Those undertaken by governments will not be disclosed until the Preparatory Committee for the review conference meets in Geneva on 8–12 April 1991.[105] In September 1990 a conference was held in the GDR on Prevention of a Biological and Toxin Arms Race and the Responsibility of Scientists.[106] The conference dealt with developmental aspects of potential new biological weapons, possible means and methods for verification of the BWC and ethical aspects of research related to biological weapons.[107] The Federation of

[101] 'Workshop on chemical weapons and missile proliferation: implications for Asia/Pacific', *Pacific Research*, vol. 3, no. 1 (Feb. 1990), p. 32.

[102] See *SIPRI Yearbook 1990* (note 6), pp. 540–41; SIPRI, *SIPRI Yearbook 1989: World Armaments and Disarmament* (Oxford University Press: Oxford, 1989), pp. 114–45.

[103] See Geissler, E., 'Contribution of confidence-building measures to greater transparency in activities directly related to the Biological Weapons Convention', S. J. Lundin (ed.), *Views on Possible Verification Measures for the Biological Weapons Convention*, SIPRI Chemical & Biological Warfare Studies, no. 12 (Oxford University Press: Oxford, 1991). The following countries participated for the first time in the information exchange: Austria, Chile, Ecuador, Greece and Portugal.

[104] Geissler, E. (ed.), *Strengthening the Biological Weapons Convention by Confidence-Building Measures*, SIPRI Chemical & Biological Warfare Studies, no. 10 (Oxford University Press: Oxford, 1990); see also Lundin (note 103).

[105] For example, the Swedish National Defence Research Institute (FOA) arranged a meeting of governmental experts from 11 countries on improving confidence-building measures on 29–30 May 1990; the results have not been published.

[106] 12th Kühlungsborn Colloquium on Philosophical and Ethical Problems of the Biosciences, *Prevention of a Biological and Toxin Arms Race and the Responsibility of Scientists*, 14-19 Sep. 1990, organized by the Society of Physical and Mathematical Biology of the GDR together with the Central Institute of Molecular Biology and the Central Institute of Philosophy, both of the Academy of Sciences of the GDR, and the GDR Committee on Scientific Questions of Peace and Disarmament.

[107] Geissler, E. and Haynes R. H. (eds), *Prevention of a Biological and Toxin Arms Race and the Responsibility of Scientists* (Academy Verlag: Berlin, 1991).

American Scientists (FAS) presented its proposals for improvement of the BWC.[108] This study was the product of a co-operative effort by FAS representatives, international scientists and disarmament experts, but the report was produced solely by the FAS. Both the conference and the FAS study strongly advocated that there should be more explicit coverage by the BWC of potential new biological weapons. It was also argued that great care should be taken so that the BWC is not weakened by well-intentioned efforts to cover developments which are currently not perceived as threats to the BWC or which have only civil application. The review conferences (or other, new decision-making bodies under the BWC) should continue to assess which new technical developments might be misused for weapon development and therefore subject to coverage by the BWC.[109] Article IV of the BWC, which calls for national control measures, could be acted upon and states parties should implement national legislation promoting compliance with the treaty.[110] The US Biological Weapons Anti-Terrorism Act of 1989, which was signed by President Bush in May 1990, is an example of such legislation.[111] Thus far 10 countries have enacted national legislation in response to Article IV. The national implementation measures required under the future CWC which have already been started to be developed by some states are also worth noting in this context.[112] It should also be mentioned that, in September 1990 at its annual Pugwash Conference, the Pugwash Council approved a statement to be submitted to the Third Review Conference.[113]

VI. International environmental agreements relevant to CBW negotiations

It is obvious that implementation of the CWC will complement or overlap with many other initiatives to limit chemical damage to mankind and the environment. Current international efforts to negotiate other agreements to protect the environment and to prevent the misuse of chemical and biological agents are particularly relevant to the CWC negotiations. The following is a brief overview of these efforts.

In May 1990, at a UN-sponsored environmental conference in Bergen, Norway, delegates repeatedly called upon the rich countries to pay incentives

[108] Federation of American Scientists, *Proposals for the Third Review Conference of the Biological Weapons Convention*, Report of the Federation of American Scientists Working Group on Biological and Toxin Weapons Verification (FAS: Washington, Oct. 1990).

[109] Lundin, S. J., 'The present state of the negotiations on chemical weapons', Geissler and Haynes (note 107), pp. 219–37.

[110] See Scott, D., 'The concept of treaty-mandated compliance legislation and the Biological Weapons Convention', Geissler and Haynes (note 107), pp. 345–67.

[111] See Institute for Defense and Disarmament Studies, *Arms Control Reporter* (IDDS: Brookline, Mass.), sheet 701.B.59, July 1990; for the text see Wright, S. (ed.), *Preventing a Biological Arms Race* (MIT Press: Cambridge, 1990), pp. 406–9.

[112] Stock, T., 'The future Chemical Weapons Convention and Article VII: the national implementation requirement for each State Party and the lessons for the further strengthening of the BWC regime', Geissler and Haynes (note 107), pp. 369–81.

[113] 'Statement of the Pugwash Council for the Third Review Conference of the Biological Weapons Convention of 1972', *Pugwash Newsletter*, vol. 28, no. 2 (Oct. 1990), pp. 99–101.

to Third World countries in order to persuade them to take the steps needed to curb pollution and to utilize environmentally safe processes.[114] A June 1990 follow-up meeting to the Montreal Protocol on Substances that Deplete the Ozone Layer[115] took place in London at which senior ministers and representatives of many non-governmental and technical bodies from some 90 countries participated. During the conference, the experts decided to ban the use of chlorofluorocarbons (CFCs), halons and carbon tetrachloride by 2000.[116] Third World countries will be given a 10-year period in which to comply with the measures. The agreement calls for phasing out methyl chloroform by 2005, and elimination of the use of hydrochlorofluorocarbons (HCFCs) no later than 2040, and earlier than 2020 if possible. India and China, both non-members of the Montreal Protocol and countries which emit large quantities of ozone-destroying chemicals into the atmosphere, are expected to become parties to the protocol. Another important result was that a fund will be set up to help diminish the costs faced by Third World countries in switching to chemicals and technologies which are less damaging to the ozone layer. The multilateral fund for 1991–93 totals $160 million, and when India and China become parties to the Montreal Protocol, the fund will reach $240 million.[117] In continuation of these efforts several expert meetings took place to discuss the effect on mankind and the environment of the accumulation of so-called 'greenhouse' gases .[118] Current scientific research shows that the Montreal Protocol's concept of 'ozone depletion potential', which compares the impact of different compounds on the stratospheric ozone layer and calls for CFC substitutes such as HCFCs, is probably insufficient to deal with the problem.[119]

At the October 1990 Second World Climate Conference in Geneva, politicians and experts from over 100 nations discussed the steps to be taken to deal with global warming, which now seems to be recognized as a real threat by governments.[120] The Geneva meeting was not constituted to negotiate a

[114] North, R., 'US defends stance on global warming aid to third world', *The Independent*, 16 May 1990, p. 6.

[115] At the time of the London Conference, 56 states parties had ratified the Montreal Protocol of Sep. 1987 on Substances that Deplete the Ozone Layer.

[116] O'Sullivan, D., 'Ozone depletion accord: plan will aid developing nations', *Chemical & Engineering News*, vol. 68, no. 28 (9 July 1990), pp. 6–7; Rothe, M., '"Eingreiftruppe" gegen Ozonkiller macht mobil', *Neues Deutschland*, 30 June 1990, p. 4; 'Die "Ozonkiller" sollen bis zum Jahr 2000 ganz abgeschafft sein', *Frankfurter Rundschau*, 2 July 1990, p. 1.

[117] Milne, R., 'Nations approach unity on measures to protect ozone', *New Scientist*, vol. 126, no. 1723 (30 June 1990), p. 33.

[118] Gamillscheg, H., 'Treibhauseffekt vertreibt Millionen Menschen', *Frankfurter Rundschau*, 29 Aug. 1990, p. 6; Gamillscheg, H., 'Lähmung angesichts des Treibhauseffekts', *Frankfurter Rundschau*, 1 Sep. 1990, p. 5; 'Halon-Stopp zum Schutze der Ozonschicht', *Neue Zürcher Zeitung*, 4 Oct. 1990, p. 9; 'EG für Verbot der Ozonkiller von 1997 an', *Frankfurter Allgemeine Zeitung*, 9 June 1990, p. 4.

[119] See 'Scientists warn of perils posed by substitutes for CFCs', *New Scientist*, vol. 126, no. 1723 (30 June 1990), p. 33.

[120] 'Comment: global warming', *New Scientist*, vol. 128, no. 1743 (17 Nov. 1990), p. 13; 'Next steps on global warming', *Nature*, vol. 348, no. 6298 (15 Nov. 1990), pp. 181–82; Simonitsch, P., 'UN verlangen rasch Massnahmen gegen den Treibhauseffekt', *Frankfurter Rundschau*, 30 Oct. 1990, p. 1; Möhring, C., 'Die Menschen müssen ihr Verhalten ändern', *Frankfurter Allgemeine Zeitung*, 5 Nov. 1990, p. 3.

treaty on greenhouse gases but to collect advice about what should be done. An agreement was reached to begin negotiations at a meeting to be held in Washington in early 1991 in the hope that this would lead to the drafting of a climate treaty for presentation at the UN Conference on Environment and Development to be held in June 1992 in Brazil. In order to keep all of the negotiating parties, including the USA, at the table, the conference agreed only that greenhouse emissions should be significantly limited, but no specific limits were mandated. The Second Climate Conference also clearly showed that some key countries are not yet ready to accept emission restrictions.

Prior to the Second Climate Conference the European Community (EC) ministers for environment and energy met in Luxembourg and agreed upon the first binding international agreement on reduction of the emission of greenhouse gases, particularly carbon dioxide, to 1990 levels by the year 2000 in the EC countries.[121] In order to accomplish this, countries will stabilize their emissions at the 1990 level. Several countries, such as France and the FRG, hope to drastically reduce their emissions by the year 2000.

A number of countries, including the FRG and the USA, also took national initiatives by adopting legislation and regulations to improve industrial technology for air cleaning and to minimize atmospheric pollution.[122]

VII. Conclusions

The following conclusions can be drawn about the bilateral and multilateral negotiation efforts on chemical and biological weapons in 1990:

1. The US–Soviet bilateral agreement on chemical weapons was an important and necessary step towards a multilateral, comprehensive, global CWC. The agreement indicates that the USA and the USSR accept the need to diminish their stocks of chemical weapons by destruction. The hesitation which they appear still to feel about complete destruction of all chemical weapons probably owes to the fact that not all countries share their doubts about the wisdom of acquiring chemical weapons. The negotiation of this agreement demonstrated very clearly how it is possible, where there is political will, to achieve a positive result and to overcome technical and political obstacles. The bilateral agreement also constitutes a basis upon which meaningful multilateral negotiations can build.

For other countries the bilateral destruction undertaking and the announced cessation of CW production constitute powerful political signals, which should also receive positive attention at the CD. However, linking the bilateral

[121] MacKenzie, D., 'Europe split on policy to stem global warming', *New Scientist*, vol. 128, no. 1741 (3 Nov. 1990), p. 19.

[122] 'The Clean Air Act Amendments of 1990', *Wireless File*, no. 221 (United States Information Service, US Embassy: Stockholm, 15 Nov. 1990), pp. 8–10; Schaffer, J., 'Congress approves major clean air bill', *Wireless File*, no. 209 (United States Information Service, US Embassy: Stockholm, 29 Oct. 1990), pp. 24–25; 'Bonn will Schadstoffausstoss mindern', *Frankfurter Allgemeine Zeitung*, 8 Oct. 1990, p. 2; 'Klimaschutz ist nicht nur eine Aufgabe "des Staates"', *Frankfurter Rundschau*, 29 Nov. 1990, pp. 36–37.

agreement with the multilateral chemical disarmament process has previously been opposed by some CD delegations. Under the bilateral agreement, the USA and the USSR began to exchange information and visits to their CW facilities on a scale that was not previously thinkable, thereby setting an example for other countries. The pledge to co-operate in solving the problems is positive even taking into account the possibility that the USA and the USSR may be unable to destroy their weapons within the agreed time-schedule owing to a variety of difficulties such as lack of adequate destruction techniques and facilities and international environmental protest.

2. The CD negotiations have not led to the final convention, and some countries have now begun to voice concern about the verification provisions of the future CWC, including the challenge inspection (inspection on request) and the proposed *ad hoc* verification mechanism. However, there has been a decisive political improvement in that the mandate for the negotiations now no longer excludes final drafting. Conclusion of the convention is probably possible once decisive political decisions have been taken. The technical work currently conducted at the CD is necessary and, once done, may shorten the time needed to conclude the CWC after political agreement.

The national trial inspections thus far performed by 24 countries have provided the negotiators with the sort of technical and political information which will become increasingly valuable as work on the final text begins. It is possible that some of this work will also be conducted by the future Preparatory Commission and may even continue after entry into force of the CWC. Several countries continued to support the negotiations by technical contributions. Data about production statistics were also provided to the CD by a number of countries. The 1990 negotiations produced some progress on the elaboration and finalization of the text of the draft convention, in particular on the order of destruction.

3. The crisis in the Persian Gulf will probably make the countries involved in the multilateral force less interested in concluding the CWC. On the other hand, if the crisis is solved under UN auspices, there should be a strong impetus to get rid of all weapons of mass destruction in that region. In that case, a binding commitment to rapidly conclude the work on the CWC would be highly desirable so that as many states as possible could be primary signatories to it.

4. The military security of many countries may to a certain extent have been influenced by whether or not they possess chemical or biological weapons. A growing number of agreements address civilian use of chemicals and such matters as the release of dangerous chemicals into the atmosphere and large industrial disasters. The national and international agreements on these problems have led to arrangements which are closely related to those that will be necessary for the national implementation of the CWC. It will be necessary for countries and their negotiating experts to harmonize the provisions of the CWC with these. Also during the Second World Climate Conference in 1990, the participating countries agreed to start negotiations on a climate treaty.

5. The primary extra-governmental preparations for the 1991 Third Review Conference on the BWC addressed future risks to the BWC resulting from inadequacies in its text. Fears were expressed that new developments in the biological sciences and biological techniques could be misused for development of biological weapons and may not be adequately covered by the current BWC. However, the review conferences have dealt with this matter by scrutinizing new developments to see whether they might fall outside the scope of the convention. It is important that concerns not be raised about perfectly legitimate research. Chemical and biological terms should also not be linked in general with the word 'weapon' to give the impression that there may soon exist new effective chemical and biological weapons. It is, of course, possible that research and development may lead to the development of biological weapons, but this is a possibility which is addressed by the BWC.

6. The information exchange about activities related to the BWC did not function significantly better in 1990, and this process needs to be discussed and further improved at the Third Review Conference. This and many other suggestions which have been made to improve the verification and compliance activities under the BWC will also need to be given adequate attention.

Appendix 14A. Agreement between the United States of America and the Union of Soviet Socialist Republics on Destruction and Non-Production of Chemical Weapons and on Measures to Facilitate the Multilateral Convention on Banning Chemical Weapons*

The United States of America and the Union of Soviet Socialist Republics, hereinafter referred to as 'the Parties',

Determined to make every effort to conclude and to bring into force at the earliest data a convention providing for a global ban on the development, production, stockpiling and use of chemical weapons and on their destruction, hereinafter referred to as 'the multilateral convention',

Aware of their special responsibility in the area of chemical weapons disarmament,

Desiring to halt the production of chemical weapons and to begin the destruction of the preponderance of their chemical weapons stockpiles, without waiting for the multilateral convention to enter into force,

Recalling the Memorandum of Understanding between the Government of the United States of America and the Government of the Union of Soviet Socialist Republics Regarding a Bilateral Verification Experiment and Data Exchange Related to Prohibition of Chemical Weapons, signed at Jackson Hole, Wyoming on 23 September 1989, hereinafter referred to as 'the Memorandum',

Recalling the bilateral commitment to co-operate with respect to the destruction of chemical weapons, contained in the joint statement on chemical weapons issued at Jackson Hole, Wyoming on 23 September 1989, and

Mindful of the efforts of each Party aimed at the destruction of chemical weapons and desiring to co-operate in this area,

Have agreed as follows:

Article 1. General Provisions and Areas of Co-operation

1. In accordance with provisions of this Agreement, the Parties undertake:

(*a*) to co-operate regarding methods and technologies for the safe and efficient destruction of chemical weapons;

(*b*) not to produce chemical weapons;

(*c*) to reduce their chemical weapons stockpiles to equal, low levels;

(*d*) to co-operate in developing, testing, and carrying out appropriate inspection procedures; and

(*e*) to adopt practical measures to encourage all chemical weapons-capable States to become parties to the multilateral convention.

2. Each Party, during its destruction of chemical weapons, shall assign the highest priority to ensuring the safety of people and to protecting the environment. Each Party shall destroy its chemical weapons in accordance with stringent national standards for safety and emissions.

Article II. Co-operation Regarding Methods and Technologies of Destruction

1. To implement their undertaking to co-operate regarding the destruction of chemical weapons, the Parties shall negotiate a specific programme of co-operation. For this purpose, the Parties may create special groups of experts, as appropriate. The programme may include matters related to: methods and specific technologies for the destruction of chemical weapons; measures to ensure safety and protection of people and the environment; construction and operation of destruction facilities; the appropriate equipment for destruction; past, current and planned destruction activities; monitoring of destruction of chemical weapons; or such other topics as the Parties may agree. Activities to implement this programme may include: exchanges of visits to relevant facilities; exchanges of documents; meetings and discussions among experts; or such other activities as the Parties may agree.

2. Each Party shall, as appropriate, cooperate with other States that request information or assistance regarding the destruction of chemical weapons. The Parties may respond jointly to such requests.

Article III. Cessation of the Production of Chemical Weapons

Upon entry into force of this Agreement and thereafter, each Party shall not produce chemical weapons.

Article IV. Destruction of Chemical Weapons

1. Each Party shall reduce and limit its chemical weapons so that, by no later than 31 December 2002, and thereafter, its aggregate quantity of chemical weapons does not exceed 5,000 agent tons. In this Agreement, 'tons' means metric tons.

2. Each Party shall begin its destruction of chemical weapons by no later than 31 December 1992.

3. By no later than 31 December 1999, each Party shall have destroyed at least 50 per cent of its aggregate quantity of chemical weapons. The aggregate quantity of chemical weapons of a Party shall be the amount of chemical weapons declared in the data exchange carried out on 29 December 1989, or declared thereafter, pursuant to the Memorandum, as updated in accordance with paragraph 6(b) of this article.

4. In the event that a Party determines that it cannot achieve an annual rate of destruction of chemical weapons of at least 1,000 agent tons during 1995, or that it cannot destroy at least 1,000 agent tons during each year after 1995, that Party shall, at the earliest possible time, notify the other Party, in accordance with paragraph 10 of this article.

5. Each Party, in its destruction of chemical weapons, shall also destroy the munitions, devices and containers from which the chemicals have been removed. Each Party shall reduce and limit its other empty munitions and devices for chemical weapons purposes so that, by no later than 31 December 2002, and thereafter, the aggregate capacity of such munitions and devices does not exceed the volume of the remaining bulk agent of that Party.

6. Thirty days after the entry into force of this Agreement, each Party shall inform the other Party of the following:

(*a*) its current general plan for the destruction of chemical weapons pursuant to this Agreement and its detailed plan for the destruction of chemical weapons during the calendar year following the year in which this Agreement enters into force. The detailed plan shall encompass all of the chemical weapons to be destroyed during the calendar year, and shall include their locations, types and quantities, the methods of their destruction, and the locations of the destruction facilities that are to be used; and

(*b*) any changes, as of the entry into force of this Agreement, in the data contained in the data exchange carried out on 29 December 1989, or provided thereafter, pursuant to the Memorandum.

7. Beginning in the calendar year following the year in which this Agreement enters into force, each Party shall inform the other Party annually, by no later than 30 November, of its detailed plan for the destruction of chemical weapons during the following calendar year.

8. Beginning in the calendar year following the year in which this Agreement enters into force, each Party shall inform the other Party annually, by no later than 15 April, of the following:

(*a*) any further changes, as of 31 December of the previous year, to the data contained in the data exchange carried out on 29 December 1989, or provided thereafter, pursuant to the Memorandum;

(*b*) the implementation during the previous calendar year of its detailed plan for the destruction of chemical weapons; and

(*c*) any update to the general and detailed plans provided pursuant to paragraphs 6(a) or 7 of this article.

9. Each Party shall limit its chemical weapons storage facilities so that, by no later than 31 December 2002, and thereafter, the number of such facilities does not exceed eight. Each Party plans to have all such facilities located on its national territory. This is without prejudice to its rights and obligations, including those under the Protocol for the Prohibition of the Use in War of Asphyxiating, Poisonous or Other Gases, and of Bacteriological Methods of Warfare, signed at Geneva on 17 June 1925.

10. If a Party experiences problems that will prevent it from destroying its chemical weapons at a rate sufficient to meet the levels specified in this article, that Party shall immediately notify the other Party and provide a full explanation. The Parties shall

promptly consult on measures necessary to resolve the problems. Under no circumstances shall the Party not experiencing problems in its destruction of chemical weapons be required to destroy its chemical weapons at a more rapid rate than the Party that has experienced such problems.

Article V. Inspection Activities

1. Each Party shall provide access to each of its chemical weapons production facilities for systematic on-site inspection to confirm that production of chemical weapons is not occurring at those facilities.

2. Each Party shall identify and provide access to each of its chemical weapons destruction facilities and the chemical weapons holding areas within these destruction facilities for systematic on-site inspection of the destruction of chemical weapons. Such inspection shall be accomplished through the continuous presence of inspectors and continuous monitoring with on-site instruments.

3. When a Party has removed all of its chemical weapons from a particular chemical weapons storage facility, it shall promptly notify the other Party. The Party receiving the notification shall have the right to conduct, promptly after its receipt of the notification, an on-site inspection to confirm that no chemical weapons are present at that facility. Each Party shall also have the right to inspect, not more than once each calendar year, subsequent to the year of the notification and until such time as the multilateral convention enters into force, each chemical weapons storage facility for which it has received a notification pursuant to this paragraph, to determine that chemical weapons are not being stored there.

4. When a Party has completed its destruction of chemical weapons pursuant to this Agreement, it shall promptly notify the other Party. In its notification, the Party shall specify the chemical weapons storage facilities where its remaining chemical weapons are located and provide a detailed inventory of the chemical weapons at each of these storage facilities. Each Party, promptly after it has received such a notification, shall have the right to inspect each of the chemical weapons storage facilities specified in the notification, to determine the quantities and types of chemical weapons at each facility.

5. Each Party shall also have the right to inspect, not more than once each calendar year, subsequent to the year in which destruction begins and until such time as the multilateral convention enters into force, each chemical weapons storage facility of the other Party that is not already subject to annual inspection pursuant to paragraph 3 of this article, to determine the quantities and types of chemical weapons that are being stored there.

6. On the basis of the reports of its inspectors and other information available to it, each Party shall determine whether the provisions of this Agreement are being satisfactorily fulfilled and shall communicate its conclusions to the other Party.

7. Detailed provisions for the implementation of the inspection measures provided for in this Article shall be set forth in the document on inspection procedures. The Parties shall work to complete this document by 31 December 1990.

Article VI. Measures to Facilitate the Multilateral Convention

The Parties shall co-operate in making every effort to conclude the multilateral convention at the earliest date and to implement it effectively. Toward those ends, the Parties agree, in addition to their other obligations in this Agreement, to the following:

1. Each Party shall reduce and limit its chemical weapons so that, by no later than the end of the eighth year after entry into force of the multilateral convention, its aggregate quantity of chemical weapons does not exceed 500 agent tons.

2. Upon signature of this Agreement, the Parties shall enter into consultations with other participants in the multilateral negotiations and shall propose that a special conference of States parties to the multilateral convention be held at the end of the eighth year after its entry into force. This special conference would, *inter alia*, determine, in accordance with agreed procedures, whether the participation in the multilateral convention is sufficient for proceeding to the total elimination of all remaining chemical weapons stocks over the subsequent two years.

3. The Parties shall intensify their co-operation with each other and with other States to ensure that all chemical weapon-capable States become parties to the multilateral convention.

4. The Parties declare their intention to be among the original parties to the multilateral convention.

5. To gain experience and thereby facilitate the elaboration and implementation of the multilateral convention, the Parties agree to conduct bilateral verification experiments involving trial challenge inspections at facilities not declared under the Memorandum or subsequently. The detailed modalities for such experiments, including the number and location of the facilities to be inspected, as well as the procedures to be used, shall be agreed between the Parties no later than six months after the signing of this Agreement.

Article VII. Consultations

The Parties, in order to resolve questions related to this Agreement that may arise, shall use normal diplomatic channels, specifically-designated representatives, or such other means as they may agree.

Article VIII. Relationship to other Documents

1. After the multilateral convention enters into force, the provisions of the multilateral convention shall take precedence over the provisions of this Agreement in cases of incompatible obligations therein. Otherwise, the provisions of this Agreement shall supplement the provisions of the multilateral convention in its operation between the Parties. After the multilateral convention is signed, the Parties to this Agreement shall consult with each other in order to resolve any questions concerning the relationship of this Agreement to the multilateral convention.

2. The chemical weapons, chemical weapons storage facilities, and chemical weapons production facilities subject to this Agreement are those that are subject to declaration under the Memorandum.

Article IX. Amendments

Each Party may propose amendments to this Agreement. Agreed amendments shall enter into force in accordance with the procedures governing the entry into force of this Agreement.

Article X. Entry into force; Duration; Withdrawal

1. This Agreement shall enter into force upon an exchange of instruments stating acceptance of the Agreement by each Party.

2. This Agreement shall be of unlimited duration, unless the Parties agree to terminate it after the entry into force of the multilateral convention.

3. Each Party shall, in exercising its national sovereignty, have the right to withdraw from this Agreement if it decides that extraordinary events related to the subject matter of this Agreement have jeopardized its supreme interests. It shall give notice of its decision to the other Party six months prior to withdrawal from the Agreement. Such notice shall include a statement of the extraordinary events the notifying Party regards as having jeopardized its supreme interests.

Done at Washington, in duplicate, this first day of June, 1990, in the English and Russian languages, each text being equally authentic.

---

* The Agreement was signed for the USA by President George Bush and for the USSR by President Mikhail Gorbachev.

15. Multilateral and bilateral efforts towards nuclear test limitations

RAGNHILD FERM

I. Introduction

In 1990 the US–Soviet 1974 Threshold Test Ban Treaty (TTBT) and 1976 Peaceful Nuclear Explosions Treaty (PNET) were finally ratified after almost three years of negotiations on new verification protocols. Parallel with these bilateral talks multilateral efforts were made towards a comprehensive test ban (CTB): decisions were taken to hold a Partial Test Ban Treaty (PTBT) amendment conference (to extend the Treaty by forbidding nuclear testing not only in the atmosphere, in outer space and under water, but also under ground) and to re-establish the *Ad Hoc* Committee on a Nuclear Test Ban at the Conference on Disarmament (CD).

II. The US–Soviet test negotiations

The TTBT and the PNET, signed in 1974 and 1976, respectively, entered into force on 11 December 1990 with the exchange of the instruments of ratification by the foreign ministers. The treaties limit the yield[1] of nuclear explosions to 150 kt. A major purpose of the Threshold Test Ban Treaty when it was negotiated was to try to slow down the development of high-yield nuclear weapons which were regarded as potential first-strike weapons. However, nuclear warheads for smaller weapons with lower yields to be deployed on more precise delivery systems were already under development, and most of the US and Soviet explosions conducted between 1969 and 1973 were below 150 kt.

So-called peaceful nuclear explosions (PNEs) cannot be distinguished from explosions conducted for military purposes. The Peaceful Nuclear Explosions Treaty was therefore negotiated to prevent high-yield PNEs being utilized for military testing purposes. It should be noted that the USA has not conducted a PNE since 1973, and the latest Soviet nuclear explosion carried out outside the usual test sites and therefore presumed to serve non-weapon purposes took place in 1988.

Since the treaties were signed no explosions with a yield above 150 kt appear to have been unambiguously detected and proven to be a violation, which means that the treaties were effectively implemented before ratification.

[1] In this chapter the term 'yield' is used to describe the size of the explosions. In appendix 2A, table 2A.1 of this volume only the body wave magnitude (m_b) is given. To be able to give a reasonably correct estimate of yield it is necessary to have detailed information, for example on the geological conditions of the area where the test is conducted. Therefore, giving the m_b figure is an unambiguous way of listing the size of an explosion.

Nevertheless, the USA refused to ratify them, claiming that the verification methods were inadequate. During the Reagan Administration the USA several times accused the USSR of violating the 150-kt threshold, but the accusations were rejected by the USSR and also questioned by many US experts, and there was no conclusive evidence. The USSR, in turn, alleged that the USA had exceeded the 150-kt limit on several occasions, which, again, was never proven.[2] To avoid suspicion it is vital that the verification method be fully approved by both parties.

The United States insisted that the hydrodynamic method, in particular its CORRTEX (Continuous Reflectometry for Radius versus Time Experiments) technique, was the preferred means of measuring the yield of nuclear explosions.[3] The USSR and many US experts considered verification by national technical means (seismological monitoring from outside the borders of the testing state), as stipulated in the original protocols, to be sufficient. Seismological techniques have improved considerably since the treaties were negotiated. Seismic equipment outside the testing country can now observe a variety of different seismic waves, and these observations can be combined to reach more accurate yield assessments.[4] In addition the increased openness in the USSR could also facilitate proper verification.

Despite differences in views on how the treaties should be verified, the USSR agreed to open talks with the USA to work out new verification protocols agreeable to both parties in the expectation that such talks could lead to real restrictions on testing. These bilateral test negotiations were announced in September 1987. A step-by-step approach was agreed by the two negotiating parties: once verification measures permitted ratification of the TTBT and the PNET, the negotiations should proceed to further limitations on nuclear testing, leading to a complete cessation of nuclear testing as part of an effective disarmament process.[5] To test and assess verification methods acceptable to both parties, a Joint Verification Experiment (JVE) was carried out in 1988: US and Soviet scientists monitored nuclear tests at each other's test sites.[6]

On 1 June 1990, at the summit meeting in Washington, the US and Soviet Presidents signed the two new verification protocols containing regulations for hydrodynamic yield measurement, on-site inspection as well as seismic monitoring. (See excerpts from the texts of the protocols, appendix 15A.) In September 1990 the US Senate gave advice and consent to a resolution on the ratification of the two treaties with their protocols. The vote for ratification was unanimous, but the Senate's Select Committee on Intelligence had warned that on-site inspection may compromise the security of sensitive US military

[2] *Department of State Bulletin,* June 1987, p. 3; and *Pravda,* 22 Aug. 1987.

[3] CORRTEX is an intrusive on-site system which requires placing a cable in or next to the shaft containing the nuclear device. By measuring the speed at which the cable is crushed the size of the test can be measured.

[4] Panofsky, W. K. H., 'Verification of the Threshold Test Ban', *Arms Control Today*, vol. 20, no. 7 (Sep. 1990), p. 36.

[5] *Wireless File*, EUR-403 (US Information Service, US Embassy: Stockholm, 17 Sep. 1987).

[6] The Joint Verification Experiments are described in Ferm, R., 'Nuclear explosions', SIPRI, *SIPRI Yearbook 1989: World Armaments and Disarmament* (Oxford University Press: Oxford, 1989), pp. 52–53, and the text of the US–Soviet JVE Agreement is given in appendix 2B.

installations.[7] In October 1990 the Supreme Soviet voted for ratification and issued a message to the parliaments of the world, reaffirming the Soviet stand in favour of a complete cessation of nuclear testing to 'enhance the nuclear proliferation regime and the elimination of nuclear weapons'.[8]

The Threshold Test Ban Treaty Protocol permits hydrodynamic measurement of each explosion associated with a nuclear weapon test (which may consist of one or more explosion) with a planned yield above 50 kt. Tests with planned yields greater than 50 kt can also be monitored using three designated seismic stations off the test site but within the territory of the testing state. The TTBT Protocol also allows on-site inspection of tests with planned yields above 35 kt (but not exceeding 50 kt if the hydrodynamic method is used), and includes special provisions for monitoring unusual cases (tests with non-standard geometries) and tests with multiple nuclear explosions with planned aggregate yields exceeding 150 kt. The main purpose of these provisions is to exchange geological data, rock samples, and so on, which can improve the yield assessment, and to check that no unusual processes are going on which might lead to misleading yield estimates.

If, in each of the first five years of the Treaty, a party does not conduct at least two tests with planned yields above 50 kt, the other party may use hydrodynamic measurement that year on two tests from among those having the highest planned yields below 50 kt. The parties shall provide advance information on 1 June every year of the number of explosions with planned yields above 35 kt and 50 kt for the subsequent 12 months. No later than 200 days prior to the planned date of any test at which the other party has the right to carry out verification procedures, the testing party shall provide information on the date, designation and location of the test and on whether the yield will exceed 35 kt or 50 kt. Within 20 days of receipt of such notification the verifying party must inform the testing party whether or not it intends to carry out verification activities and which methods it plans to use.

The Peaceful Nuclear Explosions Treaty Protocol also permits hydrodynamic measurement of explosions with planned yields over 50 kt and of each individual explosion in a test with aggregate yield exceeding 50 kt. It allows on-site inspection of the geological conditions near the emplaced nuclear device for explosions with planned yields above 35 kt (but not exceeding 50 kt if the hydrodynamic method is used) and a local seismic network for group explosions exceeding 150 kt. Notification requirements are similar to those for the TTBT.

Under both treaties the parties shall as necessary discuss implementation and verification issues. All notifications shall be transmitted through the US–Soviet Nuclear Risk Reduction Centres, established in 1987 and opened in 1988.[9]

Many experts dispute that the hydrodynamic technique is a more accurate method for verifying compliance with the yield limit than seismic monitoring.

[7] *Threshold Test Ban and Peaceful Nuclear Explosions Treaties*, Executive Report submitted by the Committee on Foreign Relations, US Senate, 101st Congress (US Government Printing Office: Washington, DC, 1990), pp. 178–85.

[8] TASS, 9 Oct. 1990.

[9] See also the discussion of the Nuclear Risk Reduction Centres in chapter 12 of this volume.

In addition, hydrodynamic techniques are regarded as not as useful for measuring explosions below about 50 kt. They would not, therefore, be particularly useful for verifying compliance with a considerably lower-yield threshold for tests, which is partly why proponents of a comprehensive test ban see the process of negotiation on verification methods as diverting attention from the real issue: a cessation of all nuclear explosions within the near future. At a Senate hearing on nuclear testing issues held in November 1989, it was claimed that the negotiations had just delayed progress.[10] On the other hand, it is argued by some that on-site presence at an adversary's test sites would shed light on the purposes of that country's test programme and reduce the chances of a technological surprise and, in the long run, this could be a far more important factor for stability than accurate knowledge of yields. CORRTEX is a very expensive method. A network of 20 seismic monitoring stations could be built and installed in the USSR for the price of a single CORRTEX measurement.[11] The increased costs of verification might force the parties to reduce the number of verification exercises; if the parties exercise all rights available to verify and the costs are to be drawn from the test programme, the costs might have the added effect of forcing a reduction in the *number* of tests that the laboratories could afford to conduct.

However, the step-by-step procedure towards a CTB, agreed at the beginning of the US–Soviet talks, has come to a halt. The US Administration announced at the beginning of January 1990 that, before further steps can be taken, a period of implementation of the TTBT and the PNET should be observed to assess the verification process and to consider additional moves from a national security standpoint.[12] In addition, the 1989 US Department of Energy Annual Report to Congress on Nuclear Testing stated that the USA would require 10 years of testing and study to make a recommendation on the acceptability of further testing limitations.[13]

III. The PTBT amendment conference

According to Article II of the 1963 multilateral Partial Test Ban Treaty, any party may propose amendments to the Treaty. If one-third or more of the parties to the Treaty so request, the depositary governments (the USA, the USSR and the UK) shall convene a conference to consider the proposed amendments. As early as 1984 the Parliamentarians for Global Action, an organization of parliamentarians from all over the world, put forward the idea

[10] *Nuclear Testing Issues*, Statement by Rear Admiral E. J. Caroll, Hearings before the Committee on Foreign Relations, US Senate, 101st Congress (US Government Printing Office: Washington, DC, 1990), pp. 12–13.

[11] *Threshold Test Ban and Peaceful Nuclear Explosion Treaties with the USSR*, Statement by G. E. van der Vink (Director of Planning, the IRIS Consortium), Hearings before the Committee on Foreign Relations, US Senate, 110th Congress, 2nd Session (US Government Printing Office: Washington, DC, 1990), pp. 73–76.

[12] *Wireless File*, EUR-301 (US Information Service, US Embassy: Stockholm, 24 Jan. 1990).

[13] US Department of Energy, Annual Report to Congress, Vol. I, *Program Status of Preparations for Further Limitations on Nuclear Testing* (US Government Printing Office: Washington, DC, Dec. 1989), p. 17.

of using this amendment provision to transform the Treaty into a comprehensive test ban treaty.

Proponents of an amendment conference argued that *if* a majority of the parties, including the three original signatories, do support a CTB it becomes binding on *all* the parties, including the so-called threshold countries—Argentina, Brazil, India, Israel, Pakistan and South Africa—states not parties to the 1968 Non-Proliferation Treaty (NPT) and suspected of developing or possessing nuclear weapons. It is interesting to note that India and Pakistan were among the states which requested the amendment conference. Thus a CTB could prevent even non-NPT members from going ahead with plans to manufacture nuclear weapons. Without being able to test, threshold states would find it very difficult to start producing nuclear weapons. Therefore, progress on a CTB is regarded by many as necessary to protect the NPT regime. (For further discussion of how a CTB could affect the NPT, see chapter 16.)

However, a CTB cannot be achieved as long as nuclear weapon states maintain that testing is necessary for the modernization and development of their nuclear weapons; and even if the nuclear weapon states decide to freeze their currently held nuclear stockpiles, or to reduce them further, they may argue that it is necessary to test the reliability and efficiency of the warheads. The US Administration has repeatedly stated that testing is necessary to ensure the reliability of the nuclear deterrent and to improve the safety, security and survivability of the weapons. However, a growing body of scientific experts testifies that explosive testing is *not* required to ensure the reliability of stockpiled weapons, which have already undergone rigorous design and production testing.[14] In the period 1970–88, only 8 of the nearly 300 US explosions were conducted in order to test older warheads in stockpiles.[15]

By 1 April 1989 more than the required number of parties—all non-nuclear states, most of them non-aligned countries—had formally supported a request for an amendment conference. According to the Treaty, any amendment must be approved by a majority of the parties, *including the three depositary states*, before it can take effect. This in fact gives the USA, the UK and the USSR each a veto over any changes or amendments. Although the USA and the UK were reluctant to convene the conference (they had declared at an early stage that they would object to any changes in the original Treaty), the three depositary states fulfilled their duty as such and arranged for the amendment conference to be held in New York on 7–18 January 1991.

In the decision adopted at the end of the conference it was stated that further work needs to be undertaken. The President of the conference was mandated to 'conduct consultations with a view to achieving progress on those issues and resuming the work of the Conference at an appropriate time'. Seventy-four countries voted in favour of the decision and two against—the USA and the

[14] Mark, J. C., 'Do we need nuclear testing?', *Arms Control Today*, vol. 20, no. 9 (Nov. 1990), pp. 12–17.

[15] Norris, R. S., Cochran, T. B. and Arkin, W. M., 'Known US nuclear tests July 1945 to 31 December 1988', *Nuclear Weapons Databook*, Working Paper no. 86-2 (Rev. 2C) (Natural Resources Defense Council: Washington, DC, Jan. 1989).

UK. The remaining NATO states taking part in the conference abstained from voting or voted in favour.[16]

IV. The Conference on Disarmament

In July 1990 the Conference on Disarmament (CD) agreed by consensus on a mandate for its *Ad Hoc* Committee on a Nuclear Test Ban. The Committee had not functioned since 1983, because of disagreement among the members on the mandate. A majority of the Conference members have been in favour of starting negotiations on a CTB, but the USA and the UK have opposed any mandate which includes such negotiations as the goal of the talks. France has refused to participate in any talks on limiting nuclear testing, but declared that it did not object to re-establishing the Committee nor did it oppose the adoption of the mandate, reiterating its earlier commitment to 'refrain from participating in any exercise which corresponds to a conception of nuclear disarmament' that it does not share.[17] China stated that it would take an active part in the work of the Committee and strive for an early CTB and effective disarmament.[18]

Under the mandate,[19] which is based on a 1988 proposal put forward by the President of the CD,[20] the Conference requests the *Ad hoc* Committee to initiate, as a first step towards achieving a comprehensive nuclear test ban treaty, substantive work on specific and interrelated test ban issues, including the structure and scope as well as verification and compliance. As only one month of work remained at the 1990 CD session, there was only time for initial discussions on the nuclear test issue.[21]

The CD *Ad Hoc* Group of Scientific Experts continued its work on a verification system for a CTB based on the international exchange of seismological data under a future treaty. The Group reported that the second phase of the Second Technical Test (GSETT-2) was conducted by 21 countries from 16 January to 6 March 1990. Twenty-five countries participated in the test carried out on 19–27 June during which four experimental international data centres operated.[22] The main phase of GSETT-2, with full-scale testing over an extended period of time, was scheduled to be conducted during the first half of 1991.

[16] *Amendment Conference of the States Parties to the Treaty Banning Nuclear Weapon Tests in the Atmosphere, in Outer Space and Under Water*, PTBT/CONF/L.1 and Draft Decision, 18 Jan. 1991. For further analysis of the amendment conference see Müller, H., Fischer, D. and Kötter, W., SIPRI, *Nuclear Disarmament and Global Order* (provisional title) (Oxford University Press: Oxford, forthcoming).

[17] Conference on Disarmament document CD/PV.565, 17 July 1990, p.13.

[18] See note 17.

[19] Conference on Disarmament document CD/1016*, 17 July 1990.

[20] Conference on Disarmament document CD/863, 23 Aug. 1988.

[21] Conference on Disarmament document CD/NTB/CRP.8, 16 Aug. 1990.

[22] Conference on Disarmament document CD/1032, 10 Aug. 1990.

Appendix 15A. Excerpts from the protocols to the Threshold Test Ban and Peaceful Nuclear Explosions treaties*

Excerpts from the Protocol to the Treaty between the USA and the USSR on the Limitation of Underground Nuclear Weapon Tests

Section I. Definitions

For the purposes of this Protocol:

1. The term 'test site' means a geographical area for the conduct of underground nuclear weapons tests, specified in paragraph 1 or in accordance with paragraph 2 of Section II of this Protocol.

2. The term 'underground nuclear weapon test,' hereinafter 'test', means either a single underground nuclear explosion conducted at a test site, or two or more underground nuclear explosions conducted at a test site within an area delineated by a circle having a diameter of two kilometers and conducted within a total period of time of 0.1 second. The yield of a test shall be the aggregate yield of all explosions in the test.

3. The term 'explosion' means the release of nuclear energy from an explosive canister.

(. . .)

7. The term 'Designated Personnel' means personnel appointed by the Verifying Party from among its nationals and included on its list of Designated Personnel, in accordance with Section IX of this Protocol, to carry out activities related to verification in accordance with this Protocol in the territory of the Testing Party.

(. . .)

10. The term 'hydrodynamic yield measurement method' means the method whereby the yield of a test is derived from on-site, direct measurement of the properties of the shock wave as a function of time during the hydrodynamic phase of the ground motion produced by the test.

11. The term 'seismic yield measurement method' means the method whereby the yield of a test is derived from measurement of parameters of elastic ground motion produced by the test.

12. The term 'on-site inspection' means activities carried out by the Verifying Party at the test site of the Testing Party, in accordance with Section VII of this Protocol, for the purposes of independently obtaining data on conditions under which the test will be conducted and for confirming the validity of data provided by the Testing Party.

13. The term 'emplacement hole' means any drill-hole, shaft, adit or tunnel in which one or more explosive canisters, associated cables, and other equipment are installed for the purposes of conducting a test.

(. . .)

15. The term 'satellite hole' means any drill-hole, shaft, adit or tunnel in which sensing elements and cables and transducers are installed by the Verifying Party for the purposes of hydrodynamic measurement of the yield of a specific test.

(. . .)

29. The term 'Bilateral Consultative Commission' means the Commission established in accordance with Section XI [Procedures for Consultation and Coordination: 5 pp] of this Protocol.

30. The term 'Coordinating Group' means a working group of the Bilateral Consultative Commission that is established for each test with respect to which activities related to verification are carried out.

31. The term 'coordinated schedule' means the schedule, including the specific times and durations for carrying out activities related to verification for a specific test, established in the Coordinating Group as specified in paragraph 12 of Section XI of this Protocol.

(. . .)

Section II. Test sites

1. The test sites for the Parties are: the Nevada Test Site, for the United States of America; and the Northern Test Site (Novaya Zemlya) and the Semipalatinsk Test Site, for the Union of Soviet Socialist Republics.

Upon entry into force of the Treaty, each Party, for each of its test sites, shall provide the other Party with:

(a) a precise written description of the boundaries; and

(b) a diagram with geographic coordinates of the boundaries to the nearest second, to a scale no smaller than 1:250,000.

2. Following entry into force of the Treaty, if a Party decides to establish a new test site or to change the boundaries of a test site specified in paragraph 1 of this Section, the description and diagram specified in paragraph 1 of this Section shall be transmitted to the other Party no less than 12 months prior to the planned date for conducting the first test at the new test site or area of expansion of a previously specified test site.

3. A test site of a Party shall be located only within its territory. All tests shall be conducted solely within test sites specified in paragraph 1 or in accordance with paragraph 2 of this Section.

4. For the purposes of the Treaty and this Protocol, all underground nuclear explosions at test sites specified in paragraph 1 or in accordance with paragraph 2 of this Section shall be considered underground nuclear weapon tests and shall be subject to all provisions of the Treaty and this Protocol.

Section III. Verification measures

1. For purposes of verification of compliance with the Treaty, in addition to using available national technical means, the Verifying Party shall have the right, with respect to tests that are conducted 200 days or more following entry into force of the Treaty:

(a) with respect to a test having a planned yield exceeding 50 kilotons, to carry out any or all of the verification activities associated with use of the hydrodynamic yield measurement method, in accordance with Section V [Hydrodynamic Yield Measurement Method: 36 pp] of this Protocol, with respect to each explosion in the test;

(b) with respect to a test having a planned yield exceeding 50 kilotons, to carry out any or all of the verification activities associated with use of the seismic yield measurement method, in accordance with Section VI [Seismic Yield Measurement Method: 6+ pp] of this Protocol; and

(c) with respect to a test having a planned yield exceeding 35 kilotons, to carry out any or all of the verification activities associated with on-site inspection, in accordance with Section VII [On-Site Inspection: 5+ pp] of this Protocol, with respect to each explosion in the test, except that such activities may be carried out with respect to a test having a planned yield exceeding 50 kilotons only if the Verifying Party does not use the hydrodynamic yield measurement method.

2. In addition to the rights specified in paragraph 1 of this Section, for the purposes of building confidence in the implementation of this Protocol and improving its national technical means of verification, the Verifying Party shall have the right:

(a) if, in each of the five calendar years immediately following entry into force of the Treaty, the Testing Party does not conduct a least two tests having a planned yield exceeding 50 kilotons, to use the hydrodynamic yield measurement method, in accordance with Section V of this Protocol, with respect to two tests from among those having the highest planned yields that the Testing Party conducts in that calendar year;

(b) if, in the sixth calendar year following entry into force of the Treaty and in each calendar year thereafter, unless the Parties otherwise agree, the Testing Party does not conduct a least one test having a planned yield exceeding 50 kilotons, to use the hydrodynamic yield measurement method, in accordance with Section V of this Protocol, with respect to one test from among those having the highest planned yield that the Testing Party conducts in that calendar year;

(c) if, in any calendar year, the Testing Party postpones a test having a planned yield of 50 kilotons or less to the following calendar year, after having been notified by the Verifying Party of its intent to use the hydrodynamic yield measurement method with respect to that test, to use such method with respect to that test in the following calendar year. This right shall be additional to the rights specified in the paragraph 1(a) of this Section and in subparagraphs (a) and (b) of this paragraph; and

(d) in addition to the rights specified in subparagraphs (a), (b), and (c) of this paragraph, if, in each of the five calendar years beginning with the conduct of the first test by the Testing Party at a new test site, the Testing Party does not conduct at least two tests having a planned yield exceeding 50 kilotons at the new test site, the Verifying Party shall have the right to use the hydrodynamic yield measurement method, in accordance with Section V of this Protocol; with respect to two tests from among those having the highest planned yields that the Testing Party conducts at the new test site in that calendar year.

3. If the Verifying Party has notified the Testing Party that it intends to use the hydrodynamic yield measurement method with respect to a specific test including more than one explosion, unless the Parties agree on verification measures with respect to such a test:

(a) the distance between the closest points of any two adjacent explosive canisters shall be no less than 50 meters; and

(b) the time of each explosion shall be established by the Testing Party so as to permit the carrying out of hydrodynamic yield measurements for each explosion for a distance of no less than 30 meters in the satellite hole closest to the emplacement hole with which it is associated.

4. If the Verifying Party has notified the Testing Party that it intends to use the hydrodynamic yield measurement method with respect to a specific test, and if that test is conducted in more than one emplacement hole, the Testing Party shall have the right to conduct that test only if no more than one emplacement hole has characteristics or contains explosive canisters having characteristics differing from those set forth in paragraph 2 or 3 of Section V of this protocol with respect to a test of standard configuration, unless the Parties agree to verification measures with respect to such a test.

5. The Testing Party shall have the right to conduct a test having a planned yield exceeding 35 kilotons within a time period of less than two seconds of any other test having a planned yield exceeding 35 kilotons only if the Parties agree on verification measures with respect to such tests. No test shall be conducted within 15 minutes prior to or following a reference test, unless the Parties otherwise agree.

6. The Testing Party shall have the right to conduct a test having a planned yield exceeding 35 kilotons in a cavity having a volume exceeding 20,000 cubic meters only if the Parties agree on verification measures with respect to such a test.

7. The Verifying Party, by notifying the Testing Party that it intends to use the hydrodynamic yield measurement method with respect to a test of non-standard configuration having a planned yield exceeding 50 kilotons, shall have the right to require a reference test for this non-standard test, in order to compare the yields measured through its national technical means for these two associated tests with the yield obtained by carrying out hydrodynamic yield measurement of the reference test. The right of the Verifying Party to a reference test shall be independent of whether or not it actually carries out hydrodynamic yield measurements of the test of non-standard configuration.

8. With respect to the requirement for a reference test:

(a) if the Testing Party, at the time it provides notification of a test, identifies that test as a reference test for a future test of non-standard configuration, and if the Verifying Party does not use the hydrodynamic yield measurement method with respect to the identified reference test, the Verifying Party shall forfeit its right to require a reference test for that test of non-standard configuration and for any subsequent test of non-standard configuration that would be associated with that reference test, if the Testing Party conducts the identified reference test;

(b) the Testing Party shall have the right to identify only one test of standard configuration as a reference test not associated with any specific test of non-standard configuration until it has conducted an associated test of non-standard configuration for which this test serves as a reference test, or unless it simultaneously provides notification of the associated test of non-standard configuration; and

(c) if the Testing Party, at the time it provides notification of a test of standard configuration, indicates that the test will satisfy a requirement for a reference test for a previously conducted test of non-standard configuration, and if the Verifying Party notifies the Testing Party of its intent not to use the hydrodynamic yield measurement method with respect to that reference test, the Verifying Party shall forfeit its right to require a reference test for the previously conducted test of non-standard configuration. In that case, the Testing Party shall have the right to cancel that reference test.

9. Following notification by the Verifying Party, in accordance with paragraph 5 of Section IV [Notifications and Information Relating to Tests: 9+ pp] of this Protocol, of whether or not it intends to carry out any of the activities related to verification for a specific test, and, if so, which activities, the Verifying Party shall forfeit its right to revise that notification unless the Testing Party changes the previously declared location of that test by more than one minute of latitude

or longitude or changes the planned yield of a test from 50 kilotons or less to a planned yield exceeding 50 kilotons. If the Testing Party makes any such change, the Verifying Party shall have the right to revise its previous notification and to carry out any of the activities specified in paragraph 1 or 2 of this Section and, if the Verifying Party notifies the Testing Party that it intends to carry out activities related to verification with respect to that test, in accordance with paragraph 20 of Section IV of this Protocol, the Testing Party shall not conduct the test less than 180 days following the date of the revised notification by the Verifying Party, unless the Parties otherwise agree.

10. Designated Personnel shall have the right to carry out activities related to verification in accordance with this Protocol, 24 hours a day, provided such activities are consistent with the safety requirements of the Testing Party at the test site or Designated Seismic Station. All operations and procedures that require the participation of Designated Personnel and personnel of the Testing Party shall be carried out in accordance with the technical operations and practices at the test site or Designated Seismic Station of the Testing Party, and in this connection:

(a) Designated Personnel:

(i) shall not interfere with activities of personnel of the Testing Party at the test site or Designated Seismic Station; and

(ii) shall be responsible for the working of their equipment, its timely installation and operation, participation in such operations, including dry runs, as the Testing Party may request, and recording of data; and

(b) the Testing Party:

(i) shall be under no obligation to delay the test because of any malfunction of the equipment of the Verifying Party or inability of Designated Personnel to carry out their functions, unless the Testing Party caused such a situation to arise; and

(ii) shall bear full responsibility for the preparation and conduct of the test and shall have exclusive control over it.

11. If the Verifying Party has notified the Testing Party that it intends to carry out activities related to verification for a specific test, the Testing Party shall have the right to make changes in the timing of its operations related to the conduct of that test, except the Testing Party shall not make changes in the timing of its operations related to the conduct of that test that would preclude Designated Personnel from carrying out their rights related to verification provided in this Protocol.

If the Testing Party notifies the Verifying Party of a change in the timing of its operations that the Verifying Party deems would either preclude or significantly limit the exercise of such rights, the Coordinating Group shall meet at the request of the Representative of the Verifying Party to the Coordinating Group, to consider the change in order to ensure that the rights of the Verifying Party are preserved. If the Coordinating Group cannot agree on a revision to the coordinated schedule that will ensure the rights of both Parties as provided in this Protocol, there shall be no advancement of events within the coordinated schedule due to such a change. Either Party may request that the Bilateral Consultative Commission consider any such change in timing of operations or in the coordinated schedule, in accordance with paragraph 15 of Section XI of this Protocol.

(. . .)

Excerpts from the Protocol to the Treaty between the USA and the USSR on Underground Nuclear Explosions for Peaceful Purposes

Section I. Definitions

(. . .)

8. The term 'local seismic network' means the array of seismic stations and the control point temporarily deployed, in accordance with this Protocol, for the purpose of identifying the number of individual explosions in a specific group explosion.

(. . .)

Section II. Explosion depth and composition

1. No explosion shall be conducted at a distance in meters from the ground surface less than 30 times the 3.4 root of the planned yield of that explosion in kilotons.

2. No group explosion shall have an aggregate yield exceeding 150 kilotons unless the Parties agree on specific procedures to implement appropriate provisions of this Protocol so as to permit identification of

each individual explosion and determination of the yield of each individual explosion in the group.

3. No explosion having a planned yield exceeding 35 kilotons shall be conducted in a cavity having a volume exceeding 20,000 cubic meters, unless the Parties agree on verification measures for such an explosion.

Section III. Verification measures

1. For the purposes for the Treaty, all underground nuclear explosions conducted outside national nuclear test sites shall be considered underground nuclear explosions for peaceful purposes subject to all the provisions of the Treaty. For purposes of verification of compliance with the Treaty, in addition to using available national technical means, the Verifying Party shall have the right:

(a) to use the hydrodynamic yield measurement method, in accordance with Section V [Hydrodynamic Yield Measurement Method: 7 typed pages] of this Protocol, to measure the yield of each explosion that the Party carrying out the explosion notifies, in accordance with paragraph 3 of Section IV [Notification and Information Relating to Explosions] of this Protocol, to have a planned yield exceeding 50 kilotons;

(b) to use the hydrodynamic yield measurement method, in accordance with Section V of this Protocol, to monitor the yield of each individual explosion in a group explosion that the Party carrying out the explosion notifies, in accordance with paragraph 3 of Section IV of this Protocol, to have a planned aggregate yield exceeding 50 kilotons;

(c) to use, in conjunction with the use of the hydrodynamic yield measurement method, a local seismic network, in accordance with Section VI [Local Seismic Network: 2 typed pages] of this Protocol, for each group explosion that the Party carrying out the explosion notifies, in accordance with paragraph 3 of Section IV of this Protocol, to have a planned aggregate yield exceeding 150 kilotons; and

(d) to carry out on-site inspection, in accordance with Section VII [On-Site Inspection: 2 typed pages] of this Protocol, with respect to any explosion that the Party carrying out the explosion notifies, in accordance with paragraph 3 of Section IV of this Protocol, to have a planned yield exceeding 35 kilotons and, with respect to any explosion having a planned yield exceeding 50 kilotons, only if the Verifying Party has decided not to use the hydrodynamic yield measurement method.

2. The Party carrying out the explosion shall bear full responsibility for, and have exclusive control over, the conduct of the explosion.

3. Designated Personnel shall be responsible for the working of their equipment, its timely installation and operation, for participating in such operations, including dry runs, as the Party carrying out the explosion may request, and for recording data at the time of the explosion. The Party carrying out the explosion shall be under no obligation to change the time of the explosion because of any malfunction of the equipment of the Verifying Party or inability of Designated Personnel to carry out their functions, unless actions of the Party carrying out the explosion have caused such a situation to arise.

(. . .)

---

* Extracts selected by Richard Kokoski.

Part IV. Special features

16. The fourth review of the Non-Proliferation Treaty

DAVID FISCHER and HARALD MÜLLER

I. Introduction

From 20 August to 14 September 1990, 84 states parties to the 1968 Non-Proliferation Treaty (NPT) met in Geneva to review how nuclear weapon states and non-nuclear weapon states parties had observed their Treaty obligations during the five preceding years. It was the fourth of the NPT Review Conferences that have taken place every five years since the Treaty entered into force in 1970,[1] and the last before the 1995 Extension Conference which must decide whether to extend the Treaty indefinitely or for a specified period. It has become the practice to seek consensus at each conference on a Final Document embodying agreed statements of policy and recommendations for action. Such a document is meant to send a powerful message to those states that are not parties to the Treaty. The achievement of a final consensus has emerged, for better or worse, as the standard against which to measure the success of each conference. Despite strenuous efforts, the fourth NPT Review Conference (like the one in 1980) failed to issue a Final Document. In the final analysis, division related to a single issue—the priority to be given to the negotiation of a comprehensive test ban treaty (CTBT).

II. Attendance and organization

The 84 participants included 6 of the 10 new states parties that had acceded since 1985, among them North Korea, Spain and Saudi Arabia. The unusually large number of observer delegations included those from China and France for the first time. Four threshold states—Argentina, Brazil, Israel and Pakistan—sent observer delegations while India and South Africa abstained. Chile, Cuba, Myanmar (formerly Burma), Oman, Tanzania, Zimbabwe and, notably, Byelorussia and Ukraine were also present. The Palestine Liberation

[1] Previous NPT Review Conferences are well documented in *SIPRI Yearbooks*. For 1975 see 'The implementation of agreements related to disarmament', SIPRI, *World Armaments and Disarmament: SIPRI Yearbook 1976* (Taylor & Francis: London, 1976), pp. 363–92, and 'Final Declaration of the Review Conference of the parties to the Treaty on the Non-Proliferation of Nuclear Weapons, 30 May 1975', appendix 9A, pp. 403–13; for 1980 see 'The Second NPT Review Conference', SIPRI, *World Armaments and Disarmament: SIPRI Yearbook 1981* (Taylor & Francis: London, 1981), chapter 10, pp. 297–338, and 'Final Document of the Second Review Conference of the parties to the Treaty on the Non-Proliferation of Nuclear Weapons', appendix 10A, pp. 339–62; for 1985 see Goldblat, J., 'The third review of the Non-Proliferation Treaty', SIPRI, *World Armaments and Disarmament: SIPRI Yearbook 1986* (Oxford University Press: Oxford, 1986), pp. 469–80, and 'Final Declaration of the third Review Conference', appendix 20A, pp. 481–94.

Organization (PLO) was listed as an 'observer', not as a state ('category A'), but under an undefined 'category B'. Intergovernmental organizations represented were the UN, the International Atomic Energy Agency (IAEA), the Agency for the Prohibition of Nuclear Weapons in Latin America (OPANAL), the Arab League and the Commission of the European Communities.

The number of non-governmental organizations (including political action groups, research units and peace organizations from all the five continents) represented was also far larger than previously.

As in 1985, the fourth Review Conference established three committees in which all delegations had the right to take part: a General Committee consisting of the conference's principal officers to help the President oversee and steer the proceedings; a Drafting Committee of 30 members whose task would be to edit and harmonize the output of the three Main Committees; and a Credentials Committee to verify the credentials of delegates.

The chief task of Main Committee 1 was to deal with questions of arms control and disarmament, including the prohibition of all nuclear tests, and negative and positive security assurances (negative security assurances are assurances given by nuclear weapon states to non-nuclear weapon states that they will not be attacked by nuclear arms nor subjected to nuclear threats; positive security assurances are undertakings by the nuclear weapon states to come to the aid of non-nuclear weapon states if they are so threatened or attacked). The committee split into three working groups, one on Articles I and II of the NPT,[2] one on security assurances and one on disarmament.

Main Committee 2 dealt with Articles III and VII of the NPT, chiefly IAEA safeguards, conditions for supply of nuclear hardware and technology, nuclear weapon-free zones (NWFZs) and steps to secure wider acceptance of the NPT.

Main Committee 3 focused on international co-operation in peaceful uses of nuclear energy and, like Main Committee 2, on wider acceptance of the Treaty.

The reports of the Main Committees were to be sent to the Drafting Committee, which would consolidate them in a single document and eliminate duplication and inconsistencies. Its report would be submitted to a plenary meeting for formal approval.

As in 1985, the President convened informally a smaller group of senior conference participants. This 'friends of the President' group met informally to discuss solutions to the various controversies that threatened consensus. Some observers regretted that this group was not convened more often during the last crucial days.

Delegations met regularly in private meetings in three caucuses: the Western group of all Organization for Economic Co-operation and Development (OECD) countries; the Eastern group, in which the GDR still participated; and the neutral and non-aligned (NNA) group, consisting *inter alia* of virtually all the Third World delegations. The groups tried to establish common positions

[2] The status of implementation of the NPT is examined article by article in Müller, H., 'Prospects for the fourth review of the Non-Proliferation Treaty', SIPRI, *SIPRI Yearbook 1990: World Armaments and Disarmament* (Oxford University Press: Oxford, 1990), pp. 553–86.

on the issues before the conference. Somewhat exceptionally, the Western group was on this occasion the most successful, despite many remaining differences.

The non-aligned countries initially presented an appearance of unity. But divergences over tactics and particular issues where interests diverged soon appeared. Because of the Gulf crisis and their actions in apparent breach of the NPT, Iraq and North Korea were sharply at odds with other members of their group. Absence of clear leadership compounded an existing lack of cohesion. Mexico, Egypt and Nigeria took independent and sometimes unco-ordinated initiatives. The only well-planned, effectively co-ordinated and executed non-aligned operation was that of Mexico and a small group of supporters (Iran, Indonesia, the Philippines, Sri Lanka and Venezuela) on a single issue, the CTBT.

The once monolithic Eastern group had lost cohesion. Several of its members were determined to demonstrate their newly won freedom of action. Hungary occasionally even attended meetings of the Western caucus. The group held two meetings during the first week and then ceased to function except for a short meeting during the last night.

Besides these three groups, two caucuses met regularly during the conference. One was composed of the 12 European Community (EC) countries that, for the first time at a Review Conference, discussed regularly the proceedings of the conference but did not take joint initiatives to move it forward.

The most efficient group of delegations consisted of Australia, Austria, Canada, Denmark, Finland, Ireland, the Netherlands, New Zealand, Norway and Sweden. As the conference progressed, Hungary and the Czech and Slovak Federal Republic worked increasingly closely with this 'group of 10'. The 10 countries had joined forces in preparing the conference and continued to co-operate throughout its course. They were responsible for some of the most important sections in the draft Final Document, for instance the recommendation that nuclear supplies should only be sent to non-nuclear weapon states if they have accepted safeguards on all their nuclear activities (full-scope safeguards), as the non-nuclear weapon states parties to the NPT are required to do.

III. The main positions in the General Debate

In both tone and substance the first days of the conference seemed to promise a successful outcome. Israel and the PLO were admitted without objection as observers, a half-hearted attempt by Iraq to evict the delegation of Kuwait was firmly quashed by President Oswaldo de Rivero of Peru,[3] and the delegations of the Association of South-East Asian Nations (ASEAN) and Viet Nam did not pursue in public their dispute over whether the Viet Nam-supported Democratic Republic of Cambodia was rightfully listed as a party. Comments on South Africa's and Israel's nuclear arsenals were relatively mild, despite

[3] NPT/CONF.IV/SR.3; NPT/CONF.IV/SR.7

the reports on Israel's nuclear weapons that have come out since 1985. No side-issue of this kind disrupted the proceedings.

In marked contrast with 1980 and 1985, most statements in the General Debate were constructive, avoided polemics and showed a good deal of understanding of the basic problems confronting the non-proliferation regime. This reflected not only more thorough preparations for the conference but also greater Third World interest in the substance and desire for the success of the NPT, which is seen increasingly as a genuine contribution to global security. Much of the improvement was also due to the dissolution of the East–West conflict and the concomitant changes in the world political climate.

The hope was frequently expressed that the first appearance at an NPT conference of Chinese and French observers might portend early accession to the Treaty. Perhaps in anticipation of such comments the French observer had circulated a statement recalling that France had committed itself in 1968 'to behave exactly like states that chose to accede to the NPT'. It referred to the Extension Conference in 1995 and to the importance of the fourth review as a prelude, underlining France's commitment to 'an equitable and stable non-proliferation regime based, in part, on the necessary balance between the non-proliferation of nuclear weapons and the development of the civil applications of the atom'.[4]

Neutral and non-aligned countries

The first signs of trouble came when the delegation of Yugoslavia presented a statement on behalf of the 'Non-Aligned and Other States',[5] giving absolute priority to Article VI of the Treaty and to the relevant paragraphs of the preamble (on disarmament and a CTBT). It relegated to second place the Treaty's role of preventing the further spread of nuclear weapons to other states. The statement took note of some progress in arms reduction talks but emphasized that 'the qualitative improvements in nuclear weapons continued unabated'. It went on to say that the possibility of a significant extension of the Treaty 'would be enhanced by the effective implementation by 1995 of obligations laid down in it relating to nuclear disarmament'. It laid the strongest emphasis on the early achievement of a CTBT. The NNA statement appears to be more the traditional long-term agenda of the non-aligned states in the Conference on Disarmament (CD) than a document specifically prepared as a serious base for negotiation at a four-week conference. In fact, most of the NNA agenda was known to be unacceptable to the present US and British governments except perhaps as a set of long-term objectives.

However, there were indications in the statements of, for example, Bangladesh, Egypt, Indonesia, Kenya, Malaysia, Morocco, Nigeria, Peru and Thailand that several delegations were by no means wedded to every demand. Indeed, except for Mexico, Uganda and Venezuela, the NNA states were

[4] NPT/CONF.IV/24/Corr.1.
[5] NPT/CONF.IV/30.

much less importunate in their individual speeches in the General Debate than in the joint NNA statement.

The eighth and last paragraph of this statement also contained a complaint about 'many unjustified restrictions and constraints imposed on developing non-nuclear weapon states parties'. It called for enhanced financial assistance, not only for nuclear energy but for other energy sources as well, and in particular for preferential treatment to the parties to the Treaty in co-operation on the use of nuclear materials for purposes other than power generation and in nuclear safety. This first formal proposal that the IAEA should extend its services to cover non-nuclear sources of energy was pressed, without much success, by the delegate of Bangladesh. The relatively mild tone of the complaint about export controls was sharpened by an NNA paper submitted later in the conference.[6] It contained five pages of complaints about the implementation of Article IV, supplier policy and discrimination against developing non-nuclear weapon states parties to the NPT. It called for a significant increase in technical assistance and the creation of a 'Trust Fund' to be managed by the IAEA to provide preferential assistance to NPT states parties. This paper reflected more the fears of 1980 than the realities of 1990 when the number of NPT states parties pressing ahead with nuclear power had shrunk to two or three (the two Korea states and Taiwan). Since many NNA states were to express very different views both in the General Debate and in the committees, and went on to support full-scope safeguards as well as the strengthening of export controls, one is bound to ask whether this statement was truly representative of NNA opinion.

East European countries

The newly independent East European countries stressed their concern with disarmament, arms control and security and disclosed an ambivalence towards nuclear power reflecting the shock of the Chernobyl accident. They emphasized, none the less, that the energy situation in their countries and the urgent need to reduce fossil fuel emissions left them with no alternative but to include nuclear power in their energy planning. At the same time they placed great weight on nuclear safety and recognized the need to co-operate with Western Europe to enhance nuclear safety standards. Their views on the more controversial issues were similar to those of most West European states, much closer in fact than to the positions taken in the NNA statement. While most East European states attached great importance to a CTBT, they were much more prepared than the NNA states to give credit to the progress achieved in arms reductions and to the steps taken to limit testing. Their statements also gave more prominence to the role of IAEA safeguards and were explicit about the need for more resources for that programme.[7]

[6] NPT/CONF.IV/L.2.
[7] See, e.g., NPT/CONF.IV/SR.2 (Poland) and NPT/CONF.IV/SR.6 (Bulgaria).

West European countries

At their summit meeting in Dublin in July 1990, the EC states had issued their first comprehensive statement on nuclear non-proliferation.[8] All EC members parties to the NPT referred to and quoted from this joint statement in the General Debate, thus underlining that there was general West European agreement on the fundamentals of a common non-proliferation policy. All EC participants were outspoken in their condemnation of the Iraqi invasion of Kuwait and in their pointed criticism of North Korea for not having completed the negotiation of its NPT safeguards agreement.

A clear majority was in support of full-scope safeguards, but Britain, Belgium and Italy (and also France) dissented. The delegations working in the 'group of 10' emphasized the need for enhancing IAEA resources but with limited support from the others and clear objections from Belgium. Belgium, Denmark, Ireland, Italy, the Netherlands and Spain all supported the wider application of IAEA safeguards in nuclear weapon states. The British, however, showed little enthusiasm for the proposal.

All European countries stressed the need to make nuclear energy safer through international co-operation. However, the emphasis in the Italian, Luxembourg, Danish, Irish and Greek statements was on the risks inherent in nuclear energy, while the German, Spanish, British and Belgian delegates placed more stress on the benefits of nuclear power and the need for greater international co-operation to strengthen its role.

On the CTBT, the EC remained divided. The UK alone (although with the unspoken French concurrence) supported the US position and declared that a CTBT 'remains a long-term goal'. For the other EC participants, a test ban was an arms control measure to be pursued at an earlier date than heretofore and with more vigour. However, one could observe a certain convergence of views between the UK and other EC states in favour of a step-by-step approach, bringing further tangible constraints on testing in the near future. The clear preference of EC participants was also to pursue the issue in the framework of the CD rather than to press for a CTBT at the Partial Test Ban Treaty (PTBT) Amendment Conference.[9]

The nuclear weapon states

The three nuclear weapon states parties to the NPT were unanimous in their determination to maintain and prolong the Treaty. Each strongly commended the unprecedented progress in arms control since 1985 and rejected any suggestion that the extension of the NPT should be made contingent on the conclusion of any other arms control agreement such as a CTBT. Their com-

[8] The Dublin summit meeting declaration on nuclear non-proliferation is reprinted in *PPNN Newsbrief*, no. 10 (summer 1990), p. 12.

[9] For a more detailed analysis of EC positions, see Fischer, D. and Müller, H., *A Treaty in Trouble. Western Europe and the Fourth NPT Review Conference*, Peace Research Institute Frankfurt (PRIF) Report, Frankfurt, 1991.

mon approach for a successful conference and a future extension of the Treaty was manifest in their joint proposals for the preparation and holding of the 1995 Extension Conference.[10]

The Soviet Union reaffirmed its readiness to enter at once into negotiations on a CTBT, pointed to its current moratorium for testing (begun in autumn 1989, terminated in autumn 1990) and affirmed its willingness to suspend all testing pending negotiations on a CTBT. It endorsed the consideration of nuclear weapon-free zones in a wide range of regions and referred pointedly to its ratification of protocols 2 and 3 of the Treaty of Rarotonga, which the USA and the UK had not accepted. The USSR informed the conference of its intention to cease production of highly enriched uranium for weapon purposes and its plan to stop the production of military plutonium by the year 2000.

In contrast, the USA and the UK reaffirmed their opposition to the early conclusion of a CTBT, remained silent on the question of new or existing nuclear weapon-free zones, and claimed that the negative security assurances that they had already given were perfectly adequate but that they were prepared to take another look at the matter. The first US statements were thus seen as defensive and inflexible. The UK put forward some new ideas about safeguards but showed little flexibility on other contentious issues.

In their acceptance of safeguards on their civilian programmes the USA and the UK had gone farther than the USSR: all US and British civilian plants are on the lists of those 'eligible' for IAEA safeguards. The USSR announced at the conference that it would extend its list of facilities open to safeguards to all its power reactors and certain additional research reactors. All these three nuclear weapon states sounded warnings about the costs the IAEA would have to bear if it were to inspect all their civilian plants.

Despite their differences on particular questions, the three nuclear weapon states were more united in purpose and appearance than at any previous Review Conference.

IV. Specific political issues in the General Debate

Germany

The presence of FRG Foreign Minister Hans-Dietrich Genscher was significant as a mark of German commitment to the Treaty. This was reinforced by his speech, in which he made an agreed statement on behalf of both German states, reaffirming their contractual and unilateral undertaking not to manufacture, possess or have control over nuclear, biological and chemical weapons, declaring that the united Germany would abide by that obligation and by the NPT and would seek 'continued validity' of the NPT beyond 1995.[11]

The statement revealed a greater openness than in the past towards the creation of NWFZs and towards the extension of security assurances. The

[10] NPT/CONF.IV/MC.2/WP.10/Rev. 1.
[11] NPT/CONF.IV/SR.4.

degree to which German security had depended on nuclear deterrence during the cold war had previously made both concepts suspect in West German eyes. Another widely noted passage in Genscher's speech was his announcement that as a condition for future significant nuclear exports Germany would require full-scope safeguards in the importing country and would renegotiate existing contracts during the next five years.

NPT parties with dubious activities

When the conference opened North Korea, in contradiction to its obligation under the NPT, had still not concluded the standard safeguards agreement with the IAEA. The country reportedly operated an unsafeguarded reactor and was building an unsafeguarded reprocessing facility. The North Korean situation was addressed by IAEA Director General Hans Blix, and indirectly even in the message from UN Secretary-General Javier Perez de Cuellar. Numerous countries from every group criticized North Korea and called upon Pjongyang to conclude its safeguards agreement without further delay. North Korea defended itself by alleging that the 'nuclear threat' posed by the presence of US nuclear weapons on and close by the peninsula, and implicit in large US–South Korean military exercises, were putting the country in a special situation. North Korea, its delegate maintained, had expected this threat to come to an end with its accession to the NPT, but instead the exercises had become more menacing. The safeguards agreement with the IAEA would be concluded if North Korea received 'juridical guarantees' about its own security. He also indicated that the beginning of North–South Korean arms control tasks with a view to reducing the number of nuclear weapons in the region could possibly improve the security situation of his country; that may open a way out of the situation in which Pjongyang had manœuvred itself.

There were several critical references to statements by Colonel Muammar Qadhafi, implying that he had nuclear weapon ambitions. Some delegates also addressed the breach of Article II of the NPT implicit in Iraqi attempts to smuggle special electronic switches with possible applications to nuclear weapons out of the USA and the UK, and in clandestine attempts to obtain gas-centrifuge enrichment technology and equipment.

Safeguards issues

IAEA safeguards

In his opening address, IAEA Director General Blix had described the growing safeguards responsibilities of the Agency and the constricting effect of the 'zero-growth' lid that had been placed upon the IAEA budget. There was almost universal praise for the safeguards operation, and many delegates pressed for ending zero-growth restrictions. On this point, however, the superpowers, most EC members and Japan were silent. Japan and Czechoslovakia recommended that the IAEA re-examine current safeguards approaches and look into the possibility of, for instance, substituting random inspections for inspections at regular intervals. This might help to stretch available resources and to extend the coverage of safeguards in the nuclear weapon states.

Canada joined the UK in recommending that the IAEA should not hesitate to make use of provisions in its NPT safeguards system[12] that permit the IAEA to make special inspections at additional locations if it has reason to believe that an NPT state party had not notified it of all the locations in which it was holding or processing nuclear material.[13] These special inspections require the consent of the state concerned, but if this is refused the IAEA may command the state to comply. If it still fails to do so the IAEA may find that it is no longer able to verify that there has been no diversion and may invoke its statutory sanctions.[14] Since the IAEA has only limited experience with special inspections the UK and Canada, no doubt with the nuclear activities of Iraq in mind, recommended that it should study what the implications would be and how it would set about doing so.

Wider safeguards coverage in the nuclear weapon states

Following a proposal by Sweden the 1985 Review Conference had recommended that a study be made of the possibility of gradually extending the application of IAEA safeguards in the nuclear weapon states parties until their entire civilian fuel cycles were covered. This would entail strict separation of civilian and military nuclear activities. Once the former were under safeguards, it would require relatively little extra effort to verify an eventual agreement to stop the production of fissile material for nuclear weapons.

The recommended study was not made but in the 1990 General Debate the concept again found support among all three main groups and was pressed once more by Sweden and by the Netherlands and Belgium.

Until the ceiling on the IAEA safeguards budget is removed and the operation can be massively expanded there is little prospect of putting the Swedish proposal into effect; thus, the nuclear weapon states wrapped their reservations against extended safeguards on their territory in warnings against the operational costs of an extension.

Full-scope safeguards

One of the most positive developments in the General Debate was the growing support for full-scope safeguards as a requirement for all nuclear exports to non-nuclear weapon states. In 1985 the Australian delegation had proposed that the conference should explicitly endorse the principle. As a result of FRG, Belgian and Swiss objections the Australian proposal was watered down to the recommendation that suppliers 'take effective steps' to gain acceptance of such a principle.

12 INFCIRC/153, paras 73 and 77.
13 See also IAEA Statute Article XII.A.6.
14 INFCIRC/153 paras 18 and 19.

Since 1985, two states have disregarded the principle. The Soviet Union has negotiated an agreement with India for the sale of two nuclear power reactors. The reactors themselves, their fuel and the fissile material they produce (and any plant in India that reprocesses their spent fuel) will be under IAEA safeguards but the USSR has not required—and probably not asked—India to bring its entire fuel cycle under safeguards. In addition the USSR has leased to India a nuclear submarine to which no IAEA safeguards can be applied.

France, as a non-party, was not under an obligation to take the 1985 recommendation into account. With President François Mitterrand's assent the French Government has made a preliminary commitment to sell Pakistan a large nuclear plant under a safeguards agreement similar in scope to that between India, the USSR and the IAEA. There are also reports that France has been negotiating with India for the sale of two power reactors, possibly as an alternative to the Soviet reactors.

The support for full-scope safeguards increased as compared to 1985. Undoubtedly the most important change was in the country that had led the opposition at the third Review Conference, the FRG. For various reasons, including the revelation of several illegal nuclear exports, German policy in this field has become much more restrictive than in the past.[15] Thus, Foreign Minister Genscher announced not only that all future nuclear exports would be subject to full-scope safeguards but also that, within five years, existing commitments would be adapted 'to the requirements of a consolidated non-proliferation policy'. This implies that Germany's existing agreements with Argentina and Brazil would have to be converted within five years into full-scope safeguards agreements, or terminated after five years.

The USA requires *de facto* full-scope safeguards[16] on its own exports. However, in 1985 it was the main supplier of nuclear fuel to Spain, not yet party to the NPT. US backing of the Australian proposal was accordingly lukewarm in 1985. In the 1990 General Debate the US delegate came out in full support of the principle. Japan, silent on the issue in 1985, also declared itself in favour. Several East European states, led by Hungary, did likewise. More surprising was the widespread support of the principle by NNA delegations. In 1985 only Peru among the Third World countries had come out clearly in favour of full-scope safeguards. It was now joined by many NNA states including Egypt, Ghana, Indonesia, Bangladesh and Thailand.

The widespread support for full-scope safeguards exposed those exporters not yet committed to the principle. In their statements in the General Debate, both the USSR and the UK supported the principle—in principle. William Waldegrave stated that the UK was ready to join with other major exporters in requiring full-scope safeguards. Belgium maintained that agreement of all

[15] Müller, H., *After The Scandals: German Nonproliferation Policy*, Peace Research Institute Frankfurt (PRIF) Report, Frankfurt, 1990.

[16] A distinction is drawn between *de facto* full-scope safeguards—IAEA safeguards in place on all nuclear fuel and plants in the importing state at the time of a new transfer but no commitment by that state to place all future nuclear activities under safeguards—and *de jure* full-scope safeguards under which, as in the case of non-nuclear weapon states parties to the NPT, there is a continuing legal obligation to accept comprehensive safeguards.

suppliers (including France and maybe China) was necessary. The Swiss statement was more opaque. Switzerland applied a case-by-case approach with full-scope safeguards 'one among several options'.

Co-operation in the peaceful uses of nuclear energy

Technical assistance

Almost every speaker in the General Debate had some encouraging words to say about the IAEA's technical co-operation activities. Most NNA states called for them to be expanded and to enhance the financial means for technical assistance especially earmarked for Third World states parties to the NPT. A couple of speakers picked up remarks made in the opening statement by IAEA Director General Blix on the need for an expanded nuclear power programme to help combat the greenhouse effect.

Nuclear safety

Nuclear safety, a neglected topic at previous Review Conferences, had been thrust into prominence by the Chernobyl accident. At the outset Ireland asked that the conference address nuclear safety as an important point on its agenda and in the Final Document. Deputy Foreign Minister Vladimir Petrovskiy spoke movingly about the inability of the USSR to cope single-handed with Chernobyl's consequences and the need for increased international co-operation to ensure the safety of nuclear energy. This theme was taken up by several East European countries, Italy, other West European states and countries from other regions. The general tenor of their remarks was praise for the post-Chernobyl activities of the IAEA and a call for them to be increased, as a significant step to implement Article IV of the NPT.

Security and disarmament issues

Nuclear weapon-free zones

Several delegations commented on the evolution of NWFZs since 1985, the extent to which it had been possible to promote them, and the obstacles and objections they had run into. Indonesia and Malaysia (but not Thailand and Singapore) strongly endorsed the proposal for a NWFZ in the ASEAN region and appealed to those who were standing in its way to withdraw their objections, an oblique reference to the USA. Nigeria, Ghana and Morocco reaffirmed support for an African NWFZ, and Morocco and Egypt reaffirmed their support for a NWFZ in the Middle East. References to Israel and South Africa were more measured than in 1985, and there were also more direct appeals to the four other threshold countries to reconsider their rejection of or absence from the NPT.

New Zealand, but not Australia, pressed the USA and the UK to sign the Protocols of the Rarotonga Treaty and deplored the continuation of French tests at Mururoa. In contrast to earlier conferences, the issue of NWFZ in Europe did not lead to an acerbic dispute between East and West.

Attack on nuclear installations

Neither Israel's attack on the Iraqi Tammuz reactor in 1981 nor the subsequent attacks by Iraq on the nuclear power plant construction site at Bushire in Iran led to a release of radioactivity as neither target was yet in operation. The Chernobyl accident, however, demonstrated what could happen if operating reactors were blown to pieces. Moved by these examples, the Hungarian, Netherlands and several other delegations from West and East European and NNA states urged the early conclusion of an international agreement to prohibit attacks on nuclear plants and requested that the matter be addressed in the Final Document. The silence of the US delegation on this point suggested that it still had reservations about giving a blanket commitment not to attack nuclear installations of the adversary in the event of war.

Security assurances

In the past the non-nuclear weapon states had been generally in favour of obtaining explicit negative security assurances from the nuclear weapon states. However, they had been somewhat wary of positive assurances, wishing to ensure that it would be the threatened state and not the assisting nuclear weapon state that would determine whether or not a threat existed and what assistance if any would be required.

Neither form of security assurance had figured prominently on the agendas of previous Review Conferences. However, Nigeria had recently taken the initiative in the CD and in the preparatory committee of the fourth Review Conference to press for a binding and explicit international treaty on negative security assurances.[17] Reflecting on a potential nuclear threat from Israel, Egypt had taken the initiative of proposing that the nuclear weapon states should commit themselves in a new Security Council resolution to more explicit and unconditional positive assurances than those they had given in 1968 (Security Council Resolution 255).[18]

The great majority of delegates from all three groups considered that the existing assurances should be reviewed, and there was widespread support for the proposals of Nigeria and Egypt. Delegates from developing countries in particular defined security assurances as an important *quid pro quo*, and there were repeated calls to limit such assurances exclusively to non-nuclear weapon states parties to the NPT. The USA and the UK, as noted, maintained that their existing security assurances were adequate but that they were none the less prepared to take another look at the matter.

[17] NPT/CONF.IV/17.
[18] NPT/CONF.IV/31.

International plutonium storage and a cut-off of fissile materials production for weapon purposes

A large-scale reduction of nuclear arsenals such as that envisaged in the impending START agreement or that might follow a second START agreement could make available large quantities of plutonium and highly enriched uranium from dismantled nuclear warheads. This prospect has revived some interest in the proposal to establish an international plutonium storage system as an extension of IAEA safeguards, as foreseen in a dormant provision of the IAEA Statute (Article XII A.5). This idea was mentioned by a few delegates, among others Italy.

Soviet Deputy Foreign Minister Petrovskiy proposed that a freeze on the production of all fissile material for military purposes be formalized by an agreement between the USA and the USSR which could become multilateral and should be verified by the IAEA.

A comprehensive test ban treaty

The General Debate showed that as in 1980 and 1985 the issue of a CTBT would be the most divisive issue before the conference. There was a wide range of views on this subject stretching from those of the USA and the UK at one extreme to those of Mexico and its supporters on the other. One could discern five broad groups in this spectrum but there were nuances between the positions of states in each group.

As at previous conferences the USA and the UK were alone in maintaining that it was necessary to continue testing: as long as they depended on nuclear deterrence for their security they would have to carry out a 'sensible' programme of underground tests to ensure the 'safety, security, survivability and reliability' of their nuclear arsenals. Both regarded a comprehensive test ban as a long-term goal. In the meantime the USA was engaged in negotiations with the USSR on a step-by-step approach to impose further limits on their testing programmes.

Most NATO states were in principle in favour of a CTBT and argued that concrete steps should be taken towards it. None the less they welcomed the modest progress already made—the agreements between the superpowers to limit the yield of their tests to a maximum of 150 kt. They also welcomed the prospect of US–Soviet negotiations on further limits and the revival of the *Ad Hoc* group on nuclear testing at the CD (not, however, empowered to negotiate a treaty) as steps in an incremental approach towards a CTBT. The differences between this group and the next were in emphasis rather than substance.

The Soviet, East European and Nordic delegates also recognized the (limited) value of the agreements already reached and the negotiations under way but wished to see more rapid progress towards a CTBT and the extension of a full negotiating mandate to the CD *Ad Hoc* committee. Finland and Sweden also introduced an argument that had not been pressed in 1985—the ecological damage caused by testing.[19] The USSR stressed that bilateral as well as multilateral negotiations were needed. The USSR was prepared to consider converting the PTBT into CTBT at the January 1991 Amendment Conference, though Petrovskiy was careful not to commit the USSR unequivocally to this solution.[20] Clearly the USSR did not wish to embarrass the USA unduly nor to stress the difference of views between the superpowers.

A substantial group of NNA delegations pressed for the early conclusion of a CTBT but laid equal or similar weight on other measures of arms control such as security assurances (Nigeria, Ghana, Cameroon and Egypt), 'deep cuts' in nuclear arsenals (Peru) and additional NWFZs (Egypt, Uganda and Senegal). Some recognized that a CTBT was 'not for tomorrow' (Bangladesh). Many in this group also recognized the utility of bilateral negotiations and a step-by-step approach (e.g., Sri Lanka). The attitudes of this group might be summed up as attaching great value to the CTBT but perceiving it as one of several desirable steps needed to supplement and strengthen the NPT, to which they attach overriding importance.[21]

The fifth group, led by Mexico, included Venezuela, Iran, Syria and, on occasion, Uganda, Indonesia and Sri Lanka. Their attitude was succinctly expressed in the statement of the Venezuelan delegate in the General Debate. In the absence of a CTBT 'the nuclear arms race would continue indefinitely'—a CTBT was an 'indispensable measure' for its cessation. He could not accept the proposition that the main obligation of the nuclear weapon states 'was being fulfilled through the conclusion of the INF Treaty and the START negotiations' nor could his delegation accept the step-by-step approach. Bilateral agreements 'would merely serve to justify the continuation of the nuclear weapons race . . . their entry into force constituted a step backwards'. Mexico added another touch: the non-aligned statement was intended to reinforce the NPT and ensure that it would remain in force after 1995. The threat was veiled but clear: without a CTBT the chances for prolongation would be poor.[22]

Extension of the NPT

Almost all delegates recalled that this conference was the last to be held before the 1995 conference which must decide whether the NPT is 'to continue in force indefinitely, or shall be extended for an additional fixed period or periods'. The delegate of Spain made the point that the NPT does not expire in 1995. He and other delegates recommended that the 1995 conference should both review the operation of the Treaty and decide on the length of its extension (i.e., instead of confining itself to the decision on extension—or holding two separate conferences).

[19] NPT/CONF.IV/SR.2, p. 22; NPT/CONF.IV/SR.5, p. 9.
[20] NPT/CONF.IV/SR.2, p. 16.
[21] E.g., NPT/CONF.IV/SR.6, p. 15; NPT/CONF.IV/SR. 11, p. 9; NPT/CONF.IV/SR.6, p. 21.
[22] NPT/CONF.IV/SR9, p. 5.

Numerous speakers, including many from the Third World (e.g., Peru, Thailand, Egypt and, to the surprise of many, Libya), considered that the Treaty was indispensable to international security and should be maintained without, however, specifying the length of the extension. Nine speakers were already prepared to commit themselves to an indefinite extension (from Ireland, the USA, Denmark, Greece, Spain, Austria, Canada, Norway and Finland). In a paper circulated well before the conference opened, all the Nordic countries had already supported an indefinite extension.[23]

V. The work of the committees

Main Committee 1

Compliance with Articles I and II

In 1985 and again in 1990 the main issue was whether the nuclear weapon states, in particular the USA and the UK, had breached their obligations under Article I not to help any non-nuclear weapon state to acquire nuclear weapons by giving such assistance to Israel and South Africa as could help them make the bomb. Since 1985 doubts had also arisen whether Iraq and North Korea were complying with their obligations under Article II of the Treaty not to seek nuclear weapons.

Despite the potentially damaging character of Israeli technician Mordechai Vanunu's claims and ensuing speculations about the size and quality of Israel's nuclear arsenal, delegations continued to ignore them in the working group. The changing political situation in South Africa, the likelihood that it would soon accede to the NPT and the fact that there have been no nuclear supplies to it since 1985 reduced the plausibility of any charge that it was receiving forbidden assistance and the incentive for making such charges.

On the problems posed by activities in Iraq and North Korea it was agreed to call for 'scrupulous and unreserved compliance with treaty obligations' by all parties. Agreement was not reached on a passage clearly directed at Libya and Iraq that would have called upon parties to eschew actions and statements that could indicate intentions of non-compliance.

On the question of how to address the threshold countries, the issue was whether Israel and South Africa should be mentioned by name while the other four remained anonymous. With the support of the Western group, the UK submitted a proposal that all six should be named but each with a different marking. Argentina and Brazil were to be commended for their bilateral confidence building, India and Pakistan to be given some credit for their agreement to refrain from attacking each other's nuclear plant, South Africa to be encouraged to make good its half-promises to accede to the NPT, but for Israel there were no kind words. Mexico, no doubt after conversations with representatives of some nuclear threshold states, rejected this proposal on the grounds

[23] NPT/CONF.IV/18.

that countries should be named only if they had already been named in previous UN resolutions. Another report had it that Mexico would accept the naming of threshold states only if the countries in the region concerned so wished.

There was some general discussion of the need for strict and uniform export controls. The committee's report affirmed 'the determination to stop the spread of nuclear explosive capabilities' and the need for all parties 'to ensure that they do not assist directly or indirectly the nuclear programmes of the threshold states'. This was a significant improvement on the 1985 final document since it placed emphasis for the first time on explosive capabilities and thus implicitly endorsed controls on the export of sensitive technologies—in other words, the London Club's Guidelines.[24]

For the rest, the committee's report on Articles I and II consisted chiefly of the corresponding paragraphs of the 1985 Final Document; however, there were some notable additions, namely: (*a*) it welcomed the improved international climate and expressed the hope that the improvement would continue; (*b*) it welcomed the 16 parties that had joined the NPT since 1985; and (*c*) it expressed satisfaction with the joint declaration of the two German states.

Security assurances

In Main Committee 1 Nigeria proposed that, since there would be insufficient time at the Review Conference to examine its proposal, a special conference should be convened 'not later than 1991' to draw up an agreement providing negative security assurances to non-nuclear weapon states parties to the NPT.[25] This suggestion, as well as the Egyptian proposal on positive security assurances, drew wide support from Third World countries.

In the past the Western nuclear weapon states have been prepared to give only very conditional and ambiguous assurances, tailored to the needs of NATO's flexible response strategy. NATO was not prepared to promise the Soviet Union's former East European allies immunity from nuclear threat.

All this has now changed, and some NATO delegations expressed greater flexibility than in the past. The USA and the UK proclaimed their willingness to examine the matter further, but there had been no change yet in their positions. All of the Western group had doubts about the feasibility of holding a conference in 1991 and of dealing with the matter solely within the framework of the NPT: the collaboration of the non-NPT nuclear weapon states, France and China, would be needed. Much the same applied to positive security assurances.

The committee's report left it to the Drafting Committee to reconcile these differing views on the 1991 conference and the need for a new Security Council resolution. But the committee did go well beyond the findings of the 1985 Review Conference. It now recognized the need for 'effective international arrangements that could be included in an international legally binding instrument' on negative security assurances but that for maximum effective-

[24] *Guidelines for Nuclear Transfers*, INFCIRC/254 (IAEA: Vienna, 1977).
[25] NPT/CONF.IV/MC.1/SR.3, p. 5.

ness such an instrument should be accepted by all the five nuclear weapon states. The committee also recognized that it might be necessary to protect non-nuclear weapon states parties to the NPT against nuclear threats that might be made by some of the threshold states. One effective way of ensuring such protection would be the establishment of new NWFZs. The committee also regretted the lack of progress in the CD on the question of negative assurances and encouraged the CD to continue its attempts to surmount the problems it faced.

In sum, Nigeria and Egypt failed to get the committee's endorsement of their proposals but nevertheless succeeded in launching a comprehensive review of the issues and in gaining support in principle for their ideas.

Disarmament and a comprehensive test ban treaty

As in the General Debate—and as in 1985—the attack on the positions of the USA and the UK was led by the delegate of Mexico. But whereas in 1985 Ambassador Garcia Robles had been able to rally the support of virtually all non-aligned and several neutral states, and thus successfully isolate the USA and Britain, the position taken by his successor was too radical for most of the delegates in the Western and Eastern groups and for many NNA states. There was of course little sympathy with the US and British views on the CTBT, but most delegations recognized that significant progress had been made between East and West since 1985 in reversing the nuclear arms race and reducing conventional weapons and that more progress was in prospect.

The delegate of Mexico, however, was reluctant to admit that there had been a significant change. In a paper that his delegation and those of Venezuela, the Philippines, Ghana and Kenya jointly submitted for inclusion in the Final Document[26] the conference was asked to note 'with regret' negative events and trends in the nuclear arms race. The paper maintained that, despite US–Soviet negotiations since 1985, 'no agreements relating to the cessation of the arms race had emerged so far'. It asked the conference 'to deplore' among other matters that 'a new environment, space, was increasingly being drawn into the arms race'. Its major conclusion was that 'the objective under article VI had not been achieved'. The paper was clearly designed to pre-empt any claim by the Western powers that, while their views on the need for nuclear testing had not changed, they were making substantial progress in other ways in carrying out their obligations under Article VI. In a similar vein the Mexican group sought to deny that progress in reducing conventional weapons through the Treaty on Conventional Armed Forces in Europe (CFE) or towards eliminating chemical weapons was relevant to Article VI. Only a CTBT would count.

In an attempt to find a compromise, Chairman B. A. Adeyemi circulated a paper containing his own appreciation of the situation. This recognized the progress that had been achieved in arms control but called upon the nuclear

[26] NPT/CONF.IV/MC.1, WP. 4.

weapon states to take further steps particularly in regard to the prohibition of nuclear tests. It emphasized the importance of a CTBT as a contribution to the NPT but stopped short of establishing a formal link between the treaties.

The paper evoked a barrage of counter-proposals, and the Chairman was asked to issue a new paper combining his own proposals as well as those of other delegates. This compendium[27] showed that for each of the Chairman's 17 paragraphs the Mexican delegation had submitted an alternative, usually framed in terms critical of the disarmament record. For its part, the USA had countered by submitting 12 paragraphs commending the progress made in arms control and making no link between the conclusion of a CTBT and the prospects for the NPT. The FRG, Finland, Greece and Hungary (an unprecedented combination of sponsors) had submitted a draft paragraph that noted the relevance of conventional disarmament to Article VI, welcomed the prospects for a CFE agreement and expressed concern about the continued growth of conventional forces in other parts of the world, a discreet reminder that Article VI applies equally to the armed forces of non-aligned Third World countries.

In a separate paper Italy and the USSR joined forces in proposing that the nuclear weapon states 'should look for a peaceful use' for fissile material withdrawn from dismantled warheads, 'part of which might be used to the benefit of developing countries',[28] an idea that found its place in the committee's report.

In another attempt at compromise the Chairman redrafted his own paper to incorporate a few of the Mexican and US proposals. Two points in the new draft caused problems for some Western delegations: a recommendation urging the CD to 'begin negotiations on the cessation of the production of weapons-grade fissionable material, production of nuclear weapons, of their delivery vehicles; and the prevention of an arms race in outer space' and a statement that a CTBT 'would enhance the universality and durability of the NPT beyond 1995'. The latter formulation implied more of a 'linkage' than the USA was prepared to accept at this point. The group led by Mexico made it clear that it was dissatisfied with the whole redraft.

Since time was running short, the committee agreed to annex the Chairman's paper to its report merely 'as a useful basis for further discussion . . . without prejudice to the position of any delegation' and pass the incomplete report to the Drafting Committee.[29]

Main Committee 2

Safeguards

The small minority of delegates that were still reluctant to make an unqualified recommendation in favour of full-scope safeguards now found themselves

[27] NPT/CONF.IV/MC.1/WG.3/CRP.5.
[28] NPT/CONF.IV/MC.1/WP 2.
[29] NPT/CONF.IV/MC.1/1.

joined by Italy. The concerns of Belgium and Switzerland were chiefly that France, not requiring the acceptance of full-scope safeguards by its non-NPT customers, would take lucrative nuclear orders away from those that did so require. Italy's concern may have been that since Italy held the presidency of the EC it should not leave France to be exposed. Soviet concerns were that it should not be embarrassed by its sales to Argentina and India. One explanation heard for the British attitude was that the UK feared that insistence on full-scope safeguards would reduce the prospect of French accession to the NPT. Whatever their motives the five proposed the formula that all suppliers were 'to join together' in requiring full-scope safeguards. The committee's report, in contrast, used the Australian formula, namely 'the conference . . . urges the nuclear supplier states to require as a necessary condition for the transfer of relevant nuclear supplies to non-nuclear weapon states under new supply arrangements . . . to accept IAEA safeguards on all peaceful nuclear activities both current and future', and noted the reservations of the five dissenters on this paragraph.

The committee's recommendations related to safeguards, in the eyes of many the main achievement of the whole conference, included:

1. Emphasis on the need to use up-to-date cost-effective technologies and to study the use of 'randomized' inspections;
2. Provision of resources adequate to ensure effective safeguards;
3. Streamlining the procedures for accrediting inspectors;
4. A call for universal acceptance of the Convention on Physical Protection and for stringent national physical protection measures especially in nuclear weapon states;
5. A proposal by Canada that export controls should be applied to non-nuclear materials that are usable in nuclear weapons, such as tritium;
6. A call for further separation of civilian and military fuel cycles and broader safeguards coverage in nuclear weapon states. The latter were also asked to offer for verification any nuclear material or plant transferred from military to civilian use;
7. A call upon all states to apply the trigger list drawn up by the Zangger Committee (defining the items whose export 'triggers' the demand for safeguards). This was the first formal endorsement of the Zangger list by a Review Conference and thus gives the list the status of an authoritative interpretation of Article III.2;
8. The recommendation to the IAEA to make full use of its right to carry out special inspections;
9. A recommendation that the IAEA should increase the transparency of its safeguards activities.

Nuclear weapon-free zones

The committee's report on NWFZs began by repeating the corresponding paragraphs of the 1985 report on the subject. It noted the 'continued successful operation' of the Treaty of Tlatelolco and the 'continued existence' of the Antarctic Treaty. It reaffirmed the exhortation of the General Assembly to France to ratify Protocol 1 to the Tlatelolco Treaty (which would apply its terms to French territories in the region). Luxembourg entered a reservation about the exhortation: why should France be singled out by name when no mention was made of Brazil's failure to bring it into force, Argentina and Chile's failure to ratify it and Cuba's failure even to sign it?

The report took note of the entry into force of the Rarotonga Treaty and of the diverse actions the nuclear weapon states have taken about the protocols. Protocol 1 would oblige France, the UK and the USA to apply the Treaty in their own territories in the region. Protocol 2 would oblige all five to give negative security assurances to the parties, and Protocol 3 would require the nuclear weapon states to refrain from carrying out nuclear tests in the region. The report noted that only China and the USSR had adhered to protocols 2 and 3. The USA had said that none of its operations in the region were inconsistent with the Treaty, the UK that it would respect the intentions of the states in the region on Protocols 1 and 3, while France had continued to test nuclear weapons and had formally decided not to adhere to the Protocols.

The report used the wording of the 1985 Final Document on the question of an African NWFZ. South Africa was called upon to accept full-scope safeguards and accede to the NPT. The USA entered a reservation about the reference to South Africa and about the wording of the reference to Egypt's proposal for the creation of a NWFZ in the Middle East containing critical reference to Israel and calling on all parties, especially the NWS, to help persuade Israel to accede to the NPT.

North Korea proposed a text supporting its contention that the US nuclear threat was the impediment to the conclusion of its safeguards agreement and that the creation of a Korean NWFZ would solve the matter. South Korea countered that a NWFZ must take account of the characteristics of the region and must be 'on the basis of arrangements freely arrived at among the states of the region concerned'.[30] The report of the committee simply noted North Korea's proposal and the objections of South Korea and other states.

On the proposal of Indonesia and the Philippines the report also noted the efforts of the South-East Asian countries to establish an ASEAN NWFZ.

Preparations for 1995

A wide gap opened in the committee concerning preparations for the 1995 conference. Led by Mexico 14 countries proposed that the 1995 preparatory committee should meet in September 1991 and be authorized to consider 'all substantive matters relating to the conference'.[31] This seemed tantamount to holding a mini-review conference each year until 1995 in order to put as much pressure as possible on the USA and the UK to negotiate a CTBT. The NNA

[30] NPT/CONF.IV/DC/I/Add.3(A), p. 12, para. 2.
[31] NPT/CONF.IV/L.3.

resolution would also have decided to 'explore ways and means to intensify the dialogue with states non-parties to the Treaty'.

The three depositaries had earlier proposed that the 1995 conference should have the task of reviewing the operation of the NPT as well as extending it, that the Preparatory Committee should meet for the first time in April/May 1993, and that, as previously, its work should be confined essentially to organizational, financial and procedural matters. Later, to meet the NNA states part of the way, the three-power resolution was amended to include in the remit of the Preparatory Committee 'any other business considered relevant by the Preparatory Committee', to adopt the proposal for a dialogue with non-parties and 'to report any significant developments to the President (of the conference'.

These different approaches went unresolved to the Drafting Committee.

Main Committee 3

Chairman Chusei Yamada got off to an early start by circulating his own proposal for a draft section of the committee's report. This consisted almost entirely of a repetition of sections of the 1985 Final Document. They had already appeared in a paper informally circulated by the UK.

The committee also had before it the lengthy NNA statement complaintive on the achievements of peaceful nuclear co-operation. The representative of the Netherlands demonstrated convincingly that it was hardly consistent with the proposals for strengthening safeguards and export controls adopted in Main Committee 2. The NNA resolution left no traces in the generally constructive report that the committee adopted without much further ado.

Supply assurances

The committee did its best to extract whatever comfort could be drawn out of the failure of two major attempts to reconcile those that insisted on cast iron assurances of supply with those that gave priority to mutually acceptable considerations of non-proliferation, the IAEA Committee on Assurances of Supply (CAS) and the UN Conference on the Promotion of International Cooperation in the Peaceful Uses of Nuclear Energy (UNCPICPUNE). It noted that CAS surveys on international nuclear markets were 'a very useful source of information' and it called for a resumption of formal discussions in CAS, *when appropriate,* not a promising request in an extremely slack nuclear market. On UNPICPUNE, the committee stated that technical papers of the conference 'could be utilized by governments in planning national programmes for the development, use and safety of nuclear energy' and recalled UNCPICPUNE's praise for the IAEA.

The committee also made an obeisance towards the Brundtland Report[32] by proposing that the Conference should note its importance and that the IAEA

[32] Brundtland, G. H., *et al.*, *Our Common Future: World Commission on Environment and Development*, rev. edn (Oxford University Press: Oxford, 1987).

had reported on its contribution to achieve 'environmentally sound and sustainable development'. An attempt by a group of pro-nuclear countries to emphasize the role of nuclear energy in combatting the greenhouse effect was blocked by Austria, Ireland, Denmark and other nuclear critics.

Nuclear safety and the management of nuclear waste

The committee's report commended the role of the IAEA following the Chernobyl accident and welcomed the intensification of international co-operation since the accident that was largely initiated by the Agency. It urged universal ratification of the two post-accident conventions on early notification and mutual assistance in the case of an accident.

The Italian delegation proposed that the IAEA take on the task of supervising compliance with its nuclear safety standards, thus making the latter mandatory and internationally verifiable. As most leading Western users of nuclear energy had already indicated in IAEA gatherings, such a major extension of the Agency's authority was unacceptable to them. The Committee's report therefore stressed the primary responsibility of individual states for the safety of their nuclear plants while recognizing the importance of international co-operation in *assisting* individual nations to make safe use of nuclear energy.

The report also urged all states to make full use of the IAEA's wide range of safety missions and the similar services it is beginning to provide for waste management. It called upon the nuclear industry to maintain the highest standards of safety through such industrial organizations as the recently founded World Association of Nuclear Operators. Finally the report commended the IAEA and the OECD Nuclear Energy Agency for their recent work on improving the international regime for liability in the case of nuclear damage.

Promotion, financing and technical assistance

On the basis of a US draft, the report commended the (very gloomy) study of a senior IAEA expert group. The subdued tone of the report is reflected in its recommendation that 'export credit agencies, supported by governments of exporting states and multilateral credit institutions, if requested, should evaluate nuclear power as one option when assessing a borrowing country's electric power programme'.[33]

The committee's report on technical assistance took over extended passages from the Final Document of the 1985 Review Conference and added some material proposed by the group of 10 countries and the USA. The report made the customary appeal for more funds for technical assistance and repeated the 1985 conference's endorsement of channelling assistance to NPT parties. The report also placed slightly more emphasis than its predecessor on regional

[33] International Atomic Energy Agency, *Promotion and Financing of Nuclear Power Programmes in Developing Countries* (IAEA: Vienna, 1987).

projects, welcoming the creation of two new regional co-operation agreements, and on assistance to the least developed NPT parties.

Other elements of the committee's report

The report dealt at greater length than in 1985 with the issue of attacks on civilian nuclear reactors; the additions to the 1985 Final Document stemmed from proposals made jointly by Hungary, the Netherlands and Sweden, and separately by Iran. They contained a call upon all states to become parties to the 1967 First Additional Protocol to the 1949 Geneva Convention and on the nuclear weapon states, when reviewing their military doctrines, to take into account the dangerous radiation that an attack on a nuclear plant might release. This aimed at US military planners who have reportedly been reluctant to grant nuclear facilities complete immunity from the consequences of attacks on military or industrial targets. Acceptance of this language was thus a concession by the USA.

In its report on Article V—peaceful nuclear explosions (PNEs)—the committee accepted a proposal by the group of 10 states in which it was noted that the potential of the technology had not been sufficiently demonstrated and that the IAEA had received no requests for services under Article V. It was also noted for the first time that no nuclear weapon state had an active PNE programme, since the USSR had terminated its PNE activities after 1985.

The committee also had more to say under Article IX—accession and membership—than in 1985. Presumably with Iraq and North Korea in view it underlined the necessity of strict compliance by existing parties with their Treaty obligations. It also stated that an informal dialogue with non-parties would help to enhance the Treaty's universality.

VI. The last week: the end of the NPT Review Conference

When the last week of the conference opened, the general mood was optimistic despite wide remaining differences on the principal issue. Chairman Carl-Magnus Hyltenius (Sweden) of the Drafting Committee decided to leave the less controversial paragraphs of the draft Final Document to be elaborated by the Drafting Committee as a whole, meeting under vice-chairwoman Ambassador Peggy Mason of Canada. He himself formed a 'splinter group' of the most interested delegations to tackle the most stubborn issues. But the drafting group got off to a bad start. The Mexican and Western groups had introduced a number of amendments or additions to Chairman Adeyemi's second paper on the work of Main Committee 1. At this point Mexican Ambassador Marin Bosch insisted that an 'integrated paper' be prepared showing each of the Adeyemi paragraphs and, at the same time, whatever amendments may have been submitted to that paragraph. Most delegates saw no need at this late stage for such a time-consuming exercise; yet the meeting had to be adjourned, and the Drafting Committee lost half a day while the Secretariat worked on the Mexican request. In the Western group, this provoked accusations of delaying tactics to increase the pressure as the conference moved towards its end.

Moves by the USA

The chance of a consensus depended not only on Mexico but also on the still unknown degree of flexibility in the US positions. The USA was known to be unenthusiastic about additional NWFZs and on additional steps to outlaw attacks on nuclear plants, but it had meanwhile accepted the majority view on both subjects. On the other two leading political issues, security assurances and the CTBT, the USA had made its position very clear.

By the middle of the final week, under pressure from some of its Western allies as well as its friends among the NNA states, the US delegation was seeking new instructions from Washington. There were indications that while Washington would not agree to a 1991 conference on security assurances it would be prepared to pursue the matter more seriously in the CD, but this concession was withheld until the issue of a CTBT could be settled. It was obvious that neither the USA nor the UK would commit itself to a moratorium on nuclear testing or to the early negotiation of a CTBT; but hints were emerging that they might be prepared explicitly to recognize the fact that the majority of states attached importance to such negotiations and considered them to be relevant to the ability of the NPT to survive. On 12 September, the USA circulated an informal paper which read: 'The Conference recognizes the significant importance placed upon both negotiations towards a comprehensive nuclear test ban treaty during the next five years and the relationship between the discontinuance of all nuclear explosions in all environments and the long term viability of the Treaty.'

The USA was thus ready to recognize the existence of a widespread perception of a long-term 'linkage', but would not agree that the fate of the NPT must or should depend on the conclusion of a CTBT before 1995. This position was close to that of many states in all groups, but it did not go far enough for the Mexican group. Iran submitted a counter proposal: 'The Conference further stressed that the discontinuance of nuclear testing would play an essential/a central role in the future of the NPT beyond 1995'. In other words, unless nuclear testing was stopped by 1995 the future of the NPT would be in doubt. In another paper, the Mexican group asked that a negotiation mandate be given to the CD.

Compromise in sight?

In the meantime, the full-scope safeguards issue was being solved, largely as result of negotiations in the Western group. Australia had been outspokenly critical of the British position, and on 12 September, a day after the conclusion of the Treaty on the final settlement with respect to Germany,[34] the FRG joined in with a forceful statement about Soviet as well as British reservations. By the next day the UK had withdrawn its reservation, and after the British move the opposition of Belgium, Switzerland, Italy and the USSR collapsed. But there was still debate whether the conference should urge 'suppliers' (Australian preference), 'all suppliers' (British/Belgian preference) or 'the suppliers' to adopt full-scope safeguards. The latter was adopted.[35]

By the evening of 13 September agreement was also in prospect on the entire section dealing with Articles I and II of the NPT. The disputed phrase in the report of Main Committee 1 (urging nuclear weapon states parties to put pressure on Israel) had been withdrawn and the USA had withdrawn its reservation about naming South Africa. Concern was to be expressed about the nuclear programmes of both countries and both would be called upon to accept full-scope safeguards and accede to the NPT. Luxembourg had also withdrawn its reservation about naming France, after a joint EC position failed because of Irish opposition. Toned-down versions of the obliquely critical references to Iraq and Libya had been accepted. NNA pressure had eliminated references to the need for stricter export controls.[36]

The last day

On Friday Ambassador Hyltenius invited the 'splinter group' to the Swedish mission. By the end of the afternoon it had become obvious that Mexico would not accept the text on the CTBT that the USA had proposed; the link with the future of the NPT must be more explicit. To increase pressure on the USA and even at the cost of disquieting some of its supporters the Mexican delegation doggedly resisted proposals commending the progress made in arms control and disarmament since 1985.

On Friday evening Chairman Hyltenius reconvened the full Drafting Committee. By now it was clear that compromises had been reached on most of the draft report on Article VI. Also, in separate Nigerian–US and Egyptian–US talks, the divergences on security assurances were resolved; negative assurances were to be taken up promptly in the CD, and the five nuclear weapon states were invited to enter consultations on extended positive assurances.

A balanced text was emerging that would take note of the meagre progress made with regard to the testing issue in the CD. The compromise text would also recognize the major improvement in US–Soviet relations, the entry into force of the INF Treaty, the prospects for an early conclusion of a START treaty, and the progress made in reducing short-range nuclear forces, in negotiating a ban on chemical weapons and in reducing conventional forces in Europe.

[34] For the text of this 'Two-plus-Four' Treaty, see Rotfeld, A. D. and Stützle, W. (eds), SIPRI, *Germany and Europe in Transition* (Oxford University Press: Oxford, 1991), pp. 183–86.

[35] NPT/CONF.IV/DC/1/Add. 3 (B).

[36] NPT/CONF.IV/DC/1/Add.3 (C).

As the evening wore on, to the surprise of many the USA now accepted a clear-cut link between the future of the NPT and the CTBT and implicitly endorsed early action on the latter. The compromise drawn up by the Chairman read:

The Conference further recognized that the discontinuance of nuclear testing would play a central role in the future of the NPT. The Conference also stressed the significant importance placed upon negotiations, multilateral and bilateral, during the next five years to conclude a CTBT. The Conference again calls for early action towards that objective, by the Conference on Disarmament, at the beginning of its first session in 1991. The Conference urges that the *Ad-hoc* Committee on a Nuclear Test Ban be given an appropriate mandate to pursue the objective of negotiations to conclude a comprehensive nuclear test ban treaty.[37]

The USA and the USSR[38] wished this to be followed by a paragraph contained in the original draft, in which the conference would also note 'the jointly declared statement of the USA and the Soviet Union to proceed with step-by-step negotiations on further intermediate limitations on nuclear testing, having the ultimate objective of the complete cessation of nuclear testing as part of an effective disarmament process'.

The eight NNA states in the Swedish mission agreed to accept the Chairman's proposal but (as he had recommended) without amendment or addition and with no reference to the US–Soviet negotiations 'which they regarded as a disclaimer that undermined the Chairman's compromise text'. The other states present (the three depositaries, Australia, Canada, New Zealand and Poland) insisted on retaining the disputed passage. Attempts at a compromise (e.g., by placing the reference to bilateral negotiations elsewhere or shortening it) were rejected by the USA; it was also not clear whether Mexico would have accepted them.[39]

Mexico and its supporters may well have argued that bilateral negotiations are simply put forward as a substitute for a CTBT, to avoid serious negotiations on an end to testing, that they engage only the superpowers, that they provide a licence for continued testing and for the qualitative improvement of nuclear weapons even if only on a reduced scale. On the other hand there is no doubt that the USA had made a major concession in accepting a clear-cut link between the prospects for the NPT and a CTBT, in implicitly recognizing that the continuation of testing was likely to impair the prospects for the NPT and in agreeing that the *Ad hoc* committee of the CD should be given the tortuous 'appropriate mandate to pursue the objective of negotiations' on a CTBT. Even the US–Soviet formula concerning bilateral talks marked a considerable advance on the attitude of the Reagan Administration: it implied a superpower commitment further to reduce the number and yields of tests and reaffirmed the ultimate objective of a CTBT. Reducing the number of tests should in itself retard the development of new arms and the pace of the arms race.

37 Quoted in the letter circulated by the Mexican delegation to the CD after the conference.
38 Press Release NPT/136, 17 Sep. 1990.
39 This account is based on a lengthy paper, circulated as a letter by the Mexican delegation (note 37).

It was understood that the other significant undecided issue, preparations for 1995, would fall into place once agreement had been reached on the testing issue. In a further small concession to the Mexican viewpoint the UK had proposed convening the first meeting of the Preparatory Committee in autumn 1992 instead of spring 1993, and the West had accepted New York, where there were more NNA delegations than in Geneva, as the venue for 1995.

After midnight: breakdown

By the early hours of Saturday morning it was clear that there was no way out of the impasse. The President convened a final plenary meeting and, in a last desperate attempt to salvage a consensus, circulated a short draft paper. It stated simply that 'the conference deeply regretted that it was unable to arrive at an agreed text in its review of Article VI and preambular paragraphs 8 to 12' and that, despite considerable efforts, 'consensus was not possible other than to note the President's brief characterization of the differences'. This was followed by a summary of the views of each of the two groups. The paper then stated that 'all States Parties continued to express their strong support for the continued viability of the Treaty in the future.' In a final paragraph the conference regretted that it was unable to reach agreement on the preparations for the 1995 conference. The text was presented on a take-it-or-leave-it basis, no change would be accepted. The President asked the three main groups to convene and gave them 20 minutes to consider his draft. When the plenary reconvened Ambassador Marin Bosch announced that the Mexican delegation was unable to accept the paper. The President then said that he had no option but to bring the conference to a close without a Final Document.

A number of delegates chose to make a final statement, full of mutual recriminations. While almost all speakers sought to minimize the impact of the night's failure and reaffirmed their strong and abiding support for the NPT, it was obvious that the final confrontation had left a bitter taste in many mouths.

VII. Analysis and assessment

The reasons for failure

As the principal architect of the Tlatelolco Treaty and an eloquent representative of the Third World in the negotiations of the NPT, Mexico has played an important role in promoting non-proliferation and in advancing the cause of nuclear disarmament. At the four Review Conferences Mexico took the lead in pressing for a CTBT and was the chief critic of the policies of the USA. Mexico's redoubtable representative, Ambassador Garcia Robles, managed to wring many diplomatic concessions from the nuclear weapon states.

Ambassador Garcia Robles none the less used to draw back at the last moment from blocking consensus. At the fourth Review Conference Mexico was prepared to go the whole way even though the inability of the conference to produce a Final Document was likely to be regarded by many, as in 1980, as a token of failure and of deep divisions that could weaken the Treaty and the regime. In its single-minded pursuit of one disarmament measure, a CTBT, it was strongly assisted by the skilful, professional advice and lobbying of a non-governmental organization, Parliamentarians for Global Action, that, like Mexico, had attended the conference with the clear intention either to extract an unlikely concession from the USA or to prevent a final consensus.

In the Western and Eastern groups, it was evident that the changed political climate and the success of arms control since 1985 had lulled the US, British, Soviet and other delegations into a false sense of security—resting too much on their laurels and assuming that their case was so convincing that it would not be possible for anybody to make a convincing case for criticizing the disarmament record. This was apparent in the seemingly inflexible opening statements of the USA in the General Debate and in Main Committee 1. To some extent the US and British assumptions were correct, the NNA states were more divided than in 1985, but a sufficiently large and well-organized group was still ready to follow the Mexican lead. In short the Western delegations may have been too complacent.

The US delegation was quite different from that of 1985; the leader and his deputy were relative newcomers to non-proliferation and lacked the vast hoard of conference experience and non-proliferation lore that the Mexican (or for that matter the Soviet) delegate could draw upon. The US delegation thus had to depend more than usual on the help of its friends and on the USSR. In dealing with the more technical issues before the fourth Review Conference the USA certainly had the support of Australia, the FRG, the Netherlands and, except on the issue of full-scope safeguards, the UK. But when it came to the politically crucial issues Western support was not sufficient. However, leading and indispensable friends of the USA among the NNA states, such as Egypt and Nigeria who had played a key role in achieving a consensus in 1985, were largely engaged in advancing their own schemes (security assurances), while some other architects of the 1985 consensus, such as Sri Lanka, had now become radical. The USSR, plagued by problems at home, played a helpful but rather passive role and was distracted by divisions within its own ranks on the question of full-scope safeguards. The UK, as co-ordinator of the West European group, was divided from most other West European states on the main political and technical issues of the conference, the CTBT and full-scope safeguards. Hence leadership on many of the issues before Main Committees 2 and 3 passed to Australia, much helped by the Netherlands and Ireland, while the UK itself was part of the problem in Main Committee 1.

After a promising start in the speech of Hans-Dietrich Genscher, the German delegation played a helpful but still rather passive role. The startling changes taking place in Germany and the realignment of its nuclear export policies were probably too topical to allow it to assume the leading role to which its status as Europe's foremost industrial and economic power and major non-nuclear weapon state would entitle it. For somewhat different reasons the role of Japan at such conferences is also still somewhat passive.

When all is said and done, however, the strait through which the conference had to sail was very narrow, with jagged reefs on either side. On one was the long-standing demand of the NNA states that the nuclear weapon states should live up to their 1963 promise, repeated in 1968, 'to seek to achieve the discontinuance of all test explosions of nuclear weapons for all time and to continue negotiations to this end'. This demand reflected a deeply felt resentment because of promises unkept and hopes deferred. The NNA states showed some flexibility in not insisting on a Final Document that would have called for the immediate negotiation and conclusion of a CTBT and in contenting themselves with a US and British admission that its absence would jeopardize the outcome of the 1995 conference.

On the other side was the equally inescapable fact that there was no prospect that an NPT Review Conference would bring about a fundamental change in US deterrence policies no matter how much 'continuance' of testing might imperil the future of the NPT. The USA and the UK showed some flexibility, however, in being prepared to admit implicitly that their nuclear testing did indeed present such a danger.

The reefs were made more hazardous, moreover, by self-serving attitudes on either side. There was a strong impression that some of the radical NNA states, aside from justified criticism, were possessed by an inexorable zeal to browbeat the USA in a forum where this could be done with little risk. The presence in the Mexican group of countries whose own disarmament record is dismal at best, such as Iran and Syria, lends credibility to this suspicion.

On the other hand the USA and, to a lesser degree, the UK are showing a disposition to wrap purely national interests (or even sub-national interests of certain circles in the Pentagon and the nuclear weapon laboratories), such as the qualitative improvement of nuclear weaponry or the development of a new generation of nuclear weapons for which not even the closest allies show any enthusiasm, in the noble coat of self-appointed stewardship for world security. The current Gulf conflict might even exacerbate this deceptive confusion of global and purely national interest and diminish the chances that a serious review of such 'sacred cows' as continued testing might be conducted.

Between these two reefs only a surer captaincy and consummate diplomatic seamanship—and perhaps a little more time—would have brought the conference safely through. The final decision lay in Mexico's hands. It had wrung from the USA a surprisingly frank admission of 'linkage'. It was clear that the USA had reached the limit of what it could concede, however. Indeed, some suggest that the US delegation may have gone beyond that limit. It would have cost Mexico little to accept the US insistence (and the Soviet desire) that there should also be a reference to their bilateral step-by-step approach, and to hold the superpowers accountable for the commitment implicit in that statement. Instead Mexico chose to show that it could command a blocking minority. In doing so it may have intended to fire a shot across the bows of the Anglo-Saxon nuclear weapon states but the shot hit the conference amidships.

Damage assessment

The absence of a Final Document means that the conference failed to endorse and give formal status to any of these recommendations. In a salvage operation and on the initiative of 23 delegations (including Mexico but not the depositaries) the agreed parts of the reports of Main Committees 2 and 3 were circulated to the IAEA's General Conference, its Board of Governors and its member states.[40] The reports were not formally placed on the agenda of either of the IAEA's governing bodies but made available for their information. None the less it is open to those bodies to follow up any of the committee's recommendations. It is also open to any member state and to the IAEA Secretariat to propose that a particular recommendation be acted upon.

The IAEA would be the proper place to follow up recommendations relating to technical co-operation, nuclear safety and safeguards. It would not, however, be the right place to follow up recommendations regarding the strengthening of export controls or to re-knit the unravelled consensus on full-scope safeguards. The representatives of states not parties to the NPT would probably be able to block any such action. It is more likely that the suppliers will convene to consider the implications of the newly emerging consensus.

The display of deep division between the parties, their inability to bridge them and the criticisms that the radical NNA states aimed at the nuclear weapon states must have given satisfaction to the critics and opponents of the Treaty in the threshold countries. While the conference's failure to reach consensus might not change the mind of any state contemplating accession it can hardly be seen to have enhanced the authority and attraction of the Treaty or have strengthened the non-proliferation regime.

The criticism to which North Korea was exposed was no doubt reported to Pyongyang, but the absence of a Final Document must have tempered its effect and pleased the North Korean delegation and likewise attenuated whatever faint pressure the conference brought to bear on Iraq and Libya.

The absence of instructions on how to prepare for 1995 leaves the preparations for the crucial Extension Conference uncertain and controversial. The danger for the continued existence of the NPT is not yet overwhelming, but it is there. There are few almost universally agreed rules that can serve as building blocks of order in the difficult period now before the world. The NPT is one, and a CTBT, no doubt, would be a welcome and useful addition. The conference has shown that some are willing to damage the NPT if they cannot get a CTBT. The danger of this tactic is that in the late 1990s it may leave the world with neither.

[40] IAEA GC (XXXIV)/INF/291, 19 Sep. 1990.

17. New security structures in Europe: concepts, proposals and decisions

ADAM DANIEL ROTFELD

I. New institutions

The heads of state and government of the 34 nations of the Conference on Security and Co-operation in Europe (CSCE), assembled in Paris on 19–21 November 1990, adopted a set of documents consolidating the profound and fundamental changes which had taken place in Europe. The five documents which were adopted are the Treaty on Conventional Armed Forces in Europe (CFE), the Joint Declaration of Twenty-Two States, the Charter of Paris for a new Europe, the Supplementary Document to give effect to certain provisions contained in the Charter, and the Vienna Document 1990 on new confidence- and security-building measures (CSBMs).[1] The Paris meeting participants reaffirmed the principles contained in the Helsinki CSCE Final Act of 1975 and declared their commitment to the rule of law, pluralistic democracy and free elections. They approved the results of the negotiations on German unification and on the military aspects of security in Europe by accepting the new decisions on CSBMs and by taking note of the CFE Treaty, signed within the framework of the CSCE by the 22 North Atlantic Treaty Organization (NATO) and Warsaw Treaty Organization (WTO) states. The Paris meeting also adopted guidelines for negotiations on military security.

The changes in Europe opened a new possibility to establish institutional arrangements in the field of political and military security. In the Paris documents, the summit meeting decided:

1. To organize every two years meetings of heads of state or government (on the occasion of CSCE follow-up meetings);
2. To establish a Council of Ministers for Foreign Affairs. The 34 ministers will meet regularly, at least once a year (the first meeting will take place in Berlin). Meetings of other ministers may also be agreed by the CSCE states;
3. To institute a Committee of Senior Officials with the aim of preparing the meetings of the Council and carrying out its decisions;
4. That the Council of Ministers for Foreign Affairs should discuss the possibility of establishing an emergency mechanism (to convene meetings of the Committee of Senior Officials in emergency situations);

[1] For the text of the CFE Treaty, see appendix 13A in this volume; for the Joint Declaration of Twenty-Two States, see appendix 17A; for the Charter of Paris, see appendix 17B; for the Supplementary Document, see excerpts in Rotfeld, A. D. and Stützle, W. (eds), SIPRI, *Germany and Europe in Transition* (Oxford University Press: Oxford, 1991); and for the Vienna Document 1990, see appendix 13C in this volume.

5. That the CSCE follow-up meetings will be held as a rule every two years and will not exceed three months;

6. To establish in Prague a permanent CSCE Secretariat in order to provide administrative support to other CSCE institutions, meetings and consultations;

7. To create a Conflict Prevention Centre (CPC) in Vienna;

8. To establish an Office for Free Elections in Warsaw;

9. To discuss and prepare the creation of a CSCE parliamentary assembly (an 'Assembly of Europe') involving members of parliaments from all 34 states (this new body could be based on the Parliamentary Assembly of the Council of Europe); and

10. To establish, in the framework of the new CSBMs, a mechanism for consultation and co-operation as regards unusual military activities and to create a communications network for the transmission of messages relating to agreed measures. Annual implementation assessment meetings are also envisaged (the CPC will serve as the forum for these meetings). The beginnings of some new structures are connected with the annual exchange of military information such as plans for the deployment of major weapon and equipment systems and military budgets.

The question is: To what extent are these and other new institutions adequate to meet the threats and challenges of post-cold war Europe?

II. New threats and challenges

The crucial factors for European security are developments in the USSR, the dissolution of the WTO and the unification of Germany.[2] A qualitatively new element is the re-emergence of the long suppressed drive of nations to self-determination. However, a number of new threats are inherent in this movement, among which the most dangerous is resurgent nationalism, which has an impact on both external relations and powerful domestic centrifugal tendencies that could weaken or even tear apart certain states composed of many nationalities. This applies above all to the Soviet Union and Yugoslavia.

The basic premise of the system taking shape at the start of the 1990s is that full-scale war in Europe between East and West is *impossible*. The reasoning is as follows. Since the WTO—an alliance of nations with not only common political interests but also a common ideology and form of government—is disintegrating, issues that divide the members are now more numerous than those which unite them. Although they may be prepared to co-operate in negotiations on arms control, they will not consent to placing their armed forces under a joint command. The unification of Germany has *de facto* removed one member of the WTO from the alliance, the former German Democratic Republic; and Hungary, Czechoslovakia and Poland have officially signalled their political determination to leave.[3] These and other factors, not least the

[2] For the text of the Treaty on the final settlement with respect to Germany, see appendix 17C in this volume.

[3] *Pravda*, 16 June 1990; *The Guardian*, 15 Jan. 1991.

real possibility of disintegration of the Soviet Union, have completely changed the political configuration of the WTO and raised questions about the point of preserving the alliance.[4]

In contrast, NATO has in this transition period managed to maintain its cohesion as well as adapt to the new circumstances and challenges. In fact, some Central and East European states view NATO and its mechanisms as elements of a future pan-European security system.

A new security situation has arisen which is something of a paradox: the reduction of the probability of war in Europe between hostile blocs to virtually nil has been accompanied by an ominous increase in the danger of nationalist conflicts erupting into local clashes and limited wars. The new security institutions and structures should be designed to prevent such conflicts and to perform a stabilizing function. In the past, the bipolar system and its institutions served as responses to the specific risks of the cold war. With its termination and the end of politico-military and ideological confrontation, a need has arisen to fashion new instruments for these tasks.

The catalogue of threats and challenges in the new Europe can be summarized in four elements:

1. In the internal sphere, the failure of economic reforms could undo the construction of democratic institutions in Central and Eastern Europe and result in totalitarian systems of the left being replaced by authoritarian regimes of the right. In early 1991 the use of force against independence movements in the Baltic republics may have jeopardized the democratization process in the USSR as a whole.

2. A resurgence of previously stifled national aspirations could in economic crisis grow into mass populist–nationalist movements which would challenge liberal–democratic élites, be xenophobically intolerant of national minorities and be hostile to neighbouring nations.

3. In the external sphere, two developments could in certain circumstances have an important destabilizing effect on Europe:

(*a*) Disintegration of the Soviet Union as well as state-building processes in the republics (Lithuania, Latvia, Estonia and Ukraine, Byelorussia, the Caucasian republics and above all Russia) may also trigger not always predictable territorial claims, while economic disaster could unleash a huge

[4] Foreign Minister Shevardnadze stated that the Warsaw Pact 'is not eternal and was concluded for a definite period' and that it has proved too inflexible in its response to change and incapable of meeting the challenges of the new political reality. 'It will last only as long as it serves the real interests of its members.' *Pravda*, 26 June 1990. In the context of the 1991 military intervention of Soviet troops in the Baltics, the Parliament of Czechoslovakia urged the acceleration of negotiations on the formal dissolution of the WTO. The Foreign Ministers of Hungary, Poland and Czechoslovakia met in Budapest to elaborate their common position in this regard. *Rzeczpospolita*, 17 Jan. and 22 Jan. 1991. In letters of 11 Feb. 1991, addressed to President Vaclav Havel of Czechoslovakia and President Lech Walesa of Poland, President Mikhail Gorbachev proposed the dissolution of the military structures of the WTO as of 1 Apr. 1991. A Declaration signed in Budapest on 25 Feb. 1991 by Foreign and Defence Ministers from the six WTO countries said that the 'member states of the Warsaw Treaty, acting as sovereign states with equal rights, decided that by March 31, 1991, they will dismantle the military organs and structures of the Treaty'.

migratory wave of labour, political refugees and fugitives from hunger and a rising tide of ethnic and racial animosities.

(*b*) The united German state is a powerful centre which will both dominate Western groupings (NATO and the European Community) and to a large extent determine real and perceived security in Central and South-Eastern Europe. German unification, or rather the ingestion of the GDR by the FRG, will also temporarily slow down integration in the European Community.

4. Security will also be influenced to an increasing degree by problems arising from a widening technological gap and ecological mishaps. The Chernobyl accident and the ecological disaster areas spanning the borders of Poland, Czechoslovakia and eastern Germany are signals of the extent of the common threats to Europe.

The prerequisite for a new European order is recognition of the interdependence of political, military, economic, technological and ecological problems.

III. Concepts and blueprints

Structures are by definition more durable than the conditions in which they came into being. Changes in existing structures and the creation of new institutions are the result of various kinds of action. First, institutions which have outlived their usefulness or are obstacles to the process of democratic change and construction of a new European order will continue to wither away. Second, some structures which, although products of different conditions, still offer participants a framework for the advancement of their interests will undergo transformations, a process which has already started. Third, there is a need to create new co-operation mechanisms. This applies in particular to institutionalization of the CSCE process.[5] The cornerstones of a new European security system will be the structures now in place or being developed within NATO, the EC and the pan-European CSCE process.

'The cold war is over'. In these words many participants of the 1990 Paris summit meeting characterized the present situation in Europe.[6] The basic task of present and prospective institutional security structures is to diminish threats and to ensure crisis control or management. It is therefore important first to define properly the current sources of insecurity in Europe: they arise today not from the likelihood of deliberate sudden attack but rather as the unforeseeable and unpredictable result of instabilities and uncertainties.

All the blueprints for a new European security system invariably postulated the necessity of synchronizing German unification with the creation of new

[5] A review of proposals for a new European regional order is presented in Sharp, J. M. O., 'Europe's new architecture: a reporter's guide to proposals for a new regional order', *Deadline: A Bulletin from the Center for War, Peace, and the News Media*, vol. 5, nos 5–6 (autumn 1990). See also Eavis, P. (ed.), *European Security, The New Agenda* (Saferworld Foundation: Bristol, UK, Nov. 1990). A valuable suggestion in this respect appears in Clesse, A. and Rühl, L. (eds), *Beyond East–West Confrontation, Searching for a New Security Structure in Europe* (Nomos: Baden-Baden, 1990).

[6] Statement by President Bush, issued in Paris, 21 Nov. 1990. See also the statement by President Gorbachev, 19 Nov. 1990, *Pravda*, 20 Nov. 1990.

structures. What was sought was a solution which would ensure both control and anchorage of Germany as an integral part of the new system. Lord Ismay, NATO Secretary-General in the 1950s, is credited with authorship of the oft-cited triple formula justifying the existence of NATO: 'To keep the Americans in, the Russians out and the Germans down'. Despite the many changes in Europe, this *raison d'être* seems, to judge by the reactions of the Federal Republic's neighbours and allies, to have remained largely intact. A report prepared by European and US specialists made the apt point that the new structures 'will be composed of organizations principally dealing with economic rather than military matters'.[7] Nevertheless, a structure diminishing the risk of war is also necessary. Arguing the need to accommodate collective action, the authors write: 'NATO at present provides the promise of collective action if Germany were attacked from Poland. The new security structure should provide the same kind of guarantee if Poland were attacked from Germany'.[8] They see the new structures as ensuring the whole of Europe the same degree of security as Western Europe is provided by NATO.

A qualitatively new element is the return to the European stage of Germany as a great power. The fact that twice in this century it has provoked world wars has inclined many authors to make pessimistic predictions,[9] which boil down to fears that in post-cold war Europe the policies of a united Germany will be the chief source of instability.[10] However, extrapolation is not the most reliable method of forecasting. Mechanical projection into the future of threats known in the past would be a gross over-simplification or even total misunderstanding. Germany at the end of the 20th century is not and will not be the Germany of the end of the 19th or the first half of the 20th century, for the following reasons.[11]

1. Germany has never existed within the borders to which it was reduced as a result of defeat in World War II, started by its invasion of Poland.

2. Over 40 years of democratic institutions have led, for the first time in German history, to the coalescence in the Federal Republic of a society which, extremist and fringe phenomena apart, has frequently confirmed its commitment to the ideal of a state based on the rule of law (*Rechtsstaat*).

3. Germany has never in the past been, as it is now, an integral part of Western political, military and economic structures (NATO, the EC, the Council of Europe, the Western European Union, etc.), to which it has like

[7] See a report compiled by the British–American Security Information Council (BASIC), 'A New Security Structure for Europe', written by Frank Blackaby in association with specialists in Europe and the USA (BASIC: London–Washington, 1990), p. 3.

[8] See Blackaby (note 7), p. 3.

[9] The fears aroused by German unification have sparked off a major debate, producing judgements as extreme as that expressed by the British Secretary of State for Industry, Nicholas Ridley. See 'Saying the unsayable about Germany', *Spectator*, 14 July 1990. In the debate, which led to Ridley's resignation, former NATO Secretary-General Joseph Luns stated: 'Ridley said out loud what many Europeans think'; see Melcher, R. and Rollnick, R., 'Axis urged to counter Bonn', *The European*, 27–29 July 1990.

[10] See Bellak, L., 'Why I fear the Germans', *New York Times*, 25 Apr. 1990.

[11] See Van Evera, S., 'Primed for peace, Europe after the cold war', *International Security*, vol. 15, no. 3 (winter 1990/91).

other members transferred some of its sovereign rights, including those pertaining to decisions with a fundamental bearing on security, economics and welfare.

4. In contrast to the past, German unification is not being effected by 'blood and iron' (Bismarck's Prussia) but is a sovereign political process consequent on the decision of the GDR Parliament to unite with the Federal Republic in accordance with Article 23 of the Basic Law.[12]

5. In the German unification process, due provision was made for the external aspects of unification and the security of neighbouring countries through the acceptance of certain obligations: definitive recognition of the frontiers of the united state, which covers the areas of the FRG and the former GDR, including Berlin; reduction of its armed forces to 370 000 troops in the course of three to four years; renunciation of the production and possession of nuclear, biological and chemical weapons and commitments to dispose of them; and continued adherence to the Non-Proliferation Treaty, of which both German states were signatories.[13]

In the Europe of the early 1990s, the cleavage is no longer demarcated by ideological conflict but by a gulf in economic development, standards of living, technology and civilization. From San Francisco to Vladivostok, 34 governments can now claim the credentials of democratic election or are seeking to do so. There is also general acceptance of the territorial *status quo* and the inviolability of frontiers, although this does not exclude the possibility of peaceful changes, practical illustrations of which are both the unification of Germany and the moves towards secession among the Soviet republics. The latter process is likely to see not only Lithuania, Latvia and Estonia breaking away and acquiring full sovereignty but other republics following suit if steps are not taken to set up a new type of community, some kind of Soviet 'Commonwealth'.[14]

The end of the cold war also spells the end of a bipolar world. However, the new European landscape will be more complex than a simple transformation of bi- into multipolarity. The fact that Soviet forces are being withdrawn from Central and Eastern Europe does not mean that US forces will eventually pull out from Western Europe.[15] Indeed, US military presence, although reduced, is

[12] A motion to this effect jointly tabled by four political parties (CDU, DSU, SPD and FDP) was carried by the Volkskammer on 23 Aug. 1990 with 294 votes in favour, 62 against and 7 abstentions. The majority was far greater than the required two-thirds.

[13] For the table of parties to major multilateral agreements, see annexe A in this volume. These and other commitments have been confirmed in bi- and multilateral agreements concluded by the FRG and the GDR with the USSR, Poland and the three Western powers and in the '2 plus 4' talks.

[14] This question is the subject of theoretical debate and practical searches for optimum solutions in which two extreme positions have emerged: a unitary state with a strong central government taking the form of a federation invested with full sovereignty, and a confederation whose components (the republics) would be sovereign states associated in a new community. A survey of the different concepts has been made by the director of the Ethnography Institute of the USSR Academy of Sciences; see Tishkov, V., 'Tupiki natsyonalnoy gosudartviennosti', *Pravda*, 4 Sep. 1990, p. 3.

[15] *Pravda*, 5 Nov. 1989, reported that Soviet forces had been reduced to 235 000 and that 7000 tanks and 700 aircraft had been withdrawn from Europe. The leader of the Soviet delegation to the CFE talks demanded a drastic cut in US forces in Europe to 70 000–80 000 men. See *International Herald Tribune*, 10 Sep. 1990.

treated both by the NATO states and the former members of the WTO as buttressing rather than undermining European stability and security.[16] The role of the Soviet Union has changed, as has the function performed in the European system by the United States. Although both the USSR and the USA have retained certain elements of domination over the other actors, particularly in the area of global strategic decisions, the role of super-referee on questions of security can only be claimed by the United States.[17]

Speaking a month after the fall of the Berlin Wall, US Secretary of State James Baker outlined a 'new architecture' for the immediate future:

> This new architecture must have a place for old foundations and structures that remain very valuable—like NATO—while recognizing that they can also serve new collective purposes. The new architecture must continue the construction of institutions—like the European Community—that can help draw together the West while also serving as an open door to the East. And the new architecture must build up frameworks—like the CSCE process—that can overcome the division of Europe and that, at the same time, can bridge the Atlantic Ocean.[18]

In Baker's opinion, the new structures must also accomplish two specific objectives: to promote peaceful unification of Germany as well as reinstatement of the whole of Berlin; and to promote linkage—military, political and economic—of the security of Europe with the security of North America, that is, the USA and Canada.

An interesting example of thinking along these lines is a study written by the US diplomat and student of international relations, James E. Goodby.[19] His objective seems to be to persuade US public opinion of the necessity of constructing a new European security system. He states that, while NATO and the EC have made a significant contribution to the idea of a 'Europe whole and free', 'those institutions can never embrace the whole of Europe'.[20] Goodby takes the view, unquestionably unpopular in the United States, that the CSCE is the only institution which can meet the increased requirements and expectations in the area of European co-operation ('there is no plausible alternative to the CSCE'). The conventional wisdom in the USA is that 'NATO and the CSCE represent different, and contradictory, ideas about the organization of Europe'.[21] Goodby argues in contrast that NATO, the EC and the CSCE are complementary structures which need not be a differentiated triad but can reinforce each other. The CSCE process can supplement both these Western

[16] Sharp, J. M. O. (ed.), SIPRI, *Europe After an American Withdrawal: Economic and Military Issues* (Oxford University Press: Oxford, 1990).

[17] Witness events in the Persian Gulf. The Iraqi aggression has for the first time led the international community to close ranks and unite politically to impose collective sanctions. The USA has effectively assumed the role reserved by the UN Charter for its own agencies and units; see chapter 18 in this volume.

[18] Secretary of State James Baker, addressing the Berlin Press Club; see Baker, J. A., 'A New Europe, a New Atlanticism: Architecture for a New Era', Press Release no. 245, 12 Dec. 1989, p. 3.

[19] Goodby, J. E., *CSCE: The Diplomacy of Europe Whole and Free*, Occasional Paper of the US Atlantic Council (Atlantic Council: Washington, DC, July 1990).

[20] See Goodby (note 19).

[21] See Goodby (note 19).

structures (NATO and the EC) by adding to security policy a constraining factor in the form not so much of countervailing power as the 'exercise of democratic control over national decisions'.[22]

It is worth noting in this context that, whatever the banes of the cold war (tension, confrontation, the arms race, and deformation of international relations and models of internal development), the bipolar system did act as a lid, stopping Europe's strains and dissensions from boiling over into armed conflict. Some authors maintain that the rules of the cold war saved Europe from hot war.[23] The search for a new security system should not only seek to eliminate new threats but also take account of the factors which have often been a source of conflict and war in Europe, including ethnic strife and resurgent nationalism. The gradually disappearing bipolar system is by definition more stable than a multipolar system, with its inherent asymmetry and multiplicity of actors and decision-making centres.

At the December 1989 Malta summit meeting the US and Soviet presidents decided to accelerate the timetable for finalization of the accords defining European security structures until the end of the century.

The announcement of forthcoming decisions in this area activated a process of deep change in NATO and the WTO. The significance of non-military—that is, politico-legal—aspects of security grew. NATO, as Baker intimated,[24] is becoming a forum for co-operation in negotiations, implementation and verification of accords and for expansion of the scope of East–West agreements.

The new structures which can ensure the external security of states will be of basic significance to the future of Europe. This applies in particular to states which do not or soon will not belong to the existing military alliances, chiefly members of the WTO. Since the dismemberment of the WTO is imminent, this has left them in a situation in which old security guarantees have become meaningless and new structures and ground-rules are not yet in place. In this context it is worth noting certain similar elements in the approach of a number of countries, both those formerly on either side of the old divide and those in the neutral and non-aligned group.

IV. Institutionalizing the CSCE: convergent proposals and views

Preparations for the Paris CSCE summit meeting gave a new impetus to the all-European debate on institutionalization of the CSCE process. The direction of change was defined by three factors:

[22] Goodby writes: 'It can help construct a security system in which power is constrained not just by countervailing power but by the exercise of democratic control over national decisions'. See Goodby (note 19), p. IV.

[23] Mearsheimer, J. J., 'Back to the future: Instability in Europe after the cold war', *International Security*, vol. 15, no. 11 (summer 1990), pp. 5–56.

[24] See Baker (note 18), p. 4.

1. Earlier decisions—that led to a more prominent place in the CSCE process for the military aspects of security, implementation of the decisions of the Stockholm Conference on CSBMs, the continuation of negotiations on a 'second generation' of CSBMs and the initiation 'within the CSCE process' of the parallel CFE Negotiation—bear witness to the momentous significance that agreement on such measures will have for developing a co-operative security system in Europe.

2. The need to take account of the systemic changes in Central and Eastern Europe and the USSR—political pluralism, liberalization of emigration policies and freedom of speech—has totally changed the complexion of the problem-area of individual freedoms, human contacts and humanitarian issues in international relations; these questions have ceased to be an area of ideological confrontation and require a machinery for routine co-operation and oversight.

3. The existing forms of the CSCE process (periodic conferences at the political level and meetings of experts) and the political rather than legal nature of its decisions no longer meet the needs of the new situation; structures will have to be adapted to new political realities.

The first outline of a new CSCE structure was presented by Polish Prime Minister Tadeusz Mazowiecki in a parliamentary exposé on 18 January 1990 in which he proposed a European Co-operation Council.[25] Details of its remit and structure were circulated to the CSCE countries through diplomatic channels.[26] Poland saw the Council as a permanent CSCE political agency performing consultative and co-ordinating functions in all areas regulated by CSCE decisions. It would ensure continuity of the CSCE process, and its tasks would embrace (*a*) review and evaluation of current problems in political, military, economic, humanitarian and environmental co-operation; and (*b*) co-ordination of the work of institutions and bodies set up by the CSCE.

The European Co-operation Council's purview would include assessment of the functioning of CSBMs, prevention of potential conflicts, and study of new proposals for co-operation and contacts. The Council would comprise representatives of the 34 CSCE states at the ambassador level. Administration would be handled by an international secretariat, which would also run a CSCE Documentation and Information Centre. Polish Foreign Minister Krzysztof Skubiszewski added in the Sejm that the Council could have three commissions: for political relations and security, economics and ecology, and humanitarian questions.[27]

At the meeting of WTO foreign ministers in Prague (17 March 1990) the new Czechoslovak Foreign Minister, Jiri Dienstbier, proposed the formation

[25] A modified version of this proposal was presented by Prime Minister Mazowiecki at the European Parliament (Strasbourg, 30 Jan. 1990). See also Freedman, L. (ed.), *Europe Transformed: Documents on the End of the Cold War: Key Treaties, Agreements, Statements, Speeches* (Tri-Service Press: London, 1990), pp. 430–35.

[26] Polish Proposal on the Outline of the Council of European Co-operation in the Framework of the CSCE Process (Warsaw, 9 Mar. 1990), in Rotfeld and Stützle (note 1), p. 134.

[27] Sejm exposé by Foreign Minister Skubiszewski (Warsaw, 26 Apr. 1990).

of a European Security Commission[28] to act as the executive organ of a regional security system in accordance with the UN Charter. Dienstbier proposed that the Commission have two chambers: a political and, subordinate to it, a military chamber.[29] The Polish Foreign Minister expressed his readiness to enter into consultations on this matter with Czechoslovakia.[30]

The democratic changes in Central and Eastern Europe have led most of the countries of this region to seek membership of *the Council of Europe*. In each case they are motivated by a need to manifest that they belong to or, as Polish Prime Minister Mazowiecki put it in Strasbourg, are returning to Europe ('Back to Europe!').[31]

The 12 members of the *European Community (EC)* believe that the best path to overcoming divisions lies in the CSCE process. The Declaration adopted by the EC foreign ministers (in Dublin, on 20 February 1990) and the recommendations of the special session of the Council (in Dublin, on 28 April 1990) concerning the further development of the CSCE stated that the Paris summit meeting ought to consider 'the development of pluralist democracy, the rule of law, human rights, better protection of minorities, human contacts, security, economic cooperation, the environment, further cooperation in the Mediterranean and cooperation in the field of culture'.[32] The EC Declaration also stated that the summit meeting should consider new institutional arrangements, 'taking also into account proposals made by the Central and Eastern European countries'. A new element was the attention given to the need to define the interaction between the CSCE process and institutions such as the Council. The document issued by its next session stated that the Paris summit meeting ought to create 'an opportunity to define the crucial role which the CSCE will play in a future architecture of Europe and in establishing a new set of relations between participating States'.[33]

Regarding the position of *the neutral and non-aligned (NNA) states*, they no longer play the role of 'honest brokers'. Their position is close to that adopted by the EC.[34] Some of them also indicated that they were willing to offer their capitals as headquarters of a permanent CSCE Secretariat.[35] The Ministers for Foreign Affairs of nine NNA countries (Austria, Cyprus, Finland, Liechtenstein, Malta, San Marino, Sweden, Switzerland and

[28] See the text in Rotfeld and Stützle (note 1), p. 138.

[29] This proposal was circulated to all the CSCE countries in the form of a detailed memorandum on 6 Apr. 1990. See Rotfeld and Stützle (note 1), p. 139.

[30] See Sejm exposé (note 27).

[31] See Freedman (note 25), p. 430.

[32] Guidelines on CSCE, Adopted by Special Meeting of the European Council, Presidency Conclusion. See the text in Rotfeld and Stützle (note 1).

[33] European Council, Presidency Conclusions, Dublin, 25–26 June 1990; see Rotfeld and Stützle (note 1).

[34] Speech of President Mauno Koivisto of Finland in the European Parliament (Strasbourg, 9 May 1990).

[35] See, for example, 'Building a more secure Europe in the 1990s', Keynote Address by Mr Sten Andersson, Swedish Minister for Foreign Affairs, at the 1990 International Conference of the Institute of East–West Security Studies, Stockholm, 7 June 1990.

Yugoslavia) met in Helsinki on 1–2 November 1990.[36] They discussed the new role of the CSCE and future co-operation within this process. As far as the negotiation on military aspects of security in Europe is concerned, the ministers 'called for the creation of a single CSCE forum' for negotiations. They voiced their support for establishing a Conflict Prevention Centre and a CSCE Secretariat. They stressed the importance of the creation of a parliamentary assembly based on that of the Council of Europe.[37] The document was the first joint political statement of the NNA states reflecting their reaction to the dramatic change in the European security environment.

Regarding the position of *the Soviet Union*, an article by Foreign Minister Eduard Shevardnadze sketched the Soviet concept of a future European security structure.[38] Taking the view that countries which are or will be outside the military blocs will require new security guarantees, he argued that they could be provided by new all-European institutions such as: (*a*) a Council (Assembly) of Great Europe, a summit-level forum of the leaders of all the CSCE states, meeting once every two years to discuss issues of cardinal importance and take basic decisions; (*b*) a Foreign Ministers Committee (Council), convening once or twice a year and charged with preparing the material for decisions to be taken at summit level and monitoring their implementation; (*c*) a permanent consultative organ, composed of 34 ambassadors, based in the capital of the country which is the seat of the CSCE executive secretariat and dealing with current problems and contacts; and (*d*) a Permanent Secretariat, composed of groups of experts and administrative staff and coming under the Council (Committee) of Foreign Ministers and Coordination Commission. It would be based in the capital of the country which is the seat of the Commission. In the Soviet concept, an important role should be played by a Centre for Military–Political Stability in Europe, with two tasks: information–verification, and conflict and crisis prevention. Shevardnadze proposed Berlin as the seat of the Centre. He also believed that the summit meeting should set up a special group to prepare recommendations for co-ordinating the work of the existing organizations concerned with economics, ecology, communications, information and human contacts.[39]

The USSR attached particular importance to synchronization of the processes of CSCE institutionalization and German unification, both of which were gaining momentum. The close substantive and formal connection between them was reflected in both the decisions finalizing the '2 plus 4' talks and the inclusion in the accords of a package of agreements approved by the CSCE foreign ministers (in New York, on 1–2 October 1990) and at the Paris summit meeting.

[36] The full text of the Communiqué of the Meeting of the Ministers for Foreign Affairs of the neutral and non-aligned countries of the CSCE (Helsinki, 1–2 Nov. 1990) is published in Rotfeld and Stützle (note 1), pp. 156–57.

[37] See note 36.

[38] 'Europe: the mission of a generation', *Izvestia*, 30 May 1990.

[39] More details were provided by Shevardnadze in 'Towards a greater Europe—the WTO and NATO in a renewing Europe', *NATO's Sixteen Nations*, June 1990, pp. 18–20.

Describing the tasks facing *NATO*, President George Bush stated: 'I share the view of those who have called for regular consultations among senior representatives of the CSCE countries. We should consider whether the new CSCE mechanisms can help mediate and settle disputes in Europe'.[40]

The most eloquent call for institutionalization among the NATO states was expressed by the Foreign Minister of the Federal Republic of Germany, Hans-Dietrich Genscher. At a conference of the Tutzing Protestant Academy in January 1990, he said:

The CSCE summit can also contribute towards an East–West partnership for stability and a peaceful European order by dealing with the establishment of European institutions, such as:

1. An institution to coordinate East–West economic cooperation. The European Development Bank must also be seen in this context.
2. A pan-European institution for the protection of human rights. The application of the Council of Europe's human rights convention to the whole of Europe suggests itself.
3. A centre for the creation of a European legal area aimed at legal harmonization.
4. A European environment agency.
5. The extension of EUREKA cooperation to the whole of Europe.
6. Collaboration between ESA and corresponding Eastern institutions.
7. A centre to develop European telecommunications.
8. A centre to develop European transport infrastructure and policy.
9. A European verification centre.
10. A European conflict-management centre.

To keep the CSCE process moving, one might also set up a council for foreign ministers of the CSCE countries, which would meet at regular intervals.[41]

A few days later, at an international conference organized jointly by SIPRI and the Institut für Internationale Politik und Wirtschaft (IPW) and held in Potsdam,[42] Genscher again listed the 10 new pan-European institutions and added that, by deepening and reinforcing the CSCE process, 'all participating States are prepared to create a framework of stability and network of security for foreseeable and unforeseeable developments in Europe'.[43]

The London Declaration adopted at the July 1990 NATO summit meeting[44] presented for the first time a NATO attitude to development of the CSCE process closely attuned to the expectations of the USSR, the Central and Eastern European countries, and the NNA states. It made the following six proposals with respect to CSCE institutionalization:

[40] President George Bush's Address at Oklahoma State University, 4 May 1990; see Rotfeld and Stützle (note 1), p. 97.

[41] *Mitteilung für die Presse des Bundesministers des Auswärtigen*, Bonn, no. 1026/90 (31 Jan. 1990), pp. 46–47. The English text was issued by the Western European Union and is published in Freedman (note 25), pp. 436–45.

[42] For an account of the SIPRI–IPW conference, see Rotfeld and Stützle (note 1).

[43] The full text is published in Rotfeld and Stützle (note 1), pp. 20–29.

[44] London Declaration on a Transformed North Atlantic Alliance, issued by the heads of state and government participating in the meeting of the North Atlantic Council, London, 5–6 July 1990; NATO

1. Regular annual consultative meetings of heads of state and government or foreign ministers suitably prepared beforehand and assured of continuation.
2. Biannual CSCE review conferences to assess progress towards a 'Europe whole and free'.
3. Formation of a small secretariat to co-ordinate these meetings.
4. Establishment of machinery for monitoring elections in all 35 countries on the basis of the Document adopted in Copenhagen by the Conference of the Human Dimension of the CSCE.
5. Formation of a Centre for Conflict Prevention which could act as a forum for exchange of military information, discussion of 'unusual military activities' and mediation in disputes.
6. Establishment of a CSCE parliamentary body, based on the existing Parliamentary Assembly of the Council of Europe in Strasbourg and embracing all the participants in the CSCE process.

According to the London Document, 'The sites of these new institutions should reflect the fact that the newly democratic countries of Central and Eastern Europe form part of the political structure of the new Europe'.[45]

V. Decisions and prospects

The CSCE committee for preparation of the Paris summit meeting met in Vienna on 10 July 1990. Its proceedings were based on officially tabled proposals, the positions of states as indicated in various forms and the NATO London Declaration. A joint proposal by Czechoslovakia, the GDR and Poland[46] combined parts of the Mazowiecki initiative (for a European Cooperation Council) and the Dienstbier initiative (for a European Security Commission), setting out the tasks, functions and structure of a Council for Security and Co-operation in Europe to meet biannually at foreign or other minister level.[47]

The CSCE ministerial session, the first CSCE meeting to be held in the United States (in New York, on 1–2 October 1990), preceded the Paris summit meeting. This was the last time the German Democratic Republic was represented as an independent country. Albania, which expressed interest in the CSCE process (in 1972–73 it declined the invitation to join), has since then gained observer status. The participation of the three Soviet Baltic republics—Lithuania, Latvia and Estonia—remained unresolved, despite their interest in

Press Service, Press Communiqué S-1(90)36. For the full text, see Rotfeld and Stützle (note 1), pp. 150–52.

[45] See the London Declaration (note 44).

[46] CSCE document CSCE/GVA. 1, Vienna, 11 July 1990.

[47] It is worth adding that on 26–27 Sep. 1990 the Parliamentary Assembly of the Council of Europe in Strasbourg (10th Sitting) expressed its unequivocal support for the principle that the Strasbourg Parliamentary Assembly should be the basis for the missing 'parliamentary dimension' of the CSCE. For the first time the European Parliament sessions were attended not only by states with special guest status (Poland among them) which are seeking admission to the European Council, but also by other CSCE states with no previous connection with the Council, such as Canada, Monaco, Romania and the USA.

being invited as independent states to participate in the CSCE work.[48] Speaking at the opening of the CSCE ministerial session, US Secretary of State Baker said that 'the time has come to strengthen CSCE politically and institutionally'. In the US view, the basis was the set of six proposals contained in the NATO London Declaration. Baker said that these proposals 'would not duplicate, but would complement the work performed by other bodies'. The US Government presented the completion of the CFE Treaty as 'an essential precondition to holding the CSCE Summit'.[49] In general, the US approach to the new CSCE institution has been subordinated to the principle that 'form follows function'. The New York meeting of CSCE Foreign Ministers approved the agenda for the Paris summit meeting.[50]

The unprecedented convergence of the negotiating positions led to agreement in a relatively brief period of time. The agreed new mechanisms and structures will be at least as important for the future of European security as the general principles laid down in the Charter of Paris for a New Europe and in the Joint Declaration of Twenty-Two States. In the latter document the heads of state and government 'pledge to work together with the other participating States to strengthen the CSCE process so that it can make an ever greater contribution to security and stability in Europe. They recognize in particular the need to enhance political consultations among CSCE participants and to develop other CSCE mechanisms'.[51] These commitments will be put to the test: the functioning of the new institutions will show whether and how they can influence the security of states and the region as a whole, or whether they join the long list of declarations full of rhetoric and empty wording.

Regarding European security, three areas are of key concern: new mechanisms of political–military consultation, new mechanisms of decision-making, and the adaptation of the existing structures and organizations to the new identity of the CSCE process. In other words, the effectiveness of the new CSCE institutions will considerably depend on changes taking place in NATO and on the process of transforming the EC, which is striving to act as a focus of stability in Europe, into a political union.[52]

The NATO states, in particular the USA, were as a rule unwilling to institutionalize the CSCE process,[53] because of misgivings in the past that the Soviet Union could gain the *droit de regard* on the Western alliance. Now their resistance to shaping pan-European structures seems to stem from a fear that security institutions created within the CSCE framework will inevitably

[48] The US official noted that 'at the present time consensus on Baltic states' membership is not possible'. *Wireless File*, no. 190 (United States Information Service, US Embassy: Stockholm, 1 Oct. 1990), p. 7.

[49] See *Wireless File* (note 48), p. 5.

[50] The full text of the CSCE Ministerial Communiqué appears in *Wireless File*, no. 191 (United States Information Service, US Embassy: Stockholm, 2 Oct. 1990), p. 4.

[51] For the full text, see appendix 17B in this volume.

[52] Document of the EC European Council, Presidency Conclusions, Rome, 14–15 Dec. 1990. For the text of the Declaration on US–EC relations, adopted in Rome on 23 Nov. 1990, see appendix 17D in this volume.

[53] See also Rotfeld, A. D., 'Follow up to the conference: forms of co-operation after the CSCE', in *CSCE, A Polish View* (PWN, Polish Scientific Publishers: Warsaw, 1976), pp. 221–70.

lead to the erosion and weakening of NATO and that the new institutions have no real *raison d'être* but will simply create a new bureaucracy.[54]

In effect, the Conflict Prevention Centre has been overshadowed by the on-going CFE and CSBMs negotiations. The main task of the CPC is to assist the Council of Foreign Ministers in reducing the risk of conflict. During its initial stage of operation the Centre's role will consist in giving support to the implementation of CSBMs such as:

- mechanism for consultation and co-operation as regards unusual military activities;
- annual exchange of military information;
- communications network;
- annual implementation assessment meetings;
- co-operation as regards hazardous incidents of a military nature.[55]

The Council of the Foreign Ministers might assign other functions to the CPC in the future. Hopes have been pinned on the Centre to open a new and decisive chapter in building a co-operative security system. However, the very fact that the main CPC organ, the Consultative Committee, is composed of the heads of delegation to the CSBM Negotiations reduces the role of this new institution to a sort of a subsidiary negotiating body. The two other institutions set up at the Paris meeting (the CSCE Secretariat in Prague and the Office of Free Elections in Warsaw) run the risk of performing only symbolic functions in the new security structure. Their work in the CSCE context can be brought to the first test by the extent to which they promote a peaceful process in the three Baltic republics and facilitate opportunities for states to explain their positions and behaviour.

The political will of states will continue to be crucial in the work of the CSCE. New institutions may play only a role that is adequate to the security interests of the 34 participants of the CSCE process. In the CSCE, consensus is the rule of decision making. Consensus decisions are based on political authority, and—more important—states therefore implement them. The arduous and lengthy procedures for reaching consensus are compensated by their effectiveness.[56] Likewise, the fact that CSCE provisions are of a political, not a legal, character does not mean that they are less obligatory per se. The will and common interests of the participants are of decisive importance. The impact of decisions is not determined by the form and nature of the commitments but by the effectiveness and efficiency of their implementation. Accordingly, a commitment by states to report and politically account for their conduct is more important than setting in motion an appropriate legal procedure. However, as in the case of arms control agreements, ratification by par-

[54] Kampelman, M. M., 'CSCE should not rush to build permanent institutions', *Wireless File*, no. 209 (United States Information Service, US Embassy: Stockholm, 29 Oct. 1990). He also warned against making big changes in the CSCE process: 'If it ain't broke don't fix it'. *Wireless File*, no. 205 (United States Information Service: US Embassy: Stockholm, 23 Oct. 1990).

[55] See excerpts of the Supplementary Document to give effect to certain provisions contained in the Charter of Paris for a New Europe in Rotfeld and Stützle (note 1), pp. 226–32.

[56] Lehne, S., 'Vom Prozess zur Institution. Zur aktuellen Debatte über die Weiterentwicklung des KSZE-Prozesses', *Europa-Archiv*, 25 Aug. 1990, p. 505.

liamentary bodies is an additional guarantee that commitments will be respected. The inclusion in the CSCE process of a body representing the 34 parliaments, the new Assembly of Europe, could strengthen its effectiveness. Much depends on how the status, mandate, rules and functions of this Assembly are determined by the CSCE parliamentarians.

In summary, there is no need either to multiply the number of existing structures and institutions in Europe or to create a new type of 'UN' body for Europe and North America. The best approach would be a pragmatic one which applies the principle of Occam's razor: entities must not be multiplied unnecessarily. The Paris summit meeting participants were anxious to prevent a situation in which the tasks of the new institutions were beyond the means prescribed for them. Expectations regarding what they can accomplish must be set according to the capacities, competences and mechanisms of the institutions. However, there still exists a risk that the new CSCE institutions will have such a limited scope of action and such modest capacities that they will find themselves on the periphery of the main European security issues.

Two conclusions emerge from an analysis of the concepts, proposals and decisions described above. Institutionalization is neither a value nor an end in itself. For example, regular reports by states on the implementation of decisions in different fields could be equally important elements. Such reports could answer specified and standardized questions; the compilation of such information by the Secretariat would, on the one hand, be a way for governments to show they had complied with their commitments and, on the other, give parliaments and non-governmental organizations access to source materials so that they could formulate their own opinions and recommendations. For instance, one of the tasks of the CPC could be the preparation of a yearbook on armaments, arms reductions and CSBMs. The Secretariat could prepare a similar publication, containing information on progress in human rights, co-operation in environmental protection, implementation of decisions on economic co-operation, and so on.

The second conclusion follows from the lesson of the experience of international organizations: that new bodies and institutions should be set up only if the solution of specific problems of security and co-operation is impossible without them. The institutional arrangements agreed in Paris seem to be limited in their ability to prevent new threats and challenges. They will not decisively reshape the security order in Europe. However, it is important that the first step was taken towards creating common security institutions involving all the European states, including the USSR, as well as the USA and Canada. In the Paris decisions, security is denationalized, and Europe is treated as a whole and as a homogeneous security area. A new European security system will be measured by both the efficiency of the new institutions and the political will of states. The creation of these new institutions should be seen as a stage in the process. The institutional arrangements decided in Paris reflect strong political consensus regarding consolidation of the positive changes occurring in Europe.

Appendix 17A. Joint Declaration of Twenty-Two States

Paris, 19 November 1990

The Heads of State or Government of Belgium, Bulgaria, Canada, the Czech and Slovak Federal Republic, Denmark, France, Germany, Greece, Hungary, Iceland, Italy, Luxembourg, the Netherlands, Norway, Poland, Portugal, Romania, Spain, Turkey, the Union of Soviet Socialist Republics, the United Kingdom and the United States of America

– greatly welcoming the historic changes in Europe

– gratified by the growing implementation throughout Europe of a common commitment to pluralist democracy, the rule of law and human rights, which are essential to lasting security on the continent,

– affirming the end of the era of division and confrontation which has lasted for more than four decades, the improvement in relations among their countries and the contribution this makes to the security of all,

– confident that the signature of the Treaty on Conventional Armed Forces in Europe represents a major contribution to the common objective of increased security and stability in Europe, and

– convinced that these developments must form part of a continuing process of co-operation in building the structures of a more united continent,

Issue the following Declaration:

1. The signatories solemnly declare that, in the new era of European relations which is beginning, they are no longer adversaries, will build new partnerships and extend to each other the hand of friendship.

2. They recall their obligations under the Charter of the United Nations and reaffirm all of their commitments under the Helsinki Final Act. They stress that all of the ten Helsinki Principles are of primary significance and that, accordingly, they will be equally and unreservedly applied, each of them being interpreted taking into account the others. In that context, they affirm their obligation and commitment to refrain from the threat or use of force against the territorial integrity or the political independence of any State, from seeking to change existing borders by threat or use of force, and from acting in any other manner inconsistent with the principles and purposes of those documents. None of their weapons will ever be used except in self-defence or otherwise in accordance with the Charter of the United Nations.

3. They recognize that security is indivisible and that the security of each of their countries is inextricably linked to the security of all States participating in the Conference on Security and Co-operation in Europe.

4. They undertake to maintain only such military capabilities as are necessary to prevent war and provide for effective defence. They will bear in mind the relationship between military capabilities and doctrines.

5. They reaffirm that every State has the right to be or not to be a party to a treaty of alliance.

6. They note with approval the intensification of political and military contacts among them to promote mutual understanding and confidence. They welcome in this context the positive responses made to recent proposals for new regular diplomatic liaison.

7. They declare their determination to contribute actively to conventional, nuclear and chemical arms control and disarmament agreements which enhance security and stability for all. In particular, they call for the early entry into force of the Treaty on Conventional Armed Forces in Europe and commit themselves to continue the process of strengthening peace in Europe through conventional arms control within the framework of the CSCE. They welcome the prospect of new negotiations between the United States and the Soviet Union on the reduction of their short-range nuclear forces.

8. They welcome the contribution that confidence- and security-building measures have made to lessening tensions and fully support the further development of such measures. They reaffirm the importance of the 'Open Skies' initiative and their determination to bring the negotiations to a successful conclusion as soon as possible.

9. They pledge to work together with the other CSCE participating States to strengthen the CSCE process so that it can make an even greater contribution to security and stability in Europe. They recognize in particular the need to enhance political consultations among CSCE participants and to develop other CSCE mechanisms. The are convinced that the Treaty on Conventional Armed Forces in Europe and agreement on a substantial new set of CSBMs, together with new patterns of co-operation in the framework of the CSCE, will lead to increased security and thus to enduring peace and stability in Europe.

10. They believe that the preceding points reflect the deep longing of their peoples for close co-operation and mutual understanding and declare that they will work steadily for the further development of their relations in accordance with the present Declaration as well as with the principles set forth in the Helsinki Final Act.

The original of this Declaration of which the English, French, German, Italian, Russian and Spanish texts are equally authentic will be transmitted to the Government of France which will retain it in its archives. The Government of France is requested to transmit the text of the Declaration to the Secretary-General of the United Nations, with a view to its circulation to all the members of the organization as an official document of the United Nations, indicating that it is not eligible for registration under Article 102 of the Charter of the United Nations. Each of the signatory States will receive from the Government of France a true copy of this Declaration.

In witness whereof the undersigned High Representatives have subscribed their signatures below.

Appendix 17B. The Charter of Paris for a new Europe

Paris, 21 November 1990

A NEW ERA OF DEMOCRACY, PEACE AND UNITY

We, the Heads of State or Government of the States participating in the Conference on Security and Co-operation in Europe, have assembled in Paris at a time of profound change and historic expectations. The era of confrontation and division of Europe has ended. We declare that henceforth our relations will be founded on respect and co-operation.

Europe is liberating itself from the legacy of the past. The courage of men and women, the strength of the will of the peoples and the power of the ideas of the Helsinki Final Act have opened a new era of democracy, peace and unity in Europe.

Ours is a time for fulfilling the hopes and expectations our peoples have cherished for decades: steadfast commitment to democracy based on human rights and fundamental freedoms; prosperity through economic liberty and social justice; and equal security for all our countries.

The Ten Principles of the Final Act will guide us towards this ambitious future, just as they have lighted our way towards better relations for the past fifteen years. Full implementation of all CSCE commitments must form the basis for the initiatives we are now taking to enable our nations to live in accordance with their aspirations.

Human rights, democracy and rule of law

We undertake to build, consolidate and strengthen democracy as the only system of government of our nations. In this endeavour, we will abide by the following:

Human rights and fundamental freedoms are the birthright of all human beings, are inalienable and are guaranteed by law. Their protection and promotion is the first responsibility of government. Respect for them is an essential safeguard against an over-mighty State. Their observance and full exercise are the foundation of freedom, justice and peace.

Democratic government is based on the will of the people, expressed regularly through free and fair elections. Democracy has as its foundation respect for the human person and the rule of law. Democracy is the best safeguard of freedom of expression, tolerance of all groups of society, and equality of opportunity for each person.

Democracy, with its representative and pluralist character, entails accountability to the electorate, the obligation of public authorities to comply with the law and justice administered impartially. No one will be above the law.

We affirm that, without discrimination, every individual has the right to:

freedom of thought, conscience and religion or belief,
freedom of expression,
freedom of association and peaceful assembly,
freedom of movement;

no one will be:

subject to arbitrary arrest or detention,
subject to torture or other cruel, inhuman or degrading treatment or punishment;

everyone also has the right:

to know and act upon his rights,
to participate in free and fair elections,
to fair and public trial if charged with an offence,
to own property alone or in association and to exercise individual enterprise,
to enjoy his economic, social and cultural rights.

We affirm that the ethnic, cultural, linguistic and religious identity of national minorities will be protected and that persons belonging to national minorities have the right freely to express, preserve and develop that identity without any discrimination and in full equality before the law.

We will ensure that everyone will enjoy recourse to effective remedies, national or international, against any violation of his rights.

Full respect for these precepts is the bedrock on which we will seek to construct the new Europe.

Our States will co-operate and support each other with the aim of making democratic gains irreversible.

Economic liberty and responsibility

Economic liberty, social justice and environmental responsibility are indispensable for prosperity.

The free will of the individual, exercised in democracy and protected by the rule of law, forms the necessary basis for successful economic and social development. We will promote economic activity which respects and upholds human dignity.

Freedom and political pluralism are necessary elements in our common objective of developing market economies towards sustainable economic growth, prosperity, social justice, expanding employment and efficient use of economic resources. The success of the transition to market economy by countries making efforts to this effect is important and in the interest of us all. It will enable us to share a higher level of prosperity which is our common objective. We will co-operate to this end.

Preservation of the environment is a shared responsibility of all our nations. While supporting national and regional efforts in this field, we must also look to the pressing need for joint action on a wider scale.

Friendly relations among participating states

Now that a new era is dawning in Europe, we are determined to expand and strengthen friendly relations and co-operation among the States of Europe, the United States of America and Canada, and to promote friendship among our peoples.

To uphold and promote democracy, peace and unity in Europe, we solemnly pledge our full commitment to the Ten Principles of the Helsinki Final Act. We affirm the continuing validity of the Ten Principles and our determination to put them into practice. All the Principles apply equally and unreservedly, each of them being interpreted taking into account the others. They form the basis for our relations.

In accordance with our obligations under the Charter of the United Nations and commitments under the Helsinki Final Act, we renew our pledge to refrain from the threat or use of force against the territorial integrity or political independence of any State, or from acting in any other manner inconsistent with the principles or purposes of those documents. We recall that non-compliance with obligations under the Charter of the United Nations constitutes a violation of international law.

We reaffirm our commitment to settle disputes by peaceful means. We decide to develop mechanisms for the prevention and resolution of conflicts among the participating States.

With the ending of the division of Europe, we will strive for a new quality in our security relations while fully respecting each other's freedom of choice in that respect. Security is indivisible and the security of every participating State is inseparably linked to that of all the others. We therefore pledge to co-operate in strengthening confidence and security among us and in promoting arms control and disarmament.

We welcome the Joint Declaration of Twenty-Two States on the improvement of their relations.

Our relations will rest on our common adherence to democratic values and to human rights and fundamental freedoms. We are convinced that in order to strengthen peace and security among our States, the advancement of democracy, and respect for and effective exercise of human rights, are indispensable. We reaffirm the equal rights of peoples and their right to self-determination in conformity with the Charter of the United Nations and with the relevant norms of international law, including those relating to territorial integrity of States.

We are determined to enhance political consultation and to widen co-operation to solve economic, social, environmental, cultural and humanitarian problems. This common resolve and our growing interdependence will help to overcome the mistrust of decades, to increase stability and to build a united Europe.

We want Europe to be a source of peace, open to dialogue and to co-operation with other countries, welcoming exchanges and involved in the search for common responses to the challenges of the future.

Security

Friendly relations among us will benefit from the consolidation of democracy and improved security.

We welcome the signature of the Treaty on Conventional Armed Forces in Europe by twenty-two participating States, which will lead to lower levels of armed forces. We endorse the adoption of a substantial new set of Confidence- and Security-building Measures which will lead to increased transparency and confidence among all participating States. These are important steps towards enhanced stability and security in Europe.

The unprecedented reduction in armed forces resulting from the Treaty on Conventional Armed Forces in Europe, together with new approaches to security and co-operation within the CSCE process, will lead to a new perception of security in Europe and a new dimension in our relations. In this context we fully recognize the freedom of States to choose their own security arrangements.

Unity

Europe whole and free is calling for a new beginning. We invite our peoples to join in this great endeavour.

We note with great satisfaction the Treaty on the Final Settlement with respect to Germany signed in Moscow on 12 September 1990 and sincerely welcome the fact that the German people have united to become one State in accordance with the principles of the Final Act of the Conference on Security and Co-operation in Europe and in full accord with their neighbours. The establishment of the national unity of Germany is an important contribution to a just and lasting order of peace for a united, democratic Europe aware of its responsibility for stability, peace and co-operation.

The participation of both North American and European States is a fundamental characteristic of the CSCE; it underlies its past achievements and is essential to the future of the CSCE process. An abiding adherence to shared values and our common heritage are the ties which bind us together. With all the rich diversity of our nations, we are united in our commitment to expand our co-operation in all fields. The challenges confronting us can only be met by common action, co-operation and solidarity.

The CSCE and the world

The destiny of our nations is linked to that of all other nations. We support fully the United Nations and the enhancement of its role in promoting international peace, security and justice. We reaffirm our commitment to the principles and purposes of the United Nations as enshrined in the Charter and condemn all violations of these principles. We recognize with satisfaction the growing role of the United Nations in world affairs and its increasing effectiveness, fostered by the improvement in relations among our States.

Aware of the dire need of a great part of the world, we commit ourselves to solidarity with all other countries. Therefore, we issue a call from Paris today to all the nations of the world. We stand ready to join with any and all States in common efforts to protect and advance the community of fundamental human values.

GUIDELINES FOR THE FUTURE

Proceeding from our firm commitment to the full implementation of all CSCE principles and provisions, we now resolve to give a new impetus to a balanced and comprehensive development of our co-operation in order to address the needs and aspirations of our peoples.

Human dimension

We declare our respect for human rights and fundamental freedoms to be irrevocable. We will fully implement and build upon the provisions relating to the human dimension of the CSCE.

Proceeding from the Document of the Copenhagen Meeting of the Conference on the Human Dimension, we will co-operate to strengthen democratic institutions and to promote the application of the rule of law. To that end, we decide to convene a seminar of experts in Oslo from 4 to 15 November 1991.

Determined to foster the rich contribution of national minorities to the life of our societies, we undertake further to improve their situation. We reaffirm our deep conviction that friendly relations among our peoples, as well as peace, justice, stability and democracy, require that the ethnic, cultural, linguistic and religious identity of national minorities be protected and conditions for the promotion of that identity be created. We declare that questions related to national minorities can only be satisfactorily resolved in a democratic political framework. We further acknowledge that the rights of persons belonging to national minorities must be fully respected as part of universal human rights. Being aware of the urgent need for increased co-operation on, as well as better protection of, national minorities, we decide to convene a meeting of experts on national minorities to be held in Geneva from 1 to 19 July 1991.

We express our determination to combat all forms of racial and ethnic hatred, anti-semitism, xenophobia and discrimination against anyone as well as persecution on religious and ideological grounds.

In accordance with our CSCE commitments, we stress that free movement and contacts among our citizens as well as the free flow of information and ideas are crucial for the maintenance and development of free societies and flourishing cultures. We welcome increased tourism and visits among our countries.

The human dimension mechanism has proved its usefulness, and we are consequently determined to expand it to include new procedures involving, *inter alia*, the services of experts or a roster of eminent persons experienced in human rights issues which could be raised under the mechanism. We shall provide, in the context of the mechanism, for individuals to be involved in the protection of their rights. Therefore, we undertake to develop further our commitments in this respect, in particular at the Moscow Meeting of the Conference on the Human Dimension, without prejudice to obligations under existing international instruments to which our States may be parties.

We recognize the important contribution of the Council of Europe to the promotion of human rights and the principles of democracy and the rule of law as well as to the development of cultural co-operation. We welcome moves by several participating States to join the Council of Europe and adhere to its European Convention on Human Rights. We welcome as well the readiness of the Council of Europe to make its experience available to the CSCE.

Security

The changing political and military environment in Europe opens new possibilities for common efforts in the field of military security. We will build on the important achievements attained in the Treaty on Conventional Armed Forces in Europe and in the Negotiations on Confidence- and Security-Building Measures. We undertake to continue the CSBM negotiations under the same mandate, and to seek to conclude them no later than the Follow-up Meeting of the CSCE to be held in Helsinki in 1992. We also welcome the decision of the participating States concerned to continue the CFE negotiation under the same mandate and to seek to conclude it no later than the Helsinki Follow-up Meeting. Following a period for national preparations, we look forward to a more structured co-operation among all participating States on security matters, and to discussions and consultations among the thirty-four participating States aimed at establishing by 1992, from the conclusion of the Helsinki Follow-Up Meeting, new negotiations on disarmament and confidence and security building open to all participating States.

We call for the earliest possible conclusion of the Convention on an effectively verifiable, global and comprehensive ban on chemical weapons, and we intend to be original signatories to it.

We reaffirm the importance of the Open Skies initiative and call for the successful conclusion of the negotiations as soon as possible.

Although the threat of conflict in Europe has diminished, other dangers threaten the stability of our societies. We are determined to co-operate in defending democratic institutions against activities which violate the independence, sovereign equality or territorial integrity of the participating States. These include illegal activities involving outside pressure, coercion and subversion.

We unreservedly condemn, as criminal, all acts, methods and practices of terrorism and express our determination to work for its eradication both bilaterally and through multilateral co-operation. We will also join together in combating illicit trafficking in drugs.

Being aware that an essential complement to the duty of States to refrain from the threat or use of force is the peaceful settlement of disputes, both being essential factors for the maintenance and consolidation of international peace and security, we will not only seek effective ways of preventing, through political means, conflicts which may emerge, but also define, in conformity with international law, appropriate mechanisms for the peaceful resolution of any disputes which may arise. Accordingly, we undertake to seek new forms of co-operation in this area, in particular a range of methods for the peaceful settlement of disputes, including mandatory third-party involvement. We stress that full use should be made in this context of the opportunity of the meeting on the peaceful settlement of disputes which will be convened in Valletta at the beginning of 1991. The Council of Ministers for Foreign Affairs will take into account the Report of the Valletta Meeting.

Economic co-operation

We stress that economic co-operation based on market economy constitutes an essential element of our relations and will be instrumental in the construction of a prosperous and united Europe. Democratic institutions and economic liberty foster economic and social progress, as recognized in the Document of the Bonn Conference on Economic Co-operation, the results of which we strongly support.

We underline that co-operation in the economic field, science and technology is now an important pillar of the CSCE. The participating States should periodically review progress and give new impulses in these fields.

We are convinced that our overall economic co-operation should be expanded, free enterprise encouraged and trade increased and diversified according to GATT rules. We will promote social justice and progress and further the welfare of our peoples. We recognize in this context the importance of effective policies to address the problem of unemployment.

We reaffirm the need to continue to support democratic countries in transition towards the establishment of market economy and the creation of the basis for self-sustained economic and social growth, as already undertaken by the Group of twenty-four countries. We further underline the necessity of their increased integration, involving the acceptance of disciplines as well as benefits, into the international economic and financial system.

We consider that increased emphasis on economic co-operation within the CSCE process should take into account the interests of developing participating States.

We recall the link between respect for and promotion of human rights and fundamental freedoms and scientific progress. Co-operation in the field of science and technology will play an essential role in economic and social development. Therefore, it must evolve towards a greater sharing of appropriate scientific and technological information and knowledge with a view to overcoming the technological gap existing among the participating States. We further encourage the participating States to work together in order to develop human potential and the spirit of free enterprise.

We are determined to give the necessary impetus to co-operation among our States in the fields of energy, transport and tourism for economic and social development. We welcome, in particular, practical steps to create optimal conditions for the economic and rational development of energy resources, with due regard for environmental considerations.

We recognize the important role of the European Community in the political and economic development of Europe. International economic organizations such as the Economic Commission for Europe of the United Nations (ECE/UN), the Bretton Woods Institutions, the Organization for Economic Co-operation and Development (OECD), the European Free Trade Association (EFTA) and the International Chamber of Commerce (ICC) also have a significant task in promoting economic co-operation, which will be further enhanced by the establishment of the European Bank for Reconstruction and Development (EBRD). In order to pursue our objectives, we stress the necessity for effective co-ordination of the activities of these organizations and emphasize the need to find methods for all our States to take part in these activities.

Environment

We recognize the urgent need to tackle the problems of the environment and the importance of individual and co-operative efforts in this area. We pledge to intensify our endeavours to protect and improve our environment in order to restore and maintain a sound ecological balance in air, water and soil. Therefore, we are determined to make full use of the CSCE as a framework for the formulation of common environmental commitments and objectives, and thus to pursue the work reflected in the Report of the Sofia Meeting on the Protection of the Environment.

We emphasize the significant role of a well-informed society in enabling the public and individuals to take initiatives to improve the environment. To this end, we commit ourselves to promote public awareness and education on the environment as well as the public reporting of the environmental impact of policies, projects and programmes.

We attach priority to the introduction of clean and low-waste technology, being aware of the need to support countries which do not yet have their own means for appropriate measures.

We underline that environmental policies should be supported by appropriate legislative measures and administrative structures to ensure their effective implementation.

We stress the need for new measures providing for systematic evaluation of compliance with the existing commitments and, moreover, for the development of more ambitious commitments with regard to notification and exchange of information about the state of the environment and potential environmental hazards. We also welcome the creation of the European Environment Agency (EEA).

We welcome the operational activities, problem-oriented studies and policy reviews in various existing international organizations engaged in the protection of the environment, such as the United Nations Environment Program (UNEP), the Economic Commission for Europe of the United Nations (ECE/UN) and the Organization for Economic Co-operation and Development (OECD). We emphasize the need for strengthening their co-operation and for their efficient co-ordination.

Culture

We recognize the essential contribution of our common European culture and our shared values in overcoming the division of the continent. Therefore, we underline our attachment to creative freedom and to the protection and promotion of our cultural and spiritual heritage, in all its richness and diversity.

In view of the recent changes in Europe, we stress the increased importance of the Cracow Symposium and we look forward to its consideration of guidelines for intensified co-operation in the field of culture. We invite the Council of Europe to contribute to this Symposium.

In order to promote greater familiarity amongst our peoples, we favour the establishment of cultural centres in cities of other participating States as well as increased co-operation in the audio-visual field and wider exchange in music, theatre, literature and the arts.

We resolve to make special efforts in our national policies to promote better understanding, in particular among young people, through cultural exchanges, co-operation in all fields of education and, more specifically, through teaching and training in the languages of other participating States. We intend to consider first results of this action at the Helsinki Follow-up Meeting in 1992.

Migrant workers

We recognize that the issues of migrant workers and their families legally residing in host countries have economic, cultural and social aspects as well as their human dimension. We reaffirm that the protection and promotion of their rights, as well as the implementation of relevant international obligations, is our common concern.

Mediterranean

We consider that the fundamental political changes that have occurred in Europe have a positive relevance to the Mediterranean region. Thus, we will continue efforts to strengthen security and co-operation in the Mediterranean as an important factor for stability in Europe. We welcome the Report of the Palma de Mallorca Meeting on the Mediterranean, the results of which we all support.

We are concerned with the continuing tensions in the region, and renew our determination to intensify efforts towards finding just, viable and lasting solutions, through peaceful means, to outstanding crucial problems, based on respect for the principles of the Final Act.

We wish to promote favourable conditions for a harmonious development and diversification of relations with the non-participating Mediterranean States. Enhanced co-operation with these States will be pursued with the aim of promoting economic and social development and thereby enhancing stability in the region. To this end, we will strive together with these countries towards a substantial narrowing of the prosperity gap between Europe and its Mediterranean neighbours.

Non-governmental organizations

We recall the major role that non-governmental organizations, religious and other groups and individuals have played in the achievement of the objectives of the CSCE and will further facilitate their activities for the implementation of the CSCE commitments by the participating States. These organizations, groups and individuals must be involved in an appropriate way in the activities and new structures of the CSCE in order to fulfil their important tasks.

NEW STRUCTURES AND INSTITUTIONS OF THE CSCE PROCESS

Our common efforts to consolidate respect for human rights, democracy and the rule of law, to strengthen peace and to promote unity in Europe require a new quality of political dialogue and co-operation and thus development of the structures of the CSCE.

The intensification of our consultations at all levels is of prime importance in shaping our future relations. To this end, we decide on the following:

We, the Heads of State or Government, shall meet next time in Helsinki on the occasion of the CSCE Follow-up Meeting 1992. Thereafter, we will meet on the occasion of subsequent follow-up meetings.

Our Ministers for Foreign Affairs will meet, as a Council, regularly and at least once a year. These meetings will provide the central forum for political consultations within the CSCE process. The Council will consider issues relevant to the Conference on Security and Co-operation in Europe and take appropriate decisions.

The first meeting of the Council will take place in Berlin.

A Committee of Senior Officials will prepare the meetings of the Council and carry out its decisions. The Committee will review current issues and may take appropriate decisions, including in the form of recommendations to the Council.

Additional meetings of the representatives of the participating States may be agreed upon to discuss questions of urgent concern.

The Council will examine the development of provisions for convening meetings of the Committee of Senior Officials in emergency situations.

Meetings of other Ministers may also be agreed by the participating States.

In order to provide administrative support for these consultations we establish a Secretariat in Prague.

Follow-up meetings of the participating States will be held, as a rule, every two years to allow the participating States to take stock of developments, review the implementation of their commitments and consider further steps in the CSCE process.

We decide to create a Conflict Prevention Centre in Vienna to assist the Council in reducing the risk of conflict.

We decide to establish an Office for Free Elections in Warsaw to facilitate contacts and the exchange of information on elections within participating States.

Recognizing the important role parliamentarians can play in the CSCE process, we call for greater parliamentary involvement in the CSCE, in particular through the creation of a CSCE parliamentary assembly, involving members of parliaments from all participating States. To this end, we urge that contacts be pursued at parliamentary level to discuss the field of activities, working methods and rules of procedure of such a CSCE parliamentary structure, drawing on existing experience and work already undertaken in this field.

We ask our Ministers for Foreign Affairs to review this matter on the occasion of their first meeting as a Council.

* * *

Procedural and organizational modalities relating to certain provisions contained in the Charter of Paris for a New Europe are set out in the Supplementary Document which is adopted together with the Charter of Paris.

We entrust to the Council the further steps which may be required to ensure the implementation of decisions contained in the present document, as well as in the Supplementary Document, and to consider further efforts for the strengthening of security and co-operation in Europe. The Council may adopt any amendment to the Supplementary Document which it may deem appropriate.

* * *

The original of the Charter of Paris for a new Europe, drawn up in English, French, German, Italian, Russian and Spanish, will be transmitted to the Government of the French Republic, which will retain it in its archives. Each of the participating States will receive from the Government of the French Republic a true copy of the Charter of Paris.

The text of the Charter of Paris will be published in each participating State, which will disseminate it and make it known as widely as possible.

The Government of the French Republic is requested to transmit to the Secretary-General of the United Nations the text of the Charter of Paris for a New Europe, which is not eligible for registration under Article 102 of the Charter of the United Nations, with a view to its circulation to all members of the Organization as an official document of the United Nations.

The Government of the French Republic is also requested to transmit the text of the Charter of Paris to all other international organizations mentioned in the text.

Wherefore, we, the undersigned High Representatives of the participating States, mindful of the high political significance we attach to the results of the Summit Meeting, and declaring our determination to act in accordance with the provisions we have adopted, have subscribed our signatures below:

Done at Paris, on 21 November 1990, in the name of (signatures) . . .

Appendix 17C. Treaty on the final settlement with respect to Germany

Moscow, 12 September 1990

The Federal Republic of Germany, the German Democratic Republic, the French Republic, the Union of Soviet Socialist Republics, the United Kingdom of Great Britain and Northern Ireland and the United States of America,

Conscious of the fact that their peoples have been living together in peace since 1945;

Mindful of the recent historic changes in Europe which make it possible to overcome the division of the continent;

Having regard to the rights and responsibilities of the Four Powers relating to Berlin and to Germany as a whole, and the corresponding wartime and post-war agreements and decisions of the Four Powers;

Resolved in accordance with their obligations under the Charter of the United Nations to develop friendly relations among nations based on respect for the principle of equal rights and self-determination of peoples, and to take other appropriate measures to strengthen universal peace;

Recalling the principles of the Final Act of the Conference on Security and Cooperation in Europe, signed in Helsinki;

Recognizing that those principles have laid firm foundations for the establishment of a just and lasting peaceful order in Europe;

Determined to take account of everyone's security interests;

Convinced of the need finally to overcome antagonism and to develop cooperation in Europe;

Confirming their readiness to reinforce security, in particular by adopting effective arms control, disarmament and confidence-building measures; their willingness not to regard each other as adversaries but to work for a relationship of trust and cooperation; and accordingly their readiness to consider positively setting up appropriate institutional arrangements within the framework of the Conference on Security and Cooperation in Europe;

Welcoming the fact that the German people, freely exercising their right of self-determination, have expressed their will to bring about the unity of Germany as a state so that they will be able to serve the peace of the world as an equal and sovereign partner in a united Europe;

Convinced that the unification of Germany as a state with definitive borders is a significant contribution to peace and stability in Europe;

Intending to conclude the final settlement with respect to Germany;

Recognizing that thereby, and with the unification of Germany as a democratic and peaceful state, the rights and responsibilities of the Four Powers relating to Berlin and to Germany as a whole lose their function;

Represented by their Ministers for Foreign Affairs who, in accordance with the Ottawa Declaration of 13 February 1990, met in Bonn on 5 May 1990, in Berlin on 22 June 1990, in Paris on 17 July 1990 with the participation of the Minister for Foreign Affairs of the Republic of Poland, and in Moscow on 12 September 1990;

Have agreed as follows:

Article 1

(1) The united Germany shall comprise the territory of the Federal Republic of Germany, the German Democratic Republic and the whole of Berlin. Its external borders shall be the borders of the Federal Republic of Germany and the German Democratic Republic and shall be definitive from the date on which the present Treaty comes into force. The confirmation of the definitive nature of the borders of the united Germany is an essential element of the peaceful order in Europe.

(2) The united Germany and the Republic of Poland shall confirm the existing border between them in a treaty that is binding under international law.

(3) The united Germany has no territorial claims whatsoever against other states and shall not assert any in the future.

(4) The Governments of the Federal Republic of Germany and the German Democratic Republic shall ensure that the constitution of the united Germany does not contain any provision incompatible with these principles. This applies accordingly to the provisions laid down in the preamble, the

second sentence of Article 23, and Article 146 of the Basic Law for the Federal Republic of Germany.

(5) The Governments of the French Republic, the Union of Soviet Socialist Republics, the United Kingdom of Great Britain and Northern Ireland and the United States of America take formal note of the corresponding commitments and declarations by the Governments of the Federal Republic of Germany and the German Democratic Republic and declare that their implementation will confirm the definitive nature of the united Germany's borders.

Article 2

The Governments of the Federal Republic of Germany and the German Democratic Republic reaffirm their declarations that only peace will emanate from German soil. According to the constitution of the united Germany, acts tending to and undertaken with the intent to disturb the peaceful relations between nations, especially to prepare for aggressive war, are unconstitutional and a punishable offence. The Governments of the Federal Republic of Germany and the German Democratic Republic declare that the united Germany will never employ any of its weapons except in accordance with its constitution and the Charter of the United Nations.

Article 3

(1) The Governments of the Federal Republic of Germany and the German Democratic Republic reaffirm their renunciation of the manufacture and possession of and control over nuclear, biological and chemical weapons. They declare that the united Germany, too, will abide by these commitments. In particular, rights and obligations arising from the Treaty on the Non-Proliferation of Nuclear Weapons of 1 July 1968 will continue to apply to the united Germany.

(2) The Government of the Federal Republic of Germany, acting in full agreement with the Government of the German Democratic Republic, made the following statement on 30 August 1990 in Vienna at the Negotiations on Conventional Armed Forces in Europe:

'The Government of the Federal Republic of Germany undertakes to reduce the personnel strength of the armed forces of the united Germany to 370, 000 (ground, air and naval forces) within three to four years. This reduction will commence on the entry into force of the first CFE agreement. Within the scope of this overall ceiling no more than 345, 000 will belong to the ground and air forces which, pursuant to the ageed mandate, alone are the subject of the Negotiations on Conventional Armed Forces in Europe. The Federal Government regards its commitment to reduce ground and air forces as a significant German contribution to the reduction of conventional armed forces in Europe. It assumes that in follow-on negotiations the other participants in the negotiations, too, will render their contribution to enhancing security and stability in Europe, including measures to limit personnel strengths'.

The Government of the German Democratic Republic has expressly associated itself with this statement.

(3) The Governments of the French Republic, the Union of Soviet Socialist Republics, the United Kingdom of Great Britain and Northern Ireland and the United States of America take note of these statements by the Governments of the Federal Republic of Germany and the German Democratic Republic.

Article 4

(1) The Governments of the Federal Republic of Germany, the German Democratic Republic and the Union of Soviet Socialist Republics state that the united Germany and the Union of Soviet Socialist Republics will settle by treaty the conditions for and the duration of the presence of Soviet armed forces on the territory of the present German Democratic Republic and of Berlin, as well as the conduct of the withdrawal of these armed forces which will be completed by the end of 1994, in connection with the implementation of the undertaking of the Federal Republic of Germany and the German Democratic Republic referred to in paragraph 2 of Article 3 of the present Treaty.

(2) The Governments of the French Republic, the United Kingdom of Great Britain and Northern Ireland and the United States of America take note of this statement.

Article 5

(1) Until the completion of the withdrawal of the Soviet armed forces from the territory of the present German Democratic Republic and of Berlin in accordance with Article 4 of

the present Treaty, only German territorial defence units which are not integrated into the alliance structures to which German armed forces in the rest of German territory are assigned will be stationed in that territory as armed forces of the united Germany. During that period and subject to the provisions of paragraph 2 of this Article, armed forces of other states will not be stationed in that territory or carry out any other military activity there.

(2) For the duration of the presence of Soviet armed forces in the territory of the present German Democratic Republic and of Berlin, armed forces of the French Republic, the United Kingdom of Great Britain and Northern Ireland and the United States of America will, upon German request, remain stationed in Berlin by agreement to this effect between the Government of the united Germany and the Governments of the states concerned. The number of troops and the amount of equipment of all non-German armed forces stationed in Berlin will not be greater than at the time of signature of the present Treaty. New categories of weapons will not be introduced there by non-German armed forces. The Government of the united Germany will conclude with the Governments of those states which have armed forces stationed in Berlin treaties with conditions which are fair taking account of the relations existing with the states concerned.

(3) Following the completion of the withdrawal of the Soviet armed forces from the territory of the present German Democratic Republic and of Berlin, units of German armed forces assigned to military alliance structures in the same way as those in the rest of German territory may also be stationed in that part of Germany, but without nuclear weapon carriers. This does not apply to conventional weapon systems which may have other capabilities in addition to conventional ones but which in that part of Germany are equipped for a conventional role and designated only for such. Foreign armed forces and nuclear weapons or their carriers will not be stationed in that part of Germany or deployed there.

Article 6

The right of the united Germany to belong to alliances, with all the rights and responsibilities arising therefrom, shall not be affected by the present Treaty.

Article 7

(1) The French Republic, the Union of Soviet Socialist Republics, the United Kingdom of Great Britain and Northern Ireland and the United States of America hereby terminate their responsibilities relating to Berlin and to Germany as a whole. As a result, the corresponding, related quadripartite agreements, decisions and practices are terminated and all related Four Power institutions are dissolved.

(2) The united Germany shall have accordingly full sovereignty over its internal and external affairs.

Article 8

(1) The present Treaty is subject to ratification or acceptance as soon as possible. On the German side it will be ratified by the united Germany. The Treaty will therefore apply to the united Germany.

(2) The instruments of ratification or acceptance shall be deposited with the Government of the united Germany. That Government shall inform the Governments of the other Contracting Parties of the deposit of each instrument of ratification or acceptance.

Article 9

The present Treaty shall enter into force for the united Germany, the French Republic, the Union of Soviet Socialist Republics, the United Kingdom of Great Britain and Northern Ireland and the United States of America on the date of deposit of the last instrument of ratification or acceptance by these states.

Article 10

The original of the present Treaty, of which the English, French, German and Russian texts are equally authentic, shall be deposited with the Government of the Federal Republic of Germany, which shall transmit certified true copies to the Governments of the other Contracting Parties.

AGREED MINUTE:

Any questions with respect to the application of the word 'deployed' as used in the last sentence of paragraph 3 of Article 5 will be decided by the Government of the united Germany in a reasonable and responsible way taking into account the security interests of each Contracting Party as set forth in the preamble.

Appendix 17D. Declaration on US–EC relations

Rome, 23 November 1990

The United States of America on one side and, on the other, the European Community and its member States,

– mindful of their common heritage and of their close historical, political, economic and cultural ties,

– guided by their faith in the values of human dignity, intellectual freedom and civil liberties, and in the democratic institutions which have evolved on both sides of the Atlantic over the centuries;

– recognizing that the transatlantic solidarity has been essential for the preservation of peace and freedom and for the development of free and prosperous economies as well as for the recent developments which have restored unity in Europe,

– determined to help consolidate the new Europe, undivided and democratic,

– resolved to strengthen security, economic co-operation and human rights in Europe in the framework of the CSCE, and in other fora,

– noting the firm commitment of the United States and the EC member states concerned to the North Atlantic Alliance and to its principles and purposes,

– acting on the basis of a pattern of co-operation proven over many decades, and convinced that by strengthening and expanding this partnership on an equal footing they will greatly contribute to continued stability, as well as to political and economic progress in Europe and in the world,

– aware of their shared responsibility, not only to further common interests but also to face transnational challenges affecting the well-being of all mankind,

– bearing in mind the accelerating process by which the European Community is acquiring its own identity in economic and monetary matters, in foreign policy and in the domain of security,

– determined to further strengthen transatlantic solidarity through the variety of their international relations, have decided to endow their relationship with long-term perspectives.

Common goals

The United States of America and the European Community and its member States solemnly reaffirm their determination further to strengthen their partnership in order to:

– support democracy, the rule of law and respect for human rights and individual liberty, and promote prosperity and social progress world-wide;

– safeguard peace and promote international security, by co-operating with other nations against aggression and coercion, by contributing to the settlement of conflicts in the world and by reinforcing the role of the United Nations and other international organizations;

– pursue policies aimed at achieving a sound world economy marked by sustained economic growth with low inflation, a high level of employment, equitable social conditions, in a framework of international stability;

– promote market principles, reject protectionism and expand, strengthen and further open the multilateral trading system;

– carry out their resolve to help developing countries by all appropriate means in their efforts towards political and economic reforms;

– provide adequate support, in co-operation with other states and organizations, to the nations of Eastern and Central Europe undertaking economic and political reforms and encourage their participation in the multilateral institutions of international trade and finance.

Principles of US–EC partnership

To achieve their common goals, the European Community and its member States and the United States of America will inform and consult each other on important matters of common interest, both political and economic, with a view to bringing their positions as close as possible, without prejudice to their respective independence. In appropriate international bodies, in particular, they will seek close co-operation.

The US–EC partnership will, moreover, greatly benefit from the mutual knowledge and understanding acquired through regular consultations as described in this declaration.

Economic co-operation

Both sides recognized the importance of strengthening the multilateral trading system. They will support further steps towards liberalization, transparency, and the implementation of GATT and OECD principles concerning both trade in goods and services and investment.

They will further develop their dialogue, which is already underway, on other matters such as technical and non-tariff barriers to industrial and agricultural trade, services, competition policy, transportation policy, standards, telecommunications, high technology and other relevant areas.

Education, scientific and cultural co-operation

The partnership between the European Community and its member States on the one hand, and the United States on the other, will be based on continuous efforts to strengthen mutual co-operation in various other fields which directly affect the present and future well-being of their citizens, such as exchanges and joint projects in science and technology, including, inter alia, research in medicine, environment protection, pollution prevention, energy, space, high-energy physics, and the safety of nuclear and other installations, as well as in education and culture, including academic and youth exchanges.

Trans-national challenges

The United States of America and the European Community and its member States will fulfill their responsibility to address trans-national challenges, in the interest of their own peoples and of the rest of the world. In particular, they will join their efforts in the following fields:

– combatting and preventing terrorism;

– putting an end to the illegal production, trafficking and consumption of narcotics and related criminal activities, such as the laundering of money;

– co-operating in the fight against international crime;

– protecting the environment, both internationally and domestically, by integrating environmental and economic goals;

– preventing the proliferation of nuclear armaments, chemical and biological weapons, and missile technology.

Institutional framework for consultation

Both sides agree that a framework is required for regular and intensive consultation. They will make full use of and further strengthen existing procedures, including those established by the President of the European Council and the President of the United States on 27th February 1990, namely:

– bi-annual consultations to be arranged in the United States and in Europe between, on the one side, the President of the European Council and the President of the Commission, and on the other side, the President of the United States;

– bi-annual consultations between the European Community Foreign Ministers, with the Commission, and the US Secretary of State, alternately on either side of the Atlantic;

– ad hoc consultations between the Presidency Foreign Minister or the Troika and the US Secretary of State;

– bi-annual consultations between the Commission and the US Government at Cabinet level;

– briefings, as currently exist, by the Presidency to US representatives on European Political Co-operation (EPC) meetings at the Ministerial level.

Both sides are resolved to develop and deepen these procedures for consultation so as to reflect the evolution of the European Community and of its relationship with the United States.

They welcome the actions taken by the European Parliament and the Congress of the United States in order to improve their dialogue and thereby bring closer together the peoples on both sides of the Atlantic.

Note: On 15 December 1990, in Rome, the European Community adopted its European Council Presidency Conclusions. Following the November CSCE summit meeting in Paris, the EC European Council proposed the gradual extension of the role of the Political Union in the area of common security, to include: arms control and disarmament, CSCE matters, UN peace-keeping operations, economic and technological co-operation in the armaments field, co-ordination of armaments export policy and non-proliferation. (See European Council document SN 424/1/90.)

18. The role of the United Nations in the Iraq–Kuwait conflict in 1990[1]

BRIAN URQUHART

I. The Iraq–Kuwait conflict

The end of the cold war has revitalized the United Nations Security Council[2] and has allowed it to begin to act in the way the authors of the UN Charter intended. In 1990 the Iraq–Kuwait conflict, an unusually clear and unambiguous case of aggression, provided the Security Council with a major challenge which is also the first full-scale test of collective action against aggression by the United Nations through a united Security Council. On this occasion the Security Council demonstrated the decisiveness and sense of urgency which had been notably absent on many previous occasions. Both the post-cold war political climate and the stark clarity of the aggression itself contributed to this reaction, which was in dramatic contrast to the Security Council's pusillanimous response to Iraq's 1980 aggression against Iran.

Between 2 August and 29 November 1990, the UN Security Council adopted 12 resolutions[3] on the Iraq–Kuwait conflict:

1. On the day of the Iraqi invasion of Kuwait, Resolution 660 (2 August 1990) condemned the invasion and demanded Iraq's immediate withdrawal from Kuwait.
2. Resolution 661 (6 August 1990) imposed mandatory sanctions and set up a committee[4] to monitor their implementation.
3. Resolution 662 (9 August 1990) declared Iraq's annexation of Kuwait[5] to be non-valid.
4. Resolution 664 (18 August 1990) demanded that Iraq permit and facilitate the immediate departure from Iraq and Kuwait of nationals of third countries, and that Iraq rescind its order for the closing of diplomatic and consular missions in Kuwait.

[1] This chapter is based on a longer piece by Brian Urquhart: *Collective Security After the Cold War—Thoughts in the Margin of the Gulf Crisis* (Ford Foundation: New York, forthcoming).

[2] The permanent members of the Security Council are the USA, the USSR, the UK, France and China. Ten other members are elected for two-year periods; in 1990 these member states were Canada, Colombia, Côte d'Ivoire (Ivory Coast), Cuba, Ethiopia, Finland, Malaysia, Romania, the Republic of Yemen and Zaire. Decisions are taken by a vote of 9 members including the 5 permanent members (Article 27.3 of Chapter V of the UN Charter).

[3] For the full texts of these UN Security Council resolutions, see appendix 18A.

[4] The committee is composed of all 15 Security Council members. It has met regularly throughout the crisis and has requested, and received, reports from member states on the application of the sanctions.

[5] Iraq declared that it had annexed Kuwait as a province of Iraq on 8 Aug. 1990.

5. Resolution 665 (25 August 1990) called on states with maritime forces in the area to halt all inward and outward shipping to verify that the sanctions imposed by Resolution 661 were being respected.[6]

6. In Resolution 666 (13 September 1990) the Security Council turned its attention to humanitarian questions relating to the conflict.

7. Resolution 667 (16 September 1990) condemned Iraq's aggressive acts against diplomatic premises and personnel in Kuwait.

8. Resolution 669 (24 September 1990) dealt with the question of assistance, under Article 50 of the Charter, to states involved in implementing the sanctions which were suffering from the application of sanctions.

9. Resolution 670 (25 September 1990) spelled out the obligation to apply sanctions on all means of transport into Iraq, including aircraft.[7]

10. Resolution 674 (29 October 1990) dealt with the taking of third-state nationals as hostages and demanded their immediate departure, and insisted on the safety and well-being of diplomatic and consular personnel and premises, asking the Secretary-General to continue his efforts on these matters.

11. Resolution 677 (28 November 1990) condemned Iraq's attempt to alter the demographic composition of the Kuwait population and to destroy civil records.

12. Finally, Resolution 678 (29 November 1990) authorized 'Member States co-operating with the Government of Kuwait, unless Iraq on or before 15 January 1991 fully implements [the Security Council's previous resolutions], to use all necessary means to uphold and implement [these resolutions] and to restore international peace and security in the area'.

These 12 resolutions constitute an unprecedented, if selective, course of action under Chapter VII (Articles 39–51) of the UN Charter (Action with Respect to Threats to the Peace, Breaches of the Peace, and Acts of Aggression).[8]

II. The UN Charter and the Security Council's actions

Regardless of their final impact, it is important to examine how the articles of Chapter VII and other provisions of the Charter have in fact been used or not used, particularly because the Council's actions in this crisis are already being hailed as setting an important precedent for the future.

On 25 August, the Security Council, in asking states with maritime forces in the Persian Gulf area to monitor shipping (Resolution 665), had already begun to depart from UN Charter Chapter VII, in which Articles 46 and 47

[6] Two hundred direct responses to resolution 661 were received from governments and published as Security Council documents. A further 83 replied to the Sanctions Committee's questionnaire. These were also published as Security Council documents.

[7] Prior to the hostilities which started on 16 Jan. 1991, many notifications of flights were received, as well as requests to the committee for guidance and authorization. These latter communications were answered by the Chairman on the advice of the Legal Counsel of the United Nations. In addition, 14 responses to the resolution itself were received.

[8] For the text of UN Charter Chapter VII, see appendix 18B.

clearly imply that enforcement measures under Chapter VII would be under the control of the Security Council and its Military Staff Committee (MSC).[9] On 29 November, Resolution 678 went much further down this divergent road in authorizing the use of force after 15 January 1991. This tendency was evident from the beginning of the conflict. While the Security Council acted quickly and forcefully in early August, it was not in a position to assure the security of other states in the area against possible Iraqi attack. Thus a parallel, international coalition operation, under Article 51 of the UN Charter (which refers to the 'inherent right of individual or collective self-defence'; see appendix 18B) and independent of Security Council decisions, was mounted under the leadership of the United States to protect Saudi Arabia. When this force deployment began it was the accepted wisdom that *sanctions* were to be the means of securing Iraq's withdrawal from Kuwait and that it might well be six months or longer before they began to have a serious impact on Iraq. Later, however, when the *defensive* arms buildup in Saudi Arabia began to acquire *offensive* capability, the view began to emerge that sanctions were too slow and that, if Iraq did not speedily withdraw, force would have to be used to drive Iraq out of Kuwait. This tendency culminated on 29 November 1990 in Resolution 678, setting the 15 January 1991 deadline for Iraq's withdrawal from Kuwait.

There are practical as well as other reasons for the divergence from the course of action set out in Chapter VII. The 40 years of the cold war meant, among many other things, that the steps foreseen in the UN Charter for providing the Security Council with standby forces to enforce its decisions have never been taken, because of the lack of agreement among the permanent members. No agreements have been concluded under Article 43 with member states to make armed forces, assistance and facilities available to the Security Council. The MSC, which was designed to assist the Security Council in the application of armed force, has conducted purely token meetings throughout the cold war period and, despite recent Soviet suggestions for revival of the MSC, is still a largely inactive body. It holds periodic meetings of a purely formal nature.

The very idea of a UN command under the Security Council, although traditionally accepted for peace-keeping operations, was not seriously considered for enforcement operations in the Persian Gulf. Suggestions that naval or other military forces in the Persian Gulf should come under the Security Council and the MSC (see Article 47.3) were apparently regarded as unrealistic or unacceptable by the main participants in the international military buildup in Saudi Arabia.[10] These same powers apparently never considered the possibility of the Security Council working out arrangements for the *command* of forces in the Persian Gulf, as is also suggested in Article 47.3 of the Charter. Although in Korea in 1950 the Security Council, in the

[9] The MSC is composed of the Chiefs of Staff of the 5 permanent members of the UN Security Council (see UN Charter, Article 47.2).

[10] See also chapter 19 in this volume for an account of the military mobilization leading up to Operation Desert Storm in January 1991.

absence of the Soviet Union,[11] designated the United States as the Unified Command, it evidently considered that in the Persian Gulf a Security Council-designated command, or even the discussion of such a thing, would prejudice operational effectiveness.

The Charter places an important condition on the ultimate use of force. Article 42 states that 'Should the Security Council consider that measures provided for in Article 41 [sanctions] would be inadequate or have proved to be inadequate, it may take such action by air, sea or land forces as may be necessary to maintain or restore international peace and security'. The goal of Chapter VII is action short of force if possible. No such determination about the inadequacy of sanctions has ever been made by the Security Council.

Nevertheless, on 29 November 1990 the Security Council authorized the use, after 15 January 1991, of 'all necessary means' (which is generally understood to mean the use of force) by 'member states co-operating with the Government of Kuwait' to enforce its previous decisions, thus marrying the resolutions of the Security Council on the Iraq–Kuwait conflict and the forces assembled in Saudi Arabia under Article 51 under the political and military leadership of the United States in a future enforcement action. A main objective of the Security Council's 29 November authorization was to legitimize future action and to preserve the coalition represented by military contingents in Saudi Arabia. It remains to be seen how effectively this objective will be achieved in the long run. What is certain is that the Security Council and the member governments need to consider urgently the most effective and appropriate means of facing future acts of aggression.

In the search for a solution to the Iraq–Kuwait conflict, there has been a growing emphasis on unconditional withdrawal and on military means for securing such withdrawal if persuasion fails. In this case collective security has turned out to mean large-scale war.[12] In the present state of military technology, this is a sobering phenomenon which points urgently to the need to develop other methods of rolling back or, preferably, preventing aggression. Without betraying the principles of the UN Charter by rewarding aggression, non-violent methods—sanctions included—for dealing with aggression, including the acceptability of face-saving mechanisms, need to be urgently studied by the UN and governments.

III. The development of the international system

The Iraq–Kuwait conflict provides a practical test of the international system as it now exists, as well as an indication of how that system might be developed in the future. The UN system—the central political organization and some 30 loosely affiliated agencies and programmes—was set up in 1945 at a time of great post-war optimism. The system never died, although one

[11] The USSR boycotted the work of the Security Council during this period over the issue of Chinese representation on the Security Council.
[12] See Hoffman, S., 'The price of war', *New York Review of Books*, 17 Jan. 1991.

might sometimes infer this from the media reports. In fact some remarkable things were accomplished in a dismal political climate, laying the foundations for the current 'renaissance'. None the less, much of the original system established by the Charter, especially on the political and security side, lay dormant during the long winter of the cold war and is only now coming to life. The United Nations is waking up in a world very different from the world of 1945 in which it was born.

It was not only the cold war that, from 1946 to 1990, inhibited the Security Council from applying the procedures of Chapter VII of the Charter to breaches of peace and acts of aggression, but also the fact that the aggressions of the 1930s (which had inspired Chapter VII) were not of the type which typically occurred in the years after World War II. In cases which did not involve the crossing of recognized borders so much as disagreement on the borders themselves, aggression and unquestioned responsibility for it were difficult to establish or to agree upon, particularly when those concerned were affiliated with one side or the other of the cold war. Thus, for over 40 years the Security Council largely operated under Chapter VI (Pacific Settlement of Disputes), relying increasingly on the Secretary-General's good offices, and using processes of mediation, conciliation and peace-keeping (an *ad hoc* extension of Chapter VI), which tended to treat the parties to a conflict 'even-handedly'. Such processes were often indecisive in settling disputes and, in the cold war period, could not be backed with a credible show of collective influence or pressure. It was in such conditions that Secretary-General Javier Pérez de Cuellar described the situation in 1982 as being 'perilously near to a new international anarchy'.[13]

It is not only the end of the cold war but also the unusual clarity of the Iraqi aggression against Kuwait that has made possible the speed and unanimity of Security Council action in the current conflict. Future challenges are unlikely to present such a clear basis for the Council's action. It is therefore urgently necessary to consider what system of collective security will be best suited to the conflicts that are likely to arise in the future.

The United Nations faces two main tasks: (*a*) to combine peace-keeping, peace-making and collective action into a reliable international security system, and (*b*) to deal with the great socio-economic problems of global interdependence. Both of these tasks have become increasingly pressing and urgent—the first because of the volatile post-cold war situation and the devastating destructiveness, and expense, of modern weapons; the second as we see the new generation of global problems themselves developing into threats to international security and even to human survival. The two tasks are mutually dependent.

The UN system needs to be put in shape for its new world-encircling mission, its much heralded renaissance. It needs a thorough overhauling. It needs to be brought up to date. It needs to be properly manned, financed and

[13] *Report of the Secretary-General on the Work of the Organization*, A/37/1 (United Nations: New York, Sep. 1982), p. 3.

supported. And its course needs to be carefully and authoritatively charted. It needs to be professionally competitive with the best in government and in the private sector. Above all its member governments need to consider what basic attitudes and policies must be changed if the United Nations system is to respond effectively to the great, and very real, problems of today and tomorrow.

Without a reliable system of international peace and security, it will be impossible to devote the necessary energy, resources and co-operation to the great global problems of our time. Article 26 of the Charter, in formulating the task of the Security Council in the 'regulation of armaments', refers to 'the establishment and maintenance of international peace and security with the least diversion for armaments of the world's human and economic resources'.

It is in the work of the UN Security Council that the end of the cold war has had the most immediate effect, in particular because of the new spirit of co-operation and partnership of the permanent members. A number of practical steps forward have already been taken—in Namibia, Afghanistan, the Iraq–Iran War, Cambodia, Central America and Western Sahara, to name a few examples. The Iraq–Kuwait conflict has evoked an unprecedentedly firm and united response, and is putting to the test, as well as raising questions about, the concept of collective security.

Hitherto, the United Nations has not provided a *system* for peace and security so much as a last resort and safety net. Sometimes it was able to mount peace-keeping operations as a sort of sheriff's posse when things had already got out of hand. The question is whether, in the new international climate, the nations of the world are capable of the effort—and expenditure—to create and maintain a *system*, based on vigilance, consensus, common interest and law. Ideally, such a system would keep a permanent watch on international peace and security around the world, pre-empt or prevent conflict, mediate disputes, assure the protection of the weak, and deal authoritatively with aggressors or would-be aggressors.

This is a very large order. It requires, first of all, a return to the provisions of the UN Charter which were the distillation of the terrible lessons of the World War II and of the events that led up to it, including the failure of the League of Nations. A first step in this direction has been taken by the Security Council in invoking Chapter VII of the Charter to vote enforcement measures against Iraq. But the creation of a reliable *system* for peace and security involves more than reacting, however forcefully, to a conflict that has already occurred. It requires both the creation of conditions in which peace can be maintained, and the capacity to anticipate and to *prevent* breaches of the peace. It requires respect for, and confidence in, the authority both of the Security Council and of international law and some capacity for enforcement. That respect has been stunted and eroded in the cold war period. It will take time and effort to restore it and to make sure that confidence in the Security Council is shared by the whole body of the UN membership.

Governments will also have to be prepared to put adequate resources and support behind both global and regional security systems. Peace-keeping, for example, will need more convincing support than the shoe-string basis it has operated on until now. Apart from dynamic diplomatic action, and an increasing effort to apply legal norms where these are relevant, two main operational functions are required to give reality to the Council's decisions. They are (*a*) peace-keeping, which may be compared to the police function in a nation state, and (*b*) enforcement, which corresponds to the military function. Until recently, popular emphasis and interest lay mainly in peace-keeping, an original creation of the United Nations. The Iraq conflict has highlighted the necessity of also maintaining the capacity for enforcement.

Obviously, the outcome of the Iraq–Kuwait conflict, whatever it is, will have a crucial impact on the international system, and especially on the future of the Security Council and the international system for peace and security. Even before we know the outcome, the conflict provides some useful lessons and pointers to the future.

It is already clear that the world we have to deal with will resemble neither the world of the 1930s on which the Charter was based, nor the world of the cold war which dominated and stunted the United Nations in its first 40 years. The removal of the constraints of the cold war has not only liberated the UN Security Council; it has also removed constraints on a wide variety of disputes and conflict situations throughout the world. We are entering a period of great instability, engendered by a mixture of causes and motivations. Old international rivalries, ethnic and religious turmoil, a vast flow of arms and military technology, domestic disintegration, poverty, economic inequities, instant and universal communication, population pressures, natural and ecological disasters, scarcity of vital resources, vast movements of population—such are the components of a highly volatile and unstable period of world history.

In such a situation, no one nation, or even a partnership of two or three powerful nations, is going to be able to assume the role of world moderator and policeman, even supposing all the others would accept it. The United Nations therefore must grow to maturity as the accepted mechanism for assuring a reasonable degree of peace, security and stability.

The reduction of the nuclear tensions of East–West rivalry has released a variety of other tensions in many parts of the world. The ensuing disputes and conflicts should *all* be a matter of concern to the United Nations if we are to talk seriously of a 'new world order'. In addition, far greater attention will have to be given to the underlying causes of conflict, some of which are mentioned above, if the new world order is to have any real meaning.

In the current conflict in the Persian Gulf, we have seen the tremendous effort and resources required to mount a convincing response to just one conflict situation, admittedly a particularly flagrant act of aggression in a particularly sensitive part of the world. A credible international security system, or 'new world order', will have to respond, either through regional or through global organizations, to the whole wide range of disputes, threats to the peace, conflicts, breaches of the peace, or even acts of aggression, which are likely to occur in the aftermath of the cold war. It is no longer acceptable that significant international action is only taken when a situation is a serious threat to the interests of the most powerful nations. A system of international peace and security which is comprehensive, universal, and in which all nations can participate must therefore be the aim. Is this a feasible proposition?

IV. The essential elements of a new international system

The basic principles of such a system are already set out in the Charter. Its necessary elements are also clear enough.

1. The mechanism for political co-ordination and consultation among governments needs to be far more effective and comprehensive than any arrangement that has gone before. The Security Council, the General Assembly, the Secretary-General, and the corps of permanent national representatives to the United Nations provide a working basis for such a mechanism. Their procedures and functioning will have to become far better informed, more active, more consistent and more universal.

2. The word 'security', in order to acquire real significance, must mean a permanent watch on developments all over the world, socio-economic as well as political and military. Special attention must be given to dangerous buildups of armaments and to potential threats, especially to the weaker nations.

3. Necessary action to pre-empt or correct dangerous situations should be taken as a matter of course by the Security Council.

4. The mechanisms for carrying out the decisions of the Security Council need to be developed and systematized. These include:

(*a*) *Pacific settlement* (sometimes called peace-making)—mediation, concerted diplomatic activity, conciliation, good offices, and so on, and legal recourse on matters of a justiciable nature. Here the Secretary-General and his senior colleagues will continue to play a major role. The International Court of Justice should be used more often and more imaginatively.[14] The activities of regional arrangements and agencies should be strengthened and co-ordinated as necessary with the work of the Security Council.

(*b*) *Conflict control* (sometimes called peace-keeping). The technique of peace-keeping needs to be put on a much stronger logistical and financial basis. The technique and its use need to be diversified. Systematic stand-by and training arrangements should be set up world-wide. Alternative methods of financing—including subventions by large interests which benefit from peace-keeping—should be urgently studied. Peace-keeping units should be regarded not as an abnormal expense but as a routine and indispensable

[14] See UN Charter, Article 36.3.

feature of the 'new world order'. They should be deployed in dangerous areas in advance of crisis or conflict.

(*c*) *Enforcement capability*. Although Chapter VII was originally considered to be the most innovative part of the Charter, less attention has been given to its implementation than to any other part of the Charter. This has become glaringly obvious during the Iraq–Kuwait conflict. The Security Council should be far better prepared for the next threat to the peace or act of aggression.

The basic task remains as the Charter outlined it in 1945—the gradual conversion of the present military set-up into a worldwide system of common security. The Military Staff Committee should therefore be instructed to embark on an extensive study of the conversion of modern military technology and establishments, including the concept and means of deterrence, to the needs of an international system of security in a highly volatile and unstable period. The MSC representation of the five permanent members is, to some extent, an anachronism; the Security Council should co-opt other states of strategic and representative political importance to take part in the MSC's work.

The Military Staff Committee was also originally supposed to advise and assist the Security Council on 'the regulation of armaments, and possible disarmament' (UN Charter, Article 47.1) which is closely related to the basic task of conversion. It is surely time that this part of the MSC mandate was actively revived.

The MSC, in studying the gradual conversion of the present world military set-up to an international security system, should give particular attention to agreements to provide forces under Article 43 of the Charter and to the question of future strategic direction and command of such forces, should enforcement measures again become necessary. The MSC should be required to study and report on the extent to which the provision of forces under Article 43 is still a feasible, and practical, option in present circumstances.

V. Conclusion

The key to the effectiveness of a future system of peace and security will be the combination and interaction of all these various elements. Thus the Security Council would be informed and activated by a consistent and constant world-wide peace and security watch. It would meet regularly to survey the *whole* fabric of international peace and security, not just particular holes and tears in it, as in the past. The Council should develop a far greater degree of consultation with states which are not members of the Council. It should provide the centre for a new, standing process of political and diplomatic consultation on the interlinked elements of global security. Regional organizations or arrangements, which should be strengthened and developed, should be regularly consulted by the Security Council, and should, in principle (Article 52 of the Charter), be the resort of first instance in regional disputes.

It should be possible in the future to prevent or pre-empt many disputes from degenerating into actual conflict. There will, however, always be places where the danger of conflict is imminent. In such areas the Council should deploy peace-keeping mechanisms of an appropriate kind to observe and report on the situation and to contain it while diplomatic and pacific solutions are being sought.

If these peace-keeping efforts fail, they should have the function of a trip-wire which activates, after suitable warnings, pre-planned enforcement action under Chapter VII of the Charter. The actual existence of a wide range of enforcement capability, through the work of the Security Council and the MSC, and the general agreement that in certain prescribed circumstances these enforcement measures will come into play, would provide a strong deterrent to aggression.

The lessons of the Iraq–Kuwait conflict would seem to point to the need to develop a system such as this. Such a system would admittedly be a giant step forward from the belated and improvised efforts to which the Security Council has so far been limited. It can only work if governments, especially the more powerful ones, genuinely accept and co-operate in the aim of converting both the present diplomatic framework and the present military framework into a system of common security. This would be the best test of a firm belief in a 'new world order'.

Appendix 18A. UN Security Council resolutions on the Iraqi invasion of Kuwait

Resolution 660 (2 August 1990)

The Security Council,

Alarmed by the invasion of Kuwait on 2 August 1990 by the military forces of Iraq,

Determining that there exists a breach of international peace and security as regards the Iraqi invasion of Kuwait,

Acting under Articles 39 and 40 of the Charter of the United Nations,

1. *Condemns* the Iraqi invasion of Kuwait;

2. *Demands* that Iraq withdraw immediately and unconditionally all its forces to the positions in which they were located on 1 August 1990;

3. *Calls upon* Iraq and Kuwait to begin immediately intensive negotiations for the resolution of their differences and supports all efforts in this regard, and especially those of the League of Arab States;

4. *Decides* to meet again as necessary to consider further steps to ensure compliance with the present resolution.

In favour 14: Canada, China, Colombia, Côte d'Ivoire, Cuba, Ethiopia, Finland, France, Malaysia, Romania, UK, USA, USSR, Zaire

The Republic of Yemen did not participate in the vote.

Resolution 661 (6 August 1990)

The Security Council,

Reaffirming its resolution 660 (1990) of 2 August 1990,

Deeply concerned that that resolution has not been implemented and that the invasion by Iraq of Kuwait continues with further loss of human life and material destruction,

Determined to bring the invasion and occupation of Kuwait by Iraq to an end and to restore the sovereignty, independence and territorial integrity of Kuwait,

Noting that the legitimate Government of Kuwait has expressed its readiness to comply with resolution 660 (1990),

Mindful of its responsibilities under the Charter of the United Nations for the maintenance of international peace and security,

Affirming the inherent right of individual or collective self-defence, in response to the armed attack by Iraq against Kuwait, in accordance with Article 51 of the Charter,

Acting under Chapter VII of the Charter of the United Nations,

1. *Determines* that Iraq so far has failed to comply with paragraph 2 of resolution 660 (1990) and has usurped the authority of the legitimate Government of Kuwait;

2. *Decides*, as a consequence, to take the following measures to secure compliance of Iraq with paragraph 2 of resolution 660 (1990) and to restore the authority of the legitimate Government of Kuwait;

3. *Decides* that all States shall prevent:

(*a*) The import into their territories of all commodities and products originating in Iraq or Kuwait exported therefrom after the date of the present resolution;

(*b*) Any activities by their nationals or in their territories which would promote or are calculated to promote the export or transshipment of any commodities or products from Iraq or Kuwait; and any dealings by their nationals or their flag vessels or in their territories in any commodities or products originating in Iraq or Kuwait and exported therefrom after the date of the present resolution, including in particular any transfer of funds to Iraq or Kuwait for the purposes of such activities or dealings;

(*c*) The sale or supply by their nationals or from their territories or using their flag vessels of any commodities or products, including weapons or any other military equipment, whether or not originating in their territories but not including supplies intended strictly for medical purposes, and, in humanitarian circumstances, foodstuffs, to any person or body in Iraq or Kuwait or to any person or body for the purposes of any business carried on in or operated from Iraq or Kuwait, and any activities by their nationals or in their territories which promote or are calculated to promote such sale or supply of such commodities or products;

4. *Decides* that all States shall not make available to the Government of Iraq or to any commercial, industrial or public utility undertaking in Iraq or Kuwait, any funds or any other financial or economic resources and shall prevent their nationals and any persons within their territories from remov-

ing from their territories or otherwise making available to that Government or to any such undertaking any such funds or resources and from remitting any other funds to persons or bodies within Iraq or Kuwait, except payments exclusively for strictly medical or humanitarian purposes and, in humanitarian circumstances, foodstuffs;

5. *Calls upon* all States, including States non-members of the United Nations, to act strictly in accordance with the provisions of the present resolution notwithstanding any contract entered into or licence granted before the date of the present resolution;

6. *Decides* to establish, in accordance with rule 28 of the provisional rules of procedure of the Security Council, a Committee of the Security Council consisting of all the members of the Council, to undertake the following tasks and to report on its work to the Council with its observations and recommendations:

(*a*) To examine the reports on the progress of the implementation of the present resolution which will be submitted by the Secretary-General;

(*b*) To seek from all States further information regarding the action taken by them concerning the effective implementation of the provisions laid down in the present resolution;

7. *Calls upon* all States to co-operate fully with the Committee in the fulfilment of its task, including supplying such information as may be sought by the Committee in pursuance of the present resolution;

8. *Requests* the Secretary-General to provide all necessary assistance to the Committee and to make the necessary arrangements in the Secretariat for the purpose;

9. *Decides* that, notwithstanding paragraphs 4 through 8 above, nothing in the present resolution shall prohibit assistance to the legitimate Government of Kuwait, and *calls upon* all States:

(*a*) To take appropriate measures to protect assets of the legitimate Government of Kuwait and its agencies;

(*b*) Not to recognize any régime set up by the occupying Power;

10. *Requests* the Secretary-General to report to the Council on the progress of the implementation of the present resolution, the first report to be submitted within thirty days;

11. *Decides* to keep this item on its agenda and to continue its efforts to put an early end to the invasion by Iraq.

In favour 13: Canada, China, Colombia, Côte d'Ivoire, Ethiopia, Finland, France, Malaysia, Romania, UK, USA, USSR, Zaire
Abstaining 2: Cuba, Republic of Yemen

Resolution 662 (9 August 1990)

The Security Council,

Recalling its resolutions 660 (1990) and 661 (1990),

Gravely alarmed by the declaration by Iraq of a 'comprehensive and eternal merger' with Kuwait,

Demanding, once again, that Iraq withdraw immediately and unconditionally all its forces to the positions in which they were located on 1 August 1990,

Determined to bring the occupation of Kuwait by Iraq to an end and to restore the sovereignty, independence and territorial integrity of Kuwait,

Determined also to restore the authority of the legitimate Government of Kuwait,

1. *Decides* that annexation of Kuwait by Iraq under any form and whatever pretext has no legal validity, and is considered null and void;

2. *Calls upon* all States, international organizations and specialized agencies not to recognize that annexation, and to refrain from any action or dealing that might be interpreted as an indirect recognition of the annexation;

3. *Further demands* that Iraq rescind its actions purporting to annex Kuwait;

4. *Decides* to keep this item on its agenda and to continue its efforts to put an early end to the occupation.

In favour 15: Canada, China, Colombia, Côte d'Ivoire, Cuba, Ethiopia, Finland, France, Malaysia, Romania, UK, USA, USSR, Republic of Yemen, Zaire

---

Resolution 664 (18 August 1990)

The Security Council,

Recalling the Iraqi invasion and purported annexation of Kuwait and resolutions 660, 661 and 662,

Deeply concerned for the safety and well being of third state nationals in Iraq and Kuwait,

Recalling the obligations of Iraq in this regard under international law,

Welcoming the efforts of the Secretary-General to pursue urgent consultations with the Government of Iraq following the concern and anxiety expressed by the members of the Council on 17 August 1990.

Acting under Chapter VII of the United Nations Charter:

1. *Demands* that Iraq permit and facilitate the immediate departure from Kuwait and Iraq of the nationals of third countries and grant immediate and continuing access of consular officials to such nationals;

2. *Further demands* that Iraq take no action to jeopardize the safety, security or health of such nationals;

3. *Reaffirms* its decision in resolution 662 (1990) that annexation of Kuwait by Iraq is null and void, and therefore demands that the government of Iraq rescind its orders for the closure of diplomatic and consular missions in Kuwait and the withdrawal of the immunity of their personnel, and refrain from any such actions in the future;

4. *Requests* the Secretary-General to report to the Council on compliance with this resolution at the earliest possible time.

In favour 15: Canada, China, Colombia, Côte d'Ivoire, Cuba, Ethiopia, Finland, France, Malaysia, Romania, UK, USA, USSR, Republic of Yemen, Zaire

---

Resolution 665 (25 August 1990)

The Security Council,

Recalling its resolutions 660 (1990), 661 (1990), 662 (1990) and 664 (1990) and demanding their full and immediate implementation,

Having decided in resolution 661 (1990) to impose economic sanctions under Chapter VII of the Charter of the United Nations,

Determined to bring an end to the occupation of Kuwait by Iraq which imperils the existence of a Member State and to restore the legitimate authority, and the sovereignty, independence and territorial integrity of Kuwait which requires the speedy implementation of the above resolutions,

Deploring the loss of innocent life stemming from the Iraqi invasion of Kuwait and determined to prevent further such losses,

Gravely alarmed that Iraq continues to refuse to comply with resolutions 660 (1990), 661 (1990), 662 (1990) and 664 (1990) and in particular at the conduct of the Government of Iraq in using Iraqi flag vessels to export oil,

1. *Calls upon* those Member States cooperating with the Government of Kuwait which are deploying maritime forces to the area to use such measures commensurate to the specific circumstances as may be necessary under the authority of the Security Council to halt all inward and outward maritime shipping in order to inspect and verify their cargoes and destinations and to ensure strict implementation of the provisions related to such shipping laid down in resolution 661 (1990);

2. *Invites* Member States accordingly to co-operate as may be necessary to ensure compliance with the provisions of resolution 661 (1990) with maximum use of political and diplomatic measures, in accordance with paragraph 1 above;

3. *Requests* all States to provide in accordance with the Charter such assistance as may be required by the States referred to in paragraph 1 of this resolution;

4. *Further requests* the States concerned to co-ordinate their actions in pursuit of the above paragraphs of this resolution using as appropriate mechanisms of the Military Staff Committee and after consultation with the Secretary-General to submit reports to the Security Council and its Committee established under resolution 661 (1990) to facilitate the monitoring of the implementation of this resolution;

5. *Decides* to remain actively seized of the matter.

In favour 13: Canada, China, Colombia, Côte d'Ivoire, Ethiopia, Finland, France, Malaysia, Romania, UK, USA, USSR, Zaire
Abstaining 2: Cuba, Republic of Yemen

Resolution 666 (13 September 1990)

The Security Council,

Recalling its resolution 661 (1990), paragraphs 3(c) and 4 of which apply, except in humanitarian circumstances, to foodstuffs,

Recognizing that circumstances may arise in which it will be necessary for foodstuffs to be supplied to the civilian population in Iraq or Kuwait in order to relieve human suffering,

Noting that in this respect the Committee established under paragraph 6 of that resolution has received communications from several Member States,

Emphasizing that it is for the Security Council, alone or acting through the

Committee, to determine whether humanitarian circumstances have arisen,

Deeply concerned that Iraq has failed to comply with its obligations under Security Council resolution 664 (1990) in respect of the safety and well-being of third State nationals, and reaffirming that Iraq retains full responsibility in this regard under international humanitarian law including, where applicable, the Fourth Geneva Convention,

Acting under Chapter VII of the Chapter of the United Nations,

1. *Decides* that in order to make the necessary determination whether or not for the purposes of paragraph 3(c) and paragraph 4 of resolution 661 (1990) humanitarian circumstances have arisen, the Committee shall keep the situation regarding foodstuffs in Iraq and Kuwait under constant review;

2. *Expects* Iraq to comply with its obligations under Security Council resolution 664 (1990) in respect of third State nationals and reaffirms that Iraq remains fully responsible for their safety and well-being in accordance with international humanitarian law including, where applicable, the Fourth Geneva Convention;

3. *Requests*, for the purposes of paragraphs 1 and 2 of this resolution, that the Secretary-General seek urgently, and on a continuing basis, information from relevant United Nations and other appropriate humanitarian agencies and all other sources on the availability of food in Iraq and Kuwait, such information to be communicated by the Secretary-General to the Committee regularly;

4. *Requests* further that in seeking and supplying such information particular attention will be paid to such categories of persons who might suffer especially, such as children under 15 years of age, expectant mothers, maternity cases, the sick and the elderly;

5. *Decides* that if the Committee, after receiving the reports from the Secretary-General, determines that circumstances have arisen in which there is an urgent humanitarian need to supply foodstuffs to Iraq or Kuwait in order to relieve human suffering, it will report promptly to the Council its decision as to how such need should be met;

6. *Directs* the Committee that in formulating its decisions it should bear in mind that foodstuffs should be provided through the United Nations in co-operation with the International Committee of the Red Cross or other appropriate humanitarian agencies and distributed by them or under their supervision in order to ensure that they reach the intended beneficiaries;

7. *Requests* the Secretary-General to use his good offices to facilitate the delivery and distribution of foodstuffs to Kuwait and Iraq in accordance with the provisions of this and other relevant resolutions;

8. *Recalls* that resolution 661 (1990) does not apply to supplies intended strictly for medical purposes, but in this connection recommends that medical supplies should be exported under the strict supervision of the Government of the exporting State or by appropriate humanitarian agencies.

In favour 13: Canada, China, Colombia, Côte d'Ivoire, Ethiopia, Finland, France, Malaysia, Romania, UK, USA, USSR, Zaire
Against 2: Cuba, Republic of Yemen

Resolution 667 (16 September 1990)

The Security Council,

Reaffirming its resolutions 660 (1990), 661 (1990), 662 (1990), 664 (1990), 665 (1990) and 666 (1990),

Recalling the Vienna Conventions of 18 April 1961 on diplomatic relations and of 24 April 1963 on consular relations, to both of which Iraq is a party,

Considering that the decision of Iraq to order the closure of diplomatic and consular missions in Kuwait and to withdraw the immunity and privileges of these missions and their personnel is contrary to the decisions of the Security Council, the international Conventions mentioned above and international law,

Deeply concerned that Iraq, notwithstanding the decisions of the Security Council and the provisions of the Conventions mentioned above, has committed acts of violence against diplomatic missions and their personnel in Kuwait,

Outraged at recent violations by Iraq of diplomatic premises in Kuwait and at the abduction of personnel enjoying diplomatic immunity and foreign nationals who were present in these premises,

Considering that the above actions by Iraq constitute aggressive acts and a flagrant violation of its international obligations which strike at the root of the conduct of international relations in accordance with the Charter of the United Nations,

Recalling that Iraq is fully responsible for any use of violence against foreign nationals or against any diplomatic or consular missions in Kuwait or its personnel,

Determined to ensure respect for its decisions and for Article 25 of the Charter of the United Nations,

Further considering that the grave nature of Iraq's actions, which constitute a new escalation of its violations of international law, obliges the Council not only to express its immediate reaction but also to consider further concrete measures to ensure Iraq's compliance with the Council's resolutions,

Acting under Chapter VII of the Charter of the United Nations,

1. *Strongly condemns* aggressive acts perpetrated by Iraq against diplomatic premises and personnel in Kuwait, including the abduction of foreign nationals who were present in those premises;

2. *Demands* the immediate release of those foreign nationals as well as nationals mentioned in resolution 664 (1990);

3. *Further demands* that Iraq immediately and fully comply with its international obligations under resolutions 660 (1990), 662 (1990) and 664 (1990) of the Security Council, the Vienna Conventions on diplomatic and consular relations and international law;

4. *Further demands* that Iraq immediately protect the safety and well-being of diplomatic and consular personnel and premises in Kuwait and in Iraq and take no action to hinder the diplomatic and consular missions in the performance of their functions, including access to their nationals and the protection of their **person** and interests;

5. *Reminds* all States that they are obliged to observe strictly resolutions 661 (1990), 662 (1990), 664 (1990), 665 (1990) and 666 (1990);

6. *Decides* to consult urgently to take further concrete measures as soon as possible, under Chapter VII of the Charter, in response to Iraq's continued violation of the Charter, of resolutions of the Council and of international law.

In favour 15: Canada, China, Colombia, Côte d'Ivoire, Cuba, Ethiopia, Finland, France, Malaysia, Republic of Yemen Romania, UK, USA, USSR, Zaire

---

Resolution 669 (24 September 1990)

The Security Council,

Recalling its resolution 661 (1990) of 6 August 1990,

Recalling also Article 50 of the Charter of the United Nations,

Conscious of the fact that an increasing number of requests for assistance have been received under the provisions of Article 50 of the Charter of the United Nations,

Entrusts the Committee established under resolution 661 (1990) concerning the situation between Iraq and Kuwait with the task of examining requests for assistance under the provisions of Article 50 of the Charter of the United Nations and making recommendations to the President of the Security Council for appropriate action.

In favour 15: Canada, China, Colombia, Côte d'Ivoire, Cuba, Ethiopia, Finland, France, Malaysia, Romania, UK, USA, USSR, Republic of Yemen, Zaire

---

Resolution 670 (25 September 1990)

The Security Council,

Reaffirming its resolutions 660 (1990), 661 (1990), 662 (1990), 664 (1990), 665 (1990), 666 (1990) and 667 (1990),

Condemning Iraq's continued occupation of Kuwait, its failure to rescind its actions and end its purported annexation and its holding of third State nationals against their will, in flagrant violation of resolutions 660 (1990), 662 (1990), 664 (1990) and 667 (1990) and of international humanitarian law,

Condemning further the treatment by Iraqi forces of Kuwaiti nationals, including measures to force them to leave their own country and mistreatment of persons and property in Kuwait in violation of international law,

Noting with grave concern the persistent attempts to evade the measures laid down in resolution 661 (1990),

Further noting that a number of States have limited the number of Iraqi diplomatic and consular officials in their countries and that others are planning to do so,

Determined to ensure by all necessary means the strict and complete application of the measures laid down in resolution 661 (1990),

Determined to ensure respect for its decisions and the provisions of Articles 25 and 48 of the Charter of the United Nations,

Affirming that any acts of the Government of Iraq which are contrary to the above-mentioned resolutions or to Articles 25 or 48 of the Charter of the United Nations, such as Decree No. 377 of the Revolution Command Council of Iraq of 16 September 1990, are null and void,

Reaffirming its determination to ensure compliance with Security Council resolutions by maximum use of political and diplomatic means,

Welcoming the Secretary-General's use of his good offices to advance a peaceful solution based on the relevant Security Council resolutions and noting with appreciation his continuing efforts to this end,

Underlining to the Government of Iraq that its continued failure to comply with the terms of resolutions 660 (1990), 661 (1990), 662 (1990), 664 (1990), 666 (1990) and 667 (1990) could lead to further serious action by the Council under the Charter of the United Nations, including under Chapter VII,

Recalling the provisions of Article 103 of the Charter of the United Nations,

Acting under Chapter VII of the Charter of the United Nations,

1. *Calls upon* all States to carry out their obligations to ensure strict and complete compliance with resolution 661 (1990) and, in particular, paragraphs 3, 4 and 5 thereof;

2. *Confirms* that resolution 661 (1990) applies to all means of transport, including aircraft;

3. *Decides* that all States, notwithstanding the existence of any rights or obligations conferred or imposed by any international agreement or any contract entered into or any licence or permit granted before the date of the present resolution, shall deny permission to any aircraft to take off from their territory if the aircraft would carry any cargo to or from Iraq or Kuwait other than food in humanitarian circumstances, subject to authorization by the Council or the Committee established by resolution 661 (1990) and in accordance with resolution 666 (1990), or supplies intended strictly for medical purposes or solely for UNIIMOG;*

4. *Decides* further that all States shall deny permission to any aircraft destined to land in Iraq or Kuwait, whatever its State of registration, to overfly its territory unless:

(*a*) The aircraft lands at an airfield designated by that State outside Iraq or Kuwait in order to permit its inspection to ensure that there is no cargo on board in violation of resolution 661 (1990) or the present resolution, and for this purpose the aircraft may be detained for as long as necessary; or

(*b*) The particular flight has been approved by the Committee established by resolution 661 (1990); or

(*c*) The flight is certified by the United Nations as solely for the purposes of UNIIMOG;

5. *Decides* that each State shall take all necessary measures to ensure that any aircraft registered in its territory or operated by an operator who has his principal place of business or permanent residence in its territory complies with the provisions of resolution 661 (1990) and the present resolution;

6. *Decides* further that all States shall notify in a timely fashion the Committee established by resolution 661 (1990) of any flight between its territory and Iraq or Kuwait to which the requirement to land in paragraph 4 above does not apply, and the purpose for such a flight;

7. *Calls upon* all States to co-operate in taking such measures as may be necessary, consistent with international law, including the Chicago Convention, to ensure the effective implementation of the provisions of resolution 661 (1990) or the present resolution;

8. *Calls upon* all States to detain any ships of Iraqi registry which enter their ports and which are being or have been used in violation of resolution 661 (1990), or to deny such ships entrance to their ports except in circumstances recognized under international law as necessary to safeguard human life;

9. *Reminds* all States of their obligations under resolution 661 (1990) with regard to the freezing of Iraqi assets, and the protection of the assets of the legitimate Government of Kuwait and its agencies, located within their territory and to report to the Committee established under resolution 661 (1990) regarding those assets;

10. *Calls upon* all States to provide to the Committee established by resolution 661 (1990) information regarding the action taken by them to implement the provisions laid down in the present resolution;

11. *Affirms* that the United Nations Organization, the specialized agencies and other international organizations in the

* United Nations Iran–Iraq Military Observer Group.

United Nations system are required to take such measures as may be necessary to give effect to the terms of resolution 661 (1990) and this resolution;

12. *Decides* to consider, in the event of evasion of the provisions of resolution 661 (1990) or of the present resolution by a State or its nationals or through its territory, measures directed at the State in question to prevent such evasion;

13. *Reaffirms* that the Fourth Geneva Convention applies to Kuwait and that as a High Contracting Party to the Convention Iraq is bound to comply fully with all its terms and, in particular, is liable under the Convention in respect of the grave breaches committed by it, as are individuals who commit or order the commission of grave breaches.

In favour 14: Canada, China, Colombia, Côte d'Ivoire, Ethiopia, Finland, France, Malaysia, Romania, UK, USA, USSR, Republic of Yemen, Zaire
Against 1: Cuba

Resolution 674 (29 October 1990)

The Security Council,

Recalling its resolutions 660 (1990), 661 (1990), 662 (1990), 664 (1990), 665 (1990), 666 (1990), 667 (1990) and 670 (1990),

Stressing the urgent need for the immediate and unconditional withdrawal of all Iraqi forces from Kuwait, for the restoration of Kuwait's sovereignty, independence and territorial integrity and of the authority of its legitimate government,

Condemning the actions by the Iraqi authorities and occupying forces to take third-State nationals hostage and to mistreat and oppress Kuwait and third-State nationals, and the other actions reported to the Security Council, such as the destruction of Kuwait demographic records, the forced departure of Kuwaitis, the relocation of population in Kuwait and the unlawful destruction and seizure of public and private property in Kuwait, including hospital supplies and equipment, in violation of the decisions of the Council, the Charter of the United Nations, the Fourth Geneva Convention, the Vienna Conventions on Diplomatic and Consular Relations and international law,

Expressing grave alarm over the situation of nationals of third States in Kuwait and Iraq, including the personnel of the diplomatic and consular missions of such States,

Reaffirming that the Fourth Geneva Convention applies to Kuwait and that as a High Contracting Party to the Convention Iraq is bound to comply fully with all its terms and in particular is liable under the Convention in respect of the grave breaches committed by it, as are individuals who commit or order the commission of grave breaches,

Recalling the efforts of the Secretary-General concerning the safety and well-being of third-State nationals in Iraq and Kuwait,

Deeply concerned at the economic cost and at the loss and suffering caused to individuals in Kuwait and Iraq as a result of the invasion and occupation of Kuwait by Iraq,

Acting under Chapter VII of the Charter of the United Nations,

Reaffirming the goal of the international community of maintaining international peace and security by seeking to resolve international disputes and conflicts through peaceful means,

Recalling the important role that the United Nations and its Secretary-General have played in the peaceful solution of disputes and conflicts in conformity with the provisions of the Charter,

Alarmed by the dangers of the present crisis caused by the Iraqi invasion and occupation of Kuwait, which directly threaten international peace and security, and seeking to avoid any further worsening of the situation,

Calling upon Iraq to comply with the relevant resolutions of the Security Council, in particular its resolutions 660 (1990), 662 (1990) and 664 (1990),

Reaffirming its determination to ensure compliance by Iraq with the Security Council resolutions by maximum use of political and diplomatic means,

A

1. *Demands* that the Iraqi authorities and occupying forces immediately cease and desist from taking third-State nationals hostage, mistreating and oppressing Kuwaiti and third-State nationals and any other actions, such as those reported to the Security Council and described above, that violate the decisions of this Council, the Charter of the United Nations, the Fourth Geneva Convention, the Vienna Conventions on Diplomatic and Consular Relations and international law;

2. *Invites* States to collate substantiated information in their possession or submitted to them on the grave breaches by Iraq as per paragraph 1 above and to make this information available to the Security Council;

3. *Reaffirms* its demand that Iraq immediately fulfil its obligations to third-State nationals in Kuwait and Iraq, including the personnel of diplomatic and consular missions, under the Charter, the Fourth Geneva Convention, the Vienna Conventions on Diplomatic and Consular Relations, general principles of international law and the relevant resolutions of the Council;

4. *Also reaffirms* its demand that Iraq permit and facilitate the immediate departure from Kuwait of those third-State nationals, including diplomatic and consular personnel, who wish to leave;

5. *Demands* that Iraq ensure the immediate access to food, water and basic services necessary to the protection and well-being of Kuwaiti nationals and of nationals of third States in Kuwait and Iraq, including the personnel of diplomatic and consular missions in Kuwait;

6. *Reaffirms* its demand that Iraq immediately protect the safety and well-being of diplomatic and consular personnel and premises in Kuwait and in Iraq, take no action to hinder these diplomatic and consular missions in the performance of their functions, including access to their nationals and the protection of their person and interests and rescind its orders for the closure of diplomatic and consular missions in Kuwait and the withdrawal of the immunity of their personnel;

7. *Requests* the Secretary-General, in the context of the continued exercise of his good offices concerning the safety and well-being of third-State nationals in Iraq and Kuwait, to seek to achieve the objectives of paragraphs 4, 5 and 6 above and in particular the provision of food, water and basic services to Kuwaiti nationals and to the diplomatic and consular missions in Kuwait and the evacuation of third-State nationals;

8. *Reminds* Iraq that under international law it is liable for any loss, damage or injury arising in regard to Kuwait and third States, and their nationals and corporations, as a result of the invasion and illegal occupation of Kuwait by Iraq;

9. *Invites* States to collect relevant information regarding their claims, and those of their nationals and corporations, for restitution or financial compensation by Iraq with a view to such arrangements as may be established in accordance with international law;

10. *Requires* that Iraq comply with the provisions of the present resolution and its previous resolutions, failing which the Security Council will need to take further measures under the Charter;

11. *Decides* to remain actively and permanently seized of the matter until Kuwait has regained its independence and peace has been restored in conformity with the relevant resolutions of the Security Council.

B

12. *Reposes* its trust in the Secretary-General to make available his good offices and, as he considers appropriate, to pursue them and to undertake diplomatic efforts in order to reach a peaceful solution to the crisis caused by the Iraqi invasion and occupation of Kuwait on the basis of Security Council resolutions 660 (1990), 662 (1990) and 664 (1990), and calls upon all States, both those in the region and others, to pursue on this basis their efforts to this end, in conformity with the Charter, in order to improve the situation and restore peace, security and stability;

13. *Requests* the Secretary-General to report to the Security Council on the results of his good offices and diplomatic efforts.

In favour 13: Canada, China, Colombia, Côte d'Ivoire, Ethiopia, Finland, France, Malaysia, Romania, UK, USA, USSR, Zaire
Abstaining 2: Cuba, Republic of Yemen

Resolution 677 (28 November 1990)

The Security Council,

Recalling its resolutions 660 (1990) of 2 August 1990, 662 (1990) of 9 August 1990 and 674 (1990) of 29 October 1990,

Reiterating its concern for the suffering caused to individuals in Kuwait as a result of the invasion and occupation of Kuwait by Iraq,

Gravely concerned at the ongoing attempt by Iraq to alter the demographic composition of the population of Kuwait and to destroy the civil records maintained by the legitimate Government of Kuwait,

Acting under Chapter VII of the Charter of the United Nations,

1. *Condemns* the attempts by Iraq to alter the demographic composition of the population of Kuwait and to destroy the civil

records maintained by the legitimate Government of Kuwait;

2. *Mandates* the Secretary-General to take custody of a copy of the population register of Kuwait, the authenticity of which has been certified by the legitimate Government of Kuwait and which covers the registration of the population up to 1 August 1990;

3. *Requests* the Secretary-General to establish, in co-operation with the legitimate Government of Kuwait, an Order of Rules and Regulations governing access to and use of the said copy of the population register.

In favour 15: Canada, China, Colombia, Côte d'Ivoire, Cuba, Ethiopia, Finland, France, Malaysia, Romania, UK, USA, USSR, Republic of Yemen, Zaire

---

Resolution 678 (29 November 1990)

The Security Council,

Recalling and reaffirming its resolutions 660 (1990), 661 (1990), 662 (1990), 664 (1990), 665 (1990), 666 (1990), 667 (1990), 669 (1990), 670 (1990) and 674 (1990),

Noting that, despite all efforts by the United Nations, Iraq refuses to comply with its obligation to implement resolution 660 (1990) and the above subsequent resolutions, in flagrant contempt of the Council,

Mindful of its duties and responsibilities under the Charter of the United Nations for the maintenance and preservation of international peace and security,

Determined to secure all compliance with its decisions,

Acting under Chapter VII of the Charter of the United Nations,

1. *Demands* that Iraq comply fully with resolution 660 (1990) and all subsequent relevant resolutions and decides, while maintaining all its decisions, to allow Iraq one final opportunity, as a pause of goodwill, to do so;

2. *Authorizes* Member States co-operating with the Government of Kuwait, unless Iraq on or before 15 January 1991 fully implements, as set forth in paragraph 1 above, the foregoing resolutions, to use all necessary means to uphold and implement Security Council resolution 660 (1990) and all subsequent relevant resolutions and to restore international peace and security in the area;

3. *Requests* all States to provide appropriate support for the actions undertaken in pursuance of paragraph 2 of this resolution;

4. *Requests* the States concerned to keep the Council regularly informed on the progress of actions undertaken pursuant to paragraphs 2 and 3 of this resolution;

5. *Decides* to remain seized of the matter.

In favour 12: Canada, Colombia, Côte d'Ivoire, Ethiopia, Finland, France, Malaysia, Romania, UK, USA, USSR, Zaire
Against 2: Cuba, Republic of Yemen
Abstaining 1: China

Appendix 18B. Chapter VII of the United Nations Charter

Chapter VII—Action with respect to threats to the peace, breaches of the peace, and acts of aggression

Article 39

The Security Council shall determine the existence of any threat to the peace, breach of the peace, or act of aggression and shall make recommendations, or decide what measures shall be taken in accordance with Articles 41 and 42, to maintain or restore international peace and security.

Article 40

In order to prevent an aggravation of the situation, the Security Council may, before making the recommendations or deciding upon the measures provided for in Article 39, call upon the parties concerned to comply with such provisional measures as it deems necessary or desirable. Such provisional measures shall be without prejudice to the rights, claims, or position of the parties concerned. The Security Council shall duly take account of failure to comply with such provisional measures.

Article 41

The Security Council may decide what measures not involving the use of armed force are to be employed to give effect to its decisions, and it may call upon the Members of the United Nations to apply such measures. These may include complete or partial interruption of economic relations and of rail, sea, air, postal, telegraphic, radio, and other means of communication, and the severance of diplomatic relations.

Article 42

Should the Security Council consider that measures provided for in Article 41 would be inadequate or have proved to be inadequate, it may take such action by air, sea, or land forces as may be necessary to maintain or restore international peace and security. Such action may include demonstrations, blockade, and other operations by air, sea, or land forces of Members of the United Nations.

Article 43

1. All Members of the United Nations, in order to contribute to the maintenance of international peace and security, undertake to make available to the Security Council, on its call and in accordance with a special agreement or agreements, armed forces, assistance, and facilities, including rights of passage, necessary for the purpose of maintaining international peace and security.
2. Such agreement or agreements shall govern the numbers and types of forces, their degree of readiness and general location, and the nature of the facilities and assistance to be provided.
3. The agreement or agreements shall be negotiated as soon as possible on the initiative of the Security Council. They shall be concluded between the Security Council and Members or between the Security Council and groups of Members and shall be subject to ratification by the signatory states in accordance with their respective constitutional processes.

Article 44

When the Security Council has decided to use force it shall, before calling upon a Member not represented on it to provide armed forces in fulfillment of the obligations assumed under Article 43, invite that Member, if the Member so desires, to participate in the decisions of the Security Council concerning the employment of contingents of that Member's armed forces.

Article 45

In order to enable the United Nations to take urgent military measures, Members shall hold immediately available national air-force contingents for combined international enforcement action. The strength and degree of readiness of these contingents and plans for their combined action shall be determined, within the limits laid down in the special agreement or agreements referred to in Article 43, by the Security Council with the assistance of the Military Staff Committee.

Article 46

Plans for the application of armed force shall be made by the Security Council with the assistance of the Military Staff Committee.

Article 47

1. There shall be established a Military Staff Committee to advise and assist the Security Council on all questions relating to the Security Council's military requirements for the maintenance of international peace and security, the employment and command of forces placed at its disposal, the regulation of armaments, and possible disarmament.

2. The Military Staff Committee shall consist of Chiefs of Staff of the permanent members of the Security Council or their representatives. Any Member of the United Nations not permanently represented on the Committee shall be invited by the Committee to be associated with it when the efficient discharge of the Committee's responsibilities requires the participation of that Member in its work.

3. The Military Staff Committee shall be responsible under the Security Council for the strategic direction of any armed forces placed at the disposal of the Security Council. Questions relating to the command of such forces shall be worked out subsequently.

4. The Military Staff Committee, with the authorization of the Security Council and after consultation with appropriate regional agencies, may establish regional subcommittees.

Article 48

1. The action required to carry out the decisions of the Security Council for the maintenance of international peace and security shall be taken by all the Members of the United Nations or by some of them, as the Security Council may determine.

2. Such decisions shall be carried out by the Members of the United Nations directly and through their action in the appropriate international agencies of which they are members.

Article 49

The Members of the United Nations shall join in affording mutual assistance in carrying out the measures decided upon by the Security Council.

Article 50

If preventive or enforcement measures against any state are taken by the Security Council, any other state, whether a Member of the United Nations or not, which finds itself confronted with special economic problems arising from the carrying out of those measures shall have the right to consult the Security Council with regard to a solution of those problems.

Article 51

Nothing in the present Charter shall impair the inherent right of individual or collective self-defense if an armed attack occurs against a Member of the United Nations, until the Security Council has taken measures necessary to maintain international peace and security. Measures taken by Members in the exercise of this right of self-defense shall be immediately reported to the Security Council and shall not in any way affect the authority and responsibility of the Security Council under the present Charter to take at any time such actions as it deems necessary in order to maintain or restore international peace and security.

19. Military mobilization in the Persian Gulf conflict

BARRY R. POSEN*

I. Introduction

Early in the morning of 2 August 1990 as many as 140 000 Iraqi soldiers crossed the border into Kuwait. They were led by units of the Republican Guard, reportedly two armoured divisions, a mechanized infantry division and a special forces division.[1] Regular Iraqi Army infantry units amounting to perhaps three divisions followed immediately behind.[2] Kuwait City was occupied within 12 hours. Although the Iraqi invasion appears to have been a surprise attack, immediate and subsequent accounts indicate that Kuwaiti air and ground units did resist, often intensively. With a total Kuwaiti ground force strength of 16 000, however, the outcome was inevitable. By the following day, 3 August, large Iraqi forces were moving towards the border with Saudi Arabia.

After meetings between Secretary of Defense Richard Cheney and King Fahd of Saudi Arabia on the evening of 6 August, the USA was invited to send forces to Saudi Arabia. The first units left the United States at 12:45 a.m. GMT on 7 August. In these consultations Secretary Cheney had assured King Fahd that the USA would move fast, 'deploy enough force to get the job

[1] Divisions are typically the major formations in any army that contain both a mix of most, if not all, weapons employed in ground warfare, as well as the command, engineering, supply and maintenance units necessary to employ that weaponry effectively. Divisions vary enormously both within and among armies. Iraqi divisions appear to be rather small, perhaps 9000–10 000 troops; US divisions are very large, 16 000–18 000 troops. Additionally, Western divisions usually contain three large subsidiary units called brigades, which can operate with considerable autonomy. Brigades usually control three to five battalions. Battalions range in size from 400 to 800 troops and are usually composed of one type of force: infantry, tank or artillery. Independent brigades are also often found in many armies. Some small armies consist only of brigades and do not organize division-sized formations. As an organizational form, the division came into widespread use during the Napoleonic wars.

[2] US Secretary of Defense Cheney gives the figure of 140 000 troops in *Statement before the Committee on Armed Services*, US House of Representatives, 14 Dec. 1990 (mimeograph), p. 1. For an estimate of the number of divisions involved, see Gordon, M., 'Iraq bolsters invasion force and moves on Saudi border', *New York Times*, 4 Aug. 1990, p. A4, which puts the number of troops at 100 000.

* This chapter was prepared with the assistance of the Massachusetts Institute of Technology Defense and Arms Control Studies Program, Conventional Forces Working Group: Nick Beldecos, Owen Cote, Eric Heginbotham, Eric Labs, Jonathan Ladinsky, Daniel Lindley, Brian Nichiporuk and Kevin Oliveau. Special thanks are due to Nick Beldecos and Laura Peters, who helped prepare the final draft. The working group sifted hundreds of sources in the preparation of the estimates contained in this chapter, and readers should understand that they *are* estimates. Space constraints prohibit a full accounting of all the sources employed. Those referenced proved particularly useful.

Figure 19.1. Map of the Persian Gulf region

done', stay as long as necessary, and leave when asked.[3] Almost immediately thereafter, Saudi Arabia also appealed for British assistance.[4] The initial deployment subsequently grew to the massive total multinational force of 680 000 men and women that ultimately, on 17 January 1991, launched Operation Desert Storm, the campaign to eject Iraq from Kuwait.[5] Although the effort was *launched* largely on the basis of bilateral understandings between the USA and Saudi Arabia, the huge military coalition that ultimately enforced the United Nations embargo, and then forcibly ejected Iraq from Kuwait, found its legitimacy and its explicit objectives in a dozen UN resolutions.[6] Total Iraqi strength in southern Iraq and Kuwait, referred to by

[3] Cheney (note 2), p. 4.

[4] 'Excerpts from news conference by Cheney and Powell at the Pentagon', *New York Times*, 9 Aug. 1990, p. A16.

[5] Queries addressed by the author to the Pentagon elicited the following list of 34 countries supporting Operation Desert Shield and Desert Storm: Afghanistan (Mujahideen), Argentina, Australia, Bahrain, Bangladesh, Belgium, Canada, China, Czechoslovakia, Denmark, Egypt, France, Germany, Greece, Hungary, Italy, Kuwait, Morocco, the Netherlands, New Zealand, Niger, Oman, Pakistan, Poland, Qatar, Saudi Arabia, Senegal, South Korea, Spain, Syria, the United Arab Emirates, Turkey, the UK and the USA. The precise nature of the contributions of some states is not given. Note that dates and times are according to Greenwich Mean Time; so, for example, while Operation Desert Storm started on 16 Jan. 1991 according to most US sources, it started on 17 Jan. according to GMT and in the time zones of the Middle East.

[6] For the UN Security Council resolutions on the embargo, see chapter 18 in this volume. The texts of all the UN Security Council resolutions relating to the Iraq–Kuwait conflict appear in appendix 18A.

US planners as the Kuwait Theatre of Operations (KTO), had reached roughly 540 000 by 16 January, with one-half of these troops stationed inside Kuwait.

This chapter describes the mobilization efforts of both sides from the outset of the crisis on 2 August 1990 to the eve of war on 16 January 1991. For three reasons the US mobilization effort is at the core of the discussion. First, the USA served as the military leader of the coalition. Second, the USA provided the bulk of the land, air and naval forces that participated in the multinational reinforcement effort. By 15 January, 53 per cent of the brigades, 80 per cent of the combat aircraft and roughly 60 per cent of the ships were from the US armed forces. The UK and France provided about 7 per cent of the brigades, with Arab countries providing 40 per cent. Saudi Arabia provided 13 per cent of the naval assets, with the other 27 per cent provided by 14 extra-regional powers. Saudi Arabia and Kuwait provided 13 per cent of the combat aircraft, with the UK, France, Canada and Italy providing a total of 7 per cent. Third, given the tremendous combat power of US ground force units, they were inevitably destined to do the bulk of the heavy work of engaging the most powerful Iraqi ground forces in the event of fighting.

The coalition reinforcement effort was designated Operation Desert Shield by the US military. From the point of view of intentions, the coalition buildup can be divided into two phases—defensive from 7 August to 7 November, and offensive thereafter.[7] For analytic purposes, however, it is useful further to divide the process into four periods, based on judgements of the military capability achieved by the coalition forces against Iraqi forces in the Kuwait Theatre of Operations at specified times. Inevitably there is some arbitrariness in the dates. Iraqi forces are discussed in reference to coalition intentions and capabilities, since it is somewhat difficult to tell what the Iraqi purposes were in any given phase of their buildup.[8]

The first period, 7–20 August, produced a 'deterrent' force that presented Iraq with the prospect of direct encounters with US ground forces if the Saudi border were crossed. The presence of non-Saudi Arab forces helped legitimate the US deployment in the Arab world. Substantial air capabilities were also achieved which would have delayed an Iraqi advance. Although there was some public confusion at the time, it is clear in retrospect that there was no intention to stop at this objective. *The second period, 21 August–30 September,* established a 'stalwart defensive' capability on the ground in Saudi Arabia and arguably an offensive capability in the air. The former was essential to give Saudi leaders the confidence to shut off Iraqi oil exports through their pipelines and keep them closed in the face of expected threats, thus supporting the UN embargo. *The third period, 1 October–7 November,* produced a ground 'counter-offensive' capability with the sizeable US armour

[7] Atkinson, R. and Woodward, B., 'The doctrine of invincible force', *Washington Post*, National Weekly Edition, 10–16 Dec. 1990, pp. 6–7.

[8] For order of battle estimates for Iraq and the coalition nations, it is useful to have at hand standard reference works such as: International Institute for Strategic Studies, *The Military Balance 1990–1991* (Brassey's: Oxford, 1990), and the *SIPRI Yearbooks*. Gazit, S. and Eytan, Z., *The Middle East Military Balance 1988–1989* (Jerusalem Post and Westview Press: Boulder, Colo., 1989) provides more detail on the regional actors.

and artillery reinforcements that reached the theatre.[9] *The fourth period, 8 November–16 January,* explicitly aimed at the development of an 'offensive option'. It was initiated by President George Bush's decision of 8 November to more than double the number of US forces in the Persian Gulf.

II. Phase I, period I (7–20 August): deterrence and delay

Following the Saudi decision on 6 August to invite US forces into the country, Secretary of Defense Cheney visited Egypt and Morocco to seek their co-operation in the enhancement of Saudi Arabia's defences. President Bush had apparently initiated this process in a number of telephone conversations with President Hosni Mubarak of Egypt and other Middle Eastern leaders. Saudi Arabia initiated its own appeals for regional assistance, which were formalized in an Arab League resolution passed in Cairo on the evening of 10 August. This resolution also called for Iraq's withdrawal from Kuwait and the support of Arab League members for Kuwaiti efforts to regain their country.[10] Iraqi forces had crossed the Saudi border during the early phases of the invasion of Kuwait, and diplomatic efforts to communicate with Iraq had been rebuffed, further increasing Saudi fears.[11]

The threat to Saudi Arabia was too great for the US or Saudi leadership to ignore. Land- and sea-based airpower was the main military asset that the USA could initially employ, but ground power was not ignored. US and Saudi leaders wanted to put at least some US ground forces close to the Saudi–Kuwaiti border very quickly—in part, it seems, simply to ensure that Iraq understood that an invasion meant war with the USA. However, US political and military leaders presumably would not have wanted to send token forces; rather, they would have wanted US forces to possess some initial capability to conduct a real military operation, if only a fighting withdrawal in the event Iraq attacked Saudi Arabia.

By 20 August, the USA had roughly one brigade positioned close to each of the two most obvious invasion routes into Saudi Arabia from Kuwait—the main roads crossing the border. US ground forces could have defended Saudi air fields from Iraqi commando attacks, and point air defences could have blunted any Iraqi bomber attacks that penetrated the fighter screen. Unaided, US ground units could not have stopped the Iraqi forces then in Kuwait from moving south. However, in combination with Saudi ground forces and with

[9] In the judgement of the author, the coalition then had at least a limited offensive capability that would have permitted ground operations to liberate southern Kuwait; see Posen, B. R., *Political Objectives and Military Options in the Persian Gulf,* Defense and Arms Control Studies Working Paper (Massachusetts Institute of Technology: Cambridge, Mass., 5 Nov. 1990).

[10] Kifner, J., 'Arabs vote to sent troops to help Saudis; boycott of Iraqi oil is reported near 100%', *New York Times,* 11 Aug. 1990, p. A1. The members of the League of Arab States, founded in 1945, are: Algeria, Bahrain, Djibouti, Egypt (membership suspended in 1979), Iraq, Jordan, Kuwait, Lebanon, Libya, Mauritania, Morocco, Oman, Palestine, Qatar, Saudi Arabia, Somalia, Sudan, Syria, Tunisia, the United Arab Emirates, and North and South Yemen (the Yemen states were united as the Republic of Yemen in May 1990).

[11] Miller, J., 'Saudis tell of Iraq hot-line drama', *New York Times,* 4 Oct. 1990, p. A15.

Table 19.1. Ground forces in the Kuwait Theatre of Operations, 20 August 1990

	Iraq	Coalition[a]
Personnel	230 000	75 000
Brigades	45	12
Tanks	1 500	300
Guns and MRLs[b]	1 000	200

[a] See the text below for national contributions.

[b] Multiple rocket launchers.

Sources: Berry, F. C., 'Massive airlift stabilized situation', *National Defense* (Dec. 1990), p. 7, indicates that from 7 Aug. to 6 Sep. more than 81 000 troops were moved by air to Saudi Arabia, which suggests about 40 000 in the first two weeks. This figure would easily have permitted the arrival of two airborne/air assault and 1 US Marine Corps brigade, totalling perhaps 15 000 troops. The USMC 7th Marine Expeditionary Brigade was reported to have unloaded its first equipment from Maritime Prepositioning Ships based at Diego Garcia on 15 Aug., although the same source reports that the brigade was not fully deployed until 28 Aug.; Naylor, S. D., 'Heavy going in the rush to the Gulf', *Army Times*, 10 Sep. 1990, p. 24. On Arab forces, see Naylor, S. D., 'Allies under the shield', *Army Times*, 22 Oct. 1990, pp. 14–15.

the support of the airpower then in the theatre, they might have delayed Iraqi forces, withdrawn in good order, and permitted devastating battlefield interdiction and close-air support operations by fighter aircraft. Whether this combination would have been sufficient to actually throw back an attack by Iraqi forces then in Kuwait is difficult to judge, although it appears plausible that such an attack could have been stalled or even stopped after gaining some ground. (The ground forces of Iraq and the coalition are given in table 19.1.)

Estimating the number of military personnel on both sides is problematic. For Iraq, personnel estimates were generated by the US intelligence community presumably on the basis of satellite imagery[12] and the electronic signatures of particular military units. These would have produced data on large Iraqi formations, such as divisions and brigades. Rules of thumb were then apparently applied to convert these units to personnel numbers.[13] The conversion factor seems to have varied between 15 000 and 20 000 per detected division. These figures roughly correspond to estimates of Soviet divisional 'slice' (division plus non-divisional support troops) employed by Western analysts for many years. Coalition personnel included four Saudi Army mechanized infantry brigades and two National Guard brigades, an Egyptian commando brigade, a Moroccan infantry battalion, the remnants of the Kuwaiti Army, a joint brigade of the Gulf Co-operation Council (GCC)[14]

[12] See chapter 3 in this volume.

[13] Smith, R. J., 'Iraqis fortify defenses in Kuwait', *Washington Post*, 21 Aug. 1990, p. 1, suggests 200 000 troops in the area, with 160 000 in Kuwait. See also Broder, J. and Healy, M., 'Iraq's Kuwait, Saudi border force grows', *Los Angeles Times*, 29 Aug. 1990.

[14] The Co-operation Council for the Arab States of the Gulf (GCC) was established in 1981 by Bahrain, Kuwait, Oman, Qatar, Saudi Arabia and the United Arab Emirates.

Table 19.2. Coalition air power in the Kuwait Theatre of Operations, 20 August 1990

Number	Type	Nation	Origin
Air Defence (AD)			
60	F-15C	Saudi Arabia	In place
24	Tornado	Saudi Arabia	In place
48	F-15C	USA	1 TFW, Langley AFB, Va.
12	Tornado	UK	RAF Coningsby, UK, on exercises at Akrotiri, Cyprus
All Weather Day/Night Ground Attack (AWD/N) & Heavy Bomber			
30	Tornado	Saudi Arabia	In place
14	F-111E	USA	20 TFW, RAF Upper Heyford, UK, on exercises at Incirlik, Turkey
22	F-117A	USA	37 TFW, Tonopah Test Range, Nev.
24	F-15E	USA	4 TFW, Seymour-Johnson AFB, N.C.
20	A6E	USA	3rd Marine Air Wing, El Toro, Calif., and 2nd Marine Air Wing, Cherry Point, N.C.
8	B-52G	USA	42 BW, Loring AFB, Maine
Multipurpose Fighter/Ground Attack (MPFGA)			
45	F-5E	Saudi Arabia	In place
20	A-4KU	Kuwait	Est. aircraft surviving Iraqi invasion
15	Mirage F-1	Kuwait	Est. aircraft surviving Iraqi invasion
44	F-16	USA	363 TFW, Shaw AFB, S.C.
48	F/A-18	USA	3rd Marine Air Wing, El Toro, Calif., and 2nd Marine Air Wing, Cherry Point, N.C.
12	Jaguar	UK	RAF Coltishall, UK
Close Air Support (CAS)			
24	Hawk	Saudi Arabia	In place
6	Hawk	Kuwait	Est. aircraft surviving Iraqi invasion
48	A-10	USA	354 TFW, Myrtle Beach AFB, S.C.
60	AV-8B	USA	3rd Marine Air Wing, El Toro, Calif., and 2nd Marine Air Wing, Cherry Point, N.C.
Suppression of Enemy Air Defences (SEAD), Early Warning (EW), C^3I, Reconnaissance (Recce)			
10	RF-5E	Saudi Arabia	In place
5	E-3 AWACS	Saudi Arabia	In place
5	E-3 AWACS	Saudi Arabia	52 AWACW, Tinker AFB, Okla.
12	F-4G	USA	35 TFW, George AFB, Calif.
6	EA-6B	USA	3rd Marine Air Wing, El Toro, Calif., and 2nd Marine Air Wing, Cherry Point, N.C.
4	Nimrod	UK	?

TFW = Tactical Fighter Wing
BW = Bomber Wing
AWACW = AWACS Wing
AFB = Air Force Base

Sources: On the Saudi air order of battle and initial reinforcements, see Janssen Lok, J., 'Defense of Saudi Arabia', *Jane's Defence Weekly*, 20 Oct. 1990, p. 757. On US Air Force deployments, see Ginovsky, J., 'Langley F-15s were the first to arrive', *Air Force Times*, 20 Aug. 1990, p. 56; Weber, S., 'Shaw deploys F-16s, support personnel', *Air Force Times*, 27 Aug. 1990, p. 12; Ginovsky, J., 'Seymour-Johnson F-15Es join Mideast force', *Air Force Times*, 27 Aug. 1990, p. 13; and on early British deployments, see 'US/UK deploy aircraft to Saudi Arabia', *Flight International*, 15–21 Aug. 1990, p. 5.

regularly based at King Khalid Military City near Hafr al Batin, two to three US airborne/air assault brigades, and one US Marine brigade.[15] It is important to note that most of this initial capability was provided by the US strategic airlift fleet, which flew more than 2000 missions in the first 30 days of Operation Desert Shield and delivered roughly 40 000 tonnes of cargo.[16]

Intelligence organizations presumably based Iraqi weapon estimates on the same technical means of intelligence gathering. After laundering to protect sources and methods, estimates were then revealed to the press. For coalition forces, standard Western tables of organization and equipment (TOE) are employed to estimate the number of weapons likely to have been associated with the units reported to have been present in the theatre. Thus, the entire Saudi order of battle[17] is not included in this estimate; instead, the forces that were plausibly available at the border are included. For example, both Saudi armoured brigades, normally in the north-western and south-western parts of the country, probably remained there to deal with possible Iraqi or Yemeni raids.

Press reports suggest that US Army ground forces were initially stationed near Hafr al Batin, south of the 'tri-border' area, and that the Marines were just north of the port of Jubail, on the coast road. These would have been the two easiest invasion routes because this is where the main roads enter Saudi Arabia from Kuwait. A US 'tripwire' was thereby established. Although one ought not to make too much of it, the coalition could have deployed a brigade every 15 km along the Saudi–Kuwaiti border. This is the extreme upper range of what has been considered an acceptable front for mechanized brigades in the European theatre. Thus, although many of the units then in Saudi Arabia were relatively lightly armed infantry formations, the front-line forces might have been able to fight a delaying action. Their heterogeneity and relative inexperience would have worked against them. However, considerable air power *was* already present to assist them.

As table 19.4 suggests, this air order of battle compares well with that of Iraq. The speedy US aerial reinforcement was aided by considerable stocks of *matériel* that were apparently prepositioned in Saudi Arabia in the mid-1980s, and by a Saudi air base structure which had been engineered with considerable excess capacity.[18] A good many of the Iraqi aircraft, in consequence of range or basing constraints, probably could not have been brought to bear on any attack on Saudi Arabia. It is unlikely that Iraq could have mustered more than about 240 deep-attack aircraft and 60 or so high-quality air-to-air fighters. These would have encountered nearly 150 of the finest air defence aircraft in the world, all operating under AWACS (airborne warning and control system) control and linked to a ground-based early-warning system as well. An undis-

[15] See the sources below table 19.1.

[16] Berry, F. C., 'Massive airlift stabilized situation', *National Defense* (Dec. 1990), p. 7, suggests 63 000 tonnes, but some of this cargo actually arrived by sea.

[17] 'Order of battle' refers to the identification, strength, command structure and disposition of the personnel, units and equipment of a military force.

[18] Fulghum, D., 'US airlift to Mideast is biggest ever mounted', *Aviation Week & Space Technology*, 20 Aug. 1990, pp. 18–21.

Table 19.3. Coalition naval strength, 20 August 1990

Nation	Force	Location
USA	Eisenhower Carrier Task Force[a]	Red Sea
	Independence Carrier Task Force	Arabian Sea
	Saratoga Carrier Task Force	Mediterranean Sea
	Cruisers *England* and *Antietam*	Persian Gulf
	5 frigates, 1 destroyer	Persian Gulf
UK	Destroyer *York*, 2 frigates	Persian Gulf
France	2 frigates	Persian Gulf
USSR	1 destroyer	Arabian Sea

[a] A typical single-carrier task force would contain an aircraft-carrier, two cruisers and four destroyers. Usually two such task forces would be supported by an underway replenishment group of four supply vessels, escorted by three frigates and a destroyer.

Source: Evans, M., 'Naval armada to enter record books', *London Times*, 18 Aug. 1990, diagram, pp. 2–3.

closed number of Patriot missiles were brought into Saudi Arabia early in the crisis to augment the already substantial Saudi surface-to-air missile inventory. The course of the air war after 17 January suggests that Iraq probably would not have been able to operate aircraft over Saudi Arabia with much success.

Iraqi ground forces would have been opposed by at least 440 land-based attack aircraft of all kinds; 118 were capable of night operations at considerable range, although they could perhaps fly only one sortie every 24 hours. The 322-odd day fighters could have flown two sorties per day, with US A-10s flying three or more. Television coverage of the war showed a great many attack helicopters in the theatre early in the deployment, but it is difficult to say how many there were. Thus, the total land-based air force could have generated nearly 800 sorties per day. Each one might have destroyed, on average, at least half an armoured vehicle. These attacks would have been augmented by missions flown from perhaps two or three carrier air wings, another 250 aircraft, although some would not have had the range to participate. Aircraft would have suffered daily attrition, but the USA could have easily replaced lost aircraft to keep the force at full strength. Indeed, the USA might have managed to bring in reinforcements during the battle. Thus, it is reasonable to conclude that the Iraqi Army might have suffered the loss of 400 armoured vehicles per day, roughly 10 per cent of their front-line strength in the KTO, if armoured personnel carriers and infantry fighting vehicles are included. This is a high daily loss rate. Given the distances in Saudi Arabia, had the ground forces been able to hold Iraqi forces to a 20-km-per-day rate of advance, the air forces would have had at least three days to attack Iraqi forces before they reached any truly valuable objectives—the first major off-shore oil field along the coast. This clearly would not have been enough time to annihilate the Iraqi force, but that force would have been in woeful shape.

These very simple calculations suggest that the coalition was reaching a level of strength that would have given an Iraqi military commander some pause, and as well suggest a degree of inevitable US involvement that might even have influenced President Saddam Hussein.[19]

Finally, the coalition began to assemble a naval force to police the UN embargo. The non-indigenous naval forces that seem to have been present by 20 August are given in table 19.3.

III. Phase I, period II (21 August–30 September): defensive, and period III (1 October–7 November): counter-offensive

As table 19.4 indicates, most of the coalition's initial air buildup was completed by the end of period II. The period II deployments were probably ample to fight and win a major aerial campaign against the entire Iraqi Air Force and air defence organization.[20] Thus, the coalition was already in an extremely strong position to wage an aerial offensive or counter-offensive against Iraq by the end of September. Theatre shortages of munitions and spare parts arising from the haste of the deployment might qualify this judgement. The principal distinction between periods II and III, then, lies in the area of ground forces—specifically coalition heavy ground forces. Another noteworthy aspect of this period is that Iraqi forces in the KTO rapidly built up through the end of September to a personnel strength estimated at 430 000, where they remained until the USA announced its offensive buildup on 8 November.

Coalition ground force personnel in period II include all Saudi forces, less 15 000 for three brigades on other fronts. The actual brigade count includes only 7 of the 10 brigades in the Saudi force structure. A plausible estimate for US forces is 95 000 personnel, including 13 Marine and Army brigades.[21] The US Army forces were under the control of the XVIII Airborne Corps, which had long been tasked with the rapid deployment mission for the US Central Command. It is apparent from the number of transport ships accumulated by the USA by 13 September that a major, sustained effort was planned. The US military had activated or already employed 127 sealift ships, including 53 roll-on/roll-off ships. Thirty-eight ships of the total were foreign

[19] Around 20 August Iraqi troops in Kuwait assumed a more defensive posture, and Republican Guard units involved in the invasion of Kuwait were redeployed to southern Iraq. Smith, R. J., 'Iraqis fortify defenses in Kuwait', *Washington Post*, 21 Aug. 1990, p. 1; and Brode, J. M. and Healy, M., 'Iraq's Kuwait, Saudi border force grows', *Los Angeles Times*, 29 Aug. 1990, p. 1.

[20] This argument is made in Posen (note 9).

[21] 'The balance of forces in the Gulf: in the region and en route', *Washington Post*, 24 Sep. 1990, p. 24, suggests 165 000 US military personnel in the region. In this chapter, estimates were made of those associated with Navy crews, and with USAF and USMC air units, and these were subtracted to arrive at the 95 000 figure. The article also suggests that most of the US 82nd Airborne, 101st Air Assault and 24th Mechanized Infantry Divisions were deployed, along with the 3rd Armored Cavalry Regiment, and the 7th, 1st and 4th Marine Expeditionary Brigades. On 17 Aug., the Civil Reserve Air Fleet was activated for the first time in its 38-year history, providing the military with 17 civilian passenger and 21 cargo planes; see Kitfield, J., 'The race by air', *Government Executive*, Nov. 1990, p. 20.

Table 19.4. Air orders of battle, fixed-wing aircraft, periods I–IV, 7 August 1990–16 January 1991

	Coalition[b]				
Forces[a]	Period: I	II[c]	III[c]	IV[c]	Iraq
Land-based					
AD	144	174	186	210	275
AWD/N	118	196	206	326	24
MPFGA	184	256	282	442	284
CAS	138	186	186	294+	60
SEAD/EW Recce/C^3I	42	84+	88+	118+	14
Subtotal	626	886+	948+	1 390+	657
Sea-based					
All types	240	240	320	480	*n.a.*
Total	**866**	**1 126+**	**1 268+**	**1 870+**	**657**

[a] For acronyms, see table 19.2.
[b] See the text for national contributions to coalition forces.
[c] A plus sign (+) indicates that aircraft in excess of the number shown are likely to have been deployed.

Sources:

Period I: See Berry, F. C., 'Massive airlift stabilized situation', *National Defense*, Dec. 1990, p. 7; and table 19.2.

Period II: Among numerous sources, see: Atkinson, R., 'US to rely on air strikes if air war erupts', *Washington Post*, 16 Sep. 1990, p. A36; on US Marine Corps aviation, see Morocco, J. D., 'USAF establishes command and control procedures, boosts support efforts', *Aviation Week & Space Technology*, 10 Sep. 1990, p. 29; Fulghum, D. A., 'Marine Corps completes Mideast deployment', *Aviation Week & Space Technology*, 17 Sep. 1990, p. 24.

Period III: French, Italian and Canadian fighter aircraft arrived; Janssen Lok, J., 'Defence of Saudi Arabia', *Jane's Defence Weekly*, 20 Oct. 1990; see also 'Air forces build-up continues in Gulf', *Flight International*, 26 Sep.–2 Oct. 1990, p. 8.

Period IV: Substantial redeployments of USAF units from Europe, including F-15s, F-16s, F-111s, F-4Gs and A-10s, were made in period IV; see Smith, R. F., 'US military cuts in Europe hastened by crisis over Iraq', *Washington Post*, 26 Dec. 1990, p. 32. Twenty-four F-16s that were deployed in Incirlik, Turkey, may be from a unit not accounted for in this tally. Another 48 aircraft were to be deployed in Incirlik in January; see 'More US aircraft for Turkey', *London Financial Times*, 16 Jan. 1991, p. 2. Whether any of these aircraft were redeployed in the Gulf is unclear.

For the Iraqi Air Force, see International Institute for Strategic Studies, *The Military Balance 1990–1991* (Brassey's: Oxford, 1990), p. 106. Aircraft are categorized by judgement using standard criteria.

Table 19.5. Ground forces in the Kuwait Theatre of Operations, periods II and III, 21 August–7 November 1990

	Coalition[a]		Iraq
	Period: II	III	II–III
Personnel	185 000	275 000	430 000
Brigades	26	31	75
Tanks	900	1 350	3 500
Guns and MRLs[b]	500	850	1 700–2 200

[a] See the text for national contributions to coalition forces.
[b] Multiple rocket launchers.

Sources: By late September, specific estimates of Iraqi strength were regularly provided to journalists by the Pentagon. Gordon, M., 'Pentagon, disputing Moscow, says 500 to 1000 Soviet advisers are in Iraq', *New York Times*, 26 Sep. 1990, p. 8, rates Iraqi strength in Kuwait and southern Iraq at 430 000 troops, 3500 tanks, 2500 armoured personnel carriers and 1700 artillery pieces. See also Healy, M., 'Cheney sees signs Hussein may lash out', *Los Angeles Times*, 27 Sep. 1990, p. 12. The estimated distribution of the forces has varied, although it appears that on the average only 50 per cent of the force has been in Kuwait.

flag charters.[22] Egypt is credited with one commando and two armoured brigades.[23] The rest of the forces include a Syrian infantry brigade, the Moroccan battalion mentioned above, remnants of the Kuwaiti Army and the Peninsula Shield Force of the Gulf Co-operation Council.

By the close of period III on 7 November, a new increment of offensive ground power arrived with the US 1st Cavalry Division, at least 350 M-1 tanks strong.[24] All three of the non-divisional artillery brigades of III Corps had also probably arrived, with a likely total strength of 72 155-mm howitzers, 72 203-mm howitzers, and 81 Multiple Launch Rocket System launcher vehicles.[25] US attack helicopters may have numbered as many as 250. Although Iraq was reported to have about 150 attack helicopters, most were not in the class of these US weapons, and it is doubtful that they were all deployed in Kuwait. Thus, coalition ground forces had a reach and a responsiveness that would have given them considerable advantages in armour

[22] Donovan, F. R., 'Surge and sustainment', *Sea Power*, Nov. 1990, pp. 39–45. By the end of Oct., these ships, plus the equivalent of about 90 US commercial liner sailings, had delivered nearly 1 million tons of dry cargo to the Middle East, according to one source. 'MSC ship to load up for 3rd trip to Mideast', *Journal of Commerce*, 24 Oct. 1990, p. 1B. (The article actually says '2 billion tons', but the figure is so preposterous that it is assumed to be a misprint and that it should have read 'pounds'.)

[23] Assembly of the Western European Union, *Consequences of the Invasion of Kuwait: Continuing Operations in the Gulf Region*, Document 1248, 7 Nov. 1990, pp. 19–20, 24. See also Naylor, S. D., 'Allies under the shield', *Army Times*, 22 Oct. 1990, pp. 14–15.

[24] Donnelly, T., 'Fort Hood's 1st Cavalry, 2nd Armored Divisions begin journey to the Gulf', *Army Times*, 17 Sep. 1990, p. 8. The 1st Cavalry Division normally has two active and one reserve brigade. For deployment to the Gulf, the 2nd Armored Division, then in the process of decommissioning, provided the 'round-out' brigade, and presumably much more. Indeed, the reinforced 1st Cavalry Division that was deployed was probably the most lavishly equipped armoured division ever fielded by the US Army.

[25] On current non-divisional artillery force structure, see Ingram, B. L. (Capt.), '3x8 and beyond: force structure changes for the field artillery of tomorrow', *Field Artillery*, Feb. 1989, pp. 19–23.

Table 19.6. Coalition naval forces, November 1990 and January 1991

	Mid-November 1990	Mid-January 1991
Argentina	2	2
Australia	3	3
Belgium	4	3
Canada	3	3
Denmark	1	1
France	11	9
Greece	1	1
Italy	4	4
Netherlands	3	3
Norway	1	1
Portugal	1	1
Saudi Arabia	25	25
Spain	4	4
UK	12	15
USA	60 (4 carriers)	111 (6 carriers)
USSR	2–4	2–4

Sources: The best single source is 'Aspin grades international performance in the Mideast', House Armed Services Committee News Release, 16 Nov. 1990, tables II and III. See also Assembly of the Western European Union, *Consequences of the Invasion of Kuwait*, Documents 1243, 1248 and 1248 Addendum; see also 'The Times Guide to the Gulf confrontation', *London Times*, 12 Nov. 1990, diagram, p. 10. Germany stationed five minesweepers in the Mediterranean but they are excluded from the table on the grounds that they could have made no conceivable contribution to the embargo or to military action. On the mid-January 1991 force, see 'The Times Guide to the Gulf confrontation', *London Times*, 16 Jan. 1991, diagram, p. 16. Naval deployments changed little from mid-Nov. 1990, with the obvious exception of those of the USA, which doubled its force.

battles with the Iraqi Army. The British Seventh Armoured Brigade with 114 Challenger tanks was lending much needed support to the US Marine Corps.[26] French 'light' forces had arrived in brigade strength, but still lacked tanks.[27] The allies had amassed a very powerful armoured striking force, backed by tremendous air power. An Iraqi attack at this point would likely have ended in disaster. Indeed, had the objectives been limited, the coalition force probably could have taken the offensive.[28]

The principal military action prior to 17 January was in the naval sphere. By 3 December, 13 000 challenges had been made to shipping in the Gulf area in support of the UN embargo, with 500 boardings. After the USA, France was the most active in this endeavour.[29]

[26] Naylor, S. D., 'New desert rats', *Army Times*, 22 Oct. 1990, p. 15.

[27] For a survey of the entire coalition ground order of battle then present in or destined for the theatre, see Assembly of the Western European Union (note 23), pp. 24–26.

[28] Posen (note 9).

[29] Assembly of the Western European Union (note 23), Document 1248 Addendum, 4 Dec. 1990, Appendix I, p. 26, lists the number of naval challenges made in support of the embargo and the identity of the challengers.

IV. Phase II, period IV (8 November–16 January): offensive

On 8 November 1990, President Bush announced his decision to deploy additional forces in Saudi Arabia. In the words of Secretary of Defense Cheney, the purpose of this buildup was to present Saddam Hussein 'with the prospect of a serious Iraqi defeat'.[30] General Colin Powell, Chairman of the Joint Chiefs of Staff, subsequently provided details regarding the forces sent.[31] These included the 1st and 3rd Armored Divisions, a forward brigade of the 2nd Armored Division, the 2nd Armored Cavalry Regiment, the HQ VII Corps, and numerous combat support and combat service support units, all from Germany.[32] The 1st Infantry Division (Mechanized), additional support units, roughly two-thirds of an additional Marine division and three carrier battle groups were sent from the USA. Fourteen fighter squadrons, two bomber squadrons, and 11 support squadrons of tankers and transports were also sent from various bases in Europe and the USA. The deployment would ultimately require the mobilization of 227 490 US National Guard and Reserve personnel. Although it is difficult to be precise, unit tables of organization and equipment suggest that the USA had perhaps 500 attack helicopters by 16 January 1991, or shortly thereafter. By 15 December 1990 an additional two dozen cargo ships had been added to the initial 123 accumulated at the outset of the crisis. Over 2.3 million measurement tons of equipment had either arrived in Saudi Arabia or were *en route*. Subsequent data show that nearly a quarter million additional personnel were associated with the development of the offensive option. The entire additional ground force dispatched had not arrived by 16 January, and indeed some small elements of the force were not even in place by early February 1991. However, most of the force does seem to have arrived by the time Operation Desert Storm was launched. Estimates of tank and gun numbers are based on inferences from standard tables of organization and equipment.[33]

Pentagon spokesman Pete Williams offered the estimates of Iraqi troops, tanks and guns shown in table 19.7.[34] Somewhat more than half of these were apparently deployed in Kuwait, with a line of as many as 15 infantry divisions (45 brigades) defending the coastline and the Saudi–Kuwaiti border.[35] Each defended perhaps 20 km of front. Three more divisions likely defended the Saudi–Iraqi border west of the Wadi al Batin, a dry river bed. Small reserves

[30] Cheney (note 2), p. 6.

[31] *Statement of General Colin L. Powell, US Chairman of the Joint Chiefs of Staff Before the Committee on Armed Services*, US House of Representatives on Operation Desert Shield, 14 Dec. 1990, (mimeograph), pp. 13–14.

[32] See also chapter 13 in this volume, particularly tables 13.14 and 13.15.

[33] See note 8.

[34] Baker, C., 'Against the wall', *Army Times*, 28 Jan. 1991, p. 24. In the 23 Jan. televised briefing from the Pentagon, Secretary of Defense Cheney offered the figure of 120 brigades, organized in 'more than 30 divisions'.

[35] In general, Iraqi Army and Republican Guard divisions seem to vary between two and four brigades in strength, with three the average. Many independent 'brigades' seem to exist, however. For a similar although not identical estimate of the number and distribution of Iraqi ground forces, see Gordon, M. R., 'Final Iraqi preparations indicate Hussein wants war, officials say', *New York Times*, 15 Jan. 1991, p. A1, and map on p. A10.

Table 19.7. Ground forces in the Kuwait Theatre of Operations, 16 January 1991

	USA	Other coalition[a]	Total coalition	Iraq
Personnel	300 000	140 000	440 000	545 000
Brigades	29	26	55	120
Tanks	2 000	1 350	3 350	4 200
Guns & MRLs	1 200	600	1 800	3 100

[a] 'Other coalition' forces include 10 Saudi brigades and five independent artillery battalions; a British armoured division of two brigades, a French light division credited with two brigades, two Egyptian armoured divisions and a commando unit, totalling seven brigades; a Syrian armoured division credited with two brigades; and Peninsular Shield Force, Kuwaiti and Pakistani forces credited with a brigade each. A certain amount of guesswork went into estimating their deployed tanks and guns. For example, as a standard, it was assumed that most non-US artillery battalions have 18 guns, except for British battalions, which are known to have 24.

Sources: On British, French, Egyptian and Syrian strength, see Western European Union, *The Gulf Crisis, Chronology of Events from 4 December 1990 to 15 January 1991*, 21 Jan. 1991. On 14 Dec. 1990, General Powell acknowledged many smaller contributions of troops or medical teams from Bangladesh, Morocco, Nepal, Niger, Senegal, Sierra Leone, Somalia and Zaire. He also noted assistance from Bulgaria, Czechoslovakia, Hungary, Poland and Romania. He put the present or planned contributions of Egypt and Syria at 50 000 and credited Saudi Arabia with 150 000 total troops; see *Statement of General Colin L. Powell, US Chairman of the Joint Chiefs of Staff Before the Committee on Armed Services*, US House of Representatives on Operation Desert Shield, 14 Dec. 1990, (mimeograph), pp. 13–14. This listing of ground forces as of 15 January gives the former a total of about 40 000, and the latter a total of about 40 000 since the purpose was to provide the best estimate of combat capability then available.

of mechanized forces were deployed immediately behind the front line, and somewhat larger reserves were said to be deployed deeper in Kuwait, for a total of perhaps six mechanized divisions (18 armoured and mechanized brigades). Three to six additional regular infantry divisions (9–18 brigades) were probably to be found between the Kuwaiti–Iraqi border and Basra.

Divisions of the Iraqi Republican Guard were reported to be deployed along the north-western Kuwaiti–Iraqi border, and in Basra. Considerable confusion has reigned regarding their true strength, but a reasonable number seems to be a maximum of six division-sized formations.[36] Three of them appear to be heavy units, controlling among them six or seven very well-equipped armoured and mechanized brigades and some infantry brigades.[37]

[36] Rottman, G. L., 'Saddam's juggernaut or armed horde?', *International Defense Review*, Nov. 1990, pp. 1240–42; *Winning in the Desert II*, Center for Army Lessons Learned (US Army Combined Arms Training Activity: Fort Leavenworth, Kan., Sep. 1990), pp. 31–33.

[37] See Jacobson, M. R., 'Armor in Desert Shield', *Infantry*, Nov.–Dec. 1990, pp. 32–37; and 'Iraqi infantry', *Infantry*, Jan.–Feb. 1991. The author suggests that each of these heavy divisions should have 3 brigades, but other sources suggest not more than about 6 heavy brigades in the Guard, along with 20-odd infantry and/or special forces brigades. See Pelletiere, S., Johnson, D. and Rosenberger, L., *Iraqi Power and US Security in the Middle East* (US Army War College: Carlisle, Penn., 1990), chap. 2, pp. 15–17.

These were on the Kuwaiti–Iraqi border. The other three were largely infantry formations. Some may have been close to Basra, although some Republican Guard units must have remained in Baghdad. As many as 20 infantry brigades were said to have been associated with the Republican Guard and were presumably distributed among all six divisional formations. However, some of these brigades were dubbed commando units, and were likely quite small.[38]

One of the most noteworthy features of the Iraqi occupation of Kuwait was the elaborate system of fortifications constructed. This effort began shortly after the invasion and apparently continued to the eve of Operation Desert Storm.[39] Some estimated the system at 175 km in length, extending from the sea along the Saudi–Kuwaiti border, and across the Wadi al Batin to cover a segment of the Iraqi–Saudi border. These fortifications, consisting of earth berms and bunkers, housed company and battalion strongpoints. Guarding the approaches to these positions was a network of obstacles, some 3000 metres deep, which included berms, ditches, razor wire and minefields.[40]

V. Conclusions

There are four main lessons to be drawn from this chronicle.

1. A very diverse collection of forces was mustered for this effort. Although many of the contributions were small, it is most unlikely that so many nations would have offered help had the operation not been organized under the auspices of the United Nations. The UN legitimated their individual contributions, even as each additional individual contribution added legitimacy to the entire coalition.

2. The special positions in international politics of Britain and France were undoubtedly reinforced. While many extra-regional powers contributed to the coalition effort, only these two states could join the USA in contributing the full panoply of air, ground and naval forces. This was as much a function of political will as military capability. Their contributions distinguish them from other middle powers whose ability to act in such serious crises is clearly much more circumscribed.

3. The bulk of the military contributions came from the USA, a superpower. One doubts that a military coalition of this size could have been assembled in the absence of superpower participation, indeed of US participation.

[38] A Republican Guard amphibious brigade is said to have participated in the initial phase of the invasion of Kuwait.

[39] Tyler, P. and Moore, M., 'Iraqis setting up strong defensive line, US says', *Washington Post*, 10 Aug. 1990, p. 1, quotes Pentagon spokesman Pete Williams on the construction of defensive earth barricades south-west of Kuwait City and tank revetments. By Oct., these had apparently grown into the elaborate prepared defences familiar to those who have followed news from the Gulf. Broder, J., 'Iraqi defenses take a cue from history', *Los Angeles Times*, 13 Oct. 1990, p. 8. For more details, see Jacobson, 'Iraqi infantry' (note 37). On the extension of these fortifications west of the Wadi al Batin, see Evans, M., 'Saddam's Maginot line can be broken', *London Times*, 24 Nov. 1990, p. 9.

[40] Baker (note 34), p. 24.

4. The combined size and speed of the US contribution deserve special comment. For nearly 15 years the USA worked on two military problems. The first was to reinforce its forces in Central Europe quickly and massively to help the NATO alliance thwart an attack by 100 Warsaw Pact mechanized divisions, expected to occur with less than 30 days' warning time. To accomplish this the USA invested heavily in airlift and prepositioning. Second, the USA and its allies also made elaborate plans to marshal shipping quickly, load that shipping with military equipment, and move much of that equipment to Europe where it would marry with US troops flown in by civilian airliners. The US military engaged in considerable planning and conducted numerous exercises in support of this mission.

Similarly, since the fall of the Shah and the Soviet invasion of Afghanistan in 1979, the US military has prepared to move very large forces to the Persian Gulf with great speed to thwart a Soviet overland invasion of Iran. Air bases were constructed in Saudi Arabia to facilitate this effort. Military equipment and supplies were prepositioned on ships at Diego Garcia and in the Pacific and the Atlantic. A great many exercises were conducted in support of this mission.

This was a $2.5 trillion effort,[41] and although the cold war had lost much of its military intensity by August 1990, very little of the capabilities so laboriously developed over the preceding 15 years had been run down. In addition, precisely because of the new US–Soviet *détente*, the USA was free to employ resources against Iraq that had long been intended for a conflict with the Soviet Union. The military circumstances could not have been less propitious for a challenge to US interests in the Persian Gulf. All of this should have been obvious to even a junior grade Iraqi military intelligence officer or a junior grade foreign ministry official. That none of these facts was sufficiently brought to the attention of President Saddam Hussein, or if they were, were ignored by him, must be counted as one of the great puzzles of this entire conflict.

[41] It is difficult to put a precise price tag on this buildup, but $2.5 trillion gives a rough order of magnitude. See also chapter 5 in this volume.

20. The SIPRI 1990 Olof Palme Memorial Lecture: 'Third World interests in the new era of East–West relations'

In October 1986, SIPRI's Governing Board decided to arrange an annual public lecture, named after the late Swedish Prime Minister Olof Palme. The lecture is to be delivered in Stockholm by a political leader of international stature or an eminent scholar in order to highlight the need for, and problems of, peace and security, in particular of arms control and disarmament. The lecture is also intended to draw attention to SIPRI's commitment to a future with fewer arms and more freedom. On 18 September 1987, Willy Brandt, former Chancellor of the Federal Republic of Germany, delivered the first annual Olof Palme Memorial Lecture. On 29 September 1988, Sergey F. Akhromeyev, Chief of General Staff, First Deputy Minister of Defence and Marshal of the Soviet Union, delivered the second lecture. On 26 September 1989, Victor F. Weisskopf, Professor Emeritus, Massachusetts Institute of Technology, USA, delivered the third lecture. On 10 September 1990, Oscar Arias Sánchez, former President of the Republic of Costa Rica and Nobel Peace Prize Laureate, delivered the fourth lecture.

OSCAR ARIAS SÁNCHEZ

There is not a single man or woman from the Third World who, upon arriving in Sweden, does not think of Olof Palme. I am from the South and feel very honoured to be participating, before so distinguished an audience, in the 1990 Olof Palme Memorial Lecture.

Olof Palme lives on in the hearts of many men and women of the Third World, that often forgotten, often neglected part of our planet. For us of the South, his continual support of our countries' struggles for dignity has permanently marked our hearts and minds. To us, his name stands for solidarity, and his memory, for the remembrance of persistent dedication to the poor.

Olof Palme was an enlightened optimist. His optimism was intimately linked to his faith in the potential of freedom and democracy; it was essentially connected to the possibilities of achieving the broadest social justice in a world threatened by political oppression, poverty, disease, war, and the great economic and social inequalities of underdevelopment.

Today, when democracy is becoming more generalized, when tyrannies are falling one by one, when the walls of misunderstanding and shame have been torn down by men and women hungry for freedom and weary of oppression, when yesterday's enemies join together to participate in the search for solutions to the problems they share, the ideas and purposes that Olof Palme left to the world have a chance to flourish.

From Olof Palme we learned that we must struggle continuously for co-operation between the world's peoples and for disarmament. Only thus can we make the next century the first of an era without violence, without injustice, without hunger and without oppression.

Development and disarmament were Palme's principal international concerns. He knew—with a vision shared only by great statesmen—that the economic prosperity of the industrialized nations was inexorably linked to the development of the poorest. He also knew that the security of the superpowers does not rest in the military advantages one bloc has over another, nor in the distribution of its arms, but in the satisfaction of the needs shared by both large and small countries.

Faith that the future will be different allows us to think that 'the discontent of rising expectations', the phrase that defined Palme's political thought, is a universal concept: at the crossroads of our time, we must choose the path taken by Palme or permit the earth to become increasingly divided between rich and poor nations. We must follow his footsteps or resign ourselves to an increasingly unsafe world.

We live on a planet that is becoming smaller—one in which countries, communities and individuals are becoming more and more interdependent. Today, every human being contributes to the fate of every other one of his fellow human beings. No leader of our era can ignore the obligation of knowing what is occurring in every part of the world, of knowing the thoughts, actions and aspirations of all cultures and of all nations.

Ours is a very complex and diverse world, in which, however, interdependence has led to an odd simplification: the great problems of our time are global, shared by all communities. Problems no longer exclusively affect one country or one region. Today, we are universally aware that the search for collective solutions must take place before we can confront our problems and threats. Never before has the human race been so convinced that it cannot delay the enforcement of a great pact of mutual defence comprising all peoples and nations.

The concept of mutual defence reminds us of the bitter aggression and hostility that continue to be a part of our daily lives. The threatening conflict in the Middle East, the internal war of El Salvador and the prolonged military conflicts of Afghanistan and Cambodia are a part of this reality. In the past, mutual defence was a pretext for war. In most modern languages, the word 'defence', a word indisputably linked to the term 'security', has always had a predominantly military connotation. A kind of accepted hypocrisy allows us to consider the institutions and resources that nations dedicate to aggression as institutions and resources of defence that guarantee security.

Let us speak, then, of national defence and security as the driving forces behind the most important events of our time. A dangerous mythology full of prejudices, of sectarianism, of wasted nationalism and of arrogance has been built around these terms. Under the pretext of guaranteeing national defence and security, the world has not only ignored the real natural threats against human life, but has created others that are equally or more complex and dangerous. Advances in science have allowed us to have a detailed and trustworthy vision of the geological history and biological limitations of our planet. This knowledge reveals that from nature itself emerge serious threats that endanger the human race and all of the animal and plant species accompanying us. Humanity's greatest efforts should be dedicated to guaranteeing defence and security before these threats.

Inequality, ignorance, disease, poverty, environmental decay: these are the enemies that require a strategy, a defence method, a security plan. Ironically, the concepts of defence and security that the leaders of the world have been proposing have not only made us neglect these enemies of humanity, but have produced more urgent ones: the arms race, militarization, poverty intensified by wasted resources, the aftermath of the oppression, destruction and death provoked by war.

On this occasion, I want to call on the memory of Olof Palme to reflect on the need for the leaders of industrialized and developing nations to come together to combat militarism and the buildup of arms, to establish the basis for a pact of security and mutual defence among all human beings.

Understandably, Europe's recent political experience has made us optimistic about disarmament. The existence and growth of that optimism can be found in the disappearance of some of the tensions that led to the escalated arms race in this part of the world. In light of the classic definitions of defence and security, there is no doubt that the European nations, members or not of the two most important military blocs, can now feel more secure and less threatened. The time has come to dismantle nuclear arsenals. The destruction of those arsenals is an immediate priority. Retaining them represents not only the enormous waste of material resources but also the continued threat of global destruction.

We must make every possible effort to persuade powers to destroy their nuclear arms and prevent others from manufacturing them. We should make that task a successful crusade before a major proliferation is made possible.

The progress observed until now in the process of *détente* allows us to foresee, for the near future, important reductions in the conventional arsenals of Europe and a significant decrease in armed forces. This prediction brings hope and satisfaction to every one of the world's inhabitants.

However, the struggle against the nuclear arms buildup and the advances in the conventional disarmament of Europe must be taken to their ultimate and logical conclusions. Today's events demonstrate that peace in Europe is not independent of events in the less developed countries of Africa, Asia and Latin America. Even if the world achieves the elimination of nuclear arms, and Europe continues to reduce its arms arsenals, peace in this continent will not be assured.

We must condemn the horror of the arms race without reservation. Nuclear, chemical, biological or conventional, the buildup of arms constitutes an aggression against life and against civilization: we condemn it in all of its forms. The potential horror of the atomic, biological and chemical arsenal is greater than any imaginable hell; we understand why its prevention is a serious concern and permanent task of Europe. But, to the millions of human beings living in the Third World, it is important for the peoples of the North to recognize that conventional arms have caused, and continue to cause, death and destruction throughout many regions of the world. The sale and use of conventional arms, manufactured in industrial countries, condemn millions of human beings to oppression, to poverty and to death.

Let us speak, then, of disarmament. Of disarmament in every sense. Let us speak of peace, of change and of development. Those are the benefits that we, the people of the Third World, are denied by the manufacture and sale of weapons. Let us speak today of the enormous profits obtained from the sale of arms—from the sale of death—by individuals and businesses. Let us speak of the social sacrifice that the purchase of these arms represents for the most impoverished peoples of the earth. Let us speak of the suffering that war adds to this sacrifice. Let us speak of the moral responsibility that all of that poverty and all of that pain gives to citizens, businessmen and leaders of those countries that base part of their prosperity on the business of death.

Let us express our fear of nuclear arms. Let us think of them when we speak of disarmament. Let us remember the holocausts of Hiroshima and Nagasaki, and let us determine to avoid their repetition. We should never forget the instant when, for the first time, a voracious flash produced the death or mutilation of hundreds of thousands of human beings. Let us continue to recall that painful memory so that our determination to eliminate the nuclear threat will prevail.

We must fear and condemn chemical weapons. Today, the possibility of entire populations dying beneath poisonous clouds, of children and old people, women and men, suffering the indiscriminate and heart-rending effects of chemical agents, is not so remote. Let us join our voices to the outcry for the dismantling of chemical arsenals and the permanent elimination of those substances whose mere possession makes men and nations the enemies of humanity.

We should condemn, then, the buildup of chemical, biological and nuclear arms. But let us not forget that the suffering and deaths that have taken place in Angola, Afghanistan, Viet Nam, Peru, Nicaragua and El Salvador have been caused by conventional weapons, primarily proceeding from industrialized nations. We should not forget that almost all of the countries of the Third World spend, in the purchase of these arms, many of the resources that should be dedicated to the betterment of their people's living conditions. Developing nations dedicate hundreds of times more resources to military expenditure than they invest in education and health.

A deep ethical contradiction between the industrialized countries' longing for peace and the prosperity that some of them achieve as producers and exporters of arms is evident. It is neither logical nor ethical that the material well-being of the wealthiest part of humanity depends on the actions of irresponsible Third World governments which, through military expenditures, submit their peoples to poverty and oppression, or subject them to the tragedies of war.

Those of us who have struggled for peace in the Third World, who armed only with the weapons of dialogue and persuasion have had to confront the obstinacy of the most war-loving spirits of the earth, know from personal experience what the trading of arms signifies. To us, it represents a business of death that reflects the hypocrisy and duplicity often abounding in international relations.

We have experienced the perplexity of dealing with individuals and states that declare war an unforgivable madness in international forums, but do not hesitate to supply the weapons that make that war possible. While proclaiming themselves the champions of peace and justice, states that manufacture and accumulate chemical weapons in great quantities have condemned a third party because it employed those kinds of weapons in a regional war and internal conflict. International opinion has deplored and condemned the use of chemical weapons against the Kurdish rebels. But we continue to ask ourselves whether that condemnation loses its moral base when some countries manufacture these weapons with the obvious purpose of eventually using them.

An enormous burden must weigh on the consciences of those nations and individuals that gather the dividends of war for their own material gain. My country has had a positive experience because of what we have called the dividends of peace. In 1948, under the inspiration and guidance of President José Figueres, the people of Costa Rica chose to demilitarize. Figueres suppressed the army, and his fellow citizens believed in the courage and viability of that decision. In 1949, that act became a constitutional principle, and now we and our children enjoy its fruits.

International development agencies recognize that Costa Rica today has a standard of living comparable to that of the industrialized countries. It is universally accepted that the extraordinary advances of my country in the fields of education, health, housing and social welfare are basically due to the fact that we do not dedicate our resources to the purchase of arms. The absence of an army has strengthened the Costa Rican democratic system, making it one of the most consolidated democracies of Latin America.

To us, these are the dividends of peace. These are the dividends that would be within the grasp of all Third World countries if they did not dedicate a very important part of their resources to the purchase of arms. These are the dividends that we hope will be obtained from an ethical re-statement of the problem of disarmament. As East–West tensions disappear, we hope that the industrialized nations of Asia, North America and Europe will dedicate more attention and resources to the resolution of the great inequalities that subsist in present North–South relations. We hope that the productive capacity of the war industry will not be maintained at the cost of the blood and well-being of our peoples. That would be neither just nor acceptable.

We are surprised and concerned by the complaints of certain sectors of the industrialized nations for whom the advent of peace seems to constitute a misfortune and not a blessing. Some argue that, with peace, disarmament will bring unemployment and poverty to many citizens of the industrialized world. The military–industrial complex is mobilizing to pressure governments against disarmament, claiming that they cannot condemn thousands of war industry workers to unemployment. We fear that powerful businesses manufacturing arms may attempt to delay disarmament and even encourage new wars.

National economies that to some degree depend on the sale of arms will have to undergo readjustment processes that will demand important short-term sacrifices. Their societies will have to subject themselves to limitations, but these will never be as serious or as painful as the ones that the societies of the Third World have suffered for decades. Poor nations have had to deter many of their countrymen from the labours of peace. That has been our sacrifice. Rich nations must be willing to make the sacrifice of deterring many of their citizens from the labours of war. They must face the challenge of turning swords into ploughshares: that must be their sacrifice.

The ingenuity and productive capacity that until now have been dedicated to destruction should be channelled towards the solution of the immense problems threatening our survival as a species: the environment, poverty and disease. This is the concept of security that must prevail as we approach the next century.

Unfortunately, what has occurred in the Middle East poses a serious threat to world peace. The leader of a totalitarian state has led his people into a war of aggression that has caused the intensification of the arms race in that region. In that escalation, the only winners will again be the dealers of death. What may emerge is the intensification of the arms race, extending from the Middle East to the rest of the Third World. Such a tragedy could eliminate all of our hopes for peace, justice, democracy and development.

We hope that all of the countries of the world will contribute to a negotiated resolution of this conflict. From personal experience, I know that dialogue and negotiation are the best tools for conflict resolution. In Central America, this belief made possible a process of pacification that involved leaders of differing ideologies. It was dialogue, and not the power of weapons, that permitted us to put an end to the war in Nicaragua and establish the basis for a future full of hope for its children and young people. Let us hope that other conflict regions of the world will be given a chance to build new paths for peace.

In our struggle for peace in Central America, I learned many things. I learned the value of humility. In conflicts, we tend to believe that the only correct perspective is our own. But that isn't necessarily true. We must be sufficiently humble to accept the possibility that we may be wrong, that other perspectives exist, and that other solutions may be more appropriate. We should listen carefully to the voices of those who are most directly affected by conflicts and work with them to achieve solutions instead of imposing them from the outside.

I also learned the value of prudence, of seeking results and not applause. May a decision never be taken just to please the gallery! But above all, in the search for peace, I learned the enormous value of patience. Before the brutal aggression of Iraq against Kuwait, the world was united in its rejection of force as a means to achieve political ends. The almost unanimous condemnation of the aggressor and his isolation are inspiring indications that the international community can constitute a powerful moral force.

This is the moment to moderate the postures and rhetoric of war. This is, for all of us, the moment to accept the moral responsibility of making peace our primary purpose.

In order to speak of disarmament and security in a language that is understandable from the different perspectives of the industrialized and developing nations, it is necessary for both parties to assume the shared responsibility taken by people who live in the same neighbourhood. Let us adopt, then, the idea that our planet has shrunk to the point that the peace of some is impossible if the peace of others is not guaranteed. Let us retake, then, the concept of security as one which demands redefinition.

In a short time we have seen, in America and Europe, the disappearance of many reasons for distrust and fear. Numerous obstacles that stood in the path towards peace and democratization have been removed. Many of the walls of misunderstanding have been torn down. Let us take advantage of the reason that seems to be reigning over international dialogue, and not allow the momentary shadows of the present to deprive us of the opportunity to renounce hate and construct a future without fear.

The developing world demands, now more than ever, the reform of the unjust structures that we have tolerated since the beginning of colonialism. We demand that the gap between our countries become smaller. Our impatience and our expectations grow. We implore change because we are certain that our demands are just. We want a world in which the war of the powerful does not have, as its stage and actors, the land and people of the Third World.

Let us redirect our thoughts to Olof Palme's ideals of solidarity. Let us recognize, once and for all, that the dignity and prosperity of the industrialized countries cannot be based on the suffering of the majority, of the poor of the world. Let us recognize, as Olof Palme did many years ago, that the well-being and security of every inhabitant of the industrialized world depend on the development of the Third World.

Annexes

Annexe A. Major multilateral arms control agreements

Annexe B. Chronology 1990

Annexe A. Major multilateral arms control agreements

RAGNHILD FERM

For the texts of the arms control agreements, see Goldblat, J., SIPRI, *Agreements for Arms Control: A Critical Survey* (Taylor & Francis: London, 1982); for the Treaty of Rarotonga, see SIPRI, *World Armaments and Disarmament: SIPRI Yearbook 1986* (Oxford University Press: Oxford, 1986), pp. 509–19.

I. Summaries of the agreements

Protocol for the prohibition of the use in war of asphyxiating, poisonous or other gases, and of bacteriological methods of warfare (Geneva Protocol)

Signed at Geneva on 17 June 1925; entered into force on 8 February 1928.

Declares that the parties agree to be bound by the above prohibition, which should be universally accepted as part of international law, binding alike the conscience and the practice of nations.

Antarctic Treaty

Signed at Washington on 1 December 1959; entered into force on 23 June 1961.

Declares the Antarctic an area to be used exclusively for peaceful purposes. Prohibits any measure of a military nature in the Antarctic, such as the establishment of military bases and fortifications, and the carrying out of military manœuvres or the testing of any type of weapon. Bans any nuclear explosion as well as the disposal of radioactive waste material in Antarctica, subject to possible future international agreements on these subjects. An international convention on the regulation of Antarctic mineral resource activities (which has not yet entered into force) was signed in Wellington, New Zealand, in 1988.

At regular intervals consultative meetings are convened to exchange information and hold consultations on matters pertaining to Antarctica, as well as to recommend to the governments measures in furtherance of the principles and objectives of the Treaty.

Treaty banning nuclear weapon tests in the atmosphere, in outer space and under water (Partial Test Ban Treaty—PTBT)

Signed at Moscow on 5 August 1963; entered into force on 10 October 1963.

Prohibits the carrying out of any nuclear weapon test explosion or any other nuclear explosion: (*a*) in the atmosphere, beyond its limits, including outer space, or under water, including territorial waters or high seas; (*b*) in any other environment if such explosion causes radioactive debris to be present outside the territorial limits of the state under whose jurisdiction or control the explosion is conducted.

Treaty on principles governing the activities of states in the exploration and use of outer space, including the moon and other celestial bodies (Outer Space Treaty)

Signed at London, Moscow and Washington on 27 January 1967; entered into force on 10 October 1967.

Prohibits the placing into orbit around the earth of any objects carrying nuclear weapons or any other kinds of weapons of mass destruction, the installation of such weapons on celestial bodies, or the stationing of them in outer space in any other manner. The establishment of military bases, installations and fortifications, the testing of any type of weapons and the conduct of military manœuvres on celestial bodies are also forbidden.

Treaty for the prohibition of nuclear weapons in Latin America (Treaty of Tlatelolco)

Signed at Mexico City on 14 February 1967; entered into force on 22 April 1968.

Prohibits the testing, use, manufacture, production or acquisition by any means, as well as the receipt, storage, installation, deployment and any form of possession of any nuclear weapons by Latin American countries.

The parties should conclude agreements with the IAEA for the application of safeguards to their nuclear activities.

Under *Additional Protocol I* the extra-continental or continental states which, *de jure* or *de facto*, are internationally responsible for territories lying within the limits of the geographical zone established by the Treaty (France, the Netherlands, the UK and the USA) undertake to apply the statute of military denuclearization, as defined in the Treaty, to such territories.

Under *Additional Protocol II* the nuclear weapon states undertake to respect the statute of military denuclearization of Latin America, as defined and delimited in the Treaty, and not to contribute to acts involving a violation of the Treaty, nor to use or threaten to use nuclear weapons against the parties to the Treaty.

Treaty on the non-proliferation of nuclear weapons (NPT)

Signed at London, Moscow and Washington on 1 July 1968; entered into force on 5 March 1970.

Prohibits the transfer by nuclear weapon states, to any recipient whatsoever, of nuclear weapons or other nuclear explosive devices or of control over them, as well as the assistance, encouragement or inducement of any non-nuclear weapon state to manufacture or otherwise acquire such weapons or devices. Prohibits the receipt by non-nuclear weapon states from any transferor whatsoever, as well as the manufacture or other acquisition by those states of nuclear weapons or other nuclear explosive devices.

Non-nuclear weapon states undertake to conclude safeguard agreements with the International Atomic Energy Agency (IAEA) with a view to preventing diversion of nuclear energy from peaceful uses to nuclear weapons or other nuclear explosive devices.

The parties undertake to facilitate the exchange of equipment, materials and scientific and technological information for the peaceful uses of nuclear energy and to ensure that potential benefits from peaceful applications of nuclear explosions will be made available to non-nuclear weapon parties to the Treaty. They also undertake to pursue negotiations in good faith on effective measures relating to cessation of the nuclear arms race at an early date and to nuclear disarmament, and on a treaty on general and complete disarmament.

Twenty-five years after the entry into force of the Treaty (1995), a conference shall be convened to decide whether the Treaty shall continue in force indefinitely or shall be extended for an additional fixed period or periods.

Treaty on the prohibition of the emplacement of nuclear weapons and other weapons of mass destruction on the sea-bed and the ocean floor and in the subsoil thereof (Sea-Bed Treaty)

Signed at London, Moscow and Washington on 11 February 1971; entered into force on 18 May 1972.

Prohibits emplanting or emplacing on the sea-bed and the ocean floor and in the subsoil thereof beyond the outer limit of a 12-mile sea-bed zone any nuclear weapons or any other types of weapons of mass destruction as well as structures, launching installations or any other facilities specifically designed for storing, testing or using such weapons.

Convention on the prohibition of the development, production and stockpiling of bacteriological (biological) and toxin weapons and on their destruction (BW Convention)

Signed at London, Moscow and Washington on 10 April 1972; entered into force on 26 March 1975.

Prohibits the development, production, stockpiling or acquisition by other means or retention of microbial or other biological agents, or toxins whatever their origin or method of production, of types and in quantities that have no justification of prophylactic, protective or other peaceful purposes, as well as weapons, equipment or means of delivery designed to use such agents or toxins for hostile purposes or in armed conflict. The destruction of the agents, toxins, weapons, equipment and means of delivery in the possession of the parties, or their diversion to peaceful purposes, should be effected not later than nine months after the entry into force of the Convention.

Convention on the prohibition of military or any other hostile use of environmental modification techniques (Enmod Convention)

Signed at Geneva on 18 May 1977; entered into force on 5 October 1978.

Prohibits military or any other hostile use of environmental modification techniques having widespread, long-lasting or severe effects as the means of destruction, damage or injury to states party to the Convention. The term 'environmental modification techniques' refers to any technique for changing—through the deliberate manipulation of natural processes—the dynamics, composition or structure of the Earth, including its biota, lithosphere, hydrosphere and atmosphere, or of outer space. The understandings reached during the negotiations, but not written into the Convention, define the terms 'widespread', 'long-lasting' and 'severe'.

Convention on the prohibitions or restrictions on the use of certain conventional weapons which may be deemed to be excessively injurious or to have indiscriminate effects ('Inhumane Weapons' Convention)

Signed at New York on 10 April 1981; entered into force on 2 December 1983.

The Convention is an 'umbrella treaty', under which specific agreements can be concluded in the form of protocols.

Protocol I prohibits the use of weapons intended to injure by fragments which are not detectable in the human body by X-rays.

Protocol II prohibits or restricts the use of mines, booby-traps and similar devices.

Protocol III prohibits or restricts the use of incendiary weapons.

South Pacific Nuclear Free Zone Treaty (Treaty of Rarotonga)

Signed at Rarotonga, Cook Islands, on 6 August 1985; entered into force on 11 December 1986.

Prohibits the manufacture or acquisition by other means of any nuclear explosive device, as well as possession or control over such device by the parties anywhere inside or outside the zone area described in an annex. The parties also undertake not to supply nuclear material or equipment unless subject to IAEA safeguards; and to prevent in their territories the stationing as well as the testing of any nuclear explosive device. Each party remains free to allow visits, as well as transit, by foreign ships and aircraft.

Under Protocol 1, France, the UK and the USA would undertake to apply the treaty prohibitions relating to the manufacture, stationing and testing of nuclear explosive devices in the territories situated within the zone, for which they are internationally responsible.

Under Protocol 2, China, France, the UK, the USA and the USSR would undertake not to use or threaten to use a nuclear explosive device against the parties to the treaty or against any territory within the zone for which a party to Protocol 1 is internationally responsible.

Under Protocol 3, China, France, the UK, the USA and the USSR would undertake not to test any nuclear explosive device anywhere within the zone.

II. Status of the implementation of the major multilateral arms control agreements, as of 1 January 1991

Number of parties

1925 Geneva Protocol	125	Sea-Bed Treaty	83
Antarctic Treaty	39	BW Convention	112
Partial Test Ban Treaty	118	Enmod Convention	54
Outer Space Treaty	92	'Inhumane Weapons' Convention	31
Treaty of Tlatelolco	23	Treaty of Rarotonga	11
Additional Protocol I	3	Protocol 1	0
Additional Protocol II	5	Protocol 2	2
Non-Proliferation Treaty	141	Protocol 3	2
NPT safeguards agreements (non-nuclear weapon states)	86		

Notes

1. The Federal Republic of Germany and the German Democratic Republic merged into one state in 1990. The dates of entry into force of the treaties listed in the table for the united Germany are the dates previously given for the FR Germany.

2. The Yemen Arab Republic and the People's Democratic Republic of Yemen merged into one state in 1990. All agreements which either state has entered into are in force for Yemen, according to a statement by the united Yemen state. Accordingly, SIPRI gives the dates of entry into force of the treaties listed in the table for Yemen as the earliest dates previously given for either of the former Yemen states.

3. The table records year of ratification, accession or succession.

4. The Partial Test Ban Treaty, the Outer Space Treaty, the Non-Proliferation Treaty, the Sea-Bed Treaty and the BW Convention provide for three depositaries—the governments of the UK, the USA and the USSR. The dates given for these agreements are the earliest date on which countries deposited their instruments of ratification, accession or succession—whether in London, Washington or Moscow. The dates given for the other agreements, for which there is only one depositary, are the dates of the deposit of the instruments of ratification, accession or succession with the depositary in question, except in the case of the 1925 Geneva Protocol, where the dates refer to the date of notification by the depositary.

5. The 1925 Geneva Protocol, the Partial Test Ban Treaty, the Outer Space Treaty, the Non-Proliferation Treaty, the Sea-Bed Treaty, the BW Convention, the Enmod Convention and the 'Inhumane Weapons' Convention are open to all states for signature.

The Antarctic Treaty is subject to ratification by the signatories and is open for accession by UN members or by other states invited to accede with the consent of all the contracting parties whose representatives are entitled to participate in the consultative meetings provided for in Article IX.

The Treaty of Tlatelolco is open for signature by all the Latin American republics; all other sovereign states situated in their entirety south of latitude 35° north in the western hemisphere; and (except for a political entity the territory of which is the subject of an international dispute) all such states which become sovereign, when they have been admitted by the General Conference; Additional Protocol I—by 'all extra-continental or continental states having *de jure* or *de facto* international responsibility for territories situated in the zone of application of the Treaty'; Additional Protocol II—by 'all powers possessing nuclear weapons', that is, the USA, the USSR, the UK, France and China.

The Treaty of Rarotonga is open for signature by members of the South Pacific Forum; Protocol 1—by France, the UK and the USA; Protocol 2—by France, China, the USSR, the UK and the USA; Protocol 3—by France, China, the USSR, the UK and the USA.

6. Key to abbreviations used in the table:

S	Signature without further action
PI, PII	Additional Protocols to the Treaty of Tlatelolco
P1, P2, P3	Protocols to the Treaty of Rarotonga
CP	Party entitled to participate in the consultative meetings provided for in Article IX of the Antarctic Treaty
SA	Nuclear safeguards agreement in force with the International Atomic Energy Agency as required by the Non-Proliferation Treaty or the Treaty of Tlatelolco, or concluded by a nuclear weapon state on a voluntary basis.

7. The footnotes are listed at the end of the table and are grouped separately under the heading for each agreement. The texts of the statements contained in the footnotes have been abridged, but the wording is close to the original version.

8. A complete list of UN member states and year of membership appears in section III.

State	Geneva Protocol	Antarctic Treaty	Partial Test Ban Treaty	Outer Space Treaty	Treaty of Tlatelolco	Non-Proliferation Treaty	Sea-Bed Treaty	BW Convention	Enmod Convention	'Inhumane Weapons' Convention	Treaty of Rarotonga
Afghanistan	1986		1964	1988		1970 SA	1971	1975	1985	S	
Albania	1989					1990					
Algeria			S								
Angola	1990[1]										
Antigua and Barbuda	1989[2]		1988[1]	1988[1]	1983[2]	1985[1]	1988[12]		1988[7]		
Argentina	1969	1961 CP	1986	1969	S[1]		1983[1]	1979	1987[1]	S	
Australia	1930[1]	1961 CP	1963	1967		1973 SA	1973	1977	1984	1983	1986
Austria	1928	1987	1964	1968		1969 SA	1972	1973[1]	1990[10]	1983	
Bahamas			1976[1]	1976[1]	1977[2]	1976[1]	1989	1986			
Bahrain	1988[1, 4]					1988[2]		1988[2]			

Bangladesh	1989[1]		1985	1986		1979 SA		1985	1979		
Barbados	1976[3]			1968	1969[2]	1980		1973			
Belgium	1928[1]	1960 CP	1966	1973		1975 SA	1972	1979	1982	S	
Belize						1985[1]		1986			
Benin	1986		1964	1986		1972	1986	1975	1986	1989[1]	
Bhutan	1979		1978			1985 SA		1978			
Bolivia	1985		1965	S	1969[2]	1970	S	1975	S		
Botswana			1968[1]	S		1969	1972	S			
Brazil	1970	1975 CP	1964	1969[2]	1968[3]		1988[2]	1973	1984		
Brunei Darussalam						1985 SA					
Bulgaria	1934[1]	1978	1963	1967		1969 SA	1971	1972	1978	1982	
Burkina Faso	1971		S	1968		1970					

State	Geneva Protocol	Antarctic Treaty	Partial Test Ban Treaty	Outer Space Treaty	Treaty of Tlatelolco	Non-Proliferation Treaty	Sea-Bed Treaty	BW Convention	Enmod Convention	'Inhumane Weapons' Convention	Treaty of Rarotonga
Burma see: Myanmar											
Burundi			S	S		1971	S	S			
Byelorussia	1970[5]		1963[2]	1967[3]			1971	1975	1978	1982	
Cambodia (Kampuchea)	1983[11]					1972	S	1983			
Cameroon	1989		S	S		1969	S				
Canada	1930[1]	1988	1964	1967		1969 SA	1972[3]	1972	1981	S	
Cape Verde			1979			1979	1979	1977	1979		
Central African Republic	1970		1964	S		1970	1981	S			
Chad			1965			1971					
Chile	1935[1]	1961 CP	1965	1981	1974[4]			1980			
China	1952[6]	1983 CP		1983	PII: 1974[5]			1984[3]		1982[2]	P2: 1989[1] P3: 1989[1]

Colombia		1989	1985	S	1972[2] SA	1986	S	1983			
Congo						1978	1978	1978			
Cook Islands											1985
Costa Rica			1967		1969[2] SA[16]	1970 SA	S	1973			
Côte d'Ivoire	1970		1965			1973 SA	1972	S			
Cuba	1966	1984		1977[4]			1977[4]	1976	1978	1987	
Cyprus	1966[2]		1965	1972		1970 SA	1971	1973	1978	1988[3]	
Czechoslovakia	1938[7]	1962	1963	1967		1969 SA	1972	1973	1978	1982	
Denmark	1930	1965	1964	1967		1969 SA	1971	1973	1978	1982	
Dominica					S	1984[1]					
Dominican Republic	1970		1964	1968	1968[2] SA[16]	1971 SA	1972	1973			

State	Geneva Protocol	Antarctic Treaty	Partial Test Ban Treaty	Outer Space Treaty	Treaty of Tlatelolco	Non-Proliferation Treaty	Sea-Bed Treaty	BW Convention	Enmod Convention	'Inhumane Weapons' Convention	Treaty of Rarotonga
Ecuador	1970	1987	1964	1969	1969[2] SA[16]	1969 SA		1975		1982	
Egypt	1928		1964	1967		1981[3] SA		S	1982	S	
El Salvador	S		1964	1969	1968[2] SA[16]	1972 SA		S			
Equatorial Guinea	1989		1989	1989		1984	S	1989			
Ethiopia	1935		S	S		1970 SA	1977	1975	S		
Fiji	1973[1, 2]		1972[1]	1972[1]		1972[1] SA		1973			1985
Finland	1929	1984 CP	1964	1967		1969 SA	1971	1974	1978	1982	
France	1926[1]	1960 CP		1970	PI: S[6] PII: 1974[7]	[4]		1984		1988[4]	
Gabon			1964			1974		S			
Gambia	1966[2]		1965[1]	S		1975 SA	S	S			

Germany*	**1929**	1979 CP	**1964**	1971		1975[5] SA	**1975**	1983[4]	**1983**	S	
Ghana	**1967**		**1963**	S		1970 SA	**1972**	1975	**1978**		
Greece	**1931**	1987	**1963**	1971		1970 SA	**1985**	1975	**1983**	S	
Grenada	**1989**[2]				**1975**[2]	1975[1]		1986			
Guatemala	**1983**		**1964**[3]		**1970**[2] **SA**[16]	1970 SA	**S**	1973	**1988**[2]	1983	
Guinea						1985	**S**				
Guinea-Bissau	**1989**		**1976**	1976		1976	**1976**	1976			
Guyana				S				S			
Haiti			**S**	S	**1969**[2]	1970		S			
Holy See (Vatican City)	**1966**			S		1971[6] SA			**S**		

* The Federal Republic of Germany and the German Democratic Republic merged into one state in 1990. The dates of entry into force of the treaties listed in this table for the united Germany are the dates previously given for FR Germany.

State	Geneva Protocol	Antarctic Treaty	Partial Test Ban Treaty	Outer Space Treaty	Treaty of Tlatelolco	Non-Proliferation Treaty	Sea-Bed Treaty	BW Convention	Enmod Convention	'Inhumane Weapons' Convention	Treaty of Rarotonga
Honduras			1964	S	1968[2] SA[16]	1973 SA	S	1979			
Hungary	1952	1984	1963	1967		1969 SA	1971	1972	1978	1982	
Iceland	1967		1964	1968		1969 SA	1972	1973	S	S	
India	1930[1]	1983 CP	1963	1982			1973[5]	1974[5]	1978	1984	
Indonesia	1971[2]		1964	S		1979[7] SA		S			
Iran	1929		1964	S		1970 SA	1971	1973	S		
Iraq	1931[1]		1964	1968		1969 SA	1972[4]	S	S		
Ireland	1930[8]		1963	1968		1968 SA	1971	1972[6]	1982	S	
Israel	1969[9]		1964	1977							

Italy	1928	1981 CP	1964	1972		1975[8] SA	1974[6]	1975	1981	S[5]	
Jamaica	1970[2]		S	1970	1969[2] SA[16]	1970 SA	1986	1975			
Japan	1970	1960 CP	1964	1967		1976[9] SA	1971	1982	1982	1982	
Jordan	1977[10]		1964	S		1970 SA	1971	1975			
Kampuchea see Cambodia											
Kenya	1970		1965	1984		1970		1976			
Kiribati						1985[1] SA					1986
Korea, Dem. People's Rep. of (North)	1988[1, 12]	1987				1985		1987	1984		
Korea, Republic of (South)	1988[1]	1986 CP	1964[3]	1967[4]		1975[10, 11] SA	1987	1987[7]	1986[3]		
Kuwait	1971[13]		1965[4]	1972[5]		1989[12]		1972[8]	1980[4]		

State	Geneva Protocol	Antarctic Treaty	Partial Test Ban Treaty	Outer Space Treaty	Treaty of Tlatelolco	Non-Proliferation Treaty	Sea-Bed Treaty	BW Convention	Enmod Convention	'Inhumane Weapons' Convention	Treaty of Rarotonga
Lao People's Dem. Republic	1989		1965	1972		1970	1971	1973	1978	1983	
Lebanon	1969		1965	1969		1970 SA	S	1975	S		
Lesotho	1972[2]			S		1970 SA	1973	1977			
Liberia	1927		1964			1970	S	S	S		
Libya	1971[14]		1968	1968		1975 SA	1990	1982			
Liechtenstein						1978[13] SA				1989	
Luxembourg	1936		1965	S		1975 SA	1982	1976	S	S	
Madagascar	1967		1965	1968[6]		1970 SA	S	S			
Malawi	1970		1964[1]			1986		S	1978		
Malaysia	1970		1964	S		1970 SA	1972	S			

Maldives	1966[2]					1970 SA					
Mali			S	1968		1970	S	S			
Malta	1964[2]		1964[1]			1970 SA	1971	1975			
Mauritania			1964								
Mauritius	1970[2]		1969[1]	1969[1]		1969 SA	1971	1972			
Mexico	1932		1963	1968	1967[2, 8] SA	1969[14] SA	1984[7]	1974[9]		1982	
Monaco	1967										
Mongolia	1968[15]		1963	1967		1969 SA	1971	1972	1978	1982	
Morocco	1970		1966	1967		1970 SA	1971	S	S	S	
Mozambique						1990					
Myanmar (formerly Burma)			1963	1970			S	S			

State	Geneva Protocol	Antarctic Treaty	Partial Test Ban Treaty	Outer Space Treaty	Treaty of Tlatelolco	Non-Proliferation Treaty	Sea-Bed Treaty	BW Convention	Enmod Convention	'Inhumane Weapons' Convention	Treaty of Rarotonga
Nauru						1982 SA					1987
Nepal	1969		1964	1967		1970 SA	1971	S			
Netherlands	1930[16]	1967 CP	1964	1969	PI: 1971[9] SA[17]	1975 SA	1976	1981	1983[5]	1987[6]	
New Zealand	1930[1]	1960 CP	1963	1968		1969 SA	1972	1972	1984[6]	S	1986
Nicaragua	1990		1965	S	1968[2, 10] SA[16]	1973 SA	1973	1975	S	S	
Niger	1967[2]		1964	1967			1971	1972			
Nigeria	1968[1]		1967	1967		1968 SA		1973		S	
Niue											1986
Norway	1932	1960 CP	1963	1969		1969 SA	1971	1973	1979	1983	
Pakistan	1960[2]		1988	1968				1974	1986	1985	

Panama	1970		1966	S	1971[2] SA	1977	1974	1974			
Papua New Guinea	1981[1, 2]	1981	1980[1]	1980[1]		1982 SA		1980	1980		1989
Paraguay	1933[17]		S		1969[2] SA[16]	1970 SA	S	1976			
Peru	1985	1981 CP	1964	1979	1969[2] SA[16]	1970 SA		1985			
Philippines	1973		1965[3]	S		1972 SA		1973		S	
Poland	1929	1961 CP	1963	1968		1969 SA	1971	1973	1978	1983	
Portugal	1930[1]		S			1977 SA	1975	1975	S	S	
Qatar	1976					1989	1974	1975			
Romania	1929[1]	1971[1]	1963	1968		1970 SA	1972	1979	1983	S[7]	
Rwanda	1964[2]		1963	S		1975	1975	1975			

State	Geneva Protocol	Antarctic Treaty	Partial Test Ban Treaty	Outer Space Treaty	Treaty of Tlatelolco	Non-Proliferation Treaty	Sea-Bed Treaty	BW Convention	Enmod Convention	'Inhumane Weapons' Convention	Treaty of Rarotonga
Saint Christopher and Nevis	1989[2]										
Saint Lucia	1988					1979[1] SA		1986[10]			
Saint Vincent and the Grenadines						1984[1]					
Samoa, Western			1965			1975 SA					1986
San Marino			1964	1968		1970[10]		1975			
Sao Tome and Principe						1983	1979	1979	1979		
Saudi Arabia	1971			1976		1988	1972	1972			
Senegal	1977		1964			1970 SA	S	1975			
Seychelles			1985	1978		1985	1985	1979			
Sierra Leone	1967		1964	1967		1975	S	1976	S	S	

Singapore			1968[1]	1976		1976 SA	1976	1975			
Solomon Islands	1981[2]					1981[1]	1981[12]	1981[10]	1981[7]		1989
Somalia			S	S		1970		S			
South Africa	1930[1]	1960 CP	1963	1968			1973	1975			
Spain	1929[18]	1982 CP	1964	1968		1987 SA	1987	1979	1978	S	
Sri Lanka	1954		1964	1986		1979 SA		1986	1978		
Sudan	1980		1966			1973 SA	S			S	
Suriname					1977[2] SA[16]	1976[1] SA					
Swaziland			1969			1969 SA	1971				
Sweden	1930	1984 CP	1963	1967		1970 SA	1972	1976	1984	1982	

State	Geneva Protocol	Antarctic Treaty	Partial Test Ban Treaty	Outer Space Treaty	Treaty of Tlatelolco	Non-Proliferation Treaty	Sea-Bed Treaty	BW Convention	Enmod Convention	'Inhumane Weapons' Convention	Treaty of Rarotonga
Switzerland	1932	1990	1964	1969		1977[13] SA	1976	1976[11]	1988[8]	1982	
Syria	1968[19]		1964	1968[7]		1969[10]		S	S		
Taiwan	1929[20]		1964	1970[8]		1970	1972[8]	1973[12]			
Tanzania	1963		1964				S	S			
Thailand	1931		1963	1968		1972 SA		1975			
Togo	1971		1964	1989		1970	1971	1976		S	
Tonga	1971[2]		1971[1]	1971[1]		1971[1]		1976			
Trinidad and Tobago	1962[2]		1964	S	1970[2]	1986					
Tunisia	1967		1965	1968		1970 SA	1971	1973	1978	1987	
Turkey	1929		1965	1968		1980[15] SA	1972	1974	S[9]	S	
Tuvalu						1979[1]					1986

Uganda	1965		1964	1968		1982			S		
UK	1930[1]	1960 CP	1963[5]	1967	PI: 1969[11] PII: 1969[11]	1968[16] SA[17]	1972[9]	1975[13]	1978	S	
Ukraine			1963[2]	1967[3]			1971	1975	1978	1982	
United Arab Emirates								S			
Uruguay	1977	1980[2] CP	1969	1970	1968[2] SA[16]	1970 SA	S	1981			
USA	1975[21]	1960 CP	1963	1967	PI: 1981[12] PII: 1971[13] SA[17]	1970 SA[18]	1972	1975	1980	S[8]	
USSR	1928[22]	1960 CP	1963	1967	PII: 1979[14]	1970 SA[19]	1972	1975	1978	1982	P2: 1988[2] P3: 1988[2]
Venezuela	1928		1965	1970	1970[2, 15] SA[16]	1975 SA		1978			
Viet Nam	1980[1]			1980		1982 SA	1980[10]	1980	1980	S	

State	Geneva Protocol	Antarctic Treaty	Partial Test Ban Treaty	Outer Space Treaty	Treaty of Tlatelolco	Non-Proliferation Treaty	Sea-Bed Treaty	BW Convention	Enmod Convention	'Inhumane Weapons' Convention	Treaty of Rarotonga
Yemen*	1971[23]		1979	1979		1979	1979	1979	1977		
Yugoslavia	1929[24]		1964	S		1970[20] SA	1973[11]	1973		1983	
Zaire			1965	S		1970 SA		1977	S		
Zambia			1865[1]	1973			1972				
Zimbabwe								1990			

* The Yemen Arab Republic and the People's Democratic Republic of Yemen merged into one state in 1990. All agreements which either state has entered into are in force for Yemen, according to a statement by the united Yemen state. Accordingly, SIPRI gives the dates of entry into force of the treaties listed in the table for Yemen as the earliest dates previously given for either of the former Yemen states.

The 1925 Geneva Protocol

[1] The Protocol is binding on this state only as regards states which have signed and ratified or acceded to it. The Protocol will cease to be binding on this state in regard to any enemy state whose armed forces or whose allies fail to respect the prohibitions laid down in it.
Australia withdrew its reservation in 1986, New Zealand in 1989.

[2] Notification of succession.

[3] In notifying its succession to the obligations contracted in 1930 by the UK, Barbados stated that as far as it was concerned the reservation made by the UK was to be considered as withdrawn.

[4] The accession of Bahrain to the Protocol shall in no way constitute recognition of Israel or be a cause for the establishment of any relations with it.

[5] In a note of 2 Mar. 1970, submitted at the UN, Byelorussia stated that 'it recognizes itself to be a party' to the Protocol. However, it has not notified the depositary.

[6] On 13 July 1952 the People's Republic of China issued a statement recognizing as binding upon it the 1929 accession to the Protocol in the name of China. China considers itself bound by the Protocol on condition of reciprocity on the part of all the other contracting and acceding powers.

[7] Czechoslovakia shall cease to be bound by this Protocol towards any state whose armed forces, or the armed forces of whose allies, fail to respect the prohibitions laid down in the Protocol. This reservation was withdrawn in 1990.

[8] Ireland does not intend to assume, by this accession, any obligation except towards the states having signed and ratified this Protocol or which shall have finally acceded thereto, and should the armed forces or the allies of an enemy state fail to respect the Protocol, the government of Ireland would cease to be bound by the said Protocol in regard to such state. In 1972, Ireland declared that it had decided to withdraw the above reservations made at the time of accession to the Protocol.

[9] The Protocol is binding on Israel only as regards states which have signed and ratified or acceded to it. The Protocol shall cease to be binding on Israel as regards any enemy state whose armed forces, or the armed forces of whose allies, or the regular or irregular forces, or groups or individuals operating from its territory, fail to respect the prohibitions which are the object of the Protocol.

[10] The accession by Jordan to the Protocol does not in any way imply recognition of Israel. Jordan undertakes to respect the obligations contained in the Protocol with regard to states which have undertaken similar commitments. It is not bound by the Protocol as regards states whose armed forces, regular or irregular, do not respect the provisions of the Protocol.

[11] The accession was made on behalf of the Coalition Government of Democratic Kampuchea (the government in exile), with a statement that the Protocol will cease to be binding on it in regard to any enemy state whose armed forces or whose allies fail to respect the prohibitions laid down in the Protocol. France declared that as a party to the Geneva Protocol (but not as the depositary) it considers this accession to have no effect. A similar statement was made by Austria, Bulgaria, Cuba, Czechoslovakia, GDR, Hungary, Mauritius, Netherlands, Poland, Romania, USSR and Viet Nam, which do not recognize the Coalition Government of Kampuchea, which in Feb. 1990 changed its official name to the National Government of Cambodia.

[12] The Dem. People's Rep. of Korea states that it will not exclude the right to exercise its sovereignty *vis-à-vis* the other contracting party which violates the Protocol in its implementation.

[13] The accession of Kuwait to the Protocol does not in any way imply recognition of Israel or the establishment of relations with the latter on the basis of the present Protocol. In case of breach of the prohibition laid down in this Protocol by any of the parties, Kuwait will not be bound, with regard to the party committing the breach, to apply the provisions of this Protocol.

[14] The accession to the Protocol does not imply recognition of Israel. The Protocol is binding on Libya only as regards states which are effectively bound by it and will cease to be binding on Libya as regards states whose armed forces, or the armed forces of whose allies, fail to respect the prohibitions which are the object of this Protocol.

[15] In the case of violation of this prohibition by any state in relation to Mongolia or its allies, Mongolia shall not consider itself bound by the obligations of the Protocol towards that state. This reservation was withdrawn in 1990.

[16] As regards the use in war of asphyxiating, poisonous or other gases and of all analogous liquids, materials or devices, this Protocol shall cease to be binding on the Netherlands with regard to any enemy state whose armed forces or whose allies fail to respect the prohibitions laid down in the Protocol.

[17] This is the date of receipt of Paraguay's instrument of accession. The date of the notification by the depositary government 'for the purpose of regularization' is 1969.

[18] Spain declared the Protocol as binding *ipso facto*, without special agreement with respect to any other member or state accepting and observing the same obligation, that is, on condition of reciprocity.

[19] The accession by Syria to the Protocol does not in any case imply recognition of Israel or lead to the establishment of relations with the latter concerning the provisions laid down in the Protocol.

[20] The Protocol, signed in 1929 in the name of China, is taken to be valid for Taiwan (the Republic of China, which is part of the People's Republic of China.) However, unlike the People's Republic of China, Taiwan has not reconfirmed its accession to the Protocol.

[21] The Protocol shall cease to be binding on the USA with respect to use in war of asphyxiating, poisonous or other gases, and of all analogous liquids, materials, or devices, in regard to an enemy state if such state or any of its allies fail to respect the prohibitions laid down in the Protocol.

[22] The Protocol only binds the USSR in relation to the states which have signed and ratified or which have definitely acceded to the Protocol. The Protocol shall cease to be binding on the USSR in regard to any enemy state whose armed forces or whose allies *de jure* or *de facto* do not respect the prohibitions which are the object of this Protocol.

[23] In case any party fails to observe the prohibition under the Protocol, the People's Democratic Republic of Yemen will consider itself free of its obligation. *Note:* SIPRI considers this reservation to be valid for the united state of Yemen until further notification is made.

[24] The Protocol shall cease to be binding on Yugoslavia in regard to any enemy state whose armed forces or whose allies fail to respect the prohibitions which are the object of the Protocol.

The Antarctic Treaty

[1] Romania stated that the provisions of Article XIII, paragraph 1 of the Treaty were not in accordance with the principle according to which multilateral treaties whose object and purposes concern the international community, as a whole, should be open for universal participation.

[2] In acceding to the Treaty, Uruguay proposed the establishment of a general and definitive statute on Antarctica in which the interests of all states involved and of the international community as a whole would be considered equitably. It also declared that it reserved its rights in Antarctica in accordance with international law.

The Partial Test Ban Treaty

[1] Notification of succession.

[2] The USA considers that Byelorussia and Ukraine are already covered by the signature and ratification by the USSR.

[3] With a statement that this does not imply the recognition of any territory or regime not recognized by this state.

[4] Kuwait stated that its signature and ratification of the Treaty do not in any way imply its recognition of Israel nor oblige it to apply the provisions of the Treaty in respect of the said country.

[5] The UK stated its view that if a regime is not recognized as the government of a state, neither signature nor the deposit of any instrument by it, nor notification of any of those acts, will bring about recognition of that regime by any other state.

The Outer Space Treaty

[1] Notification of succession.

[2] Brazil interprets Article X of the Treaty as a specific recognition that the granting of tracking facilities by the parties to the Treaty shall be subject to agreement between the states concerned.

[3] The USA considers that Byelorussia and Ukraine are already covered by the signature and ratification by the USSR.

[4] This does not imply the recognition of any territory or regime not recognized by this state.

[5] Kuwait acceded to the Treaty with the understanding that this does not in any way imply its recognition of Israel and does not oblige it to apply the provisions of the Treaty in respect of the said country.

[6] Madagascar acceded to the Treaty with the understanding that under Article X of the Treaty the state shall retain its freedom of decision with respect to the possible installation of foreign observation bases in its territory and shall continue to possess the right to fix, in each case, the conditions for such installation.

[7] Syria acceded to the Treaty with the understanding that this should not mean in any way the recognition of Israel, nor should it lead to any relationship with Israel that could arise from the Treaty.

[8] China declared as illegal and null and void the signature and ratification of the Outer Space Treaty by the Taiwan authorities.

The Treaty of Tlatelolco

[1] On signing the Treaty, Argentina stated that it understands Article 18 as recognizing the rights of parties to carry out, by their own means or in association with third parties, explosions of nuclear devices

for peaceful purposes, including explosions which involve devices similar to those used in nuclear weapons.

[2] The Treaty is in force for this country due to a declaration, annexed to the instrument of ratification in accordance with Article 28, paragraph 2, which waived the requirements for the entry into force of the Treaty, specified in paragraph 1 of that Article: namely, that all states in the region deposit the instruments of ratification; that Protocol I and Protocol II be signed and ratified by those states to which they apply; and that agreements on safeguards be concluded with the IAEA. (Colombia made this declaration subsequent to the deposit of ratification, as did Nicaragua and Trinidad and Tobago.)

[3] On signing the Treaty, Brazil stated that, according to its interpretation, Article 18 of the Treaty gives the signatories the right to carry out, by their own means or in association with third parties, nuclear explosions for peaceful purposes, including explosions which involve devices similar to those used in nuclear weapons. This statement was reiterated at the ratification. Brazil also stated that it did not waive the requirements for the entry into force of the Treaty laid down in Article 28. *The Treaty is therefore not yet in force for Brazil.*

[4] Chile has not waived the requirements for the entry into force of the Treaty laid down in Article 28. *The Treaty is therefore not yet in force for Chile.*

[5] On signing Protocol II, China stated, *inter alia*: China will never use or threaten to use nuclear weapons against non-nuclear Latin American countries and the Latin American nuclear weapon-free zone; nor will China test, manufacture, produce, stockpile, install or deploy nuclear weapons in these countries or in this zone, or send its means of transportation and delivery carrying nuclear weapons to cross the territory, territorial sea or airspace of Latin American countries. The signing of the Protocol does not imply any change whatsoever in China's stand on the disarmament and nuclear weapons issue and, in particular, does not affect its stand against the Non-Proliferation Treaty and the Partial Test Ban Treaty.

China holds that, in order that Latin America may truly become a nuclear weapon-free zone, all nuclear countries, and particularly the superpowers, must undertake not to use or threaten to use nuclear weapons against the Latin American countries and the Latin American nuclear weapon-free zone, and implement the following undertakings: (1) dismantle all foreign military bases in Latin America and refrain from establishing new bases there, and (2) prohibit the passage of any means of transportation and delivery carrying nuclear weapons through Latin American territory, territorial sea or airspace.

[6] On signing Protocol I, France made the following reservations and interpretative statements: The Protocol, as well as the provisions of the Treaty to which it refers, will not affect the right of self-defence under Article 51 of the UN Charter; the application of the legislation referred to in Article 3 of the Treaty relates to legislation which is consistent with international law; the obligations under the Protocol shall not apply to transit across the territories of the French Republic situated in the zone of the Treaty, and destined to other territories of the French Republic; the Protocol shall not limit, in any way, the participation of the populations of the French territories in the activities mentioned in Article 1 of the Treaty, and in efforts connected with the national defence of France; the provisions of Articles 1 and 2 of the Protocol apply to the text of the Treaty as it stands at the time when the Protocol is signed by France, and consequently no amendment to the Treaty that might come into force under Article 29 thereof would be binding on the government of France without the latter's express consent.

[7] On signing Protocol II, France stated that it interprets the undertaking contained in Article 3 of the Protocol to mean that it presents no obstacle to the full exercise of the right of self-defence enshrined in Article 51 of the UN Charter; it takes note of the interpretation of the Treaty given by the Preparatory Commission for the Denuclearization of Latin America and reproduced in the Final Act, according to which the Treaty does not apply to transit, the granting or denying of which lies within the exclusive competence of each state party in accordance with the pertinent principles and rules of international law; it considers that the application of the legislation referred to in Article 3 of the Treaty relates to legislation which is consistent with international law. The provisions of Articles 1 and 2 of the Protocol apply to the text of the Treaty as it stands at the time when the Protocol is signed by France. Consequently, no amendment to the Treaty that might come into force under the provision of Article 29 would be binding on the government of France without the latter's express consent. If this declaration of interpretation is contested in part or in whole by one or more contracting parties to the Treaty or to Protocol II, these instruments would be null and void as far as relations between France and the contesting state or states are concerned. On depositing its instrument of ratification of Protocol II, France stated that it did so subject to the statement made on signing the Protocol. On 15 Apr. 1974, France made a supplementary statement to the effect that it was prepared to consider its obligations under Protocol II as applying not only to the signatories of the Treaty, but also to the territories for which the statute of denuclearization was in force in conformity with Article 1 of Protocol I.

[8] On signing the Treaty, Mexico said that if technological progress makes it possible to differentiate between nuclear weapons and nuclear devices for peaceful purposes, it will be necessary to amend the relevant provisions of the Treaty, according to the procedures established therein.

[9] The Netherlands stated that Protocol I shall not be interpreted as prejudicing the position of the Netherlands as regards its recognition or non-recognition of the rights or of claims to sovereignty of the parties to the Treaty, or of the grounds on which such claims are made.

[10] Nicaragua stated that it reserved the right to use nuclear energy for peaceful purposes such as the removal of earth for the construction of canals, irrigation works, power plants, and so on, as well as to allow the transit of atomic material through its territory.

[11] When signing and ratifying Protocol I and Protocol II, the UK made the following declarations of understanding: In connection with Article 3 of the Treaty, defining the term 'territory' as including the territorial sea, airspace and any other space over which the state exercises sovereignty in accordance with 'its own legislation', the UK does not regard its signing or ratification of the Protocols as implying recognition of any legislation which does not, in its view, comply with the relevant rules of international law.

The Treaty does not permit the parties to carry out explosions of nuclear devices for peaceful purposes unless and until advances in technology have made possible the development of devices for such explosions which are not capable of being used for weapon purposes.

The signing and ratification by the UK could not be regarded as affecting in any way the legal status of any territory for the international relations of which the UK is responsible, lying within the limits of the geographical zone established by the Treaty.

Should any party to the Treaty carry out any act of aggression with the support of a nuclear weapon state, the UK would be free to reconsider the extent to which it could be regarded as committed by the provisions of Protocol II.

In addition, the UK declared that its undertaking under Article 3 of Protocol II not to use or threaten to use nuclear weapons against the parties to the Treaty extends also to territories in respect of which the undertaking under Article I of Protocol I becomes effective.

[12] The USA ratified Protocol I with the following understandings: The provisions of the Treaty made applicable by this Protocol do not affect the exclusive power and legal competence under international law of a state adhering to this Protocol to grant or deny transit and transport privileges to its own or any other vessels or aircraft irrespective of cargo or armaments; the provisions of the Treaty made applicable by this Protocol do not affect rights under international law of a state adhering to this Protocol regarding the exercise of the freedom of the seas, or regarding passage through or over waters subject to the sovereignty of a state, and the declarations attached by the United States to its ratification of Protocol II apply also to its ratification of Protocol I.

[13] The USA signed and ratified Protocol II with the following declarations and understandings: In connection with Article 3 of the Treaty, defining the term 'territory' as including the territorial sea, airspace and any other space over which the state exercises sovereignty in accordance with 'its own legislation', the ratification of the Protocol could not be regarded as implying recognition of any legislation which does not, in the view of the USA, comply with the relevant rules of international law.

Each of the parties retains exclusive power and legal competence, unaffected by the terms of the Treaty, to grant or deny non-parties transit and transport privileges.

As regards the undertaking not to use or threaten to use nuclear weapons against the parties, the USA would consider that an armed attack by a party, in which it was assisted by a nuclear weapon state, would be incompatible with the party's obligations under Article 1 of the Treaty.

The definition contained in Article 5 of the Treaty is understood as encompassing all nuclear explosive devices; Articles 1 and 5 of the Treaty restrict accordingly the activities of the parties under paragraph 1 of Article 18.

Article 18, paragraph 4 permits, and US adherence to Protocol II will not prevent, collaboration by the USA with the parties to the Treaty for the purpose of carrying out explosions of nuclear devices for peaceful purposes in a manner consistent with a policy of not contributing to the proliferation of nuclear weapon capabilities.

The USA will act with respect to such territories of Protocol I adherents, as are within the geographical area defined in Article 4, paragraph 2 of the Treaty, in the same manner as Protocol II requires it to act with respect to the territories of the Parties.

[14] The USSR signed and ratified Protocol II with the following statement:

The USSR proceeds from the assumption that the effect of Article 1 of the Treaty extends, as specified in Article 5 of the Treaty, to any nuclear explosive device and that, accordingly, the carrying out by any party to the Treaty of explosions of nuclear devices for peaceful purposes would be a violation of its obligations under Article 1 and would be incompatible with its non-nuclear status. For states parties to the Treaty, a solution to the problem of peaceful nuclear explosions can be found in accordance with the provisions of Article V of the Non-Proliferation Treaty and within the framework of the international procedures of the IAEA. The signing of the Protocol by the USSR does not in any way signify recognition of the possibility of the force of the Treaty being extended beyond the territories of the states parties to the Treaty, including airspace and territorial waters as defined in accordance with international law. With regard to the reference in Article 3 of the Treaty to 'its own legislation' in connection with the territorial waters, airspace and any other space over which the states parties to the Treaty exercise sovereignty, the signing of the Protocol by the USSR does not signify recognition of their claims to the exercise of sovereignty which are contrary to generally accepted standards of international law. The USSR takes note of the interpretation of the Treaty given in the Final Act of the Preparatory Commission for the Denuclearization of Latin America to the effect that the transport of nuclear weapons by the parties to the Treaty is covered by the prohibitions in Article 1 of the Treaty. The USSR reaffirms its position that authorizing the transit of nuclear weapons in any form would be contrary to the objectives of the Treaty, according to which, as specially mentioned in the preamble, Latin America must be completely free from nuclear weapons, and that it would be incompatible with the non-nuclear status of the states parties to the Treaty and with their obligations as laid down in Article 1 thereof.

Any actions undertaken by a state or states parties to the Treaty which are not compatible with their non-nuclear status, and also the commission by one or more states parties to the Treaty of an act of aggression with the support of a state which is in possession of nuclear weapons or together with such a state, will be regarded by the USSR as incompatible with the obligations of those countries under the Treaty. In such cases the USSR reserves the right to reconsider its obligations under Protocol II. It further reserves the right to reconsider its attitude to this Protocol in the event of any actions on the part of other states possessing nuclear weapons which are incompatible with their obligations under the said Protocol. The provisions of the articles of Protocol II are applicable to the text of the Treaty of Tlatelolco in the wording of the Treaty at the time of the signing of the Protocol by the Soviet Union, due account being taken of the position of the USSR as set out in the present statement. Any amendment to the Treaty entering into force in accordance with the provisions of Articles 29 and 6 of the Treaty without the clearly expressed approval of the USSR shall have no force as far as the USSR is concerned.

In addition, the USSR proceeds from the assumption that the obligations under Protocol II also apply to the territories for which the status of the denuclearized zone is in force in conformity with Protocol I of the Treaty.

[15] Venezuela stated that in view of the existing controversy between Venezuela on the one hand and the UK and Guyana on the other, Article 25, paragraph 2 of the Treaty should apply to Guyana. This paragraph provides that no political entity should be admitted, part or all of whose territory is the subject of a dispute or claim between an extra-continental country and one or more Latin American states, so long as the dispute has not been settled by peaceful means.

[16] Safeguards agreements under the Non-Proliferation Treaty cover the Treaty of Tlatelolco.

[17] Safeguards agreements under Protocol I.

The Non-Proliferation Treaty

[1] Notification of succession.

[2] Bahrain declared that its accession to the Treaty shall in no way constitute recognition of Israel or be a cause of establishment of any relations of any kind therewith.

[3] On the occasion of the deposit of the instrument of ratification, Egypt stated that since it was embarking on the construction of nuclear power reactors, it expected assistance and support from industrialized nations with a developed nuclear industry. It called upon nuclear weapon states to promote research and development of peaceful applications of nuclear explosions in order to overcome all the difficulties at present involved therein. Egypt also appealed to these states to exert their efforts to conclude an agreement prohibiting the use or threat of use of nuclear weapons against any state, and expressed the view that the Middle East should remain completely free of nuclear weapons.

[4] France, not party to the Treaty, declared that it would behave like a state adhering to the Treaty and that it would follow a policy of strengthening appropriate safeguards relating to nuclear equipment, material and technology. In 1981 an agreement between France, the European Atomic Energy Community (Euratom) and the IAEA for the application of safeguards in France entered into force. The agreement covers nuclear material and facilities notified to the IAEA by France.

[5] On depositing the instrument of ratification, FR Germany reiterated the declaration made at the time of signing: it reaffirmed its expectation that the nuclear weapon states would intensify their efforts in accordance with the undertakings under Article VI of the Treaty, as well as its understanding that the security of FR Germany continued to be ensured by NATO; it stated that no provision of the Treaty may be interpreted in such a way as to hamper further development of European unification; that research, development and use of nuclear energy for peaceful purposes, as well as international and multinational co-operation in this field, must not be prejudiced by the Treaty; that the application of the Treaty, including the implementation of safeguards, must not lead to discrimination of the nuclear industry of FR Germany in international competition; and that it attached vital importance to the undertaking given

by the USA and the UK concerning the application of safeguards to their peaceful nuclear facilities, hoping that other nuclear weapon states would assume similar obligations.

[6] On acceding to the Treaty, the Holy See stated, *inter alia*, that the Treaty will attain in full the objectives of security and peace and justify the limitations to which the states party to the Treaty submit, only if it is fully executed in every clause and with all its implications. This concerns not only the obligations to be applied immediately but also those which envisage a process of ulterior commitments. Among the latter, the Holy See considers it suitable to point out the following: (*a*) The adoption of appropriate measures to ensure, on a basis of equality, that all non-nuclear weapon states party to the Treaty will have available to them the benefits deriving from peaceful applications of nuclear technology. (*b*) The pursuit of negotiations in good faith of effective measures relating to cessation of the nuclear arms race at an early date and to nuclear disarmament, and on a treaty on general and complete disarmament under strict and effective control.

[7] On signing the Treaty, Indonesia stated, *inter alia*, that it attaches great importance to the declarations of the USA, the UK and the USSR affirming their intention to provide immediate assistance to any non-nuclear weapon state party to the Treaty that is a victim of an act of aggression in which nuclear weapons are used. Of utmost importance, however, is not the action *after* a nuclear attack has been committed but the guarantees to prevent such an attack. Indonesia trusts that the nuclear weapon states will study further this question of effective measures to ensure the security of the non-nuclear weapon states. On depositing the instrument of ratification, Indonesia expressed the hope that the nuclear countries would be prepared to co-operate with non-nuclear countries in the use of nuclear energy for peaceful purposes and implement the provisions of Article IV of the Treaty without discrimination. It also stated the view that the nuclear weapon states would observe the provisions of Article VI of the Treaty relating to the cessation of the nuclear arms race.

[8] Italy stated that in its belief nothing in the Treaty was an obstacle to the unification of the countries of Western Europe; it noted full compatibility of the Treaty with the existing security agreements; it noted further that when technological progress would allow the development of peaceful explosive devices different from nuclear weapons, the prohibition relating to their manufacture and use shall no longer apply; it interpreted the provisions of Article IX, paragraph 3 of the Treaty, concerning the definition of a nuclear weapon state, in the sense that it referred exclusively to the five countries which had manufactured and exploded a nuclear weapon or other nuclear explosive device prior to 1 Jan. 1967, and stressed that under no circumstance would a claim of pertaining to such category be recognized by Italy for any other state.

[9] On depositing the instrument of ratification, Japan expressed the hope that France and China would accede to the Treaty; it urged a reduction of nuclear armaments and a comprehensive ban on nuclear testing; appealed to all states to refrain from the threat or use of force involving either nuclear or non-nuclear weapons; expressed the view that peaceful nuclear activities in non-nuclear weapon states party to the Treaty should not be hampered and that Japan should not be discriminated against in favour of other parties in any aspect of such activities. It also urged all nuclear weapon states to accept IAEA safeguards on their peaceful nuclear activities.

[10] A statement was made containing a disclaimer regarding the recognition of states party to the Treaty.

[11] On depositing the instrument of ratification, the Republic of Korea took note of the fact that the depositary governments of the three nuclear weapon states had made declarations in June 1968 to take immediate and effective measures to safeguard any non-nuclear weapon state which is a victim of an act or an object of a threat of aggression in which nuclear weapons are used. It recalled that the UN Security Council adopted a resolution to the same effect on 19 June 1968.

[12] On depositing the instruments of ratification, Kuwait declared that the ratification of the Treaty does not mean in any way a recognition of Israel. No treaty relation will arise between Kuwait and Israel.

[13] On depositing the instruments of accession and ratification, Liechtenstein and Switzerland stated that activities not prohibited under Articles I and II of the Treaty include, in particular, the whole field of energy production and related operations, research and technology concerning future generations of nuclear reactors based on fission or fusion, as well as production of isotopes. Liechtenstein and Switzerland define the term 'source or special fissionable material' in Article III of the Treaty as being in accordance with Article XX of the IAEA Statute, and a modification of this interpretation requires their formal consent; they will accept only such interpretations and definitions of the terms 'equipment or material especially designed or prepared for the processing, use or production of special fissionable material', as mentioned in Article III of the Treaty, that they will expressly approve; and they understand that the application of the Treaty, especially of the control measures, will not lead to discrimination of their industry in international competition.

[14] On signing the Treaty, Mexico stated, *inter alia*, that none of the provisions of the Treaty shall be interpreted as affecting in any way whatsoever the rights and obligations of Mexico as a state party to the Treaty of Tlatelolco.

It is the understanding of Mexico that at the present time any nuclear explosive device is capable of being used as a nuclear weapon and that there is no indication that in the near future it will be possible to manufacture nuclear explosive devices that are not potentially nuclear weapons. However, if technological advances modify this situation, it will be necessary to amend the relevant provisions of the Treaty in accordance with the procedure established therein.

[15] The ratification was accompanied by a statement in which Turkey underlined the non-proliferation obligations of the nuclear weapon states, adding that measures must be taken to meet adequately the security requirements of non-nuclear weapon states. Turkey also stated that measures developed or to be developed at national and international levels to ensure the non-proliferation of nuclear weapons should in no case restrict the non-nuclear weapon states in their option for the application of nuclear energy for peaceful purposes.

[16] The UK recalled its view that if a regime is not recognized as the government of a state, neither signature nor the deposit of any instrument by it, nor notification of any of those acts, will bring about recognition of that regime by any other state.

[17] This agreement, signed by the UK, Euratom and the IAEA, provides for the submission of British non-military nuclear installations to safeguards under IAEA supervision.

[18] This agreement provides for safeguards on fissionable material in all facilities within the USA, excluding those associated with activities of direct national security significance.

[19] The agreement provides for the application of IAEA safeguards in Soviet peaceful nuclear facilities designated by the USSR.

[20] In connection with the ratification of the Treaty, Yugoslavia stated, *inter alia*, that it considered a ban on the development, manufacture and use of nuclear weapons and the destruction of all stockpiles of these weapons to be indispensable for the maintenance of a stable peace and international security; it held the view that the chief responsibility for progress in this direction rested with the nuclear weapon powers, and expected these powers to undertake not to use nuclear weapons against the countries which have renounced them as well as against non-nuclear weapon states in general, and to refrain from the threat to use them. It also emphasized the significance it attached to the universality of the efforts relating to the realization of the Non-Proliferation Treaty.

The Sea-Bed Treaty

[1] On signing and ratifying the Treaty, Argentina stated that it interprets the references to the freedom of the high seas as in no way implying a pronouncement of judgement on the different positions relating to questions connected with international maritime law. It understands that the reference to the rights of exploration and exploitation by coastal states over their continental shelves was included solely because those could be the rights most frequently affected by verification procedures. Argentina precludes any possibility of strengthening, through this Treaty, certain positions concerning continental shelves to the detriment of others based on different criteria.

[2] On signing the Treaty, Brazil stated that nothing in the Treaty shall be interpreted as prejudicing in any way the sovereign rights of Brazil in the area of the sea, the sea-bed and the subsoil thereof adjacent to its coasts. It is the understanding of Brazil that the word 'observation', as it appears in paragraph 1 of Article III of the Treaty, refers only to observation that is incidental to the normal course of navigation in accordance with international law. This statement was repeated at the time of ratification. The USA declared, in 1989, that under customary international law and Article III of the Treaty, these observations may be undertaken whether or not they are incidental to a so-called 'normal course of navigation,' and that such activity is not subject to unilateral coastal state restriction. The USSR and the FRG also stated that they did not agree with Brazil's interpretation of the term 'observation'.

[3] In depositing the instrument of ratification, Canada declared: Article I, paragraph 1, cannot be interpreted as indicating that any state has a right to implant or emplace any weapons not prohibited under Article I, paragraph 1, on the sea-bed and ocean floor, and in the subsoil thereof, beyond the limits of national jurisdiction, or as constituting any limitation on the principle that this area of the sea-bed and ocean floor and the subsoil thereof shall be reserved for exclusively peaceful purposes. Articles I, II and III cannot be interpreted as indicating that any state but the coastal state has any right to implant or emplace any weapon not prohibited under Article I, paragraph 1 on the continental shelf, or the subsoil thereof, appertaining to that coastal state, beyond the outer limit of the sea-bed zone referred to in Article I and defined in Article II. Article III cannot be interpreted as indicating any restrictions or limitation upon the rights of the coastal state, consistent with its exclusive sovereign rights with respect to the continental shelf, to verify, inspect or effect the removal of any weapon, structure, installation, facility or device implanted or emplaced on the continental shelf, or the subsoil thereof, appertaining to that coastal

state, beyond the outer limit of the sea-bed zone referred to in Article I and defined in Article II. On 12 Apr. 1976, FR Germany stated that the declaration by Canada is not of a nature to confer on the government of this country more far-reaching rights than those to which it is entitled under current international law, and that all rights existing under current international law which are not covered by the prohibitions are left intact by the Treaty.

[4] A statement was made containing a disclaimer regarding recognition of states party to the Treaty.

[5] On the occasion of its accession to the Treaty, the government of India stated that as a coastal state, India has, and always has had, full and exclusive rights over the continental shelf adjoining its territory and beyond its territorial waters and the subsoil thereof. It is the considered view of India that other countries cannot use its continental shelf for military purposes. There cannot, therefore, be any restriction on, or limitation of, the sovereign right of India as a coastal state to verify, inspect, remove or destroy any weapon, device, structure, installation or facility, which might be implanted or emplaced on or beneath its continental shelf by any other country, or to take such other steps as may be considered necessary to safeguard its security. The accession by the government of India to the Treaty is based on this position. In response to the Indian statement, the USA expressed the view that, under existing international law, the rights of coastal states over their continental shelves are exclusive only for the purposes of exploration and exploitation of natural resources, and are otherwise limited by the 1958 Convention on the Continental Shelf and other principles of international law. On 12 Apr. 1976, FR Germany stated that the declaration by India is not of a nature to confer on the government of this country more far-reaching rights than those to which it is entitled under current international law, and that all rights existing under current law which are not covered by the prohibitions are left intact by the Treaty.

[6] On signing the Treaty, Italy stated, *inter alia*, that in the case of agreements on further measures in the field of disarmament to prevent an arms race on the sea-bed and ocean floor and in their subsoil, the question of the delimitation of the area within which these measures would find application shall have to be examined and solved in each instance in accordance with the nature of the measures to be adopted. The statement was repeated at the time of ratification.

[7] Mexico declared that in its view no provision of the Treaty can be interpreted to mean that a state has the right to emplace nuclear weapons or other weapons of mass destruction, or arms or military equipment of any type, on the continental shelf of Mexico. It reserves the right to verify, inspect, remove or destroy any weapon, structure, installation, device or equipment placed on its continental shelf, including nuclear weapons or other weapons of mass destruction.

[8] Ratification of the Treaty by Taiwan is considered by Romania as null and void.

[9] The UK recalled its view that if a regime is not recognized as the government of a state neither signature nor the deposit of any instrument by it, nor notification of any of those acts, will bring about recognition of that regime by any other state.

[10] Viet Nam stated that no provision of the Treaty should be interpreted in a way that would contradict the rights of the coastal states with regard to their continental shelf, including the right to take measures to ensure their security.

[11] On 25 Feb. 1974, the Ambassador of Yugoslavia transmitted to the US Secretary of State a note stating that in the view of the Yugoslav Government, Article III, paragraph 1, of the Treaty should be interpreted in such a way that a state exercising its right under this Article shall be obliged to notify in advance the coastal state, in so far as its observations are to be carried out 'within the stretch of the sea extending above the continental shelf of the said state'. On 16 Jan. 1975 the US Secretary of State presented the view of the USA concerning the Yugoslav note, as follows: In so far as the note is intended to be interpretative of the Treaty, the USA cannot accept it as a valid interpretation. In addition, the USA does not consider that it can have any effect on the existing law of the sea. In so far as the note was intended to be a reservation to the Treaty, the USA placed on record its formal objection to it on the grounds that it was incompatible with the object and purpose of the Treaty. The USA also drew attention to the fact that the note was submitted too late to be legally effective as a reservation. A similar exchange of notes took place between Yugoslavia and the UK on 12 Apr. 1976. FR Germany stated that the declaration by Yugoslavia is not of a nature to confer on the government of this country more far-reaching rights than those to which it is entitled under current international law, and that all rights existing under current international law which are not covered by the prohibitions are left intact by the Treaty.

[12] Notification of succession.

The BW Convention

[1] Considering the obligations resulting from its status as a permanently neutral state, Austria declares a reservation to the effect that its co-operation within the framework of this Convention cannot exceed the limits determined by the status of permanent neutrality and membership of the UN.

[2] Bahrain declared that its accession to the Convention shall in no way constitute recognition of Israel or be a cause of establishment of any relations of any kind with it.

[3] China stated that the BW Convention has the following defects: it fails explicitly to prohibit the use of biological weapons; it does not provide for 'concrete and effective' measures of supervision and verification; and it lacks measures of sanctions in case of violation of the Convention. China hopes that these defects will be corrected at an appropriate time, and also that a convention for complete prohibition of chemical weapons will soon be concluded. The signature and ratification of the Convention by the Taiwan authorities in the name of China are considered illegal and null and void.

[4] On depositing its instrument of ratification, FR Germany stated that a major shortcoming of the BW Convention is that it does not contain any provisions for verifying compliance with its essential obligations. The Federal Government considers the right to lodge a complaint with the UN Security Council to be an inadequate arrangement. It would welcome the establishment of an independent international committee of experts able to carry out impartial investigations when doubts arise as to whether the Convention is being complied with.

[5] In a statement made on the occasion of the signature of the Convention, India reiterated its understanding that the objective of the Convention is to eliminate biological and toxin weapons, thereby excluding completely the possibility of their use, and that the exemption with regard to biological agents or toxins, which would be permitted for prophylactic, protective or other peaceful purposes, would not in any way create a loophole in regard to the production or retention of biological and toxin weapons. Also any assistance which might be furnished under the terms of the Convention would be of a medical or humanitarian nature and in conformity with the UN Charter. The statement was repeated at the time of the deposit of the instrument of ratification.

[6] Ireland considers that the Convention could be undermined if the reservations made by the parties to the 1925 Geneva Protocol were allowed to stand, as the prohibition of possession is incompatible with the right to retaliate, and that there should be an absolute and universal prohibition of the use of the weapons in question. Ireland notified the depositary government for the Geneva Protocol of the withdrawal of its reservations to the Protocol, made at the time of accession in 1930. The withdrawal applies to chemical as well as to bacteriological (biological) and toxin agents of warfare.

[7] The Republic of Korea stated that the signing and ratification of the Convention does not in any way mean or imply the recognition of any territory or regime which has not been recognized by the Republic of Korea as a state or government.

[8] In the understanding of Kuwait, its ratification of the Convention does not in any way imply its recognition of Israel, nor does it oblige it to apply the provisions of the Convention in respect of the said country.

[9] Mexico considers that the Convention is only a first step towards an agreement prohibiting also the development, production and stockpiling of all chemical weapons, and notes the fact that the Convention contains an express commitment to continue negotiations in good faith with the aim of arriving at such an agreement.

[10] Notification of succession.

[11] The ratification by Switzerland contains the following reservations:

1. Owing to the fact that the Convention also applies to weapons, equipment or means of delivery designed to use biological agents or toxins, the delimitation of its scope of application can cause difficulties since there are scarcely any weapons, equipment or means of delivery peculiar to such use; therefore, Switzerland reserves the right to decide for itself what auxiliary means fall within that definition.

2. By reason of the obligations resulting from its status as a perpetually neutral state, Switzerland is bound to make the general reservation that its collaboration within the framework of this Convention cannot go beyond the terms prescribed by that status. This reservation refers especially to Article VII of the Convention as well as to any similar clause that could replace or supplement that provision of the Convention.

In a note of 18 Aug. 1976, addressed to the Swiss Ambassador, the US Secretary of State stated the following view of the USA with regard to the first reservation: The prohibition would apply only to (*a*) weapons, equipment and means of delivery, the design of which indicated that they could have no other use than that specified, and (*b*) weapons, equipment and means of delivery, the design of which indicated that they were specifically intended to be capable of the use specified. The USA shares the view of Switzerland that there are few weapons, equipment or means of delivery peculiar to the uses referred to. It does not, however, believe that it would be appropriate, on this ground alone, for states to reserve unilaterally the right to decide which weapons, equipment or means of delivery fell within the definition. Therefore, while acknowledging the entry into force of the Convention between itself and Switzerland, the USA enters its objection to this reservation.

[12] The deposit of the instrument of ratification by Taiwan is considered by the Soviet Union as an illegal act because the government of the People's Republic of China is regarded by the USSR as the sole representative of China.

[13] The UK recalled its view that if a regime is not recognized as the government of a state, neither signature nor the deposit of any instrument by it nor notification of any of those acts will bring about recognition of that regime by any other state.

The Enmod Convention

[1] Argentina interprets the terms 'widespread, long-lasting or severe effects' in Article I, paragraph 1, of the Convention in accordance with the definition agreed upon in the understanding on that article. It likewise interprets Articles II, III and VIII in accordance with the relevant understandings.

[2] Guatemala accepts the text of Article III on condition that the use of environmental techniques for peaceful purposes does not adversely affect its territory or the use of its natural resources.

[3] It is the understanding of the Republic of Korea that any technique for deliberately changing the natural state of rivers falls within the meaning of the term 'environmental modification techniques' as defined in Article II of the Convention. It is further understood that military or any other hostile use of such techniques, which could cause flooding, inundation, reduction in the water-level, drying up, destruction of hydrotechnical installations or other harmful consequences, comes within the scope of the Convention, provided it meets the criteria set out in Article I thereof.

[4] Kuwait made the following reservations and understanding: This Convention binds Kuwait only towards states parties thereto; its obligatory character shall *ipso facto* terminate with respect to any hostile state which does not abide by the prohibition contained therein. It is understood that accession to this Convention does not mean in any way recognition of Israel by Kuwait; furthermore, no treaty relation will arise between Kuwait and Israel.

On 23 June 1980, the UN Secretary-General, the depositary of the Convention, received from the government of Israel a communication stating that Israel would adopt towards Kuwait an attitude of complete reciprocity.

[5] The Netherlands accepts the obligation laid down in Article I of the Enmod Convention as extending to states which are not party to the Convention and which act in conformity with Article I of this Convention.

[6] New Zealand declared that, in its interpretation, nothing in the Convention detracts from or limits the obligations of states to refrain from military or any other hostile use of environmental modification techniques which are contrary to international law.

[7] Notification of succession.

[8] Because of its obligation incumbent upon it by virtue of its status of perpetual neutrality, Switzerland made a general reservation specifying that its co-operation in the framework of this Convention cannot go beyond the limits imposed by this status. This reservation refers, in particular, to article V, paragraph 5, of the Convention, and to any similar clause which may replace or supplement this provision in the Convention (or in any other arrangement).

[9] On signing the Convention, Turkey declared that the terms 'widespread', 'long-lasting' and 'severe effects' contained in the Convention need to be more clearly defined, and that so long as this clarification was not made, Turkey would be compelled to interpret for itself the terms in question and, consequently, reserved the right to do so as and when required. Turkey also stated its belief that the difference between 'military or any other hostile purposes' and 'peaceful purposes' should be more clearly defined so as to prevent subjective evaluations.

[10] Austria's instrument of accession contains the following reservation: 'Considering the obligations resulting from its status as a permanently neutral state, the Republic of Austria declares a reservation to the effect that its co-operation within the framework of this Convention cannot exceed the limits determined by the status of permanent neutrality and membership with the United Nations'.

The 'Inhumane Weapons' Convention

[1] The accession of Benin refers only to Protocols I and III of the Convention.

[2] Upon signature, China stated that the Convention fails to provide for supervision or verification of any violation of its clauses, thus weakening its binding force. The Protocol on mines, booby traps and other devices fails to lay down strict restrictions on the use of such weapons by the aggressor on the territory of the victim and to provide adequately for the right of a state victim of an aggression to defend itself by all necessary means. The Protocol on incendiary weapons does not stipulate restrictions on the use of such weapons against combat personnel.

[3] Cyprus declared that the provisions of Article 7, paragraph 3b, and Article 8 of Protocol II of the Convention will be interpreted in such a way that neither the status of peace-keeping forces or missions of the UN in Cyprus will be affected nor will additional rights be, *ipso jure*, granted to them.

[4] France ratified only Protocols I and II. On signing the Convention France stated that it regretted that it had not been possible to reach agreement on the provisions concerning the verification of facts which might be alleged and which might constitute violations of the undertakings subscribed to. It therefore reserved the right to submit, possibly in association with other states, proposals aimed at filling that gap at the first conference to be held pursuant to Article 8 of the Convention and to utilize, as appropriate, procedures that would make it possible to bring before the international community facts and information which, if verified, could constitute violations of the provisions of the Convention and the Protocols annexed thereto. Reservation: Not being bound by the 1977 Additional Protocol I to the Geneva Conventions of 1949, France considers that the fourth paragraph of the preamble to the Convention on prohibitions or restrictions on the use of certain conventional weapons, which reproduces the provisions of Article 35, paragraph 3, of Additional Protocol I, applies only to states parties to that Protocol. France will apply the provisions of the Convention and its three Protocols to all the armed conflicts referred to in Articles 2 and 3 common to the Geneva Conventions of 1949.

[5] Italy stated its regret that no agreement had been reached on provisions that would ensure respect for the obligations under the Convention. Italy intends to undertake efforts to ensure that the problem of the establishment of a mechanism that would make it possible to fill this gap in the Convention is taken up again at the earliest opportunity in every competent forum.

[6] The Netherlands made the following statements of understanding: A specific area of land may also be a military objective if, because of its location or other reasons specified in Article 2, paragraph 4, of Protocol II and in Article I, paragraph 3, of Protocol III, its total or partial destruction, capture, or neutralization in the prevailing circumstances offers a definitive military advantage; military advantage mentioned in Article 3, paragraph 3 under c, or Protocol II, refers to the advantage anticipated from the attack considered as a whole and not only from isolated or particular parts of the attack; in Article 8, paragraph 1, of Protocol II, the words 'as far as it is able' mean 'as far as it is technically able'.

[7] Romania stated that the provisions of the Convention and its Protocols have a restricted character and do not ensure adequate protection either to the civilian population or to the combatants as the fundamental principles of international humanitarian law require.

[8] The USA stated that it had strongly supported proposals by other countries to include special procedures for dealing with compliance matters, and reserved the right to propose at a later date additional procedures and remedies, should this prove necessary, to deal with such problems.

The Treaty of Rarotonga

[1] In signing Protocols 2 and 3 China declared that it respected the status of the South Pacific nuclear-free zone and would neither use nor threaten to use nuclear weapons against the zone nor test nuclear weapons in the region. However, China reserved its right to reconsider its obligations under the Protocols if other nuclear weapon states or the contracting Parties to the Treaty took any action in 'gross' violation of the Treaty and the Protocols, thus changing the status of the zone and endangering the security interests of China.

[2] In signing Protocols 2 and 3 the USSR stated the view that admission of transit of nuclear weapons or other nuclear explosive devices by any means, as well as of visits by foreign military ships and aircraft with nuclear explosive devices on board, to the ports and airfields within the nuclear-free zone would contradict the aims of the Treaty of Rarotonga and would be inconsistent with the status of the zone. It also warned that in case of action taken by a party or parties violating their major commitments connected with the nuclear-free status of the zone, as well as in case of aggression committed by one or several parties to the Treaty, supported by a nuclear-weapon state, or together with it, with the use by such a state of the territory, airspace, territorial sea or archipelagic waters of the parties for visits by nuclear weapon-carrying ships and aircraft or for transit of nuclear weapons, the USSR will have the right to consider itself free of its non-use commitments assumed under Protocol 2.

The Soviet Union ratified Protocols 2 and 3 to the Treaty without reference to the conditions included in its statement made at the time of signature. It expressed the hope that all states members of the South Pacific Forum would join the Treaty, and called upon the nuclear powers, which had not done so, to sign and ratify the relevant Protocols.

III. UN member states and year of membership

In the following list of the 159 UN member states in 1990, the countries marked with an asterisk are also members of the Geneva-based Conference on Disarmament (CD).

Afghanistan, 1946
Albania, 1955
*Algeria, 1962
Angola, 1976
Antigua and Barbuda, 1981
*Argentina, 1945
*Australia, 1945
Austria, 1955
Bahamas, 1973
Bahrain, 1971
Bangladesh, 1974
Barbados, 1966
*Belgium, 1945
Belize, 1981
Benin, 1960
Bhutan, 1971
Bolivia, 1945
Botswana, 1966
*Brazil, 1945
Brunei Darussalam, 1984
*Bulgaria, 1955
Burkina Faso, 1960
Burma (see Myanmar)
Burundi, 1962
Byelorussia, 1945
Cambodia (Kampuchea), 1955
Cameroon, 1960
*Canada, 1945
Cape Verde, 1975
Central African Republic, 1960
Chad, 1960
Chile, 1945
*China, 1945
Colombia, 1945
Comoros, 1975
Congo, 1960
Costa Rica, 1945
Côte d'Ivoire, 1960
*Cuba, 1945
Cyprus, 1960
*Czechoslovakia, 1945
Denmark, 1945
Djibouti, 1977
Dominica, 1978
Dominican Republic, 1945
Ecuador, 1945
*Egypt, 1945
El Salvador, 1945
Equatorial Guinea, 1968
*Ethiopia, 1945
Fiji, 1970
Finland, 1955
*France, 1945
Gabon, 1960
Gambia, 1965
*Germany, 1973
Ghana, 1957
Greece, 1945
Grenada, 1974
Guatemala, 1945
Guinea, 1958
Guinea-Bissau, 1974
Guyana, 1966
Haiti, 1945
Honduras, 1945
*Hungary, 1955
Iceland, 1946
*India, 1945
*Indonesia, 1950
*Iran, 1945
Iraq, 1945
Ireland, 1955
Israel, 1949
*Italy, 1955
Ivory Coast (see Côte d'Ivoire)
Jamaica, 1962
*Japan, 1956
Jordan, 1955
Kampuchea (see Cambodia)
*Kenya, 1963
Kuwait, 1963
Lao People's Democratic Republic, 1955
Lebanon, 1945
Lesotho, 1966
Liberia, 1945
Libya, 1955
Liechtenstein, 1990
Luxembourg, 1945
Madagascar, 1960
Malawi, 1964
Malaysia, 1957
Maldives, 1965
Mali, 1960
Malta, 1964
Mauritania, 1961
Mauritius, 1968
*Mexico, 1945
*Mongolia, 1961
*Morocco, 1956
Mozambique, 1975
*Myanmar *(formerly Burma)*, 1948
Namibia, 1990
Nepal, 1955
*Netherlands, 1945
New Zealand, 1945
Nicaragua, 1945
Niger, 1960
*Nigeria, 1960
Norway, 1945
Oman, 1971
*Pakistan, 1947
Panama, 1945
Papua New Guinea, 1975
Paraguay, 1945
*Peru, 1945
Philippines, 1945
*Poland, 1945
Portugal, 1955
Qatar, 1971
*Romania, 1955
Rwanda, 1962
Saint Christopher (Kitts) and Nevis, 1983
Saint Lucia, 1979
Saint Vincent and the Grenadines, 1980
Samoa, Western, 1976
Sao Tome and Principe, 1975
Saudi Arabia, 1945
Senegal, 1960
Seychelles, 1976
Sierra Leone, 1961
Singapore, 1965
Solomon Islands, 1978
Somalia, 1960
South Africa, 1945
Spain, 1955
*Sri Lanka, 1955
Sudan, 1956
Suriname, 1975
Swaziland, 1968
*Sweden, 1946
Syria, 1945
Tanzania, 1961
Thailand, 1946
Togo, 1960
Trinidad and Tobago, 1962
Tunisia, 1956
Turkey, 1945
Uganda, 1962
*UK, 1945
Ukraine, 1945
United Arab Emirates, 1971
Uruguay, 1945
*USA, 1945
*USSR, 1945
Vanuatu, 1981
*Venezuela, 1945
Viet Nam, 1977
Yemen, 1947[a]
*Yugoslavia, 1945
*Zaire, 1960
Zambia, 1964
Zimbabwe, 1980

[a] Yemen Arab Rep., 1947; People's Dem. Rep. of Yemen, 1967.

Annexe B. Chronology 1990

RAGNHILD FERM

For the convenience of the reader, key words are indicated in the right-hand column, opposite each entry. They refer to the subject-areas covered in the entry. Definitions of the acronyms can be found on page xvi.

3 Jan.	The leader of Panama, General Noriega, surrenders to the US forces. (The USA invaded Panama on 20 Dec. 1989.)	Panama/USA
6 Jan.	The GDR Communist Party leader proposes that both German states reduce their armies by half in 1990. (NATO rejects the proposal a few days later.)	GDR/FRG; Force reductions
9 Jan.	Czechoslovakia calls for the withdrawal, by the end of 1990, of all Soviet troops stationed in Czechoslovakia.	Czechoslovakia/ USSR; Withdrawals
15 Jan.	The Supreme Soviet of the USSR issues a decree that army, navy and KGB units should be sent to Azerbaijan to try to halt violence between the ethnic groups.	USSR
16 Jan.	The UN Security Council unanimously approves that the UN administer Cambodia during an agreed transition to free elections.	Cambodia; UN
16 Jan.– 5 Feb.	The military chiefs and diplomats of the CSCE states hold a seminar, in Vienna, to discuss military doctrines in relation to the posture, structure and activities of conventional forces in the ATTU zone. The seminar is an integral part of the CSBM Negotiations.	CSCE; CSBM
17 Jan.	The FRG Chancellor says that German unification would not involve any change in existing frontiers with other states.	FRG; German unification
18 Jan.	The Soviet Foreign Ministry announces that the USSR has started reducing its presence at the Cam Ranh Bay naval base in Viet Nam.	Withdrawals; USSR/Viet Nam
18 Jan.	Hungary calls for the withdrawal, in 1990 or at the latest during 1991, of all Soviet troops stationed in Hungary.	Hungary/ USSR; Withdrawals
18 Jan.	At a meeting with the Soviet ambassador to Poland the leader of Solidarnosc, Lech Walesa, points out necessary steps to be taken to improve relations between Poland and the USSR, including the withdrawal of all Soviet troops stationed in Poland by the end of 1990.	Poland/ USSR; Withdrawals

20 Jan.	At talks in Dublin the foreign ministers of the EC states express support for President Gorbachev's proposal (30 Nov. 1989) for a summit meeting of all CSCE member states.	CSCE; EC
22 Jan.	As a step towards a START treaty the USA and the USSR reach an agreement to conduct trial inspections of each other's ballistic missile re-entry vehicles (RVs). The agreement will allow each side to demonstrate methods of verifying the number of RVs on one ICBM and one SLBM from each country.	START
29 Jan.	As a step towards a START treaty the USA and the USSR reach an agreement to conduct reciprocal exhibitions of certain strategic (heavy) bombers. The agreement will allow each side to demonstrate its approach to distinguishing strategic bombers.	START
31 Jan.	In his State of the Union address President Bush proposes that US and Soviet troops in Central Europe be reduced to 195 000 on each side.	Force reductions; USA; USSR; Europe
31 Jan.	The FRG Foreign Minister proposes at the Tutzing Conference to establish CSCE verification and conflict management centres.	CSCE
1 Feb.	The Prime Minister of the GDR presents a programme for a united Germany, based on neutrality, with the seat of government in Berlin. (The Chancellor of the FRG rejects the proposal.)	German unification; GDR;
2 Feb.	The President of South Africa announces that the Government will lift the ban on the African National Congress (ANC) and on the South African Communist Party. (Nelson Mandela, the long-imprisoned leader of the ANC, is released on 11 Feb.)	South Africa; ANC
7–9 Feb.	At a meeting in Moscow between the US Secretary of State and the Soviet Foreign Minister, the Soviet side states that it would no longer insist on having the right to withdraw from a START treaty if it thought the ABM Treaty was violated. A preliminary agreement on a CW convention, which would abolish all CW throughout the world, is also reached. The ministers agree that a meeting of all heads of state or government of the CSCE members should be held in 1990.	USA/USSR; START; ABM; CSCE; CW
8 Feb.	At the CFE Negotiation the NATO states propose that neither the USA nor the USSR shall station, within the area comprising Belgium, Czechoslovakia, Denmark, the FRG, the GDR, Hungary, Luxembourg, the Netherlands and Poland, more than 195 000 ground and air force personnel each. The troops withdrawn shall be demobilized.	CFE; NATO

10 Feb.	Meeting the Chancellor of the FRG in Moscow, President Gorbachev says that the USSR will respect the two German states' decision on their future. However, unification cannot take place against the legitimate interests of neighbouring states and allies.	German unification; FRG; USSR
11 Feb.	The Soviet Government states that, without waiting for a CFE accord, the WTO will unilaterally reduce its armed forces and make them incapable of attack. The Soviet Government confirms that all Soviet troops will be withdrawn from the territories of other European states by 1995–96 and all military bases on foreign territories will be eliminated by the year 2000.	Force reductions; Withdrawals; USSR
12–14 Feb.	The foreign ministers of the NATO and WTO states hold a conference in Ottawa to discuss the 'Open Skies' plan (proposed by President Eisenhower in 1955 and revived by President Bush in 1989). A basic plan is agreed under which unarmed military or civilian reconnaissance aircraft from each alliance would be allowed to fly legally over the territory of the other. (Expert delegations continue negotiations until 24 Feb.)	Open Skies
13 Feb.	At the Ottawa Open Skies conference a US–Soviet agreement is reached on troop levels in Europe: each side would be allowed to keep 195 000 troops in Central Europe. However, the USA would be permitted to station an additional 30 000 troops in the UK, Italy, Turkey and Greece.	Force reductions; USA; USSR; Europe
13 Feb.	At the Ottawa Open Skies conference the four victorious World War II powers (the USA, the USSR, the UK and France) and the GDR and the FRG state that they will start negotiations on the unification of the two German states (the 'Two-plus-Four' talks).	German unification
15 Feb.	Diplomatic relations between the UK and Argentina are restored after the 8-year break following the 1982 Falklands/Malvinas conflict.	UK/Argentina
15 Feb.	At the CD the *Ad Hoc* Committee on Chemical Weapons is formally re-established for 1990, for the first time with a full negotiating mandate.	CW; CD
19 Feb.	The Foreign Minister and the Defence Minister of the FRG declare that neither FRG nor NATO troops should be stationed in what is now GDR territory once the two German states unite.	German unification; FRG; NATO
22 Feb.	At the CFE Negotiation the NATO states present a detailed proposal for the maximum number of inspections each state would be required to accept. The inspections would verify that the arsenals of each state correspond to the numbers declared annually by each state.	CFE; NATO

24–25 Feb.	Meeting at Camp David, USA, President Bush and the FRG Chancellor agree that a united Germany should be a full member of NATO and be included in NATO's military command structure. However, no NATO troops will be stationed on former GDR territory, which should have a special military status to take account of the security interests of other states, including the USSR.	German unification; FRG; USA; NATO
26–27 Feb.	The President of Czechoslovakia pays an official visit to Moscow. The foreign ministers of the two states sign an agreement on the complete withdrawal of Soviet troops from Czechoslovakia by 1 July 1991.	Czechoslovakia/ USSR; Withdrawals
3 Mar.	After talks in Ulan Bator, the USSR and Mongolia agree on the full withdrawal of Soviet troops from Mongolian territory during 1991–92.	Mongolia/ USSR; Withdrawals
7 Mar.	The USA and the FRG announce a detailed plan for the removal of US CW stationed on FRG territory.	USA/FRG; CW
10 Mar.	In a note sent to all CSCE member states Poland proposes to establish a Council of European Co-operation in the framework of the CSCE process.	Poland; CSCE
11 Mar.	An agreement is signed in Moscow between the USSR and Hungary on the withdrawal of Soviet troops from Hungarian territory. The withdrawal will start on 12 Mar. and be completed by 30 June 1991.	Hungary/ USSR; Withdrawals
11 Mar.	Lithuania declares itself an independent state.	Lithuania
14 Mar.	A 'Two-plus-Four' meeting, in Bonn, agrees that Poland will be invited as soon as border questions are discussed.	German unification; Poland
15 Mar.	At the CFE Negotiation the NATO states present a proposal on the destruction of surplus weapons and introduce a draft Protocol on notification and exchange of information.	CFE; NATO
15 Mar.	At the CFE Negotiation the WTO states present a proposal on the reduction of troop levels in Europe to around 700 000 for each military bloc.	CFE; WTO
15 Mar.	General Secretary Gorbachev is elected President of the USSR by the Congress of People's Deputies.	USSR
17 Mar.	At a meeting of foreign ministers of the WTO in Prague the participants agree that the unification of Germany and the creation of new European security systems should be 'synchronized' but disagree over whether a united Germany should belong to NATO or be neutral. (Poland, Czechoslovakia and Hungary, in particular, reject the possibility of German neutrality.) The Czechoslovak Foreign	German unification; WTO; NATO; CSCE; UN

	Minister proposes the formation of a European Security Commission to act as the executive organ of a regional security system in accordance with the UN Charter.	
21 Mar.	Namibia becomes an independent state. (On 23 Apr. the UN General Assembly admits Namibia as a member of the UN.)	Namibia; UN
30 Mar.	Estonia adopts a decree that the 1920 legal instrument on independence, signed by the USSR and Estonia, is still valid. The 1940 document on Estonia's entry into the USSR is declared legally invalid.	USSR/Estonia
1 Apr.	In a speech Iraqi President Saddam Hussein states that Iraq does not need an atomic bomb since it possesses binary chemical weapons.	Iraq; Nuclear weapons; CW
4 Apr.	The South Korean Ministry of National Defense announces that 7000 US troops will be withdrawn from South Korea and that the Combined Field Army will be disbanded by 1993.	South Korea/USA; Withdrawals
4–6 Apr.	At a meeting in Washington, DC, between the US Secretary of State and the Soviet Foreign Minister the USSR withdraws its demand that a unified Germany must be neutral.	German unification; USSR
12 Apr.	The Volkskammer of the GDR affirms the inviolability of the Oder–Neisse frontier between the GDR and Poland.	GDR: Poland
19 Apr.	The US Government announces plans to withdraw up to 15 000 US troops from Asia within three years.	USA; Withdrawals
19 Apr.	The President of France and the Chancellor of the FRG, meeting in Paris, issue a joint message to the EC European Council calling for work to begin on an inter-governmental conference on a European political and monetary union.	France/FRG; EC
19 Apr.	The outgoing Sandinista Government in Nicaragua signs an agreement with the Contra leaders that an effective and definitive cease-fire will take immediate effect.	Nicaragua
23–26 Apr.	The Chinese Prime Minister pays an official visit to the USSR. (This is the first visit of a Chinese head of government since 1964.) A Sino-Soviet agreement is signed on the principles under which the large military forces stationed on their common border could be reduced.	USSR/China
27 Apr.	The defence ministers of the FRG and the GDR declare that their goal is a united Germany as a NATO member without NATO's military structures or equipment being extended to the territory of the GDR.	German unification; NATO

28 Apr.	At the special session of the EC European Council held in Dublin, a majority of the member states expresses that the EC should pursue closer political and possibly even security co-operation.	EC
3 May	President Bush announces the decision that the USA will postpone indefinitely the production of new short-range missiles to replace the Lance missiles in the FRG and cancel further modernization of US nuclear artillery shells deployed in Europe. The USA will be ready to open negotiations on all ground-based short-range nuclear forces in Europe as soon as a CFE treaty is concluded.	SNF; USA/FRG; Europe
4 May	At the annual British–French summit meeting it is announced that the UK and France will create closer links in security and defence matters.	UK/France
4 May	Latvia declares itself an independent state.	Latvia/ USSR
4 May	The South African Government and the ANC announce that they have reached an important breakthrough by agreeing to work together to end political violence in South Africa.	South Africa; ANC
5 May	The first meeting of the foreign ministers of the FRG, the GDR, the USA, the USSR, the UK and France ('Two plus Four') on German unification is held in Bonn.	German unification
8 May	Albania announces that it will apply to join the CSCE. (Albania rejected the invitation issued by the Finnish Government in 1972.)	CSCE; Albania
10 May	The NATO Nuclear Planning Group, meeting in Alberta, Canada, states that, given the political and military changes in Central and Eastern Europe together with the prospect of an early CFE treaty, there is a diminishing need for short-range nuclear systems. A review of NATO's future nuclear force posture will provide guidance for the military authorities in their further work on future nuclear requirements.	NATO; SNF; Europe
12 May	A declaration is signed, in Tallin, by Estonia, Latvia and Lithuania, to renew the Agreement and Declaration on Accord and Co-operation, which was concluded by the three states in 1934. The new agreement provides for a Baltic States Council which will seek restoration of independence of the three republics.	USSR/Estonia; Latvia; Lithuania
16–19 May	At a meeting in Moscow between the US Secretary of State and the Soviet Foreign Minister agreement is reached on the control under a START treaty of nuclear-armed ALCMs with a range exceeding 600 km. Nuclear-	USA/USSR; START; CW

armed SLCMs would be covered by a separate arrangement, committing each side not to deploy more than 880 missiles. The two sides also agree that the USA would immediately cease CW production. Both sides would then reduce stockpiles to *circa* 5000 tonnes by the year 2002.

18 May	The finance ministers of the GDR and the FRG sign a state treaty on the creation of a monetary, economic and social union between the two countries.	German unification
18 May	At the CSBM Negotiations the NATO states present a proposal for the establishment of a mechanism for requiring explanation of unusual military activities.	CSCE; CSBM
21 May	The Parliaments of the Yemen Arab Republic (North Yemen) and the Democratic People's Republic of Yemen (South Yemen) approve the unification of the two states as the Republic of Yemen.	Yemen
23 May	The NATO Defence Planning Committee, meeting in ministerial session in Brussels, states that the presence of significant US conventional and nuclear forces in Europe will continue to be crucial for building a lasting peaceful order in Europe. However, a review of NATO's military strategy will be undertaken and the operational concepts and doctrines adjusted so that they continue to meet the security requirements of the member states. The general target, first agreed in 1977, of annual 3% real increases in the defence expenditures of member states is no longer appropriate, and after a CFE treaty some reduction in overall defence expenditure can be expected.	NATO; Military spending; Europe
25 May	Speaking before the UN Security Council, meeting in Geneva, the Chairman of the Executive Committee of the PLO puts forward a plan for the solution of the Israel/Palestine conflict.	Israel/ Palestine; UN
29 May–8 June	A preparatory meeting for the conference for the amending of the 1963 Partial Test Ban Treaty is held in New York. (The amendment conference will be held in New York, 7–18 Jan. 1991.)	Nuclear tests
30 May–3 June	A US–Soviet summit meeting is held in Washington, DC. The two sides agree to pursue talks on strategic arms and the relationship between strategic offensive and defensive weapons once a START treaty is signed. Emphasis should be placed on removing incentives for a nuclear first strike. President Gorbachev affirms the Soviet support for the Missile Technology Control Regime (MTCR), concluded in 1987, limiting the transfer of certain missile technology. A Joint Statement on Non-proliferation covering nuclear, missile and CW proliferation is signed.	USA/USSR; START; CW; Missile proliferation; Nuclear tests

	An Agreement is signed on the Destruction and Non-Production of Chemical Weapons and on Measures to Facilitate the Multilateral Convention on Banning Chemical Weapons. Each side will reduce its CW stockpiles to 5000 tonnes by the end of 2002, starting by the end of 1992. Each side agrees to halve its CW stockpile by 1999. The production of CW will cease as soon as the bilateral agreement is ratified. At the meeting new verification protocols for the 1974 Threshold Test Ban Treaty and the 1976 Peaceful Nuclear Explosions Treaty are signed.	
4 June	The South Korean President meets with President Gorbachev in San Francisco, the first meeting of the presidents of the two states. On 30 Sep. Soviet–South Korean diplomatic relations are established.	USSR/South Korea
5–29 June	At the Conference on the Human Dimension of the CSCE, held in Copenhagen, a declaration is adopted committing all CSCE states to free elections, multi-party systems and the rule of law. Albania takes part in the Conference as an observer.	CSCE; Albania
5 June	In a speech at the Conference on the Human Dimension of the CSCE the Soviet Foreign Minister says that by the end of 1990 the USSR will unilaterally withdraw 60 tactical missile launchers, more than 250 nuclear artillery pieces and 1500 nuclear warheads from Central Europe.	Withdrawals; USSR; CSCE; SNF
7 June	The Political Consultative Committee of the WTO, meeting in Moscow, states that since development in Europe creates conditions for overcoming a bloc security model it is necessary to reconsider the character and functions of the WTO and its transformation into a treaty of sovereign states with equal rights, formed on a democratic basis. The meeting reiterates that it wants constructive co-operation with NATO and neutral and non-aligned states and that the institutionalization of the Helsinki process is important.	WTO
7–8 June	The North Atlantic Council, meeting in ministerial session at Turnberry, UK, endorses the US President's proposal to start negotiations with the USSR on short-range nuclear weapons in Europe as soon as a CFE treaty is concluded (see *3 May*). It also states that the CSCE process will be an important framework for reforms and stability and for the construction of a new Europe along with other European institutions, including the Alliance itself.	NATO; CSCE
9 June	The Defence Minister of Hungary says that Hungary will not take part in the WTO military exercises this year and wants to leave the WTO by late 1991.	Hungary/ WTO

14 June	At the CFE Negotiation France and Poland jointly present a compromise proposal for limits on the number of tanks and armoured combat vehicles. The number of tanks should be limited to 20 000 and that of armoured transport vehicles to 30 000 for each side. The minimum unladen weight for both tracked and wheeled tanks is 16.5 tonnes. The proposal is accepted by both sides.	France; Poland; CFE
14–15 June	The Committee of the Defence Ministers of the WTO, meeting in Berlin, considers imparting a strictly defensive character to the WTO forces, measures to reduce the armed forces and conventional arms and withdrawal of Soviet troops from the territories of the allied states. It was reaffirmed that because of the growing level of trust between the WTO and NATO structures of a European security system are emerging.	WTO
15 June	The President of Nicaragua declares that the army of her country will be halved by August 1990.	Nicaragua; Force reductions
26 June	The Hungarian Parliament votes unanimously in favour of withdrawing from the Warsaw Treaty Organization.	Hungary/WTO
27 June	At the CFE Negotiation the participating states agree on a package of arms limitations including definitions for different categories of armoured vehicles.	CFE
27 June	The UN Security Council adopts Resolution No. 658 approving a plan for a settlement of the Western Sahara question under which the UN would supervise a cease-fire between Morocco and the Polisario and conduct a referendum in which the people of Western Sahara would choose between independence and integration with Morocco.	Morocco/ Western Sahara; UN
1 July	The Treaty on a monetary, economic and social union between the two German states takes effect. (See *18 May*.)	German unification
6 July	The NATO heads of state and government, participating in the North Atlantic Council meeting in London, issue a Declaration on a transformed North Atlantic Alliance. The NATO states propose to the WTO a joint declaration that the organizations are no longer adversaries and would refrain from the threat or use of force. President Gorbachev, on behalf of the USSR, and other East European leaders are invited to address the North Atlantic Council. As Soviet troops leave Eastern Europe and a CFE treaty is implemented the NATO military structure will change, moving to smaller, more mobile units with lower levels of readiness. NATO will cut the numbers of nuclear arms in Europe and adopt a strategy making nuclear forces truly weapons of last resort. Once negotiations begin on short-range nuclear forces NATO will propose the elimination of all its artillery shells from Europe. The CSCE	NATO/Use of force; No-first-use of nuclear weapons; SNF; CSCE

	should be more prominent in Europe's future, and new institutions within the CSCE are proposed, such as a Centre for Conflict Prevention and a CSCE parliamentary body.	
8 July	Greece and the USA sign a defence co-operation agreement.	Greece/USA; Foreign bases
15 July	In a memorandum to the Arab League Secretary the Iraqi Foreign Minister states that since the outbreak of the Iraq–Iran War in 1980 Kuwait has stolen vast amounts of oil from Iraq and that Iraq has the right to retrieve these funds. He also accuses Kuwait of having encroached upon Iraqi territory.	Iraq/Kuwait
16 July	President Gorbachev and the FRG Chancellor, meeting in Zheleznovodsk, USSR, announce that the USSR no longer opposes membership of a united Germany in NATO. However, NATO structures will not be extended to the territory of the GDR. After unification the USSR will withdraw the rest of its (350 000) troops from the GDR in 3–4 years' time. The armed forces of the united Germany will be brought down to 370 000 troops during the same period of time.	German unification; USSR; FRG
17 July	At the 'Two-plus-Four' talks it is agreed that a united Germany will include the FRG, the GDR and all of Berlin. The Polish western border (along the Oder–Neisse) will remain as it is on the day when the final arrangements come into force.	German unification; Poland
17 July	The Conference on Disarmament agrees on a mandate for the *Ad Hoc* Committee on a Nuclear Test Ban. (The Committee has not functioned since 1983.)	Nuclear tests; CD
18 July	In a reaction to the Iraqi memorandum of 15 July the Kuwaiti Foreign Minister states to the Arab League that the Iraqi charges have no factual basis. The Kuwaiti Army is placed on a state of alert.	Iraq/Kuwait
18 July	The US Foreign Secretary says that to prevent a return to power of the Khmer Rouge in Cambodia, the US Administration will open a dialogue with Viet Nam and will be prepared to enhance humanitarian assistance to Cambodia.	USA/Viet Nam; USA/ Cambodia
25 July	The Hungarian Prime Minister announces that Hungary will withdraw from the WTO by the end of 1990.	Hungary/ WTO
25 July	The British Defence Secretary proposes that half of the British troops deployed in the FRG be withdrawn by 1995. The overall British military forces will be cut by 18%.	Force reductions; UK
26 July– 22 Sep.	The USA removes its CW from depots in the FRG for further transport to a chemical weapons destruction facility on the Johnston Atoll in the Pacific.	CW; USA/FRG

28 July	The GDR calls for the withdrawal of all Soviet troops stationed in the GDR.	GDR/USSR; Withdrawals
30 July	Albania and the USSR resume diplomatic relations (which were cut off in 1961).	Albania/ USSR
1–2 Aug.	The US Secretary of State and the Soviet Foreign Minister meet in Irkutsk, USSR. A Soviet pledge to stop producing rail-mounted strategic missiles from the beginning of 1991 is made at the meeting.	USA/USSR; START
2 Aug.	Iraqi troops cross the border of Kuwait and occupy the country. Kuwait's emir flees to Saudi Arabia. The Iraqi President accuses Kuwait of stealing oil from Iraq and forcing down oil prices through overproduction.	Iraq/Kuwait
2 Aug.	The UN Security Council unanimously adopts Resolution No. 660 condemning Iraq's invasion of Kuwait. It demands Iraq to withdraw immediately and unconditionally. (Yemen does not participate in the vote.)	Iraq/Kuwait; UN
5 Aug.	The Islamic Conference Organization (ICO), meeting in Cairo, issues a communiqué condemning Iraqi aggression against Kuwait. It demands the withdrawal of Iraqi forces from Kuwaiti territory.	ICO; Iraq/Kuwait
6 Aug.	The UN Security Council adopts Resolution No. 661 calling upon all states to take appropriate measures to protect the legitimate Government of Kuwait. (Cuba and Yemen abstain from voting.)	Iraq/Kuwait; UN
6 Aug.	An agreement is signed by the President of South Africa and the leader of the ANC suspending armed struggle and setting a stage for peaceful settlement. The South African Government states that it has agreed to the release of some 1300 political prisoners and the return of 22 000 anti-government exiles.	South Africa; ANC
8 Aug.	Iraq declares that Kuwait is part of Iraq and declares full unity between the two countries.	Iraq/Kuwait
8 Aug.	The US President announces that he has ordered US military units to take up defensive positions in Saudi Arabia to assist the Saudi Government in the wake of Iraq's invasion of Kuwait. The UK sends additional air and naval units to Saudi Arabia.	Iraq/Kuwait: USA; UK
9 Aug.	The UN Security Council unanimously adopts Resolution No. 662 declaring Iraq's annexation of Kuwait null and void.	Iraq/Kuwait; UN
9 Aug.	After a meeting of leaders from Gambia, Ghana, Guinea, Nigeria, Sierra Leone and the OAU, as well as representatives from Mali and Togo, it is decided that the Economic	Liberia/ ECOWAS

	Community of West African States (ECOWAS) should deploy a peace-keeping force in Monrovia, Liberia, to restore peace in the region.	
10 Aug.	An emergency summit meeting of the Arab League is held in Cairo. A decision is taken to send an Arab expeditionary force to Saudi Arabia (21 members vote in favour, 3 vote against, 3 have reservations, 2 abstain and 1 is absent).	Iraq/Kuwait; Arab League
14 Aug.	In a letter to the President of Iran, Iraq accepts the terms for a comprehensive peace plan based on the implementation of UN Security Council Resolution No. 598 of 1987.	Iran/Iraq War
15 Aug.	In a letter to the UN Secretary-General the Soviet Foreign Minister suggests limiting international sales and supplies of conventional weapons.	Arms transfers; USSR; UN
18 Aug.	The UN Security Council unanimously adopts Resolution No. 664 demanding that Iraq permit and facilitate the immediate departure from Iraq and Kuwait of nationals of third countries and that Iraq rescind its order for the closing of diplomatic and consular missions in Kuwait.	Iraq/Kuwait; UN
20 Aug.– 15 Sep.	The fourth Non-Proliferation Treaty Review Conference is held in Geneva. The Conference fails to reach agreement on a final declaration.	NPT
22 Aug.	The GDR and the FRG pledge that a united Germany will not possess nuclear, biological or chemical weapons.	German unification; GDR; FRG; Nuclear weapons; BW; CW
23 Aug.	The Volkskammer of the GDR votes to dissolve the state, to unify with the FRG on 3 Oct. and to hold general elections on 2 Dec.	German unification; GDR
25 Aug.	The UN Security Council adopts Resolution No. 665 giving states the right to enforce the economic embargo against Iraq by allowing them to halt shipping to and from Iraq. (Cuba and Yemen abstain from voting.)	Iraq/Kuwait; UN
25 Aug.	Peace-keeping forces of the Economic Community of West African States (ECOWAS) arrive in Liberia to oversee a cease-fire and bring to an end the civil war. (See *9 Aug.*)	Liberia; ECOWAS
28 Aug.	The five permanent members of the UN Security Council agree on a framework document for a solution to the conflict in Cambodia, calling for a cease-fire, elections and an interim administration under UN auspices.	Cambodia

30 Aug.	At the CFE Negotiation the NATO states propose that a CFE treaty contain a commitment to CFE follow-on talks. The negotiating states should agree not to increase manpower while follow-on talks are under way.	CFE; NATO
30 Aug.	At the CFE Negotiation the Foreign Minister of the FRG announces that the FRG and the GDR have agreed that in four years' time the army of a united Germany would not exceed 370 000 troops. (This represents an almost 50% reduction of the total number possessed by the two states as of today.) No more than 345 000 troops will belong to the ground and air forces included in the CFE Negotiation. (See *6 Jan.* and *16 July*.)	German unification; Force reduction; FRG; GDR
31 Aug.	The GDR and the FRG sign a Treaty to harmonize their political and legal systems.	German unification
4–6 Sep.	The Prime Ministers of North and South Korea meet, in Seoul, for the first time since the establishment of the North and South Korean governments in 1948.	North Korea/South Korea
7 Sep.	In a resolution adopted by the Polish Senate, Poland proposes modifying the WTO into a treaty of a more consultative and defensive character. The principle of withdrawal of military structures from Poland is emphasized.	WTO/Poland
9 Sep.	President Bush and President Gorbachev, meeting in Helsinki, declare that nothing short of the complete implementation of the UN Security Council's resolutions on Iraq is acceptable to them or to the world community.	Iraq/Kuwait; USA; USSR; UN
9–10 Sep.	At a meeting in Jakarta, convened by France and Indonesia, the four Cambodian parties accept a framework document as the basis for settling the Cambodia conflict. (See *28 Aug.*) A Supreme National Council is created.	Cambodia
10 Sep.	The US Secretary of State says that the USA considers further cuts in US forces in Europe as part of an arms control agreement expected to be signed in 1990. The earlier agreed limit of 195 000 troops has been overtaken by events.	Withdrawals; USA; Europe
12 Sep.	The four World War II victorious powers and the GDR and the FRG sign the Treaty on the final settlement with respect to Germany in Moscow, sanctioning German unification and heralding the return of their full sovereignty. The Treaty will suspend their remaining occupation rights on 3 Oct., the day that the two German states will merge.	German unification
13 Sep.	The FRG and the USSR approve a Treaty on good-neighbourliness, partnership and co-operation. (The Treaty is signed by the Soviet President and the German Chancellor on 9 Nov.)	USSR/ Germany

13 Sep.	The UN Security Council adopts Resolution No. 666 reaffirming that Iraq remains fully responsible for the safety and well-being of third-state nationals in accordance with international humanitarian law. (Cuba and Yemen vote against.)	Iraq/Kuwait; UN
16 Sep.	The UN Security Council unanimously adopts Resolution No. 667 condemning Iraq's aggressive acts against diplomatic premises and personnel in Kuwait.	Iraq/Kuwait; UN
18 Sep.	The President of France and the Chancellor of the FRG, meeting at the Franco-German Security Council in Munich, declare that France will withdraw half of its troops stationed in the FRG within the next two to three years. 15 000–20 000 troops will remain in the united Germany.	France/FRG; Withdrawals
18 Sep.	The US Defense Department announces that 151 of the US military installations abroad (in 10 European and Asian countries) will be closed within the next few years.	Withdrawals; USA; Asia; Europe
20 Sep.	The Volkskammer of the GDR and the Bundestag of the FRG approve the Treaty on the final settlement with respect to Germany. (See *12 Sep.*)	German unification; GDR; FRG
24 Sep.	The GDR officially withdraws from the WTO.	WTO/GDR
24 Sep.	The UN Security Council unanimously adopts Resolution No. 669 on the task of examining requests for assistance under Article 50 of the UN Charter to states involved in implementing the sanctions which are suffering from their application.	Iraq/Kuwait; UN
25 Sep.	The UN Security Council adopts Resolution No. 670 spelling out the obligation to apply sanctions on all means of transport into Iraq, including aircraft. (Cuba votes against.)	Iraq/Kuwait; UN
26 Sep.	The US Defense Department announces that 40 000 US troops will be withdrawn from Europe over the next 12 months in anticipation of a CFE treaty.	Withdrawals; USA; Europe
27 Sep.	Addressing the UN General Assembly the President of Brazil states that Brazil discards the idea of any experiments that might involve nuclear explosions, even if only for peaceful purposes.	Brazil; Nuclear tests
1 Oct.	The USA, the USSR, the UK and France sign in New York a declaration suspending their rights and responsibilities in Berlin and all of Germany. The governments of the GDR and the FRG take note of this declaration.	German unification
1 Oct.	At a ceremony in New York the 'Two-plus-Four' foreign ministers sign a declaration suspending the four World War II victorious powers' rights in Berlin and Germany.	German unification

3 Oct.	The GDR and the FRG officially become one state. The German Chancellor addresses to all heads of government with whom the united Germany has diplomatic relations a letter stating the political intentions of the united Germany.	GDR/FRG
3 Oct.	The US Secretary of State and the Soviet Foreign Minister announce that a US–Soviet agreement in principle is reached on ceilings on conventional weapons in Europe. According to the agreement NATO will withdraw some 20 000 tanks and the WTO will withdraw 40 000 tanks, more than 51 000 artillery pieces and more than 40 000 armoured personnel carriers.	CFE; USA; USSR; NATO; WTO
9 Oct.	The USSR and Germany sign a Treaty on the costs connected with the stationing and withdrawal of Soviet troops on the territory of the former GDR.	USSR/Germany
12 Oct.	The USSR and Germany sign a Treaty on the conditions connected with the stationing and modalities of the withdrawal of Soviet troops from the territory of the former GDR.	USSR/Germany
12 Oct.	The USSR submits for the first time to the UN a detailed account of the Soviet military potential, including 1989 Soviet military expenditures in standardized form and figures on Soviet armed forces as of 1 Jan. 1990.	USSR; Armed forces; Military spending; UN
13 Oct.	In a letter to the US Secretary of State the Soviet Foreign Minister explains that between July 1988 and Aug. 1990 the USSR withdrew 10 000 tanks, 25 480 armoured combat vehicles and 24 100 artillery pieces from the CFE treaty area of application. An unspecified number of tanks were exported.	CFE; USSR
18 Oct.	At a meeting with the ambassadors of Denmark, Norway, Sweden, Finland, Poland and Germany, the Soviet Deputy Foreign Minister announces that the USSR will not deploy, in peacetime, nuclear arms on Soviet ships in the Baltic or on aircraft above that area.	USSR; Nuclear-free Baltic
29 Oct.	A Treaty of entente and co-operation between the USSR and France is signed in Paris by the two Presidents. According to the Treaty the two states pledge to consult with each other on common policies during international crises.	USSR/France
29 Oct.	The UN Security Council adopts Resolution No. 674 demanding that the Iraqi authorities and occupying forces immediately cease and desist from the taking of third-state national hostages and ensure the immediate access to food, water and basic services necessary to the protection and well-being of Kuwaiti nationals and of nationals of third	Iraq/Kuwait; UN

	states in Kuwait and Iraq, including the personnel of diplomatic and consular missions in Kuwait. (Cuba and Yemen abstain.)	
5 Nov.	The two Shia-Muslim militias in Lebanon sign a peace agreement. The two groups promise to help the Lebanese army to control certain areas in southern Lebanon.	Lebanon
6 Nov.	Hungary is the first WTO country to become a member of the Council of Europe.	Hungary; Council of Europe
7 Nov.	The USA announces that it will withdraw all combat fighting aircraft from the Philippines in 1991. In addition, more than 1800 US troops will leave the Philippines.	USA/Philippines; Withdrawals; Foreign bases
8 Nov.	President Bush orders a doubling of the US troops deployed in the Persian Gulf area.	Iraq/Kuwait; USA
9–10 Nov.	President Gorbachev visits the FRG. A 20-year Treaty on good-neighbourliness, partnership and co-operation between the two states is signed by the Soviet President and the Chancellor of the FRG.	USSR/FRG
14 Nov.	Poland and Germany sign the Treaty of confirmation of the existing border between Poland and the FRG.	Poland/ Germany
17 Nov.	The CSCE member states adopt, in Vienna, a document on Confidence- and Security-Building Measures. The document builds on the 1986 Stockholm Document, stipulating a strengthened verification regime, a greater amount of information, improved conditions for military observers and lower thresholds for the number of troops in exercises that require advance notice. The document also establishes a communications network and the Conflict Prevention Centre. It is included in the set of documents approved at the CSCE summit meeting in Paris.	CSCE; CSBM
19–21 Nov.	A CSCE summit meeting is held in Paris. On 19 Nov., the 22 NATO and WTO states sign the Treaty on Conventional Armed Forces in Europe (CFE Treaty), reducing each side's heavy tanks to 20 000, armoured vehicles to 30 000, artillery pieces to 20 000, combat aircraft to 6800 and attack helicopters to 2000. The Treaty is accompanied by 3 declarations on the limitation of: land-based naval aviation, military personnel and the army of the united Germany (see *16 July* and *30 Aug.*). On the same day the 22 states also sign a Joint Declaration, marking the end of the era of confrontation of the cold war and promising not to use force against each other.	CSCE; CFE

On 21 Nov. the CSCE heads of government or state sign the Charter of Paris for a new Europe which reaffirms the values of the CSCE and prescribes new structures and institutions for the CSCE process. They decide to provide regular summit meetings and meetings of foreign ministers with the aim of establishing by 1992 new negotiations on disarmament and confidence- and security-building measures, open to all participating states. A Supplementary Document to the Charter is also signed which deals with institutional arrangements such as the establishment of a CSCE Secretariat, an Office of Free Elections and a Conflict Prevention Centre.

23 Nov.	The EC countries and the USA sign, in Rome, a declaration on the principles of the US–EC partnership.	USA/EC
26 Nov.	The follow-on to the CFE Negotiation, CFE IA, opens in Vienna.	CFE
28 Nov.	The Presidents of Argentina and Brazil sign, at Foz do Iguacu, Brazil, a Declaration on Common Nuclear Policy. The two states will jointly start negotiations with the IAEA for the conclusion of safeguards agreements. When such agreements are reached measures will be taken leading to entry into force for both states of the Treaty of Tlatelolco, including action aimed at updating and improving the text.	Argentina/ Brazil; NPT; NWFZ: Latin America; IAEA
28 Nov.	The UN Security Council adopts Resolution No. 677 condemning Iraq's attempt to alter the demographic composition of the Kuwaiti population and to destroy civil records. (Cuba and Yemen vote against; China abstains from voting.)	Iraq/Kuwait; UN
28 Nov.	The contending armed factions inside Liberia sign a cease-fire agreement. (On 21 Dec. the parties agree to form an interim government.)	Liberia
28 Nov.	The first plenary session of the follow-on CSBM Negotiations is held.	CSCE; CSBM
29 Nov.	The UN Security Council adopts Resolution No. 678, allowing Iraq a final opportunity to withdraw from Kuwait and restore Kuwait's legitimate government by 15 Jan. 1991, after which 'all necessary means' may be used to force Iraqi compliance with UN resolutions. (Cuba and Yemen vote against; China abstains.)	Iraq/Kuwait; UN
4 Dec.	At the Western European Union parliamentary assembly, meeting in Paris, the French Defence Minister stresses that the WEU should be the framework for the joint creation of a European defence in line with European interests. European defence should not depend on 'imported and inadequate concepts'.	WEU

6 Dec.	In a letter to the EC President, Presidents Kohl and Mitterrand propose an EC common foreign policy, also including security aspects.	EC; Germany; France
6–7 Dec.	The NATO Defence Planning Committee and Nuclear Planning Committee, meeting in Brussels, state that the future NATO force posture will be based on smaller, more mobile and flexible forces. The new nuclear force structure will reflect the reduced reliance on nuclear weapons.	NATO; Armed forces
8 Dec.	The heads of governments of Estonia, Latvia and Lithuania, meeting in Tallin, sign a document stating that the Baltic states will not possess nuclear weapons under the jurisdiction of the republics, but will pursue a policy of non-proliferation of nuclear weapons and of the establishment of nuclear weapon-free zones.	Estonia; Latvia; Lithuania; nuclear weapons; NWFZ
11 Dec.	The 1974 Threshold Test Ban Treaty and the 1976 Peaceful Nuclear Explosions Treaty enter into force.	Nuclear tests; USA/USSR
14–15 Dec.	The EC European Council, meeting in Rome, proposes considering a gradual extension of the role of an EC political union in the area of common security. Issues such as economic and technological co-operation in the armaments field, co-ordination of armaments export policy, and non-proliferation as well as defence matters should be dealt with by the union. The importance of strengthening ties within NATO is emphasized. (See *28 Apr.*)	EC; NATO
18 Dec.	The UN General Assembly adopts a resolution condemning Iraq's invasion of Kuwait and its serious violation of human rights against the Kuwaiti people and third-state nationals, in violation of the UN Charter and the international Covenants on Human Rights.	Iraq/Kuwait; UN
20 Dec.	The Soviet Foreign Minister, Eduard Shevardnadze, resigns from his post.	USSR

About the contributors

Dr Ian Anthony (United Kingdom) is a Researcher on the SIPRI arms transfers and arms production project. He is the author of the SIPRI monograph *The Naval Arms Trade* (1990) and a co-author of the SIPRI Research Report *West European Arms Production: Structural Changes in the New Political Environment* (1990). He is the author of chapters in the *SIPRI Yearbooks 1988, 1989* and *1990.*

Dr Oscar Arias Sánchez (Costa Rica) is former President of the Republic of Costa Rica (1986–90) and Nobel Peace Prize Laureate (1987).

William M. Arkin (United States) is Director of the Nuclear Information Unit of Greenpeace USA. In 1981–89 he was Director of the National Security Program of the Institute for Policy Studies, Washington, DC. He is co-editor of the Natural Resources Defense Council *Nuclear Weapons Databook* series, and co-author of *Volume IV, Soviet Nuclear Weapons* (1989) and chapters in the *SIPRI Yearbooks* since 1985. He is a co-author of *Encyclopedia of the US Military* (1990) and a member of the editorial board of the *Bulletin of the Atomic Scientists.*

Dr Thomas B. Cochran (United States) is Senior Staff Scientist and Co-director of the Nuclear Program and the *Nuclear Weapons Databook* project of the Natural Resources Defense Council, Washington, DC. He has initiated a series of nuclear weapon verification projects with the Soviet Academy of Sciences. He is a co-author of several NRDC volumes of the *Nuclear Weapons Databook*, and of chapters in the *SIPRI Yearbooks* since 1985. His publications include articles on nuclear weapon research and production, arms control, nuclear weapon proliferation, safeguards, seismic verification, national energy R&D policy and radiation exposure standards.

Agnès Courades Allebeck (France) is a Research Assistant on the SIPRI arms transfers and arms production project. She was formerly a Research Assistant at the European Parliament, Luxembourg (1985) and has done research on the external relations of the European Community. She is a co-author of the SIPRI Research Report *West European Arms Production: Structural Changes in the New Political Environment* (1990), the author of a chapter in the *SIPRI Yearbook 1989* and co-author of a chapter in the *SIPRI Yearbook 1990.*

Dr Regina Cowen Karp (Germany) is a Researcher and Leader of the SIPRI project on Security Without Nuclear Weapons?. She is the editor of the SIPRI volume *Security with Nuclear Weapons? Different Perspectives on National Security* (1991) and co-editor (with W. Stützle and B. Jasani) of *The ABM Treaty: To Defend or Not to Defend?* (1987). She is the author of a contribution in G. M. Dillon (ed.), *Defence Policy Making, A Comparative Analysis* (1988) and of chapters in the *SIPRI Yearbooks 1988, 1989* and *1990.*

Dr Saadet Deger (Turkey) is Senior Researcher and Leader of the SIPRI world military expenditure project. She was formerly a Research Fellow at Birkbeck College, University of London (1977–87). She is a member of the UNIDIR Working Group of Experts for the Economics of Disarmament. She is an expert adviser to the United Nations and consultant to the World Bank. She is a member of the Executive Board of the International Defence Economics Association and of the Editorial Board of the journal *Defence Economics: The Political Economy of Defence Disarmament and Peace*. She is co-author (with S. Sen) of the SIPRI volume *Military Expenditure: The Political Economy of International Security* (1990), a co-editor of *Defence, Security and Development* (1987) and the author of *Military Expenditure in Third World Countries: The Economic Effects* (1986). She has published many articles for international journals on disarmament and development, the economics of security in the Third World and the arms trade, and is the author of chapters in the *SIPRI Yearbooks 1988*, *1989* and *1990*.

Ragnhild Ferm (Sweden) is Researcher on the SIPRI arms control and disarmament project. She has published chapters on nuclear explosions, a comprehensive test ban and arms control agreements and the chronologies of arms control events in the *SIPRI Yearbooks* since 1982. She is the author of the annual SIPRI publication *Rustning eller Nedrustning?* (in Swedish) [*Armaments or Disarmament?*].

Richard W. Fieldhouse (United States) is Senior Research Associate at the Natural Resources Defense Council, Washington, DC, and co-author of the forthcoming NRDC publication *Nuclear Weapons Databook, Volume V, British, French and Chinese Nuclear Weapons and Nuclear Weapons Proliferation.* In 1985–89 he was a Researcher at SIPRI and leader of the project on naval forces and arms control. He is the editor of the SIPRI volume *Security at Sea: Naval Forces and Arms Control* (1990) and co-author (with S. Taoka) of the companion volume *Superpowers at Sea: An Assessment of the Naval Arms Race* (1989). He is a co-author of chapters in the *SIPRI Yearbooks* since 1985.

David Fischer (United Kingdom) is a consultant on nuclear proliferation issues. He was Special Consultant to the Director-General of the International Atomic Energy Agency in 1981–82 and Assistant Director-General of the IAEA for external relations in 1976–81. He was IAEA Director External Relations in 1957–76. He is the author of *The International Nuclear Non-Proliferation Regime* (1987) and co-author (with P. Szasz) of the SIPRI volume *Safeguarding the Atom* (1985). His forthcoming publications include *Stopping the Spread of Nuclear Weapons: The Record, the Prospects* and *The International Nuclear Non-Proliferation Regime.*

Stephen Iwan Griffiths (United Kingdom) is a Research Fellow on the SIPRI project on Nationalism in Eastern Europe: Issues of European Security. He was formerly a Teaching Assistant at the Department of Politics, University of Aberdeen. He is the author of a chapter in the *SIPRI Yearbook 1990.*

Gerd Hagmeyer-Gaverus (Germany) is Researcher on the SIPRI arms transfers and arms production project. He was formerly a Researcher at the Centre for Social Science Research at the Free University, Berlin, where he co-authored several research reports. He is a co-author of chapters on world military expenditure in the *SIPRI Yearbooks 1985, 1987* and *1988* and on arms production in the *SIPRI Yearbook 1990.*

Birger Heldt (Sweden) is a Research Assistant on the armed conflicts data project at the Department of Peace and Conflict Research, Uppsala University, Sweden. He is co-author of a chapter in *States in Armed Conflict 1989* (1991).

Aaron Karp (United States) is a consultant to the UN Under Secretary-General for Disarmament Affairs. In 1987–89 he was the Leader of the SIPRI arms trade project and is the author of chapters in the *SIPRI Yearbooks 1988, 1989* and *1990*. He is the principal author of the *Report of the Secretary-General on South Africa's Nuclear Tipped Ballistic Missile Capability* (1990) and has published articles on military industry and arms procurement.

Dr Axel Krohn (Germany) is a Researcher on the SIPRI project on the security system in Europe. He is the author of *Nuklearwaffenfreie Zone: Regionales Disengagement unter der Rahmenbedingung globaler Großmachtinteressen, Das Fallbeispiel Nordeuropa* (1989). He is the co-editor of the SIPRI Research Report *Stability and Arms Control in Europe: The Role of Military Forces within a European Security System* (1989), and author of accounts of the implementation of the 1986 Stockholm Document in the *SIPRI Yearbooks 1989* and *1990.*

Karin Lindgren (Sweden) is Leader of the armed conflicts data project at the Department of Peace and Conflict Research, Uppsala University, Sweden, and Information Secretary of NORDSAM (Nordic Co-operation Committee for International Politics), Stockholm. She is the editor of *States in Armed Conflict 1989* (1991) and author of *Världens Krig* (in Swedish, 1990) and *Participation of Women in Decision-making for Peace, Case Study on Sweden* (1989). She is co-author of a contribution in P. Wallensteen (ed.), *States in Armed Conflict 1988* (1989) and of chapters in the *SIPRI Yearbooks 1989* and *1990.*

Phitsamone Ljungqvist-Souvannavong (Laos) is a Research Assistant on the SIPRI military expenditure project. She was formerly Research Assistant at the United Nations University, Tokyo (1983–86). She is currently doing post-graduate research on Japanese security policy.

Dr S. J. Lundin (Sweden) is Senior Researcher and Head of the SIPRI chemical and biological warfare programme. He is on leave of absence as Director of Research at the Swedish Defence Research Institute (FOA), and was formerly Scientific Adviser to the Swedish Disarmament Delegation (1969–85) and to the Swedish Supreme Commander (1985–87). He is the author of chapters in the *SIPRI Yearbooks 1988, 1989* and *1990* and has written extensively in the field of chemical and biological weapons.

Paolo Miggiano (Italy) is a Research Assistant on the SIPRI arms transfers and arms production project. In 1989–90 he was Senior Visiting Fellow at the Pugwash Office, Rome. He is the author of chapters on civil defence and the defence postures of Italy and neutral states, and on Italian military policy in the yearbook of the Istituto Affari Internazionali. He is a co-editor of *L'Italia e la corsa al riarmo* (1987).

Dr Harald Müller (Germany) is Senior Researcher and Director of International Programs at the Peace Research Institute Frankfurt (PRIF). His most recent publications include *Falling into Line? France and the NPT* (1990) and *Germany and the Bomb: Nuclear Policies in the Two German States, and the United Germany's Nonproliferation Commitments* (1990). He is the editor of *Deciding on the Atom: How Western European Nuclear Policy is Made* (1990), co-author (with R. Kokoski) of the SIPRI Research Report *The Non-Proliferation Treaty: Political and Technology Prospects and Dangers in 1990* (1990) and author of a chapter in the *SIPRI Yearbook 1990.*

Kjell-Åke Nordquist (Sweden) is Research Associate at the Department of Peace and Conflict Research, Uppsala University, Sweden. He is a co-author (with C. Ahlström) of *Casualties of Conflict* (1991), prepared for the Red Cross/Red Crescent Movements' World Campaign for the Protection of Victims of War, and author of *Conflicting Peace Proposals: Four Peace Proposals in the Palestine Conflict Appraised* (1985) and of a contribution in C. Alger and M. Stohl (eds), *A Just Peace Through Transformation* (1988). He is co-author of a chapter in the *SIPRI Yearbook 1990.*

Dr Robert S. Norris (United States) is Senior Staff Analyst at the Natural Resources Defense Council, Washington, DC. He was formerly Senior Research Analyst for the Center for Defense Information, Washington, DC. He is a co-editor of the 1987 and 1989 volumes of the NRDC *Nuclear Weapons Databook,* and a co-author of chapters in the *SIPRI Yearbooks* since 1985. He is a co-author of a column for the *Bulletin of the Atomic Scientists* in his field of nuclear weapon research and production, arms control and nuclear weapon testing.

John Pike (United States) is the Director of the Space Policy Project at the Federation of American Scientists, Washington, DC. He co-ordinates FAS research and public education on space policy. A former political consultant and science writer, he is the author of over 130 studies and articles on space and national security, and co-author of *The Impact of US and Soviet Ballistic Missile Defense Programs on the ABM Treaty* (1985). He is a member of the American Institute of Aeronautics and Astronautics, the American Astronautical Society and the Council on Foreign Relations. He is the author of chapters in the *SIPRI Yearbooks 1989* and *1990* on the military use of outer space.

Professor Barry R. Posen (United States) is Professor of Political Science, Massachusetts Institute of Technology. He is the author of *Political Objectives and Military Options in the Persian Gulf* (DACS Working Paper, 1990), *The Sources of Military Doctrine: France, Britain, and Germany Between the World Wars* (1984), and of the forthcoming publication *Inadvertent Escalation: Conventional War and Nuclear Risks* (1991). He has written several articles and chapters in edited volumes on defence and security policy.

Dr Adam Daniel Rotfeld (Poland) is Senior Researcher and Leader of the SIPRI project on Building a Co-operative Security System in and for Europe. He has been head of the European Security Department in the Polish Institute of International Affairs, Warsaw since 1978. He was a member of the Polish delegation to the Second Stage of the Conference on Security and Co-operation in Europe in Geneva (1973–75) and to the CSCE follow-up meetings in Belgrade (1977–78) and Madrid (1980–83), and was Deputy Head of Delegation to the Vienna follow-up meeting (1986–88). He is co-editor (with W. Stützle) of the SIPRI volume *Germany and Europe in Transition* (1991). He has published and edited over 20 monographs and studies on European security and the CSCE process, most recently *The European Security System in Statu Nascendi* (in Polish, 1990) and *CSBMs and European Security* (in Polish, 1990).

Dr Somnath Sen (India) is a Senior Researcher on the SIPRI world military expenditure project. He is currently on leave from a Senior Lecturership at the University of Birmingham. He is co-author (with S. Deger) of the SIPRI volume *Military Expenditure: The Political Economy of International Security* (1990). His other publications include *Protectionism, Exchange Rates and the Macroeconomy* (1985), articles on disarmament and development and macroeconomics, and a chapter in the *SIPRI Yearbook 1990.*

Jane M. O. Sharp (United Kingdom) is a Senior Researcher Fellow at the Centre for Defence Studies, Kings College, London. She was a Senior Researcher at SIPRI in 1987–91. She was previously a researcher at Harvard and Cornell Universities, and National Director of the Council for a Livable World, Washington, DC. She is the editor of the SIPRI volume *Europe After an American Withdrawal: Economic and Military Issues* (1990), author of the forthcoming SIPRI book *The Negotiations on Conventional Armed Forces in Europe: History, Analysis and Evolution* and co-editor (with D. Holloway) of *The Warsaw Pact: Alliance in Transition?* (1985). She is the author of chapters in the *SIPRI Yearbooks 1988, 1989* and *1990* and numerous articles and chapters on European security, East–West arms control and alliance politics. She serves on the board of directors of the Council for a Livable World, the Institute for Defense and Disarmament, Cambridge, Massachusetts, and the Center for War, Peace and the Media of New York University.

Dr Thomas Stock (Germany) is Researcher on the SIPRI chemical and biological warfare programme. He was formerly a Researcher at the Researcher Unit for Chemical Toxicology of the former Academy of Sciences of the GDR. He is a co-author of articles in the SIPRI volumes *Verification of Conventional Arms Control in Europe* (1990) and *Non-Production by Industry of Chemical-Warfare Agents: Technical Verification under a Chemical Weapons Convention* (1988) and co-editor of *National Implementation of the Future Chemical Weapons Convention* (1990).

Dr Walther Stützle (Germany) is the Director of SIPRI (1986–91). In 1983–86 he was International Security and Defence Correspondent for the *Stuttgarter Zeitung*, and in 1976–82 Head of the Planning Staff and Under-Secretary for Defence Planning and Policy of the FRG Ministry of Defence. He is co-editor (with A. D. Rotfeld) of the SIPRI volumes *Germany and Europe in Transition* (1991) and (with B. Jasani and R. Cowen) *The ABM Treaty: To Defend or Not to Defend?* (1987). He is the author of *Adenauer und Kennedy in der Berlin Krise 1961/62* (1973) and *Politik und Kräfteverhältnis* (1983), and a co-editor of *Abschied von der alten Ordnung: Europas neue Sicherheit* (1990) and *Europe's Futures—Europe's Choices* (1966). He has directed work for the *SIPRI Yearbooks 1987, 1988, 1989* and *1990*, and is the author of the introductory chapters and of numerous articles on international security.

Sir Brian Urquhart (United Kingdom) is Scholar-in-Residence at the Ford Foundation, New York, and a Member of the SIPRI Governing Board. He served the United Nations from 1945 and was Under Secretary-General for Special Political Affairs in 1974–86. He served in various capacities relating to peace-keeping operations. He is co-author (with E. Childers) of *A World in Need of Leadership: Tomorrow's United Nations* (1990), and the author of *Decolonization and World Peace* (1989), *A Life in Peace and War* (1986), *Hammarskjold* (1972) and a chapter in the *SIPRI Yearbook 1988*.

Professor Peter Wallensteen (Sweden) holds the Dag Hammarskjöld Chair in Peace and Conflict Research and is the Head of the Department of Peace and Conflict Research, Uppsala University, Sweden. His publications include *States in Armed Conflict 1988* (1989), *Peace Research: Achievements and Challenges* (1988), *Global Militarization* (1985), *Dilemmas of Economic Coercion* (1983) and *Structure and War* (1973). He is a co-author of chapters in the *SIPRI Yearbooks 1988, 1989* and *1990*.

Dr Herbert Wulf (Germany) is Senior Researcher and Leader of the SIPRI project on arms transfers and arms production. He is on leave of absence from the Hamburg University Institute for Peace Research and Security Policy. Prior to his work in peace research, he was Director of the German Volunteer Service in India. He is a co-author of the SIPRI Research Report *West European Arms Production: Structural Changes in the New Political Environment* (1990) and *Alternative Produktion statt Rüstung* (1987), and the author of *Rüstungsexport aus Deutschland* (1989). He has published books and articles on security policy, arms production, arms transfers, conversion, and disarmament and development. He is the author of a chapter in the *SIPRI Yearbook 1985* and co-author of two chapters in the *SIPRI Yearbook 1990*.

SIPRI Yearbook 1991: World Armaments and Disarmament

Oxford University Press, Oxford, 1991, 741 pp.
(Stockholm International Peace Research Institute)
ISBN 0-19-829145-0

ABSTRACTS

ARKIN, W. M., COCHRAN, T. B., FIELDHOUSE, R. W. and NORRIS, R. S., 'Nuclear weapons', in *SIPRI Yearbook 1991*, pp. 3–40.

In 1990 the USA, the USSR, the UK, France and China began the process of transition to the new political, economic and military circumstances. The USSR and the USA moved towards a less nuclear-armed future as they approached completion of a START treaty, descended from their peak nuclear deployments of the late 1980s, and took steps (especially the USSR) towards major military reductions and restructuring. Like the USA and the USSR, the UK conducted a review of military force structure and strategy to consider possible changes, reductions and reorganization. France, like the other nuclear weapon powers, was forced to deliberate over how the changed global situation would affect its defence budget. The Chinese Government provided very little public information about its current nuclear forces.

FERM, R., 'Nuclear explosions', in *SIPRI Yearbook 1991*, pp. 41–47.

In 1990, 18 nuclear tests were conducted, the lowest number for 30 years. The USSR conducted 1 test, and the USA and France carried out fewer tests than in previous years. China conducted 2 tests; this was the first year in which China conducted more than 1 test since 1983. Soviet protests against nuclear testing for environmental and disarmament reasons led to a decision by Soviet authorities to wind down activities at the Semipalatinsk test site. In January 1991 the Soviet Council of Ministers announced a unilateral moratorium on tests for the next 4 months. Research has again shown that French nuclear testing in the South Pacific is not so harmless to mankind and the environment as the French Government claims.

PIKE, J., 'Military use of outer space', in *SIPRI Yearbook 1991*, pp. 49–84.

US military space and strategic defence systems, focused for 3 decades on the USSR, underwent a profound reorientation in 1990 towards Third World contingencies. The war with Iraq displayed an unprecedented integration of military space systems in support of terrestrial operations, amounting to a revolution in the conduct of warfare. Military space systems supported a broader range of forces, in a more direct and timely manner and over a more extended period of time, than in any previous conflict. The Gulf conflict also occasioned a transformation of the US Strategic Defense Initiative programme, which was reoriented towards defending against limited missile strikes from the Third World and Soviet missile attacks.

LUNDIN, S. J. and STOCK, T., 'Chemical and biological warfare: developments in 1990', in *SIPRI Yearbook 1991*, pp. 85–112.

The 1990 US–Soviet agreement to dispose of most US and Soviet chemical weapons, cease production and co-operate in destruction was a step towards abolishing chemical weapons. Neither country's destruction programme is on schedule because of the lack of environmentally safe destruction technology and public protest against the construction of destruction facilities. US chemical weapons were removed from the FRG and transported to the US facility on Johnston Atoll for destruction. Efforts to stop CW proliferation continued, and some countries enacted legislation. Iraq continued efforts to acquire chemical weapons, and there was fear that chemical or biological weapons might be used in the Persian Gulf conflict. Uncertainty prevailed about Iraq's chemical weapon capability, especially to launch Scud missiles armed with chemical warheads.

DEGER, S., 'World military expenditure', in *SIPRI Yearbook 1991*, pp. 115–180.

The phenomenal increase of military expenditure in the 1980s has given way to sustained reductions. The modest decline in world military expenditure that was observed during the past 2 years accelerated in 1990. Soviet and US military spending fell significantly, although from very high levels, and this contributed to a decline of over 5% in aggregate world military spending. The direct impact of arms control agreements was modest, but changes in political perceptions meant that large-scale defence acquisitions could not be justified. The level of military research and development remained high and could increase in the future. Nevertheless, a trend decline in defence spending is now more certain. Systemic and structural causes, emanating from technological and economic factors, have contributed to the 1990 decline. However, the potential for armed conflicts remains, and the decline in military expenditure will not necessarily contribute to demilitarization.

SEN, S., 'Debt, financial flows and international security', in *SIPRI Yearbook 1991*, pp. 181–195.

Economic security is increasingly becoming more important for the South as well as the countries of Eastern and Central Europe. The debt crisis can be utilized to analyse developmental failures which affect stability and international security in the post-cold war era. Data and other information on debt, development assistance and defence spending show that the potential for an international disarmament dividend remains high. However, the actual resource transfers are low relative to the huge amounts still spent on the military. The economic costs of the Persian Gulf conflict show that the multidimensional aspects of security will become increasingly dominant.

ANTHONY, I., COURADES ALLEBECK, A., HAGMEYER-GAVERUS, G., MIGGIANO, P. and WULF, H., 'The trade in major conventional weapons', in *SIPRI Yearbook 1991*, pp. 197–279.

The 1990 value of the trade in major conventional weapons was $21.7 billion. This represents a decrease of 35% from the 1989 value and the acceleration of a trend. In the total, the share of deliveries to Third World recipients was 55%—the same as recorded for 1989. The USSR and the USA remained the largest exporters of major weapon systems in 1990, together accounting for 69% of the total. In contrast to the previous 5 years, US arms deliveries exceeded the value of Soviet exports. The European Community countries accounted for one-fifth of the arms supplies, with a high concentration in France, the UK and Germany. Third World exports of major conventional weapons dropped to c. 1% of total exports. The largest arms-importing country was Saudi Arabia, followed by Japan and India. There is an increasing emphasis on retrofitting or upgrading weapon systems instead of investing in development of new platforms. Several international initiatives for arms transfers control in 1990 have not yet led to quantifiable results.

WULF, H., 'Arms production', in *SIPRI Yearbook 1991*, pp. 281–316.

The global trend of a gradual reduction of arms production continued in 1990, with the exception of a few countries. The promising signs that the 1990s would be a decade of reduced arms production are still visible despite a set-back from both the Gulf conflict and the severe economic and political difficulties in the USSR. Arms sales figures for the largest 100 companies declined from 1988 to 1989 by 4%. Companies are reacting with different strategies, including mergers and takeovers, diversification into civil production and dismissal of employees. In 1990 Soviet conversion of factories in the military–industrial sector to civilian production proved to be more difficult than predicted by planners. Conversion was caught in the economic turmoil of changing from a planned to a market economy.

KARP, A., 'Ballistic missile proliferation', in *SIPRI Yearbook 1991*, pp. 317–343.

Iraq's use of ballistic missiles against Israel and Saudi Arabia, starting in January 1991, demonstrated how ballistic missile proliferation has become a major international problem. Even when its military significance is limited, its political impact is great. However, few countries continue to make rapid progress. While Iraq and North Korea continue to develop their ballistic missile forces, most other regional powers have been forced to slow down their programmes. Events in 1990 support the conclusion that for most of the world ballistic missile proliferation has stopped or slowed. Controls on the spread of missile technology under the Missile Technology Control Regime grew stronger in 1990. The USSR joined international control efforts as did several smaller countries. Many other countries reinforced their own export control policies. Yet serious imperfections mar control efforts. China remains the most important supplier outside the MTCR control system.

LINDGREN, K., HELDT, B., NORDQUIST, K.-Å. and WALLENSTEEN, P., 'Major armed conflicts in 1990, in *SIPRI Yearbook 1991*, pp. 345–380.

A total of 31 major armed conflicts were waged in 1990. This is a slight decline from the 1980s. Some conflicts were resolved in 1990, notably in Namibia, and in Nicaragua a peace process led to the end of the conflict. A new conflict was that in Liberia, leading to over 10 000 battle-related casualties. Other devastating conflicts were waged in Ethiopia, India, Sri Lanka, Chad, South Africa, Peru and Lebanon. Most conflicts were fought over control of government and ambitions for autonomy or independence. The UN played a central role in the Namibia conflict, but in many other conflicts international and regional organizations were not able to have a significant impact. The 1980s in retrospect was a decade with many conflicts relating to religious and/or ethnic aspirations. Conflict resolution was aided by the establishment of democratic forms of government in a number of countries. South-East Asia witnessed a number of solutions to internal and autonomy conflicts.

COWEN KARP, R., 'US–Soviet nuclear arms control' in *SIPRI Yearbook 1991*, pp. 383–402.

In 1990 most of the problems towards a strategic arms reduction treaty were resolved. Agreement was reached on nuclear-armed ALCMs, nuclear-armed SLCMs, deployed mobile ICBMs, non-deployed mobile ballistic missiles, US–British collaboration, the Soviet Backfire bomber and SS-18 modernization. Four issues await settlement: a sub-ceiling for ICBM warheads, access to ballistic missile telemetry data, verification of strategic bombers and monitoring of mobile ICBM production sites. Little progress was made at the Defence and Space Talks: there was no agreement on the future of the ABM Treaty and the role of strategic defences. At the June summit meeting both sides agreed to continue the DST after a START agreement. In order to comply with START sub-limits, the USA would have to cut its ballistic missile warheads by 36% and the USSR by 50%. These cuts are offset by gains for both sides due to liberal bomber warhead counting rules and the exclusion of SLCMs. The vulnerability of US silo-based ICBMs will not be reduced. Because START counting rules do not reflect the real force sizes, post-START nuclear forces are likely to exceed START-accountable nuclear forces.

GRIFFITHS, S. I., 'The implementation of the INF Treaty', in *SIPRI Yearbook 1991*, pp. 403–406.

During 1990, the INF Treaty implementation process continued satisfactorily. By the end of the year, the USSR had eliminated all but 66 SS-20 missiles, and the USA all but 128 GLCMs and 53 Pershing II missiles. The inspection process ran smoothly, with both sides fulfilling their obligations. Despite a few problems, the INF institutions have fulfilled their missions and demonstrated their worth as models for future arms control and disarmament measures. Despite the relative lack of attention, implementation of the 1987 INF Treaty has continued to demonstrate the strength of the new US–Soviet relationship, and that disarmament treaties can be successfully implemented.

SHARP, J. M. O., 'Conventional arms control in Europe', in *SIPRI Yearbook 1991*, pp. 407–511.

The 1990 Treaty on Conventional Armed Forces in Europe established numerical parity between NATO and the WTO in 5 weapon categories. The Treaty was negotiated in record time, but ratification cannot be taken for granted in the uncertain political climate of 1991. The next phase of CFE, with the same groups of states, aims to achieve manpower ceilings and aerial inspections. Despite having welcomed Soviet unilateral cuts when announced in 1988, in 1990 non-Soviet parties to the Treaty claimed that the transfer of 80 000 pieces of Soviet treaty-limited equipment east of the Urals in 1989–90 was an evasion of the spirit of the Treaty. The Joint Consultative Group sought to resolve this and other discrepancies in data exchanged. The Vienna Document 1990 on Confidence- and Security-Building Measures, also adopted by the CSCE at the Paris summit meeting, expanded the parameters of the 1986 Stockholm Document and generated new measures.

LUNDIN, S. J. and STOCK, T., 'Multilateral and bilateral talks on chemical and biological weapons', in *SIPRI Yearbook 1991*, pp. 513–539.

The 1990 US–Soviet agreement to dispose of the majority of US and Soviet chemical weapons and stop production was an important step towards a Chemical Weapons Convention. The negotiating parties at the Conference on Disarmament expressed concern about a provision which would permit retention of some CW during the first 8 years after entry into force. The CD was given a mandate to begin work on the final text of the CWC Little progress was made with political problems, but work on technical questions advanced. The Persian Gulf conflict did not strongly influence the negotiations. Preparation was made for the 1991 Review Conference of the Biological Weapons Convention with particular attention to strengthening the BWC and ensuring compliance. Future CW and BW agreements will need to be co-ordinated with environmental legislation.

FERM, R., 'Multilateral and bilateral efforts towards nuclear test limitation', in *SIPRI Yearbook 1991*, pp. 541–551.

The 1974 US–Soviet Threshold Test Ban Treaty and 1976 Peaceful Nuclear Explosions Treaty, limiting the yields of nuclear explosions to 150 kt, entered into force in 1990. New verification protocols had been negotiated between the 2 parties and made US ratification possible. The protocols permit different types of verification measures to check that the agreed yield of the explosions is not exceeded. A conference convened by parties to the 1963 Partial Test Ban Treaty was planned in 1990 and took place in January 1991. The proponents of the conference wished to convert the Treaty into a comprehensive test ban treaty, but the final decision of the Conference only recommended further work on test limitation issues. In 1990 the Conference on Disarmament agreed on a mandate for its *Ad Hoc* Committee on a Nuclear Test Ban. The Committee had not functioned since 1983, because of disagreement on the mandate. The CD *Ad Hoc* Group of Scientific Experts continued its work on a verification system for a future comprehensive test ban.

FISCHER, D. and MÜLLER, H., 'The fourth review of the Non-Proliferation Treaty', in *SIPRI Yearbook 1991*, pp. 555–584.

The fourth NPT review conference ended without a consensus document. This stood in contrast to the unprecedented unanimity with which all countries of the CSCE region approached the conference, and to the strong endorsement the NPT received from many developing countries. A small group led by Mexico insisted on making a comprehensive test ban treaty the litmus test for the NPT's survival, while the USA and the UK refused to commit themselves to the prompt conclusion of a test ban. By failing to issue a final document, the conference lost many important agreements reached, notably a call on nuclear exporters to supply nuclear hardware and technologies only to countries applying international safeguards and an invitation to the International Atomic Energy Agency to study the possibility of 'special inspections' of undeclared facilities in non-nuclear weapon states.

ROTFELD, A. D., 'New security structures in Europe: concepts, proposals and decisions', in *SIPRI Yearbook 1991*, pp. 585–615.

New institutional arrangements were established in the decisions of the Paris CSCE summit meeting in November 1990. The extent to which newly created CSCE institutions, like the Conflict Prevention Centre, an emergency mechanism or a CSBMs communications network are adequate to meet the threats and challenges of post-cold war Europe is examined. Institutionalization is neither a value nor an end in itself. New security structures should be set up only when the solution of emerging problems is impossible without them. The transformation of the European security system now under way is not subordinated to any specific concept or theoretical model, but it should correspond to the new challenges and practical needs.

URQUHART, B., 'The role of the United Nations in the Iraq–Kuwait conflict in 1990', in *SIPRI Yearbook 1991*, pp. 617–637.

The Iraq–Kuwait conflict provided the first test of collective action against aggression by the United Nations through a united Security Council. The end of the cold war and unusual clarity of the Iraqi aggression made possible the speed and unanimity of action. A 'new world order' will have to address challenges to international peace and security as necessary. Improvements are urgently required in the UN system for peace and security. Mechanisms for co-ordination and consultation among governments need to be more effective and comprehensive. There must be a permanent watch on socio-economic, political and military developments, with Security Council action to pre-empt or correct dangerous situations. Mechanisms for carrying out Council decisions must be developed, including mechanisms for peace-making, peace-keeping and enforcement. The Military Staff Committee should embark on a study of the gradual conversion of the present world military set-up to an international security system.

POSEN, B., 'Military mobilization in the Persian Gulf conflict', in *SIPRI Yearbook 1991*, pp. 639–654.

Iraq's seizure of Kuwait precipitated an international military response of massive proportions. Led by the USA, a 34-nation coalition mounted a deployment unrivalled in military history for its combination of size and speed projected at such a great distance. The coalition effort is divided into four periods, governed by the military capabilities achieved: deterrence and delay, defensive, counter-offensive and offensive. The air, ground and naval forces available to the coalition at each stage are given in detail. Estimates of Iraqi forces in southern Iraq and Kuwait during each phase are also presented.

ARIAS SÁNCHEZ, A., 'The SIPRI 1990 Olof Palme Memorial Lecture: "Third World interests in the new era of East–West relations"', in *SIPRI Yearbook 1991*, pp. 655–661.

Under the pretext of guaranteeing national defence and security, the world has ignored the real natural threats against human life and has created others that are equally or more complex and dangerous. A security plan is needed to combat inequality, ignorance, disease, poverty and environmental decay. The concepts of defence and security proposed by world leaders have not only made us neglect these threats but have produced more urgent ones: the arms race, militarization, poverty intensified by wasted resources and the oppression, destruction and death provoked by war. Leaders of the industrialized and developing nations must come together to combat militarism and the buildup of arms, to establish the basis for a pact of security and mutual defence among all human beings. The destruction of nuclear arsenals is an immediate priority because they represent an enormous waste of material resources and the continued threat of global destruction. Advances in conventional arms control in Europe are welcomed, but recent events demonstrate that peace in Europe is not independent of events in the less developed countries of the Third World.

Errata

SIPRI Yearbook 1990: World Armaments and Disarmament

Page 16, table 1.3, under SLBMs:	A bracket was omitted by the Stingray, SS-N-18 Mod. 1/3 and Mod. 2, pointing to the figure 224; and at the end of the row a bracket was omitted by the two 'Warhead x yield' figures for the Stingray models, pointing to the figure 1 568.
Page 117, line 27, should read:	'US Defense Intelligence Agency had allegedly lowered its estimate of the'.
Page 117, footnote 62, should read:	'News chronology, September through December 1989', *Chemical Weapons Convention Bulletin*, no. 7 (Feb. 1990), item for 7 Nov., p. 15.
Page 251, table 7A.1, the figures in the Total row for 1977 should read:	'A: 22 477' (not '2 477'), and 'B: 19 436' (not '9 436').
Page 396, table 10.1, Romania:	The figure for total deaths (incl. 1989) has subsequently been revised to 1 036, according to official sources.
Page 400, table 10.1, Afghanistan, fifth line of the Comments should read:	'based in Peshawar. Forces of different anti-Govt groups initiated an unsuccessful attack . . .'
Page 406, table 10.1, Indonesia/East Timor:	The figure for No. of troops in 1989 for Fretilin should read '200–400'.
Page 413, table 10.1, Uganda, third line from bottom of the Comments, the full sentence should read:	'The HSM includes former UPDA troops.'
Page 553, chapter 16, the first footnote should read:	'In drafting this paper, I have enjoyed the collaboration and advice of Martin Kalinowski of the IANUS group at the Technical University of Darmstadt. Martin Kalinowski, a physicist who specializes in proliferation issues, has advised me particularly with regard to the technical issues involved.'

INDEX